Benchmark Papers in Electrical Engineering and Computer Science

Series Editor: John B. Thomas
Princeton University

SYSTEM SENSITIVITY ANALYSIS / *J. B. Cruz, Jr.*
RANDOM PROCESSES, PART I: Multiplicity Theory and Canonical Decompositions / *Anthony Ephremides and John B. Thomas*
ALGEBRAIC CODING THEORY: History and Development / *Ian F. Blake*
PATTERN RECOGNITION: Introduction and Foundations / *Jack Sklansky*
COMPUTER-AIDED CIRCUIT DESIGN: Simulation and Optimization / *S. W. Director*
ENVIRONMENTAL MODELING: Analysis and Management / *Douglas Daetz and Richard H. Pantell*
AUTOMATIC CONTROL: Classical Linear Theory / *George J. Thaler*
CIRCUIT THEORY: Foundations and Classical Contributions / *M. E. Van Valkenburg*
DATA COMMUNICATION: Fundamentals of Baseband Transmission / *L. E. Franks*
NONLINEAR SYSTEMS: Processing of Random Signals—Classical Analysis / *A. H. Haddad*
RANDOM PROCESSES, PART II: Poisson and Jump-Point Processes / *Anthony Ephremides*
DIGITAL FILTERS AND THE FAST FOURIER TRANSFORM / *Bede Liu*
DETECTION AND ESTIMATION: Applications to Radar / *Simon S. Haykin*
DATA COMPRESSION / *Lee D. Davisson and Robert M. Gray*
ACTIVE RC FILTERS / *L. P. Huelsman*
NONLINEAR SYSTEMS: Stability Analysis / J. K. Aggarwal and M. Vidyasagar
LINEAR ESTIMATION / *Thomas Kailath*
OPTICAL COMMUNICATION THEORY / *Robert O. Harger*
ERGODIC AND INFORMATION THEORY / *Robert M. Gray and Lee D. Davisson*
TWO-DIMENSIONAL DIGITAL SIGNAL PROCESSING / *S. K. Mitra and M. P. Ekstrom*

Benchmark Papers in Electrical Engineering and Computer Science / 20

A BENCHMARK ® Books Series

TWO-DIMENSIONAL DIGITAL SIGNAL PROCESSING

Edited by

SANJIT K. MITRA

University of California, Santa Barbara

and

MICHAEL P. EKSTROM

Lawrence Livermore Laboratory

Dowden, Hutchinson & Ross, Inc.

STROUDSBURG, PENNSYLVANIA

Benchmark Papers in Electrical Engineering and Computer Science, Volume 20
Library of Congress Catalog Card Number: 77-25337
ISBN: 0-87933-320-0

80 79 78 1 2 3 4 5
Manufactured in the United States of America.

LIBRARY OF CONGRESS CATALOGING IN PUBLICATION DATA

Main entry under title:
Two-dimensional digital signal processing.
(Benchmark papers in electrical engineering and computer science ; v. 20)
Includes index.
1. Signal processing—Addresses, essays, lectures. I. Mitra, Sanjit Kumar. II. Ekstrom, Michael P.
TK5102.5.T92 621.38'043 77-25337
ISBN 0-87933-320-0

Distributed world wide by Academic Press,
a subsidiary of Harcourt Brace Jovanovich,
Publishers.

SERIES EDITOR'S FOREWORD

This Benchmark Series in Electrical Engineering and Computer Science is aimed at sifting, organizing, and making readily accessible to the reader the vast literature that has accumulated. Although the series is not intended as a complete substitute for a study of this literature, it will serve at least three major critical purposes. In the first place, it provides a practical point of entry into a given area of research. Each volume offers an expert's selection of the critical papers on a given topic as well as his views on its structure, development, and present status. In the second place, the series provides a convenient and time-saving means for study in areas related to but not contiguous with one's principal interests. Last, but by no means least, the series allows the collection, in a particularly compact and convenient form, of the major works on which present research activities and interests are based.

Each volume in the series has been collected, organized, and edited by an authority in the area to which it pertains. In order to present a unified view of the area, the volume editor has prepared an introduction to the subject, has included his comments on each article, and has provided a subject index to facilitate access to the papers.

We believe that this series will provide a manageable working library of the most important technical articles in electrical engineering and computer science. We hope that it will be equally valuable to students, teachers, and researchers.

This volume, *Two-Dimensional Digital Signal Processing,* has been edited by S. K. Mitra of the University of California at Santa Barbara and M. P. Ekstrom of Lawrence Livermore Laboratory. It consists of forty-three papers together with the editors' comments on the two-dimensional processing of digital signals. This book should serve as a particularly timely condensation and clarification of this important new field.

JOHN B. THOMAS

PREFACE

Two-dimensional digital signal processing is a relatively new field. Almost all of the work reported in this area has occurred during the last ten years. In spite of its recent origin, significant progress has been made in the analysis, design, and implementation of two-dimensional digital signal processors. This volume, organized into three parts, is devoted to describing these advances in this important and growing field. Each part of this collection includes editorial comments to provide a sense of relationship between the various papers of the section and the total volume, the major contributions in the subject area, and some pertinent materials and references that may be of interest.

The editors would like to acknowledge the assistance and advice of many of their friends in the preparation of this volume. They are particularly indebted to Drs. J. K. Aggarwal, S. Chakrabarti, M. M. Fahmy, T. S. Huang, E. I. Jury, R. M. Mersereau, L. M. Silverman, R. E. Twogood, and J. W. Woods for reviewing the list of selected papers and making numerous constructive criticisms. It is also a pleasure to acknowledge the encouragement and support provided by Dr. L. Lynn Cleland of the Lawrence Livermore Laboratory and Mr. E. Schutzman of the National Science Foundation. The editors would also like to express their gratitude to Dr. J. B. Thomas for his encouragement and guidance during the preparation and finalization of the material for this volume. This work was performed under the auspices of the U.S. Department of Energy under contract #W-7405-ENG-48 and supported in part by a National Science Foundation Grant ENG 75-15027.

The order of the editors was determined by the flip of a coin.

S. K. MITRA
M. P. EKSTROM

CONTENTS

CONTENTS BY AUTHOR

TWO-DIMENSIONAL DIGITAL SIGNAL PROCESSING

INTRODUCTION

Digital processing of an input sequence to generate an output sequence in accordance with a prescribed computational algorithm is commonly known as digital signal processing. This type of signal processing is implemented by means of either a special purpose or general purpose digital computer. Phenomenal advances in digital integrated circuit technology during the last ten years have made the digital processing approach economically practical, and often a more convenient method of signal processing. As a consequence of this continuing decrease in the cost of implementation, and other advantages such as flexibility and reliability, there has been a continual increase in the application of digital signal processing in many diverse fields.

Depending on the type of the input and output sequences, digital signal processing can be classified into two broad groups: one-dimensional digital signal processing and multi-dimensional digital signal processing. In the first case, the data are given as a function of a single integer variable, such as that obtained by sampling a time-function. In the latter case, the data are a function of several integer variables such as that obtained by sampling a two-dimensional picture. There has been a considerable amount of research and development concerning the theory and design of one-dimensional digital signal processors.* However in most cases, the one-dimensional techniques cannot be routinely extended to deal with the multi-dimensional problems. As a result, digital processing of multi-dimensional data has required separate treat-

*A companion volume in the Benchmark series describes the developments in the one-dimensional digital signal processing field. See B. Liu (ed.), *Digital Filters and the Fast Fourier Transform,* Dowden, Hutchinson & Ross, Inc., 1975.

ment leading to the development of its separate theory and design techniques. Major emphasis in this effort has been directed to the processing of two-dimensional data because of its widespread applications in space imagery, biomedicine, geophysics, projective radiography, and radio astronomy.

This book is an accumulation of the important developments in the field of two-dimensional (2-D) digital signal processing. Here the pertinent signals are discrete and functions of two integer variables. Such signals are usually represented with the integer variables either as arguments or as subscripts, e.g., $x(m,n)$ or $x_{m,n}$, and referred to as 2-D arrays or bi-sequences. In contrast to 1-D signals, 2-D signals are almost always of finite dimensions. Typically, these arrays are of the order of 128 X 128 to 1024 X 1024 or larger, for most practical applications.

As can be seen, all the papers in this volume treat 2-D digital signal processing within a formal mathematical framework. Most have been published in the last ten years, with more than three-fourths of these appearing in the last five years. This indicates the quite recent development of 2-D digital signal processing formalisms. Because of the contemporary nature of the subject and the nonavailability of its unified treatment, it seems desirable to provide both newcomers and veterans to the field with a book containing the key contributions. This volume is intended to serve this need.

The papers in this volume have been separated into three parts. The first part deals with deterministic signal processing, the second part with statistical signal processing, and the last part concentrates on implementation techniques. The first two parts constitute the papers dealing with two-dimensional, signal processing formalisms. In Part I, "Deterministic Two-Dimensional Digital Signal Processing," Papers 1 and 2 are essentially review papers providing an introduction to the area. Papers 3, 4, and 5 are concerned with the spectral factorization problem. Next, Papers 6, 7, 8, 9, and 10 deal with stability properties of 2-D digital filters and outline some procedures for testing stability. Papers 11 through 19 are devoted to the design of recursive and nonrecursive 2-D digital filters.

The papers in Part II are in the area of statistical 2-D digital signal processing and are divided into two sections: the first section dealing with the design of non-recursive 2-D processors (Papers 20 through 23), and the second concerning the design of recursive 2-D processors (Papers 24 through 28). Our inclusion of both the deterministic and statistical approaches to two-dimensional signal processing is unusual. While many important aspects (point-of-view, error criteria, modeling, desired and a priori information, etc.) are different in the two approaches, it will become apparent from the papers presented that substantive similarities do exist in the processing formalisms and research problems associated with their multi-dimensional nature. It is our viewpoint that an important contribution of this book will be to inform the researchers in one field of developments in the other.

The last part of the volume involves two-dimensional transform methods and developments associated with implementing two-dimensional processing on both high performance scientific computers and specialized digital processors. Ultimately, the usefulness of 2-D signal processing will depend on its ability to successfully solve practical applications problems. This will in turn be dependent on both design and implementation capabilities. Part III, "Implementation Techniques," is divided into four sections. In the first section, Papers 29, 30, and 31 consider the possibility of implementing a 2-D digital filter using 1-D digital filter implementation techniques. In the second, Papers 32 through 37 are concerned with the implementation of 2-D digital filters using Fast Fourier Transform (FFT) techniques. The third section, Papers 38 through 41, describes various algorithms for realizing non-recursive and recursive digital transfer functions. Finally, in the last section, Papers 42 and 43 address themselves to the analysis of finite wordlength effects on the performance of a 2-D recursive digital filter.

Because important contributions in the 2-D signal processing field are too numerous to be included in a single collection, the present volume is limited to those publications which, in our opinion, provide a reasonable, comprehensive picture of the whole field. However, there are other interesting works, which regretfully could not be included because of size limitations. Most of these are listed as supplementary references following the editor's comments in each section.

Part I

DETERMINISTIC, TWO-DIMENSIONAL DIGITAL SIGNAL PROCESSING

Editors' Comments on Papers 1 and 2

1 **SHANKS, TREITEL, and JUSTICE**
Stability and Synthesis of Two-Dimensional Recursive Filters

2 **MERSEREAU and DUDGEON**
Two-Dimensional Digital Filtering

OVERVIEW

Two-dimensional (2-D) digital signal processing is concerned with the processing of discrete signals that are functions of two integer variables. Such signals are usually referred to as 2-D sequences. A 2-D digital filter produces a 2-D output sequence $\{y_{m,n}\}$ by operating on a 2-D input sequence $\{x_{m,n}\}$ in accordance with a specified computational algorithm. The characterization of 2-D digital filters is analogous to that used for 1-D filtering.

We consider in this section, linear, shift-invariant digital filters for which the input and output sequences are related by a linear constant coefficient difference equation of the form

$$\sum_k \sum_\ell b_{k,\ell} y_{m-k,n-\ell} = \sum_k \sum_\ell a_{k,\ell} x_{m-k,n-\ell} \tag{1.1}$$

or equivalently, in the frequency domain by the digital transfer function $H(z_1,z_2)$ defined as

$$H(z_1,z_2) = \frac{Z\{y_{m,n}\}}{Z\{x_{m,n}\}} = \frac{\sum_k \sum_\ell a_{k,\ell} z_1^{-k} z_2^{-\ell}}{\sum_k \sum_\ell b_{k,\ell} z_1^{-k} z_2^{-\ell}}$$

$$= \frac{A(z_1,z_2)}{B(z_1,z_2)} \tag{1.2}$$

where $Z\{y_{m,n}\}$ and $Z\{x_{m,n}\}$ denote the 2-D Z-transforms of the output and the input sequences respectively. The inverse Z-transform of $H(z_1, z_2)$ is a 2-D sequence $\{h_{m,n}\}$ known as the *unit sample or impulse response* of the 2-D filter. The output sequence can also be expressed as a *convolution* of the input sequence with the impulse response sequence:

$$y_{m,n} = \sum_{k=-\infty}^{\infty} \sum_{\ell=-\infty}^{\infty} h_{k,\ell} x_{m-k,n-\ell}. \tag{1.3}$$

If the impulse response has only a finite number of non-zero samples, the corresponding filter is called a *finite impulse response* (FIR), 2-D filter. Otherwise, it is known as an *infinite impulse response* (IIR), 2-D filter. In the case of a FIR 2-D filter, the output sequence can be computed from the input sequence using either the direct convolution sum of Eq. (1.3) or *Discrete Fourier Transform* techniques. As a result, such filters are also known as *nonrecursive* 2-D filters. For sinusoidal input sequences of the form $x_{m,n} = e^{j\omega_1 m} e^{j\omega_2 n}$, the output $y_{m,n}$ is given as $H(e^{j\omega_1}, e^{j\omega_2})\, e^{j\omega_1 m} e^{j\omega_2 n}$ where $H(e^{j\omega_1}, e^{j\omega_2})$ is known as the *frequency response* of the 2-D digital filter.

In the case of an IIR 2-D filter, the output has to be computed recursively from the input and a set of initial conditions by using explicit forms of Eq. (1.1). As a consequence, this second type of filter has also been called *recursive.* The most commonly considered 2-D recursive filter is of the form

$$y_{m,n} = \sum_{k=0}^{M_a} \sum_{\ell=0}^{N_a} a_{k,\ell} x_{m-k,n-\ell} - \sum_{\substack{k=0 \\ k+\ell\neq 0}}^{M_b} \sum_{\ell=0}^{N_b} b_{k,\ell} y_{m-k,n-\ell}, \tag{1.4}$$

where $b_{0,0} = 1$. This is the so-called "causal" or quarter plane form of the filter in that it is nonanticipatory in both variables. The general 2-D recursive filter is the so-called half-plane recursive filter, one form of which is given by

$$y_{m,n} = \sum_{k=0}^{M_a} \sum_{\ell=0}^{N_a} a_{k,\ell} x_{m-k,n-\ell}$$

$$- \sum\sum_{k,\ell \in R} b_{k,\ell} y_{m-k,n-\ell}, \qquad (1.5)$$

where $R \triangleq \{m \geqslant 0, n \geqslant 0\} \; U \; \{m > 0, n < 0\} - (0,0)$. Notice that this so-called half-plane filter is nonanticipatory in only one variable, yet is recursively computable in that it only uses output values which can be previously computed or are given as initial conditions.

There are basically two steps in the development of a 2-D digital filter: *design* and *implementation.* The design step involves the determination of a transfer function $H(z_1,z_2)$ meeting the specified, deterministic space domain or frequency domain requirements within an acceptable error bound. If an FIR filter is desired, $H(z_1,z_2)$ has to be a finite degree polynomial in z_1^{-1} and z_2^{-1}. On the other hand, if an IIR filter is desired, $H(z_1,z_2)$ has to be expressed as a ratio of two polynomials, $A(z_1,z_2)$ and $B(z_1,z_2)$, with real coefficients. An additional problem associated with the design of an IIR filter is that of stability. This involves the necessity of insuring that $H(z_1,z_2)$ when implemented, will lead to a stable filter for which the output sequence will remain bounded for all possible bounded input sequences. Note that FIR filters by construction are always stable (provided their coefficients are bounded). After a suitable transfer function has been determined, the corresponding 2-D filter is usually implemented on a general purpose computer. This implementation is based on the recursive difference equation of Eq. (1.1) for an IIR filter or the convolution sum of Eq. (1.3) for an FIR filter.

Some of the initial work on two-dimensional digital filtering was performed by researchers in the geophysical industries for processing of seismic, gravitic, and magnetic data. In particular, a number of early contributions were advanced by J. L. Shanks, S. Treitel, and their colleagues. Many of their results on 2-D recursive filters are summarized in Paper 1 of this volume. This paper outlines the basic concepts of 2-D recursive digital filtering, stability conditions, and design algorithms for IIR filters having some desirable frequency domain or space domain responses. It also describes an implementation algorithm. While significant improvements in the characterization and design of 2-D recursive processors have appeared since its publication, this paper offered the first comprehensive treatment of some of the basic problems in 2-D filtering and is considered by many as the initial benchmark paper in the field. A more comprehensive, up-to-date, and unified review of the complete field of 2-D filtering is presented in Paper 2 by Mersereau and Dudgeon. Their discussion covers all aspects of design and implementation of both FIR and IIR filters. For anyone contemplating working in the field, this article provides an excellent introduction and is highly recommended. The list of references should also be of considerable value.

1

Reprinted from *IEEE Trans. Audio Electroacoust.* **AU-20**(2):115–128 (1972)

Stability and Synthesis of Two-Dimensional Recursive Filters

JOHN L. SHANKS, Member, IEEE
SVEN TREITEL
Res. Cen.
Amoco Production Co.
Tulsa, Okla. 74102

JAMES H. JUSTICE
Dep. Math.
Univ. Tulsa
Tulsa, Okla.

Abstract

Two-dimensional recursive filters are conveniently described in terms of two-dimensional *z* transforms. The designer of these filters faces two fundamental problems, their stability and their synthesis. Stability is determined by the location of the zero-valued region of the filter's denominator polynomial. A conjecture based on a one-dimensional stability theorem leads to a useful empirical stabilization procedure. Two-dimensional recursive filters can be synthesized to approximate large varieties of desired two-dimensional pulse responses. A conformal transformation yields two-dimensional recursive bandpass filters from appropriately specified one-dimensional filters.

I. Introduction

There are many problems that call for digital filtering of sampled two-dimensional data. For example, the geophysical industry uses two-dimensional filtering for processing seismic records [1], [2] and gravity and magnetic data [3]. Two-dimensional filtering can also be used for the enhancement of photographic data such as weather photos, air photos, and medical X-rays [4].

Conventional implementation has made use of the two-dimensional convolution algorithm and, more recently, of the two-dimensional fast Fourier transform [28]. It is possible to extend another one-dimensional technique into two dimensions, namely recursive filtering [5], [6]. In this technique the filter algorithm uses previously computed output values as well as the input values. One-dimensional recursive filtering has often proven more efficient than either pure convolution in the time domain, or, alternatively, filtering in the frequency domain. Thus it should be interesting to study the properties of two-dimensional recursive filters.

Two problem areas of two-dimensional recursive filtering are stability and synthesis. Since a portion of the output values are used by the recursive algorithm, it is possible for the output values to become arbitrarily large independent of the size of the input values. A filter of this type is said to be unstable, a condition which is undesirable. Thus we need to know what constraints to put on the recursive filter coefficients so that the filtering operation will be stable. This problem is similar to the stability problem for one-dimensional filters, except that the added dimension materially increases the complexity of the analysis. However, we can construct stability diagrams for any particular two-dimensional recursive filter. By inspecting these diagrams we can determine whether or not the filter is stable. We shall now introduce a technique for stabilizing an unstable two-dimensional recursive filter.

In order to stabilize an unstable filter, we shall use a conjecture stating that the planar least squares (Wiener) inverse of an array is minimum phase. A rigorous proof for this property exists only for the one-dimensional case. However, we present a number of examples which uphold this conjecture.

Synthesis is concerned with the design of two-dimensional recursive filters that perform some desirable filtering operations. In this paper we show two ways to synthesize recursive filters. In the first of these we assume that the impulse response for some desired filtering operation is known. We then find the coefficients of a two-dimensional recursive filter that approximates that desired impulse response in a least mean square error sense. The second synthesis method concerns the design of two-dimensional recursive filters that have some desirable two-dimensional frequency response. In the technique presented we show how to take one-dimensional filters and convert them into two-dimensional filters with arbitrary directivity in a two-dimensional frequency response plane. These techniques have been presented briefly in the past [7]–[9]. In this paper we hope to present the complete arguments, with as much background as is practical.

The Two-Dimensional *z* Transform

Just as in one-dimensional digital filtering, the *z* transform can be used to characterize two-dimensional digital filters and two-dimensional digital filtering operations. Consider an array of data values

$$D = \begin{bmatrix} d_{11} & d_{12} & d_{13} & d_{14} \\ d_{21} & d_{22} & d_{23} & d_{24} \\ d_{31} & d_{32} & d_{33} & d_{34} \end{bmatrix}.$$

We can write D as a polynomial in two variables z_1 and z_2

$$D(z_1, z_2) = \begin{bmatrix} d_{11} + d_{12}z_2 + d_{13}z_2^2 + d_{14}z_2^3 \\ + d_{21}z_1 + d_{22}z_1z_2 + d_{23}z_1z_2^2 + d_{24}z_1z_2^3 \\ + d_{31}z_1^2 + d_{32}z_1^2z_2 + d_{33}z_1^2z_2^2 + d_{34}z_1^2z_2^3 \end{bmatrix}.$$

The power of the z_1 and z_2 variables indicates the position of the associated filter coefficient in the array.

In general, the z transform for an $M\times N$ array is

$$D(z_1, z_2) = \sum_{i=1}^{M}\sum_{j=1}^{N} d_{ij}z_1^{i-1}z_2^{j-1}.$$

II. Two-Dimensional Recursive Filters

It is possible to describe a two-dimensional filter as a ratio of 2 two-dimensional polynomials in (z_1, z_2). Consider the filter

$$F(z_1, z_2) = \frac{A(z_1, z_2)}{B(z_1, z_2)} = \frac{\sum\sum_{i,j=1}^{M_a,N_a} a_{ij}z_1^{i-1}z_2^{j-1}}{\sum\sum_{k,l=1}^{M_b,N_b} b_{kl}z_1^{k-1}z_2^{l-1}}. \quad (1)$$

This filter is applied to a data array by use of the two-dimensional recursive algorithm. Given an input data array $D(z_1, z_2)$, the output of the filter $F(z_1, z_2)$ is

$$R(z_1, z_2) = F(z_1, z_2)\cdot D(z_1, z_2).$$

Using (1) and clearing fractions, we get

$$B(z_1, z_2)R(z_1, z_2) = A(z_1, z_2)D(z_1, z_2) \quad (2)$$

or

$$\left(1 + \sum_{\substack{k=1\\k\cdot l\neq 1}}^{M_b}\sum_{l=1}^{N_b} b_{kl}'z_1^{k-1}z_2^{l-1}\right)R(z_1, z_2) = \left(\sum_{i=1}^{M_a}\sum_{j=1}^{N_a} a_{ij}'z_1^{i-1}z_2^{j-1}\right)D(z_1, z_2) \quad (3)$$

where

$$b_{kl}' = b_{k,l}/b_{11}$$
$$a_{kl}' = a_{k,l}/b_{11}$$
$$b_{11} \neq 0.$$

Solving (3) for $R(z_1, z_2)$ we obtain

$$R(z_1, z_2) = \left[\sum_{i=1}^{M_a}\sum_{j=1}^{N_a} a_{ij}'z_1^{i-1}z_2^{j-1}\right]D(z_1, z_2) - \left[\sum_{\substack{k=1\\k\cdot l\neq 1}}^{M_b}\sum_{l=1}^{N_b} b_{kl}'z_1^{k-1}z_2^{l-1}\right]R(z_1, z_2). \quad (4)$$

From (4) we deduce the recursive algorithm for computing r_{mn}, the output value at the array grid point (m, n)

$$r_{mn} = \sum_{i=1}^{M_a}\sum_{j=1}^{N_a} a_{ij}'d_{m-i+1,n-j+1} - \sum_{\substack{k=1\\k\cdot l\neq 1}}^{M_b}\sum_{l=1}^{N_b} b_{kl}'r_{m-k+1,n-l+1}. \quad (5)$$

Here we assume either that all output values $r_{m-k+1,n-l+1}$ have been computed previously or are equal to zero.

Let us consider a particular filter

$$F(z_1, z_2) = \frac{a_{11} + a_{12}z_2 + a_{21}z_1 + a_{22}z_1z_2}{1.0 + b_{12}z_2 + b_{13}z_2^2 + b_{21}z_1 + b_{22}z_1z_2 + b_{23}z_1z_2^2}.$$

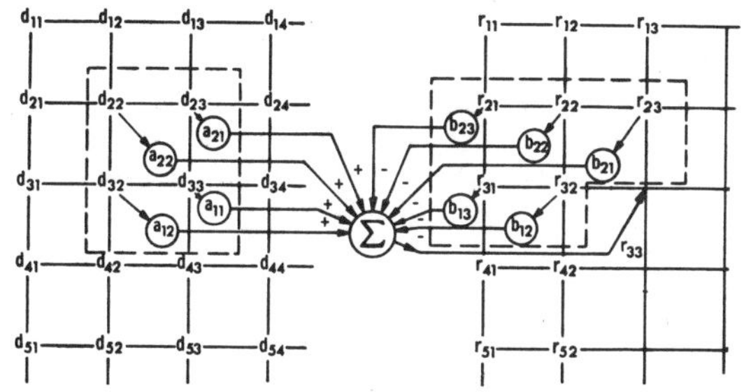

Fig. 1. Recursive filtering of an input array D to produce the output array R.

From (5), the recursive algorithm is

$$\begin{aligned} r_{mn} = {} & a_{11}d_{m,n} + a_{12}d_{m,n-1} + a_{21}d_{m-1,n} + a_{22}d_{m-1,n-1} \\ & - b_{12}r_{m,n-1} - b_{13}r_{m,n-2} - b_{21}r_{m-1,n} - b_{22}r_{m-1,n-1} \\ & - b_{23}r_{m-1,n-2}. \end{aligned} \quad (6)$$

The algorithm is illustrated in Fig. 1. The input array D is being filtered to produce the output array R. The recursive algorithm is used in Fig. 1 to compute the output value r_{33}. The input values d_{22}, d_{23}, d_{32}, and d_{33} and the previously computed output values r_{21}, r_{22}, r_{23}, r_{31}, and r_{32} are used to compute r_{33}. After r_{33} has been computed, r_{43} can be computed, assuming that r_{41} and r_{42} have been previously computed. A convenient implementation of this algorithm would be to compute the output array one column at a time. Alternately, the output could be computed one row at a time.

III. Stability

Just as in the case of one-dimensional digital feedback filters, it is possible for a two-dimensional recursive filter to be unstable. That is, given a bounded input array, the output array can be unbounded. We would like to be able to place constraints on the coefficients of the two-dimensional recursive filter to assure its stability.

The stability of the two-dimensional recursive filter is determined by the coefficients of the denominator $B(z_1, z_2)$ [see (1)]. The general criterion for stability is given by the following theorem. The proof of this theorem is being prepared for publication [26].

Theorem 1: Given that $B(z_1, z_2)$ is a polynomial in (z_1, z_2), for the coefficients of expansion of $1/B(z_1, z_2)$ in a positive powers of z_1 and z_2 to converge absolutely, it is necessary and sufficient that $B(z_1, z_2)$ not be zero for $|z_1|$ and $|z_2|$ simultaneously less than or equal to one.

Stated another way, the theorem says that if there are any values (real or complex) of z_1 and z_2 for which $B(z_1, z_2)$ is zero, and for which z_1 and z_2 are simultaneously less than or equal to one in magnitude, then the filter $1/B(z_1, z_2)$ will be unstable. If there are not, then $1/B(z_1, z_2)$ will be stable.

It is much more difficult to determine the stability of two-dimensional filters than of one-dimensional filters. In one dimension, it is only necessary to locate a finite set of roots in the z plane [10], [11]. In order to mechanize the theorem, we must in general find the continuum of (z_1, z_2) values for which $B(z_1, z_2)=0$. The authors know of no techniques for locating the continuum of zeros for a general two-dimensional polynomial in (z_1, z_2). However, for a low-order denominator, we can estimate the region of zeros. We can do this by assigning values to the variable z_1, and finding the roots in terms of z_2. For stability, all the roots of z_2 must be greater than one in magnitude if $|z_1|$ is less than one in magnitude.[1]

For example, consider the filter $F(z_1, z_2)=1/B(z_1, z_2)$, where

$$B(z_1, z_2) = 1 - 1.5z_1 + 0.6z_1^2 - 1.2z_2 + 1.8z_1z_2 - 0.72z_1^2z_2 + 0.5z_2^2 - 0.75z_1z_2^2 + 0.25z_1^2z_2^2.$$

Letting $z_1=\hat{z}_1$, we get

$$B(\hat{z}_1, z_2) = (1 - 1.5\hat{z}_1 + 0.6\hat{z}_1^2) + (-1.2 + 1.8\hat{z}_1 - 0.72\hat{z}_1^2)z_2 + (0.5 - 0.75\hat{z}_1 + 0.25\hat{z}_1^2)z_2^2. \quad (7)$$

Equation (7) is a quadratic in z_2 with complex coefficients. Thus we can compute the roots of (7) and plot them as a function of their location in the z_1 plane. Fig. 2 shows a contour map of the magnitude of one of the z_2 roots as a function of complex $z_1=u+iv$. The z_2 root chosen is the one with the minimum magnitude, since we are interested in finding those roots for which $|z_2|\leq 1.0$. The contour interval is 0.1, and the darkest contour corresponds to $|z_2|=1.0$. The shaded area is for $|z_2|<1.0$ and is therefore the mapping of the interior of the z_2 unit circle onto the z_1 plane for $B(z_1, z_2)=0$. In this case the filter is unstable, since the shaded area intersects the z_1 unit circle.

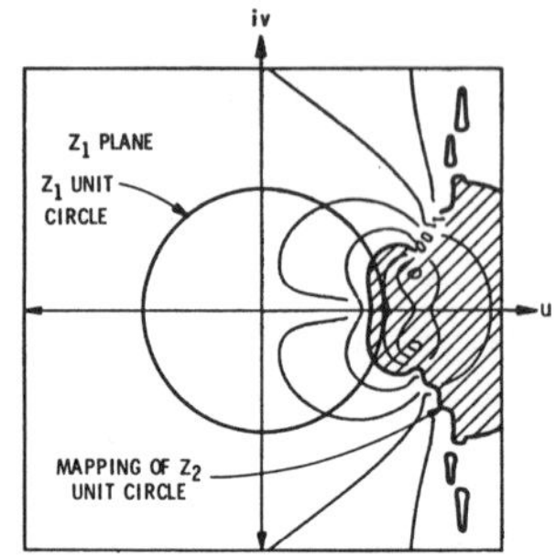

Fig. 2. Mapping of z_2 unit circle onto z_1 plane for unstable filter.

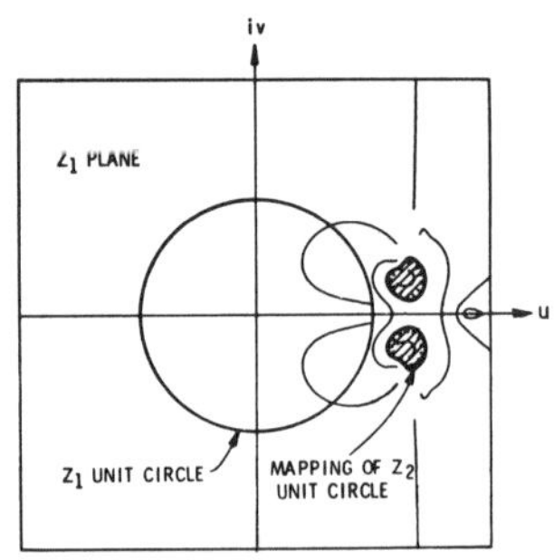

Fig. 3. Mapping of z_2 unit circle onto z_1 plane for stable filter.

Fig. 3 shows the mapping of the z_2 unit circle into the z_1 plane for the polynomial

$$B(z_1, z_2) = 1 - 1.2z_2 + 0.5z_2^2 - 1.5z_1 + 1.8z_1z_2 - 0.75z_1z_2^2 + 0.6z_1^2 - 0.72z_1^2z_2 + 0.29z_1^2z_2^2 = 0.$$

As before, the shaded area represents the mapping of the z_2 unit circle onto the z_1 plane. In this case the z_2 unit circle does not intersect the z_1 unit circle and the filter $F(z_1, z_2) = 1/B(z_1, z_2)$ is stable.[2]

In the special case that the denominator polynomial is first order in both directions, the z_1 plane$\leftrightarrow z_2$ plane transformation becomes bilinear [15], [16] and we can set some specific limits on the filter coefficients. Let $F(z_1, z_2)=1/B(z_1, z_2)$, where $B(z_1,z_2)=1+b_{21}z_1+b_{12}z_2+b_{22}z_1z_2$. The filter $F(z_1, z_2)$ will be stable if and only if the following conditions hold.

Condition 1:

$$\left|\frac{1}{b_{12}}\right| > 1.$$

Condition 2:

$$\left|\frac{1+b_{12}}{b_{21}+b_{22}}\right| > 1.$$

Condition 3:

$$\left|\frac{1+b_{12}}{b_{21}-b_{22}}\right| > 1. \quad (8)$$

The derivation of these conditions is being prepared for future publication.

For example, let us consider the filter

$$F(z_1, z_2) = 1/(1 - 0.7z_1 - 0.5z_2 + 0.3z_1, z_2). \quad (9)$$

[1] Weinberg [12] suggests a stability test involving an adaptation of a two-variable Hurwitz test as described by Ansell [13]. Huang [14] has suggested a test involving only the unit circle in one z plane and the origin of the other z plane.

[2] Figs. 2 and 3 illustrate the fact that the root maps are symmetric with respect to the u axis. The remaining root maps show only the upper half z plane.

TABLE I

Unit Pulse Response
$F(z_1, z_2) = 1/(1-0.7z_1-0.5z_2+0.3z_1z_2)$

m \ n	1	2	3	4	5	6	7	8	9	10	11
1	1.0	0.5	0.25	0.125	0.063	0.031	0.016	0.008	0.004	0.002	0.001
2	0.7	0.4	0.225	0.125	0.069	0.038	0.02	0.011	0.006	0.003	0.002
3	0.49	0.315	0.195	0.117	0.069	0.04	0.023	0.013	0.007	0.004	0.002
4	0.343	0.245	0.165	0.106	0.066	0.041	0.024	0.014	0.008	0.005	0.003
5	0.24	0.189	0.136	0.093	0.061	0.039	0.024	0.015	0.009	0.005	0.003
6	0.168	0.144	0.111	0.08	0.055	0.036	0.024	0.015	0.009	0.006	0.004
7	0.117	0.109	0.090	0.067	0.048	0.033	0.022	0.014	0.009	0.006	0.004
8	0.082	0.082	0.082	0.055	0.041	0.029	0.02	0.014	0.009	0.006	0.004
9	0.058	0.062	0.062	0.045	0.035	0.026	0.018	0.013	0.008	0.005	0.004
10	0.04	0.046	0.046	0.037	0.029	0.022	0.016	0.011	0.008	0.005	0.004
11	0.028	0.034	0.034	0.03	0.024	0.019	0.014	0.01	0.007	0.004	0.003

TABLE II

Unit Pulse Response
$F(z_1, z_2) = 1/(1-0.95z_1-0.9z_2+0.5z_1z_2)$

m \ n	1	2	3	4	5	6	7	8	9	10	11
1	1.0	0.9	0.81	0.73	0.66	0.59	0.53	0.48	0.43	0.39	0.35
2	0.95	1.2	1.4	1.5	1.6	1.7	1.7	1.8	1.8	1.7	1.7
3	0.90	1.5	2.1	2.6	3.1	3.7	4.2	4.5	4.8	5.1	5.4
4	0.86	1.7	2.8	4.0	5.3	6.6	8.0	9.5	10.	12.	14.
5	0.81	1.9	3.5	5.6	8.0	11.	14.	18.	22.	26.	30.
6	0.77	2.1	4.3	7.4	12.	17.	23.	30.	39.	49.	59.
7	0.74	2.3	5.1	9.5	15.	24.	35.	49.	66.	86.	110.
8	0.70	2.5	5.9	12.	21.	34.	52.	76.	106.	145.	191.
9	0.66	2.6	6.7	14.	27.	46.	74.	112.	164.	232.	318.
10	0.63	2.7	7.4	17.	33.	60.	101.	161.	245.	339.	509.
11	0.60	2.8	8.3	19.	41.	78.	136.	225.	355.	536.	788.

The parameters given in (9) meet all of the conditions of (8). Therefore, the filter of (9) is stable.

It would be interesting to look at the unit pulse response of the filter of (9). The unit pulse response is the output of the filter when the input is

$$U = \begin{bmatrix} 1.0 & 0 & 0 & 0 & \cdots \\ 0 & 0 & 0 & 0 & \cdots \\ 0 & 0 & 0 & 0 & \cdots \\ 0 & 0 & 0 & 0 & \cdots \\ \cdot & \cdot & \cdot & \cdot & \cdots \\ \cdot & \cdot & \cdot & \cdot & \cdots \\ \cdot & \cdot & \cdot & \cdot & \cdots \end{bmatrix}. \quad (10)$$

The size of the unit pulse array is arbitrary and depends on how many output values are desired from the filter.

The first (11×11) unit pulse response values for the filter in (9) are shown in Table I. It is characteristic of a stable two-dimensional recursive filter that its unit pulse response values tend to decrease at points away from the origin (upper left-hand corner). As can be seen in Table I, the unit pulse response values tend to decrease as m and n become large.

As an example of an unstable filter, consider the filter

$$F(z_1, z_2) = 1/(1 - 0.95z_2 - 0.95z_1 + 0.5z_1z_2). \quad (11)$$

This filter satisfies Conditions 1 and 3 of (8), but does not meet Condition 2. We can see in Table II that the first 11×11 values of the unit pulse response tend to get larger as m and n increase. This is characteristic of an unstable filter.[3]

IV. Minimum Phase in Two Dimensions

It would be useful to be able to distinguish between those denominators which produce unstable filters and those which produce stable filters. For this purpose, it is useful to introduce the concept of a two-dimensional minimum phase array.

Definition: We shall denote a minimum phase array A as one that satisfies the following two conditions.

[3] The stability of a filter cannot be determined by just looking at a finite section of the unit pulse response. However, these two examples do illustrate the decaying and the expanding characteristics of stable and unstable filters.

Condition 1: The two-dimensional array A, with the spectrum $A(f_1, f_2)$ evaluated at any real frequency $\hat{f}_1$, shall be a one-dimensionally minimum phase function of frequency f_2.

Condition 2: The two-dimensional array A, with the spectrum $A(f_1, f_2)$ evaluated at any real frequency $\hat{f}_2$, shall be a one-dimensionally minimum phase function of f_1.

The spectrum $A(f_1, f_2)$ is the two-dimensional Fourier transform of the array A. The function $A(f_1, f_2)$ can also be thought of as the z transform polynomial $A(z_1, z_2)$ evaluated along the z_1 and z_2 unit circles. Any particular frequency point (f_1, f_2) corresponds to the point in the z_1 and z_2 planes, $(z_1, z_2)=(\exp[-i\pi f_1/f_{N1}], \exp[-i\pi f_2/f_{N2}])$, where f_{N1} and f_{N2} are the Nyquist frequencies and $i=\sqrt{-1}$.[4]

The statement that a discrete sequence is one-dimensionally minimum phase is equivalent to the statement that its z transform has no zeros inside the z plane unit circle [17]. Therefore, since we can relate the frequencies f_1 and f_2 to unit circle points z_1 and z_2, we may also state Conditions 1 and 2 in the following way.

1) The two-dimensional z polynomial $A(\hat{z}_1, z_2)=0$, evaluated at any $|\hat{z}_1|=1.0$, has no zeros inside or on the z_2 unit circle.

2) The two-dimensional z polynomial $A(z_1, \hat{z}_2)=0$ evaluated at any $|\hat{z}_2|=1.0$, has no zeros inside or on the z_1 unit circle.

Having defined two-dimensional minimum phase in this way, we are ready to state a second important theorem.

Theorem 2: An array A is two-dimensionally minimum phase if and only if the polynomial $A(z_1, z_2)=0$ has no zeros for which $|z_1|\leq 1$ and $|z_2|\leq 1$, simultaneously. The proof of this theorem is being prepared for future publication.

From Theorems 1 and 2 we see that the filter $F(z_1, z_2)=1/A(z_1, z_2)$ is stable if $A(z_1, z_2)$ is minimum phase. Conversely, if $A(z_1, z_2)$ is not minimum phase, then $F(z_1, z_2)=1/A(z_1, z_2)$ will not be stable. For example, the array from (9)

$$B_1 = \begin{bmatrix} 1 & -0.5 \\ -0.7 & 0.3 \end{bmatrix}$$

is a minimum phase array, since the filter $F(z_1, z_2)=1/B_1(z_1, z_2)$ is stable. On the other hand, the array from (11)

$$B_2 = \begin{bmatrix} 1 & -0.9 \\ -0.95 & 0.5 \end{bmatrix} \tag{12}$$

is a nonminimum phase array since it produces an unstable filter.

V. The Planar Least Squares Inverse Filter

The design of a stable filter is facilitated by an understanding of the properties of the planar least squares inverse filter. Let us suppose that we have some given array C. We would like to find an array P such that C convolved with P approximates the unit pulse array U. That is

$$C * P \approx U \tag{13}$$

where the symbol $*$ denotes two-dimensional convolution. In general it will not be possible to make $C * P$ exactly equal to U. In actuality, $C * P$ will be equal to some other array, G. If we choose P such that the sum of the squares of the elements of $U-G$ is minimized, we call P a *planar least squares inverse* (PLSI) of C.

The size of the P array is arbitrary. The U and G arrays must have the same sizes, which depend on the size of the C and P arrays. If C is a $(m\times n)$ array, and P is a $(k\times l)$ array, then G and U must be $(m+k-1)$ by $(n+l-1)$ arrays. Since the size of P is arbitrary, there are many PLSI's of the array C, one for each possible set of dimensions of P. However, once the dimensions k and l of the matrix P are fixed, there is one and only one array P that minimizes the mean square difference between G and U. This array P is determined from the matrices U and C by a two-dimensional Wiener technique described by Wiggins [18].

As a group, PLSI's have interesting properties. One property in particular which we shall use is described by the following conjecture.

Conjecture 1: Given an arbitrary real finite array C, any planar least squares inverse of C is minimum phase. This is an important conjecture because it implies that the filter $F(z_1, z_2)=1/P(z_1, z_2)$ must be stable if P is a PLSI.

Unfortunately we do not have a proof for this conjecture, which is an extension of a one-dimensional theorem [25]. However, we have tested the conjecture on a large number of numerical cases, and have failed to find any counterexamples.

Numerical Examples

In order to test Conjecture 1, we chose a number of arbitrary arrays for which PLSI's were determined. We restricted the size of the PLSI's to (2×2) and (3×3), so that they could be easily tested for the minimum phase property. The PLSI's of the various test arrays were obtained with a planar Wiener filter program written for the Univac 1108 computer.

Table III shows a set of three examples for which 2×2 PLSI's were computed. The first column gives the input array C. Using the planar Wiener inverse filter program a 2×2 PLSI for C was computed. This result is shown in the second column. The P arrays were all normalized so that the upper left coefficient p_{11} is equal to 1.0. Finally, each P array was examined for stability by use of the three tests given in (8). All three tests must result in magnitudes greater than 1.0 in order for P to be minimum phase. As can be seen in Table III, every test was satisfied for each of the three samples. Therefore, each of the three P arrays are minimum phase, and the filter $F(z_1, z_2)=1/P(z_1, z_2)$ would be stable in each example.

We also computed a number of (3×3) PLSI's for various arrays. We tested the stability of the (3×3) PLSI's by plotting a root map for each PLSI. In these root maps we show the mapping of the z_2 unit circle and its interior (unit

[4] Following the usual definition of the z transform variable, we have that $z_1=\exp[-i2\pi f_1\Delta x]=\exp[-i\pi f_1/f_{N1}]$, where $f_{N1}=1/2\Delta x$ is the Nyquist frequency, and Δx is the uniform sampling increment.

Table III

Stability Tests on 2×2 PLSI's

Example	C	P Normalized so that $p_{11}=1.0$	Stability Tests on P
1	$\begin{bmatrix} 1. & 3. \\ 2. & 4. \end{bmatrix}$	$\begin{bmatrix} 1.0 & -0.3719 \\ -0.4703 & 0.2127 \end{bmatrix}$	I. 2.69 II. 2.44 III. 2.01
2	$\begin{bmatrix} 1. & 4. & 7. \\ 2. & 5. & 8. \\ 3. & 6. & 9. \end{bmatrix}$	$\begin{bmatrix} 1.0 & -0.5437 \\ -0.5862 & 0.3793 \end{bmatrix}$	I. 1.84 II. 2.21 III. 1.60
3	$\begin{bmatrix} 4. & 1. & 9. \\ 2. & 4. & 1. \\ 3. & 1. & 0. \\ 6. & 2. & 1. \\ 7. & 6. & 2. \end{bmatrix}$	$\begin{bmatrix} 1.0 & -0.2633 \\ -0.3254 & -0.2515 \end{bmatrix}$	I. 3.80 II. 2.10 III. 4.21

$$\begin{bmatrix} 1. & 4. & 7. \\ 2. & 5. & 8. \\ 3. & 6. & 9. \end{bmatrix}$$

(A)

$$\begin{bmatrix} 1.0 & -.63405 & .16389 \\ -.78183 & .52113 & -.13153 \\ .19267 & -.13399 & .03399 \end{bmatrix}$$

(B)

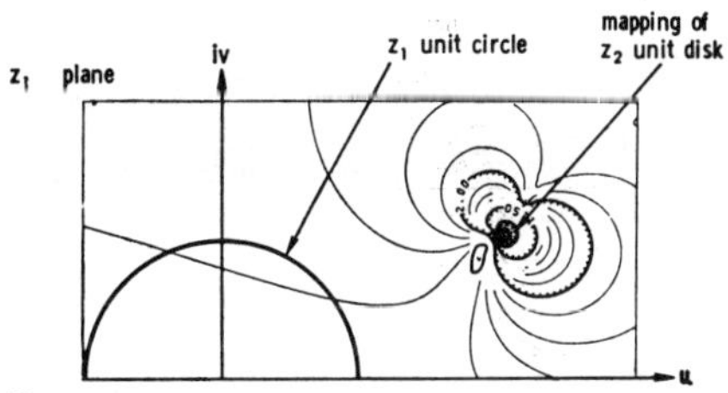

(C)

Fig. 4. Minimum phase test of (3×3) PLSI. (A) Input matrix A. (B) (3×3) PLSI of A. (C) Root map of PLSI.

disk) onto the z_1 plane. If the mapping of the z_2 unit disk does not intersect the z_1 unit disk, the array is minimum phase. For example, Figs. 4–6 show the input arrays, their (3×3) PLSI's, and the z plane root maps for the PLSI's. The symbols u and v denote the real and imaginary parts of the complex variable z_1, i.e., $z_1=u+iv$. The PLSI's have been normalized so that the upper-left coefficient is 1.0. The dark area on each root map is the mapping of the z_2 unit disk onto the z_1 plane. These root maps show that all of the PLSI's are minimum phase.

$$\begin{bmatrix} 1.0 & -.75 & .9 \\ 1.5 & -1.2 & 1.3 \\ 1.2 & .9 & .5 \end{bmatrix}$$

(A)

$$\begin{bmatrix} 1.0 & .20189 & -.17283 \\ -.47681 & -.09545 & .0036 \\ .11856 & -.00605 & -.02144 \end{bmatrix}$$

(B)

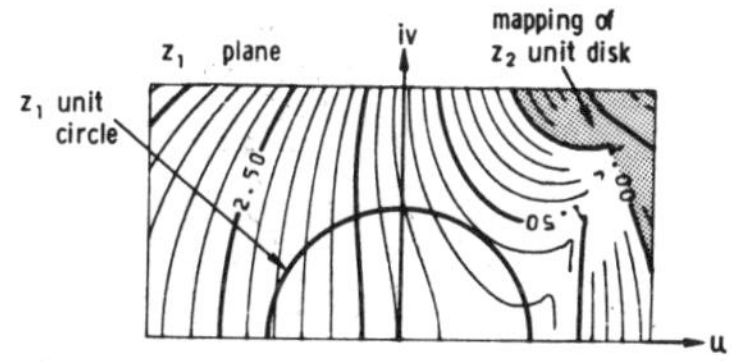

(C)

Fig. 5. Minimum phase test of (3×3) PLSI. (A) Input matrix A. (B) (3×3) PLSI of A. (C) Root map of (3×3) PLSI.

$$\begin{bmatrix} 1.0 & .5 \\ -.5 & 2. \end{bmatrix}$$

(A)

$$\begin{bmatrix} 1.0 & .17151 & .00427 \\ -.17151 & -.48877 & -.11991 \\ .00427 & .11991 & .19953 \end{bmatrix}$$

(B)

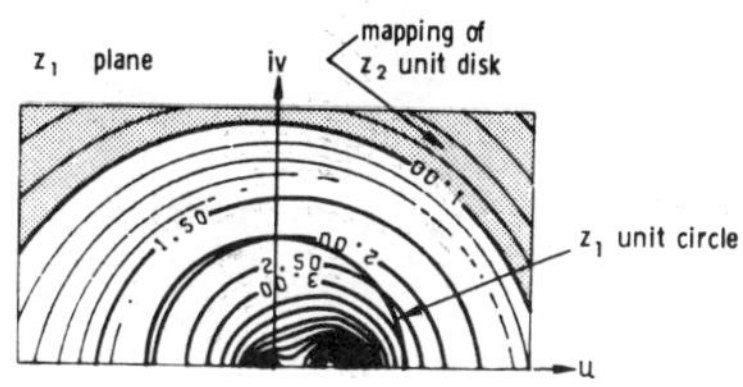

(C)

Fig. 6. Minimum phase test of (3×3) PLSI. (A) Input matrix A. (B) (3×3) PLSI of A. (C) Root map of PLSI.

Notice in particular Fig. 4. This shows that the (3×3) inverse of the array

$$A=\begin{bmatrix} 1 & 4 & 7 \\ 2 & 5 & 8 \\ 3 & 6 & 9 \end{bmatrix}$$

is minimum phase. It was also shown in Table III, Example 2, that the (2×2) PLSI of this same array is minimum phase. According to Conjecture 1 all PLSI's of an array should be minimum phase. It would be interesting to plot root maps for PLSI's larger than (3×3), but our current root map program is limited to (3×3) arrays. The examples in this section illustrate, but do not prove Conjecture 1. However, we have yet to find a counterexample, and present indications are that Conjecture 1 is true.

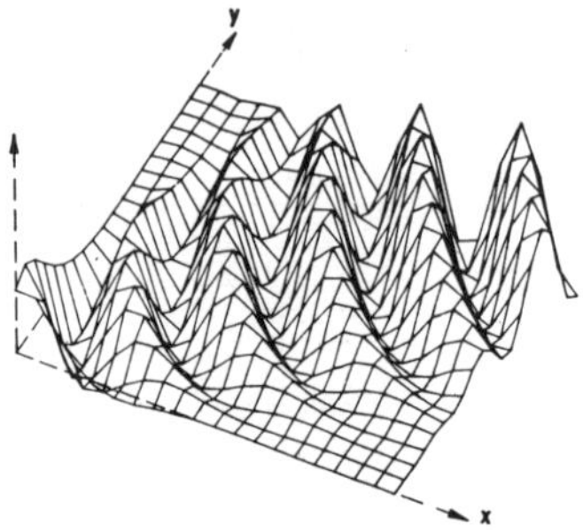

Fig. 7. Pulse response of $F(z_1, z_2) = 1/B(z_1, z_2)$.

VI. The Stabilization of an Unstable Two-Dimensional Recursive Filter

Suppose that we have designed a two-dimensional recursive filter $F(z_1, z_2) = 1/B(z_1, z_2)$, only to find that it is unstable. How can we alter the coefficients in order to produce a stable filter?

Let us denote the denominator array of this unstable filter by $\boldsymbol{B}$. We can compute a planar least squares inverse $\boldsymbol{B}'$ of $\boldsymbol{B}$. Suppose that we compute a PLSI of $\boldsymbol{B}'$, and call it $\hat{\boldsymbol{B}}$. Now $\hat{\boldsymbol{B}}$ is the inverse of the inverse, or the "double inverse" of $\boldsymbol{B}$. One intuitively feels that $\hat{\boldsymbol{B}}$ and $\boldsymbol{B}$ have some characteristics in common. Moreover, $\hat{\boldsymbol{B}}$ is itself a PLSI. Hence it is minimum phase, while $\boldsymbol{B}$ is not. Therefore, the filter $\hat{F}(z_1, z_2) = 1/\hat{B}(z_1, z_2)$ is stable. For example, consider the filter with the denominator array

$$\boldsymbol{B} = \begin{bmatrix} 1.0 & -1.20002759 & 0.40002239 \\ -1.00003018 & 1.70007079 & -0.65005088 \\ 0.40002035 & -0.7005488 & 0.25004387 \end{bmatrix}. \quad (14)$$

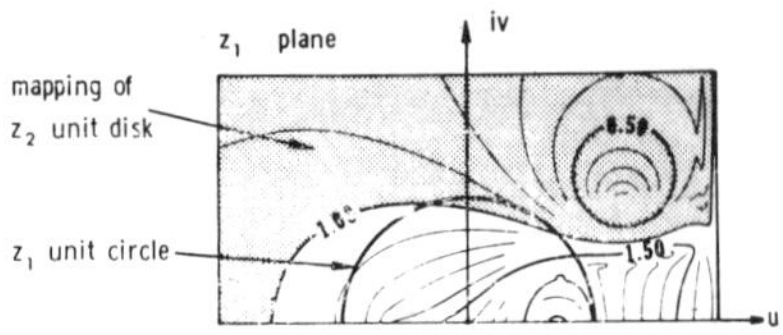

Fig. 8. Root map of B (14).

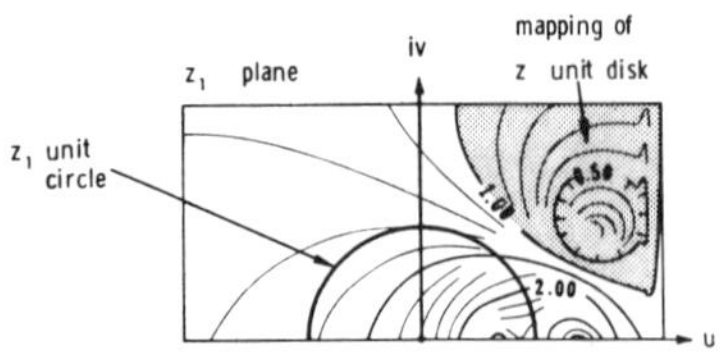

Fig. 9. Root map of $\hat{B}$ (15).

Fig. 7 shows a perspective view of the unit pulse response of $F(z_1, z_2)$. Note that the unit pulse response tends to grow larger and larger with increasing values of the spatial coordinates x and y. We can verify this instability by plotting the root map of $\boldsymbol{B}$, which is shown in Fig. 8. Since a portion of the dark area (mapping of the z_2 unit disk) falls inside the z_1 unit circle, $\boldsymbol{B}$ is not minimum phase.

In order to produce a stable filter, we first computed an (8×8) PLSI, $\boldsymbol{B}'$, of $\boldsymbol{B}$. We then computed a (3×3) PLSI of $\boldsymbol{B}'$, which we shall call $\hat{\boldsymbol{B}}$. The evaluation of $\hat{\boldsymbol{B}}$ yielded

$$\hat{\boldsymbol{B}} = \begin{bmatrix} 1.11567 & -1.23469 & 0.38431 \\ -1.0932 & 1.68765 & -0.66821 \\ 0.37577 & -0.65178 & 0.29767 \end{bmatrix}. \quad (15)$$

The root map of $\hat{\boldsymbol{B}}$ is shown in Fig. 9, which shows that $\hat{\boldsymbol{B}}$ is minimum phase. Hence $\hat{F}(z_1, z_2) = 1/\hat{B}(z_1, z_2)$ should be stable. Fig. 10 shows the unit pulse response of $\hat{F}(z_1, z_2)$, which is clearly seen to decay as x and y increase.

In what sense are $\boldsymbol{B}$ and $\hat{\boldsymbol{B}}$ similar? We have already seen that $\hat{\boldsymbol{B}}$ is minimum phase, while $\boldsymbol{B}$ does not necessarily have this property. On the other hand, $\hat{\boldsymbol{B}}$ is the double PLSI of $\boldsymbol{B}$. Hence we conjecture that $\hat{\boldsymbol{B}}$ must in fact be an approximate

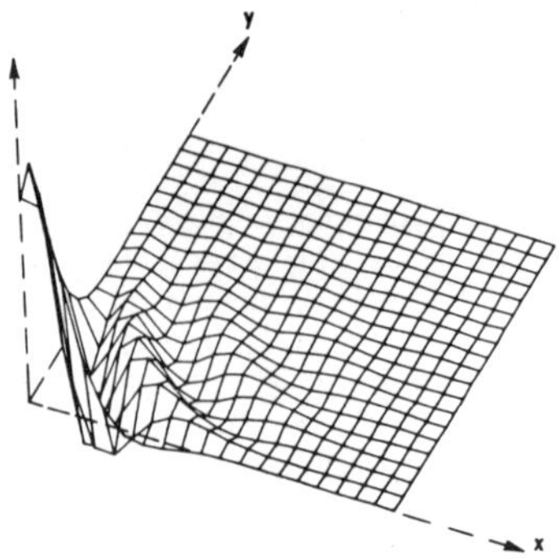

Fig. 10. Pulse response of $\hat{F}(z_1, z_2) = 1/\hat{B}(z_1, z_2)$.

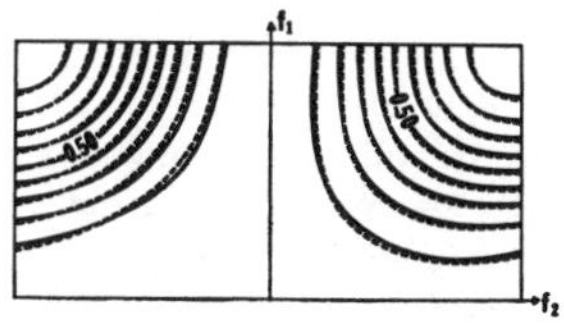

Fig. 11. Amplitude spectra of $\hat{B}$ (——) and B (- - - - - -).

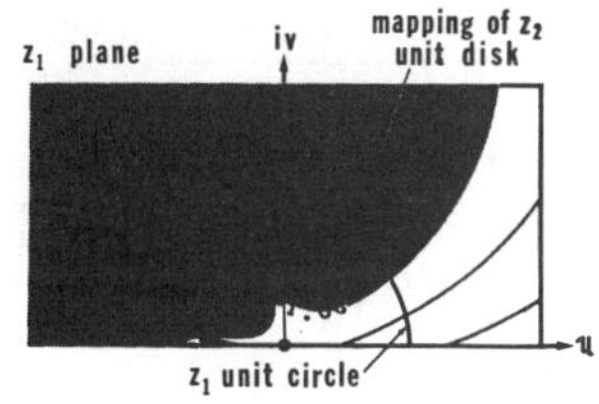

Fig. 13. Root map for array A (Fig. 5).

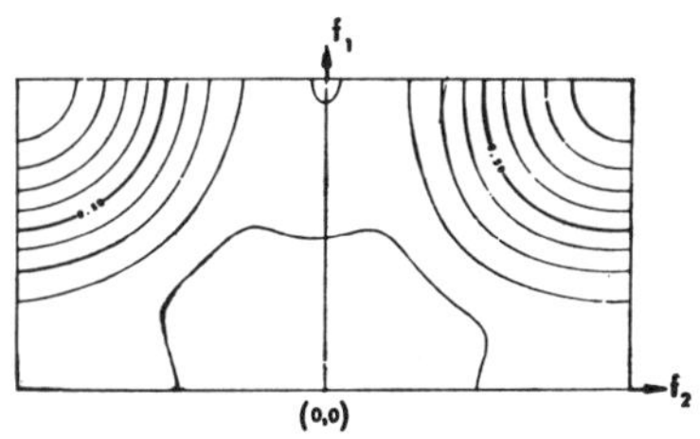

Fig. 12. Amplitude spectrum of $\hat{B}$ from (3×3) B'.

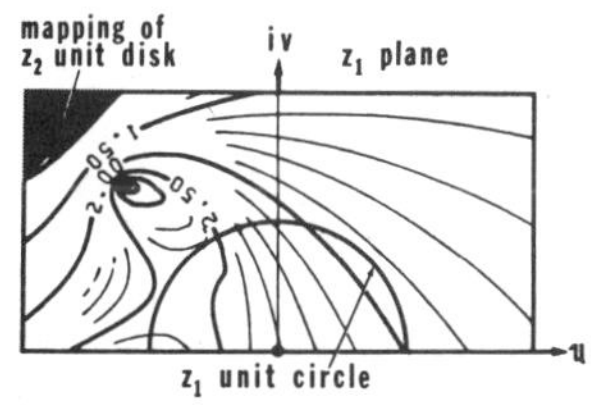

Fig. 14. Root map for array $\hat{A}$ (16) from (3×3) A'.

minimum phase version of B. In other words, we would expect the amplitude spectra of B and $\hat{B}$ to be roughly equal. Fig. 11 shows the two-dimensional amplitude spectra of B and $\hat{B}$ in the form of contour plots.[5] We observe that these spectra are similar, but not identical.

One of the factors that appears to govern the quality of the approximation is the size of the intermediate array B'. The larger the size of B', the better is the resemblance of $\hat{B}$ to the minimum phase version array B. In order to test this assertion, we computed a B' array of size (3×3) as compared to the (8×8) B' array of the previous example. We then obtained the (3×3) PLSI of this smaller B' and evaluated the amplitude spectrum of the resulting new $\hat{B}$. This spectrum is shown in Fig. 12; as expected, the agreement with the spectrum of B depicted in Fig. 11 is now much poorer.

As a second example, let us consider the filter $G(z_1, z_2) = 1/A(z_1, z_2)$, where A is the array shown in Fig. 5. The root map for A is shown in Fig. 13. As can be seen, A is not minimum phase, and the filter $F(z_1, z_2) = 1/A(z_1, z_2)$ will be unstable. We computed A', a (3×3) PLSI of A. The array A' and its root map are shown in Fig. 5. We then computed $\hat{A}$, the (3×3) PLSI of A' to get

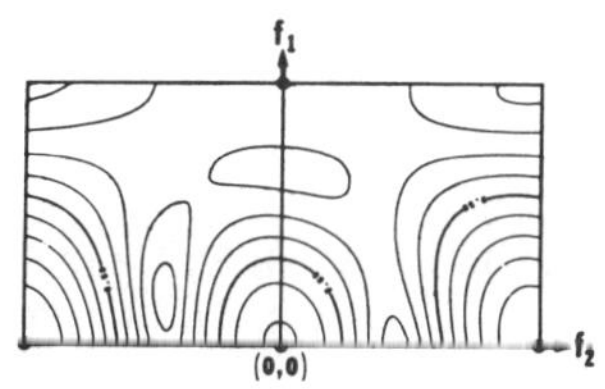

Fig. 15. Amplitude spectrum of array A (Fig. 5).

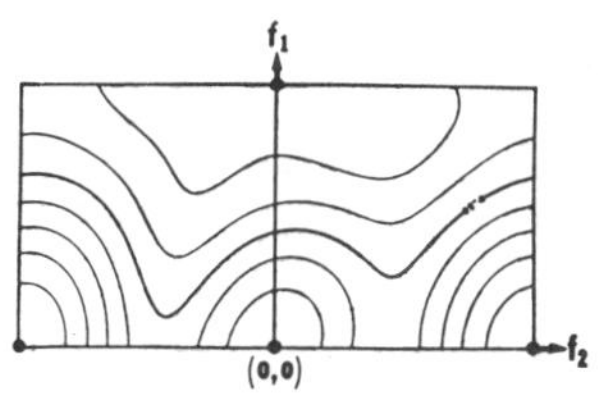

Fig. 16. Amplitude spectrum at array $\hat{A}$ from (3×3) A'.

$$\hat{A} = \begin{bmatrix} 6.761 & -1.248 & 1.258 \\ 3.156 & -0.602 & 1.027 \\ 0.706 & 0.209 & 0.368 \end{bmatrix}. \qquad (16)$$

$\hat{A}$ is minimum phase, as shown by its root map in Fig. 14. We can compare the amplitude spectra of A and $\hat{A}$ in Figs. 15 and 16. The two spectra are generally similar, but there are differences in detail.

To get a better approximation, we computed an (8×8) A' and a (15×15) A'. From these arrays the corresponding

[5] The two-dimensional amplitude spectrum must be radially symmetric about the (0,0)-frequency point. Therefore, we show only two quadrants of the two-dimensional spectrum.

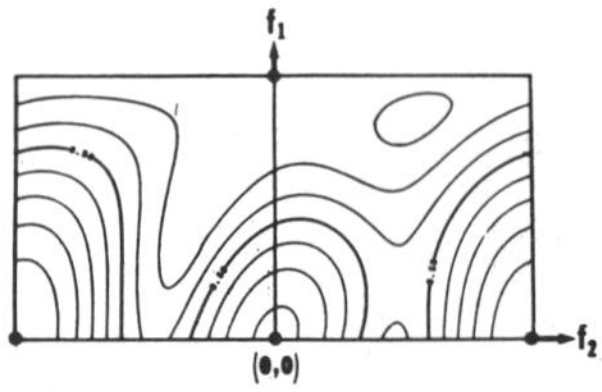

Fig. 17. Amplitude spectrum of array $\hat{A}$ from (8×8) A'.

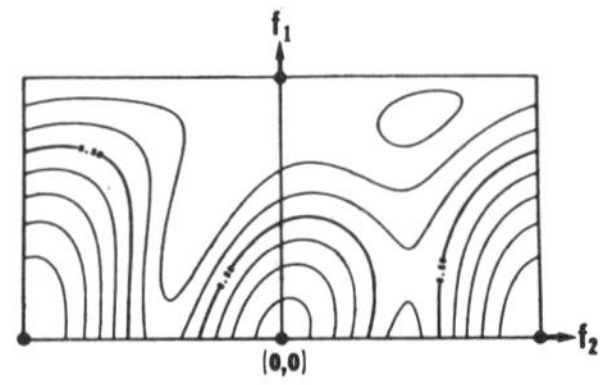

Fig. 18. Amplitude spectrum of array $\hat{A}$ from (15×15) A'.

(3×3) $\hat{A}$ arrays were obtained. Figs. 17 and 18 show the amplitude spectra of the $\hat{A}$ arrays. The amplitude spectrum of $\hat{A}$ obtained via the (8×8) A' array (Fig. 17) is in better agreement with the true amplitude spectrum of A (Fig. 15) than is the case for the (3×3) A' array employed previously (Fig. 16). However, Fig. 18 shows that very little improvement is obtained by increasing the A' array size from (8×8) to (15×15).

VII. Synthesis

We would like to be able to synthesize a two-dimensional recursive filter to perform some desired filtering operation. In this section we shall present two synthesis techniques, one in the space (measurement) domain, and one in the frequency domain.

In the special case that the filter is separable, then the synthesis becomes one-dimensional. In that case, we can express $F(z_1, z_2)$ as

$$F(z_1, z_2) = F_1(z_1)F_2(z_2)$$

and the two-dimensional filter becomes two one-dimensional filters in cascade [19]. If $F(z_1, z_2)$ is a convolutional filter, nonseparable filters can be realized by sums of separable filters [20]. In the remainder of this paper, we shall concern ourselves with the design of nonseparable two-dimensional recursive filters.

Synthesis in the Space Domain

Suppose we know the two-dimensional unit pulse response of the discrete filter which we wish to use on the data. We could use two-dimensional convolution or the two-dimensional Fourier transform to apply this filter. However, it might also be possible to approximate the impulse response with a two-dimensional recursive filter. In some cases, the recursive filter can be more efficient than straight convolution or filtering in the frequency domain.

The following is an extension of a synthesis technique in one dimension [6], [21], [22]. Suppose the desired impulse response is d_{ij}, for $i=1, 2, 3, \cdots, M_d$, and $j=1, 2, 3, \cdots, N_d$. That is,

$$D(z_1, z_2) = \sum_{i=1}^{M_d} \sum_{j=1}^{Nd} d_{ij} z_1^{i-1} z_2^{j-1}. \tag{17}$$

Let the approximating two-dimensional recursive filter be

$$F(z_1, z_2) = \frac{A(z_1, z_2)}{B(z_1, z_2)} \tag{18}$$

where

$$A(z_1, z_2) = \sum_{i=1}^{M_a} \sum_{j=1}^{N_a} a_{ij} z_1^{i-1} z_2^{j-1}$$

$$B(z_1, z_2) = \sum_{i=1}^{M_b} \sum_{j=1}^{N_b} b_{ij} z_1^{i-1} z_2^{j-1}$$

$$b_{11} = 1.0$$

$$M_a, N_a, M_b, N_b = \text{arbitrary (but fixed) parameters.}$$

From (18), $F(z_1, z_2)B(z_1, z_2) = A(z_1, z_2)$. Since multiplication of the z polynomials is equivalent to convolution of the arrays

$$a_{mn} = \sum_{i=1}^{M_b} \sum_{j=1}^{N_b} b_{ij} f_{m-i+1, n-j+1}. \tag{19}$$

The coefficients a_{mn} are defined by (19) over the integers $m=1, 2, \cdots, M_a$, $n=1, 2, \cdots, N_a$. Outside this region, the a_{mn} are zero. Let us define the set of integer pairs

$$I_a = \{(i, j): 1 \le i \le M_a,\ 1 \le j \le N_a\}.$$

Furthermore, let us define $\hat{I}_a$ as the set of all other values of (m, n) *greater than zero*.

$$I_a = \{(m, n): m > 0,\ n > 0,\ (m, n) \notin I_a\}.$$

Thus, for $(m, n) \in I_a$, the values of a_{mn} are yet to be determined. For $(m, n) \in \hat{I}_a$, $a_{mn} \equiv 0$. Therefore, using (19) and the fact that $b_{11} = 1.0$, we get

$$f_{mn} = -\sum_{\substack{i=1 \\ i\cdot j \ne 1}}^{M_b} \sum_{j=1}^{N_b} b_{ij} f_{m-i+1, n-j+1}, \qquad \text{for } (m, n) \in I_a.$$

If we judiciously choose the b_{ij}, the f_{mn} will approximate the desired response d_{mn}. Thus we may write

$$d_{mn} \approx -\sum_{\substack{i=1 \\ i\cdot j \ne 1}}^{M_b} \sum_{j=1}^{N_b} b_{ij} d_{m-i+1, n-j+1}, \quad \text{for } (m, n) \in \{\hat{I}_a \cap I_d\} \tag{20}$$

where (m, n) must be a member of the intersection of the set $\hat{I}_a$ and the set

$$I_d = \{(m, n): 1 \le m \le M_d,\ 1 \le n \le N_d\}.$$

Let us define an error e_{mn} that can be added to the right side of (20) to produce equality.

$$d_{mn} = e_{mn} - \sum_{\substack{i=1 \\ i\cdot j \neq 1}}^{M_b} \sum_{j=1}^{N_b} b_{ij} d_{m-i+1,n-j+1}, \quad \text{for } (m, n) \in \{\hat{I}_a \cap I_d\}.$$

Therefore, we can put d_{mn} inside the summation to get

$$e_{mn} = \sum_{i=1}^{M_b} \sum_{j=1}^{N_b} b_{ij} d_{m-i+1,n-j+1}, \quad \text{for } (m, n) \in \{\hat{I}_a \cap I_d\}.$$

Now that we have defined the error, let us choose the b_{ij} such as to minimize the mean square error,

$$\overline{e^2} = \sum_{(m,n)} \sum \left[\sum_{(i,j)} b_{ij} d_{m-i+1,n-j+1} \right]^2,$$

$$\text{for } (i, j) \in I_b, \qquad (m, n) \in \{\hat{I}_a \cap I_d\}. \quad (21)$$

Except for b_{11}, which is 1.0, the b_{ij} are not yet specified. By differentiating (21) with respect to the b_{ij} and setting the resultant equations equal to zero, we can find those b_{ij} that minimize the mean square error. This will give us $(M_b \cdot N_b - 1)$ equations of the type

$$\sum_{\substack{i=1 \\ i\cdot j \neq 1}}^{M_b} \sum_{j=1}^{N_b} b_{ij} \Phi_{klij} = \phi_{kl}, \qquad \text{for } k = 1, 2, 3, \cdots, M_b, \quad l = 1, 2, 3, \cdots, N_b, \quad \text{but } k \cdot l \neq 1 \quad (22)$$

where

$$\Phi_{klij} = \sum_{(m,n)} d_{m-i+1,n-j+1} d_{m-k+1,n-l+1} \quad (23)$$

$$\phi_{kl} = - \sum_{(m,n)} d_{mn} d_{m-k+1,n-j+1}, \qquad (m, n) \in \{\hat{I}_a \cap I_d\}. \quad (24)$$

Thus, (22) describes a set of $(M_b \cdot N_b - 1)$ simultaneous linear equations with $(M_b \cdot N_b - 1)$ unknowns. The solutions to (22) gives us the b_{ij} that minimizes the mean square error.

Having computed the denominator $B(z_1, z_2)$ of the filter, we now compute the numerator $A(z_1, z_2)$. One way to compute $A(z_1, z_2)$ is to compute those coefficients that minimize the mean square difference between the coefficients of $F(z_1, z_2) = A(z_1, z_2)/B(z_1, z_2)$ and the coefficients of the desired response $D(z_1, z_2)$. This is a discrete Wiener filtering problem in two dimensions [18]. It consists of finding the optimum filter $A(z_1, z_2)$ given an input $1/B(z_1, z_2)$ and a desired output $D(z_1, z_2)$.

In a simpler but less accurate method we can convolve the B array with the D array to get the A array. Since $A(z_1, z_2)/B(z_1, z_2) \approx D(z_1, z_2)$, we compute the coefficients a_{mn} from $A(z_1, z_2) = B(z_1, z_2) \cdot D(z_1, z_2)$ for $(m, n) \in I_a$. If $B(z_1, z_2)$ has been chosen well, the coefficients of $B(z_1, z_2) \cdot D(z_1, z_2)$ will be comparatively small for $(m, n) \in \hat{I}_a$.

A computer program was written to obtain a two-dimensional recursive filter using the above technique. In order to test the program, we computed the impulse response of a two-dimensional recursive filter, $D(z_1, z_2) = A'(z_1, z_2)/B'(z_1, z_2)$. Both A' and B' were 3×3 arrays. The coefficients of the A' and B' arrays are shown in Table IV, under columns A' and B'. The first 20×20 values of the impulse response were used as inputs to the computer program that finds the two-dimensional recursive filter that approximates that response. M_a, M_b, N_a, and N_b were set at three. On a CDC 3300 computer, the program yielded the coefficients given by (23) and (24), and solved (22) to find the denominator coefficients. The numerator coefficients were found by two-dimensional convolution of $B(z_1, z_2)$ with $D(z_1, z_2)$. The recursive coefficients computed by the program are shown in Table IV under Columns A and B.

TABLE IV

Test of Space Domain Synthesis

Term	A'	A	B'	B
Constant	1.0	1.0	1.0	1.0
z_1	3.0	3.0000005	−1.2	−1.1999995
z_1^2	2.0	2.0000016	0.5	0.4999996
z_2	2.0	2.0000002	−1.5	−1.4999998
$z_1 z_2$	4.0	4.0000014	1.8	1.7999989
$z_1^2 z_2$	−1.0	−0.9999976	−0.75	−0.7499992
z_2^2	−1.0	−0.9999995	0.6	0.5999998
$z_1 z_2^2$	2.0	2.0000002	−0.72	−0.7199994
$z_1^2 z_2^2$	1.0	0.9999998	0.29	0.2899996

Note: A', B' are the actual coefficients. A, B are the estimated coefficients.

Thus we see that this technique is fairly good if the desired response can be approximated by a recursive type filter.

Synthesis in the Frequency Domain

Many frequency domain recursive filter design techniques are by now available for the one-dimensional case [5], [23]. Much of the work in one-dimensional frequency domain synthesis has been in the design of the low-pass and bandpass type filter. We shall discuss a technique for designing such filters in two dimensions.

In one dimension, the design parameters are often the cutoff frequencies, or the points of transition between the "passband" and the "stopband." In two dimensions the transition is a boundary separating the pass and reject regions. The shape of the boundary depends on the designer's filter requirements and the number of coefficients he is willing to have in the filter.

Suppose we design a low-pass or bandpass filter in one dimension. Such a filter can also be viewed as a two-dimensional filter that varies in one dimension only. In terms of the two-dimensional Laplace variables [27] (s_1, s_2) such a filter would be written

$$F(s_1, s_2) = F_2(s_2).$$

The frequency response of this filter can be found by setting $s_1 = i2\pi f_1$ and $s_2 = i2\pi f_2$. Thus the response varies only in the f_2 frequency direction. However, suppose we were to perform a transformation, or mapping, of the (s_1, s_2) plane into an (s_1', s_2') such that the passband of $F(s_1, s_2)$ maps into a suitable region in the (s_1', s_2') plane. In particular, let us consider rotating the axes of the (s_1, s_2) plane.

We can rotate the (s_1, s_2) axes by an angle β by means of the transformation

$$s_1 = s_1' \cos\beta + s_2' \sin\beta \tag{25}$$

$$s_2 = s_2' \cos\beta - s_1' \sin\beta. \tag{26}$$

Rotating the (s_1, s_2) axes also causes the two-dimensional frequency axes (f_1, f_2) to be rotated. Consider a filter with the transfer function $F(s_1, s_2)$. Using (25) and (26), we can rotate the (f_1, f_2) axes by an angle β to a new position (f_1', f_2'). Conversely, this has the effect of rotating the "pass" region by an angle $(-\beta)$ with respect to the new frequency axes.

For example, consider the filter

$$F(s_1, s_2) = \frac{1}{s_2 + a}\cdot \tag{27}$$

The amplitude response of this varies in the f_2 direction only. Applying the transformation equations (25) and (26) we get

$$F'(s_1', s_2') = \frac{1}{s_2' \cos\beta - s_1' \sin\beta + a}\cdot \tag{28}$$

This filter varies in both frequency directions.

We can design two-dimensional bandpass filters whose response pattern has a particular angle with respect to the (f_1, f_2) frequency axes. First we would design a suitable one-dimensional low-pass or bandpass filter in the one-dimensional s plane. Expressing this filter as a function of either s_1 or s_2, we would use either (25) or (26) to rotate this filter some desired angle into the (s_1', s_2') plane. This resultant form $F'(s_1', s_2')$ describes a continuous two-dimensional filter.

It is still necessary to produce the equivalent two-dimensional discrete filter. For this purpose we can use the bilinear z transform [23]. We have

$$s_1' = \frac{2}{\Delta_1}\frac{1 - z_1}{1 + z_1} \tag{29}$$

$$s_2' = \frac{2}{\Delta_2}\frac{1 - z_2}{1 + z_2} \tag{30}$$

where Δ_1 and Δ_2 are the sample intervals. Applying the bilinear equations to the $F'(s_1', s_2')$ given in (28), we get

$$F(z_1, z_2) = \frac{\Delta_1\Delta_2}{2}\,\frac{(1 + z_1)(1 + z_2)}{1 + b_{21}z_1 + b_{12}z_2 + b_{22}z_1z_2}, \tag{31}$$

where

$$b_{21} = \left(\frac{a\Delta_1\Delta_2}{2} + \Delta_1\cos\beta + \Delta_2\sin\beta\right)\Big/ D \tag{32}$$

$$b_{12} = \left(\frac{a\Delta_1\Delta_2}{2} - \Delta_1\cos\beta - \Delta_2\sin\beta\right)\Big/ D \tag{33}$$

$$b_{22} = \left(\frac{a\Delta_1\Delta_2}{2} - \Delta_1\cos\beta + \Delta_2\sin\beta\right)\Big/ D \tag{34}$$

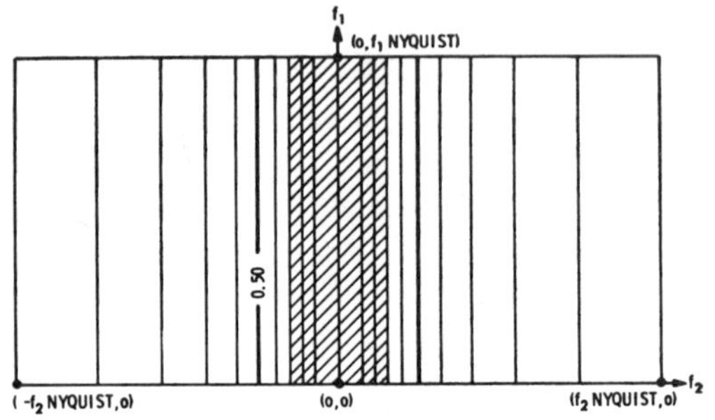

Fig. 19. Two-dimensional amplitude response of $(1+z_2)/(1-0.6186\,z_2)$.

and

$$D = \frac{a\Delta_1\Delta_2}{2} + \Delta_1\cos\beta - \Delta_2\sin\beta.$$

These equations can be simplified by letting $\Delta_1 = \Delta_2 = \Delta$. In equation (27) the constant a represents the cutoff frequency (3 dB down) in rad/unit[6] [24]. Let us set

$$a = 2\pi f_c f_N$$

where f_c is cutoff frequency as a fraction of the Nyquist frequency, and f_N is the Nyquist frequency in cycles per unit. Since $f_N = 1/2\Delta$,

$$a = \frac{\pi f_c}{\Delta}\cdot$$

Thus we get

$$b_{21} = \{\pi f_c + 2(\cos\beta + \sin\beta)\}/D$$
$$b_{12} = \{\pi f_c - 2(\cos\beta + \sin\beta)\}/D$$
$$b_{22} = \{\pi f_c - 2(\cos\beta - \sin\beta)\}/D$$
$$D = \{\pi f_c + 2(\cos\beta - \sin\beta)\}.$$

For example, let us consider a filter $F(s) = 1/(s+a)$, where a is chosen to give a cutoff frequency at 0.15 Nyquist. Thus, $f_c = 0.15$ and for $\beta = 0°$, we get

$$F(z_1, z_2) = \frac{\Delta^2}{2}\,\frac{(1 + z_1)(1 + z_2)}{\begin{bmatrix} 1 - 0.6186z_2 \\ +z_1 - 0.6186z_1z_2 \end{bmatrix}} = \frac{\Delta^2}{2}\,\frac{1 + z_2}{1 - 0.6186z_2}\cdot$$

Fig. 19 shows a contour map of the amplitude response of this filter, normalized to a peak response of 1.0. The contour interval is 0.1. The pass region, for which the amplitude response is greater than 0.7, is shaded. Since $\beta = 0$, this response does not vary along the f_1 frequency axis. Along

[6] The unit depends on the problem. In the case of seismic records, the unit in one direction would be feet, and in the other direction, time. In the case of a photograph, both units would be spatial.

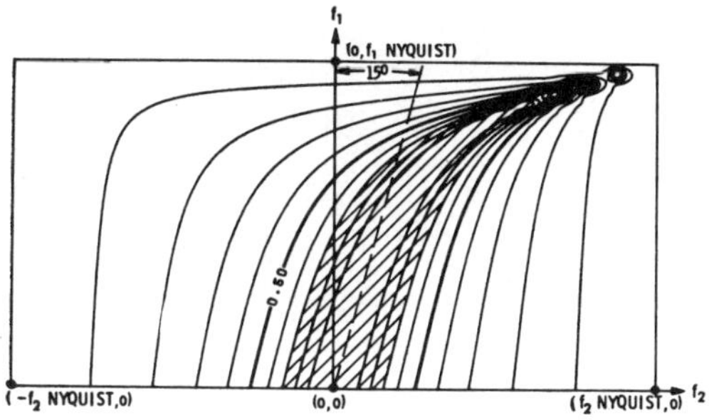

Fig. 20. Amplitude response of filter rotated 15°.

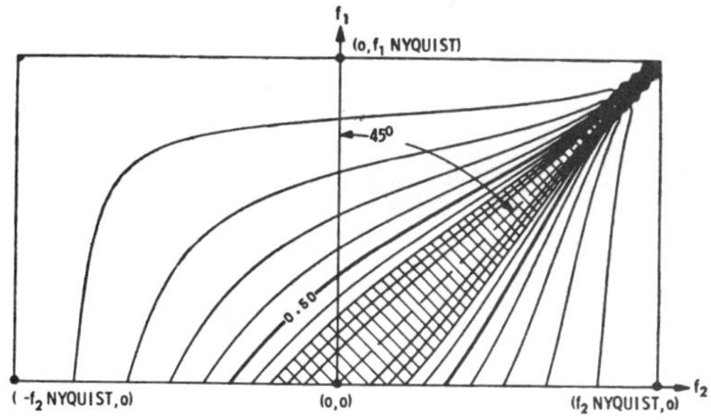

Fig. 21. Amplitude response of filter rotated 45°.

the f_2 frequency axis, the filter is a low-pass filter. Note that the cutoff frequency is 15 percent of the Nyquist frequency, which agrees with the design parameter $f_c=0.15$.

For $\beta=15°$, we get the filter

$$F(z_1, z_2) = \frac{\Delta^2}{2} \frac{(1+z_1)(1+z_2)}{\begin{bmatrix} 1 - 0.3229z_2 \\ +0.6455z_1 - 0.6773z_1z_2 \end{bmatrix}}.$$

Fig. 20 shows a contour map of the amplitude response of this filter. The contour interval is again 0.1. As required, the amplitude response has been rotated by 15° from the f_1 axis. However, the contour lines are no longer straight, and tend to be distorted at the higher (f_1, f_2) frequencies. This effect is caused by the bilinear (z_1, z_2) transform. When the bilinear equations (29) and (30) are applied to (28), the numerator of $F(z_1, z_2)$ will contain the factors $(1+z_1)(1+z_2)$. The factor $(1+z_1)$ causes $F(z_1, z_2)$ to be zero for all frequencies for which f_1 is equal to the Nyquist. Similarly, the factor $(1+z_2)$ causes the $F(z_1, z_2)$ to go to zero at the f_2 Nyquist frequencies.

Fig. 21 shows the amplitude response of a filter rotated to 45°. The filter equation is

$$F(z_1, z_2) = \frac{(1+z_1)(1+z_2)}{\begin{bmatrix} 1 + 0.1428z_2 \\ +0.1428z_1 - 0.7144z_1z_2 \end{bmatrix}}.$$

We can also "rotate" the bandpass type of filter. Fig. 22 shows the amplitude response of a bandpass filter. This is a bilinear transform of an eight-order Butterworth filter [6, pp. 37–41]. Fig. 23 shows the response of the same filter rotated 45°, again using the bilinear transform.

In order to obtain a rotation through a negative angle, we can filter the data in a different manner. Suppose $D(z_1, z_2)$ is an $M \times N$ input and $F(z_1, z_2)$ is the filter which produces the response shown in Fig. 23. Instead of applying $F(z_1, z_2)$ in the normal way, as described by equation (5), let us start filtering at the input point $d_{1,N}$, and filter backwards in the y direction. This is equivalent to applying the filter

$$G(z_1, z_2) = F(z_1, 1/z_2).$$

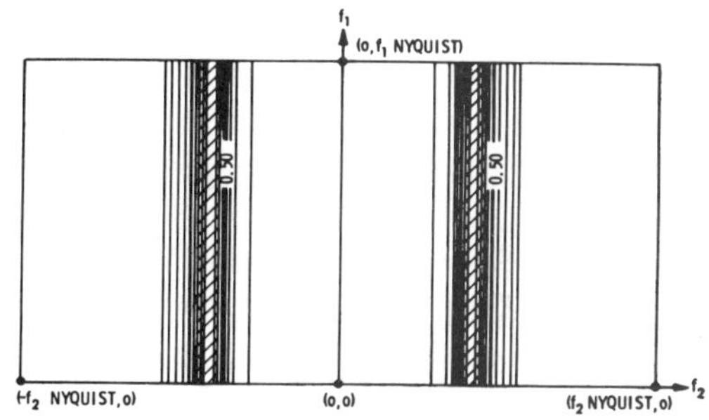

Fig. 22. Amplitude response of bandpass filter.

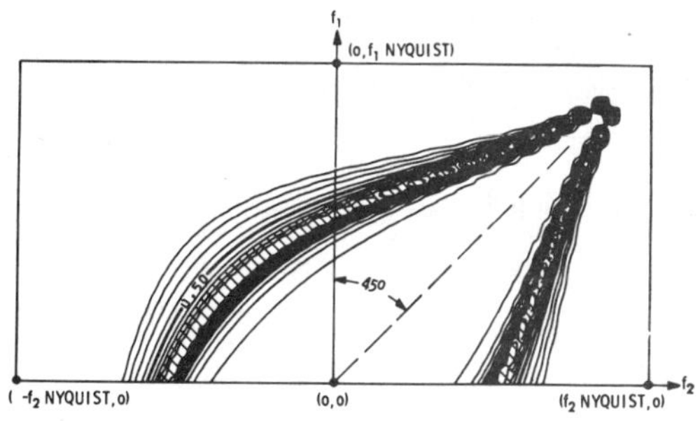

Fig. 23. Amplitude response of bandpass filter rotated 45°.

This filter has the amplitude response curve shown in Fig. 24, with the bandpass characteristic rotated by −45°.

We can also combine these rotated filters to obtain other filter shapes. Consider the filter with an amplitude response shown in Fig. 25. This is a low-pass filter rotated 45°. Suppose we filter the data with this filter in cascade with the filter described by Fig. 24. The overall amplitude response will be

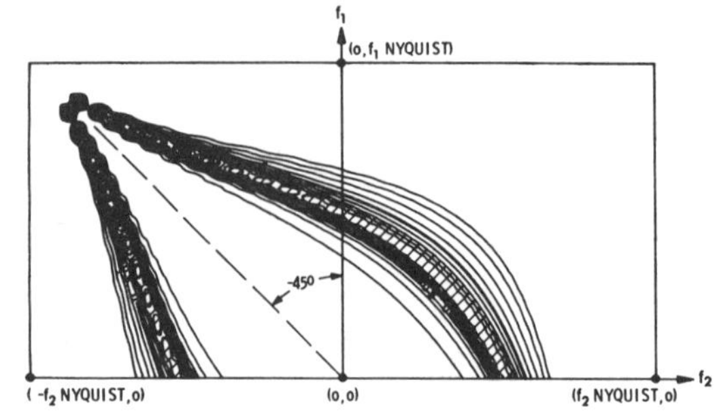

Fig. 24. Amplitude response of bandpass filter rotated −45°.

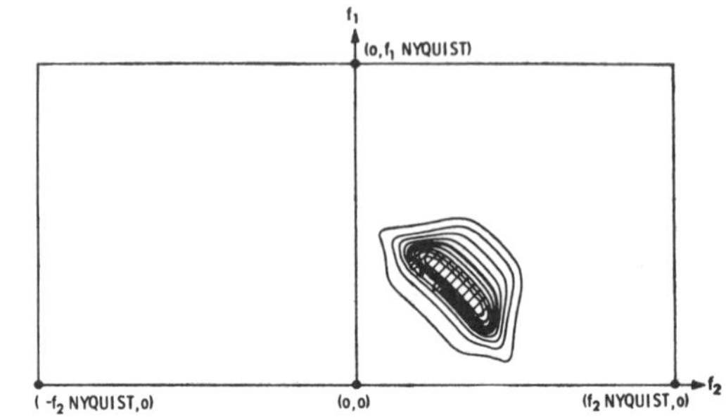

Fig. 26. Amplitude response of cascade response of filters from Figs. 24 and 25.

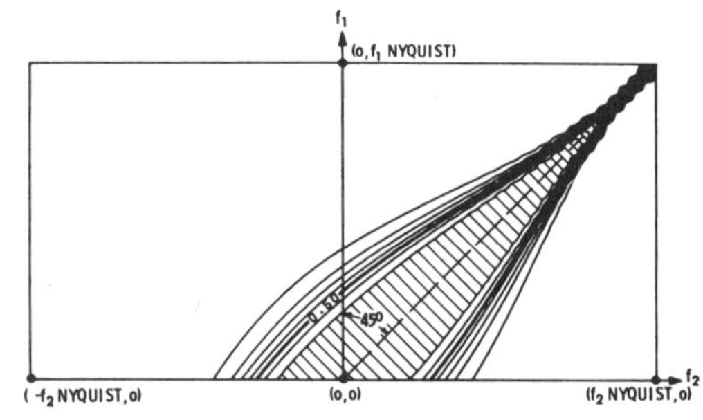

Fig. 25. Amplitude response of low-pass filter rotated 45°.

the product of the two amplitude responses. The amplitude response of the cascaded filters is shown in Fig. 26.

Phase Response of Two-Dimensional Recursive Filters

The recursive filtering algorithm of (5) produces a filter with a "one-quadrant" impulse response. If the input to the filter is a unit pulse at some grid point, the filter responds in the positive m and n directions only. Due to its one-quadrant response, the recursive filter cannot have a zero-phase response. However, by the proper combination of two or more recursive filters, we can produce a filtering operation which has zero phase or other useful symmetries.

A zero-phase two-dimensional filter has the particular property that its impulse response is symmetric about the (x, y) origin along any radius through the origin. We can perform zero-phase filtering with recursive filters by first filtering an input array with a recursive filter, starting at one corner of the input array. We then take the resultant array and, starting at the diagonally opposite corner, filter the result in the opposite direction with the same recursive filter.[7] This is equivalent to filtering with the filter

$$G(z_1, z_2) = F(z_1, z_2)\cdot F(1/z_1, 1/z_2) \qquad (35)$$

where $F(z_1, z_2)$ is the one-quadrant recursive filter.

Setting $z_1 = e^{i\omega_1}$ and $z_2 = e^{i\omega_2}$ in (35) gives us the two-dimensional frequency response of $G(z_1, z_2)$. Since $F(e^{i\omega_1}, e^{i\omega_2})$ is the complex conjugate of $F(e^{-i\omega_1}, e^{-i\omega_2})$, $G(e^{i\omega_1}, e^{i\omega_2})$ will be real for any value of (ω_1, ω_2). Therefore, the filter $G(z_1, z_2)$ has zero-phase response. The amplitude response of $G(z_1, z_2)$ will be the square of the amplitude response of $F(z_1, z_2)$. That is, $|G(\omega_1, \omega_2)| = |F(\omega_1, \omega_2)|^2$.

We can also get zero-phase response by filtering an input array forward and backward with the same recursive filter, and adding the results. This is equivalent to filtering with

$$H(z_1, z_2) = F(z_1, z_2) + F(1/z_1, 1/z_2).$$

Like $G(z_1, z_2)$, $H(z_1, z_2)$ has zero-phase response. However, the amplitude responses are different. The amplitude response of $H(z_1, z_2)$ is

$$|H(\omega_1, \omega_2)| = 2 \,\mathrm{Re}\,\{F(\omega_1, \omega_2)\}.$$

A filtering operation can be made symmetric with respect to the x axis (or y axis) by filtering forward and reverse in the x (or y) direction. For example, if we relate z_1 with the x axis and z_2 with the y axis, $G(z_1, z_2) = F(z_1, z_2)\cdot F(1/z_1, z_2)$ and $H(z_1, z_2) = F(z_1, z_2) + F(1/z_1, z_2)$ are symmetric about the x axis.

Finally, the filtering operations

$$G(z_1, z_2) = F(z_1, z_2)\cdot F(1/z_1, z_2)\cdot F(z_1, 1/z_2)\cdot F(1/z_1, 1/z_2)$$

and

$$H(z_1, z_2) = F(z_1, z_2) + F(1/z_1, z_2) + F(z_1, 1/z_2) + F(1/z_1, 1/z_2)$$

are symmetric with respect to both the x and y axes, as well as having zero-phase response.

VIII. Summary

We can perform two-dimensional filtering with the two-dimensional recursive filter algorithm. The primary prob-

[7] Use of this technique requires that the computer output array be sufficiently large to allow the output values to decay to reasonably small numbers.

lems are stability and synthesis. We have given the general theorem for stability and shown that the stability can be computed for a two by two array. For larger denominator arrays, the stability question is more difficult. We can estimate the stability by using the mapping technique as illustrated by two examples.

We have demonstrated that the denominator array of a two-dimensional recursive filter must be minimum phase if this recursive filter is to be stable. A given nonminimum phase denominator array can be stabilized by computing two successive planar least squares inverse filters. This method is based on a conjecture which states that the planar least squares inverse of any matrix is minimum phase. This conjecture appears to be a reasonable extrapolation from a theorem in the one-dimensional case. We have also presented a number of examples, all of which uphold the validity of our conjecture.

We have shown two techniques for designing or synthesizing two-dimensional recursive filters. There are several problems where the space domain synthesis might be useful. More work needs to be done on frequency domain synthesis to give the designer more control of the pass-reject boundaries. Some mappings other than the simple rotation might be useful.

We need to study further the theory and characteristics of two-dimensional recursive filters, and to find applications that will make efficient use of these techniques.

Acknowledgment

The authors wish to express their appreciation to Amoco Production Company for allowing them to publish this material.

References

[1] J. P. Fail and G. Grau, "Les filtres en éventail," *Geophys. Prospect.*, vol. 11, pp. 131–163, June 1963.
[2] S. Treitel, J. L. Shanks, and C. W. Frasier, "Some aspects of fan filtering," *Geophysics*, vol. 32, pp. 789–800, Oct. 1967.
[3] E. G. Zurflueh, "Applications of two-dimensional linear wavelength filtering," *Geophysics*, vol. 32, pp. 1015–1035, Dec. 1967.
[4] R. H. Selzer, "Improving biomedical image quality with computers," NASA Tech. Rep. 32-1336, 1968.
[5] C. M. Rader and B. Gold, "Digital filter design techniques in the frequency domain," *Proc. IEEE*, vol. 55, pp. 149–171, Feb. 1967.
[6] J. L. Shanks, "Recursion filters for digital processing," *Geophysics*, vol. 32, pp. 33–51, Feb. 1967.
[7] ——, "Two-dimensional recursive filters," in *SWIEEECO Rec.*, Apr. 1969, pp. 19E1–19E8.
[8] ——, "The design of stable two-dimensional recursive filters," in *Proc. Univ. Missouri—Mervin J. Kelly Commun. Conf.*, Oct. 1970, pp. 15-2-1–15-2-5.
[9] ——, "Two planar digital filtering algorithms," presented at the 5th Princeton Conf. Inform. Sci. Syst., Mar. 1971.
[10] J. T. Tou, *Digital and Sampled-Data Control Systems.* New York: McGraw-Hill, 1959, pp. 230–249.
[11] S. Treitel and E. A. Robinson, "The stability of digital filters," *IEEE Trans. Geosci. Electron.*, vol. GE-2, pp. 6–18, Nov. 1964.
[12] L. Weinberg, "Approximation and stability—Test procedures suitable for digital filters," presented at the Arden House Workshop on Digital Filtering, Harriman, N. Y., Jan. 1970.
[13] H. G. Ansell, "On certain two-variable generalizations of circuit theory, with applications to networks of transmission lines and lumped reactances," *IEEE Trans. Circuit Theory*, vol. CT-11, pp. 214–223, June 1964.
[14] T. S. Huang, private communication, Mar. 1971.
[15] H. Kober, *Dictionary of Conformal Representations.* New York: Dover, 1952, pp. 3–8.
[16] R. V. Churchill, *Introduction to Complex Variables and Applications.* New York: McGraw-Hill, 1948, pp. 57–63.
[17] S. Treitel and E. A. Robinson, "The stability of digital filters," *IEEE Trans. Geosci. Electron.*, vol. GE-2, pp. 15–16, Nov. 1964.
[18] R. A. Wiggins, "On factoring the correlations of discrete multivariable stochastic processes," Mass. Inst. Tech., Cambridge, Sci. Rep. 9 of Contract AF 19 (604) 7378, pp. 127–152, 1965.
[19] W. D. Fryer and G. E. Richmond, "Two-dimensional spatial filtering and computers," in *Proc. Nat. Electron. Conf.*, vol. 18, 1962, pp. 529–534.
[20] S. Treitel and J. L. Shanks, "The design of multistage separable planar filters," *IEEE Trans. Geosci. Electron.*, vol. GE-9, pp. 10–27, Jan. 1971.
[21] C. S. Burris and T. W. Parks, "Time domain design of recursive digital filters," *IEEE Trans. Audio Electroacoust.*, vol. AU-18, pp. 137–141, June 1970.
[22] T. W. Parks and C. S. Burris, "Applications and extensions of Prony's method to parameter identification and digital filtering," presented at the 5th Princeton Conf. Inform. Sci. Syst., Mar. 1971.
[23] R. M. Golden and J. F. Kaiser, "Design of wideband sampled-data filters," *Bell Syst. Tech. J.*, pp. 1533–1546, July 1964.
[24] R. G. Brown and J. W. Nilsson, *Introduction to Linear Systems Analysis.* New York: Wiley, 1962, pp. 121–125.
[25] E. A. Robinson, *Statistical Communication and Detection.* New York: Hafner, 1967, pp. 173–174.
[26] J. H. Justice and J. L. Shanks, "Stability criterion for *N*-dimensional digital filters," unpublished.
[27] V. A. Ditkin and A. P. Prudnikov, *Operational Calculus in Two Variables* (English translation by D. M. G. Wishart). New York: Pergamon, 1962.
[28] W. T. Cochran *et al.*, "What is the fast Fourier transform?" *Proc. IEEE*, vol. 55, pp. 1667–1674, Oct. 1967.

2

Reprinted from *IEEE Proc.* 63(4):610–623 (1975)

Two-Dimensional Digital Filtering

RUSSELL M. MERSEREAU, MEMBER, IEEE, AND DAN E. DUDGEON, MEMBER, IEEE

Invited Paper

***Abstract*—The problems of designing and implementing LSI systems for the processing of 2-*D* digital data, such as images or geophone arrays, are reviewed and discussed. This discussion encompasses both FIR and IIR digital filters and with respect to the latter, the issues of stability testing and filter stabilization are also considered. Techniques are also presented whereby such filtering can be accomplished using either 1 or 2-*D* LSI systems.**

I. Introduction

OUR OBJECTIVE in this paper is to review the mathematical framework underlying the two-dimensional (2-*D*) digital filtering problem and to explore available techniques for the design and implementation of 2-*D* linear shift-invariant (LSI) digital filters. In the course of this discussion, the many similarities to and differences from the one-dimensional (1-*D*) digital filtering problem will hopefully become apparent. Restricting our attention to studying LSI systems allows us to use the powerful techniques of Fourier analysis which have proved their value in numerous problems of practical interest in both 1 and 2 dimensions.

Other papers in this special issue [1]–[3] have touched upon the need for 2-*D* digital filters within the larger context of 2-*D* digital signal processing. Such filters are central to many image and array processing applications—such as X-ray enhancement, image deblurring, scene analysis, weather predictions, seismic analysis, and the processing of radar and sonar arrays, to name just a few. The discussions in this paper, while applicable to any of these problems, will not be specifically directed toward any one of them.

1-*D* LSI systems represent a special case of two and higher dimensional systems. Therefore, many 2-*D* concepts will look vaguely familiar. For this reason, those 2-*D* results which are straightforward extensions of 1-*D* results will be presented without extensive discussions. On the other hand, there are many properties of 1-*D* LSI systems which cannot be easily generalized, which is why 2-*D* digital signal processing remains a challenging and interesting field of study. These difficulties are almost always related to the fact that there is no fundamental theorem of algebra for polynomials in two independent variables [4]. The reader will recall that it is this theorem which allows us to factor a 1-*D* polynomial of degree n into a product of n polynomial factors of first degree, thereby allowing us to find the roots of polynomials, to check the stability of a filter by finding the locations of the poles of its system function, and to realize digital filters as cascade or parallel structures [5].

Because there are many operations which are more easily performed using 1-*D* mathematics, it is perhaps worthwhile to use 1-*D* filters to perform 2-*D* tasks. Interestingly this can be done, and some techniques will be presented later in this paper. While the ultimate value of such an approach remains to be established, it is the authors' belief that this approach will prove to be quite useful both conceptually as well as in practice. Inasmuch as this material is fairly new, it has received a rather heavy emphasis in this paper.

The remainder of this paper is divided into four parts. In Section II, a number of definitions are presented, primarily to establish notation and to serve as a reference for later sections. The representation of 2-*D* arrays by 1-*D* sequences is also discussed here. In Section III, we discuss the related issues of stability testing and the stabilization of unstable

Manuscript received August 14, 1974; revised October 7, 1974.

This work was supported in part by the Advanced Research Projects Agency monitored by ONR under Contract N-00014-67-A-0204-0064 and in part by the National Science Foundation under Grant GK-31353.

R. M. Mersereau is with the Department of Electrical Engineering and Research Laboratory of Electronics, Massachusetts Institute of Technology, Cambridge, Mass. 02139.

D. E. Dudgeon was with the Department of Electrical Engineering and Research Laboratory of Electronics, Massachusetts Institute of Technology, Cambridge. He is now with Bolt, Beranek, and Newman, Inc., Cambridge, Mass. 02138.

filters while in Section IV, we discuss the implementation of 2-*D* digital systems, FIR and IIR, using both 1-*D* and 2-*D* digital filters. In Section V, we shall review design techniques for 2-*D* digital filters.

II. Definitions and Generalizations from One Dimension

In this section, we will examine the underlying mathematical basis for 2-*D* LSI digital filtering. For the most part, these results are straightforward extensions of familiar 1-*D* definitions, but there are some important differences which will be discussed as they arise.

The signals with which we shall concern ourselves are discrete, that is they are functions of integer variables. 1-*D* signals are functions of a single integer variable and 2-*D* signals are functions of two integer variables. Consider the signal $x(m, n)$, shown in Fig. 1. One interpretation of x is as a function which assigns a (generally complex) number to each integer ordered pair (m, n). The function x is not defined when either or both of its arguments (m, n) is not an integer. Such a signal will be interchangeably referred to as either a 2-*D* array, a 2-*D* sequence or simply as a 2-*D* signal. In considering the sequence $x(m, n)$ as an array, referring to Fig. 1, it will be noticed that m thus corresponds to a column index and n corresponds to a row index. This is the opposite of the convention generally used in linear algebra for describing 2-*D* arrays.

An important issue which we will not consider in this paper is the problem of extracting a discrete 2-*D* array from a continuous 2-*D* signal. Clearly, in any practical digital signal processing application this must be done. For completeness, let us, therefore, state that 2-*D* sampling theorems do exist which are conceptually similar to the familiar Nyquist sampling ideas of 1-*D* bandlimited signals. The interested reader is referred to Goodman [6], to the classic paper by Peterson and Middleton [7], or to the accompanying paper by Hunt [2]. In addition, there are discrete representations of continuous signals which are not bandlimited [8]. These ideas can be readily extended to two dimensions and may be helpful for particular 2-*D* digital signal processing problems.

A system which is characterized by the operator $T[\cdot]$ is said to be *linear* if, for arbitrary inputs $x_1(m, n)$ and $x_2(m, n)$ and for an arbitrary complex constant c, the following relation is true:

$$T[x_1(m, n) + cx_2(m, n)] = T[x_1(m, n)] + cT[x_2(m, n)]. \quad (1)$$

If $y(m, n) = T[x(m, n)]$, then the system characterized by $T[\cdot]$ is said to be *shift-invariant* if and only if

$$y(m - m_0, n - n_0) = T[x(m - m_0, n - n_0)] \quad (2)$$

for all x and for all arbitrary integers m_0 and n_0. Linearity and shift-invariance are independent properties of a system in that neither property implies the other. Thus the system

$$T[x(m, n)] = h(m, n)\, x(m, n)$$

is linear but not shift invariant and the system

$$T[x(m, n)] = x^2(m, n)$$

is shift-invariant but not linear.

Any LSI system is easily shown to be completely specified by its impulse response $h(m, n)$. The impulse response (called in image processing the point spread function) is the output

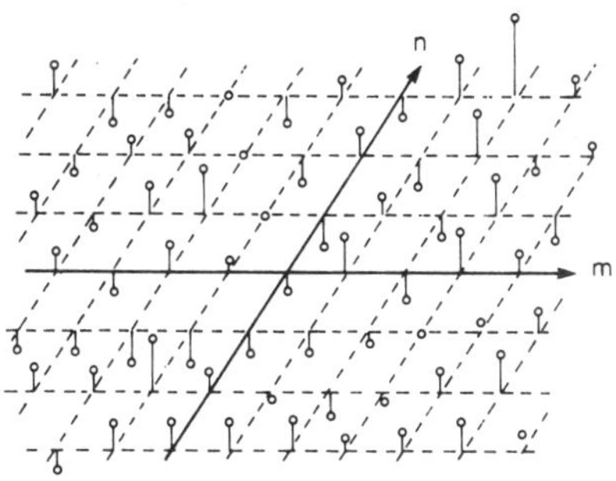

Fig. 1. A pictorial representation for a 2-*D* discrete array.

of the system when the input is a 2-*D* unit sample $\delta(m, n)$ where

$$\delta(m, n) = \begin{cases} 1, & m = n = 0 \\ 0, & \text{otherwise.} \end{cases} \quad (3)$$

Thus

$$h(m, n) = T[\delta(m, n)].$$

As with 1-*D* LSI systems, the output $y(m, n)$ of a 2-*D* LSI system is the *convolution* of the input sequence $x(m, n)$ with the impulse response $h(m, n)$. That is,

$$y(m, n) = \sum_{k=-\infty}^{\infty} \sum_{l=-\infty}^{\infty} x(k, l)h(m - k, n - l)$$

$$= x(m, n) ** h(m, n) \quad (4)$$

where the asterisks are a shorthand representation of the convolution sum of (4).

It is straightforward to verify that complex exponentials of the form $x(m, n) = w^m z^n$ (w and z are arbitrary complex numbers) are eigenfunctions of LSI systems. The associated eigenvalue is

$$H(w, z) = \sum_{k=-\infty}^{\infty} \sum_{l=-\infty}^{\infty} h(k, l)w^{-k}z^{-l} \quad (5)$$

where h is the impulse response. For such an exponential input the output would be

$$y(m, n) = w^m z^n H(w, z).$$

$H(w, z)$ is called the *system function* for the LSI system. Like the impulse response h, it too completely characterizes the system.

Let us formally define the 2-*D* *z transform* of a discrete array x as

$$X(w, z) = \sum_{m=-\infty}^{\infty} \sum_{n=-\infty}^{\infty} x(m, n)w^{-m}z^{-n}. \quad (6)$$

The system function is thus seen to be the z transform of the impulse response. We shall adopt the convention of using upper case letters to denote z transforms of arrays which are denoted by the corresponding lower case letter. The sum in (6) may not converge for all (or any) values of the complex variables w and z. Those values of w and z for which the sum does converge constitute a *region of convergence* in the (w, z) hyperplane.

A 2-*D* z transform can be inverted using the inversion formula

$$x(m,n) = \left(\frac{1}{2\pi j}\right)^2 \oint_{c_1} \oint_{c_2(z)} X(w,z) w^{m-1} z^{n-1} \, dw \, dz. \quad (7)$$

The contours for both integrals must lie completely within the region of convergence in the (w, z) hyperplane; they must be closed and they must encircle the origin counterclockwise in the plane of their respective variables.

At this point, an example might be beneficial. Consider the 2-*D* z transform

$$H(w,z) = \frac{1}{1 - aw^{-1} - bz^{-1}}$$

where $|a| + |b| < 1$ and the region of convergence includes the surface $|w| = |z| = 1$. Then

$$h(m,n) = \left(\frac{1}{2\pi j}\right)^2 \oint\oint \frac{w^m z^n}{wz - az - bw} \, dw \, dz$$

$$= \left(\frac{1}{2\pi j}\right)^2 \oint\oint \frac{\dfrac{1}{(z-b)} w^m z^n}{w - \dfrac{az}{z-b}} \, dw \, dz.$$

Considering z to be a parameter, the integrand of the w integration considered as a function of w has a simple pole at $az/(z - b)$. If we take the contours of integration to be $|z| = 1$ and $|w| = 1$, then it can be shown that $|az/(z - b)| < 1$ and hence that the pole is inside the contour $|w| = 1$. Applying Cauchy's residue theorem to perform the w integration, we get

$$h(m,n) = \begin{cases} \dfrac{1}{2\pi j} \oint \dfrac{a^m z^{m+n}}{(z-b)^{m+1}} \, dz, & \text{for } m \geq 0 \\ 0, & \text{for } m < 0. \end{cases}$$

Now we have an $(m + 1)$st-order pole at $z = b$, which lies inside the contour $|z| = 1$. Applying the residue theorem once again yields the final result

$$h(m,n) = \begin{cases} a^m b^n \dfrac{m!n!}{(m+n)!}, & \text{for } m \geq 0, n \geq 0 \\ 0, & \text{otherwise.} \end{cases}$$

If a 2-*D* sequence is absolutely summable, it can be represented by its 2-*D* Fourier transform which is a representation of the sequence as a linear combination of sinusoids

$$X(e^{j\mu}, e^{j\nu}) = \sum_{m=-\infty}^{\infty} \sum_{n=-\infty}^{\infty} x(m,n) \exp\left[-j(\mu m + \nu m)\right]. \quad (8)$$

The sequence can be recovered from $X(e^{j\mu}, e^{j\nu})$ by the inverse Fourier transform

$$x(m,n) = \frac{1}{4\pi^2} \int_{-\pi}^{\pi} \int_{-\pi}^{\pi} X(e^{j\mu}, e^{j\nu}) \exp\left[j(\mu m, \nu n)\right] d\mu \, d\nu. \quad (9)$$

The notation for the Fourier transform, although cumbersome, is used to emphasize the fact that the Fourier transform is equal to the z transform evaluated on the surface $w = e^{j\mu}$, $z = e^{j\nu}$. It also serves to remind us that X is periodic in both μ and ν with period 2π in each.

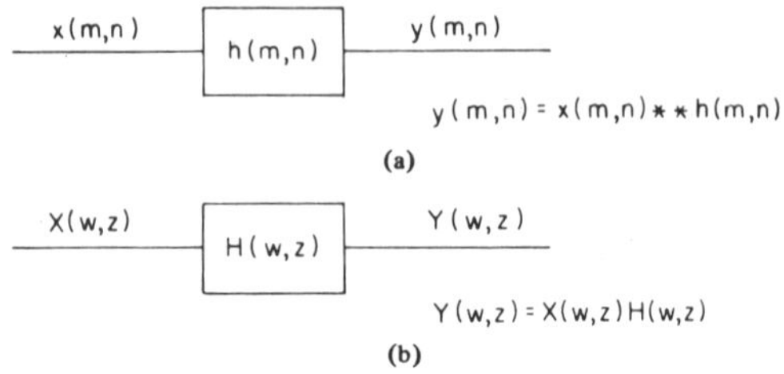

Fig. 2. Two representations of a 2-*D* LSI digital filter. (a) Signal space representation. (b) Transform space representation.

Using (4), (6), and (8), it is straightforward to show that if

$$y(m,n) = x(m,n) * h(m,n)$$

then

$$Y(w,z) = X(w,z) \, H(w,z) \quad (10)$$

and

$$Y(e^{j\mu}, e^{j\nu}) = X(e^{j\mu}, e^{j\nu}) \, H(e^{j\mu}, e^{j\nu}). \quad (11)$$

An LSI digital filter is often drawn in input–output fashion, as in Fig. 2. The variables can be specified as sequences or as transforms. In the former case it is understood that y is the convolution of x and h, and in the latter, Y is the product of X and H.

The interpretation of the Fourier transform is the same for both 1-*D* and 2-*D* signals. It is a measure of the frequency content of a signal. Slowly varying signals will have Fourier transforms which are mostly concentrated near the origin of the (μ, ν) plane and rapidly varying sequences have more frequency content for μ or ν near π.

The z transform and the Fourier transform both suffer from the computational disadvantage that they are functions of continuous variables. As in 1-*D*, the *discrete Fourier transform* (DFT) has thus proven to be an extremely useful tool for examining spectra and for implementing digital filters, particularly since the discovery of a particularly efficient algorithm (the fast Fourier transform (FFT) algorithm) for its evaluation [9].

A 2-*D* DFT may be defined by a straightforward extension of the 1-*D* definition. If we assume that the sequence $x(m, n)$ is nonzero only for $0 \leq m \leq M - 1$ and $0 \leq n \leq N - 1$, then we can define $X(k, l)$, the 2-*D* DFT of $x(m, n)$, to be

$$X(k,l) = \sum_{m=0}^{M-1} \sum_{n=0}^{N-1} x(m,n) \exp\left[-j\left[\frac{2\pi km}{M} + \frac{2\pi ln}{N}\right]\right]. \quad (12)$$

As a notational convenience, the substitution

$$W_M = \exp\left[-j\frac{2\pi}{M}\right]$$

is often made yielding

$$X(k,l) = \sum_{m=0}^{M-1} \sum_{n=0}^{N-1} x(m,n) \, W_M^{km} \, W_N^{ln}. \quad (13)$$

The inverse DFT is similar in form

$$x(m,n) = \frac{1}{MN} \sum_{k=0}^{M-1} \sum_{l=0}^{N-1} X(k,l) \, W_M^{-km} \, W_N^{-ln}. \quad (14)$$

The careful reader will notice that the DFT is equal to the Fourier transform $X(e^{j\mu}, e^{j\nu})$ evaluated at $\mu = 2\pi k/M, \nu = 2\pi l/N$ for $0 \leqslant k \leqslant M - 1$, $0 \leqslant l \leqslant N - 1$. Thus our notation of $X(k, l)$ for the DFT is ambiguous, although this double usage is quite common. Generally, it will be clear from the context whether we are referring to the z transform or the DFT. If the arguments are restricted to integer values we will always be referring to the latter.

If M and N are highly composite numbers, a 1-*D* FFT algorithm can be used efficiently to evaluate the 2-*D* DFT. This is due to the separability of the DFT kernel $W_M^{km} W_N^{ln}$. (A function x is *separable* if $x(m,n) = x_1(m)x_2(n)$). This allows us to evaluate (13) as follows: for each column of the array x (m = constant) perform an N-point DFT on $x(m,n)$ to get an intermediate array $r(m,l)$. Then perform an M-point DFT separately on each row (l = constant) of $r(m,l)$. The result is $X(k,l)$. If M and N are powers of two, the total number of real multiplications required is

$$2MN \log_2 N + 2MN \log_2 M = 2MN \log_2 MN.$$

Note that this is the same number of multiplies which would have to be performed to find the 1-*D* DFT of an MN point sequence.

The normal objective of digital filtering is to alter the frequency spectrum $X(e^{j\mu}, e^{j\nu})$ of some signal by multiplying it by some frequency response function $H(e^{j\mu}, e^{j\nu})$. The digital filtering problem consists of designing impulse responses $h(m,n)$, which simultaneously have desirable frequency responses and which can be efficiently implemented. In general, the latter constraint implies that ideal frequency responses can only by approximated by realizable systems. One approach to the approximation problem is to assume that the system function $H(w,z)$ must be a polynomial of finite degree in w and z (or w^{-1} and z^{-1}). From the inverse z transform relationship (7) we see that such an impulse response $h(m,n)$ must have only a finite number of nonzero samples. Such filters are referred to as finite-extent impulse response (FIR) filters. With FIR filters we can compute the output directly using the convolution sum (4) or by using the DFT (see Section IV).

Another approach is to assume that the system function is the ratio of finite-degree polynomials in w and z.

$$H(w,z) = \sum_{m=-\infty}^{\infty} \sum_{n=-\infty}^{\infty} h(m,n) w^{-m} z^{-n} = \frac{A(w,z)}{B(w,z)} \tag{15a}$$

where

$$A(w,z) = \sum_{k=0}^{M_1} \sum_{l=0}^{N_1} a(k,l) w^{-k} z^{-l} \tag{15b}$$

$$B(w,z) = \sum_{k=0}^{M_2} \sum_{l=0}^{N_2} b(k,l) w^{-k} z^{-l}.$$

If $X(e^{j\mu}, e^{j\nu})$ is the Fourier transform of the input and $Y(e^{j\mu}, e^{j\nu})$ is the Fourier transform of the output of the filter, then

$$Y(e^{j\mu}, e^{j\nu}) = H(e^{j\mu}, e^{j\nu}) X(e^{j\mu}, e^{j\nu}) = \frac{A(e^{j\mu}, e^{j\nu})}{B(e^{j\mu}, e^{j\nu})} X(e^{j\mu}, e^{j\nu})$$

or

$$B(e^{j\mu}, e^{j\nu})\, Y(e^{j\mu}, e^{j\nu}) = A(e^{j\mu}, e^{j\nu})\, X(e^{j\mu}, e^{j\nu}). \tag{16}$$

Equation (16) implies

$$\sum_{k,l} b(k,l) y(m-k, n-l) = \sum_{k,l} a(k,l) x(m-k, n-l). \tag{17}$$

Equation (17) is an example of a 2-*D difference equation* which relates weighted sums of shifted versions of the input and output. The output y can be computed recursively from the input x and a set of initial conditions. This can be done by rewriting (17) as

$$y(m-i, n-j) = \frac{1}{b(i,j)} \sum_{k,l} a(k,l) x(m-k, n-l)$$

$$- \frac{1}{b(i,j)} \sum_{k,l}{}' b(k,l) y(m-k, n-l). \tag{18}$$

The prime by the second summation indicates that the sum is for values of the integer ordered pair (k,l) excluding the value (i,j). Such a filter is referred to as recursive or infinite-duration impulse response (IIR). The latter term is somewhat of a misnomer since not all infinite-length impulse responses can be realized using difference equations, only those with rational transfer functions.

It is assumed in (18) that all values of $y(m-k, n-l)$ needed on the right-hand side of that equation are known either from previous calculations or from the initial conditions. If it is possible to increment the values of the indices (m,n) in such a fashion that all of the values of y can be computed in turn from a given set of initial conditions, the difference equation will be called *recursible*. In general, the recursibility of (18) will depend upon which term $y(m-i, n-j)$ is deleted from the right-hand side summation, i.e., it depends upon the value of (i,j). Recursibility will be studied in greater detail in Section IV.

In several interesting cases, 2-*D* arrays can be represented exactly by 1-*D* sequences [10]. This approach has great potential for 2-*D* digital signal processing since the resulting sequences can be processed by familiar 1-*D* techniques and the results can then be remapped to two dimensions.

A 2-*D* array of finite duration

$$\{0 \leqslant m \leqslant M-1, 0 \leqslant n \leqslant N-1\}$$

can be mapped to a 1-*D* sequence $g(p)$ by means of the mapping

$$g(Nm+n) = f(m,n), \quad 0 \leqslant m \leqslant M-1, \quad 0 \leqslant n \leqslant N-1. \tag{19}$$

The samples of $g(p)$ are thus chosen by concatenating the columns of f. This is sometimes referred to as a lexicographic ordering of the 2-*D* sequence into a 1-*D* sequence. Due to the limited extent of f, the mapping of (19) is readily seen to be invertible. If this mapping is considered in the z plane, then

$$G(z) = F(z^N, z). \tag{20}$$

Thus the z transform of g corresponds to the z transform of f evaluated along the contour $w = z^N$. The function $G(z)$ is referred to as a *slice* of $F(w,z)$ and $g(p)$ is called a *projection* of $f(m,n)$.

Restricting our attention to the Fourier plane, (20) becomes

$$G(e^{j\omega}) = F(e^{jN\omega}, e^{j\omega}). \tag{21}$$

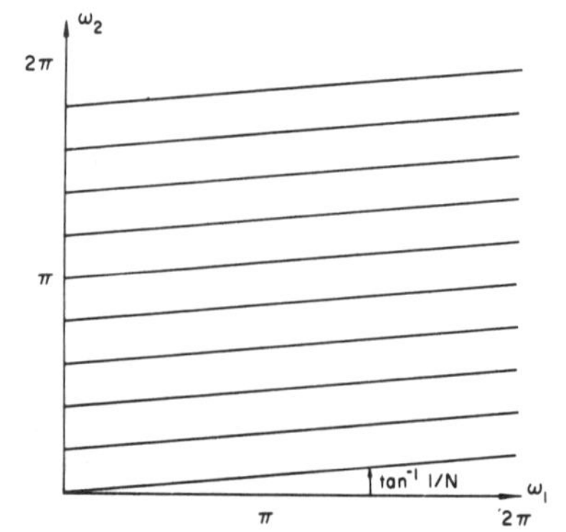

Fig. 3. Lines in the 2-D Fourier plane along which $F(\exp(j\mu), \exp(j\nu))$ and $G \exp(j\omega)$ are numerically equal for $N = 10$.

Since $F(e^{j\mu}, e^{j\nu})$ is doubly periodic in μ and ν with period 2π in each, $G(e^{j\omega})$ corresponds to the evaluation of $F(e^{j\mu}, e^{j\nu})$ along a series of N parallel lines in the 2-D Fourier plane. Each of these N lines forms an angle $\theta = \tan^{-1} 1/N$ with the μ axis, as illustrated in Fig. 3. $F(e^{j\mu}, e^{j\nu})$ can be recovered from $G(e^{j\omega})$ using the interpolation formula

$$F(e^{j\mu}, e^{j\nu}) = \sum_{k=0}^{N-1} G \exp\left(j\left(\frac{\mu + 2\pi k}{N}\right)\right) \frac{\sin \frac{N}{2}\left(\frac{\mu + 2\pi k}{N} - \nu\right)}{\sin \frac{1}{2}\left(\frac{\mu + 2\pi k}{N} - \nu\right)} \times \exp\left[-j\frac{N-1}{2}\left(\frac{\mu + 2\pi k}{N} - \nu\right)\right]. \quad (22)$$

If we are given a rational 2-D frequency response

$$H(e^{j\mu}, e^{j\nu}) = \frac{A(e^{j\mu}, e^{j\nu})}{B(e^{j\mu}, e^{j\nu})} \quad (23)$$

there exist slices from which $H(e^{j\mu}, e^{j\nu})$ can be recovered [10]. If A is a 2-D polynomial of degree N_1 in $\exp[j\nu]$ and if B is a polynomial of degree N_2 in $\exp[j\nu]$, then letting $N = \max[N_1, N_2]$, H can be recovered from the slice at angle $\tan^{-1} 1/N$. If $\hat{h}(p)$ denotes the projection associated with this slice, then

$$\hat{h}(p) = \sum_{m=-\infty}^{\infty} h(m, p - Nm). \quad (24)$$

The sequences $\hat{a}(p)$ and $\hat{b}(p)$, which are the inverse transforms of $A(z^N, z)$ and $B(z^N, z)$, respectively, can be recovered from $\hat{h}(p)$ by Prony's method, Shanks' method, or any similar technique which estimates the poles and zeros of a 1-D rational system function from samples of its impulse response [11], [12]. Then $\hat{a}(p)$ and $\hat{b}(p)$ can be separately back-projected, as in the finite duration sequence case, to give the coefficients of the numerator and denominator polynomials $a(m, n)$ and $b(m, n)$ from which $H(e^{j\mu}, e^{j\nu})$ can be reconstructed or $h(m, n)$ determined.

III. Stability of IIR Filters

An important issue in the design of digital filters is their stability. If a filter is unstable, any input, including computational noise, can cause the filter output to grow without bound, obliterating the desired output of the filter. Thus the condition is generally imposed that the output of the digital filter must remain bounded for any bounded input. This is sometimes referred to as bounded-input-bounded output (BIBO) stability. It can be easily shown [13] that BIBO stability is equivalent to requiring that the impulse response of the filter $h(m, n)$ be absolutely summable

$$\sum_{m=-\infty}^{\infty} \sum_{n=-\infty}^{\infty} |h(m, n)| < \infty. \quad (25)$$

For FIR filters, this condition is always satisfied, since $h(m, n)$ has only a finite number of nonzero terms. For IIR filters, on the other hand, stability is a very important design constraint.

In the subsequent discussion, we shall restrict ourselves to impulse responses which are nonzero only for nonnegative values of m and n (first quadrant sequences) and to system functions whose numerator and denominator polynomials are the z transforms of first quadrant sequences. (That is, the numerator and denominator of the system function are polynomials in w^{-1} and z^{-1}.) This restriction is commonly imposed upon IIR filters so that they have recursible implementations, although there is a slightly more general class of 2-D IIR filters which are also recursible [14].

The basic theorem for guaranteeing BIBO stability is due to Shanks [13], [15] and has been generalized by Farmer and Bednar [16]. Restated in our notation, Shanks' theorem states the following.

Theorem (Shanks): A recursive (IIR) filter

$$H(w, z) = \frac{A(w, z)}{B(w, z)} = \frac{\sum_{m=0}^{M} \sum_{n=0}^{N} a(m, n) w^{-m} z^{-n}}{\sum_{k=0}^{K} \sum_{l=0}^{L} b(k, l) w^{-k} z^{-l}}$$

is BIBO stable if and only if there are no values of w and z such that $B(w, z) = 0$ for $|w| \geq 1$ and $|z| \geq 1$ simultaneously.

This is analogous in 1-D to requiring that the system function have no poles outside the unit circle. One important difference is that 2-D singularities defined by $B(w, z) = 0$ are not isolated points as poles are, but are rather multidimensional surfaces or manifolds.

To apply the Shanks test is conceptually straightforward but computationally involved. The region of the z plane, $|z| \geq 1$ must be mapped into the w plane by the algebraic mapping $B(w, z) = 0$. If the image of that map lies completely within the circle $|w| = 1$ the filter is stable, otherwise, it is not.

A simpler stability test to apply is based on a theorem due to Huang [17].

Theorem (Huang): A recursive filter

$$H(w, z) = \frac{A(w, z)}{B(w, z)}$$

is stable if and only if the image of the unit circle $|z| = 1$ mapped into the w plane by $B(w, z) = 0$ does not intersect the region $|w| \geq 1$ and in addition, no point in the region $|w| \geq 1$ maps into the point $z = \infty$.

To apply Huang's test, the z unit circle must be mapped into the w plane by $B(w, z) = 0$. For the filter to be stable, its image must not intersect $|w| \leq 1$. In addition, we must solve $B(w, z) = 0$ for w, setting $z^{-1} = 0$. The magnitude of all of the roots of this equation must be less than one. Both this test and the preceding one require the mapping of a continuous region of (w, z) hyperspace and thus, in theory, both require an infinite amount of computation. A fair amount of effort has thus been expended to find implementations of Huang's test which require only a finite amount of computation. The first such attempt was by Huang himself in the same paper. He showed that his test could be reformulated in a form which only required a finite number of computational steps using a test of Ansell's [18] for continuous systems. Anderson and Jury [19] and Maria and Fahmy [20] have also proposed more efficient implementations of Huang's test. The former showed that the main component of Huang's test could be accomplished using a Schur-Cohn matrix test. This test has also been generalized for checking the stability of N-dimensional systems [21], [22]. Although the Anderson-Jury test is probably the most efficient stability testing algorithm, the testing of stability remains a major computational effort.

Any stability test, however simple, is not sufficient for designing 2-*D* recursive filters. What is required is a procedure to stabilize an unstable filter without perturbing the magnitude of its frequency response, or a design technique which will guarantee a stable filter. This is easily accomplished in one dimension since the poles of the system function which are located outside the unit circle, if any, which account for the instability, can be replaced by poles in the conjugate reciprocal locations, which are necessarily within the unit circle. This is equivalent to cascading the unstable filter with a digital all-pass filter which has no effect upon the magnitude of the frequency response. This procedure can either be performed algebraically or using discrete Hilbert transform (DHT) techniques. In 2-*D* there are difficulties with both approaches. The algebraic approach fails due to the inability to factor 2-*D* polynomials. There are also difficulties associated with the DHT.

Read and Treitel [23] have defined a 2-*D* DHT to be used for the stabilization of recursive filters. Their idea was to obtain the log-magnitude function of the denominator polynomial and use the 2-*D* DHT to calculate the minimum phase function associated with that log-magnitude function. A new denominator polynomial is then constructed by a complex exponentiation. This technique was used on examples with denominator polynomials up to degree 5×5 with moderate success. Their difficulties arose from nonexact computations of the DHT and the fact that in many cases the reconstructed minimum phase denominator polynomial is not of finite degree.

A similar attempt at stabilization utilizing the notion of a 2-*D* complex cepstrum was performed by Pistor [24] and Dudgeon [25]. As in the 1-*D* case a *complex cepstrum* can be defined as the inverse Fourier transform of the complex logarithm of the Fourier transform of a sequence. Thus, if two sequences are convolved, their cepstra add. Dudgeon [14], [26] has shown that cepstra exist for 2-*D* rational functions

$$H(w, z) = \frac{A(w, z)}{B(w, z)}$$

provided that:

1) $A(w, z)$ and $B(w, z)$ are never zero at any point (w, z) such that $|w| = 1$ and $|z| = 1$;
2) any linear phase components are removed from the Fourier transform by an appropriate linear shift of the sequence $h(m, n)$.

As an aside, it might be mentioned that we can also talk about the 1-*D* cepstrum of an image by considering the cepstrum of the projection of the image. This is also the projection of the cepstrum and like the standard 2-*D* cepstrum, it has the property that the 1-*D* cepstra add when images are convolved. The 1-*D* cepstra of 2-*D* rational functions are invertible and are defined whenever the standard 2-*D* cepstrum is defined. More importantly, they can be defined in many cases when standard cepstra cannot!

Both Pistor and Dudgeon used 2-*D* cepstra to decompose magnitude-squared frequency responses to get stable, recursible 2-*D* filters. Pistor decomposed the magnitude-squared function into a cascade of four stable recursive one-quadrant filters, each of whose impulse responses was defined on a different quadrant. Dudgeon's approach was similar in spirit. He decomposed the magnitude-squared frequency response into a cascade of two stable, recursible, half-plane filters. (The half-plane filter is one for which the impulse response is required to be zero over half the (m, n) plane rather than over three quadrants, as in the one-quadrant filter case). The filters from both techniques, having rational transfer functions, could then be implemented using difference equations [14], [27].

Both Pistor's and Dudgeon's techniques suffer from the serious difficulty also encountered by Read and Treitel [23] that an exact decomposition yields a denominator polynomial which is generally not of finite degree, in spite of the fact that the denominator of the original magnitude-squared function is of finite degree. Both Pistor and Dudgeon were forced to approximate their infinite extent results by finite extent results in order to get a practical implementation. This procedure alters the frequency response of the filter. In some cases the effect is tolerable, and in others it is quite serious.

A completely unrelated stabilization procedure is based on a conjecture due to Shanks [15]. His conjecture states that the planar least squares inverse (PLSI) of an array is a minimum phase array. To illustrate his procedure, assume that we have a possibly unstable filter

$$H(w, z) = \frac{A(w, z)}{B(w, z)}$$

which we want to stabilize. Letting $b(m, n)$ denote the coefficients of the denominator polynomial, we seek an inverse to b, denoted by p such that

$$\delta(m, n) \simeq b(m, n) ** p(m, n) \tag{26}$$

where ** denotes 2-*D* convolution. In general, the equality in (26) will not be exact and p must be chosen to minimize the error between the left and right-hand sides according to some norm. If p is chosen to minimize the mean-squared error

$$E = \sum_{m,n} (\delta(m, n) - b(m, n) ** p(m, n))^2$$

then it is referred to as the PLSI of $b(m, n)$. By the conjecture, $p(m, n)$ is a minimum phase array. Shanks further states that if the array of denominator coefficients of a rational transfer function is a minimum phase array, then the filter is stable. (An array $x(m, n)$ is minimum phase if the 1-*D* Fourier transform $X(e^{j\mu_0}, e^{j\nu})$ is minimum phase for all fixed μ_0 and all real ν and if $X(e^{j\mu}, e^{j\nu_0})$ is a 1-*D* minimum phase function for fixed ν_0 and all real μ.)

To stabilize an unstable filter, Shanks *et al.* [15] proposed taking the double PLSI of the denominator array. This double inversion, by the hypothesis, will yield a stable filter and the frequency response of the final result will hopefully approximate the original frequency response. The final frequency response will only be an approximation, however, and should be measured to see that the approximation is sufficiently good. Although the hypothesis has never been proven, it has been proven in the 1-*D* case and has been used in the design of several hundred 2-*D* filters. A counterexample has never appeared. It is perhaps reasonable, however, to test the stability of a filter obtained by this procedure.

IV. Digital Filter Implementation

As in the 1-*D* case, 2-*D* digital filters fall into two classes—those whose impulse responses contain a finite number of samples and whose transfer functions are polynomials, the FIR filters, and those whose transfer functions are rational functions, the IIR filters. Both questions of design and questions of implementation are quite different for the two classes of filters and for this reason in this section and the next, the two classes will be treated separately. FIR filters can either be implemented directly or they can be implemented using high speed convolution techniques (using either DFT or number theoretic transforms [28]). 2-*D* IIR filters are invariably implemented recursively using difference equations.

The choice between using an FIR filter or an IIR filter depends upon the application, although most applications are probably better served by FIR designs. They are always stable and they can be designed to have either zero or linear phase, which is an important consideration in image processing applications. 1-*D* design methods are generally extendable and filters of high order can be designed relatively quickly. The primary advantage of IIR filters is that fewer arithmetic operations can often be employed than by using comparable FIR filters, however these filters can be unstable. IIR filters are particularly useful in applications such as seismic data processing where volumes of data must be processed very quickly and where the extra time required to design the filter is easily justified.

FIR Filters

Referring to Fig. 2, let $x(m, n)$ and $y(m, n)$ denote the input and output respectively of a digital filter characterized by the impulse response $h(m, n)$. Since $h(m, n)$ has an impulse response which is of finite extent, we can say with no loss of generality that

$$h(m, n) = 0, \quad (m, n) \notin I_{MN}$$

$$I_{MN} = \{(m, n): 0 \leq m \leq M - 1, 0 \leq n \leq N - 1\}.$$

M is thus the horizontal extent of the array and N is its vertical extent. The output y is then the convolution of x and h

$$y(m, n) = \sum_{k=0}^{M-1} \sum_{l=0}^{N-1} x(k, l) h(m - k, n - l). \tag{27}$$

Equation (27) represents one realization of the filter.

As a tutorial device it may be instructive to define the concept of an input mask as in Fig. 4. To produce the value $y(m, n)$ the input mask is placed over the appropriate samples of the input array. Each sample is then multiplied by the impulse response coefficient associated with that position in the mask and the products are summed to give $y(m, n)$. This is repeated for all (m, n). The (m, n) can be incremented by rows, by columns, or by any other ordering since the values

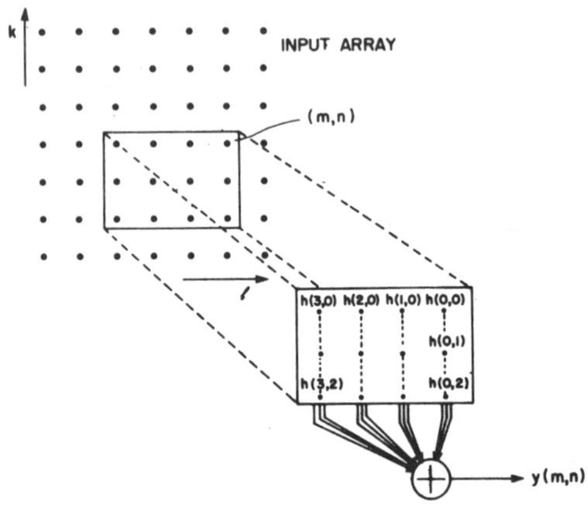

Fig. 4. An example of an input mask for a 3 × 3 FIR filter.

of the output sequence do not depend on the order in which those values are computed.

The realization of (27) has the additional advantage shared by the corresponding realization for 1-*D* FIR digital filters that there is little arithmetic quantization error. If the input samples are quantized to b_1 bits and the impulse response samples are quantized to b_2 bits, the full $b_1 + b_2$ bit products can be accumulated and only the final sum needs to be quantized.

This realization also allows savings for those filters where the impulse response possesses symmetries. If several samples of the impulse response have the same value then by applying the arithmetic distributive law and adding the appropriate input samples before multiplying by their common coefficient value, a savings in the number of multiplications can be realized. As an example, consider an impulse response which is $N_H \times N_H$ points which is used to filter an $N_X \times N_X$ input. In the absence of any exploitable symmetries in the impulse response, using (27) approximately N_H^2 multiplies and adds must be performed for each sample of the output. This results in a total of $N_X^2 N_H^2$ multiplies. In the case of octagonal symmetry

$$h(m, n) = h(-m, n) = h(m, -n) = h(n, m)$$

which is common for zero phase filters which approximate circularly symmetric frequency responses, the number of distinct coefficients is $[(N_H + 1)/2][(N_H + 3)/2]/2$, where $[\cdot]$ denotes the greatest integer function. Filtering the same image with such a filter requires $N_X^2[(N_H + 1)/2][(N_H + 3)/2]/2$ multiplies and $N_X^2 N_H^2$ adds. This is a reduction in the number of multiplies by roughly a factor of eight.

FIR filters can also be implemented using the FFT. It can be readily shown by paralleling the 1-*D* derivation that if

$$x(m, n) ** h(m, n) = y(m, n)$$

then

$$X(k, l) H(k, l) = Y(k, l) \tag{28}$$

where $X(k, l)$, $H(k, l)$, and $Y(k, l)$ are the 2-*D* DFT's of x, h, and y, respectively. Equation (28) is true for discrete Fourier transforms of any size, although to be useful for implementing filters $y(m, n)$ must be recoverable from $Y(k, l)$. Thus, if

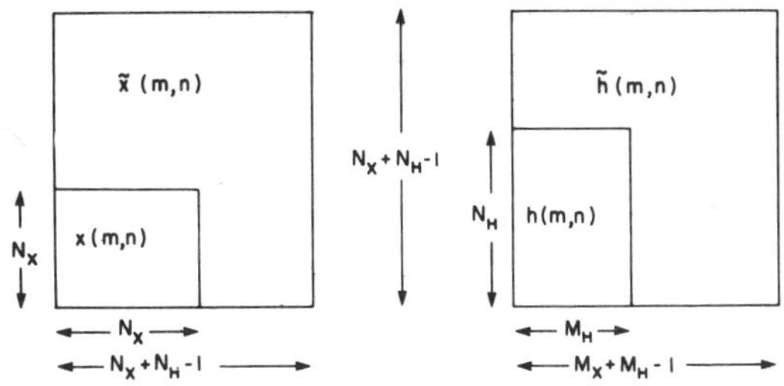

Fig. 5. This figure illustrates padding with zeros. The inverse DFT of the product of the DFT's of $\tilde{x}$ and $\tilde{h}$ corresponds to the linear convolution of x and h.

$x(m, n)$ is an $M_X \times N_X$ array and $h(m, n)$ is an $M_H \times N_H$ array so that $y(m, n)$ is an $M_Y \times N_Y$ array where

$$M_Y = M_X + M_H - 1$$

$$N_Y = N_X + N_H - 1$$

then the 2-*D* DFT's must all contain at least $M_Y \times N_Y$ points. (They can, of course, be larger in either or both dimensions.) To take the $M_Y \times N_Y$ point DFT's of the arrays x and h we can simply replace x and h by $\tilde{x}$ and $\tilde{h}$, respectively, where

$$\tilde{x}(m, n) = \begin{cases} x(m, n), & (m, n) \in I_{M_X N_X} \\ 0, & (m, n) \in I_{M_Y N_Y},\ (m, n) \notin I_{M_X N_X} \end{cases} \tag{29}$$

$$\tilde{h}(m, n) = \begin{cases} h(m, n), & (m, n) \in I_{M_H N_H} \\ 0, & (m, n) \in I_{M_Y N_Y},\ \notin I_{M_H N_H}. \end{cases} \tag{30}$$

Such an operation, commonly referred to as padding with zeros, is represented in Fig. 5.

If such an operation is performed on real sequences x and h (and the DFT of h has been precomputed and stored), then the convolution y can be performed using approximately

$$2\, M_Y N_Y \log_2 M_Y N_Y + 2\, M_Y N_Y$$

real multiplications and real additions if M_Y and N_Y are both powers of two. If $h(m, n)$ has octagonal symmetry, there are essentially no additional savings since this fact can only be used to compute $H(k, l)$, which by assumption, was precomputed.

For filtering very large inputs when machine storage is a limiting factor or when the size of the input is considerably larger than the size of the impulse response, the ideas of block convolution using either an overlap-add or overlap-save approach [29] can be applied as in the 1-*D* case. The extension to 2-*D* is straightforward. The idea is to partition the input into blocks and to filter each of the blocks separately and then appropriately combine the results. The size of the blocks used is generally limited by machine considerations and the efficiency of the whole procedure varies greatly with the block-size chosen. Hunt [30] has analyzed this problem in some depth.

For short impulse responses, the direct convolution results in a faster implementation than the one which uses the FFT, whereas for longer impulse responses the FFT approach is faster. To get a feeling for where the crossover point lies, let us assume that the total number of multiplications involved is a figure of merit. Then the FFT approach will be faster if

$$M_X N_X M_H N_H > 2\, M_Y N_Y \log_2 M_Y N_Y + 2\, M_Y N_Y. \tag{31}$$

Assuming that $M_H = N_H = M_X = N_X$, (31) is true for $N > 8$! If the filter possesses octagonal symmetry the FFT would be favored for impulse responses larger than about 24 × 24. If $M_X = N_X = 256$ and $M_H = N_H$ then (31) is true for $N_H > 6$ without octagonal symmetry and for $N_H > 18$ with it. Thus for FIR filters of reasonable size the FFT should nearly always be used in implementation.

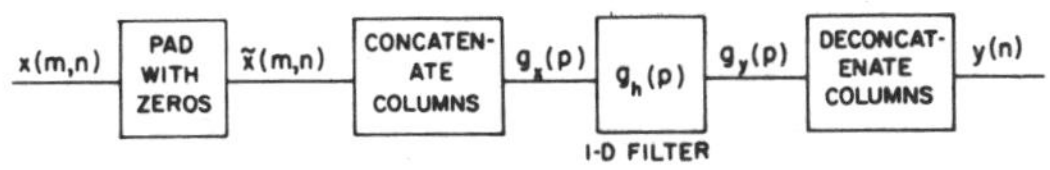

Fig. 6. A representation of a 2-*D* filter using a single 1-*D* filter using projections.

Instead of using the Fourier transform to perform the convolution, the number theoretic transform [28] could be used, with perhaps even greater computational efficiency. In fact, it has been suggested that due to the lengths of the transforms involved, the number of multiplications involved etc., that these transforms are ideally suited for implementing 2-*D* FIR filters.

2-*D* FIR filters can also be implemented as 1-*D* FIR filters using projections. To see how this can be done, if

$$y(m, n) = x(m, n) ** h(m, n)$$

then from the properties of z transforms, we know that

$$Y(w, z) = X(w, z) H(w, z)$$

and thus

$$Y(z^N, z) = X(z^N, z) H(z^N, z). \tag{32}$$

From the relations between projections and their z transforms which we developed in Section II, it follows that

$$g_y(p) = g_x(p) * g_h(p) \tag{33}$$

where g_y, g_x, and g_h are the projections of y, x, and h, respectively, and the convolution is 1-*D*. In order for (33) to be useful in a filter implementation, $y(m, n)$ must be recoverable from $g_y(p)$. This implies that the parameter N which appears in (32), which defines the projection, must be greater than or equal to the vertical extent of $y(m, n)$. To obtain projections of x and h, based on this same parameter, we merely lexicographically order the arrays x and h which were defined in (29) and (30). The final filter implementation appears in Fig. 6. The filter $g_h(p)$ is a 1-*D* FIR filter and it can be realized in any of the classical forms for these filters, including using the FFT. It can be readily seen that the number of multiplications and the amount of storage involved are identical to the corresponding 2-*D* realizations discussed here. This type of an implementation has also been treated by MacAdam [31] and is the reverse of the problem treated by Agarwal and Burrus [32] who used higher dimensional convolutions to implement 1-*D* convolutions.

Another implementation using 1-*D* filters is due to Treitel and Shanks [33] and Tretiak [34]. They have shown that any $N_H \times N_H$ point filter can be realized as a sum of separable

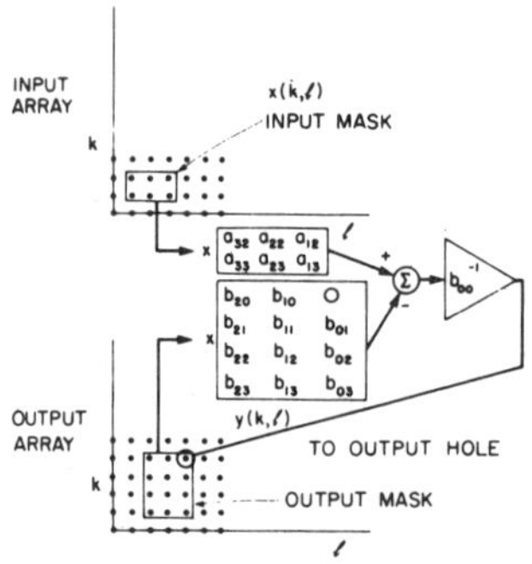

Fig. 7. An example of input and output masks to calculate $y(4, 4)$.

filters

$$h(m, n) = \sum_{i=1}^{N_H} \lambda_i f_i(m) g_i(n). \quad (34)$$

Separable filters are particularly simple to implement. Each filter in the sum can be realized by filtering each row by the 1-*D* filter $f_i(m)$ (which requires $N_H N_X^2$ multiplications) and then filtering each column of the result by $g_i(n)$ (which requires $N_X N_H (N_X + N_H - 1)$ multiplications). The complete filter thus requires $N_X N_H^2 (2N_X + N_H - 1)$ multiplications, which is roughly two times the number needed for a direct implementation. On the other hand, if a number of the eigenvalues $\{\lambda_i\}$ are sufficiently small so that only k stages of the filter are needed to approximate h, then computational speed advantages can still be realized if

$$k(2N_X + N_H)(N_X N_H) < N_X^2 N_H^2$$

or if

$$k < \frac{N_X N_H}{2N_X + N_H}.$$

For details on how the λ_i, $f_i(m)$, and $g_i(n)$ are computed, the reader is referred to [33]. These results can be generalized by using the singular value decomposition of an $M_H \times N_H$ matrix to matrices which are not square.

IIR Filters

2-*D* IIR digital filters are generally specified by 2-*D* difference equations of the general form

$$\sum_{k=0}^{M_1} \sum_{l=0}^{N_1} b(k, l) y(m-k, n-l) = \sum_{k=0}^{M_2} \sum_{l=0}^{N_2} a(k, l) x(m-k, n-l). \quad (35)$$

If $b(0, 0) \neq 0$, this can be rewritten as

$$y(m, n) = \frac{1}{b(0, 0)} \left\{ -\sum_{\substack{k=0 \\ k \cdot l \neq 0}}^{M_1} \sum_{l=0}^{N_1} b(k, l) y(m-k, n-l) + \sum_{k=0}^{M_2} \sum_{l=0}^{N_2} a(k, l) x(m-k, n-l) \right\} \quad (36)$$

which expresses $y(m, n)$ as a linear combination of input samples and previously computed values of the output y. Generalizing the concept of an input mask which we used before, we can define two masks, an input mask and an output mask, as depicted in Fig. 7. The input mask is the same as in the FIR case. It sweeps over the input array $x(m, n)$ and multiplies the appropriate values of x by the $(M_2 + 1) \times (N_2 + 1)$ array of coefficients $a(k, l)$. The output mask is similar to the input mask except that it possesses a hole at the location of the output value presently being computed. The mask corresponding to (37) thus has a hole location in its upper right (NE) hand corner. Other masks can be chosen corresponding to the difference equation of (35). In particular, masks could be developed with holes in the NW, SW, or SE corners in addition to the NE. The impulse responses corresponding to these different masks will be different for each of the four and they will not all be necessarily stable. In fact, it has been shown [24] that at most one of the four masks can correspond to a stable filter for a particular set of denominator coefficients $b(k, l)$. Generally, the choice of a mask, given a set of coefficients $b(k, l)$ is thus determined by stability considerations. An IIR filter is said to be *recursible* if for every output point (m, n) the output mask covers only points which have been previously computed. Thus the output mask in Fig. 7 corresponds to a recursible realization of (36) since if the values of the output are computed in the order $\cdots (m, n), (m, n+1), (m, n+2), \cdots, (m+1, n-1), (m+1, n), (m+1, n+1), \cdots$ the output mask only covers points which have been previously computed. Dudgeon [27] has shown that the output mask in Fig. 7 is not the most general form for a recursible output mask but is rather a special case of the mask in Fig. 8 which is columnwise recursible. It is common practice in the field, however, to deal with the masks similar to that in Fig. 7, and this we shall continue to do. This is the form, for example, assumed by most stability tests.

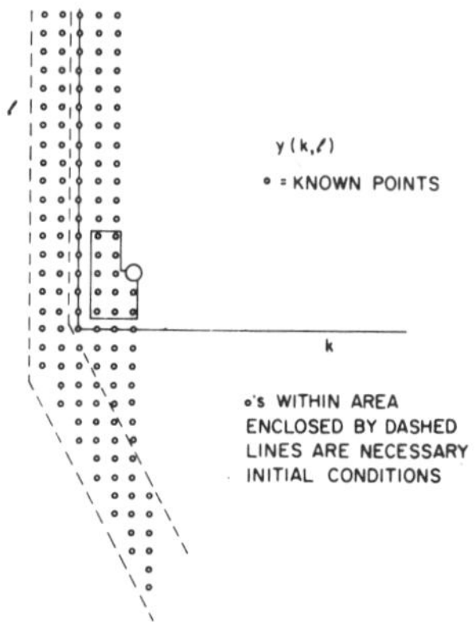

Fig. 8. A more general form for a recursible output mask. This filter is columnwise recursible if the appropriate initial conditions are given.

Implicit in the notion of recursive implementations of digital filters is the requirement that a set of initial conditions be specified. It can be shown that these initial conditions must be zero in order that the resulting system be linear, although for many problems strict linearity may not be a requirement. For the realization of (36), which recurses in the NE direction, a sufficient set of initial conditions is shown in Fig. 9. Here we are assumed to know the initial values of y in an L-shaped band which contains M_1 columns and N_1 rows. Other sets of sufficient initial conditions can be generated by linearly displacing this set of locations in either the m or n directions or both.

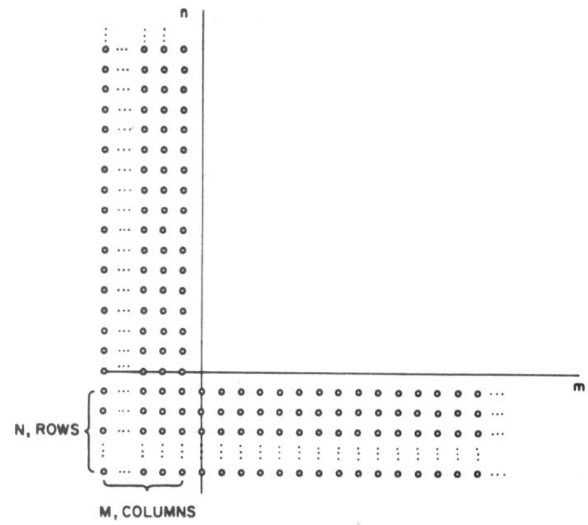

Fig. 9. A sufficient set of initial conditions for the filter of (36).

As in 1-*D*, other realizations of the filter in (35) can be obtained by rearranging the computations involved. This is most often easily done by manipulating the z transform. The realization which we have been discussing, the direct form, combines together the terms corresponding to delayed values of the input and delayed values of the output. Alternatively by writing the z transform in the form

$$H(w, z) = \left\{ \sum\sum b(k, l) w^{-k} z^{-l} \right\}^{-1} \left\{ \sum\sum a(k, l) w^{-k} z^{-l} \right\} \tag{37}$$

we can consider the same filter as a cascade of two systems—one which uses only past values of its output and one which has a finite length impulse response. This is the so-called canonic form and it corresponds to rewriting (36) as

$$u(m, n) = \frac{1}{b(0, 0)} \left\{ - \sum_{\substack{k=0 \\ k \cdot l \neq 0}}^{M_1} \sum_{l=0}^{N_1} b(k, l) u(m - k, n - l) + x(m, n) \right\}$$

$$y(m, n) = \sum_{k=0}^{M_2} \sum_{l=0}^{N_2} a(k, l) u(m - k, n - l). \tag{38}$$

Although the two realizations use the same coefficients and have the same frequency response, the two realizations will perform differently with respect to the effect of the noise which is generated as a result of arithmetic roundoff using finite precision arithmetic. Ni and Aggarwal [35] have considered the arithmetic roundoff question and while they have shown that the structures will perform differently, no definite conclusions could be made concerning which structure to prefer.

Other 1-*D* structures such as the cascade, parallel, and continued fraction structures do not generalize to two dimensions in the general case, due to the inability to factor 2-*D* polynomials. Rational z transforms which are factorable can be profitably implemented using a cascade realization, however. Such a realization is also attractive for purposes of guaranteeing stability. However, since in general the factorability property will not be present it must be imposed into the design algorithm for the filter. This complicates the design algorithm.

IIR filters can also be implemented using projections much as FIR filters can, as seen using the system in Fig. 10. The difficulty with the system is that the back projection operator is complicated. In fact, the only known technique for performing this operation is to use the technique mentioned in Section II. This is sufficiently complicated as to make this type of implementation impractical. Dudgeon [14] has suggested a simpler scheme where the backprojection operation is replaced by the simple deconcatenator of Fig. 6. The idea is as follows. A 1-*D* input sequence is formed by concatenating the length N columns of the input array. This sequence is then processed by a 1-*D* IIR filter, which can be implemented in any standard configuration. The resulting 1-*D* output sequence is then used to form a 2-*D* array by taking the first N points as the first column, the next N points as the second column, etc. The overall system is linear but shift varying with the result that only periodic impulse responses of the form $h(m, n) = h(m + 1, n - N)$ can be realized exactly by this technique. However, it is possible to approximate other impulse responses in this manner.

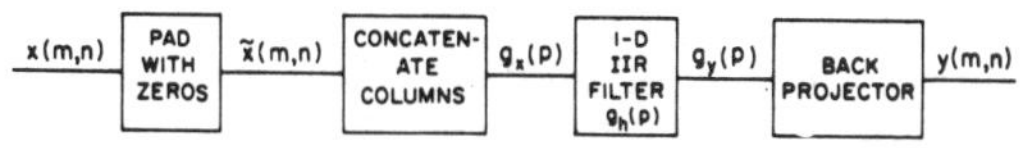

Fig. 10. A system for implementing 2-*D* IIR filters using a 1-*D* IIR filter.

This technique can be refined by considering a 2-*D* array to be a superposition of subarrays each of which contains a single row, padded with appropriate zeros. Each subarray can be filtered separately by the preceding technique and the outputs superimposed to give a final output array. Now the net operation corresponds to a LSI operation which requires doing N 1-*D* filtering operations. However, the N filtering operations can each be done with different 1-*D* filters which will give an overall system which is shift-varying in a controlled manner. Manry and Aggarwal [36] have also explored a similar approach to realizing 2-*D* IIR filters with 1-*D* IIR filters. By exploring mappings of the 2-*D* index (m, n) to the integers, they develop realizations which approximate the exact 2-*D* filter very closely (in terms of perturbation of the output for the same noise-like input) and offer considerable savings in storage. Their work offers an intriguing first step in trying to simplify realizations of 2-*D* digital recursive filters.

With FIR filters it is necessary to store a number of input rows (columns). With IIR filters it is also necessary to store a number of previously computed rows (columns) of output as well as a number of initial conditions. This is complicated by the fact that in general the output rows are infinitely long. Fortunately, in most applications we only require a limited number of samples of the output, but the fact remains that the number of rows of output that we desire effects the amount of storage required by the filter. This is a problem not encountered in 1-*D* implementations.

V. Two-Dimensional Digital Filter Design

In this section, we will discuss various methods by which 2-*D* digital filters may be designed to approximate an ideal 2-*D* frequency response. The FIR and IIR design problems will be considered separately. Our objective is to give an overview of the various design techniques. References will be provided so that the interested reader can consult the original sources if he is interested in the details of any particular method.

A. FIR Filters

The first FIR design technique, which is also the simplest, is windowing which is easily extendable from 1-*D* [29]. Assume that we have an ideal frequency response $I(e^{j\mu}, e^{j\nu})$ which we wish to approximate by a FIR filter whose impulse response is

$h(m, n)$. We can calculate the ideal impulse response $i(m, n)$ by performing an inverse 2-*D* Fourier transform either exactly using analytical techniques or approximately by taking a large inverse 2-*D* DFT. (The size of the DFT should be several times as large as the ultimate impulse response $h(m, n)$ so as to minimize aliasing effects in the space domain.) In general $i(m, n)$ will be of infinite extent. We can get a finite extent impulse response by multiplying the ideal by a window function $w(m, n)$ which is of finite extent

$$h(m, n) = i(m, n)\, w(m, n). \tag{39}$$

$H(e^{j\mu}, e^{j\nu})$ is then related to $I(e^{j\mu}, e^{j\nu})$ by a convolution in the frequency domain

$$H(e^{j\mu}, e^{j\nu}) = \frac{1}{(2\pi)^2} \int_{-\pi}^{\pi} \int_{-\pi}^{\pi} I(e^{j\alpha}, e^{j\beta}) \cdot W(e^{j(\mu-\alpha)}, e^{j(\nu-\beta)})\, d\alpha\, d\beta. \tag{40}$$

For H to approximate I, $W(e^{j\mu}, e^{j\nu})$ should approximate an impulse function. A number of finite extent windows have been proposed which approximate an impulse in their frequency response. Huang [37] has studied the use of continuous 2-*D* circularly symmetric windows taken from 1-*D* continuous windows and Rabiner and Gold [38] give examples of digital filters designed with discrete versions of these windows. Separable windows of the form

$$w(m, n) = w_1(m)\, w_2(n) \tag{41}$$

where $w_1(m)$ and $w_2(n)$ are good 1-*D* windows, are also frequently used. Designs based on the window method can be obtained very quickly but for a filter of a given size the approximation to the ideal frequency response will not be as good in general as with other techniques.

FIR filters can also be designed using frequency sampling as in 1-*D*. Here the filter is specified by determining its DFT. The impulse response of the filter can be determined at the end by performing a simple inverse DFT. The values of the DFT at selected frequencies is fixed and at other frequencies it is left variable. (Typically samples in the passband of a filter will be set equal to one, values in the stopband will be set to zero and values in a transition band, which is located between the two, will be left unspecified.) The error in the resulting approximation over the entire frequency plane is then minimized by varying the values of the unspecified DFT samples. If the filter being designed is symmetric through the origin so that its frequency response is real, linear programming can be used to perform the minimization. Hu and Rabiner [39] have used this technique for designing octagonally symmetric filters as large as 25 X 25. The primary drawback with the method is that it is computationally quite expensive.

Both Hu and Rabiner [40] and Fiasconaro [41] have designed filters by linear programming. This can be considered as an extension of frequency sampling where all of the frequency samples are considered as variables. Since the frequency response $H(e^{j\mu}, e^{j\nu})$ is linear in both its DFT samples and also its impulse response coefficients, the linear program can be applied to manipulate either. If $I(e^{j\mu}, e^{j\nu})$ is the ideal frequency response, then the constraints

$$H(e^{j\mu}, e^{j\nu}) \leq I(e^{j\mu}, e^{j\nu}) + \delta \tag{42}$$

$$H(e^{j\mu}, e^{j\nu}) \geq I(e^{j\mu}, e^{j\nu}) - \delta \tag{43}$$

can be applied to a dense grid of points in the (μ, ν) frequency plane. Linear programming then adjusts $h(m, n)$ while minimizing δ and satisfying the above constraints much as in 1-*D* [1]. The resulting filters are optimal in the minimax or Chebyshev sense.

Unfortunately this brute strength approach is even more computationally expensive than frequency sampling. Design times of an hour on high-powered computers are not uncommon. Fiasconaro [40] has designed a more efficient algorithm for designing zero-phase FIR filters. His algorithm is in the same spirit as the 1-*D* Remez second exchange algorithm used for 1-*D* filter design [1], in that it is iterative: at each iteration a linear programming minimization is performed on a small number of constraint points (typically 100) and this set of points varies from iteration to iteration. The constraint points for a given iteration include those located at the extrema of the error function from the previous iteration.

Recently, Kamp and Thiran [41] and Hersey and Mersereau [42], [43] have proposed similar algorithms which used an exchange algorithm instead of the linear programming of Fiasconaro's algorithm. Some speedup in computational time was obtained but the results were not particularly dramatic since the bulk of the time is consumed, not in the minimization but rather in the evaluation of the error function over the entire Fourier plane which all of these algorithms must perform many times. The largest optimal filter which has been designed by any of these methods was 17 X 17 and that filter possessed octagonal symmetry. The use of optimal techniques is thus confined to relatively small impulse responses.

Several methods exist for designing 2-*D* FIR filters from 1-*D* FIR designs. These methods offer the advantage of using relatively little computer time but in general the resulting filters are not optimal, although they can be very good.

The first technique is for approximating ideal frequency responses which are separable

$$I(e^{j\mu}, e^{j\nu}) = I_1(e^{j\mu}) I_2(e^{j\nu}). \tag{44}$$

Here I_1 and I_2 can be designed using any 1-*D* algorithm. If the 1-*D* filters are optimal, the 2-*D* filter will be optimal. This sort of an approach can also be used for filters which are sums of separable filters

$$I(e^{j\mu}, e^{j\nu}) = \sum_{i=1}^{N} I_{1i}(e^{j\mu}) I_{2i}(e^{j\nu}). \tag{45}$$

McClellan (see [1]) has proposed an ingenious approach for converting 1-*D* linear phase designs into 2-*D* linear phase designs. His algorithm converts a 1-*D* design $H(e^{j\omega})$ into a 2-*D* design $H(e^{j\mu}, e^{j\nu})$ by means of the substitution

$$\cos \omega = A \cos \mu + B \cos \nu + C \cos \mu \cos \nu + D. \tag{46}$$

This technique will not be elaborated here since it is discussed elsewhere in this issue [1] but it is an extremely useful approach for designing many common types of 2-*D* filters. (Circularly symmetric filters can be approximated by $A = B = C = -D = \frac{1}{2}$.) High order filters (e.g., 63 X 63 or 127 X 127) can be designed in seconds.

Projections may also be useful for some 2-*D* FIR filter design problems. To design a filter which approximates an ideal 2-*D* function, we can first evaluate the ideal on the appropriate slice to obtain a 1-*D* ideal function. Then using any 1-*D* FIR filter design algorithm, we may obtain a 1-*D* ideal. The unit sample response of this filter can then be back-projected to give a 2-*D* unit sample response, which hopefully

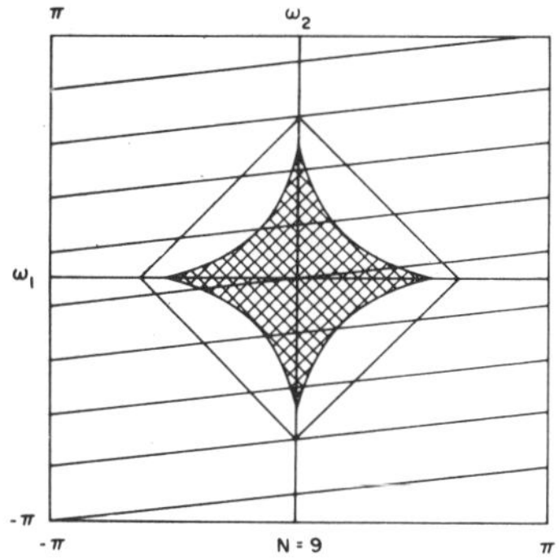

Fig. 11. An ideal filter characteristic which we want to approximate by a 9 X 9 point FIR filter. The cross hatched region represents a passband, the area outside the diamond is the stopband, and the area between the two is the transition band. The $N = 9$ slice is superimposed.

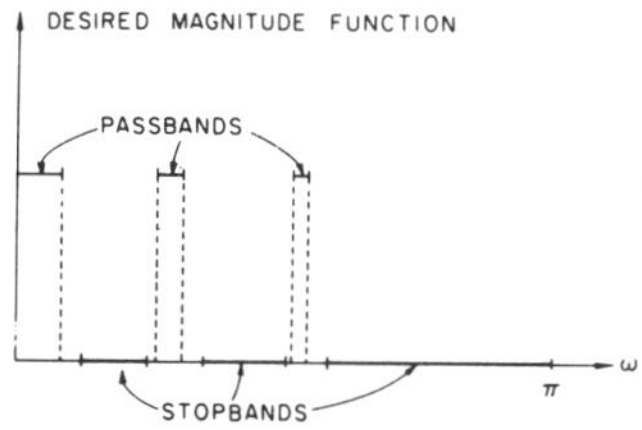

Fig. 12. The frequency response of an 81 point one-dimensional filter which approximates the $N = 9$ slice of the ideal.

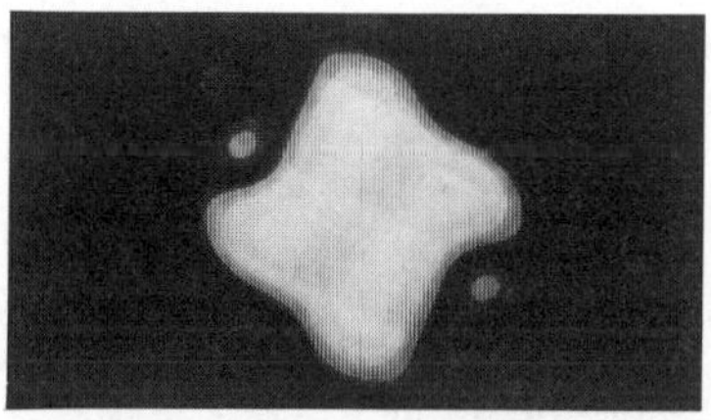

Fig. 13. An intensity plot of the log-magnitude of the resulting 2-*D* filter.

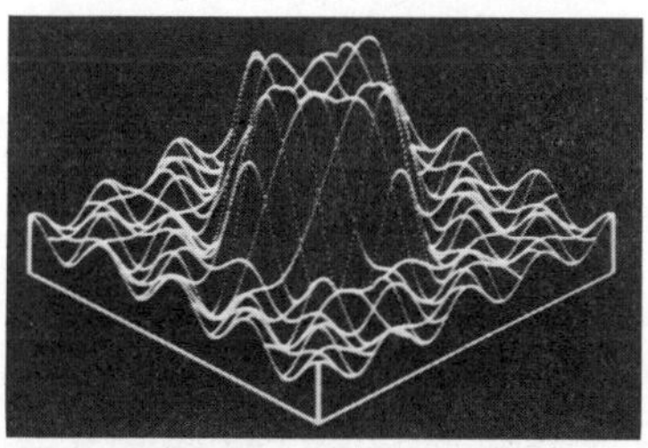

Fig. 14. A surface drawing of the filter of Fig. 13, which is an approximation to the filter of Fig. 11.

approximates the 2-*D* ideal. This technique is outlined in Figs. 11–14. In Fig. 11 is the 2-*D* ideal. It is of the form of a low-pass filter with a cusp-shaped passband and the stopband is the exterior of a diamond. The region between the pass and stop bands is a transition band where the ideal response is not specified.

In Fig. 12 is shown an approximation to the 1-*D* ideal. The approximation was obtained by using a Parks–McClellan equiripple design [21] with an 81 point impulse response. The frequency response of the 9 X 9 2-*D* filter, which results from back projecting the filter of Fig. 12, is shown in Figs. 13 and 14. Fig. 13 is an intensity plot of the log magnitude of the frequency response, and Fig. 14 is a surface drawing of the frequency response. Note that the approximation does not possess all of the symmetry of the ideal.

Recently, Hersey and Mersereau [43] have been working on ways to impose symmetry for FIR designs on a slice. This is easily done by requiring that the 2-*D* approximating function have certain coefficients equal to each other. However, this means the set of basis functions no longer satisfies the Haar condition, and degeneracies may occur if Remez' second exchange algorithm is used (as in a Parks–McClellan design). Hersey and Mersereau avoid this problem by using Remez' first exchange algorithm, which does not require the Haar condition, to perform the necessary minimization. They have successfully designed slice filters with 8-fold symmetry which closely approximate the optimal minimax designs.

Of all the design techniques, these authors feel that the most valuable are windowing, the transformation approach of McClellan and for very small filters, the optimum minimax designs of Kamp and Thiran or Hersey and Mersereau. The former two techniques are quite efficient computationally. The method of windows is the only one which can be used to approximate a completely arbitrary complex frequency response.

B. IIR Filters

In 2-*D* filter design, much less work has been done on recursive filters than on FIR filters. Recursive filters have the potential of being able to realize a given frequency response in fewer multiplications than in IIR filter. However, the recursive filter's frequency response is not linear in the design parameters, so the linear minimization techniques so successful for designing FIR filters are not directly applicable to the recursive case.

Shanks *et al.* [15] have given us two methods for designing recursive filters. The first one we will consider is the space domain approximation. Assume we are given an ideal impulse response $i(m, n)$ which we wish to approximate. (It may have come from an ideal frequency response by inverse transforming.) If we design a recursive filter with impulse response $h(m, n)$, a numerator polynomial with coefficients $a(m, n)$ and a denominator polynomial with coefficients $b(m, n)$, we know that

$$b(m, n) ** h(m, n) = a(m, n). \tag{47}$$

Both b and a are finite extent arrays. So for certain values of (m, n) in a region R we will have

$$b(m, n) ** h(m, n) = 0, \quad (m, n) \in R. \tag{48}$$

If $h(m, n)$ is a good approximation to $i(m, n)$ we can write

$$b(m, n) ** i(m, n) = e(m, n) \simeq 0, \quad (m, n) \in R \tag{49}$$

where $e(m, n)$ is an error term. If we minimize

$$\sum_{m,n} |e(m, n)|^2, \quad (m, n) \in R \tag{50}$$

by differentiating with respect to all the coefficients $b(m, n)$ and setting the derivatives equal to zero, we will get a set of linear equations for the $b(m, n)$ which are easily solved. The coefficients $a(m, n)$ may be computed by minimizing the mean-square error between $h(m, n)$ and $i(m, n)$, or less accurately, by convolving

$$b(m, n) ** i(m, n) \simeq a(m, n).$$

Burrus and Parks [44] give an excellent discussion of this technique for 1-*D* recursive filter design, and most of their results may be extended to two dimensions.

Shanks also gives a conceptually simple method of filter design in the frequency domain. In 1-*D*, recursive filters may be designed by applying the bilinear transformation to an analog filter (e.g., Butterworth, Chebyshev, or elliptic). This is not as easily done in 2-*D*, since very little design work has been done for 2-*D* analog filters. However, for certain applications, we can use 1-*D* analog filters as 2-*D* analog filters. Let $H(s)$ be a classical analog filter's Laplace transform. Then define a 2-*D* analog filter by

$$H(s_1, s_2) = H(s_1). \tag{51}$$

That is, it is constant as a function of s_2, and some function of s_1. By applying the transformation

$$s_1 = s_1' \cos\theta + s_2' \sin\theta$$

$$s_2 = s_2' \cos\theta - s_1' \sin\theta$$

we can rotate the filter response in the 2-*D* Fourier plane by the angle θ. If we now apply the bilinear transformation to each variable

$$s_1' = \frac{w-1}{w+1} \qquad s_2' = \frac{z-1}{z+1}$$

we will have a 2-*D* recursive filter. By cascading filters of different rotations and different 1-*D* prototypes, many filters can be synthesized. They gave an example of a 2-*D* filter with a pass region in the middle of the first quadrant by cascading a low-pass filter rotated at 45° with a bandpass filter rotated at -45°.

The primary drawback to this method is that a trial and error method may be needed to synthesize a desired filter. In addition, if the number of sections which must be cascaded is high, it may prove more efficient (in terms of a realization) to use an optimum design technique.

Costa and Venetsanopoulos [45] have used this design method to design good approximations to circularly symmetric filters by cascading rotated filters. In addition they have proved that if the angle θ lies within the region $270° < \theta < 360°$ the rotated filter will be stable. To implement filters with angles of rotation not in this range, the input data is simply rotated by the appropriate multiple of 90°.

Recently several workers [13], [47], [48] began exploring optimum design techniques for 2-*D* recursive filters. These methods are based on the 2-*D* differential correction method for rational approximation [46] which has been used successfully for 1-*D* recursive filter design [49], [50].

Basically the algorithm is as follows. We are given an ideal magnitude-squared frequency response $I(e^{j\mu}, e^{j\nu})$ which we wish to approximate with a ratio of 2-*D* polynomials.

$$I(e^{j\mu}, e^{j\nu}) \simeq R(e^{j\mu}, e^{j\nu}) = \frac{P(\mu, \nu)}{Q(\mu, \nu)}$$

$$= |H(e^{j\mu}, e^{j\nu})|^2 = \frac{|A(e^{j\mu}, e^{j\nu})|^2}{|B(e^{j\mu}, e^{j\nu})|^2}.$$

P and Q are 2-*D* cosine polynomials of the form

$$P(\mu, \nu) = p(0, 0) + 2 \sum_{m,n} 2p(m, n) \cos(m\mu + n\nu).$$

For the error criterion we will use the maximum difference between the ideal and the design functions

$$\Delta = \max_{\mu,\nu} |I(e^{j\mu}, e^{j\nu}) - R(e^{j\mu}, e^{j\nu})|.$$

Thus we seek to minimize Δ. The algorithm is iterative and at the kth iteration we have an approximation $R_k(e^{j\mu}, e^{j\nu})$:

1) Calculate the error Δ_k for $R_k(e^{j\mu}, e^{j\nu})$.
2) Next form the functional
$$\delta_k(R) = \max_{\mu,\nu} \left\{ \frac{|IQ - P| - \Delta_k Q}{Q_k} \right\}$$
where $R = P/Q$ and Q_k is the denominator of $R_k = P_k/Q_k$.
3) Find R_{k+1} which minimized δ_k by varying the coefficients of P and Q.
4) If $\min \delta_k \geqslant 0$ stop. R_k is a best approximation to $I(\mu, \nu)$. If not, go to step 1 and start over with R_{k+1}.

The minimization of δ_k can be structured into a linear programming problem, thus using linear programming in an iterative fashion to solve a nonlinear optimization problem.

There are two serious drawbacks to this method. Computation time is large. (Times on the order of 1 hour are not uncommon for filters with 42 coefficients and 544 constraint points [14].) Also when the design is finished one has the cosine polynomials $P(\mu, \nu)$ and $Q(\mu, \nu)$. However, it remains to solve for $A(e^{j\mu}, e^{j\nu})$ and $B(e^{j\mu}, e^{j\nu})$. This can be done by using a 2-*D* DHT as indicated in section III, but an infinite extent answer may result which then may have to be truncated in some fashion.

Bednar and Farmer [47] took a slightly different approach. They set up the minimization of Δ in terms of A and B directly, using them to approximate an ideal magnitude (rather than magnitude-square) function. In order to do this, however, the minimization of Δ becomes a nonlinear optimization problem. This leads to the usual numerical problems with nonlinear optimization, namely very large amounts of computation for minimizations with many variables, sensitivity to starting points, and the possibility of converging to a local rather than global minimum. In addition, it is necessary to check the stability of the designed filter at each iteration of the algorithm.

VI. Summary

We have tried to present a number of recently developed results for the design and implementation of 2-*D* digital filters. Although the rapidly changing nature of the subject is such that no such compendium could ever be complete, we have hopefully presented the flavor of what types of techniques are available and perhaps more importantly, which are still needed.

References

[1] L. R. Rabiner, J. H. McClellan, and T. W. Parks, "FIR digital filter design techniques using weighted Chebyshev approximation," this issue, pp. 595-610.
[2] B. R. Hunt, "Digital image processing," this issue, pp. 693-708.
[3] L. C. Wood and S. Treitel, "Seismic signal processing," this issue, pp. 649-661.
[4] T. S. Huang, W. F. Schreiber, and O. Tretiak, "Image processing," *Proc. IEEE*, vol. 59, pp. 1586-1609, Nov. 1971.

[5] R. E. Crochiere and A. V. Oppenheim, "Analysis of linear digital networks," this issue, pp. 581–595.

[6] J. W. Goodman, *Introduction to Fourier Optics*. New York: McGraw-Hill, 1968.

[7] D. P. Peterson and D. Middleton, "Sampling and reconstruction of wave number limited functions in N-dimensional Euclidean spaces," *Inform. Contr.*, vol. 5, pp. 279–323, 1962.

[8] A. V. Oppenheim and D. H. Johnson, "Discrete representation of signals," *Proc. IEEE*, vol. 60, pp. 681–691, June 1972.

[9] G. D. Bergland, "A guided tour of the fast Fourier transform," *IEEE Spectrum*, vol. 6, pp. 41–52, July 1969.

[10] R. M. Mersereau and D. E. Dudgeon, "The representation of two-dimensional sequences as one-dimensional sequences," *IEEE Trans. Acoust., Speech, and Sig. Proc.*, vol. ASSP-22, No. 5, pp. 320–325, Oct. 1974.

[11] J. L. Shanks, "Recursive filters for digital processing," *Geophysics*, vol. 32, pp. 33–51, 1967.

[12] C. S. Burrus and T. W. Parks, "Time domain design of recursive digital filters," *IEEE Trans. Audio Electroacoust.*, vol. AU-18, pp. 137–141, June 1970.

[13] J. L. Shanks, "Two-dimensional recursive filters," in *1969 SWIEEECO Rec.*, pp. 19E1–19E8.

[14] D. E. Dudgeon, "Two-dimensional recursive filtering," Sc.D. dissertation Dep. Elec. Eng., Massachusetts Inst. Technol., Cambridge, 1974.

[15] J. L. Shanks, S. Treitel, and J. H. Justice, "Stability and synthesis of two-dimensional recursive filters," *IEEE Trans. Audio Electroacoust.*, vol. AU-20, pp. 115–128, June 1972.

[16] C. Farmer and J. B. Bednar, "Stability of spatial digital filters," *Math. Biosci*, vol. 14, pp. 113–119, 1972.

[17] T. S. Huang, "Stability of two-dimensional recursive filters," *IEEE Trans. Audio Electroacoust.*, vol. AU-20, pp. 158–163, June 1972.

[18] H. G. Ansell, "On certain two-variable generalizations of circuit theory, with applications to networks of transmission lines and lumped reactances," *IEEE Trans. Circuit Theory*, vol. CT-11, pp. 214–223, June 1964.

[19] B. D. O. Anderson and E. I. Jury, "Stability test for two-dimensional recursive filters," *IEEE Trans. Audio Electroacoust.*, vol. AU-21, pp. 366–372, Aug. 1973.

[20] G. A. Maria and M. M. Fahmy, "On the stability of two-dimensional digital filters," *IEEE Trans. Audio Electroacoust.*, vol. AU-21, pp. 470–472, Oct. 1973.

[21] J. H. Justice and J. L. Shanks, "Stability criterion for N-dimensional digital filters," *IEEE Trans. Autom. Contr.*, vol. AC-18, pp. 284–286, June 1973.

[22] B. D. O. Anderson and E. I. Jury, "Stability of multidimensional digital filters," *IEEE Trans. Circuit and Syst.*, vol. CAS-21, pp. 300–304, Mar. 1974.

[23] R. R. Read and S. Treitel, "The stabilization of two-dimensional recursive filters via the discrete Hilbert transform," *IEEE Trans. Geoscience Elec.*, vol. GE-11, pp. 153–160, July 1973.

[24] P. Pistor, "Stability criterion for recursive filters," *IBM J. Res. Dev.*, vol. 18, pp. 59–71, Jan. 1974.

[25] D. E. Dudgeon, "Using a two-dimensional discrete Hilbert transform to factor two-dimensional polynomials," M.I.T. Res. Lab. Electron., Quart. Progr. Rep. 113, pp. 143–150, Apr. 1974.

[26] D. E. Dudgeon, "Existence of Cepstra for Two-Dimensional Rational Polynomials," M.I.T. Res. Lab. Electron., Quart. Progr. Rep. 114, pp. 153–157, July 15, 1974.

[27] D. E. Dudgeon, "Recursibility of two-dimensional difference equations," M.I.T. Res. Lab. of Electron., Quart. Progr. Rep. 113, pp. 151–158, Apr. 1974.

[28] R. C. Agarwal and C. S. Burrus, "Number theoretic transforms to implement fast digital convolution, this issue, pp. 550–560.

[29] A. V. Oppenheim and R. W. Schafer, *Digital Signal Processing*. Englewood Cliffs, N.J.: Prentice-Hall, 1974.

[30] B. R. Hunt, "Minimizing the computation time for using the technique of sectioning for digital filtering of pictures," *IEEE Trans. Comp.*, vol. C-21, pp. 1219–1222, Nov. 1972.

[31] D. P. MacAdam, "Digital image restoration by constrained deconvolution," *J. Opt. Soc. Amer.*, vol. 60, pp. 1617–1627, Oct. 1970.

[32] R. C. Agarwal and C. S. Burrus, "Fast one-dimensional digital convolution by multidimensional techniques," *IEEE Trans. Acoust., Speech, Sig. Proc.*, vol. ASSP-22, pp. 1–10, Feb. 1974.

[33] S. Treitel and J. L. Shanks, "The design of multistage separable planar filters," *IEEE Trans. Geosci. Elec.*, vol. GE-9, pp. 10–27, 1971.

[34] O. J. Tretiak, "Approximation of a matrix by a sum of products," M.I.T. Res. Lab. of Electron., Quart. Progr. Rep. 98, pp. 157–169, July 1970.

[35] M. D. Ni and J. K. Aggarwal, "Two-dimensional digital filtering and its error analysis," Electron. Res. Cent., Univ. of Texas, Austin, *Inform. Syst. Research Labor. Tech. Rep. 157*, Mar. 1974. Also in *IEEE Trans. Comp.*, vol. C-23, pp. 942–953, Sept. 1974.

[36] M. T. Manry and J. K. Aggarwal, "Picture processing using one-dimensional implementations of discrete planar filters," *IEEE Trans. Acoust. Speech, Sig. Proc.*, vol. ASSP-22, pp. 164–173, June 1974.

[37] T. S. Huang, "Two-dimensional windows," *IEEE Trans. Audio Electroacoust.*, vol. AU-20, pp. 88–90, Mar. 1972.

[38] L. R. Rabiner and B. Gold, *Theory and Applications of Digital Signal Processing*. Englewood Cliffs, N.J.: Prentice-Hall, Inc., New York, 1974.

[39] J. V. Hu and L. R. Rabiner, "Design techniques for two-dimensional digital filters," *IEEE Trans. Audio Electroacoust.*, vol. AU-20, pp. 249–257, Oct. 1972.

[40] J. G. Fiasconaro, "Two-Dimensional Nonrecursive Digital Filters," M.I.T. Lincoln Lab., Lexington, Mass., Tech. Rep. 505.

[41] Y. Kamp and J. P. Thiran, "Chebyshev approximation for two-dimensional nonrecursive digital filters," MBLE Res. Lab. Rep. R245, Mar. 1974. Also to be published in *IEEE Trans. Circuits and Syst.*

[42] H. S. Hersey and R. M. Mersereau, "An algorithm to perform minimax approximation in the absence of the Haar condition," M.I.T. Res. Lab. Electron., Quart. Progr. Rep. 114, Cambridge, Mass., pp. 157–160, July 15, 1974.

[43] ——, "Design techniques for two-dimensional FIR circularly symmetric digital filters," M.I.T. Res. Lab. Electron., Quart. Progr. Rep. 114, pp. 160–171, July 15, 1974.

[44] C. S. Burrus and T. W. Parks, "Time domain design of recursive digital filters," *IEEE Trans. Audio Electroacoust.*, vol. AU-18, pp. 137–141, June 1970.

[45] J. M. Costa and A. N. Venetsanopoulos, "Design of circularly symmetric two-dimensional recursive filters," *IEEE Trans. Acoust., Speech, Sig. Proc.*, vol. ASSP-22, pp. 432–443, Dec. 1974.

[46] I. Barrodale, M. J. D. Powell, and F. D. K. Roberts, "The differential correction algorithm for rational l_∞ approximation," *SIAM J. Numer. Anal.*, vol. 9, pp. 493–504, Sept. 1972.

[47] J. B. Bednar and C. Farmer, "Implementation of Chebyshev digital filter design," to be published in *IEEE Trans. Circuits and Syst.*

[48] E. H. Kaufman, Jr., and C. D. Taylor, "An application of linear programming to rational approximation," to be published in *Rocky Mountain J. Math.*

[49] D. E. Dudgeon, "Recursive filter design using differential correction," *IEEE Trans. Acoust., Speech, Sig. Proc.*, vol. ASSP-22, pp. 443–448, Dec. 1974.

[50] F. Brophy and A. Salazar, "Synthesis of spectrum shaping digital filters of recursive design," in *Rec Int. Symp. Circuits and Systems*, Apr. 1974. Also to be published in *IEEE Trans. Circuits and Syst.*

Editors' Comments on Papers 3, 4, and 5

3 **PISTOR**
Stability Criterion for Recursive Filters

4 **EKSTROM and WOODS**
Two-Dimensional Spectral Factorization with Applications in Recursive Digital Filtering

5 **DUDGEON**
The Existence of Cepstra for Two-Dimensional Rational Polynomials

SPECTRAL FACTORIZATION

One of the fundamental problems in the direct generalization of 1-D digital filtering techniques to two-dimensions is the difficulty in factoring an arbitrary polynomial of two independent variables. Factorization of 1-D polynomials into products of first order terms plays an important role in almost all aspects of 1-D digital filtering. This includes the development of stability tests, filter design algorithms, and implementation schemes and structures. The existence of the 1-D factorization is guaranteed by the Fundamental Theorem of Algebra. These applications generally involve the factorization of rational transfer functions into terms associated with prescribed regions of analyticity in the complex plane. The factorization is generally accomplished using numerical root-finding algorithms.

While 2-D polynomial factorization is equally important in the development of 2-D digital filters, the classical root-finding approach is not viable because of the lack of a 2-D equivalent to the Fundamental Theorem of Algebra. An alternate technique has recently been devised, however, which, in analogy with the 1-D case, allows the factorization of 2-D transfer functions into terms with specified regions of analyticity. This approach is developed in the papers by Pistor (Paper 3) and Ekstrom and Woods (Paper 4). Here the authors consider the factorization of a special type of two-variable polynomial $S(z_1,z_2)$ which is of zero phase and is nonnegative for all values of z_1 and z_2. This type of polynomials is obtained by analytic continuation from the square of the magnitude of an arbitrary polynomial evaluated for $|z_1| = 1$ and $|z_2| = 1$ and as a result, is often called the spectral density function.

The factorization presented by Pistor (Paper 3) is a direct extension of the 1-D spectral factorization into two dimensions. It decomposes the spectral density into four factors, each of which can be associated with a "causal" filter of appropriate direction of recursion. One such filter form is given in Eq. (1.4).

Ekstrom and Woods (Paper 4) generalized this concept of spectral factorization, presenting a family of canonical factorizations. As with Pistor's technique, their approach is based on decompositions of the complex cepstrum. Their decompositions are chosen to provide factors having prescribed regions of analyticity in the z_1, z_2 space. Because of the added dimension in the 2-D case, there is a large number of possible choices and subsequent multitude of factorizations. Specific examples included are two-, four-, and eight-term factorizations. The two-term result is a particularly important one in that the factors can be associated with half-plane recursions of the form in Eq. (1.5).

These factorizations have been restricted to decomposing spectral density functions to avoid possible problems with existence of the cepstra. Dudgeon (Paper 5) has shown that 2-D cepstra can be defined for 2-D rational polynomials, provided essentially that the Fourier transform is nonzero and bounded at all finite frequencies. A procedure for computing this 2-D cepstrum is given in a later contribution (Paper 10). Thus, it is possible to directly extend the factorization scheme in Paper 4 to this more general class of functions. In contrast to the one-dimensional case, a basic problem associated with the spectral decomposition of two-variable polynomials is that the decomposed factors are, in general, not finite-degree polynomials. Further factorization studies are continuing in the direction of obtaining approximate decompositions involving a product of finite order one-quadrant or half-plane terms (Ekstrom and Woods, 1976 and Ekstrom and Twogood, 1977).

REFERENCES

Ekstrom, M. P. and J. W. Woods. 1976. "Some Results on the Design of Two-Dimensional, Half-Plane Recursive Filters," *Proc. 1976 IEEE Internat. Symp. Circuits Syst.*, 373.

Ekstrom, M. P. and R. E. Twogood. 1977. "Finite Order, Recursive Models for Two-Dimensional Random Fields," *Proc. 20th Midwest Symp. on Circuits Syst.*, 188–189.

3

Reprinted from *IBM J. Res. Dev.* **18**(1):59–71 (1974)

Stability Criterion for Recursive Filters

P. Pistor

Abstract: A new criterion is derived that relates the stability of two-dimensional recursive filters to the properties of its cepstrum. It provides a procedure for the decomposition of unstable recursive filters having nonzero, nonimaginary frequency response into stable recursive filters. The optimal solution of the decomposition problem is discussed, including numerical implementation and nonrecursive solutions. Several numerical examples show the potentialities and limitations of the rules for decomposition and for truncation of the operators.

Contents

Introduction

In the development of two-dimensional recursive filters the two problem areas of main interest are the stability criteria and the design of stable recursive filters. Except for filters with very few samples, existing stability criteria [1–4] are difficult to apply and thus it has been virtually impossible to translate these criteria into a practical design procedure. Moreover, the criteria could not be used to justify a design procedure [2] that appears to provide stable recursive filters.

The present paper proposes a criterion that is based on the relationship of the stability of recursive filters to the absolute summability of certain operators called *cepstra* [5]. The z transform of these operators is given by the logarithm of the z transform of the recursive filters. By means of this criterion we treat a special design problem [4] concerning the decomposition of unstable recursive filters having nonzero and nonimaginary frequency response into stable recursive filters. For the case of one-dimensional filters this problem was solved by the decomposition into minimum-delay and maximum-delay operators [6, 7]. Our stability criterion leads to a proof of the existence of a similar solution for the two-dimensional case.

This paper is organized into five main sections: a compilation, for reference purposes, of z transform relations; a demonstration of the need for a decomposition procedure for zero-phase filters; a proof of the stability criterion; a solution of the decomposition problem, including methods of numerical implementation and related nonrecursive solutions; and some numerical examples.

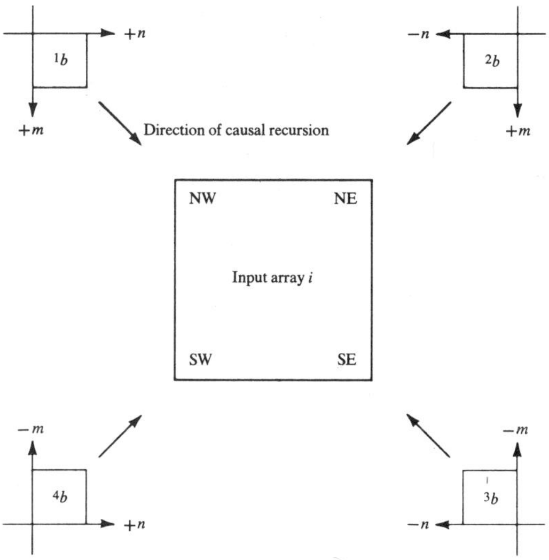

Figure 1 Four directions of recursion from the four corners of an input array. The input array is to be understood as a first-, second-, third-, or fourth-quadrant function, depending on the direction of the recursion.

Space domain operations and their z transforms

The z transform is a convenient means for characterizing certain operations involving sampled functions. Some relations [1–4, 8] needed for subsequent discussions are summarized in this section.

• *Z transform of two-dimensional functions*

The z transform of a periodically sampled, two-dimensional function

$$b = \{b_{m,n}\} \tag{1}$$

is defined by

$$B(z_1, z_2) = \sum_{m=-\infty}^{\infty} \sum_{n=-\infty}^{\infty} b_{m,n} z_1^{\,m} z_2^{\,n}, \tag{2}$$

where z_1 and z_2 denote complex variables, and $b_{m,n}$ is understood to be zero for subscript pairs (m,n) that do not belong to the definition set of b.

If b is defined for nonnegative subscripts only, it is called a *first-quadrant function* [2] and is denoted by a left superscript as

$${}^1b = \{{}^1b_{m,n}\}_{\substack{m\geq 0\\ n\geq 0}} \text{ (first-quadrant function).} \tag{3}$$

Similarly, second-, third- and fourth-quadrant functions can be defined successively as

$${}^2b = \{{}^2b_{m,n}\}_{\substack{m\geq 0\\ n\leq 0}}, \tag{4}$$

$${}^3b = \{{}^3b_{m,n}\}_{\substack{m\leq 0\\ n\leq 0}}, \tag{5}$$

$${}^4b = \{{}^4b_{m,n}\}_{\substack{m\leq 0\\ n\geq 0}} \text{ (see [9]).} \tag{6}$$

The z transform pairs in Eqs. (7)–(12), denoted by $\leftrightarrow$, relate some basic spatial operations to operations in the z transform domain [2, 8]. These relations can readily be improved by comparing terms of the same degree in z_1 and z_2.

Basic pair:

$$\{b_{m,n}\} \quad \leftrightarrow B(z_1, z_2)$$

Translation:

$$\{b_{m,n}'\} = \{b_{m-\alpha,n-\beta}\} \leftrightarrow B'(z_1, z_2) = z_1^{\,\alpha} z_2^{\,\beta} B(z_1, z_2), \tag{7}$$

Rotation through 180° around (0,0):

$$\{b_{m,n}'\} = \{b_{-m,-n}\} \quad \leftrightarrow B'(z_1, z_2) = B(1/z_1, 1/z_2), \tag{8}$$

Reflection at axis $m = 0$ (column reversion):

$$\{b_{m,n}'\} = \{b_{-m,n}\} \quad \leftrightarrow B'(z_1, z_2) = B(1/z_1, z_2), \tag{9}$$

Reflection at axis $n = 0$ (row reversion):

$$\{b_{m,n}'\} = \{b_{m,-n}\} \quad \leftrightarrow B'(z_1, z_2) = B(z_1, 1/z_2), \tag{10}$$

Weighting by row number:

$$\{b_{m,n}'\} = \{mb_{m,n}\} \quad \leftrightarrow B'(z_1, z_2) = z_1(\partial/\partial z_1) B(z_1, z_2), \tag{11}$$

Weighting by column number:

$$\{b_{m,n}'\} = \{nb_{m,n}\} \quad \leftrightarrow B'(z_1, z_2) = z_2(\partial/\partial z_2) B(z_1, z_2). \tag{12}$$

• *Convolutional filters*

The input-output relation of a convolutional filter f,

$$o = i * f, \tag{13}$$

where

$$o_{p,q} = \sum_{m=-\infty}^{\infty} \sum_{n=-\infty}^{\infty} i_{m,n} f_{p-m,q-n}, \tag{14}$$

is written in z transform notation as the product

$$O(z_1, z_2) = I(z_1, z_2)\, F(z_1, z_2). \tag{15}$$

A formal proof of this equation may be found by the same method as for Eqs. (7)–(12).

Definition 1

A filter is stable if and only if its response to any bounded input is bounded.

For a convolutional filter the stability criterion [3,4], Theorem 1, can be proved.

Theorem 1
A convolutional filter is stable if and only if

$$\sum_{m=-\infty}^{\infty} \sum_{n=-\infty}^{\infty} |f_{m,n}| < \infty. \qquad (16)$$

Condition (16) means that the z transform of f is absolutely convergent for all $(z_1, z_2) \in R$,

$$R = \{(z_1, z_2): |z_1| = 1 \wedge |z_2| = 1\}. \qquad (17)$$

• *Recursive filters*
The convolution of two discrete functions corresponds to a multiplication of their z transforms, whereas the input-output relation of recursive filters is characterized in the z transform domain by the quotient [10]

$$O(z_1, z_2) = I(z_1, z_2)/B(z_1, z_2) = I(z_1, z_2)F(z_1, z_2). \qquad (18)$$

If O, I, and B are replaced by first-quadrant functions ${}^1o$, ${}^1i$, and ${}^1b$ with

$${}^1i = \{{}^1i_{m,n}\}_{\substack{0 \le m \le M_i,\\ 0 \le n \le N_i}} \qquad (19)$$

$${}^1b = \{{}^1b_{m,n}\}_{\substack{0 \le m \le M_b,\\ 0 \le n \le N_b}} \qquad (20)$$

the space domain algorithm corresponding to Eq. (18) is

$${}^1o_{0,0} = {}^1i_{0,0}/{}^1b_{0,0},$$

$${}^1o_{p,q} = (1/{}^1b_{0,0})\left({}^1i_{p,q} - \sum_{m=0}^{\min(p,M_b)} \sum_{n=0}^{\min(q,N_b)} {}^1b_{m,n}\,{}^1o_{p-m,q-n}\right),$$
$$(m,n) \ne (0,0)$$
$$(p, q) \in \{(p, q): (p, q) \ne (0, 0) \wedge p \le M_i \wedge p \le N_i\}$$

$${}^1o_{p,q} = (1/{}^1b_{0,0})\left(- \sum_{m=0}^{\min(p,M_b)} \sum_{n=0}^{\min(q,N_b)} {}^1b_{m,n}\,{}^1o_{p-m,q-n}\right),$$
$$(m, n) \ne (0, 0).$$
$$(p, q) \in \{(p, q): (p, q) \ne (0, 0) \wedge (p > M_i \vee q > N_i)\}. \qquad (21)$$

These formulas strongly resemble the convolution algorithm. In fact, they can be derived from Eq. (14) by replacing o, i, and f by ${}^1i$, ${}^1b$, and ${}^1o$, respectively, and solving for ${}^1o_{p,q}$.

To illustrate the use of Eqs. (21), consider the input

$${}^1i = \begin{bmatrix} {}^1i_{0,0} & {}^1i_{0,1} & {}^1i_{0,2} \\ {}^1i_{1,0} & {}^1i_{1,1} & {}^1i_{1,2} \\ {}^1i_{2,0} & {}^1i_{2,1} & {}^1i_{2,2} \end{bmatrix} = \begin{bmatrix} 3 & 4 & 1 \\ 4 & 5 & 1 \\ 1 & 1 & 0 \end{bmatrix},$$

which is to be recursively filtered by

$${}^1b = \begin{bmatrix} {}^1b_{0,0} & {}^1b_{0,1} \\ {}^1b_{1,0} & {}^1b_{1,1} \end{bmatrix} = \begin{bmatrix} 3 & 1 \\ 1 & 0 \end{bmatrix}.$$

The output value to be determined first is ${}^1o_{0,0}$:

$${}^1o_{0,0} = {}^1i_{0,0}/{}^1b_{0,0} = 3/3 = 1.$$

To understand the operation of the last two formulas of Eqs. (21), we form the mask from ${}^1b$,

0 *1*
1 *3*,

and draw it across the matrices ${}^1i$ and ${}^1o$ synchronously. In the output arrays ${}^1o$ of the following examples, the symbol * designates elements not yet determined. The numeral *3* refers to the value just computed. The other entries of the mask form a dot product with the output values to which they refer. These values must be known, of course, from former steps. The result of the dot product is subtracted from the input value indicated by *3*, and this result is then divided by ${}^1b_{0,0}$.

The examples that follow demonstrate the computation of ${}^1o_{1,0}$, ${}^1o_{2,0}$, ${}^1o_{3,0}$ and ${}^1o_{1,1}$.

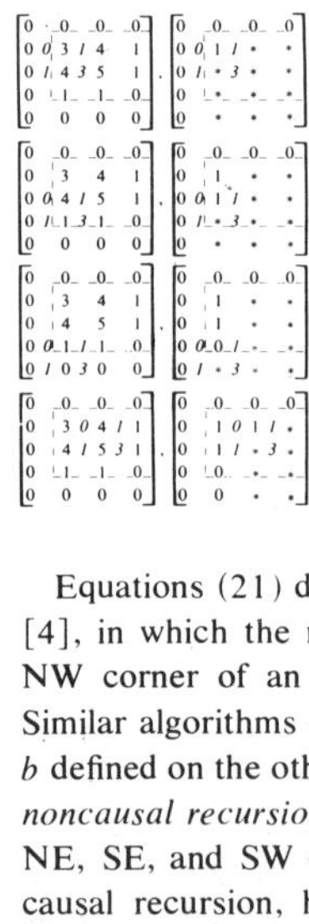

$${}^1o_{1,0} = \tfrac{1}{3}(4 - 1 \cdot 1 - 1 \cdot 0 - 0 \cdot 0) = 1;$$

$${}^1o_{2,0} = \tfrac{1}{3}(1 - 1 \cdot 1 - 1 \cdot 0 - 0 \cdot 0) = 0;$$

$${}^1o_{3,0} = \tfrac{1}{3}(0 - 1 \cdot 0 - 1 \cdot 0 - 0 \cdot 0) = 0;$$

$${}^1o_{1,1} = \tfrac{1}{3}(5 - 1 \cdot 1 - 1 \cdot 1 - 1 \cdot 0) = 1.$$

Equations (21) describe the so-called *causal recursion* [4], in which the recursive operator ${}^1b$ starts from the NW corner of an input array, as indicated in Fig. 1. Similar algorithms can be derived for functions i, o, and b defined on the other quadrants. The equations describe *noncausal recursions*, i.e., recursions that start from the NE, SE, and SW corners of an input array. Any noncausal recursion, however, can be transformed to the causal recursion. For instance, the noncausal recursive filtering of an input ${}^3i$ by an operator ${}^3b$ is transformed to a recursion with an operator ${}^1b$ simply by rotation of functions ${}^3i$ and ${}^3b$ through 180°. Because of Eq. (8), such rotations correspond to a replacement of (z_1, z_2) by $(1/z_1, 1/z_2)$. The transformation of ${}^3i$ and ${}^3b$ thus corresponds to the transition from

$$O_3(z_1, z_2) = I_3(z_1, z_2)/B_3(z_1, z_2) \qquad (22)$$

to

$$O_3(1/z_1, 1/z_2) = I_3(1/z_1, 1/z_2)/B_3(1/z_1, 1/z_2) \qquad (23)$$

in the z transform domain. This input-ouput relation is equivalent to

$$O_1(z_1, z_2) = I_1(z_1, z_2) / B_1(z_1, z_2). \quad (24)$$

For illustration, the input

$$^3i = \begin{bmatrix} ^3i_{-2,-2} & ^3i_{-2,-1} & ^3i_{-2,0} \\ ^3i_{-1,-2} & ^3i_{-1,-1} & ^3i_{-1,0} \\ ^3i_{0,-2} & ^3i_{0,-1} & ^3i_{0,0} \end{bmatrix} = \begin{bmatrix} 0 & 1 & 1 \\ 1 & 5 & 4 \\ 1 & 4 & 3 \end{bmatrix}$$

is recursively filtered by

$$^3b = \begin{bmatrix} ^3b_{-1,-1} & ^3b_{-1,0} \\ ^3b_{0,-1} & ^3b_{0,0} \end{bmatrix} = \begin{bmatrix} 0 & 1 \\ 1 & 3 \end{bmatrix}.$$

Instead of performing the appropriate noncausal recursion, which would start from the lower-right input element $^3i_{0,0}$, we rotate ^3i and ^3b, obtaining the functions ^1i and ^1b discussed in the preceding example. From the intermediate result,

$$^1o = \begin{bmatrix} ^1o_{0,0} & ^1o_{0,1} & ^1o_{0,2} & ^1o_{0,3} \cdots \\ ^1o_{1,0} & ^1o_{1,1} & ^1o_{1,2} & ^1o_{1,3} \cdots \\ ^1o_{2,0} & ^1o_{2,1} & ^1o_{2,2} & ^1o_{2,3} \cdots \\ ^1o_{3,0} & ^1o_{3,1} & ^1o_{3,2} & ^1o_{3,3} \cdots \\ \cdot & \cdot & \cdot & \cdot \\ \cdot & \cdot & \cdot & \cdot \end{bmatrix} = \begin{bmatrix} 1 & 1 & 0 & 0 & . & . \\ 1 & 1 & 0 & 0 & . & . \\ 0 & 0 & 0 & 0 & . & . \\ 0 & 0 & 0 & 0 & . & . \\ . & . & . & . & . & . \\ . & . & . & . & . & . \end{bmatrix},$$

we obtain the final result by inverting the above rotations as

$$^3o = \begin{bmatrix} . & . & . & . & . & . \\ . & . & . & . & . & . \\ . & . & . & 0 & 0 & 0 \\ . & . & . & 0 & 1 & 1 \\ . & . & . & 0 & 1 & 1 \end{bmatrix}.$$

Similarly, the replacement of z_1 by $1/z_2$, or z_2 by $1/z_2$, respectively, corresponds to the transition of the other noncausal recursions to the causal recursion.

If we put

$$^1i_{m,n} = \delta_{m,n} = \begin{cases} 1, & m = n = 0; \\ 0, & \text{else}, \end{cases} \quad (25)$$

the recursive algorithm (21) yields the convolutional filter ^1f, equivalent to the recursive filter $1/B_1(z_1, z_2) = F_1(z_1, z_2)$. According to theorem 1, the stability of these filters may be related as indicated in theorems 2, 3, and the subsequent discussion.

Theorem 2
A causal recursive filter $1/B_1(z_1, z_2)$ is stable if and only if there exists a stable filter ^1f such that $^1f * {}^1b = \delta$.

Farmer and Bednar [3] have given a stability criterion for recursive filters. This criterion comprises that of Shanks [2], which is confined to operators ^1b with a finite number of sample points.

Theorem 3
The causal recursive filter

$$F(z_1, z_2) = 1/B_1(z_1, z_2) = 1/\Big(\sum_{m=0}^{\infty} \sum_{n=0}^{\infty} {}^1b_{m,n} z_1^{\,m} z_2^{\,n}\Big), \quad (26)$$

where

$$\sum_{m=0}^{\infty} \sum_{n=0}^{\infty} |{}^1b_{m,n}| < \infty \quad (27)$$

is stable if and only if

$$B_1(z_1, z_2) \neq 0 \quad (28)$$

for all $(z_1, z_2) \in D_1$, where

$$D_1 = \{(z_1, z_2) \colon |z_1| \leq 1 \wedge |z_2| \leq 1\}. \quad (29)$$

The foregoing discussion shows how noncausal recursions can be transformed to the causal recursion. The following corollary is thus evident.

Corollary 1
Let the lth-quadrant function $^lb(l = 2, 3, 4)$ be a stable convolutional filter. Then the noncausal recursive filters $1/B_1(z_1, z_2)$ are stable if and only if $B_2(z_1, 1/z_2)$, $B_3(1/z_1, 1/z_2)$, and $B_4(1/z_1, z_2)$ are nonzero for all $(z_1, z_2) \in D_1$.

The functions that meet the conditions of Theorem 3 and Corollary 1 are characterized by Definition 2.

Definition 2
A one-quadrant function ^lb is called recursively stable if and only if ^lb is a stable convolutional filter and $1/B_l(z_1, z_2)$ is a stable recursive filter, either causal or noncausal.

• *Two-dimensional cepstrum*
We consider once more a function b and its z transform:

$$\{b_{m,n}\} \leftrightarrow B(z_1, z_2). \quad (30)$$

By formally taking the logarithm of B we obtain

$$\hat{B}(z_1, z_2) = \ln B(z_1, z_2). \quad (30')$$

The cepstrum of b, i.e., $\hat{b}$, is a function whose z transform is given by $\hat{B}(z_1, z_2)$:

$$\{\hat{b}_{m,n}\} \leftrightarrow \hat{B}(z_1, z_2) = \ln B(z_1, z_2). \quad (30'')$$

According to Eqs. (30), (30′), (30″), (13), and (15), we find, for the input-output relation $o = i * f$ of a convolutional filter,

$$\hat{O}(z_1, z_2) = \hat{I}(z_1, z_2) + \hat{F}(z_1, z_2),$$

which in turn leads to the additive relation in the cepstrum domain,

$$\hat{o} = \hat{i} + \hat{f}. \tag{31}$$

Similarly, the input-output relation (18) of a recursive filter is given by a difference in the cepstrum domain,

$$\hat{o} = \hat{i} - \hat{b}. \tag{32}$$

The problem stated in the next section is solved by means of cepstra.

Decomposition of unstable, recursive, zero-phase filters into stable recursive filters

We consider a real-valued discrete function c with a limited number of sample points, in which

$$c = \{c_{m,n}\}_{\substack{|m| \le 2\, M_c = \alpha \\ |n| \le 2\, N_c = \beta}}. \tag{33}$$

Its z transform is of zero phase and non-negative for all $(z_1, z_2) \in R$:

$$\text{Im}\,[C'(u, v)] \equiv 0, \tag{34}$$

$$C'(u, v) > 0, \tag{35}$$

where

$$C'(u, v) = \sum_{m=-\alpha}^{\alpha} \sum_{n=-\beta}^{\beta} c_{m,n} e^{-2\pi jum} e^{-2\pi jvn} \qquad j = \sqrt{-1}$$

$$= C(z_1, z_2)_{(z_1, z_2) \in R}. \tag{36}$$

Equations (34) and (36) imply central symmetry of c:

$$c_{m,n} = c_{-m,-n}. \tag{37}$$

Functions as specified in Eq. (33)ff. are found with recursive filters $F(z_1, z_2) = A(z_1, z_2)/C(z_1, z_2)$ [11] for the enhancement of images degraded by noise and by linear shift-invariant geometric distortion.

Because of Eq. (37), c is not a one-quadrant function. However, because c has a limited number of sample points, it can be transformed by translation to any quadrant function. Thus the term $1/C(z_1, z_2)$ could be associated with four different recursive filters. None, however, would be stable.

Proof By translation, the following z transforms originate from c [see Eq. (7)]:

$$C_1(z_1, z_2) = z_1^{\alpha} z_2^{\beta}\, C(z_1, z_2), \qquad \alpha = 2M_c,\ \beta = 2N_c \tag{38}$$

$$C_2(z_1, z_2) = z_1^{\alpha} z_2^{-\beta}\, C(z_1, z_2), \tag{39}$$

$$C_3(z_1, z_2) = z_1^{-\alpha} z_2^{-\beta}\, C(z_1, z_2), \tag{40}$$

$$C_4(z_1, z_2) = z_1^{-\alpha} z_2^{\beta}\, C(z_1, z_2). \tag{41}$$

According to Theorem 3 and Corollary 1, the recursive filters associated with these z transforms are unstable if a pair $(z_1, z_2) \in D_1$ exists such that

$$C_1(z_1, z_2) = C_2(z_1, 1/z_2)$$

$$= C_3(1/z_1, 1/z_2) = C_4(1/z_1, z_2) = 0. \tag{42}$$

By defining $z_2 = 1$, the existence of such a pair is proved, if a $z_1 = w$, $|w| \le 1$, can be found, which solves simultaneously the equations

$$z_1^{\alpha} C(z_1, 1) = \sum_{m=-\alpha}^{\alpha} \sum_{n=-\beta}^{\beta} c_{m,n} z_1^{\alpha+m} = z_1^{\alpha} G(z_1) = 0, \tag{43}$$

$$z_1^{\alpha} C(1/z_1, 1) = \sum_{m=-\alpha}^{\alpha} \sum_{n=-\beta}^{\beta} c_{m,n} z_1^{\alpha-m}$$

$$= z_1^{\alpha} G(1/z_1) = 0. \tag{44}$$

Since it can be concluded from the central symmetry of c that

$$G(z_1) = \sum_{m=-\alpha}^{\alpha} \sum_{n=-\beta}^{\beta} c_{m,n} z_1^{m} = G(1/z_1), \tag{45}$$

Eqs. (43) and (44) are identical. If they are solved by $z_1 = 0$, the sought pair is (0, 1). If not, Eq. (45) shows that every root $z_1 = w$ has a corresponding root $z_1 = 1/w$. Thus a pair $(z_1, z_2) \in D_1$ can always be found, solving Eqs. (42), and therefore no recursively stable operator can be obtained from c by translation.

In the one-dimensional case, unstable recursive filters $1/B(z_1)$, with no poles on the unit circle, can be decomposed into stable filters that recurse in opposite directions. Therefore the question arises whether, analogously, the unstable filter $1/C(z_1, z_2)$ can be decomposed into four stable filters that recurse in four different directions. More precisely, we ask for recursively stable one-quadrant functions ${}^1k$, ${}^2k$, ${}^3k$, and ${}^4k$ such that

$$1/C(z_1, z_2) = \prod_{l=1}^{4} 1/K_l^{-1}(z_1, z_2). \tag{46}$$

In a way similar to the solution for the one-dimensional case [12], we find a solution to this problem by transforming Eq. (46) to the cepstrum domain [viz., Eq. (30)],

$$\hat{c} = \sum_{l=1}^{4} \widehat{{}^l k}. \tag{47}$$

The results of the next section indicate how to decompose $\hat{c}$ and how to use the resulting functions to compute the operators that are sought.

Stability criterion for recursive filters based on two-dimensional cepstra

Let us assume that ${}^1b$ is a first-quadrant function that is recursively stable. This assumption implies, from Theorems 1, 2, and 3 and Definition 2, that

$$\sum_{m=0}^{\infty} \sum_{n=0}^{\infty} |{}^1f_{m,n}| < \infty, \tag{48}$$

$$^1b * {}^1f = \{\delta_{m,n}\}, \tag{49}$$

$$B_1(z_1, z_2) \neq 0, \ (z_1, z_2) \in D_1, \tag{50}$$

$$\sum_{m=0}^{\infty} \sum_{n=0}^{\infty} |^1b_{m,n}| < \infty. \tag{51}$$

From these facts we conclude that $\hat{B}_1(z_1, z_2) = \ln B_1(z_1, z_2)$ can be uniquely expanded into the power series

$$\sum_{m=0}^{\infty} \sum_{n=0}^{\infty} {}^1\hat{b}_{m,n} z_1^m z_2^n \tag{52}$$

for all $(z_1, z_2) \in D_1$.

First we consider the case $(z_1, z_2) \in D_2$,

$$D_2 = \{(z_1, z_2): |z_1| < 1 \wedge |z_2| < 1\}. \tag{53}$$

On this set $B_1(z_1, z_2)$ is regular [13], due to Eq. (51), and consequently $\hat{B}_1(z_1, z_2)$ is regular, too, because $B_1(z_1, z_2)$ is nonzero [condition (50)]. Thus a unique expansion (52) of $\hat{B}_1$ exists. The coefficient $^1\hat{b}_{0,0}$ of this power series is readily determined by evaluating

$$\hat{B}_1(z_1, z_2) = \ln B_1(z_1, z_2) \tag{54}$$

for $(z_1, z_2) = (0, 0)$:

$$^1\hat{b}_{0,0} = \ln {}^1b_{0,0}. \tag{55}$$

To find the other coefficients we differentiate Eq. (54) with respect to z_1 and z_2 and replace $1/B_1(z_1, z_2)$ by $F_1(z_1, z_2)$:

$$z_1(\partial/\partial z_1)\hat{B}_1(z_1, z_2) = F_1(z_1, z_2)\ z_1(\partial/\partial z_1)B_1(z_1, z_2), \tag{56}$$

$$z_2(\partial/\partial z_2)\hat{B}_1(z_1, z_2) = F_1(z_1, z_2)\ z_2(\partial/\partial z_2)B_1(z_1, z_2). \tag{57}$$

According to Eqs. (11), (12), (13), and (15) these relations correspond to

$$\{m^1\hat{b}_{m,n}\} = \{^1f_{m,n}\} * \{m^1b_{m,n}\} \tag{58}$$

and

$$\{n^1\hat{b}_{m,n}\} = \{^1f_{m,n}\} * \{n^1b_{m,n}\}, \tag{59}$$

or, explicitly written,

$$^1\hat{b}_{p,q} = \sum_{m=0}^{p} \sum_{n=0}^{q} (m/p)\,{}^1b_{m,n}\,{}^1f_{p-m,\,q-n}, \qquad p \neq 0 \tag{60}$$

$$^1\hat{b}_{p,q} = \sum_{m=0}^{p} \sum_{n=0}^{q} (n/q)\,{}^1b_{m,n}\,{}^1f_{p-m,\,q-n}, \qquad q \neq 0. \tag{61}$$

We prove now that the power series defined by Eqs. (52), (55), (60), and (61) is absolutely convergent and equal to $\hat{B}_1(z_1, z_2)$ for all $(z_1, z_2) \in D_1 - D_2$. From Eqs. (60) and (61) we find that, for all $(z_1, z_2) \in D_1$,

$$\sum_{p=0}^{N} \sum_{q=0}^{N} |^1\hat{b}_{p,q}|\ |z_1^p|\ |z_2^q| \leq \sum_{q=1}^{N} \sum_{n=0}^{q} (n/q)\ |^1b_{0,n}|\ |^1f_{0,q-n}|$$

$(p, q) \neq (0, 0)$,

$$+ \sum_{p=1}^{N} \sum_{q=0}^{N} \sum_{m=0}^{p} \sum_{n=0}^{q} (m/p)\ |^1b_{m,n}|\ |^1f_{p-m,q-n}|$$

$$\leq \sum_{p=0}^{N} \sum_{q=0}^{N} \sum_{m=0}^{p} \sum_{n=0}^{q} |^1b_{m,n}|\ |^1f_{p-m,q-n}|$$

$$\leq \sum_{m=0}^{N} \sum_{n=0}^{N} |^1b_{m,n}| \sum_{p=0}^{N} \sum_{q=0}^{N} |^1f_{p,q}|,$$

and thus

$$\sum_{p=0}^{\infty} \sum_{q=0}^{\infty} |^1\hat{b}_{p,q}|\ |z_1^p|\ |z_2^q| \leq \sum_{m=0}^{\infty} \sum_{n=0}^{\infty} |^1b_{m,n}| \sum_{p=0}^{\infty} \sum_{q=0}^{\infty} |^1f_{p,q}| < \infty;$$

$(p, q) \neq (0, 0)$, $(z_1, z_2) \in D_1$.
(for left-hand part only). (62)

In Eq. (62) we have used the stability of ^1f [Eq. (48)] and ^1b [Eq. (51)].

With the intermediate notation

$$E(z_1, z_2) = \sum_{m=0}^{\infty} \sum_{n=0}^{\infty} {}^1\hat{b}_{m,n}\, z_1^m\, z_2^n, \tag{63}$$

it remains to be proved that

$$E(z_1, z_2) = \hat{B}_1(z_1, z_2)$$

for all $(z_1, z_2) \in D_1 - D_2$. Because $E(z_1, z_2)$ and $B_1(z_1, z_2)$ are given by power series that are absolutely convergent for all $(z_1, z_2) \in D_1$, it follows from Abel's second theorem [14] that

$$\lim_{\rho \to 1-0} [B_1(e^{-2\pi ju}, z_2) - B_1(\rho e^{-2\pi ju}, z_2)] = 0 \tag{64}$$

and

$$\lim_{\rho \to 1-0} [E(e^{-2\pi ju}, z_2) - E(\rho e^{-2\pi ju}, z_2)] = 0 \tag{65}$$

for any fixed (u, z_2) with

$(u, z_2) \in \{(u, z_2): 0 \leq u \leq 1 \wedge |z_2| < 1\} = Q$.

From (64) we conclude [referring to condition (50) and the continuity of the logarithm] that

$$\lim_{\rho \to 1-0} [\hat{B}_1(e^{-2\pi ju}, z_2) - \hat{B}_1(\rho e^{-2\pi ju}, z_2)] = 0 \tag{66}$$

for any $(u, z_2) \in Q$. When (66) is subtracted from (65),

$$\hat{B}_1(z_1, z_2) = E(z_1, z_2) \tag{67}$$

for all $(z_1, z_2) \in \{(z_1, z_2): |z_1| \leq 1 \wedge |z_2| < 1\}$.

Repeating the argument for $z_1 = \exp(-2\pi ju)$ and $z_2 = \rho \exp(-2\pi jv)$, where u and v are any fixed values, we finally find

$$\hat{B}_1(z_1, z_2) = E(z_1, z_2), \ (z_1, z_2) \in D_1. \tag{68}$$

Thus the convolutional stability of ${}^1\hat{b}$ is a necessary condition for ${}^1b$ to be recursively stable. We prove now that this property is also a sufficient condition. Again, we first consider the set D_2. On D_2, $\hat{B}_1(z_1, z_2)$ is regular because of Eq. (62), and because $B_1(z_1, z_2) = \exp[\hat{B}_1(z_1, z_2)]$ is regular, too, the identities

$$z_1(\partial/\partial z_1)\, B_1(z_1, z_2) = B_1(z_1, z_2)\, z_1(\partial/\partial z_1)\, \hat{B}_1(z_1, z_2) \tag{69}$$

and

$$z_2(\partial/\partial z_2)\, B_1(z_1, z_2) = B_1(z_1, z_2)\, z_2(\partial/\partial z_2)\, \hat{B}_1(z_1, z_2) \tag{70}$$

relate the absolutely convergent power series. These identities yield relations similar to Eqs. (60) and (61):

$${}^1b_{p,q} = \sum_{m=1}^{p} \sum_{n=0}^{q} (m/p)\, {}^1\hat{b}_{m,n}\, {}^1b_{p-m,q-n}, \qquad p \neq 0, \tag{71}$$

$${}^1b_{p,q} = \sum_{m=0}^{p} \sum_{n=1}^{q} (n/q)\, {}^1\hat{b}_{m,n}\, {}^1b_{p-m,q-n}, \qquad q \neq 0. \tag{72}$$

For $(p, q) = (0, 0)$ we find, from Eq. (55),

$${}^1b_{0,0} = \exp\,({}^1\hat{b}_{0,0}). \tag{73}$$

With ${}^1b$ as defined by Eqs. (71)-(73), we have the

$$B_1(z_1, z_2) = \exp\,[\hat{B}_1(z_1, z_2)] = \sum_{m=0}^{\infty} \sum_{n=0}^{\infty} {}^1b_{m,n} z_1^{\,m} z_2^{\,n} \tag{74}$$

identity for all $(z_1, z_2) \in D_2$. We prove now that (74) is even valid for all $(z_1, z_2) \in D_1$. To do this it is sufficient to find a power series that is equal to $B_1(z_1, z_2)$ and absolutely convergent for all $(z_1, z_2) \in D_1$.

Since $\hat{B}_1(z_1, z_2)$ is given as a power series that is, by assumption, absolutely convergent for all $(z_1, z_2) \in D_1$, and the expansion of $\exp\,(x)$ is absolutely convergent for all $|x| < \infty$, the expansion

$$\exp[\hat{B}_1(z_1, z_2)] = \sum_{l=0}^{\infty} \frac{1}{l!}\, [\hat{B}_1(z_1, z_2)]^l$$
$$= \sum_{l=0}^{\infty} (1/l!) \left(\sum_{m=0}^{\infty} \sum_{n=0}^{\infty} {}^1\hat{b}_{m,n} z_1^{\,m} z_2^{\,n} \right)^l \tag{75}$$

is also absolutely convergent for all $(z_1, z_2) \in D_1$. Thus Eq. (75) can be rearranged as

$$\exp[\hat{B}_1(z_1, z_2)] = \sum_{m=0}^{\infty} \sum_{n=0}^{\infty} a_{m,n} z_1^{\,m} z_2^{\,n}, \tag{76}$$

where

$$a_{m,n} = \sum_{l=0}^{\infty} a_{m,n}^{\ (l)}, \tag{77}$$

$$a_{m,n}^{\ (l)} = (1/l) \sum_{p=0}^{m} \sum_{q=0}^{n} a_{p,q}^{\ (l-1)}\, {}^1\hat{b}_{m-p,n-q}, \qquad l = 1, 2, \cdots, \tag{78}$$

$$a_{m,n}^{\ (0)} = \delta_{m,n}. \tag{79}$$

Therefore the expansion (74) is valid and absolutely convergent for all $(z_1, z_2) \in D_1$. It is evident from the definition that $B_1(z_1, z_2)$ is nonzero on D_1. Theorem 4 summarizes the previous results.

Theorem 4. The quantity

$$\{{}^1b_{m,n}\}_{\substack{m \geq 0 \\ n \geq 0}}$$

is recursively stable if and only if there exists a power series

$$\sum_{m=0}^{\infty} \sum_{n=0}^{\infty} {}^1\hat{b}_{m,n} z_1^{\,m} z_2^{\,n} \tag{80}$$

that is absolutely convergent and equal to $\ln B_1(z_1, z_2)$ for all $(z_1, z_2) \in D_1$.

Theorem 4 remains applicable to noncausal recursive operators ${}^lb$, $l = 2, 3, 4$, if they are transformed to a first-quadrant function by the operations (10), (8), or (9), respectively. This leads to Corollary 2.

Corollary 2
The lth-quadrant function ${}^lb$, in which $l = 2, 3, 4$, is recursively stable if and only if $\ln\,[B_2(z_1, 1/z_2)]$, $\ln\,[B_3(1/z_1, 1/z_2)]$, or $\ln\,[B_4(1/z_1, z_2)]$, respectively, is equal to a power series of the form (80) that is absolutely convergent for all $(z_1, z_2) \in D_1$.

From Theorem 4 and Corollary 2 it is evident that the decomposition problem in the previous section can be solved if the cepstrum $\hat{c}$ of c can be decomposed [see Eq. (47)] into one-quadrant functions ${}^l\hat{k}$, in which $l = 1, 2, 3, 4$. These are stable convolutional filters. In the next sections it is proved that the decomposition is possible because of the properties of c.

Solution of the decomposition problem

- *Existence of a set of solutions*

The cepstrum $\hat{c}$ is readily obtained by evaluating the z-transform of c on R. We first have

$$\hat{C}'(u, v) = \ln \left[\sum_{m=-2M_c}^{2M_c} \sum_{n=-2N_c}^{2N_c} c_{m,n}\, e^{-2\pi j(um+vn)} \right]. \tag{81}$$

Since $\hat{C}'$ is real and positive [Eqs. (34) and (35)], the logarithm in Eq. (81) presents no problem. It is easily verified that all partial derivatives of $\hat{C}'(u, v)$ are continuous. Thus, for all u and v, $\hat{C}'(u, v)$ is given by the Fourier series expansion [15]

$$\hat{C}'(u, v) = \sum_{m=-\infty}^{\infty} \sum_{n=-\infty}^{\infty} \hat{c}_{m,n}\, e^{-2\pi j(um+vn)} \tag{82}$$

The coefficients $\hat{c}_{m,n}$ that make up the cepstrum $\hat{c}$ are obtained from

$$\hat{c}_{m,n} = \int_0^1\int \hat{C}'(u, v)\, e^{2\pi j(um+vn)}\, dudv. \tag{83}$$

From the properties of $C'(u, v)$ and from Eq. (83) it follows that

$$\hat{c}_{m,n} = \hat{c}_{-m,-n}. \tag{84}$$

Repeated integration by parts permits proof of the inequalities

$$|\hat{c}_{0,0}| \leq \alpha < \infty, \tag{85}$$

$$|\hat{c}_{0,n}| \leq \alpha/n^2, \qquad n \neq 0, \tag{86}$$

$$|\hat{c}_{m,0}| \leq \alpha/m^2, \qquad m \neq 0, \tag{87}$$

$$|\hat{c}_{m,n}| \leq \alpha/(m^2n^2), \qquad m \neq 0 \wedge n \neq 0. \tag{88}$$

Thus any decomposition

$$\hat{c} = {}^1\hat{k} + {}^2\hat{k} + {}^3\hat{k} + {}^4\hat{k} \tag{89}$$

of $\hat{c}$ into one-quadrant functions yields functions ${}^l\hat{k}$ that meet the conditions of Theorem 4 or Corollary 2. Consequently each of these decompositions corresponds to a set $\{{}^lk: l = 1, 2, 3, 4\}$ of recursively stable functions.

• *Problem of an optimal solution*

As Eqs. (71) and (72) show, the final solution $\{{}^lk: l = 1, 2, 3, 4\}$ may consist of functions with an infinite number of nonzero sample values. It would be advantageous to find such a decomposition of $\hat{c}$ which guarantees functions ${}^lk$ with a minimum finite number of nonzero sample values such that

$$1/C(z_1, z_2) = \prod_{l=1}^{4} [1/K_l(z_1, z_2)] \tag{90}$$

is true, at least in a good approximation. This is also discussed in the following section on the truncation problem. The author does not have such a decomposition procedure to offer for $\hat{c}$. Some recommendations, however, seem reasonable. First, because $\hat{c}$ is centrally symmetric, the decomposition should be such that

$${}^1\hat{k}_{m,n} = {}^3\hat{k}_{-m,-n} \tag{91}$$

and

$${}^2\hat{k}_{m,n} = {}^4\hat{k}_{-m,-n}. \tag{92}$$

If, in addition, c (and consequently $\hat{c}$) is symmetric with respect to the axes $m = 0$ and $n = 0$, the following relations should hold:

$${}^1\hat{k}_{0,n} = {}^4\hat{k}_{0,n}, \qquad n \geq 0, \tag{93}$$

$${}^2\hat{k}_{0,n} = {}^3\hat{k}_{0,n}, \qquad n \leq 0, \tag{94}$$

$${}^1\hat{k}_{m,0} = {}^2\hat{k}_{m,0}, \qquad m \geq 0, \tag{95}$$

$${}^3\hat{k}_{m,0} = {}^4\hat{k}_{m,0}, \qquad m \leq 0. \tag{96}$$

It might be of interest that the problem of optimal decomposition of $\hat{c}$ would become simple if

$$\hat{c}_{m,n} \equiv 0, \qquad m \neq 0 \wedge n \neq 0. \tag{97}$$

Such cepstra occur if and only if c is separable, that is, if

$$c_{m,n} = s_m h_n. \tag{98}$$

Equation (98) means that the two-dimensional decomposition problem is reduced to the decomposition of a column vector s and a row vector h into minimum and maximum delay operators [6, 12]. These operators correspond to the cepstra

$$\begin{aligned}&\{\hat{c}_{0,0}, \hat{c}_{1,0}, \hat{c}_{2,0}, \cdots\},\\&\{\cdots, \hat{c}_{-2,0}, \hat{c}_{-1,0}, 0\}\end{aligned} \tag{99}$$

and

$$\begin{aligned}&\{0, \hat{c}_{0,1}, \hat{c}_{0,2}, \cdots\},\\&\{\cdots, \hat{c}_{0,-2}, \hat{c}_{0,-1}, 0\}, \text{ respectively.}\end{aligned} \tag{100}$$

• *Truncation problem*

As pointed out in the previous section, the decomposition of c may result in operators ${}^lk$ having an infinite number of sample points. If these operators are to be implemented numerically, some truncation of the operators becomes mandatory. This truncation means not only that the decomposition (90) becomes an approximate one but also that the recursive stability of the operators ${}^lk$ may be affected. Theorem 5 guarantees that there exist truncations for recursively stable operators ${}^lk$ such that the resulting operators ${}^lk$ remain recursively stable.

Theorem 5

We assume that the power series

$$B_1(z_1, z_2) = \sum_{m=0}^{\infty}\sum_{n=0}^{\infty} {}^1b_{m,n} z_1^m z_2^n \tag{101}$$

is absolutely convergent and nonzero for all $(z_1, z_2) \in D_1$. There then exists a pair M_1, N_1 such that

$$|\dot{B}_1(z_1, z_2)| = \sum_{m=0}^{m_1}\sum_{n=0}^{n_1} {}^1b_{m,n} z_1^m z_2^n| > 0 \tag{102}$$

for all $(m_1, n_1) \in \{(m_1, n_1): m_1 > M_1 \wedge n_1 > N_1\}$ and for all $(z_1, z_2) \in D_1$.

The proof of this theorem is based on the uniform convergence of the power series (101). The extension of the theorem to functions other than those in the first quadrant is straightforward.

To find constraints on the truncation parameters to be applied on ${}^lk$, in which $l = 1, 2, 3, 4$, we consider $(M_1 + 1) \times (N_1 + 1)$ subarrays of ${}^1k$ and ${}^3k$; $(M_2 + 1) \times (N_2 + 1)$ subarrays of ${}^2k$ and ${}^4k$; and a $(4M_c + 1) \times (4N_c + 1)$ array c. The cascaded convolution of the subarrays yields an array with $[2(M_1 + M_2) + 1] \times [2(N_1 + N_2) + 1]$ sample points. It contains an array as large as c, if

$$(M_1 + M_2) \geq 2M_c \tag{103}$$

and

$$N_1 + N_2 \geq 2N_c. \tag{104}$$

For arrays c symmetric with respect to the axes $m = 0$ and $n = 0$, the relations (93) – (96) apply. In this case the constraints on the truncation parameters are

$$M_1 = M_2 \geq M_c, \tag{105}$$

$$N_1 = N_2 \geq N_c. \tag{106}$$

It is our experience [11] that the truncation conditions

$$M_1 = M_c, \qquad N_1 = N_c \tag{107}$$

may give satisfactory results. If Eqs. (107) hold, the decomposition of c means a saving of filter coefficients which ranges from 36 percent ($M_c = N_c = 1$) to 75 percent ($M_c \rightarrow \infty \wedge N_c \rightarrow \infty$).

• *Numerical implementation*

This section comments on the computation of $\hat{c}$, on the decomposition procedure, and on the application of the decomposed recursive filter.

To determine $\hat{c}$, Eqs. (36), (81), and (83) must be evaluated. This is done by Fast Fourier Transform (FFT) techniques [16] as indicated in Fig. 2. The functions $\{C_{m,n}\}$ and $\{\hat{C}_{m,n}\}$ are sampled versions of $C'(u, v)$ and $\hat{C}'(u, v)$, whereas $\hat{\zeta}$ is an aliased version of the desired result $\hat{c}$ [12]. The degree of aliasing can be controlled by the rate at which $C'(u, v)$ is sampled.

The function $\hat{\zeta}$, truncated at the Nyquist subscripts $\pm m_{\mathrm{Ny}}$ and $\pm n_{\mathrm{Ny}}$, is decomposed according to

$$\hat{\zeta}_{\mathrm{tr}} = {}^1\hat{\kappa} + {}^2\hat{\kappa} + {}^3\hat{\kappa} + {}^4\hat{\kappa}, \tag{108}$$

where $\dot{\kappa}$ are approximations of operators k.

Since $\hat{\zeta}_{\mathrm{tr}}$ exhibits the same symmetries as $\hat{c}$, the decomposition should be performed according to (91) and (92). This decomposition has the advantage that only ${}^1\dot{\kappa}$ and ${}^4\dot{\kappa}$ have to be determined. The values ${}^3\dot{\kappa}$ and ${}^2\dot{\kappa}$ are obtained by rotating the subscript planes through 180°.

The computation of an $(M_1 + 1) \times (N_1 + 1)$ array ${}^1\dot{\kappa}$ from ${}^1\hat{\kappa}$ should be performed in three steps. First ${}^1\dot{\kappa}_{0,0}$ is obtained by Eq. (73). Then the row vector

$$\{{}^1\dot{\kappa}_{0,n}\}_{1 \leq n \leq N_1}$$

is determined by Eq. (72), and finally the submatrix

$$\{{}^1\dot{\kappa}_{m,n}\}_{\substack{1 \leq m \leq M_1 \\ 0 \leq n \leq N_1}}$$

is obtained by Eq. (71). The same procedure is applied to

$$\{{}^4\hat{\kappa}_{-m,n}\}_{\substack{m \leq 0 \\ n \leq 0}}$$

to obtain

$$\{{}^4\dot{\kappa}_{-m,n}\}_{\substack{m \leq 0 \\ n \geq 0}}.$$

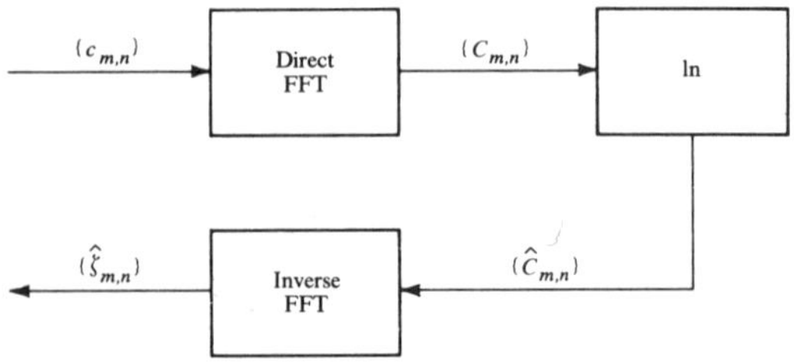

Figure 2 Block diagram of the approximate cepstrum transform for evaluation of Eqs. (36), (81), and (83). The transformation in the first step of the decomposition procedure.

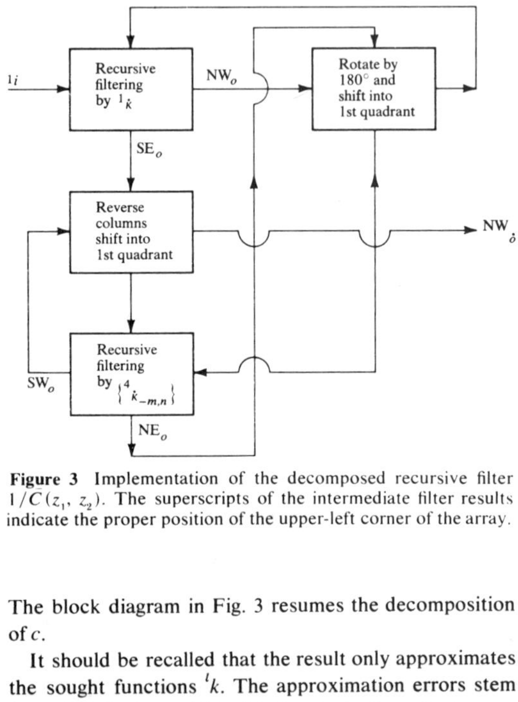

Figure 3 Implementation of the decomposed recursive filter $1/\bar{C}(z_1, z_2)$. The superscripts of the intermediate filter results indicate the proper position of the upper-left corner of the array.

The block diagram in Fig. 3 resumes the decomposition of c.

It should be recalled that the result only approximates the sought functions ${}^l k$. The approximation errors stem from the aliasing exhibited in ζ and also from the truncation applied to ${}^l\kappa$, $l = 1, 2, 3, 4$.

According to the discussions of Eqs. (22), (23), and (24), the input-output relation

$$\begin{aligned} O(z_1, z_2) &= I_1(z_1, z_2)/C(z_1, z_2) \\ &\simeq (\{[I_1(z_1, z_2)/\dot{\kappa}_1(z_1, z_2)]/\dot{\kappa}_2 \\ &\quad \times (z_1, z_2)\}/\dot{\kappa}_3(z_1, z_2))/\dot{\kappa}_4(z_1, z_2) \end{aligned} \tag{109}$$

can be evaluated only by implementing the causal recursive algorithm. Figure 4 demonstrates how this can be done. It is clear (see [2]) that the computation of the intermediate results is to be stopped after the output values have decayed to reasonably small numbers.

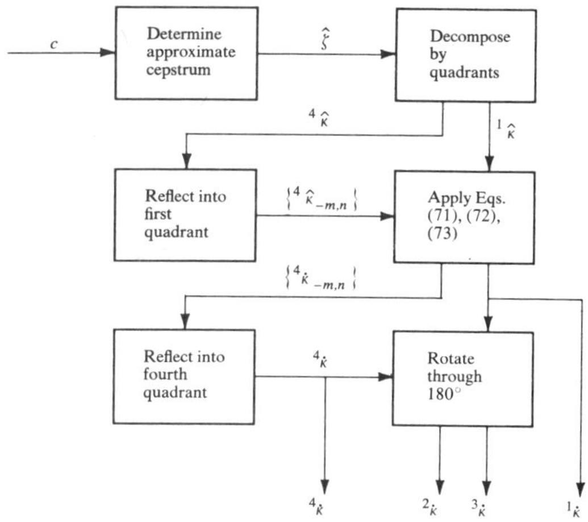

Figure 4 Block diagram of the decomposition procedure.

• *Nonrecursive solutions*

Instead of decomposing $1/C(z_1, z_2)$ into a cascade of recursive filters [see Eq. (90)], a decomposition into convolutional filters might be preferred:

$$1/C(z_1, z_2) = F_1(z_1, z_2)F_2(z_1, z_2)F_3(z_1, z_2)F_4(z_1, z_2). \tag{110}$$

The corresponding functions ${}^lf$ can also be determined from the cepstra ${}^l\hat{k}$, $l = 1, 2, 3, 4$. It is sufficient to demonstrate this for ${}^1k$ and ${}^1f$. The operator ${}^1k$ is assumed to be recursively stable. From

$$\ln (1/K_1(z_1, z_2)) = \ln F_1(z_1, z_2) \tag{111}$$

we have

$$-{}^1\hat{k} = {}^1\hat{f}. \tag{112}$$

Thus, replacing ${}^1\hat{b}$ by $(-{}^1\hat{k})$ and ${}^1b$ by ${}^1f$ in Eqs. (71), (72), and (73), we find

$${}^1f_{0,0} = \exp(-{}^1\hat{k}_{0,0}), \tag{113}$$

$${}^1f_{p,q} = -\sum_{m=1}^{p}\sum_{n=0}^{q} (m/p)\hat{k}_{m,n}f_{p-m,\,q-n}, \qquad p \neq 0, \tag{114}$$

$${}^1f_{p,q} = -\sum_{m=0}^{p}\sum_{n=1}^{q} (n/q)\hat{k}_{m,n}f_{p-m,q-n}, \qquad q \neq 0. \tag{115}$$

The decomposition (110) is preferred over (90) if it requires a lower number of filter weights for the same accuracy.

Alternatively to Eqs. (113), (114), and (115), the least-squares approximations of ${}^lf$ could be determined from ${}^lk$, $l = 1, 2, 3, 4$, by the Wiggins algorithm [2].

Numerical examples

As pointed out in the previous section, the problem of the optimal decomposition of a recursive operator c consists in finding an appropriate decomposition of $\hat{c}$ and an appropriate truncation of the resulting operators. In this section it is demonstrated that the procedure characterized by the decomposition rules (93) and (96) and the truncation rules (105)–(107) may be the optimal decomposition procedure. We also discuss examples of unsatisfactory results obtained by these rules.

The operators c that we discuss are synthesized from one-quadrant operators with 2×2 samples. The reasons are that

1. The optimal decomposition is, a priori, known exactly in two cases.
2. Since the occurring one-quadrant operators have only a few samples (2×2, 3×3, 4×4), their recursive stability can readily be checked.

An example of practical interest could also have been discussed, namely the computation of recursive filters for the enhancement of radioscintigraphic images [11, 17]. The decomposition rules mentioned here were able to specify a near-optimal solution for this problem. However, this example is not well suited to point out the potential and shortcomings of these rules.

• *Example 1*

Consider a function c,

$$c = {}^1k * {}^2k * {}^3k * {}^4k, \tag{116}$$

where

$${}^1k_{m,n} = {}^2k_{m,-n} = {}^3k_{-m,-n} = {}^4k_{-m,n}, \qquad m \geq 0,\ n \geq 0. \tag{117}$$

With the recursively stable function [4]

$${}^1k = \begin{bmatrix} 1.00 & 0.90 \\ 0.90 & 0.85 \end{bmatrix} \leftarrow, \tag{118}$$
$$\uparrow$$

we have

$$c = \begin{bmatrix} 0.68850 & 2.7639 & 4.15082 & 2.7639 & 0.68850 \\ 2.76390 & 11.0930 & 16.65830 & 11.0930 & 2.76390 \\ 4.15082 & 16.6583 & 25.01520 & 16.6583 & 4.15082 \\ 2.76390 & 11.0930 & 16.65830 & 11.0930 & 2.76390 \\ 0.68850 & 2.7639 & 4.15082 & 2.7639 & 0.68850 \end{bmatrix} \leftarrow. \tag{119}$$
$$\uparrow$$

In both (118) and (119) the vertical and horizontal arrows indicate the axes $m = 0$ and $n = 0$, respectively.

From the construction of c it is clear that $\hat{c}$ is to be decomposed according to Eqs. (93)–(96). In addition, the 3×3 result

$$
{}^{1}\dot{\kappa} = \begin{bmatrix} 1.0000 & 0.9000 & -0.9\text{E}{-6} \\ 0.9000 & 0.8500 & -0.6\text{E}{-6} \\ -0.4\text{E}{-6} & 0.2\text{E}{-6} & -0.8\text{E}{-7} \end{bmatrix} \leftarrow
$$
$$
\uparrow \qquad (120)
$$

shows that the truncation rules (105)–(107) are applicable, since ${}^{1}\dot{\kappa}$ is identical to ${}^{1}k$ except for entries in the third row and column, which may be neglected.

• *Example 2*

We modify the operator (119) by adding the spike function 0.011 δ. That means $c_{0,0} = 25.0152$ is replaced by 25.2903. Contrary to example 1, no decomposition of c is now known a priori. However, because this c exhibits the same symmetries as the function c in the first example, it can be expected that Eqs. (93)–(96) describe the optimal decomposition of $\hat{c}$. Similar to Example 1 the decomposition can be characterized by one operator, e.g., ${}^{1}\dot{\kappa}$. As a 4 × 4 result we find

$$
{}^{1}\dot{\kappa} = \begin{bmatrix} 1.4037 & 0.7090 & -0.0033 & -0.0069 \\ 0.7090 & 0.7916 & 0.0305 & -0.0183 \\ -0.0033 & 0.0305 & 0.0210 & -0.0075 \\ -0.0069 & -0.0183 & -0.0075 & 0.0024 \end{bmatrix} \leftarrow
$$
$$
\uparrow \qquad (121)
$$

The operator ${}^{1}\dot{\kappa}$ is recursively stable since it can be shown that the z transform of (121) is nonzero for all $(z_1, z_2) \in D_1$. The proof is based on the properties of the upper left 2 × 2 subarray of ${}^{1}\dot{\kappa}$.

Equation (121) does not have such a distinct cutoff beyond $m = 1$, $n = 1$ as does (120). Thus the truncation rules (105)–(107) cannot be expected to yield operators ${}^{l}\dot{\kappa}$, $l = 1, 2, 3, 4$, which, when convolved in a cascade, would result in a good approximation c_{appr} of c. Even with 4 × 4 operators, the approximation is not quite satisfactory, as a comparison of c and a subarray of $e = c_{\text{appr}} - c$ demonstrates [See Eq. (122)].

Because the operators ${}^{l}\dot{\kappa}$ are to be used for recursive filtering, the approximation error function e might be inappropriate to characterize the quality of the decomposition of c. Alternatively, c can be recursively filtered by these operators. If the decomposition is a good one, the result $\dot{o}$ should resemble a delta function. As the submatrix (123) demonstrates, the solution given by Eq. (121) does this job fairly well.

• *Example 3*

In this example a 5 × 5 function c is presented which can be decomposed into four 2 × 2 operators. However, the

$$
e_{\text{sub}} = \begin{bmatrix}
0.159264\text{E-}02 & 0.105697\text{E-}02 & 0.239717\text{E-}03 & -0.403976\text{E-}04 & -0.189802\text{E-}04 & 0.851596\text{E-}07 & 0.160910\text{E-}06 \\
-0.264569\text{E-}02 & -0.143818\text{E-}02 & 0.357122\text{E-}04 & 0.171492\text{E-}03 & 0.145821\text{E-}04 & -0.189372\text{E-}05 & 0.851477\text{E-}07 \\
-0.147880\text{E } 00 & -0.104905\text{E } 00 & -0.312095\text{E-}01 & 0.193911\text{E-}02 & 0.219917\text{E-}02 & 0.145812\text{E-}04 & -0.189804\text{E-}04 \\
-0.200212\text{E } 00 & -0.145703\text{E } 00 & -0.509612\text{E-}01 & -0.423922\text{E-}02 & 0.193903\text{E-}02 & 0.171491\text{E-}03 & -0.403971\text{E-}04 \\
-0.749245\text{E-}01 & -0.556574\text{E-}01 & -0.397092\text{E-}01 & -0.509626\text{E-}01 & -0.312099\text{E-}01 & 0.357172\text{E-}04 & 0.239724\text{E-}03 \\
0.172424\text{E-}02 & -0.724792\text{E-}04 & -0.556641\text{E-}01 & -0.145707\text{E } 00 & -0.104906\text{E } 00 & -0.143816\text{E-}02 & 0.105699\text{E-}02 \\
0.474548\text{E-}02 & 0.170898\text{E-}02 & -0.749350\text{E-}01 & -0.200217\text{E } 00 & -0.147882\text{E } 00 & -0.264567\text{E-}02 & 0.159267\text{E-}02 \\
0.175476\text{E-}02 & -0.734329\text{E-}04 & -0.556622\text{E-}01 & -0.145707\text{E } 00 & -0.104906\text{E } 00 & -0.143817\text{E-}02 & 0.105699\text{E-}02 \\
-0.749216\text{E-}01 & -0.556555\text{E-}01 & -0.397084\text{E-}01 & -0.509624\text{E-}01 & -0.312098\text{E-}01 & 0.357162\text{E-}04 & 0.239723\text{E-}03 \\
-0.200212\text{E } 00 & -0.145703\text{E } 00 & -0.509611\text{E-}01 & -0.423913\text{E-}02 & 0.193906\text{E-}02 & 0.171490\text{E-}03 & -0.403977\text{E-}04 \\
-0.147880\text{E } 00 & -0.104905\text{E } 00 & -0.312094\text{E-}01 & 0.193914\text{E-}02 & 0.219918\text{E-}02 & 0.145810\text{E-}04 & -0.189804\text{E-}04 \\
-0.264569\text{E-}02 & -0.143818\text{E-}02 & 0.357112\text{E-}04 & 0.171492\text{E-}03 & 0.145820\text{E-}04 & -0.189371\text{E-}05 & 0.851517\text{E-}07 \\
0.159264\text{E-}02 & 0.105697\text{E-}02 & 0.239716\text{E-}03 & -0.403980\text{E-}04 & -0.189803\text{E-}04 & 0.851621\text{E-}07 & 0.160911\text{E-}06
\end{bmatrix} \leftarrow
$$
$$
\uparrow \qquad (122)
$$

$$
\dot{o}_{\text{sub}} = \begin{bmatrix}
-0.613363\text{E-}02 & -0.474950\text{E-}02 & -0.159342\text{E-}02 & 0.143817\text{E-}03 & 0.139719\text{E-}03 & -0.713078\text{E-}04 & 0.121408\text{E-}04 \\
0.864866\text{E-}02 & 0.617515\text{E-}02 & 0.160198\text{E-}02 & -0.367183\text{E-}03 & -0.787302\text{E-}04 & 0.994637\text{E-}04 & -0.522314\text{E-}04 \\
-0.113625\text{E-}01 & -0.759114\text{E-}02 & -0.138257\text{E-}02 & 0.597621\text{E-}03 & -0.646202\text{E-}04 & -0.825182\text{E-}04 & 0.954837\text{E-}04 \\
0.137227\text{E-}01 & 0.861436\text{E-}02 & 0.873944\text{E-}03 & -0.743399\text{E-}03 & 0.291746\text{E-}03 & -0.321593\text{E-}04 & -0.957433\text{E-}04 \\
-0.143766\text{E-}01 & -0.852288\text{E-}02 & -0.121410\text{E-}03 & 0.654047\text{E-}03 & -0.524485\text{E-}03 & 0.280924\text{E-}03 & -0.319643\text{E-}04 \\
0.111293\text{E-}01 & 0.641693\text{E-}02 & -0.559615\text{E-}03 & -0.187813\text{E-}03 & 0.497826\text{E-}03 & -0.529216\text{E-}03 & 0.295783\text{E-}03 \\
-0.826582\text{E-}03 & -0.602300\text{E-}03 & -0.240482\text{E-}03 & 0.724608\text{E-}06 & -0.176470\text{E-}03 & 0.648627\text{E-}03 & -0.738702\text{E-}03 \\
0.130282\text{E-}02 & 0.475957\text{E-}03 & 0.530685\text{E-}03 & -0.218473\text{E-}03 & -0.609808\text{E-}03 & -0.676552\text{E-}04 & 0.817899\text{E-}03 \\
-0.292059\text{E-}03 & -0.172962\text{E-}02 & 0.405336\text{E-}03 & -0.462968\text{E-}03 & 0.624410\text{E-}02 & -0.829096\text{E-}02 & 0.832244\text{E-}02 \\
0.100263\text{E } 01 & -0.246092\text{E-}03 & 0.113352\text{E-}02 & -0.619821\text{E-}03 & 0.107933\text{E-}01 & -0.139117\text{E-}01 & 0.131143\text{E-}01 \\
-0.178754\text{E-}03 & -0.158190\text{E-}02 & 0.431565\text{E-}03 & -0.276147\text{E-}03 & 0.601419\text{E-}02 & -0.792788\text{E-}02 & 0.776092\text{E-}02 \\
0.110793\text{E-}02 & 0.248553\text{E-}03 & 0.222650\text{E-}03 & -0.137941\text{E-}03 & -0.891139\text{E-}03 & 0.356984\text{E-}03 & 0.237607\text{E-}03 \\
-0.458554\text{E-}03 & -0.128206\text{E-}03 & 0.348056\text{E-}04 & 0.243648\text{E-}03 & -0.321219\text{E-}03 & 0.814399\text{E-}03 & -0.966814\text{E-}03 \\
0.105644\text{E-}01 & 0.573408\text{E-}02 & -0.107241\text{E-}02 & -0.347991\text{E-}03 & 0.494597\text{E-}03 & -0.497626\text{E-}03 & 0.315906\text{E-}03 \\
-0.135230\text{E-}01 & -0.750595\text{E-}02 & 0.575967\text{E-}03 & 0.846770\text{E-}03 & -0.538528\text{E-}03 & 0.219587\text{E-}03 & 0.955981\text{E-}05 \\
0.124684\text{E-}01 & 0.715635\text{E-}02 & -0.334558\text{E-}04 & -0.965414\text{E-}03 & 0.365733\text{E-}03 & -0.199897\text{E-}05 & -0.117244\text{E-}03 \\
-0.956713\text{E-}02 & -0.558962\text{E-}02 & -0.232891\text{E-}03 & 0.800152\text{E-}03 & -0.181203\text{E-}03 & -0.965190\text{E-}04 & 0.113540\text{E-}03 \\
0.615787\text{E-}02 & 0.355146\text{E-}02 & 0.229146\text{E-}03 & -0.515932\text{E-}03 & 0.734335\text{E-}04 & 0.954887\text{E-}04 & -0.740222\text{E-}04 \\
-0.286662\text{E-}02 & -0.155308\text{E-}02 & -0.903530\text{E-}04 & 0.221738\text{E-}03 & -0.299074\text{E-}04 & -0.449883\text{E-}04 & 0.321043\text{E-}04
\end{bmatrix} \leftarrow
$$
$$
\uparrow \qquad (123)
$$

decomposition procedure (93) – (96) is inadequate to provide these operators. Let

$$c = \begin{bmatrix} 0.72250 & 2.8305 & 4.14922 & 2.6973 & 0.65610 \\ 2.83050 & 11.2267 & 16.65830 & 10.9593 & 2.69730 \\ 4.14922 & 16.6583 & 25.01840 & 16.6583 & 4.14922 \\ 2.69730 & 10.9593 & 16.65830 & 11.2267 & 2.83050 \\ 0.65610 & 2.6973 & 4.14922 & 2.8305 & 0.72250 \end{bmatrix} \begin{matrix} \\ \\ \leftarrow . \\ \\ \\ \end{matrix}$$
$$\begin{matrix} \quad\quad\quad\quad\quad \uparrow \end{matrix} \tag{124}$$

The function c has been constructed by the cascaded convolution

$$c = {}^1k * {}^1k * {}^3k * {}^3k, \tag{125}$$

where

$${}^1k_{m,n} = {}^3k_{-m,-n}, \qquad m \geq 0,\ n \geq 0, \tag{126}$$

and ${}^1k$ is given by Eq. (118).

From Eq. (125) it is evident that our procedure is not adequate to find the optimal decomposition of c. In the cepstrum domain this is reflected by the fact that $\hat{c}$ is identically zero for all

$$(m, n) \in \{(m, n)\colon m > 0 \wedge n < 0 \vee m < 0 \wedge n > 0\}. \tag{127}$$

For illustration a subarray of the approximate cepstrum $\hat{\xi}$ is given here.

$\hat{\xi}_{\text{sub}} =$

$$\begin{bmatrix} -0.3280506 & 0.0000012 & -0.0000008 & -0.0000002 & 0.0000004 \\ 0.4860004 & -0.0000013 & 0.0000008 & 0.0000001 & -0.0000003 \\ -0.8100015 & 0.0000028 & -0.0000022 & 0.0000013 & -0.0000011 \\ 1.7999992 & -0.0000015 & 0.0000008 & 0.0000001 & -0.0000003 \\ -0.0000005 & 1.7999954 & -0.8099970 & 0.4859967 & -0.3280469 \\ 1.7999992 & 0.0799982 & -0.0719988 & 0.0647995 & -0.0583197 \\ -0.8100015 & -0.0179969 & 0.0631976 & -0.0554384 & 0.0485987 \\ 0.4860004 & 0.0647983 & -0.0554389 & 0.0473463 & -0.0403558 \\ -0.3280506 & -0.0583183 & 0.0485989 & -0.0403555 & 0.0333769 \end{bmatrix} \begin{matrix} \\ \\ \\ \\ \leftarrow \\ \\ \\ \\ \\ \end{matrix}$$
$$\begin{matrix} \uparrow \end{matrix} \tag{128}$$

In spite of the aforementioned inadequacy it is interesting to decompose c using Eqs. (93) – (96). This means that we look for a decomposition according to

$$c = {}^1k' * {}^2k' * {}^3k' \quad {}^4k' \tag{129}$$

rather than according to Eq. (125). It can be concluded from (125) and (127) that ${}^2k'$ and ${}^4k'$ are 2×2 operators, where

$${}^2k_{m,n}{}' = {}^1k_{m,0}\,{}^3k_{0,n} \tag{130}$$

and

$${}^4k_{m,n}{}' = {}^2k_{-m,-n}{}'. \tag{131}$$

The operators ${}^1k'$ and ${}^3k'$, however, may be of infinite size since ${}^1k'$ could be obtained by recursively filtering ${}^1k * {}^1k$ with

$$\{{}^2k_{m,-n}{}'\}_{\substack{m \geq 0 \\ n \leq 0}}. \tag{132}$$

The numerical results are in accordance with these facts. Operator ${}^2\dot{\kappa}$ has a distinct cutoff beyond $m = 1$, $n = -1$, whereas ${}^1\dot{\kappa}$ slowly decays in the lower right submatrix, thus making a larger matrix desirable:

$${}^1\dot{\kappa} = \begin{bmatrix} 1.000000 & 0.899998 & -0.536E-6 & -0.238E-6 \\ 0.899999 & 0.889996 & -0.113E-5 & 0.477E-6 \\ -0.113E-5 & 0.358E-6 & 0.001600 & -0.001440 \\ -0.477E-6 & 0.596E-6 & -0.001440 & 0.001296 \end{bmatrix} \begin{matrix} \leftarrow \\ \\ , \\ \\ \end{matrix}$$
$$\begin{matrix} \uparrow \end{matrix} \tag{133}$$

$${}^2\dot{\kappa} = \begin{bmatrix} -0.238E-6 & -0.536E-6 & 0.899998 & 1.000000 \\ 0.617E-6 & -0.100E-5 & 0.809996 & 0.899999 \\ 0.772E-7 & -0.185E-6 & 0.458E-6 & -0.113E-5 \\ 0.249E-6 & -0.442E-7 & 0.812E-6 & -0.477E-6 \end{bmatrix} \begin{matrix} \leftarrow \\ \\ . \\ \\ \end{matrix}$$
$$\begin{matrix} \quad\quad\quad\quad\quad\quad\quad\quad \uparrow \end{matrix} \tag{134}$$

Nevertheless, the performance of this solution is not so far from the optimum. This property can be seen from the arrays e_{sub} [Eq. (135)] and $\dot{o}_{\text{sub}}$ [Eq. (136)], which have the same meaning as in Example 2.

$e_{\text{sub}} =$

$$\begin{bmatrix} -0.726528E-09 & -0.300764E-09 & -0.322484E-09 & -0.620874E-13 & -0.106655E-12 & 0.486984E-19 & 0.282767E-19 \\ -0.814738E-09 & -0.734651E-10 & -0.352007E-09 & -0.437013E-12 & -0.488424E-12 & 0.381983E-18 & 0.448300E-19 \\ -0.104863E-02 & 0.102956E-05 & 0.804396E-06 & 0.568578E-06 & 0.201465E-06 & -0.211685E-12 & -0.531997E-13 \\ -0.198347E-02 & 0.459036E-06 & 0.162494E-05 & 0.170333E-05 & 0.666931E-06 & -0.520017E-12 & -0.167008E-13 \\ -0.987053E-03 & -0.324249E-04 & -0.649691E-05 & 0.149511E-05 & 0.111582E-05 & -0.194287E-09 & -0.322311E-09 \\ -0.183105E-03 & -0.120163E-03 & -0.400543E-04 & -0.105484E-05 & 0.149877E-05 & -0.218226E-10 & -0.458015E-09 \\ -0.244141E-03 & -0.183105E-03 & -0.998497E-03 & -0.198635E-02 & -0.104823E-02 & 0.102559E-08 & -0.649578E-09 \\ -0.167847E-03 & -0.123978E-03 & -0.202656E-02 & -0.430578E-02 & -0.233192E-02 & 0.178572E-08 & -0.208588E-09 \\ -0.988007E-03 & -0.202274E-02 & -0.135297E-02 & -0.727915E-03 & -0.445602E-03 & 0.304066E-09 & 0.108114E-08 \\ -0.198347E-02 & -0.430326E-02 & -0.726894E-03 & 0.348490E-02 & 0.188940E-02 & -0.143340E-08 & 0.469635E-09 \\ -0.104864E-02 & -0.233199E-02 & -0.445418E-03 & 0.188950E-02 & 0.104969E-02 & -0.817241E-09 & -0.278076E-09 \\ -0.814738E-09 & 0.542428E-09 & 0.299420E-08 & -0.391476E-09 & -0.170395E-08 & 0.192376E-14 & -0.626138E-16 \\ -0.726528E-09 & -0.307082E-09 & 0.148689E-08 & 0.446240E-09 & -0.556149E-09 & 0.495778E-16 & 0.147329E-15 \end{bmatrix} \begin{matrix} \\ \\ \\ \\ \\ \\ \leftarrow \\ \\ \\ \\ \\ \\ \\ \end{matrix}$$
$$\begin{matrix} \uparrow \end{matrix} \tag{135}$$

$$\dot{o}_{\mathrm{sub}} = \begin{bmatrix}
0.280967E-02 & 0.504769E-05 & 0.624132E-05 & -0.154875E-03 & 0.714838E-03 & -0.105378E-02 & 0.119510E-02 \\
-0.312973E-02 & 0.308633E-04 & -0.490645E-04 & 0.217502E-03 & -0.835808E-03 & 0.120684E-02 & -0.135821E-02 \\
0.341518E-02 & -0.385113E-04 & 0.772769E-04 & -0.276761E-03 & 0.961070E-03 & -0.136687E-02 & 0.152817E-02 \\
-0.354451E-02 & -0.892836E-04 & 0.404467E-05 & 0.247829E-03 & -0.101446E-02 & 0.146519E-02 & -0.164298E-02 \\
0.332886E-02 & 0.519752E-03 & -0.343769E-03 & 0.272589E-05 & 0.875956E-03 & -0.139354E-02 & 0.160498E-02 \\
-0.250447E-02 & -0.148699E-02 & 0.115089E-02 & -0.662563E-03 & -0.376550E-03 & 0.999450E-03 & -0.127663E-02 \\
0.697545E-03 & 0.332353E-02 & -0.272302E-02 & 0.199929E-02 & -0.724934E-03 & -0.653226E-04 & 0.461661E-03 \\
0.164124E-02 & -0.561980E-02 & 0.469004E-02 & -0.368012E-02 & 0.212960E-02 & -0.114058E-02 & 0.599022E-03 \\
-0.525562E-02 & 0.904012E-02 & -0.764826E-02 & 0.624158E-02 & -0.432133E-02 & 0.305477E-02 & -0.229894E-02 \\
0.100776E\ 01 & -0.114254E-01 & 0.965816E-02 & -0.794142E-02 & 0.573437E-02 & -0.426973E-02 & 0.337229E-02 \\
-0.109169E-01 & 0.143818E-01 & -0.121632E-01 & 0.100813E-01 & -0.754331E-02 & 0.584474E-02 & -0.477361E-02 \\
0.861964E-02 & -0.114250E-01 & 0.837923E-02 & -0.561539E-02 & 0.377813E-02 & -0.267009E-02 & 0.207850E-02 \\
-0.649787E-02 & 0.874132E-02 & -0.503256E-02 & 0.179719E-02 & -0.643940E-03 & 0.827004E-04 & 0.863849E-04 \\
0.438516E-02 & -0.614329E-02 & 0.321013E-02 & -0.719524E-03 & -0.411731E-05 & 0.267322E-03 & -0.248437E-03 \\
-0.298584E-02 & 0.437045E-02 & -0.212724E-02 & 0.266846E-03 & 0.143739E-03 & -0.210926E-03 & 0.884780E-04 \\
0.208619E-02 & -0.316373E-02 & 0.151765E-02 & -0.178681E-03 & -0.225013E-04 & -0.193318E-04 & 0.178318E-03 \\
-0.149566E-02 & 0.229929E-02 & -0.116076E-02 & 0.247772E-03 & -0.170123E-03 & 0.251862E-03 & -0.398289E-03 \\
0.102591E-02 & -0.156502E-02 & 0.855297E-03 & -0.290920E-03 & 0.272346E-03 & -0.347115E-03 & 0.452698E-03 \\
-0.544717E-03 & 0.816697E-03 & -0.477380E-03 & 0.208324E-03 & -0.208049E-03 & 0.249972E-03 & -0.303273E-03
\end{bmatrix} \qquad (136)$$

Concluding remarks

An unstable recursive filter with a finite number of coefficients always has an infinite number of decompositions into four stable recursive filters, provided it has a nonzero frequency response with zero phase. The decomposibility is proved in this paper by means of a new criterion that relates the stability of a recursive filter to the properties of its cepstrum. One might say that this stability criterion, compared to that of Shanks, has the advantage of being constructive, i.e., it permits the specified filter to be decomposed as exactly as is desired.

Two questions remain unanswered:

1. Under what suppositions is it possible to decompose exactly an unstable recursive filter into stable recursive filters that have a minimum finite number of coefficients; and what is the number of these coefficients?
2. If such a decomposition exists, how can it be found?

In spite of these unanswered questions, the results given here are of practical value, as can be seen from the application of two-dimensional recursive filters to radioscintigraphic images [11, 17]. These filters have their theoretical foundation in the results of this paper.

Acknowledgments

The author thanks W. Kattwinkel, manager of the IBM Deutschland Heidelberg Scientific Center when this work was being done, for his support and encouragement. He also gratefully acknowledges helpful discussions with H. G. Meder and G. Jaeschke. Especially worthy of mention is the gratifying cooperation of Mrs. M. Gamp in preparing the manuscript.

References and notes

1. J. L. Shanks, "Two-dimensional Recursive Filters," in *SWIEECO Record of Technical Papers* **21**, 19E1 (1969).
2. J. L. Shanks and J. H. Justice, "Stability and Synthesis of Two-dimensional Recursive Filters," *IEEE Trans. Audio and Electroacoustics* **AU-20**, 115 (1972).
3. C. Farmer and J. B. Bednar, "Stability of Spatial Digital Filters," *Math. Biosci.* **14**, 113 (1972).
4. T. S. Huang, "Stability of Two-dimensional Recursive Filters," *IEEE Trans. Audio and Electroacoustics* **AU-20**, 158 (1972).
5. The original form of the cepstrum transformation was the power spectrum of the logarithm of a power spectrum. See B. P. Bogert, M. J. B. Healy, and J. W. Tukey, "The Quefrency Alanysis of Time Series for Echoes," in *Proc. Symp. on Time Series Analysis*, M. Rosenblatt, ed., John Wiley and Sons, Inc., New York, 1963, Ch. 15, p. 209. Throughout the present paper the (complex) cepstrum transformation is used as introduced by Oppenheim et al. [12].
6. J. L. Shanks, "Recursion Filters for Digital Processing," *Geophysics* **32**, 33 (1967).
7. E. A. Robinson, *Statistical Communication and Detection*, Hafner Press, New York, 1967, Chs. 6 and 7.
8. *Ibid.*, App. 2.
9. The positive direction of the m axis is downward, and the n axis points to the right. Compare with Fig. 1. This choice has been made following the usual subscripting of matrix elements (see also [1, 2]). In another convention [4] the m axis points to the right and the n axis is directed downward.
10. Usually, recursive filters are defined as $A(z_1, z_2)/B(z_1, z_2)$. However, because $I(z_1, z_2) \cdot A(z_1, z_2)$ can be interpreted as the input to a recursive filter $F(z_1, z_2) = 1/B(z_1, z_2)$, $A(z_1, z_2) = 1$ may be assumed without loss of generality.
11. P. Pistor, "Digital Processing of Scintigraphic Images by Two-dimensional Recursive Wiener Filters," Report 70.12.006, IBM Deutschland Heidelberg Scientific Center, 1970.
12. A. V. Oppenheim, R. W. Schafer, and T. G. Stockham, Jr., "Nonlinear Filtering of Multiplied and Convolved Signals," *Proc. IEEE* **56**, 1264 (1968).
13. W. I. Smirnow, "Lehrgang der höheren Mathematik, Teil III, 2," *VEB Deutscher Verlag der Wissenschaften*, Berlin, 1967, Ch. 4.
14. *Ibid.*, Ch. 1.
15. G. P. Tolstov, *Fourier Series*, Prentice Hall, Inc., Englewood Cliffs, N.J., 1962, p. 178.
16. W. T. Cochran et al., "What is the Fast Fourier Transform?" *Proc. IEEE* **55**, 1664 (1967).
17. P. Pistor et al., "Digital Image Processing in Nuclear Medicine," *Kerntechnik, Isotopentechnik und -Chemie* **14**, 299–306 and 353–359 (1972).

Received February 5, 1973

The author is located at the IBM Deutschland Scientific Center, Tiergartenstrasse 15, 6900 Heidelberg, Germany.

4

Reprinted from *IEEE Trans. Acoust., Speech, Signal Process.*
ASSP-24(2):115-128 (1976)

Two-Dimensional Spectral Factorization with Applications in Recursive Digital Filtering

MICHAEL P. EKSTROM AND JOHN W. WOODS, MEMBER, IEEE

***Abstract*—The concept of spectral factorization is extended to two dimensions in such a way as to preserve the analytic characteristics of the factors. The factorization makes use of a homomorphic transform procedure due to Wiener. The resulting factors are shown to be recursively computable and stable in agreement with one-dimensional (1-D) spectral factorization. The factors are not generally two-dimensional (2-D) polynomials, but can be approximated as such. These results are applied to 2-D recursive filtering, filter design, and a computationally attractive stability test for recursive filters.**

I. Introduction

A PROBLEM of interest in the digital processing of two-dimensional (2-D) signals is that of factoring 2-D discrete field, spectral density functions. Such factorizations occur, for example, in the representation of monochromatic image fields [1], and their enhancement [2] and restoration [3] by recursive filtering methods. Difficulties with factoring 2-D spectra have limited the generalization of many one-dimensional (1-D) signal processing formalisms to 2-D applications. This has led to the reformulation of many 2-D problems in 1-D characterizations in order to utilize the extensive set of 1-D analytical design formalisms [4], [5].

Factorization plays a central role in much of 1-D recursive filter design. A first step in most filter design procedures involves approximating a prescribed magnitude function either analytically or numerically by a ratio of two cosine polynomials. To realize this filter recursively, the polynominals are factored into components which are recursively stable. This factorization is an important step because it allows a desired 1-D magnitude specification to be realized (i.e., approximately) using a stable, recursive digital filter. While such factorizations are of equal interest in the recursive realizations of 2-D filters, this step has not generally been possible because of an inability to perform the required 2-D factorization.

The classical methodology for performing the 1-D spectral factorization was first introduced by Wiener in the early 1940's [6]. In the discrete case, this involves factoring the spectral density $S(z)$ into its so-called minimum phase factor $S^+(z)$ and maximum phase factor $S^-(z)$ where $S^+(z)$ has all its singularities and zeros in $\{|z| < 1\}$ and those of $S^-(z)$ are in $\{|z| > 1\}$. The inverse transforms of $S^+(z)$ are $S^-(z)$ are stable, causal, and anticausal sequences, respectively. The method originally proposed by Wiener [6] for performing the factorization involves a direct determination of the poles and zeros of $S(z)$ and their assignment to $S^+(z)$ or $S^-(z)$ depending on their magnitude. For rational z spectra, applying this root extraction process is an exact analytic means of performing the desired factorization. It can be performed computationally with a variety of root-finding routines, some of which are quite efficient [7].

In the 2-D case, the spectral factorization cannot proceed analogously because of the lack of a fundamental theorem of algebra for 2-D polynomials [8], [9]. Thus, it has been concluded that the 2-D spectral factorization of rational densities into rational factors is not generally possible [1], [9]. For a counterexample which proves this, see [1] or [10]. We thus conclude that the root-finding approach to 2-D spectral factorization is not viable.

Although Wiener apparently preferred this now classical root extraction method, he did make passing reference in his original work to an alternate technique, one based on the analytic properties of $S^+(z)$ and $S^-(z)$. Doob [15] subsequently elaborated the technique, but it has received little attention since that time until quite recently [11], [12]. In this paper we apply the fundamental principles of the Wiener–Doob technique to generalize spectral factorization to two dimensions. The generalization we propose extends the concept of spectral factorization to two dimensions so as to emulate the analytic properties associated with the 1-D factorization. Because of the added dimension, this leads to a family of canonical factorizations. The canonical factorizations result in factors which have prescribed regions of analyticity and are recursively computable as is the case in 1-D spectral factorization. We consider in detail two-, four-, and eight-factor canonical factorizations, followed by a consideration of their numerical implementation. We then present an implementation of this algorithm, whose required calculations are quite well-conditioned and easily performed using a 2-D fast Fourier transform (FFT) algorithm. Finally, we apply these factorizations to a number of topics of interest in 2-D recursive filtering. We include in this section a new filter design algorithm and a novel stability test for recursive filters. Numerical examples of each are presented.

II. Two-Dimensional Spectral Factorization

Two-dimensional linear filtering can be described in terms of the 2-D z transform. For a 2-D function $f(m, n)$ (defined on I^2, the 2-D orthogonal lattice), we denote its z transform as $F(z_1, z_2)$ given by

Manuscript received February 26, 1975; revised October 28, 1975. This work was performed under the auspices of the U.S. Energy Research and Development Administration under Contract W-7504-ENG-48 and supported in part by the Defense Advanced Research Projects Agency of the Department of Defense under ARPA Order 2436.

The authors are with the Lawrence Livermore Laboratory, University of California, Livermore, CA 94550.

$$F(z_1, z_2) = \sum_{m,n}\sum f(m,n)\, z_1^{-m} z_2^{-n}, \quad (z_1, z_2) \in \mathbb{C}^2 \tag{1}$$

where $\mathbb{C}^2$ is the space of 2-D complex variables. Corresponding to an autocorrelation or nonnegative definite function $r(m, n)$, we have the spectral function $S(z_1, z_2)$. By 2-D spectral factorization, we mean a decomposition of S into factors which are free of zeros and singularities in certain distinguished regions of $\mathbb{C}^2$. For a particular set of distinguished regions, a factorization results which is best described in terms of a generalization of the 1-D concept of minimum phase.

A. Minimum Phase in Two Dimensions

In one dimension, a complex function is said to be minimum (min) phase if all its poles and zeros are inside the unit circle. Similarly, if all the poles and zeros are outside the unit circle, the function is said to be maximum (max) phase. Now, in two dimensions there are two complex variables and hence two unit circles of concern in the region of analyticity of the z transform. It turns out to be convenient to define the following four cases.

Definition 1(a): $F(z_1, z_2)$ will be defined to be *min-min* phase if none of its singularities or zeros lie on $\{|z_1| \geq 1, |z_2| \geq 1\}$.

Definition 1(b): $F(z_1, z_2)$ will be defined to be *max-max* phase if none of its singularities or zeros lie on $\{|z_1| \leq 1, |z_2| \leq 1\}$.

Definition 1(c): $F(z_1, z_2)$ will be defined to be *min-max* phase if none of its singularities or zeros lie on $\{|z_1| \geq 1, |z_2| \leq 1\}$.

Definition 1(d): $F(z_1, z_2)$ will be defined to be *max-min* phase if none of its singularities or zeros lie on $\{|z_1| \leq 1, |z_2| \geq 1\}$.

These concepts as well as some others are convenient in characterizing the analytic properties of 2-D factors as will be shown later. However, in general there is not the fundamental importance of minimum phase as in one dimension, because factors other than these have equal if not greater practical importance. We next discuss a homomorphic transform briefly, prior to presenting our general factorization scheme.

B. Homomorphic Transform

The 2-D homomorphic transform we will be concerned with here is given in operator notation as

$$\mathcal{H} = Z^{-1} Ln Z \tag{2}$$

where Z is the 2-D z transform operator and Ln is the pointwise natural logarithm operator. We will define this operation for absolutely summable bisequences, i.e., l^1. The inverse z transform is taken on the distinguished border of the unit polydisk in $\mathbb{C}^2$, i.e., $\{|z_1| = 1, |z_2| = 1\}$. The z transform and inverse are thus seen to be equivalent to the Fourier transform ($\mathfrak{F}$) defined for l^1 functions on this discrete 2-D space. This Fourier transform is really just the familar Fourier series with the roles of transform and inverse reversed. The block diagram for this $\mathcal{H}$ transform is seen in Fig. 1(a). Its inverse

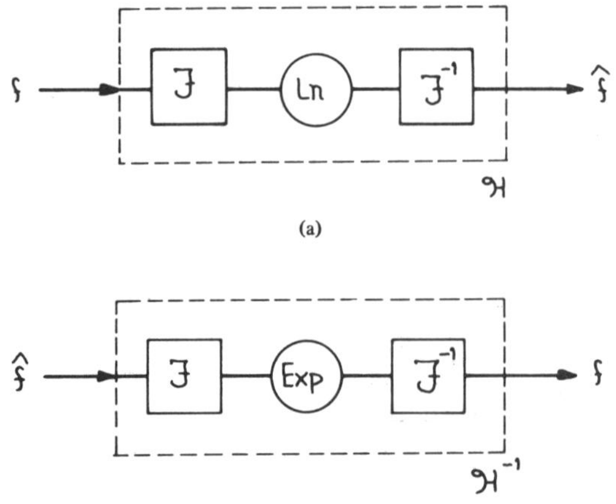

Fig. 1. Block diagram of the (a) homomorphic transform and (b) inverse.

$$\mathcal{H}^{-1} = Z^{-1} \exp Z \tag{3}$$

is shown in Fig. 1(b).

One-dimensional $\mathcal{H}$ transforms and their inverses have been studied by Oppenheim *et al.* [14]. Interestingly, it turns out that the 1-D $\mathcal{H}$ transform above is the operator used in the Wiener-Doob spectral factorization [6], [15]. If we start out with a positive definite or autocorrelation array, then its homomorphic transform is the z transform of the logarithm of its spectrum. This function is sometimes called the cepstrum. We will be interested in the case when the cepstrum is in l^1. Such functions are also said to be convolutionally stable.

The question "when is the cepstrum in l^1?" was answered in 1-D in 1933 by Wiener and Lévy [16], [17]. Their result and method of proof are not changed for two dimensions. They are repeated here for convenience.

Theorem (Wiener-Lévy): If a continuous function $S(u, v) > 0$ has an absolutely convergent Fourier series $r(m, n)$ then $Ln\, S(u, v)$ has an absolutely convergent Fourier series.[1]

We now establish some notation.

1) $\mathcal{L} \triangleq [-0.5, +0.5] \times [-0.5\ +0.5] \subset \mathbb{R}^2$.
2) $T^2 \triangleq \{|z_1| = 1, |z_2| = 1\} \subset \mathbb{C}^2$.
3) $\hat{b} = \mathcal{H}(b)$ when it exists.
4) $B(u, v) = Z(b)\big|_{z_1 = e^{+j2\pi u},\, z_2 = e^{+j2\pi v}}$.

We can restate the Wiener-Lévy Theorem in our notation.

Theorem 1: Let $b \in l^1$ with continuous $B(u, v) > 0$, then $\hat{b} = \mathcal{H}(b)$ exists and is in l^1.

In [13] Pistor derived this result for the case where the array b was of finite support, but here we see that the result is considerably more general. The importance of this theorem lies in the fact that b in l^1 will lead to appropriate analyticity conditions on projections of $\hat{b}$ onto subspaces of l^2. These

[1] The proofs of all theorems are to be found in the Appendix.

projections will turn out to be central to our spectral factorization method.

Alternately one may derive conditions on the spectral function B, which are based entirely in the frequency domain.

Theorem 2: Let $B \in L^2$ with continuous $B > 0$ on $\mathcal{L}$. Then if the mixed partial derivatives of B exist and are bounded out to second order in u and v, there exists $\hat{b} = \mathcal{H}(b) \in l^1$.

The importance of this homomorphic transform to 2-D spectral factorization can now be discussed. The logarithmic transformation in the frequency domain changes products into sums so that one may factor a given positive definite function by additively decomposing its homomorphic or $\mathcal{H}$ transform. The l^1 condition will give us the needed analytic characteristics of the factors.

C. General Factorization

The basis of Wiener and Doob's 1-D factorization method is the logarithmic mapping of poles and zeros into singularities [12]. Hence the resulting cepstrum has singularities at the poles and zeros of its inverse homomorphic ($\mathcal{H}$) transform. It follows that the causal component of the cepstrum contains the poles and zeros with magnitude less than one. Similarly the anticausal component of the cepstrum contains the poles and zeros with magnitude greater than one. Thus the inverse $\mathcal{H}$ transform of each of these components or projections (with the origin equally divided) yields a factorization into minimum and maximum phase sequences.[2] Our 2-D recursive factorization is a generalization of the Wiener–Doob method to two dimensions. We determine the factors by sectoring the range of $\mathcal{H}$. We offer the following definitions.

Definition: A *nonsymmetric half plane* is a region of the form $\{m \geqslant 0, n \geqslant 0\} \cup \{m > 0, n < 0\}$ or $\{m \geqslant 0, n \leqslant 0\} \cup \{m > 0, n > 0\}$ or their rotations.

For example, see Fig. 2. "Rotation" is conducted continuously, then the lattice points inside are taken to constitute the "rotated" figure. The term sector will be taken in the same way, i.e., the lattice points inside a continuous sector.

Definition: An *admissible region* is the set theoretic product of sectors and a nonsymmetric half-plane.

For examples, see Fig. 3. In these terms we can define a general 2-D spectral factorization which reduces to the normal definition in one dimension.

Definition: A *2-D spectral factorization* is determined by a decomposition of the range of $\mathcal{H}$ into admissible regions (possibly overlapping at the borders) along with a corresponding set of projection operators whose sum is the identity. The factors are then the inverse transform of the projections onto these regions.

The fact that the projected $\hat{b}$'s sum to $\hat{b}$ yields, after inverse $\mathcal{H}$ transformation, a convolution product factorization. It will be shown below that this factorization results in factors which have analytic properties which generalize those of the one-dimensional factors.

We next state a lemma and theorem on admissible regions.

Lemma 1: Project $b \in l^1$ onto a sector less than or equal to 180° with projection operator P_s, then $b_s = \mathcal{H}^{-1} P_s(\hat{b})$ has support on this sector.

[2] This analytic property of the cepstrum has been expressed in terms of homomorphic transforms by Oppenheim *et al.* [14].

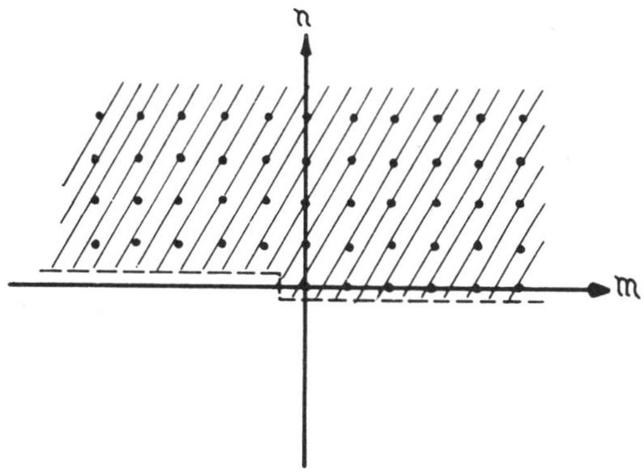

Fig. 2. Nonsymmetric half-plane (shaded region).

Theorem 3: Let $b \in l^1$ and let P be a projection operator which projects onto an admissible region. Then $b = \mathcal{H}^{-1} P(\hat{b})$ is recursively computable and stable.

Theorem 3 shows that the class of admissible regions generates factors which are recursively computable and stable in analogy with 1-D spectral factorization. Clearly a difference between the 2-D case and the more familar 1-D case is that the admissible regions cover a plane rather than a line. This leads to two main distinctions: first, the regions are not unique in two dimensions, i.e., there is much (infinitely) more variety and second, the regions can overlap along their edges. This last distinction leads to further variety in the specification of the projection mapping on the edge.

III. Canonical Factorization

In this section we consider in detail some useful sets of admissible regions and the analytic properties of the corresponding factors. For each sectoring of the $\mathcal{H}$ plane into admissible regions, we obtain a spectral factorization. The two most important cases are the 90° sectors which lead to the min-min phase four-factor decomposition, and the nonsymmetric half-plane which leads to a two-factor decomposition which will be classified as min-mix or mix-min phase. We call these canonical factorizations. Advantages and disadvantages of the various types of factorizations will be discussed. We will treat the application of these ideas to recursive filtering more fully in Section V.

A. Four-Factor Case

We define the following regions in the $\mathcal{H}$ plane.

Definition:

$$\mathcal{R}_{++} \triangleq \{m \geqslant 0, n \geqslant 0\}, \qquad \mathcal{R}_{-+} \triangleq \{m \leqslant 0, n \geqslant 0\},$$

$$\mathcal{R}_{+-} \triangleq \{m \geqslant 0, n \leqslant 0\}, \text{ and } \mathcal{R}_{--} \triangleq \{m \leqslant 0, n \leqslant 0\}.$$

These regions are shown in Fig. 4. Let $\hat{b}$ be the homomorphic transform of a positive definite or autocorrelation function $r \in l^1$.

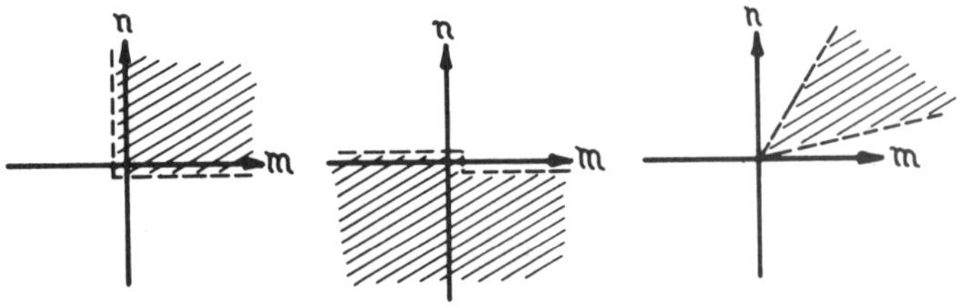

Fig. 3. Examples of admissible regions.

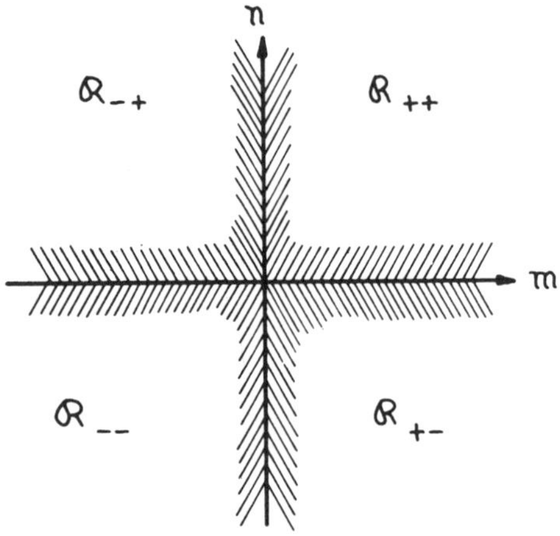

Fig. 4. Quarter-plane regions of the $\mathcal{H}$-plane. (Note: overlap occurs on both m and n coordinate axes.)

Denote by $\hat{b}_{..}$ the projection of $\hat{b}$ onto $\mathcal{R}_{..}$ where $\hat{b}$ on the overlapping boundaries is divided up arbitrarily, consistent with the $b_{..}$ being in l^1. We note, however, that each different apportionment of the boundary values provides a distinct canonical factorization. Denote by $b_{..}$ the inverse $\mathcal{H}$ transform of $\hat{b}_{..}$; we then have the factorization

$$b = b_{++} * b_{-+} * b_{+-} * b_{--}\text{[3]} \tag{4}$$

where

the factor b_{++} is min-min phase,
the factor b_{-+} is max-min phase,
the factor b_{+-} is min-max phase, and
the factor b_{--} is max-max phase.

These analytic properties of the factors are proved in [13] for finite support centro symmetric r; however, the Wiener-Lévy Theorem and Theorem 2 allows us to generalize to a much broader class, $r \in l^1$.

We now give an example of this factorization. The spectral density function of a 2-D Markov random field [1] will be factored into four factors analytic in various regions of $\mathbb{C}^2$. We take the field to be almost white with spectrum of the form

[3] In this notation * denotes convolution.

$$S(u, v) = c_0/(1 - \sum_{\mathbb{D}_\rho} a(k, l) \exp - j2\pi(ku + lv)) \tag{5}$$

where

1) $\mathbb{D}_\rho \triangleq \{(k, l) \mid k^2 + l^2 \leqslant \rho^2 \text{ and } (k, l) \neq (0, 0)\}$
2) $a(k, l) = a(-k, -l)$
3) $\sum_{kl} |a(k, l)| << 1.$

Because of 3), using $\ln(1 + x) \cong x$ for $|x| << 1$, we obtain

$$\hat{r} \triangleq \mathcal{H}(r) \cong \begin{cases} \ln(c_0) & (m, n) = 0 \\ a(m, n) & (m, n) \in \mathbb{D}_\rho. \end{cases}$$

Projecting onto the four quadrants, we obtain the decomposition

$$\hat{r} \cong \hat{r}_{++} + \hat{r}_{+-} + \hat{r}_{-+} + \hat{r}_{--}.$$

If we divide up the axis values ($m = 0$ or $n = 0$) equally between the two factors sharing the axis we obtain

$$\hat{r}_{++} = \begin{cases} 1/4 \ln c_0 & (m, n) = 0 \\ 1/2\, a(m, n) & (m = 0 \text{ or } n = 0) \cap \{(m, n) \neq (0, 0)\} \\ a(m, n) & (m > 0, n > 0) \\ 0 & \text{elsewhere} \end{cases}$$

and

$$\hat{r}_{-+} = \begin{cases} 1/4 \ln c_0 & (m, n) = 0 \\ 1/2\, a(m, n) & (m = 0 \text{ or } n = 0) \cap \{(m, n) \neq (0, 0)\} \\ a(m, n) & m < 0, n > 0 \\ 0 & \text{elsewhere.} \end{cases}$$

Upon inverse transforming, making use of $e^x \cong 1 + x \cong \frac{1}{1 - x}$ for $|x| << 1$, we obtain

$$r = r_{++} * r_{+-} * r_{-+} * r_{--}$$

where

$$R_{..}(u, v) = c_0^{1/4}\left(1 - \sum_{\mathbb{D}_\rho} \hat{r}_{..}(m, n) \exp - j2\pi(mu + nv)\right)$$

for

$$(..) \in \{(++), (-+), (+-), (--)\}.$$

If we identify + with min and − with max, then we see that this is the min-min phase decomposition discussed previously. That the ++ factor is min-min phase, i.e., has no singularities or zeros in $\{|z_1| \geq 1, |z_2| \geq 1\}$, follows from the fact that $\hat{r}$, and hence its projection is in l^1.

A difficulty with this four-factor decomposition is "nonuniqueness" due to the overlapping boundaries of the regions. This could be "solved" by defining the regions to not overlap. However, this would prevent simple factorizations that might otherwise occur by say, dividing by 2 on the overlap regions. We solve this problem by including the projection on the boundary as part of the definition of the canonical factorization. The four-factor decomposition was first presented by Pistor in 1970 in an unpublished report [18]. The work was subsequently published by Bednar and Farmer in 1972 [19] and by Pistor in 1974 [13].

B. Eight-Factor Case

If we define regions

$$\mathcal{R}_{+o} \triangleq \{m \geq 0, n = 0\}, \mathcal{R}_{o+} = \{m = 0, n \geq 0\},$$

$$\mathcal{R}_{-o} \triangleq \{m \leq 0, n = 0\}, \mathcal{R}_{o-} = \{m = 0, n \leq 0\},$$

$$\mathcal{R}'_{..} \triangleq \mathcal{R}_{..} \cap \{m \neq 0 \text{ and } n \neq 0\}$$

shown in Fig. 5, we can obtain the eight-factor decomposition

$$\hat{b} = \hat{b}'_{++} + \hat{b}'_{+-} + \hat{b}'_{-+} + \hat{b}'_{--} + \hat{b}_{+o} + \hat{b}_{o+} + \hat{b}_{-o} + \hat{b}_{o-}$$

or

$$b = b'_{++} * b'_{+-} * b'_{-+} * b'_{--} * b_{+o} * b_{o+} * b_{-o} * b_{o-}. \quad (6)$$

This eight-factor decomposition is interesting from a theoretical standpoint. First, its projection operators are unique up to a scalar multiplier since the regions only overlap at the origin of the $\mathcal{H}$ plane. Second, it provides a decomposition or factorization of the min min phase factor

$$b_{++} = b'_{++} * b_{+o} * b_{o+}.$$

This can be written as

$$b_{++} = b'_{++} * [b_{+o}(m, 0)\, b_{o+}(0, n)] \quad (7)$$

because the $+o$ and the $o+$ factors are one dimensional.

In fact it can be shown that when $b(m, n)$ is separable, i.e.,

$$b(m, n) = {}_1b(m)\, {}_2b(n),$$

then

$$b_{+o}(m, 0) = {}_1b_+(m)$$

and

$$b_{o+}(0, n) = {}_2b_+(m)$$

where ${}_1b_+$ and ${}_2b_+$ are the minimum phase factors of ${}_1b$ and ${}_2b$, respectively.

The factor b'_{++} in (7) represents the nonseparable component of b_{++}. These results are equivalent to the fact that the $\mathcal{H}$ transform of a separable function is zero off the m and n axes. It follows that for separable functions, the four- and eight-factor decompositions are identical and result in a Cartesian product of the factors arising from the 1-D Wiener-Doob factorization method [15].

It is also easy to show that b'_{++} is of the form

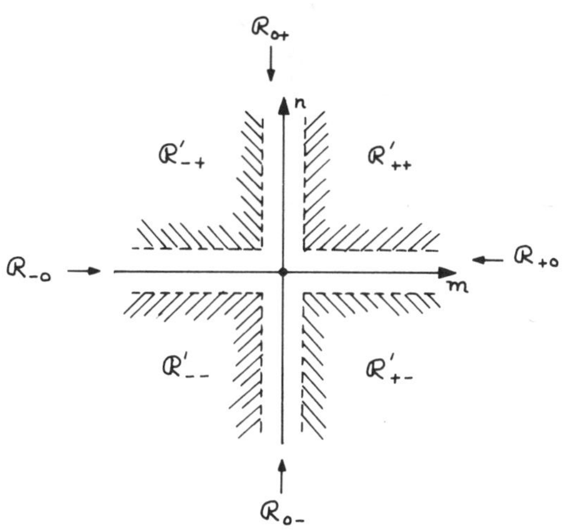

Fig. 5. Regions of support for eight-factor decomposition. (Note: no overlap for this decomposition except for the 1-D factors at the origin.)

$$b'_{++} = \begin{cases} 1 & m = n = 0 \\ 0 & (m = 0, n \neq 0) \text{ or } (m \neq 0, n = 0) \\ b'_{++} & \text{otherwise} \end{cases}$$

which justifies the term nonseparable component. The factors $b'_{..}$ are still analytic in the appropriate regions including T^2, i.e., min-min phase, etc., however, they have had their separable component removed. Huang [20] has shown that, in our notation, for a factor b with support $\mathcal{R}_{++}$, the min-min phase property is equivalent to

$$B(z_1, \infty) \neq 0 \text{ on } \{|z_1| \geq 1\} \quad (8)$$

$$B(e^{+j2\pi u}, z_2) \neq 0 \text{ on } \{|z_2| \geq 1\} \text{ for all } u \in [-\tfrac{1}{2}, +\tfrac{1}{2}]. \quad (9)$$

Rewriting (7) as

$$b_{++} = (b'_{++} * b_{o+}) * b_{+o}$$

we see that (8) is the requirement that b_{+o} be 1-D minimum phase and (9) is the condition that $(b'_{++} * b_{o+})$ be min-min phase. Obviously analogous decompositions into separable and nonseparable components hold for the other factors b_{-+}, b_{+-}, and b_{--}. We note in passing that the factor $(b_{++} * b_{o+})$ has support on the region $\{m \geq 0, n > 0\}$.

C. Two-Factor Case

Here we consider projections onto nonsymmetric half-planes. As a result we obtain two term canonical factorizations. We first define a set of regions.

Definition:

1) $\mathcal{R}_{\oplus +} \triangleq \mathcal{R}_{++} \cup \mathcal{R}'_{-+}$
2) $\mathcal{R}_{\ominus -} \triangleq \mathcal{R}_{--} \cup \mathcal{R}'_{+-}$.

These regions are shown in Fig. 6.

Similarly one defines $\mathcal{R}_{+\oplus}$, $\mathcal{R}_{-\oplus}$, $\mathcal{R}_{\ominus +}$, etc., by noting that the two signs determine the $\mathcal{R}_{..}$ part and the "o" determines

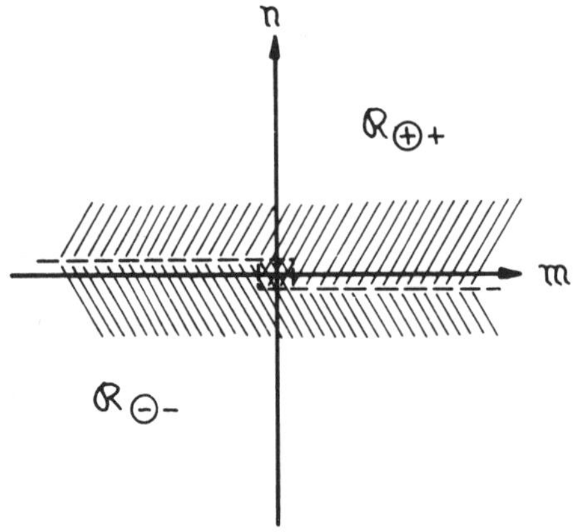

Fig. 6. Regions of support for two-factor decomposition. (Note: overlap occurs only at the origin, i.e., $m = n = 0$.)

the $\mathcal{R}'_{..}$ part by a change of the corresponding sign. By this method we catalog the eight nonsymmetric half-planes which have coordinate axes as edges. Then for a cepstrum $\hat{b}$ we have

$$\hat{b} = \hat{b}_{\oplus+} + \hat{b}_{\ominus-}$$

or

$$b = b_{\oplus+} * b_{\ominus-}. \tag{10}$$

Because of the definition of $\mathcal{R}_{\oplus+}$ we note that

$$b_{\oplus+} = b_{++} * b'_{-+}. \tag{11}$$

Now b_{++} is min-min phase and b'_{-+} is max-min phase, thus $b_{\oplus+}$ is mix-min phase, i.e., is analytic in an open region including $\{|z_1| = 1, |z_2| \geqslant 1\}$. It is convenient here to extend our 2-D minimum phase characterizations to include this case of functions. As before the definitions are classified according to the root locations of $F(z_1, z_2)$.

Definition:

1) $F(z_1, z_2)$ is said to be *min-mix* phase if none of its singularities or zeros lies on $\{|z_1| \geqslant 1, |z_2| = 1\}$.

2) $F(z_1, z_2)$ is said to be *max-mix* phase if none of its singularities or zeros lies on $\{|z_1| \leqslant 1, |z_2| = 1\}$.

3) $F(z_1, z_2)$ is said to be *mix-min* phase if none of its singularities or zeros lies on $\{|z_1| = 1, |z_2| \geqslant 1\}$.

4) $F(z_1, z_2)$ is said to be *mix-max* phase if none of its singularities or zeros lies on $\{|z_1| = 1, |z_2| \leqslant 1\}$.

While these definitions provide useful terminology, they do not fully describe the two-factor case, unlike the situation in the four-factor canonical factorization mentioned earlier. This can be realized by noting that there are eight possible canonical factorizations corresponding to the nonsymmetric half-planes defined above. Yet there are only four combinations of min-mix, etc. The additional analyticity of $b_{\oplus+}$, for example, is that $b_{\oplus+}(m, 0)$ is minimum phase in one dimension. We formalize this in a theorem.

Theorem 4: The $\oplus+$ canonical factorization (10) is characterized by the following analytical properties of factor $b_{\oplus+}$:

1) $b_{\oplus+}$ is mix-min phase, and

2) $b_{\oplus+}(m, 0)$ is 1-D min phase.

If we had defined the symmetric half-planes, e.g., $\{n \geqslant 0\}$, instead of nonsymmetric half-planes, the mixed phase notation would characterize the factors. Then they would have the stable inverses considered by Justice and Shanks [21]. However, since those factors have support on half-planes not nonsymmetric half-planes, they are not recursively computable. Thus, the nonsymmetric half-plane support constraint leads to the 1-D min phase constraint on $b_{\oplus+}$ and vice versa.

D. A Complication

At this point we elaborate a complication, alluded to in the Introduction, which arises in 2-D factorization and not in the 1-D problem. This complication is that the factors of a finite order, 2-D polynomial are *generally* not finite-order polynomials. In 1-D factorization we know that a polynomial spectrum of order $2N$ will factor into minimum and maximum phase polynomials of order N. While 2-D factorizations having appropriately generalized analytic properties have been defined above, the factors obtained are generally infinite order. This is simply a restatement of the fact that 2-D spectra are almost always irreducible.

The fact that the decomposition of almost all polymonial spectra leads to infinite-order factors, does not preclude the useful application of our factorization in engineering problems. It turns out that for many problems in image processing, for example, the spectra to be factored are nonrational. Thus as in the case of dealing with nonrational 1-D spectra, approximation methods must be applied. In designing 2-D recursive filters and other 2-D applications, one is necessarily concerned with approximation with the class of polynomials which is factorable into polynomials. (We will give examples of this in a subsequent section.)

For the reducible class of polynomial spectra, our factorization algorithm generates the appropriate factors. However, it should be noted that the canonical form of the factorization must also be known. This is because a finite-order factorization for one form will generally not be finite order for a different canonical form.

IV. Numerical Implementation

We now discuss some of the questions related to the numerical implementation of these factorizations using the discrete Fourier transform (DFT). The block diagram for this approximate $\tilde{\mathcal{H}}$ transform is seen in Fig. 7.

We see that the only difference is the substitution of the DFT and inverse DFT (IDFT) for the Fourier transform and inverse operators. As a result of this substitution, errors occur due to the approximate relation of the DFT to the Fourier transform ($\mathcal{F}$) over all of the discrete space. To gain an appreciation of this error we will consider a periodic repetition of the input autocorrelation function with period M' on the m axis and N' on the n axis.

$$r_p(m, n) = \sum_{m', n'} r(m + m'M', n + n'N'). \tag{12}$$

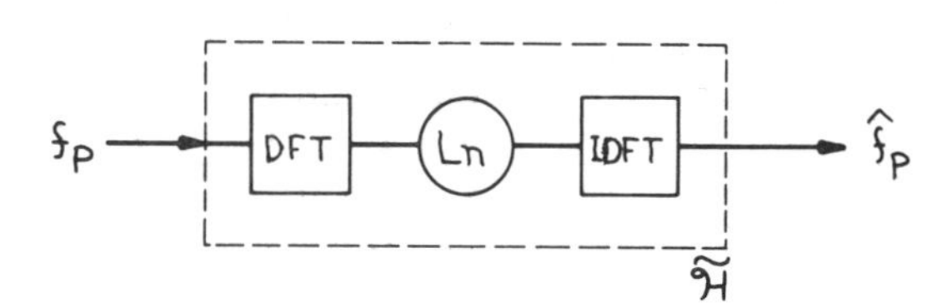

Fig. 7. Approximate homomorphic transform using FFT.

We will now consider factoring the spectrum of this periodic autocorrelation. It is well known that the $\mathcal{F}$ transform of a periodic function will consist of impulse generalized functions. We extend the notion of logarithm to the impulse generalized function as follows:

$$\ln [S(u,v)\delta(u,v)] \triangleq \ln [S(u,v)]\, \delta(u,v).$$

Also, one can show that the DFT coefficients evaluated for block size $M' \times N'$ are exactly the areas of the impulses obtained from the periodic construct r_p. It then turns out that the cepstrum obtained from the use of the DFT is exactly a period $(M' \times N')$ of the cepstrum of the periodic autocorrelation. Thus, any errors must be due solely to the overlap caused by the periodic repetition in the spatial and cepstral domains.

In the spatial domain, for absolutely summable r, it is easy to choose M' and N' large enough so that the source of the error arising from overlap in the spatial domain is small. The case of an $M \times N$ autocorrelation array is particularly simple. If one takes $M' > M$ and $N' > N$, then there will be no error from this source. The problem of overlap in the cepstral domain is more subtle and defies analytical solution. However, empirically it has been determined that for most problems of practical interest, a choice of $M' \gg 2M$ and $N' \gg 2N$ should result in errors which are satisfactorily small, where in this case M and N may correspond to a truncation of the autocorrelation function resulting in acceptable errors from that source.

The important point to be noted is that because the errors are directly associated with the size of the DFT, the numerical accuracy of the factorization can be improved by simply increasing M' and N'. The computational burden involved with increasing the DFT sizes is generally quite modest. For example, we are able to compute a 1024×1024 FFT on our CDC 7600 in only 6.5 s. For almost all factorizations of practical interest, this size FFT would provide accuracies well within the numerical noise of the FFT itself. As an example of a more typical situation which illustrates the error behavior, we have factored small correlation arrays, 9×9, using 32×32 FFT's and obtained errors $0(10^{-13})$. A more detailed characterization of the error behavior may be developed along the same lines as Rino [11] used in the 1-D case.

One concluding point concerning an additional effect of the DFT's periodicity may be of interest. The periodic nature of r generated from $\tilde{\mathcal{H}}$ leads to possible additional overlap of admissible regions. For example, for the four-factor decomposition, the regions corresponding to $\mathcal{R}_{++}$ and $\mathcal{R}_{-+}$ would become

$$\tilde{\mathcal{R}}_{++} \triangleq \{0 \leqslant m \leqslant M'/2, 0 \leqslant n \leqslant N'/2\}$$

and

$$\tilde{\mathcal{R}}_{-+} \triangleq \{m = 0, 0 \leqslant n \leqslant N'/2\} \cup \{M'/2 \leqslant m \leqslant M' - 1, 0 \leqslant n \leqslant N'/2\}.$$

These regions overlap at both $m = 0$ and $m = M'/2$. Similarly, including $\tilde{\mathcal{R}}_{+-}$ and $\tilde{\mathcal{R}}_{--}$ defined analogously, there will be pairwise overlap at $n = 0$ and $n = N'/2$. Four-region overlap will occur at $(m, n) \in \{(0, 0), (0, N'/2), (M'/2, 0), (M'/2, N'/2)\}$. In this last case an additional degree of freedom exists in the factorization in apportioning values on the regions of overlap to the various factors.

V. Applications in 2-D Recursive Filtering

In this section we will consider the application of our spectral factorization algorithm to a number of problems in recursive filtering. We first describe a new class of recursive digital filters and state a stability theorem applicable to this class. We then turn to the more classical problems of designing stable, 2-D recursive filters and testing their stability. In each of these applications the factorization plays a fundamental role.

Prior to considering these topics, we now establish some notation concerning recursive filters. Consider the so-called "causal" recursive filter [20]

$$y(m,n) = \sum\sum_{\mathcal{R}_{++}} a(k,l)\, x(m-k, n-l) - \sum\sum_{\mathcal{R}_{++} - (0,0)} b(k,l)\, y(m-k, n-l) \tag{13}$$

where $b(0,0) = 1$. The implicit rule for solving this equation, with reference to Fig. 4, is that one must always use output values below and to the left in computing the solution. This convention determines the solution and a region of analyticity for the z transform

$$H(z_1, z_2) = \frac{A(z_1, z_2)}{B(z_1, z_2)}.$$

This region is $\{|z_1| \geqslant 1, |z_2| \geqslant 1\}$. The filter is known to be stable when $B(z_1, z_2)$ has no zeros in this region, i.e., when $B(z_1, z_2)$ is min-min phase according to our Definition 1(a). We call the recursive filter (13) a ++ recursive filter. Thus we can say the ++ recursive filter is stable if and only if the denominator array is min-min phase. In an analogous manner one defines −+, +−, and −− recursive filters. It is known that they are recursively stable if they are (in our notation) max-min, min-max, and max-max phase, respectively [20].

The general application of 2-D spectral factorization to recursive filtering arises because the factors in the canonical factorizations are recursively stable. For example, as we mentioned in the Introduction, if one wanted a filter which approximated a given ideal magnitude function, one could use any of the canonical spectral factorizations to factor the given magnitude function $H(e^{ju}, e^{jv})$ into recursively stable factors. One could do this because a magnitude function fulfills the requirements of a spectral density function. (To satisfy existence conditions, the magnitude function would also need to be positive everywhere). While there are difficulties with the order of the factors, as we mentioned above, they can usually be handled satisfactorily using approximation methods.

Our spectral factorization algorithm and its multiplicity of allowable factorizations also afford insight into general recursive filter formulations. Examples of this are as follows. We define the term *type of filter* as the canonical name of a factor, such as ++, -+, +-, etc. Now this effectively delimits the possible ways that the output can be computed recursively. We agree to call each possible way a *sequence of computation*, i.e., the order in which the output points are computed. We can now establish a lemma which points up the relationship and significance of these two terms.

Lemma 2: For all the sequences of computation by which (13) can be computed, the answers are identical and agree with the inverse z transform solution computed on a closed curve encircling the origin in or on $\{|z_1| \geq 1, |z_2| \geq 1\}$.

The preceeding lemma has the following implication. In the literature on recursive filtering, one often finds the phrase "direction of recursion" used to describe a type of filter. However, the above lemma tells us that the "direction" in which the data are processed is unimportant! The important thing is that the chosen sequence of computation have the property that each output point makes use of allowable inputs as determined by the type of filter. For these important reasons, it seems to the authors that the phrase "direction of recursion" is somewhat inappropriate. There is no one "direction of recursion," but rather a countable number of sequences of computation which are consistent with the type of filter. We will see below that this point of view can be of considerable utility in understanding 2-D recursive filtering.

Next consider the use of the factorization to recursively realize a filter function as discussed above. To obtain a zero-phase recursive implementation, it would be necessary to use all the factors from the decomposition. This requires multiple recursions, each of which would be performed in a different sequence of computation (to ensure stability). However, if a magnitude-only approximation were desired, then one could omit the factors with conjugate symmetry. For example, in the four factor decomposition a zero phase approximation would require the ++, +-, -+, and -- factors. A magnitude only approximation would require one of (++, --) *and* one of (+-, -+). Thus contrary to some recent research reported in the literature [22], [23], it is apparent that the classical ++ recursive filter *cannot* be used to realize a general magnitude function. It is capable of realizing only the restrictive class of magnitude functions which factor into min-min and max-max phase terms. With this in mind we consider a new class of recursive filters.

A. Nonsymmetric Half-Plane Recursive Filters

In analogy with the quarter-plane recursive filter discussed above, one can factor a 2-D magnitude function into nonsymmetric half-plane factors and thus generate a new class of recursive filters. For example, if we factor $H(z_1, z_2)$ into ⊕+ and ⊖- factors, we can generate a ⊕+ recursive filter. It is characterized by the following recursive equation:

$$y(m,n) = \sum_{\mathcal{R}_{\oplus+}} \sum a(k,l)x(m-k,n-l) - \sum_{\mathcal{R}_{\oplus+}-(0,0)} \sum b(k,l)y(m-k,n-l) \tag{14}$$

where $b(0, 0) = 1$. Again the implicit rule for solving these equations makes the answer unique. That is, the equation cannot be rewritten and solved for other than $y(m, n)$ because this would produce a different answer than the one we want. Now from Theorem 3, we know that the ⊕+ recursive filters satisfying the conditions 1) and 2) of Theorem 4,

1) $b_{\oplus+}$ is the mix-min phase and
2) $b_{\oplus+}(m, 0)$ is 1-D min phase,

are recursively computable and stable. We emphasize that the recursive answer agrees with the inverse z transform of $A(z_1, z_2)/B(z_1, z_2)$ computed on T^2. To see this we note that the recursive solution will have support on $\mathcal{R}_{\oplus+}$. Note by 1) and 2) this must also be the support of the inverse z transform solution (by two applications of the residue theorem). Now the uniqueness of the inverse z transform computed on T^2 guarantees equality. The necessity of conditions 1) and 2) for the stability of ⊕+ recursive filters is easily seen and will be omitted. Thus, we have Theorem 5.

Theorem 5: A ⊕+ recursive filter is recursively stable iff conditions 1) and 2) of Theorem 4 hold.

It is interesting to note that the stability of $b_{\oplus+}$ can also be established from known results concerning its factors

b_{++} and b'_{-+}

via

$$b_{\oplus+} = b_{++} * b'_{-+}$$

First, we note that recursion with $b_{\oplus+}$ is equivalent to recursion with b_{++} followed by recursion with b'_{-+}. This is clear since they can be performed with the same sequence of computation, namely left to right in each row from bottom to top. The fact that the first row of b'_{-+} is zero permits this sequence of computation. Lemma 2 above assures us that the answer is the same as for a more conventional sequence of computation of a -+ recursive filter, i.e., right to left, bottom to top. Now from Theorem 4, we see that conditions 1) and 2) are necessary and sufficient for the stability of the output of the cascade of b_{++} and b'_{-+}, but this is the output of $b_{\oplus+}$, hence we are done.

The reader should note the importance of distinguishing sequence of computation from the type of recursive filter in clarifying this argument. Obviously, these results extend to the other seven nonsymmetric half-plane factors mentioned earlier (for example, see Theorem 6).

Theorem 6: A ⊖+ recursive filter is recursively stable iff:

1) $b_{\ominus+}(m, n)$ is mix-min phase; and
2) $b_{\ominus+}(m, 0)$ is 1-D max phase.

An important advantage of this class of filters is implicit from the factorization. This is that the nonsymmetric half-plane filter can be used to realize the general class of positive-definite magnitude functions. This property follows from the unusual combination of having each factor min or max phase in only one variable and also recursively realizable. From an application viewpoint, it would appear that this class of filters would find much broader use than the classical ++ filters for example, and deserves much more attention. We will present some examples of these filters in the next section.

B. Filter Design

Here we present some preliminary design work involving the use of window functions to produce all-pole recursive filter designs. The design procedure/rationale is as follows. Given a prescribed 2-D magnitude response function, we form and spectrally factor its inverse. We know that the factors are recursively computable and stable by Theorem 3, however, we also know they are generally infinite dimensional.

For the recursive filter to be of practical value, it must be of finite order. Interestingly, we observe that the factors essentially have their energy concentrated in a region near the origin. This suggests the possibility of truncating or, more generally, windowing the factors to obtain a finite-order approximation. The windowing we apply involves both a truncation and a smoothing and is used in the following manner. Each infinite-order factor is first truncated onto a specific support. The size of this support is determined either from a specification on the order of the filter or it is chosen to be sufficiently large to meet a prescribed frequency domain design criterion. A weighting is then applied to the truncated array. This weighting smooths out some perturbations in the frequency spectra introduced by the truncation and tends to stabilize the truncated factors.

This last effect of the windowing we apply is a crucial one. Although the use of window functions in both 1-D and 2-D filter design is well known [24], [25], it is almost exclusively used for nonrecursive filter design. The reason for the nonapplication of window functions to recursive filter design, an area where they can be considerably useful, is in part due to the lack of a theorem guaranteeing that the windowed recursive filter array would be recursively stable. Thus, while the truncation-type windowing is a direct means of generating finite order factors, it should be clear that uncertainties exist concerning the stability of the truncated factors. We have employed two specific weighting sequences to smooth the factors:

1) $w(m,n) = \max\left\{\left(1 - \frac{|m|}{M}\right)\left(1 - \frac{|n|}{N}\right), 0\right\}$; and

2) $w(m,n) = \exp - (\alpha|m| + \beta|n|), \quad (\alpha,\beta) \geq 0.$

The application of these smoothing sequences has a stabilizing effect on the truncated recursive factors. We will present both an intuitive and a rigorous argument as to why this is so. First the intuitive one. The truncation operator creates rapid oscillations in $B(z_1, z_2)$ which can lead to zeros in the required zero free regions, e.g., $\{|z_1| \geq 1, |z_2| \geq 1\}$ in the case of the ++ factor. The subsequent application of the smoothing sequence reduces these oscillations so as to remove any zeros that the truncation inserted.

With regard to the exponential smoothing sequence, we can be rigorous. Letting b_w be the result of multiplying b by the exponential smoothing sequence w, we obtain its z transform

$$B_w(z_1, z_2) = \sum_{m,n} b(m,n)(z_1 e^{\alpha})^{-m}(z_2 e^{\beta})^{-n}$$

for the ++ factor. Thus

$$B_w(z_1, z_2) = B(z_1 e^{\alpha}, z_2 e^{\beta}).$$

This accomplishes a mapping of the zero loci according to

$$(z_1, z_2) \longrightarrow (z_1 e^{-\alpha}, z_2 e^{-\beta}),$$

thus moving the zeros out of the region $\{|z_1| \geq 1, |z_2| \geq 1\}$ which is required to be zero free for recursive stability of this ++ factor. With appropriate modification these results are equally applicable to the nonsymmetric half-plane factors.

We now present two examples of using this recursive filter design method. In both examples, truncation of the infinite dimensional factors resulted in finite factors which were recursively unstable. Both of the above window functions were then used to obtain stable and suitably approximate filters. The appropriate parameters for the windows (i.e., M, N and α, β) were determined iteratively.

Example 1–Low-Pass Filter: Let the filter magnitude response be given by

$$|H(u,v)| = \begin{cases} 1 & \sqrt{u^2+v^2} \leq 0.2 \\ 0.02 & \text{elsewhere.} \end{cases}$$

After factorization of $|H|^{-1}$ using the four-factor canonical decomposition with the overlaps equally divided in the $\mathcal{H}$ plane, we have

$$|H(u,v)|^{-1} = |B_{++}(u,v)B_{-+}(u,v)|^2$$

where we use the fact that $B_{++} = \overline{B_{--}}$ and $B_{+-} = \overline{B_{-+}}$ (the overbar indicates conjugation). The factors b_{++} and b_{-+} were truncated to the third order, i.e., 4 X 4 arrays and multipled by the triangular smoothing sequence 1) with $M = N = 4$. The resulting demonimator arrays were

b_{++} (n ↑ vertical, $0 \longrightarrow m$ horizontal)

0.0170	0.0206	0.0025	-0.0043
-0.0127	0.0233	0.0353	0.0025
-0.1987	-0.1610	0.0233	0.0206
0.7965	-0.1987	-0.0127	0.0170

and

b_{-+} (n ↑ vertical, $0 \longrightarrow -m$ horizontal)

0.0170	0.0206	0.0024	-0.0043
-0.0127	0.0233	0.0353	0.0024
-0.1988	-0.1609	0.0233	0.0206
0.7963	-0.1988	-0.0127	0.0170.

The nonidenticalness of the factors in the fourth digit is due to the numerical error induced from the 32 × 32 FFT used in the factorization. The magnitude response of the resulting filter is shown in Fig. 8. It is a reasonably good approximation to the desired characteristic. The 3 dB point occurs at a radius of 0.15 and the filter reaches 0.02 of the peak value at approximately 0.34. A somewhat sharper cutoff was obtained by using the exponential weighting. Here the values $\alpha = \beta = 0.13$ resulted in the "best" response, which is shown in Fig. 9. This filter has a 3 dB point at approximately 0.12 and is down to 0.02 of the peak at approximately 0.25.

Example 2–Fan Filter: In this example, the filter magnitude function is taken to be a fan filter. Its spectrum is given by

$$|H(u, v)| = \begin{cases} 1 & |v| \leqslant |u| \\ 0.02 & \text{elsewhere.} \end{cases}$$

The nonsymmetric half-plane canonic factorization was used to generate factors $b_{+\oplus}$ and $b_{-\ominus}$, so that

$$|H(u, v)|^{-1} = |B_{+\oplus}|^2.$$

The factor was truncated to fourth order and exponentially smoothed ($\alpha = \beta = 0.2$), yielding the denominator array below,

$$b_{+\oplus}: \quad \begin{array}{ccccc} 0.001 & -0.001 & 0 & 0 & 0 \\ -0.010 & 0.004 & -0.025 & -0.002 & 0 \\ 0.051 & 0.002 & 0.025 & -0.002 & 0 \\ -0.227 & -0.152 & 0.004 & 0.003 & 0.001 \\ 0.355 & 0.218 & 0.026 & 0.007 & 0.001 \\ & 0.040 & 0.050 & 0.005 & 0.001 \\ & -0.037 & -0.009 & 0.005 & 0 \\ & 0.003 & -0.004 & -0.002 & 0 \\ & -0.001 & 0 & 0 & 0 \end{array}$$

(axes: $n \uparrow$, $0 \longrightarrow m$)

and the frequency response in Fig. 10. Note that because the filter is +⊕, not ++, we have 5 × 5 array on the ++ quadrant and a 4 × 4 array in the +− quadrant, so that the total number of coefficients in the fourth-order +⊕ factor is 25 + 16 = 41. As one can see, we have a reasonable approximation to a fan filter that would be acceptable in many applications.

We wish to emphasize that these all-pole filters were designed with simple window function techniques and are not optimal in any formal sense. We have simply attempted here to demonstrate the feasibility of our design procedure in approximating a general class of response functions. All three designs were performed under a constraint on the order of the filter. Having performed the designs, the next step would involve a detailed comparison of the filter response with the desired function. If it was found to be unsatisfactory, the design would be iterated using successively larger filter orders until the criterion was satisfied.

C. Stability Test

As we have indicated, the stability of a recursive filter is determined by the region of analyticity of its system function, $H(z_1, z_2)$. Inasmuch as the fundamental property of our spectral factorization algorithm is its ability to separate a 2-D function into components with prescribed regions of analyticity, it seems reasonable to consider its use in testing the stability of 2-D recursive filters. We present stability tests for both the quarter-plane and nonsymmetric half-plane classes of filters discussed above. While the tests are identical in spirit and most detail, we present them separately to ensure clarity. Because the filter stability is determined by the pole loci of $A(z_1, z_2)/B(z_1, z_2)$, we conduct our tests on $b(m, n)$.

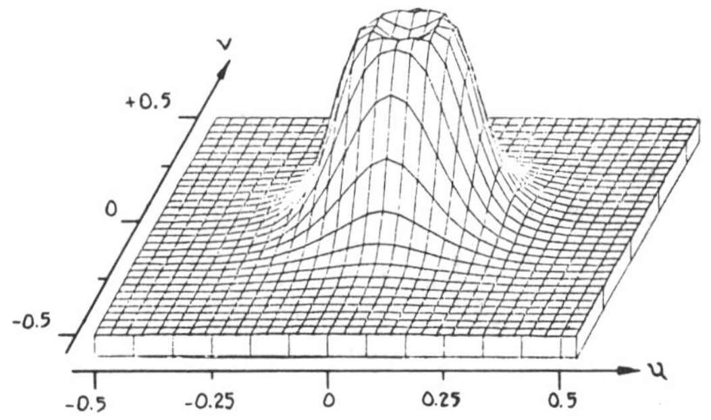

Fig. 8. Low-pass filter designed with triangular smoothing sequence.

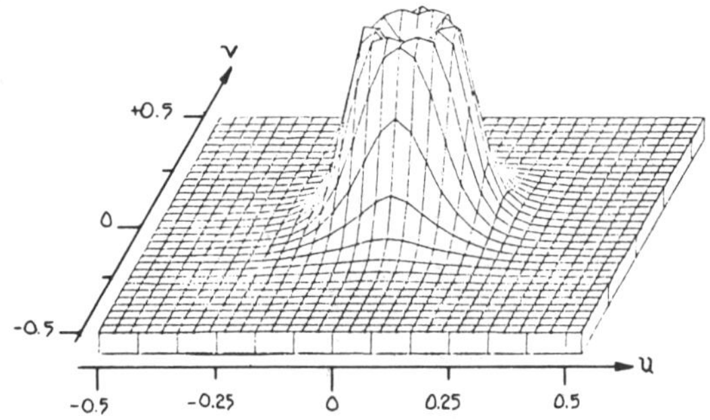

Fig. 9. Low-pass filter designed with exponential smoothing sequence.

Quarter-Plane Stability Test

Step 1: Autocorrelate $b(m, n)$ with itself, $r \triangleq b \circledast b$.

Step 2: Compute $\hat{r} = \mathcal{H}(r)$.

Step 3: Project $\hat{r}$ into quadrant with max energy (++ or −+).

Step 4: Compute $r_{..} = \mathcal{H}^{-1}(r_{..})$.

Step 5: Compare $r_{..}$ with b within a translation and conclude b is

a) (++) stable if $(..) = (++)$ and $r_{++}(m, n) = b(m, n)$,
b) (−−) stable if $(..) = (++)$ and $r_{++}(m, n) = b(-m, -n)$,
c) (−+) stable if $(..) = (-+)$ and $r_{-+}(m, n) = b(m, n)$,
d) (+−) stable if $(..) = (-+)$ and $r_{-+}(m, n) = b(-m, -n)$,
e) otherwise, $b(m, n)$ is an unstable quarter-plane filter.

Notice that in Step 5 we have determined whether a filter is any one of the four types of stable quarter-plane filters. This is possible because the supports of these four types are related by simple translation. This is not the situation for nonsymmetric half-plane filters, hence our test for these factors will deal with one type filter at a time (the one that its support matches). This test will be stated in general terms for a $(\cdot\cdot)$ recursive filter where $(\cdot\cdot) \in \{\oplus+, \oplus-, \ominus-, \ominus+, +\oplus, +\ominus, -\ominus, -\oplus\}$.

Nonsymmetric Half-Plane Stability Test

Step 1: Autocorrelate $b(m, n)$ with itself, $r \triangleq b @ b$.
Step 2: Compute $\hat{r} = \mathcal{H}(r)$.
Step 3: Project $\hat{r}$ onto $\mathcal{R}_{\cdot\cdot}$.
Step 4: Compute $r_{\cdot\cdot} = \mathcal{H}^{-1}(\hat{r}_{\cdot\cdot})$.
Step 5: Compare $r_{\cdot\cdot}$ with b and conclude b is

a) $(\cdot\,\cdot)$ stable if $r_{\cdot\cdot}(m, n) = b(m, n)$,
b) otherwise $b(m, n)$ is an unstable nonsymmetric half-plane filter.

At this point a brief comment on the use of the autocorrelation in the two tests may be of interest. It is introduced in Step 1 to simplify $\mathcal{H}$. Because $r = b @ b$ is a real, centrosymmetric function, its Fourier transform is real and the unwinding of phase associated with the complex logarithm can be avoided. In effect, we have generated a spectrum from b, so that spectral factorization can be applied. However, the autocorrelation introduces additional factors (in the quarter-plane test (++ and --) or (-+ and +-)) where there had originally been only one. This causes the slight complication in the decision process of Step 5 of the quarter-plane test. In both tests, the autocorrelation is actually implemented as a magnitude square operation after the DFT in the $\tilde{\mathcal{H}}$ transform. Some examples of application are now presented.

Example 1–Quarter-Plane Filter: This example is a test of a filter examined by Huang [20], that is

$$B(z_1, z_2) = 1 + \tfrac{1}{2}z_1^{-1} + \tfrac{1}{2}z_2^{-1} + \tfrac{1}{4}z_1^{-1}z_2^{-1} + \tfrac{1}{4}z_1^{-2} + \tfrac{1}{4}z_2^{-2}, \quad (15)$$

or

$$b(m,n) \quad \begin{array}{l|lll} n\uparrow & \tfrac{1}{4} & & \\ & \tfrac{1}{2} & \tfrac{1}{4} & \\ & 1 & \tfrac{1}{2} & \tfrac{1}{4} \\ \hline 0 & & & \rightarrow m \end{array}.$$

We tested this array and obtained the following output array for an 8 × 8 FFT:

$r_{++}(m, n)$ (n upward, m rightward from 0):

8.875E-04	3.853E-04	4.616E-03	-1.408E-03	-1.950E-03
6.348E-03	9.502E-04	-1.522E-03	4.4.5E-03	-1.408E-03
2.452E-01	7.521E-04	-2.168E-03	1.522E-03	4.616E-03
4.978E-01	2.474E-01	7.521E-04	9.502E-04	3.853E-04
1.003E+00	4.978E-01	2.452E-01	6.348E-03	8.815E-04.

With a 16 × 16 FFT we obtained

$r_{++}(m, n)$ (n upward, m rightward from 0):

-7.745E-05	5.093E-05	-7.220E-05	-3.300E-05	1.461E-04
-1.679E-05	-3.706E-05	8.402E-05	-6.472E-05	-3.300E-05
2.501E-01	-2.325E-05	-3.380E-05	8.420E-05	-7.220E-05
5.000E-01	2.501E-01	-2.325E-05	-3.706E-05	5.093E-05
1.000E+00	5.000E-01	2.501E-01	-1.679E-05	-7.745E-05.

It is apparent in both cases that $\cdot\cdot = ++$ and $r_{++} = b$ where the largest error in the 8 × 8 case is $9(10^{-3})$, and for 16 × 16 the largest error is $O(10^{-5})$. Thus we have achieved reasonably accurate results with quite modest sized FFT's and conclude that (15) is (++) stable.

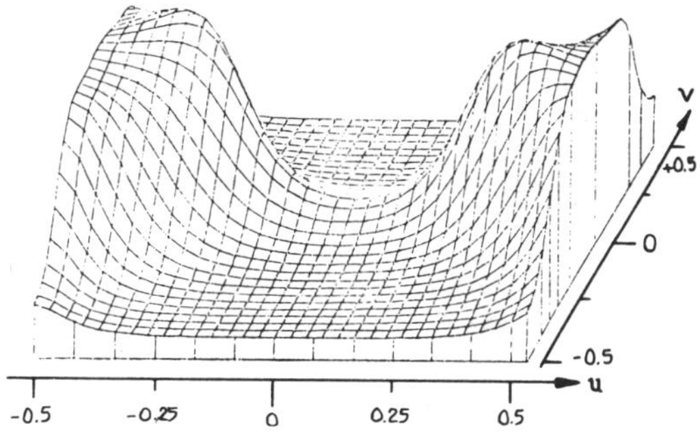

Fig. 10. Fan-filter designed with exponential smoothing sequence.

Example 2–Nonsymmetric Half-Plane Filter: In this example we test the stability of a +⊖ array using a 32 × 32 FFT. The array to be tested is

$b_{+\oplus}$ (m rightward from 0, n downward):

-1.40E-02	-1.91E-02	-7.89E-03
3.88E-02	-3.56E-03	-2.91E-02
2.31E-01	1.62E-01	3.93E-03
4.38E-01	2.22E-01	1.41E-02
	2.25E-02	-2.28E-02
	-3.43E-02	-1.90E-02
	3.10E-05	-1.72E-05.

The resulting output $r_{+\ominus}$ of Step 5 of the stability test is

(m rightward from 0, n downward):

-1.97E-09	1.21E-09	-1.62E-09	2.63E-09
-1.40E-02	-1.91E-02	-7.89E-03	3.70E-09
3.88E-02	-3.56E-03	-2.91E-02	4.62E-09
2.31E-01	1.62E-01	3.93E-03	5.15E-09
4.38E-01	2.22E-01	1.41E-02	5.15E-09
-1.44E-10	2.25E-02	-2.28E-02	4.63E-09
-1.21E-10	-3.43E-02	-1.90E-02	3.73E-09
-9.24E-11	3.10E-05	-1.72E-05	2.67E-09
-6.35E-11	1.49E-09	-1.19E-09	1.68E-09

and we conclude that $b(m, n)$ is +⊖ stable. Here, the largest error is $O(10^{-9})$.

Our stability tests compare quite favorably with previously suggested procedures [20], [26] for testing quarter-plane filters. No comparative tests exist, to the authors' knowledge, for testing nonsymmetric half-plane filters. The calculations required are of lower order variation (with respect to the order of the filter) than these earlier tests. For example, the test of Maria and Fahmy [26] requires computations of $0(N^2 2^{2(N-1)})$ for an $N \times N$ filter while our test requires $0(N^2 \log N)$. In addition, their test is extremely difficult to implement for higher order filters ($N >> 2$). An equally important comparison is based on programming considerations. Our algorithm is substantially easier to program, the only computation of any involvement being the 2-D FFT. We have shown that suitable accuracies are obtained using quite reasonably sized FFT's. The only modification of our routine to test higher order filters is simply an increase in FFT size. While we have stated tests for two specific classes of filters, it should be clear that the underlying concept easily generalizes to all other classes of recursive filters corresponding to the countable infinity of canonical factorizations.

VI. Summary and Conclusions

We have generalized the concept of spectral factorization to two dimensions. We found a family of canonical factorizations which are obtainable through the homomorphic transform and have analytic properties analogous to those of one-dimensional factors. The four-, eight-, and two-factor decompositions were treated in detail and their analytic properties determined. The numerical implementation via the FFT was discussed in some detail where it was suggested (and verified in the examples) that the computation is quite well conditioned and only modest FFT sizes need be used. These results were then applied in recursive filtering. A new class of recursive filters having support on nonsymmetric half-planes was defined. Their z transform stability theorem was stated and proven. Recursive filter design was discussed briefly where a window method was introduced. Finally, a new stability test for 2-D recursive filters was presented. This test is very easy to implement and offers the only computationally feasible way to test large recursive filters.

Appendix

Proof of Theorem (Wiener-Lévy): Let

$$S(u, v) = \sum_{m,n} r(m, n) \exp - j2\pi(mu + nv).$$

Then let

$$A\{S\} = \sum_{m,n} |r(m, n)| < \infty.$$

Let

$$\Phi(x) \triangleq Ln(x).$$

Then

$$\Phi(x) = \sum_{k=0}^{\infty} \phi_k(x - c)^k, 0 < x \leqslant 2c$$

where

$$c \triangleq A\{S\}.$$

So

$$\begin{aligned}\Phi(S)(u, v) &= \sum_{k=0}^{\infty} \phi_k(S(u, v) - c)^k \\ &= \sum_k \phi_k \left[\sum_{m,n} r(m, n) \exp - j2\pi(mu + nv) - c \right]^k \\ &\leqslant \sum_k |\phi_k| c^k \\ &= \sum_k |\phi_k| (A\{S\})^k < \infty\end{aligned}$$

because $S > 0$ and Φ is analytic on twice the range of S.

Proof of Theorem 2: $B \in L^2$ and continuous $B > 0 \Rightarrow \hat{b}$ exists. Using the well-known derivative relation for Fourier series, we have

$$m^2 n^2 \hat{b}(m, n) \overset{\mathcal{F}}{\longleftrightarrow} \frac{\partial^4}{\partial u^2 \partial v^2} \ln (B(u, v)). \tag{16}$$

Thus, if the right-hand side is in $L^2(\mathcal{L})$, the left-hand side is its inverse Fourier transform and is in l^2 by Parseval's theorem. Computing the right-hand side we obtain

$\dfrac{\partial^4 LnB(u, v)}{\partial u^2 \partial v^2}$ = linear combination of finite number of products of B^{-k} and the mixed partials of B out to order 2.

Thus it is in L^2 so the left-hand side of (16) is in l^2. Hence

$$\exists M < \infty \text{ such that } |m^2 n^2 \hat{b}(m, n)| \leqslant M \text{ for all } m, n$$

so

$$|\hat{b}(m, n)| \leqslant m^{-2} n^{-2} M \Longrightarrow \hat{b} \in l^1.$$

Proof of Lemma 1: Let $\hat{b}_s = P_s(\hat{b})$, then $\hat{b}_s$ has support on the sector S, see Fig. 11, bounded by straight lines through the origin at angles $\theta_2 > \theta_1$. Since $b_s = \mathcal{H}^{-1}(\hat{b}_s)$, we have $B_s = \exp (\hat{B}_s)$ on $\mathcal{L}$ so that using the well-known expansion for the exponential

$$\exp (x) = \sum_{k=0}^{\infty} x^k/k!, \qquad |x| < \infty,$$

we have

$$B_s = \sum_{k=0}^{\infty} \hat{B}_s^k/k!$$

Since the series converges absolutely, we may compute an inverse Fourier transform term by term to yield

$$b_s = \sum_{k=0}^{\infty} \hat{b}_s^{*k}/k!$$

where $*k$ denotes a k-fold convolution.

Now with reference to Fig. 11, it is seen that two functions with support on a sector will have the same support for their convolution. Because $\hat{b}_s$ has support S, it follows that $\hat{b}_s^{*k}$ has

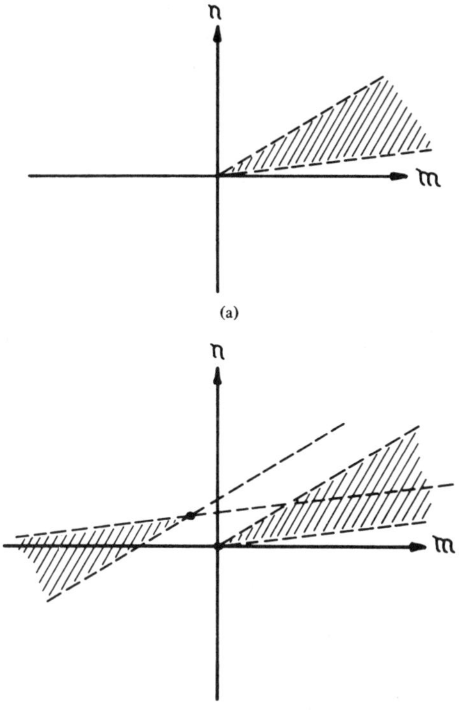

Fig. 11. Diagrams used in proof of Lemma 1. (a) Sector S and (b) convolution of two functions with support S.

support S. Finally, since b_s is a sum of such terms, it also has support S.

Proof of Theorem 3: By the definition of admissible region we see that the support of $P(\hat{b})$ will be contained in a nonsymmetric half-plane. Therefore, by the reasoning in the proof of Lemma 1, we have that the support of b will be contained in a nonsymmetric half-plane. Hence it is recursively computable.

To prove recursive stability, we first note, letting $\hat{b}_p = P(\hat{b})$, that $\hat{b}_p \in l^1$. This means that $\hat{B}_p$ is analytic in a region including T^2. Hence B_p is free of zeros in this same region, so that $b_p \triangleq \mathcal{H}^{-1}(\hat{b}_p)$ is in l^1. Also the inverse of the z transform and recursive filter solution must agree by the uniqueness of the inverse z transform because they have the same support.

Proof of Theorem 4: This can be shown to hold by the results of the four-factor case. Using (11) and Huang's simplification of the min-min and max-min phase properties (8) and (9), we obtain the following:

1) $B_{++}(z_1, \infty)$ on $\{|z_1| \geqslant 1\}$,
2) $B'_{-+}(z_1, \infty) \neq 0$ on $\{|z_1| \leqslant 1\}$,
3) $B_{++}(z_1, z_2) \neq 0$ on $\{|z_1| = 1, |z_2| \geqslant 1\}$, and
4) $B'_{-+}(z_1, z_2) \neq 0$ on $\{|z_1| = 1, |z_2| \geqslant 1\}$.

Now 3) and 4) are the same requirement on $B_{\oplus +}$, i.e., that $b_{\oplus +}$ is mix-min phase. 2) is trival since $B'_{-+}(z_1, \infty) = 1$. Then 1) is the statement that $b_{\oplus +}(m, 0)$ is 1-D min phase.

Proof of Lemma 2: First, it is clear that the answer is the same for all possible sequences of computation because each filter operation is the same, thus composite operations to yield any output point are the same. Hence, they must agree. Agreement with the inverse z transform on T^2 follows from, in fact is equivalent to, the uniqueness of the inverse z transform in its region of holomorphy.

Acknowledgment

The authors would like to acknowledge helpful comments from Prof. W. Rudin of the University of Wisconsin, Madison and Prof. S. Mitra of the University of California, Davis.

References

[1] J. W. Woods, "Two-dimensional discrete Markovian fields," *IEEE Trans. Inform. Theory*, vol. IT-18, pp. 232–240, Mar. 1972.
[2] A. Habibi, "Two-dimensional Bayesian estimates of images," *Proc. IEEE (Special Issue on Digital Picture Processing)*, vol. 60, pp. 878–883, July 1972.
[3] M. P. Ekstrom, "Numerical restoration of random images," Lawrence Livermore Lab., Univ. California, Livermore, Rep. UCRL-51129, Oct. 1971.
[4] J. H. McClellan and T. W. Parks, "Equiripple approx. of fan filters," *Geophysics*, vol. 37, pp. 573–583, Aug. 1972.
[5] R. M. Mersereau and D. E. Dudgeon, "The representation of two-dimensional sequences as one-dimensional sequences," *IEEE Trans. Acoust., Speech, Signal Processing*, vol. ASSP-22, pp. 320–325, Oct. 1974.
[6] N. Wiener, *Extrapolation, Interpolation and Smoothing of Stationary Time Series*. New York: Wiley, 1949.
[7] G. Wilson, "Factorization of the covariance generating function of a pure moving average process," *SIAM J. Numer. Anal.*, vol. 6, pp. 1–7, Mar. 1969.
[8] R. A. Wiggins, "On factoring the correlations of discrete multivariable stochastic processes," Mass. Inst. Technol., Cambridge, Rep. AFCRL-65-207, Feb. 1965.
[9] T. S. Huang, W. F. Schrieber, and O. J. Tretiak, "Image processing," *Proc. IEEE*, vol. 59, pp. 1586–1609, Nov. 1971.
[10] W. Rudin, *Fourier Analysis on Groups*. New York: Wiley-Interscience, 1967, pp. 208–209.
[11] C. L. Rino, "Factorization of spectra by discrete Fourier transforms," *IEEE Trans. Inform. Theory* (Corresp.), vol. IT-16, pp. 484–485, July 1970.
[12] M. P. Ekstrom and R. L. Rhoads, "Discrete-process spectral factorization by generalized linear filtering," Lawrence Livermore Lab., Univ. California, Livermore, Rep. UCRL-51012, Feb. 1971.
[13] P. Pistor, "Stability criterion for recursive filters," *IBM J. Res. Develop.*, pp. 59–71, Jan. 1974.
[14] A. V. Oppenheim, R. W. Schafer, and T. G. Stockham, Jr., "The nonlinear filtering of multiplied and convolved signals," *Proc. IEEE*, vol. 56, pp. 1264–1291, Aug. 1968.
[15] J. L. Doob, *Stochastic Processes*. New York: Wiley, 1953.
[16] R. E. A. C. Paley and N. Wiener, "Fourier transforms," *AMS Trans.*, p. 791, Oct. 1933.
[17] P. Lévy, "Sur la convergence absolue des series de Fourier," *Paris Compte Rendus*, vol. 196, p. 463, 1933.
[18] P. Pistor, "Digital processing of scintigraphic images by Two-dimensional Wiener filters," IBM Deutschland Heidelberg Scientific Center, Rep. 70.12.006, 1970.
[19] J. B. Bednar and C. H. Farmer, "Factorization of recursive filters," in *Proc. Image Processing Symp.*, University of Missouri, Columbia, Aug. 1972.
[20] T. S. Huang, "Stability of two-dimensional recursive filters," *IEEE Trans. Audio Electroacoust.*, vol. AU-20, pp. 158–163, June 1972.
[21] J. H. Justice and J. L. Shanks, "Stability criterion for n-dimensional digital filters," *IEEE Trans. Automat. Contr.* (Short Papers), vol. AC-18, pp. 284–286, June 1973.
[22] J. L. Shanks, S. Treitel, and J. H. Justice, "Stability and synthesis of two-dimensional recursive filters," *IEEE Trans. Audio Electroacoust.*, vol. AU-20, pp. 115–128, June 1972.
[23] R. R. Read and S. Treitel, "The stabilization of two-dimensional

recursive filters via the discrete Hilbert transform," *IEEE Trans. Geosci. Electron.*, vol. GE-12, pp. 153–160, July 1974.

[24] J. F. Kaiser, "Non-recursive digital filter design using the I_0-sinh window function," *Proc. 1974 IEEE Int. Symp. Circuits and Systems*, Apr. 1974, pp. 20–23.

[25] T. S. Huang, "Two-dimensional windows," *IEEE Trans. Audio Electroacoust.* (Corresp.), vol. AU-20, pp. 88–89, Mar. 1972.

[26] G. A. Maria and M. M. Fahmy, "On the stability of two-dimensional digital filters," *IEEE Trans. Audio Electroacoust.* (Corresp.), vol. AU-21, pp. 470–472, Oct. 1973.

5

Reprinted from *IEEE Trans. Acoust., Speech, Signal Process.*
ASSP-23(2):242–243 (1975)

THE EXISTENCE OF CEPSTRA FOR TWO-DIMENSIONAL RATIONAL POLYNOMIALS

D. E. Dudgeon

***Abstract*—The use of cepstral analysis is helpful for some problems where two one-dimensional signals are combined by convolution [1]. In such problems it is important to ensure that the phase function associated with the resultant signal may be defined so that it is a continuous, odd, and periodic function of frequency [2], [3]. One class of one-dimensional signals which have this property is the class whose z-transforms are rational polynomials [2]. In this correspondence, we shall show that these results are extendible to two dimensions, and that 2-D cepstra can be defined for 2-D rational polynomials.**

INTRODUCTION

In signal processing problems where two signals are combined by convolution, homomorphic filtering techniques [1] may be useful. For example, such a technique may be useful for inverse filtering problems or problems where a signal and noise are combined by convolution. To make the definition of the complex cepstrum mathematically tractable, it is important to ensure that the phase function associated with a signal is continuous, odd, and periodic. Defining the phase by using the imaginary part of the principal part of the complex logarithm or the arctangent is usually not satisfactory [3]. Therefore phase must be defined as an integral [1] to ensure continuity. If the phase so defined is not periodic, it may be made so by subtracting off a linear phase component. This simply corresponds to a shift in the time origin of the original signal.

At this juncture, it should be noted that we must restrict the set of signals for which cepstra can be defined to exclude signals whose Fourier transforms go to zero or infinity at some frequency. This corresponds to their z-transforms having a pole or a zero on the unit circle, resulting in a phase discontinuity which cannot be removed by the integral formulation. Equivalently the log of their Fourier transforms become infinite at such frequencies.

One class of signals for which cepstra may be defined is the class whose z-transforms are rational polynomials and are analytic and nonzero on the unit circle. For these signals the cepstra also have z-transforms which are analytic on the unit circle [1].

Manuscript received June 10, 1974; revised October 25, 1974. This work was supported in part by the Advanced Research Projects Agency of the Department of Defense under Contract N0014-67-A-0204-0064 at the M.I.T. Research Laboratory of Electronics, and in part by the M.I.T. Lincoln Laboratory (sponsored by the U. S. Air Force).

The author was with the M.I.T. Research Laboratory of Electronics, Digital Signal Processing Group, Cambridge, Mass. He is now with the Computer Systems Division, BBN Inc., Cambridge, Mass. 02138.

TWO-DIMENSIONAL CEPSTRUM

In an exactly analogous manner, we can demonstrate that any 2-D array having a rational z-transform will also have a well-defined 2-D complex cepstrum provided 1) the Fourier transform is not equal to zero or infinity at any frequency, and 2) we are careful to eliminate any linear phase components by an appropriate shift of the original array.

This is most easily seen by first considering finite extent 2-D arrays. These arrays have Fourier transforms which are 2-D polynomials in $\exp[j\mu]$ and $\exp[j\nu]$. We shall now show that if a finite extent array $b(m,n)$ has a Fourier transform $B(\mu,\nu)$ which is nonzero for all μ and ν, then the phase function associated with $B(\mu,\nu)$ will be the sum of a linear component plus a continuous, odd, and periodic component. First, let $z = \exp[j\nu]$, and consider

$$B_\mu(z) = \sum_n [\sum_m b(m,n) \exp(-j\mu m)]z^{-n}$$

as a one-dimensional polynomial in z with a parameter μ. Since $b(m,n)$ is of finite extent, $B_\mu(z)$ will have poles inside the unit circle only at $z = 0$. Now let us define the phase function $\phi(\mu,\nu)$ as a contour integral

$$\phi(\mu,\nu) = \text{Im}\left\{\oint \frac{B_\mu'(z)}{B_\mu(z)}\,dz\right\} + \phi(\mu,0) \qquad (1)$$

where the prime denotes differentation with respect to z. The contour of integration starts at $z = 1$ and proceeds around the unit circle to $z = \exp[j\nu]$. It is necessary to define the constant $\phi(\mu,0)$. This is done by considering $\phi(\mu,0)$ to be the phase as a function of μ for $\nu = 0$, and may be derived as in Oppenheim and Schafer [3].

By constructing $\phi(\mu,\nu)$ in this manner we are assured that $\phi(\mu,\nu)$ is continuous and odd; that is, $\phi(\mu,\nu) = -\phi(-\mu,-\nu)$.

When $\nu = 2\pi$, the contour of integration in (1) is a closed curve. Using Cauchy's residue theorem and [4, theorems 1 and 2] we have

$$\phi(\mu,2\pi) = 2\pi r - 2\pi N$$

where r is the number of roots (including multiplicities) of $B_\mu(z)$ inside the unit circle and N is the number of poles at $z = 0$. If we let $k_\nu = r - N$, then it is clear that

$$\phi(\mu,\nu + 2\pi) = \phi(\mu,\nu) + 2\pi k_\nu.$$

Similarly we can derive the relation

$$\phi(\mu + 2\pi,\nu) = \phi(\mu,\nu) + 2\pi k_\mu.$$

What remains to be shown is that k_ν is not a function of μ, and that k_μ is not a function of ν. If we examine the roots of $B_\mu(z)$ as we continuously vary the parameter μ from zero to 2π, we discover that the roots move about in a continuous manner [4, p. 3]. Thus, for a root to move from inside to outside the unit circle (or vice versa), it must lie on the unit circle for some value of μ. This, however, violates the hypothesis that $B(\mu,\nu) \neq 0$. Therefore the number of roots r inside the unit circle is not a function of μ, and hence k_ν is not a function of μ. A similar argument can be made to show that k_μ is not a function of ν.

Given a continuous, odd phase function $\phi(\mu,\nu)$ we can subtract the linear phase component

$$\phi_L(\mu,\nu) = k_\nu \nu + k_\mu \mu$$

to give a term $\phi_A(\mu,\nu)$ which is easily shown to be continuous, odd, and periodic.

This result is easily extended to arrays with rational Fourier transforms. The phase function for such an array can be defined as the phase function of the numerator polynomial minus the phase function of the denominator polynomial. Clearly this difference can also be written as a linear phase component plus a continuous, odd, and periodic component.

It is easy to verify that subtracting off the linear phase component is equivalent to a shift of the array. Therefore, if we are given an array $h(m,n)$ whose Fourier transform is a ratio of polynomials, we can form a new sequence

$$g(m,n) = h(m - k_\mu, n - k_\nu)$$

whose phase function will be $\phi_A(\mu,\nu)$ which is continuous, odd, and periodic. We may then form the function

$$\hat{G}(\mu,\nu) = \ln |H(\mu,\nu)| + j\phi_A(\mu,\nu)$$

which has a real inverse Fourier transform denoted by $\hat{g}(m,n)$ and called the cepstrum of $g(m,n)$. Since $H(\mu,\nu)$ is a ratio of polynomials, the properties of $\hat{g}(\mu,\nu)$ imply that the z-transform of $\hat{g}(m,n)$ is an analytic function for $|w| = |z| = 1$. This may be seen by considering $\hat{G}(\mu,\nu)$ to be a one-dimensional Fourier transform in ν with a parameter μ. Then the z-transform is analytic for all values of μ and $|z| = 1$. Similarly the z-transform is analytic for all values of ν and $|w| = 1$. Therefore it is analytic for $|w| = |z| = 1$.

SUMMARY

From this derivation two conclusions can be drawn. First, any real 2-*D* array whose Fourier transform has a log-magnitude which is continuous, even, and periodic, and whose phase is continuous, odd, and periodic over the frequency plane will have a well-defined 2-*D* cepstrum which is real. Second, one class of 2-*D* arrays with well-defined cepstra is the class whose z-transforms are the ratio of two 2-*D* polynomials (with the restriction that the z-transforms are not zero or infinite for $|w| = |z| = 1$), care having been taken first to shift the arrays to eliminate any linear phase component. In addition, for this class of 2-*D* arrays we can argue that the logarithm of the z-transform is analytic on the surface $|w| = |z| = 1$.

ACKNOWLEDGMENT

The author would like to express his thanks to Prof. A. V. Oppenheim of M.I.T. for his suggestions and encouragement during the course of this work.

REFERENCES

[1] A. V. Oppenheim, R. W. Schafer, and T. G. Stockham, Jr., "Nonlinear filtering of multiplied and convolved signals," *Proc. IEEE*, vol. 56, pp. 1264–1291, Aug. 1968.

[2] B. Gold and C. M. Rader, *Digital Processing of Signals*. New York: McGraw-Hill, 1969.

[3] A. V. Oppenheim and R. W. Schafer, *Digital Signal Processing*. Englewood Cliffs, N. J.: Prentice-Hall, 1975.

[4] M. Marden, *The Geometry of the Zeros of a Polynomial in a Complex Variable*. New York: Amer. Math. Soc., 1949.

Editors' Comments on Papers 6 Through 10

6 HUANG
Stability of Two-Dimensional Recursive Filters

7 ANDERSON and JURY
Stability Test for Two-Dimensional Recursive Filters

8 SILJAK
Stability Criteria for Two-Variable Polynomials

9 GOODMAN
Some Stability Properties of Two-Dimensional Linear Shift-Invariant Digital Filters

10 EKSTROM and TWOGOOD
A Stability Test for 2-D Recursive Digital Filters Using the Complex Cepstrum

STABILITY CONSIDERATIONS

A major concern in the design of a 2-D digital filter is ensuring the stability of the filter. If the filter is unstable, the output sequence generated due to any input (including the computational roundoff noise) can become arbitrarily large, masking the desired output. A widely used stability criterion is the so-called BIBO stability condition which requires that the output 2-D sequence of the filter remain bounded for all possible, bounded input sequences. The BIBO stability is ensured if the impulse response coefficients $h_{m,n}$ of the filter are absolutely summable, i.e.,

$$\sum_{m=-\infty}^{\infty} \sum_{n=-\infty}^{\infty} |h_{m,n}| < \infty \tag{1.6}$$

This condition is automatically satisfied by a bounded 2-D FIR filter, as it has only a finite number of nonzero impulse response coefficients.

On the other hand, it is substantially more difficult to ensure that the above condition holds for a 2-D IIR filter.

Considerable effort has been spent by a number of authors in developing stability theorems for IIR filters and formulating practical tests based on these theorems. To this end, most efforts have been directed toward 2-D IIR filters characterized by impulse responses which are first quadrant or quarter-plane sequences, i.e., for which $h_{m,n}$ is nonzero only for $m \geqslant 0, n \geqslant 0$. This is the "causal" recursive filter form of Eq. (1.4). For this case

$$B(z_1,z_2) = \sum_{k=0}^{M_b} \sum_{\ell=0}^{N_b} b_{k,\ell} z_1^{-k} z_2^{-\ell}. \tag{1.7}$$

The earliest statement of a theorem for the BIBO stability of such filters is due to Shanks (1969), who showed that a 2-D IIR quarter-plane filter characterized by a real rational transfer function $H(z_1,z_2) = 1/B(z_1,z_2)$ (of the form of Eq. (1.2)) is stable if and only if $B(z_1,z_2) \neq 0$ on $\{|z_1| \geqslant 1, |z_2| \geqslant 1\}$. A detailed explanation of this stability theorem will be found in Paper 1 by Shanks et al. Direct implementation of Shanks' theorem requires a determination of the zeros of $B(z_1,z_2)$. Because of the fundamental difficulties involving the factorization of arbitrary 2-D polynomials discussed in the previous section, this determination is computationally impractical.

An alternate (and equivalent) stability theorem which offers the potential of more practical implementation was stated by Huang (Paper 6). According to his theorem, a 2-D IIR filter is BIBO stable if and only if the map of the unit circle $|z_1| = 1$ in the z_2-plane, according to the relation $B(z_1,z_2) = 0$, lies outside the region $|z_2| \geqslant 1$, and no point on or outside the unit circle $|z_2| = 1$ maps into the point $z_1 = \infty$ by the relation $B(z_1,z_2) = 0$. Recent contributions by Davis (1976) and Goodman (1976) have completed the proof of this theorem. Huang also proposed an algorithm to implement his stability theorem (Paper 6). A number of other authors have attempted to make Huang's stability test computationally more efficient. To this end, notable contributions have been made by Anderson and Jury (Paper 7), and Maria and Fahmy (1973). These latter contributions are essentially based on a generalization of the Schur-Cohn test (Jury, 1964) or the Marden test developed for checking stability of one-dimensional digital filters, and are computationally superior to Huang's test. A more improved stability testing algorithm which can be performed by a finite number of algebraic manipulations has been recently proposed by Siljak (Paper 8). Here, stability of a two-variable polynomial is tested by applying two times an equivalent of the Marden test (Marden, 1966) developed for the one-variable polynomial root location testing. The algebraic stability testing procedures referred to above still require a substantial amount of computer time even for 2-D filters of small-to-moderate complexity.

For a general, real rational 2-D transfer function, the stability conditions of Shanks (or the equivalent ones by Huang) are sufficient but no longer necessary due to the presence of the numerator term. This fact has been recently pointed out by Goodman (Paper 9), who also proved that if a real-rational transfer function $H(z_1,z_2)$ with a numerator polynomial of degree greater than or equal to one is BIBO stable, then $H(z_1,z_2)$ has no poles in the closed unit bidisc ($|z_1| \geqslant 1$, $|z_2| \geqslant 1$) and no nonessential singularities of the second kind on the closed unit bidisc except possibly on its distinguished boundary, $|z_1| = 1$, $|z_2| = 1$. Interestingly, he has shown that BIBO stability is not guaranteed by the condition that the impulse response be square summable.

The Shanks and Huang stability theorems are not applicable to the more general class of half-plane IIR filters (see Eq. (1.5)). A stability theorem for this class was presented by Ekstrom and Woods (Paper 4). Consider the case of

$$B(z_1,z_2) = \sum_{k,\ell \in R}\sum b_{k,\ell} z_1^{-k} z_2^{-\ell} \tag{1.8}$$

where $R \triangleq \{k \geqslant 0, \ell \geqslant 0\} \cup \{k > 0, \ell < 0\}$. Their theorem states that $1/B(z_1,z_2)$ is BIBO stable if and only if $B(z_1,z_2) \neq 0$ on $\{|z_1| = 1, |z_2| \geqslant 1\}$ and $B(z_1,\infty) \neq 0$ on $\{|z_1| = 1\}$. The full implications of this theorem are discussed in detail in their paper.

A stability test based on this theorem and utilizing their spectral factorization procedure is given in Paper 4. An alternate test, also based on the factorization and the mapping properties of the 2-D complex cepstrum has been presented by Ekstrom and Twogood (Paper 10). Both tests are quite simple in structure and are derived from the basic property of the factorization to decompose the transfer function into terms with prescribed regions of anlyticity. These regions, of course, are chosen to correspond to those of the stability theorem. Because of their generality, the tests can be easily adapted for use with "causal," quarter-plane filters as well as the half-plane filters. In contrast to the previous tests, they are quite suitable for practical implementation, even for high-order filters. The most demanding calculation required is 2-D FFT.

REFERENCES

Davis, D. L. 1976. "A Correct Proof of Huang's Theorem on Stability," *IEEE Trans. Acoust., Speech, Signal Process.* **ASSP-24**:425–426.

Goodman, D. 1976. "An Alternate Proof of Huang's Stability Theorem," *IEEE Trans. Acoust., Speech, Signal Process.* **ASSP-24**:426–427.
Jury, E. I. 1964. *Theory and Application of the Z-Transform Method.* New York: Wiley Interscience.
Marden, M. 1966. *Geometry of Polynomials.* Providence, R.I.: American Math. Soc.
Maria, G. A., and M. M. Fahmy. 1973. "On the Stability of Two-Dimensional Digital Filters," *IEEE Trans. Audio Electroacoust.* **AU-21**:470–472.
Shanks, J. L. 1969. "Two-Dimensional Recursive Filter," *SWIEEECO Record,* 19E1–19E8.

6

Reprinted from *IEEE Trans. Audio Electroacoust.* **AU-20**(2):158-163 (1972)

Stability of Two-Dimensional Recursive Filters

THOMAS S. HUANG, Member, IEEE
Dep. Elec. Eng.
Res. Lab. Electron.
and Lincoln Lab.
Mass. Inst. Tech.
Cambridge, Mass. 02173

Abstract

We discuss some aspects of the stability problem in two-dimensional recursive filtering. In particular, we derive a simplified version of a stability theorem due to Shanks and show that it is equivalent to some results of Ansell. We also give several examples of stability tests and pose a few unsolved problems.

Manuscript received November 23, 1971.

The work reported in this paper was performed at Lincoln Laboratory, a center for research operated by the Massachusetts Institute of Technology, Cambridge, Mass., with the support of the Department of the Air Force.

I. Introduction

In digital image processing, we often want to do linear filtering. The use of recursive instead of nonrecursive filters has the potential of saving computation time. A two-dimensional digital recursive filter is characterized by the two-dimensional z transform

$$H(z_1, z_2) = \frac{\sum_{m=0}^{n} \sum_{n=0}^{q} a_{mn} z_1{}^m z_2{}^n}{\sum_{m=0}^{p} \sum_{n=0}^{q} b_{mn} z_1{}^m z_2{}^n} \tag{1}$$

where a_{mn} and b_{mn} are constants, and without loss of generality we can set $b_{oo} = 1$. The degrees of the numerator and the denominator polynomials do not have to be equal, since some of the coefficients a_{mn} and b_{mn} can be zero. The variables z_1 and z_2 are defined as

$$z_1 = e^{-s_1 A}$$
$$z_2 = e^{-s_2 B} \tag{2}$$

where s_1 and s_2 are, respectively, the horizontal and vertical complex spatial frequency variables and A and B are constants (sampling periods in the horizontal and vertical directions, respectively).

Let $f(m, n)$ and $g(m, n)$ be the input and output, respectively, of the filter. Then the spatial-domain difference equation corresponding to (1) is

$$\sum_{m=0}^{p}\sum_{n=0}^{q} b_{mn}g(M-m, N-n) = \sum_{m=0}^{p}\sum_{n=0}^{q} a_{mn}f(M-m, N-n). \quad (3)$$

We can express $g(M, N)$ in terms of the rest, and thus obtain a recursive filter that recurses in the $(+m, +n)$ direction. We can also express $g(M-p, N-q)$ in terms of the rest, and obtain a recursive filter that recurses in the $(-m, -n)$ direction. Similarly, we can get from (3) recursive filters recursing in the $(+m, -n)$ and the $(-m, +n)$ directions. (See Fig. 1.) We shall call the recursive filter that recurses in the $(+m, +n)$ direction *causal*. Thus, the z transform of (1) can be associated with four different recursive filters. However, unless otherwise specified, we shall associate it with the causal one.

To start recursing, we need initial conditions. For the causal filter, we need to know the values of the output g in the shaded region in Fig. 2.

There are two major problems in the design of a recursive filter: approximation and stability. The approximation problem consists of determining the coefficients a_{mn} and b_{mn} in (1) so that $H(e^{-juA}, e^{-jvB})$ approximates a given frequency responses $H_1(ju, jv)$, where u and v are, respectively, the horizontal and vertical spatial frequencies. In order for the resulting filter $H(z_1, z_2)$ to be useful, we have to require that it be stable, i.e., if we expand $H(z_1, z_2)$ into a power series

$$H(z_1, z_2) = \sum_{m=0}^{\infty}\sum_{n=0}^{\infty} h_{mn}z_1{}^m z_2{}^m \quad (4)$$

the coefficients h_{mn}, which represent the impulse response of the filter, should satisfy

$$\sum_{m=0}^{\infty}\sum_{n=0}^{\infty} |h_{mn}| < \infty. \quad (5)$$

If the filter is unstable, any noise (including the roundoff errors in computation) will propagate through the output and possibly be amplified. Furthermore, in many important applications, such as inverse filtering for image restoration, the initial conditions are usually not known. We, therefore, desire that the result of the recursive filtering be roughly independent of the initial conditions. This will be true if the impulse response $|h_{mn}| \approx 0$ for $m, n > A$, where A is a constant that is much smaller than the linear dimensions of the input and output images. Therefore, in some cases, stability itself is not sufficient; we require in addition that the impulse response die out fast enough.

In this paper we shall discuss some aspects of the stability problem of two-dimensional recursive filters. In particular, we shall derive a simplified version of the Shanks stability theorem [1], [2] and show that it is equivalent to the Ansell stability theorem [3], [4].

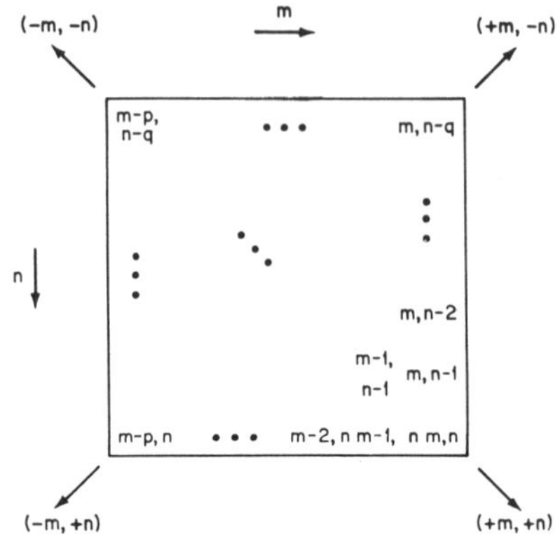

Fig. 1. Four directions of recursion.

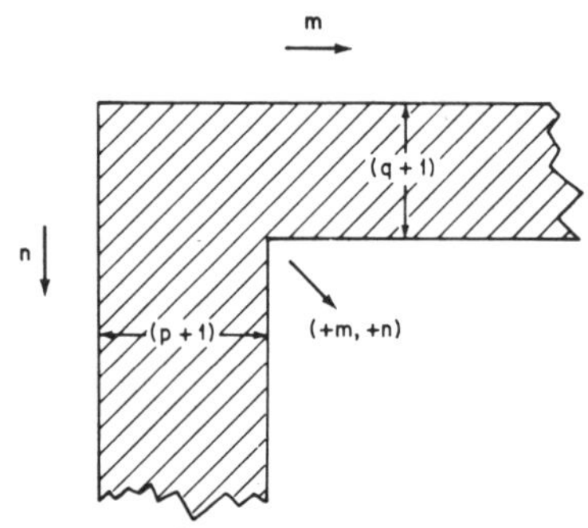

Fig. 2. Initial conditions for a causal filter.

II. Shanks Stability Theorem

Both the theorems of Shanks and Ansell tell us stability conditions in the frequency domain. In this section we shall state and prove the Shanks stability theorem and discuss its implications.

A. Statement of the Theorem

Theorem 1 (Shanks): A causal recursive filter with the z transform $H(z_1, z_2) = A(z_1, z_2)/B(z_1, z_2)$, where A and B are polynomials in z_1 and z_2, is stable if and only if there are no values of z_1 and z_2 such that $B(z_1, z_2) = 0$ and $|z_1| \leq 1$ and $|z_2| \leq 1$.

Proof: Let

$$H(z_1, z_2) = \frac{A(z_1, z_2)}{B(z_1, z_2)} = \sum_{m=0}^{\infty}\sum_{n=0}^{\infty} h_{mn}z_1{}^m z_2{}^n. \quad (6)$$

What we want to show is that

$$\sum_{m=0}^{\infty}\sum_{n=0}^{\infty} |h_{mn}| < \infty$$

if and only if $H(z_1, z_2)$ is analytic in the region $D = \{(z_1, z_2); |z_1| \leq 1 \cap |z_2| \leq 1\}$.

The "if" part: If $H(z_1, z_2)$ is analytic in D, we can find $\epsilon>0$ such that $H(z_1, z_2)$ is analytic in $D_1=\{(z_1, z_2); |z_1|<1+\epsilon \cap |z_2|<1+\epsilon\}$, which implies that

$$\sum_m \sum_n h_{mn} z_1^m z_2^n$$

is absolutely convergent in D_1. Therefore

$$\sum_m \sum_n |h_{mn}| < \infty.$$

The "only if" part: If

$$\sum_m \sum_n |h_{mn}| < \infty,$$

then by the M test,

$$\sum_m \sum_n h_{mn} z_1^m z_2^n$$

is absolutely convergent in D, which implies in turn that $H(z_1, z_2)$ is analytic in D. Q.E.D.

Independent proofs of this theorem have been given by Shanks *et al.* [5] and by Farmer [6].

B. Test Procedure

Testing stability using the Shanks theorem is hard work. One way is to map the unit disk $d_1 \equiv (z_1, |z_1| \leq 1)$ in the z_1 plane into the z_2 plane by the implicit mapping relation $B(z_1, z_2)=0$. The filter is stable if and only if the image of d_1 in the z_2 plane does not overlap the unit disk $d_2 \equiv (z_2, |z_2| \leq 1)$. This is hard work because for each particular value of $z_1 = z_1^0 \epsilon d_1$, we have to solve the equation $B(z_1^0, z_2)=0$ for z_2. And we have to do that for all (in practice, a large number of points) in d_1.

C. An Implication

We emphasize that the Shanks theorem was stated for causal filters, that is, filters that recurse in the $(+m, +n)$ direction. Let us consider the filter associated with $H(z_1, z_2) = A(z_1, z_2)/B(z_1, z_2)$ that recurses in the $(-m, -n)$ direction. We assume that the orders of A in both z_1 and z_2 are smaller than the corresponding orders of B. Then the stability condition becomes: the image of $(z_1, |z_1| \geq 1)$ in the z_2 plane, according to $B(z_1, z_2)=0$, does not overlap $(z_2; |z_2| \geq 1.)$

This can be shown by expanding $H(z_1, z_2)$ in terms of z_1^{-1} and z_2^{-1}:

$$H(z_1, z_2) = \frac{A(z_1, z_2)}{B(z_1, z_2)} = \sum_{m=0}^{\infty} \sum_{n=0}^{\infty} h_{mn}{}^1 z_1^{-m} z_2^{-n}. \tag{7}$$

Making the change of variables

$$w_1 = z_1^{-1}$$
$$w_2 = z_2^{-1} \tag{8}$$

we get

$$H(w_1^{-1}, w_2^{-1}) = \frac{A(w_1^{-1}, w_2^{-1})}{B(w_1^{-1}, w_2^{-1})} = \sum_m \sum_n h_{mn}{}^1 w_1^m w_2^n. \tag{9}$$

Now

$$\frac{A(w_1^{-1}, w_2^{-1})}{B(w_1^{-1}, w_2^{-1})} = w_1^a w_2^b \frac{A_1(w_1, w_2)}{B_1(w_1, w_2)} \tag{10}$$

where a and b are nonnegative integers, and $B_1(w_1, w_2)=0$ if and only if $B(w_1^{-1}, w_2^{-1})=0$. Applying the Shanks theorem to

$$\frac{w_1^a w_2^b A_1(w_1, w_2)}{B_1(w_1, w_2)} = \sum_m \sum_n h_{mn}{}^1 w_1^m w_2^n \tag{11}$$

we know that

$$\sum_m \sum_n |h_{mn}{}^1| < \infty$$

if and only if the image of $(w_1, |w_1| \leq 1)$ in the w_2 plane, according to $B_1(w_1, w_2)=0$, does not overlap $(w_2; |w_2| \leq 1)$. But $w_1 = z_1^{-1}$, and $w_2 = z_2^{-1}$. So our assertion has been proved.

Similarly, we can get the stability conditions for the filters associated with $H(z_1, z_2)$ that recurse in the $(+m, -n)$ and the $(-m, +n)$ directions. We can easily convince ourselves that at most one of the four stability conditions, corresponding to the four recursing directions, can be satisfied. Therefore, we have the following.

Theorem 2: Among the four recursive filters we can associate with the z transform of (1), or equivalently the difference equation of (3), at most one can be stable.

III. A Simplified Version of the Shanks Theorem

Shanks' result as stated in Theorem 1 is very tedious to apply. We now show that his stability conditions can be simplified considerably.

A. The Result

Theorem 3: A causal filter with a z transform $H(z_1, z_2) = A(z_1, z_2)/B(z_1, z_2)$, where A and B are polynomials, is stable if and only if: 1) the map of $\partial d_1 \equiv (z_1; |z_1|=1)$ in the z_2 plane, according to $B(z_1, z_2)=0$, lies outside $d_2 \equiv (z_2; |z_2| \leq 1)$; and 2) no point in $d_1 = (z_1; |z_1| \leq 1)$ maps into the point $z_2=0$ by the relation $B(z_1, z_2)=0$.

Proof: We want to establish that the stability conditions of Theorems 1 and 3 are equivalent. It is obvious that the stability conditions of Theorem 1 imply those of Theorem 3. So we proceed to show the implication in the other direction.

The two-variable polynomial $B(z_1, z_2)=0$ defines an algebraic function [7] $z_2 = f(z_1)$. We first modify the unit-circle contour in the z_1 plane to exclude any singular points of f inside the contour, resulting in a modified contour $\partial d_1'$, as shown in Fig. 3. We use d_1' to denote the closed region enclosed by $\partial d_1'$. A point $z_1 = z_1^0$ is called a singular point of $z_2 = f(z_1)$, if $B(z_1^0, z_2)=0$, considered as an equation in z_2, has multiple (finite or infinite) roots.

According to the theory of algebraic functions, in d_1' the function $z_2 = f(z_1)$ has a number of branches, each of which is holomorphic. Therefore, from the maximum-modulus theorem, the maximum of $|f(z_1)|$ over d_1' occurs on $\partial d_1'$, and the minimum of $|f(z_1)|$ over d_1' can occur in the interior only if the minimum is zero. However, condition 2) of Theorem

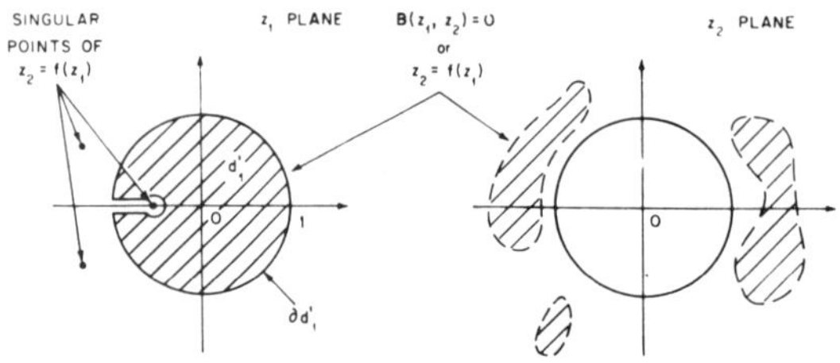

Fig. 3. Mapping $z_2 = f(z_1)$.

3 says $f(z_1)$ is never zero in d_1'. Therefore, the minimum of $f(z_1)$ occurs on $\partial d_1'$, i.e.,

$$|f(d_1')| \geq \min |f(\partial d_1')| \tag{12}$$

which implies that if $|f(\partial d_1')| > 1$, then $|f(d_1')| > 1$, i.e., to ensure that $f(d_1')$ lies outside the unit disk $d_2 = (z_2, |z_2| \leq 1)$, it is sufficient to ensure that $f(\partial d_1')$ lies outside d_2.

We are almost there, but not quite. What we really want to show is that if $|f(\partial d_1)| > 1$, then $|f(d_1)| > 1$, where $d_1 \equiv (z_1, |z_1| \leq 1)$ and $\partial d_1 \equiv (z_1; |z_1| = 1)$. Since the detour in $\partial d_1'$ can be any path leading from ∂d_1 to the singular point, what is left to show is simply that $|f(s)| > 1$, where s is the singular point. But since each branch of $z_2 = f(z_1)$ is continuous at $z_1 = s$, and since $|f(s + \epsilon e^{j\theta})| > 1$ for arbitrarily small ϵ and any θ, we have $|f(s)| > 1$. Q.E.D.

B. Test Procedure

To test stability using Theorem 3, we have to map $\partial d_1 = (z_1; |z_1| = 1)$ into the z_2 plane according to $B(z_1, z_2) = 0$, and see whether the image lies outside $d_2 \equiv (z_2; |z_2| \leq 1)$. Also, we have to solve $B(z_1, 0) = 0$ to see whether there is any root with magnitude smaller than 1.

IV. Ansell Stability Theorem

To test the stability of a two-dimensional recursive filter using Theorem 3 is much simpler than using Shanks' original Theorem 1. However, the procedure is still infinite in the sense that in principle we have to map every point on the continuum ∂d into the z_2 plane. In this section we show that Theorem 3 can be reduced to a result due to Ansell. We shall see that Ansell's result enables us to test stability in a finite number of steps, which, unfortunately, can still be very tedious.

Make the change of variables

$$p_1 = \frac{1 - z_1}{1 + z_1} \tag{13a}$$

and

$$p_2 = \frac{1 - z_2}{1 + z_2} \tag{13b}$$

and let

$$H(z_1, z_2) = \frac{E(p_1, p_2)}{F(p_1, p_2)} \tag{14}$$

where E and F are polynomials in p_1 and p_2. We can restate Theorem 3 as follows.

Theorem 4 (Ansell): The causal recursive filter $H(z_1, z_2)$ is stable if and only if: 1) for all real finite w, the complex polynomial in p_2, $F(jw, p_2)$ has no zeros in Re $p_2 \geq 0$; and 2) the real polynomial in p_1, $F(p_1, 1)$ has no zeros in Re $p_1 \geq 0$.

This theorem is essentially the same as [3, Lemma 6]. The nice part about this theorem is that condition 1) can be tested using standard techniques of circuit theory [8].

Theorem 5 (Ansell): Condition 1) of Theorem 4 is equivalent to the following. Let us express $F(jw, j\Omega)$, where w and Ω are real, as

$$F(jw, j\Omega) = b_0(w)\Omega^n + b_1(w)\Omega^{n-1} + \cdots + b_n(w) + j[a_0(w)\Omega^n + a_1(w)\Omega^{n-1} + \cdots + a_n(w)] \tag{15}$$

where $a_i(w)$ and $b_i(w)$ are real polynomials in w and neither $a_0(w)$ nor $b_0(w)$ is identically zero. Let $C_{r,s}(w)$ be defined as

$$C_{r,s} = a_r b_s - a_s b_r \tag{16}$$

for $0 \leq r, s \leq n$ [using zero for a's and b's not present in $F(jw, j\Omega)$]. Let $D(w)$ denote the $n \times n$ symmetrical polynomial matrix whose typical element $D_{ij}(w)(1 \leq i, j \leq n)$ is the sum of all those $C_{r,s}(w)(0 \leq r, s \leq n)$ for which both

$$s + r = i + j - 1 \tag{17a}$$

and

$$s - r > |i - j| \tag{17b}$$

are satisfied. Then, the n successive principal minors of $D(w)$ must be positive for all real w.

This is part of Theorem 2 of Ansell [3], [9]. We note that Sturm's method [8] can be used to test whether each minor of $D(w)$ is positive for all real w.

V. Examples

We now give some examples illustrating the use of Theorems 3, 4, and 5. Ideally, we would like to determine what relations the coefficients a_{ij} and b_{ij} in (1) must satisfy to ensure the stability of a causal recursive filter $H(z_1, z_2)$. Unfortunately, this does not seem possible except for very simple filters.

A. First-Order Filters

Let us consider the first-order filter

$$H(z_1, z_2) = \frac{1}{1 + az_1 + bz_2} \tag{18}$$

where a and b are constants. In this simple case, we can easily determine the stability conditions in terms of a and b using Theorem 1.

Let

$$B(z_1, z_2) = 1 + az_1 + bz_2 = 0.$$

We have

$$z_2 = -\frac{1}{b} - \frac{a}{b} z_1. \tag{19}$$

Therefore, $B=0$ maps the unit disk in the z_1 plane into a disk in the z_2 plane with center at $z_2=-1/b$ and radius $|a/b|$. This image will not overlap the unit disk in the z_2 plane if and only if

$$\left|\frac{1}{b}\right| - \left|\frac{a}{b}\right| > 1 \tag{20}$$

or equivalently

$$|a| + |b| < 1. \tag{21}$$

We now try to establish the same result by using Theorems 4 and 5. Making the change of variables according to (13), we get

$$H(z_1, z_2) = \frac{E(p_1, p_2)}{F(p_1, p_2)} = \frac{1 + p_1 + p_2 + p_1p_2}{(1+a+b) + (1-a+b)p_1 + (1+a-b)p_2 + (1-a-b)p_1p_2}. \tag{22}$$

Condition 2) of Theorem 4 dictates that $F(p_1, 1)=2(1+a)+2(1-a)p_1=0$ has no root in Re $p_1 \geq 0$. Therefore

$$|a| < 1. \tag{23}$$

Now we tackle condition 1) of Theorem 4 with the help of Theorem 5. We have

$$F(jw, j\Omega) = (1+a+b) - (1-a-b)w\Omega + j[(1-a+b)w + (1+a-b)\Omega]$$

and we want

$$C_{0,1} = (1+a-b)(1+a+b) + (1-a+b)(1-a-b)w^2 > 0, \qquad \text{for all real } w. \tag{24}$$

Let $A_1=1+a-b$, $A_2=1+a+b$, $A_3=1-a+b$, and $A_4=1-a-b$. Then inequality (24) is valid if either of the following four conditions is satisfied: 1) $A_1>0$, $A_2>0$, $A_3>0$, $A_4>0$; 2) $A_1>0$, $A_2>0$, $A_3<0$, $A_4<0$; 3) $A_1<0$, $A_2<0$, $A_3>0$, $A_4>0$; 4) $A_1<0$, $A_2<0$, $A_3<0$, $A_4<0$. It is easy to show that 1) is equivalent to inequality (21), while 2)–4) are incompatible with inequality (23).

B. A Special Class of Second-Order Filters

We next consider the filter

$$H(z_1, z_2) = \frac{1}{1 + az_1 + bz_2 + cz_1z_2} \tag{25}$$

which we shall call a bilinear filter. We first establish stability conditions in terms of the coefficients a, b, and c, using Theorem 3.

Let

$$B(z_1, z_2) = 1 + az_1 + bz_2 + cz_1z_2 = 0.$$

We have

$$z_2 = -\frac{1 + az_1}{b + cz_1} \tag{26}$$

which is a bilinear transformation mapping circles into circles. The image of the unit circle $\partial d_1 \equiv (z_1;\ |z_1| = 1)$ in the z_2 plane is then a circle. From (26), the center of this image circle is on the real axis, and it intersects the real axis at

$$z_2 = -\frac{1-a}{b-c} \quad \text{and} \quad z_2 = -\frac{1+a}{b+c}.$$

It is easy to see that condition 1) of Theorem 3 is satisfied if

$$\left|\frac{1-a}{b-c}\right| > 1 \tag{27a}$$

and

$$\left|\frac{1+a}{b+c}\right| > 1. \tag{27b}$$

Finally, condition 2) of Theorem 3 is equivalent to

$$|a| < 1. \tag{27c}$$

Conditions (27) agree with those of Shanks [1].

We now try to establish the stability conditions using Theorems 4 and 5. Making the change of variables according to (13), we get

$$H(z_1, z_2) = \frac{E(p_1, p_2)}{F(p_1, p_2)} = \frac{1 + p_1 + p_2 + p_1p_2}{B_1 + B_2p_1 + B_3p_2 + B_4p_1p_2} \tag{28}$$

where

$$\begin{aligned} B_1 &= 1 + a + b + c \\ B_2 &= 1 - a + b - c \\ B_3 &= 1 + a - b - c \\ B_4 &= 1 - a - b + c. \end{aligned} \tag{29}$$

For condition 2) of Theorem 4, we form

$$F(p_1, 1) = 2(1+a) + 2(1-a)p_1 = 0$$

where Re $p_1 > 0$ means

$$|a| < 1. \tag{30}$$

Now we tackle condition 1) of Theorem 4. We have

$$F(jw, j\Omega) = B_1 - B_4w\Omega + j[(B_2w + B_3\Omega)].$$

According to Theorem 5, we want

$$C_{0,1} = B_3B_1 + B_2B_4w^2 > 0, \qquad \text{for all real } w \tag{31}$$

which is valid if either of the following four conditions is satisfied: 1) $B_1>0$, $B_3>0$, $B_2>0$, $B_4>0$; 2) $B_1>0$, $B_3>0$, $B_2<0$, $B_4<0$; 3) $B_1<0$, $B_3<0$, $B_2>0$, $B_4>0$; 4) $B_1<0$, $B_3<0$, $B_2<0$, $B_4<0$.

It is easy to show that 1) is equivalent to

$$|a+b| - 1 < c < 1 - |a-b| \tag{32}$$

and 2)–4) are ruled out by inequality (30). It can be verified readily that (32) is equivalent to (27).

C. A Numerical Example

Unfortunately, filters described by (18) and (25) are the only types for which we were able to derive stability conditions in terms of the coefficients in the z transforms. For more complicated filters, the application of Theorem 3 becomes unwieldy. However, for any filter with numerical coefficients, we can always apply Theorems 4 and 5 and determine whether or not it is stable in a finite number of steps. We illustrate by an example.

Consider the filter

$$H(z_1, z_2) = \frac{1}{1 + \frac{1}{2}z_1 + \frac{1}{2}z_2 + \frac{1}{4}z_1z_2 + \frac{1}{4}z_1^2 + \frac{1}{4}z_2^2}. \tag{33}$$

Making the change of variables according to (13), and proceeding as before, we find that

$$F(p_1, 1) = 3p_1^2 + 6p_1 + 7$$

whose zeros have negative real parts. Therefore, condition 2) of Theorem 4 is satisfied.

For condition 1) of Theorem 4, we form

$$C_{0,1}(w) = \frac{3}{4}w^4 - \frac{3}{2}w^2 - \frac{15}{4}$$

which certainly can be negative for some real values of w. Therefore, condition 1) is violated. The filter of (33) is unstable.

VI. Concluding Remarks

We have discussed in this paper how to test the stability of two-dimensional recursive filters in the frequency domain. In conclusion, we would like to point out that what we have discussed is only part of the stability problem. Furthermore, it is not the most important part. The more important questions we want to ask are: How do we design filters that are guaranteed to be stable? And, if a given filter is unstable, how do we stabilize it without changing its frequency response?

One approach to attacking the first question is to study what types of frequency responses we can get from classes of simple filters whose stability we know how to control.

With respect to the second question, we might mention a conjecture of Shanks [2]. Let $H(z_1, z_2)=1/B(z_1, z_2)$ be an unstable filter. We first determine a least mean-square inverse of $B(z_1, z_2)$, which we call

$$C(z_1, z_2) = \sum_{m=0}^{a}\sum_{n=0}^{b} c_{mn}z_1^m z_2^n \tag{34}$$

where a and b are positive integers.

Let

$$B(z_1, z_2)C(z_1, z_2) = 1 + d_{10}z_1 + d_{01}z_2 + d_{11}z_1z_2 \cdots.$$

Then among all polynomials of order (a, b), C is the least mean-square inverse of B if the coefficients of C are such that $(d_{10}^2+d_{01}^2+d_{11}^2+\cdots)$ is minimized. We next determine a least mean-square inverse of $C(z_1, z_2)$, which we call $\hat{B}(z_1, z_2)$. Then Shanks' conjecture states that the filter $\hat{H}(z_1, z_2) = 1/\hat{B}(z_1, z_2)$ is stable and that the magnitude of the frequency response of $\hat{H}$ is approximately equal to that of H. The approximation becomes better when we use larger values for a and b in (34). It was proven by Robinson [10], [11] that a one-dimensional version of Shanks' procedure does yield stable (one-dimensional) filters. However, whether the procedure yields stable filters in two dimensions is still an open question. Also, no analysis is available on how close the magnitude of the frequency response of $\hat{H}$ is to that of H.

We have yet another open question with respect to question 2. In the one-dimensional case, if we are doing non-real-time filtering, then we can always decompose an unstable filter with no poles on the unit circle into two stable ones recursing in opposite directions. We associate the poles of the original filter outside the unit circle with a filter recursing in the positive direction, and the poles inside the unit circle with a filter recursing in the negative direction. Is there an analogous procedure in two dimension?

References

[1] J. L. Shanks, "Two-dimensional recursive filters," in *1969 SWIEECO Rec.*, pp. 19E1–19E8.

[2] ——, "The design of stable two-dimensional recursive filters," in *Proc. UMR—M. J. Kelley Commun. Conf.*, Univ. Missouri, Rolla, Oct. 1970.

[3] H. G. Ansell, "On certain two-variable generalizations of circuit theory, with applications to networks of transmission lines and lumped reactances," *IEEE Trans. Circuit Theory*, vol. CT-11, pp. 214–223, June 1964.

[4] L. Weinberg, "Approximation and stability-test procedures suitable for digital filters," presented at the Arden House Workshop on Digital Filtering, Harriman, N. Y., Jan. 1970.

[5] J. L. Shanks, S. Treitel, and J. H. Justice, "Stability and synthesis of two-dimensional recursive filters," this issue, pp. 115–128.

[6] C. Farmer, Tulsa Univ., Tulsa, Okla., private communication, May 1971.

[7] G. A. Bliss, *Algebraic Functions.* New York: Amer. Math. Soc., 1933.

[8] E. A. Guillemin, *Synthesis of Passive Networks.* New York: Wiley, 1957.

[9] F. R. Gantmacher, *Applications of the Theory of Matrices.* New York: Interscience, 1959, pp. 295–296.

[10] E. A. Robinson, *Statistical Communication and Detection.* New York: Hafner, 1967.

[11] ——, "Structural properties of stationary stochastic processes with applications," ch. 12 in *Time Series Analysis*, M. Rosenblatt, Ed. New York: Wiley, 1963.

7

Reprinted from *IEEE Trans. Audio Electroacoust.* **AU-21**(4):366-372 (1973)

Stability Test for Two-Dimensional Recursive Filters

BRIAN D. O. ANDERSON and ELIAHU I. JURY

Abstract—For deciding the stability of a two-dimensional filter, it is of interest to determine whether or not a prescribed polynomial in the variables z_1 and z_2 is nonzero in the region $|z_1| \leqslant 1 \cap |z_2| \leqslant 1$. A new procedure for testing for this property is given, which does not involve the use of bilinear tranformations. Key parts of the test involve the construction of a Schur-Cohn matrix and the checking for positivity on the unit circle of a set of self-inversive polynomials.

I. Introduction

A two-dimensional digital recursive linear filter can be defined by its two-dimensional z-transform

Manuscript received November 9, 1972. This work was supported in part by the Australian Research Grants Committee.

B. D. O. Anderson is with the Department of Electrical Engineering, University of Newcastle, New South Wales 2308, Australia.
E. I. Jury is with the Department of Electrical Engineering and Computer Science, University of California, Berkeley, Calif. 94720.

$$G(z_1, z_2) = \frac{\sum_{i=0}^{m} \sum_{j=0}^{n} a_{ij} z_1^i z_2^j}{\sum_{i=0}^{p} \sum_{j=0}^{q} b_{ij} z_1^i z_2^j} \tag{1}$$

where the quantities a_{ij} and b_{ij} are real constants. The filter is stable if and only if (see [1]-[3])

$$B(z_1, z_2) = \sum_{i=0}^{p} \sum_{j=0}^{q} b_{ij} z_1^i z_2^j \neq 0,$$

$$|z_1| \leqslant 1 \cap |z_2| \leqslant 1. \tag{2}$$

The object of this paper is to present a procedure for checking the stability condition; the procedure is believed to be simpler than those of [2] and [3]. In fact, the procedure of [2] is not finite in the sense that the procedure requires the construction of a theoretically infinite number of mappings. The procedure of [3], though finite, requires application of two bilinear transformations to pose the problem in a form solved by Ansell [4]. In essence, Ansell's main contribution is to couple the use of a Hermite test for checking stability [5] with a series of Sturm tests [6] checking positivity.

Our procedure, like that of [3], is finite. We require no bilinear transformation and we replace the Hermite test component of the main part of Ansell's procedure by a Schur-Cohn matrix test [7]-[9]. Then we allow either a series of Sturm tests, or, what turns out to be equivalent, a series of tests for establishing the root distribution of a polynomial.

The plan of the paper is as follows. In Section II, we recall the initial simplification to the stability problems made by Huang [3] (which is equivalent to a similar simplification made by Ansell [4] for a related problem). Then we indicate the structure of our test. In Sections III-V, we indicate the test in detail. Section VI contains concluding remarks.

II. Outline Solution of the Problem

Huang's simplification of the stability problem rests on the following observation: $B(z_1, z_2) \neq 0$ for $|z_1| \leqslant 1 \cap |z_2| \leqslant 1$ if and only if the following two conditions hold:

$$B(z_1, 0) \neq 0, \qquad |z_1| \leqslant 1 \tag{3}$$

$$B(z_1, z_2) \neq 0, \qquad |z_1| = 1, |z_2| \leqslant 1. \tag{4}$$

Let us now indicate in outline how these conditions may be checked.

First, the checking of (3) is straightforward because $B(z_1, 0)$ is a single variable polynomial and there are a number of tests for determining whether or not its zeros all lie outside the unit circle. One group of tests relies on replacement of $B(z_1, 0)$ by another polynomial through a bilinear transformation which has to be checked for Hurwitz character. (A polynomial is Hurwitz if all its zeros have negative real parts.) There are, of course, many tests for checking the Hurwitz property (see, e.g., [5]). We shall omit further discussion of this approach, this being the philosophy used by Huang.

A second test involves use of the Schur-Cohn matrix [7]-[10]; this matrix is square, Hermitian, of size equal to the degree of $B(z_1, 0)$, and with elements which are simple functions of the coefficients of $B(z_1, 0)$. The matrix is negative definite if and only if $B(z_1, 0)$ has all its zeros in $|z_1| > 1$. Of course, the negative definiteness can be established by examining the signs of the leading principal minors of the matrix.

Other tests may be found in [9]-[12]. These tests include a computationally efficient one involving the recursive construction of a finite set of polynomials, each member of the set having lower degree than the preceding, and with $B(z_1, 0)$ being the first member of the set. Examination of some of the coefficients of these polynomials quickly yields the information as to whether (3) holds.

To avoid burdening the reader with too much detail, we shall restrict consideration in this paper to use of the Schur-Cohn test and the test based on recursive construction of a set of polynomials. We remark that although some simplification in the computations involved in applying the two tests to real polynomials is possible (see [9], [12], and [13]), we avoid stating these simplifications, again to avoid giving too much detail. The tests are given in Section III, together with examples illustrating their application to the checking of condition (3).

Now we comment on how condition (4) is to be checked. We omit further discussion of the approach of Huang, based on a bilinear transformation in both variables. Broadly, the checking falls into two distinct phases. First, we apply a Schur-Cohn test, after a fashion. Second, we check the positivity of a number of polynomials on $|z_1| = 1$.

Application of the Schur-Cohn test is described in Section IV. The idea is to think of z_1 as a parameter, so that $B(z_1, z_2)$ is a polynomial in z_2, whose coefficients are functions of a parameter z_1. For stability, the polynomial cannot be zero inside $|z_2| \leqslant 1$; this means that the associated Schur-Cohn matrix must be negative definite.

We have already noted that the entries of the Schur-Cohn associated with a polynomial are simple functions of the polynomial coefficients. Accordingly, here the entries of the Schur-Cohn matrix become simple functions of z_1. In fact, the entries are polynomial in z_1 and/or z_1^*, as we shall see.

To check negative definiteness of the Schur-Cohn matrix requires determination of the principal minors, which must have certain signs. These minors are again polynomial in z_1 and z_1^* and are real because the Schur-Cohn matrix is Hermitian.

Accordingly, the condition for (4) to hold becomes one requiring the sign definiteness on $|z_1| = 1$ of a set of functions which take real values and are polynomial in z_1 and z_1^*. Bearing in mind that on $|z_1| = 1$, we have $z_1^* = z_1^{-1}$, it follows that the functions are of the form

$$f_i(z_1) = \sum_{j=1}^{N_i} c_j z_1^j + 2c_0 + \sum_{j=1}^{N_i'} d_j z_1^{-j} \tag{5}$$

on $|z_1| = 1$. Because $f_i(z_1)$ is real for all $|z_1| = 1$, one must have c_0 real, $c_j = d_j^*$, and $N_i = N_i'$. In actual fact, because the original $B(z_1, z_2)$ has real coefficients, one finds that c_j is real so that

$$f_i(z_1) = \sum_{j=1}^{N_i} (c_j z_1^j + c_j z_1^{-j}) + 2c_0 = \sum_{j=0}^{N_i} c_j (z_1^j + z_1^{-j}) \tag{6}$$

on $|z_1| = 1$. One still requires a test for this polynomial to be positive or negative as required for all z_1 on $|z_1| = 1$, and this is the second phase in the checking of condition (4).

Before describing this second phase, we deal with one further issue associated with the first phase. Since the zero distribution of a one variable polynomial can be obtained by constructing a sequence of polynomials, one might expect this procedure to be possible for $B(z_1, z_2)$, treating z_1 as a parameter. Indeed, this is so; however, organization of the calculations appears more complex than organization of the calculations based on use of the Schur–Cohn matrix, at least for other than very simple $B(z_1, z_2)$. Therefore, we prefer use only of the Schur–Cohn ideas for the first phase.

Turning to the second phase, we note first that there is more than one approach to checking sign definiteness, covered in detail in Section V. First, one may use a Sturm test; this comes about as follows: Setting $z_1 = \exp(i\theta)$ in (6), where $i = \sqrt{-1}$, one finds, following [16]

$$f_i(\exp i\theta) = f_i(\cos\theta) = \sum_{j=0}^{N_i} 2c_j \cos j\theta. \tag{7}$$

Then one may set $x = \cos\theta$ and use the Chebyshev polynomials to obtain a real polynomial $g_1(x)$ in x which has constant sign on $(-1, 1)$ if and only if $f_i(z_1)$ in (6) has constant sign on $|z_1| = 1$; the sign of $g_i(x)$ can be checked with a Sturm test.

A second approach is based on examining the zero distributions of $z_1^{N_i} f_i(z_1)$. Observe first that if $f_i(z_1)$ has constant sign on $|z_1| = 1$, it has no zero there, and all zeros are in $|z_1| < 1$ or $|z_1| > 1$. If z_1^0 is a zero, so is $(z_1^0)^{-1}$, as is easily seen, irrespective of sign definiteness. So the number of zeros of $z_1^{N_i} f_i(z_1)$ in $|z_1| < 1$ is N_i and in $|z_1| > 1$ is N_i, if and only if $f_i(z_1)$ has constant sign on $|z_1| = 1$. We shall describe in Section V the variant on the zero distribution tests of Section III required for checking positivity. One such variant has been used in [14].

III Zero Distribution of Single Variable Polynomials and the Checking of Condition (3)

We shall look first at the Schur–Cohn criterion (see [7]–[10]):

Schur-Cohn Criterion

Suppose that

$$f(z) = \sum_{i=0}^{n} a_i z^i \qquad (a_n \neq 0) \tag{8}$$

and associate with it the $n \times n$ Hermitian matrix $C = (\gamma_{ij})$ defined by

$$\gamma_{ij} = \sum_{p=1}^{i} (a_{n-i+p} a^*_{n-j+p} - a^*_{i-p} a_{j-p}), \qquad i \leqslant j. \tag{9}$$

Then the number of zeros z_i of $f(z)$ for which $|z_i| < 1$ and for which z_i^{-1} is not also a zero is the number of positive eigenvalues of C; the number of zeros z_i for which $|z_i| > 1$ and for which z_i^{-1} is not also a zero is the number of negative eigenvalues of C; the number of zeros z_i for which either $|z_i| = 1$ or z_i^{-1} is also a zero (or both) is the nullity of C.

Let us now see how this criterion can be used in checking condition (3).

First, the polynomial $B(z_1, 0)$ will be real and so will the matrix C. For (3) to hold, all eigenvalues of C must be negative definite. This will hold if and only if odd-order leading principal minors are negative and even-order leading principal minors are positive.

Example 1: Consider

$$B(z_1, z_2) = 1 + az_1 + bz_2 + cz_1 z_2. \tag{10}$$

Evidently,

$$B(z_1, 0) = 1 + az_1$$

and it is trivial to see that we require $|a| < 1$ to satisfy (3).

Example 2: Consider

$$B(z_1, z_2) = 12 + 10z_1 + 6z_2 + 5z_1 z_2 + 2z_1^2 + z_1^2 z_2. \tag{11}$$

Then

$$B(z_1, 0) = 12 + 10z_1 + 2z_1^2.$$

We could simply observe by direct calculation that the roots of this polynomial lie outside $|z_1| = 1$. However, the Schur–Cohn matrix is

$$C = \begin{bmatrix} a_2^2 - a_0^2 & a_1 a_2 - a_1 a_0 \\ a_1 a_2 - a_1 a_0 & a_2^2 - a_0^2 \end{bmatrix} = \begin{bmatrix} -140 & -100 \\ -100 & -140 \end{bmatrix}$$

which is easily seen to have a negative 1×1 leading principal minor and a positive 2×2 leading principal minor. Thus condition (3) is satisfied for $B(z_1, z_2)$ in (11).

Next, we consider a procedure based on forming a sequence of polynomials. To provide us with a result needed for later sections as well as this, we state a theorem of Jury [12], which represents an extension of the original theorem of Cohn [7].

Theorem: Suppose that

$$f(z) = \sum_{i=0}^{n} a_i z^i \tag{8}$$

and define a sequence of polynomials

$$F_0(z) = f(z), \quad F_1(z) = \sum_{i=0}^{n-1} a_i^{(1)} z^i,$$

$$F_2(z) = \sum_{i=0}^{n-2} a_i^{(2)} z^i, \cdots \tag{12}$$

by

$$F_{j+1}(z) = a_0^{*(j)} F_j(z) - a_{n-j}^{(j)} F_j^*(z) \tag{13}$$

where

$$F_j^*(z) = a_0^{*(j)} z^{n-j} + a_1^{*(j)} z^{n-j-1} + \cdots + a_{n-j}^*. \tag{14}$$

(Thus $F_j^*(z)$ is obtained from $F_j(z)$ by coefficient reversal and conjugation.) Set $\delta_j = F_j(0)$ and $P_j = \delta_1 \delta_2 \cdots \delta_j$. Then

1) all zeros of $f(z)$ lie inside $|z| < 1$ if and only if $P_j < 0$ for all j;
2) all zeros of $f(z)$ lie outside $|z| \leqslant 1$ if and only if $P_j > 0$ for all j, or, equivalently, $\delta_j > 0$ for all j;
3) if all P_j are nonzero, the number of negative P_j is the number of zeros of $f(z)$ inside $|z| < 1$ and the number of positive P_j is the number of zeros outside $|z| \leqslant 1$.

We comment that statement 2 above covers the requirement for checking the zero distribution of $B(z_1, 0)$. The theorem, however, is incomplete in the sense that it does not discuss what happens when some P_j are zero; note though that this cannot happen if all zeros of $f(z)$ are in $|z| < 1$ or outside $|z| \leqslant 1$.

Example 1 (continued): We have

$$B(z_1, 0) = 1 + az_1 = F_0(z).$$

Then

$$F_1(z) = (1 + az_1) - a(a + z_1) = 1 - a^2.$$

Then $P_1 = F_1(0) = 1 - a^2$, and $P_1 > 0$ if $|a| < 1$. This is the required condition.

Example 2 (continued): We have

$$B(z_1, 0) = 12 + 10z_1 + 2z_1^2.$$

Then

$$F_1(z) = 12(12 + 10z_1 + 2z_1^2) - 2(2 + 10z_1 + 12z_1^2) = 140 + 100z_1$$

$$F_2(z) = 140(140 + 100z_1) - 100(100 + 140z_1) = 9600.$$

We have $\delta_1 = 140$, $\delta_2 = 9600$. The positivity indicates satisfaction of condition (3).

IV. Application of the Schur-Cohn Test to the Checking of Condition (4)

The Schur-Cohn test associates with the polynomial

$$f(z) = \sum_{i=0}^{n} a_i z^i \tag{8}$$

the Hermitian matrix $C = (\gamma_{ij})$ with

$$\gamma_{ij} = \sum_{p=1}^{i} (a_{n-i+p}a_{n-j+p}^* - a_{i-p}^* a_{j-p}), \qquad i \leqslant j. \tag{9}$$

To check condition (4), we replace $f(z)$ by $B(z_1, z_2)$, written as a polynomial in z_2:

$$B(z_1, z_2) = \sum_{j=0}^{q} \left(\sum_{i=0}^{p} b_{ij} z_1^i \right) z_2^j. \tag{15}$$

The coefficient a_j is therefore $\Sigma_{i=0}^{p} b_{ij} z_1^i$ and the matrix C is $q \times q$, with entries which from (9) are seen to be polynomial in z_1 and z_1^* with real coefficients.

The condition $B(z_1, z_2) \neq 0$ for $|z_1| = 1$, $|z_2| \leqslant 1$ holds if and only if for all $|z_1| = 1$, C is negative definite, i.e., if and only if the leading principal minors of C have appropriate signs. These principal minors being linear combinations of products of the γ_{ij} are themselves polynomial in z_1 and z_1^* with real coefficients; they are also real, since C is Hermitian. On $|z_1| = 1$, we may set $z_1^* = z_1^{-1}$ and the polynomials have the form $\Sigma_{j=0}^{N} c_j(z_1^j + z_1^{-j})$. Such polynomials are termed self-inversive; if $z_1 = z_{1\alpha}$ is a zero, so is $z_1 = z_{1\alpha}^{-1}$.

Example 1 (continued): We write

$$B(z_1, z_2) = (1 + az_1) + (b + cz_1)z_2.$$

This is a degree one polynomial in z_2 and, accordingly, the Schur-Cohn matrix is of size 1×1; it is simply

$$C = (b + cz_1^*)(b + cz_1) - (1 + az_1^*)(1 + az_1).$$

For it to be negative definite on $|z_1| = 1$, we require

$$(1 + az_1^*)(1 + az_1) - (b + cz_1^*)(b + cz_1) > 0, \qquad |z_1| = 1$$

or, recognizing that $z_1^* = z_1^{-1}$ on $|z_1| = 1$, we have

$$(a - bc)z_1 + (1 + a^2 - b^2 - c^2) + (a - bc)\frac{1}{z_1} > 0.$$

Notice that, as claimed, the coefficients of this expression are real. Conditions guaranteeing the inequality are easy to find in this case because of the simplicity of $B(z_1, z_2)$. General methods for checking the inequality are described in Section V.

Example 2 (continued): In this case, we have

$$B(z_1, z_2) = (12 + 10z_1 + 2z_1^2) + (6 + 5z_1 + z_1^2)z_2.$$

This leads to a 1×1 Schur-Cohn matrix:

$$C = [(6 + 5z_1 + z_1^2)(6 + 5z_1^{-1} + z_1^{-2}) - (12 + 10z_1 + 2z_1^2)(12 + 10z_1^{-1} + 2z_1^{-2})] = [-18z_1^2 - 105z_1 - 186 - 105z_1^{-1} - 18z_1^{-2}], \qquad |z_1| = 1$$

and C is negative definite if

$$18z_1^2 + 105z_1 + 106 + 105z_1^{-1} + 10z_1^{-2} > 0, \qquad |z_1| = 1.$$

The checking of this condition is discussed in Section V.

Note that, in general, the Schur-Cohn matrix will be larger than 1×1; in case it has dimension q, there will be q positivity conditions to be checked, one associated with each leading principal minor of C.

V. Checking Positivity of Self-Inversive Polynomials

In this section, we discuss two techniques for checking the positivity on $|z_1| = 1$ of a self inversive polynomial:

$$f(z_1) = \sum_{j=0}^{N} c_j (z_1^j + z_1^{-j}). \tag{16}$$

Here, the c_j are real constants. Of course, one has immediate necessary conditions, obtained by putting $z_1 = 1, -1$:

$$\sum_{j=0}^{N} c_j > 0 \qquad \sum_{j=0}^{N} (-1)^j c_j > 0. \tag{17}$$

(Equally simply conditions can be obtained by setting $z_1 = \sqrt{-1}$ also. See [17].)

The first approach requires us to set $z_1 = \exp(i\theta)$; then (16) becomes

$$f(\cos\theta) = 2 \sum_{j=0}^{N} c_j \cos j\theta. \tag{18}$$

Positivity is to be checked for $0 \leqslant \theta \leqslant 2\pi$ (or $-\pi \leqslant 0 \leqslant \pi$). Since $\cos j\theta = \cos(-j\theta)$, it is enough to check positivity for $0 \leqslant \theta \leqslant \pi$. Let us set

$$x = \cos\theta \qquad T_k(x) = \cos k\theta \tag{19}$$

where $T_k(x)$ is the kth Chebyshev polynomial of the first kind, defined recursively by

$$T_{k+1}(x) = 2x\, T_k(x) \quad T_{k-1}(x), \qquad T_1(x) = x$$
$$T_0(x) = 1. \tag{20}$$

Then (dropping the unessential factor 2) we require

$$g(x) = \sum_{j=0}^{N} c_j T_j(x) > 0 \tag{21}$$

for $-1 \leqslant x \leqslant 1$. Observe that $T_j(x)$ is a polynomial in x of degree j; therefore, $g(x)$ has the form

$$g(x) = \sum_{j=0}^{N} d_j x^j \tag{22}$$

for some real d_j. Positivity can be checked by forming a Sturm chain, see, e.g., [6]. Set $f_0(x) = g(x)$, $f_1(x) = g'(x)$ and define $f_2(x), f_3(x), \cdots$ by

$$f_0(x) = q_1(x) f_1(x) - f_2(x)$$
$$f_1(x) = q_2(x) f_2(x) - f_3(x)$$

with $f_{i+2}(x)$ the remainder obtained when dividing $f_i(x)$ by $f_{i+1}(x)$, and with $f_{i+2}(x)$ of lower degree than $f_{i+1}(x)$. Assuming $f_0(x)$ and $f_1(x)$ have no common factors, the algorithm terminates with $f_N(x)$ equal to a constant, and $f_0(x)$ is positive on $[-1, 1]$ if and only if $f_0(0) > 0$ and

$$V\{f_0(-1), f_1(-1), \cdots, f_N(-1)\} = V\{f_0(1), f_1(1), \cdots, f_N(1)\}$$

where $V\{\cdots\}$ denotes the number of variations in sign of the numbers inside $\{\cdots\}$. In case $f_0(x)$ and $f_1(x)$ have a common zero (i.e., $g(x)$ has repeated zeros), some modification is necessary (see [6]).

Example 1 (continued): We are required to check that

$$(a - bc) z_1 + (1 + a^2 - b^2 - c^2) + (a - bc) \frac{1}{z_1} > 0$$

on $|z_1| = 1$, or, equivalently,

$$(a - bc) T_1(x) + \tfrac{1}{2}(1 + a^2 - b^2 - c^2) > 0, \qquad -1 \leqslant x \leqslant 1$$

or

$$(a - bc) x + \tfrac{1}{2}(1 + a^2 - b^2 - c^2) > 0, \qquad -1 \leqslant x \leqslant 1.$$

The conditions for this are obvious;

$$|a - bc| < \tfrac{1}{2}(1 + a^2 - b^2 - c^2).$$

This condition is easily rewritten in the form stated by Huang [3]:

$$\left|\frac{1-a}{b-c}\right| > 1 \qquad \left|\frac{1+a}{b+c}\right| > 1.$$

Example 2 (continued): We are required to check the inequality

$$18z_1^2 + 105z_1 + 186 + 105z_1^{-1} + 18z_1^{-2} > 0$$

or

$$18\, T_2(x) + 105\, T_1(x) + 93 > 0, \qquad -1 \leqslant x \leqslant 1$$

or

$$36x^2 + 87x + 93 > 0, \qquad -1 \leqslant x \leqslant 1.$$

We have

$$f_0(x) = 36x^2 + 87x + 93$$
$$f_1(x) = 72x + 87$$
$$f_2(x) = \frac{93 \times 144 - 87 \times 87}{144}.$$

Then $V\{f_0(-1), f_1(-1), f_2(-1)\} = 0 = V\{f_0(1), f_1(1), f_2(1)\}$. This establishes that condition (4) is satisfied.

The second approach to checking positivity is based on the determination of the zero distribution of $f(z_1)$ in (16). Because $f(z_1)$ is self-inversive, there are as many zeros of $z_1^N f(z_1)$ inside $|z_1| < 1$ as outside. Therefore, $f(z_1)$ is positive on $|z_1| = 1$ if and only if $f(1) > 0$ [or $f(z_1)$ is positive at any one point of $|z_1| = 1$] and $z_1^N f(z_1)$ has N zeros inside $|z_1| < 1$.

(For then $z_1^N f(z_1)$ has N zeros outside, and thus no zeros on $|z_1| = 1$. Hence $f(z_1)$ has constant sign on $|z_1| = 1$.)

Unfortunately, the Schur–Cohn matrix for a self-inversive polynomial is zero, and the other procedure given in Section III based on setting up a sequence of polynomials leads to a zero polynomial at the first recursion. Hence, neither procedure, as it stands, is of help. However, the following result is of assistance:

Theorem [7], [12]: Let $f(z_1)$ be as in (16). The number of zeros of $g(z_1) = z_1^N f(z_1)$ in $|z_1| < 1$ is the same as the number of zeros of $z_1^{2N-1} g'(1/z_1)$ in $|z_1| < 1$, where $g'(1/z_1)$ is obtained by differentiating $g(z_1)$ with respect to z_1 and then substituting z_1^{-1} for z_1.

Assuming that $z_1^{2N-1} g'(1/z_1)$ has neither zeros reciprocal with respect to $|z_1| = 1$ nor zeros on $|z_1| = 1$, the Schur–Cohn criterion will yield the number of zeros inside $|z_1| < 1$ and the procedure based on generating of a recursive set of polynomials may also. If, on the other hand, $z_1^{2N-1} g'(1/z_1)$ does have reciprocal zeros or zeros on the unit circle, the situation must be analyzed further. In view of the easily established relation

$$z_1 g'(z_1) + z_1^{2N-1} g'\left(\frac{1}{z_1}\right) = f(z_1) \tag{23}$$

it follows that such zeros of $z_1^{2N-1} g'(1/z_1)$ are also zeros of $f(z_1)$. Accordingly, the highest common factor of $z_1^{2N-1} g'(1/z_1)$ and $f(z_1)$ can be found, and then its zero properties can in turn be studied, it too being self-inversive. Proceeding in this way, we can determine ultimately the number of zeros of $z_1^{2N-1} g'(1/z_1)$ and therefore $z_1^N f(z_1)$ inside $|z_1| < 1$, to conclude whether or not $f(z_1)$ is positive.

Example 1 (continued): We have

$$z_1 f(z_1) = (a - bc) z_1^2 + (1 + a^2 - b^2 - c^2) z_1 + (a - bc)$$

and this has the same number of zeros inside $|z_1| = 1$ as does

$$2(a - bc) + (1 + a^2 - b^2 - c^2) z_1.$$

Evidently, this has a zero inside $|z_1| = 1$ if and only if

$$|2(a - bc)| < |1 + a^2 - b^2 - c^2|.$$

Further, for $f(z_1)$ to be positive we require $f(1) > 0$ or

$$2(a - bc) + (1 + a^2 - b^2 - c^2) > 0.$$

Example 2 (continued): We have

$$g(z_1) = z_1^2 f(z_1) = 18z_1^4 + 105z_1^3 + 186z_1^2 + 105z_1 + 18.$$

This has the same number of zeros inside $|z_1| = 1$ as does

$$z_1^3 g'\left(\frac{1}{z_1}\right) = z_1^3 [72z_1^{-3} + 315z_1^{-2} + 372z_1^{-1} + 105]$$

or

$$105z_1^3 + 372z_1^2 + 315z_1 + 72$$

or

$$h(z_1) = 35z_1^3 + 124z_1^2 + 105z_1 + 24.$$

The Schur–Cohn matrix is

$$C = \begin{bmatrix} a_3^2 - a_0^2 & a_2a_3 - a_0a_1 & a_1a_3 - a_0a_2 \\ a_2a_3 - a_0a_1 & a_3^2 + a_2^2 - a_0^2 - a_1^2 & a_2a_3 - a_0a_1 \\ a_1a_3 - a_0a_2 & a_2a_3 - a_0a_1 & a_3^2 - a_0^2 \end{bmatrix} = \begin{bmatrix} 649 & 1820 & 699 \\ 1820 & 5000 & 1820 \\ 699 & 1820 & 649 \end{bmatrix}.$$

The signs of the leading principal minors of C are positive, negative, negative. Jacobi's theorem [15] states that the number of positive eigenvalues of an $n \times n$ Hermitian C is the number of permanences of sign of $\{1, D_1, D_2, \cdots D_n\}$ where D_i is the ith leading principal minor, provided all D_i are nonzero. In our case, C therefore has two positive eigenvalues and one negative eigenvalue. By the Schur–Cohn theorem, the polynomial $h(z_1)$ has two zeros inside $|z_1| = 1$; hence $g(z_1)$ has two zeros inside $|z_1| = 1$. Because $f(1) > 0$, it follows that $f(z_1) > 0$ for all $|z_1| = 1$.

Finally, we comment that conditions (3) and (4) can obviously be replaced by conditions involving interchange of z_1 and z_2 which may be simpler:

$$B(0, z_2) \neq 0, \qquad |z_2| \leq 1 \tag{24}$$

$$B(z_1, z_2) \neq 0, \qquad |z_1| \leq 1, \quad |z_2| = 1. \tag{25}$$

As a further illustration of the techniques presented, which of course apply to (24) and (25), we shall verify that (24) and (25) hold for the same polynomial in z_1 and z_2 as appeared in Example 2. This time we shall use shortcuts.

Example 3:

$$B(z_1, z_2) = 12 + 10z_1 + 6z_2 + 5z_1z_2 + 2z_1^2 + z_1^2 z_2. \tag{11}$$

We have

$$B(0, z_2) = 12 + 6z_2.$$

Obviously, condition (24) holds. Next,

$$B(z_1, z_2) = (12 + 6z_2) + (10 + 5z_2) z_1 + (2 + z_2) z_1^2.$$

The Schur–Cohn matrix obtained by regarding this as a polynomial in z_1 and by setting $z_2^{-1} = z_2^*$ is

$$C = \begin{bmatrix} |a_2|^2 - |a_0|^2 & a_2a_1^* - a_0^*a_1 \\ a_2^*a_1 - a_0a_1^* & |a_2^2| - |a_0|^2 \end{bmatrix} = \begin{bmatrix} -175 - 70z_2 - 70z_2^{-1} & -125 - 50z_2 - 50z_2^{-1} \\ -125 - 50z_2 - 50z_2^{-1} & -175 - 70z_2 - 70z_2^{-1} \end{bmatrix}.$$

For (25) to hold, the 1,1 entry of C must be negative definite and C must have positive determinant, i.e.,

$$175 + 70z_2 + 70z_2^{-1} > 0, \quad |z_2| = 1$$

$$(175 + 70z_2 + 70z_2^{-1})^2 - (125 + 50z_2 + 50z_2^{-1})^2 > 0, \quad |z_2| = 1.$$

The first inequality is obviously satisfied. To check the second, set $x = \cos\theta$. The inequality which has to be satisfied becomes

$$(175 + 140x)^2 - (125 + 100x)^2 > 0, \qquad -1 \leqslant x \leqslant 1$$

i.e.,

$$(300 + 240x)(50 + 40x) > 0, \qquad -1 \leqslant x \leqslant 1$$

and clearly this inequality holds.

VI. Conclusion

Our method is really to be compared against that of Huang. By and large, the methods involve the same sort of calculations, save that we avoid the bilinear transformation component of Huang's method. Undoubtedly, this represents a substantial computational load for any but the simplest two-variable polynomials. On the other hand, as soon as the two-variable polynomials under test become at all complex, presumably computers will be used to do the checking, and it might well prove the case that programming considerations determine which is the better method. Nevertheless, for those who have some knowledge of the Schur-Cohn and related theory, our method would probably be conceptually preferable. For those whose knowledge encompasses Hurwitz testing but not the Schur-Cohn material, Huang's method might be conceptually preferable.

Of course, in programming either algorithm, one would want to take full advantage of the fact that a number of the component tests involve real polynomials, and some simplification may then be possible.

References

[1] C. Farmer and J. B. Bednar, "Stability of spatial digital filters," *Math. Biosci.*, vol. 14, pp. 113-119, 1972.

[2] J. L. Shanks, S. Treitel, and J. H. Justice, "Stability and synthesis of two-dimensional recursive filters," *IEEE Trans. Audio Electroacoust.*, vol. AU-20, pp. 115-128, June 1972.

[3] T. S. Huang, "Stability of two-dimensional recursive filters," *IEEE Trans. Audio Electroacoust.*, vol. AU-20, pp. 158-163, June 1972.

[4] H. G. Ansell, "On certain two-variable generalizations of circuit theory, with applications to networks of transmission lines and lumped reactances," *IEEE Trans. Circuit Theory*, vol. CT-11, pp. 214-223, June 1964.

[5] S. H. Lehnigk, *Stability Theorems for Linear Motions, with an Introduction to Lyapunov's Direct Method.* Englewood Cliffs, N.J.: Prentice-Hall, 1966.

[6] E. A. Guillemin, *Synthesis of Passive Networks.* New York: Wiley, 1957.

[7] A. Cohn, "Über die Anzahl der Wurzeln einer algebraischen Gleichung in einer Kreise," *Math. J.*, vol. 14, pp. 110-148, 1922.

[8] M. Fujiwara, "Über die Algebraischen Gleichungen, deren Wurzeln in einem Kreise oder in einer Halbebene liegen," *Math. J.*, vol. 24, pp. 160-169, 1926.

[9] E. I. Jury, *Theory and Application of the z-Transform Method.* New York: Wiley, 1964.

[10] S. Barnett, *Matrices in Control Theory.* London, England: Van Nostrand Reinhold, 1971.

[11] K. J. Åstrom, *Introduction to Stochastic Control Theory.* New York: Academic, 1970.

[12] E. I. Jury, "A note on the reciprocal zeros of a real polynomial with respect to the unit circle," *IEEE Trans. Circuit Theory*, vol. CT-11, pp. 292-294, June 1964.

[13] B. D. O. Anderson and E. I. Jury, "A simplified Schur-Cohn test," *IEEE Trans. Automat. Contr.*, vol. AC-18, pp. 157-163, Apr. 1973.

[14] D. D. Siljak, "Algebraic criteria for positive realness on the unit circle," in Proc. 2nd Int. Symp. Network Theory, Yugoslavia, July 1972.

[15] F. R. Gantmacher, *The Theory of Matrices.* New York: Chelsea, 1959.

[16] H. A. Nour Eldin, "Ein neues Stabilitätskriterium für abgetaste Regelsysteme," *Regelungstechnik*, vol. 7, pp. 301-307, 1971.

[17] E. I. Jury and G. Gupta, "Alternate coefficients constraints for stability test," *Proc. IEEE* (Lett.), vol. 55, pp. 1769-1770, Oct. 1967.

8

Reprinted from *IEEE Trans. Circuits Syst.* **CAS-22**(3):185–189 (1975)

Stability Criteria for Two-Variable Polynomials

D. D. ŠILJAK, MEMBER, IEEE

***Abstract*—Recursive algebraic algorithms are developed for testing various stability properties of two-variable polynomials in a finite number of steps. Stability is tested with respect to either the half-plane or the unit circle applying only two times the equivalent of the Routh or Marden test, regardless of the degree of the polynomial.**

I. Introduction

IN CLASSICAL stability theory of linear systems [1], the problem is to determine the location of zeros of a real or complex polynomial relative to the half-plane or the unit circle. A large number of solutions to the stability problem as a set of algebraic inequalities involving only polynomial coefficients is available as variations of the Hermite and Routh–Hutwitz criteria for the half-plane, and Schur–Cohn and Marden–Jury criteria for the unit circle.

Theory of electric networks brought forward as a fundamental concept the notion of positive real functions and matrices [1]. The positive real property has two essential aspects, one being stability of a real polynomial with respect to the open half-plane or the unit disc, the other being nonnegativity of a real polynomial on the imaginary axis or the unit circle. It was shown only recently [2], [3] how nonnegativity of polynomials and polynomial matrices can be resolved in the context of the algebraic stability algorithms, and test the positive realness of real rational functions and matrices by a single recursive scheme in finite number of steps.

When the positive real concept in network theory was extended to multivariable functions by Ozaki and Kasami [4], the stability problem of multivariable polynomials arose in a natural way. For two-variable polynomials, the stability problem was reduced by Ansell [5] to testing stability of a real polynomial in the open half-plane and nonnegativity of a set of real polynomials on the imaginary axis. The set of polynomials was generated by the principal minors of the corresponding Hermite matrix, and their number was equal to the order of the matrix. Subsequently, the stability problem of two-dimensional digital filters considered by Shanks *et al.* [6], was resolved by Huang [7] using the solution mechanism of Ansell [5] and the bilinear transformation. Once the direct methods [3] were available for testing nonnegativity of polynomials and polynomial matrices on the unit circle, it was possible to avoid bilinear transformation in this context and use the Schur–Cohn criterion [8]. As an alternative to the Schur–Cohn criterion, the Marden–Jury table can be used as shown in [9]. In [9], however, the Sturm's test was recommended for testing positivity of scalar polynomials on the finite interval, which is inferior to the direct methods [3] both conceptually and algorithmically.

The purpose of this paper is to combine the classical stability conditions for one-variable polynomials [1] with algebraic criteria for positivity of polynomial matrices [2], [3] developed in the framework of positive reality, and produce direct recursive algorithms for testing stability properties of two-variable polynomials. The algorithms can be readily extended to polynomials in more than two variables and, in turn, used to verify positive real characterestics of multivariable rational functions and matrices. The outstanding simplicity of the algorithms is achieved by utilizing the fact [3], [10] that positivity of a polynomial matrix is equivalent to positivity of the matrix at a single value of the independent variable and positivity of its determinant. Stability of two-variable polynomials is tested with respect to either the open half-plane or the unit disk by executing only two times the equivalent of the Routh or Marden tests regardless of the degree of the polynomials involved. The proposed algorithms are completely recursive and are performed by finite number of algebraic calculations, which makes them suitable for computer implementation [11].

II. Positivity of Matrix Polynomials

Let us consider a regular, real matrix polynomial[1]

$$F(z) \equiv \sum_{k=0}^{N} A_k z^k \tag{1}$$

where A_k are real, constant $m \times m$ coefficient matrices, $A_N \neq 0$, and $z = \sigma + i\omega$ is the complex variable. By following [2], [3], we will outline the necessary and sufficient conditions for the Hermitian part

$$H(z) = \tfrac{1}{2}[F(z) + F^*(z)] \tag{2}$$

of $F(z)$ to be positive either on the imaginary axis $\boldsymbol{I} = \{z\colon \operatorname{Re} z = 0\}$ axis positivity; or positive on the unit circle $\boldsymbol{C} = \{z\colon |z| = 1\}$ circle positivity. That is, for all constant complex vectors $U \neq 0$ of a unitary m-dimensional vector space, we have either

$$U^*H(z)U > 0, \qquad \forall z \in \boldsymbol{I} \tag{3}$$

or

$$U^*H(z)U > 0, \qquad \forall z \in \boldsymbol{C}. \tag{4}$$

Let us first consider the *axis positivity* of $H(z)$. We replace z by $i\omega$ in $H(z)$ and use $f(\omega^2) \equiv \det H(i\omega)$, which

Manuscript received March 15, 1974; revised August 21, 1974. This work was supported by NASA under Grant 05-017-010.
The author is with the Department of Electrical Engineering and Computer Science, University of Santa Clara, Santa Clara, Calif. 95053.

[1] With some obvious exceptions, capital light-face letters denote matrices and vectors, lower case and Greek letters denote scalars, and capital bold face letters denote sets.

is a real even scalar polynomial

$$f(\omega^2) \equiv \sum_{k=0}^{n} a_{2k}\omega^{2k} \tag{5}$$

where $a_{2n} \neq 0$. By forming the modified Routh array [2] corresponding to $f(\omega^2)$,

$$\begin{array}{lllll} r_0 = & (-1)^n a_{2n} & (-1)^{n-1} a_{2n-2} & \cdots -a_2 & a_0 \\ r_1 = & (-1)^n n a_{2n} & (-1)^{n-1}(n-1) a_{2n-2} & \cdots -a_2 & \\ \vdots & & \vdots & & \\ r_{2n} = & . a_0 & & & \end{array} \tag{6}$$

we conclude that $f(\omega^2) \not\equiv a_0$ is positive for all ω,

$$f(\omega^2) > 0, \qquad \forall \omega \tag{7}$$

if and only if $v(f) = v[r_0, r_1, \cdots, r_{2n}] = n$ and $a_0 > 0$, where $v(f)$ is the number of sign changes among the Routh numbers $r_0, r_1, \cdots, r_{2n}$; or $f(\omega^2) \equiv a_0 > 0$ [2].

We establish the following theorem.

Theorem 1

A Hermitian polynomial $m \times m$ matrix $H(z)$ is axis positive if and only if the polynomial $f(\omega^2) \equiv \det H(i\omega)$ satisfies

$$v(f) = n, \qquad a_0 > 0 \tag{8}$$

and $H(0) > 0$.

Proof: Since (8) is equivalent to $\det H(i\omega) > 0$, $\forall \omega$, it remains to show that

$$H(z) > 0, \qquad \forall z \in \boldsymbol{I} \tag{9}$$

implies and is implied by

$$\det H(i\omega) > 0, \qquad \forall \omega \tag{10a}$$

$$H(0) > 0. \tag{10b}$$

Obviously, (9) implies (10). Let (10) be satisfied. Then, from (10a), $\det H(i\omega) \neq 0$, $\forall \omega$; that is, the eigenvalues $\lambda_k(\omega)$, $\forall k = 1,2,\cdots,m$, of $H(i\omega)$ are different from zero for all ω. Since $\lambda_k(\omega)$ are real and continuous functions of ω, and for $\omega = 0$ they are all positive due to (10b), they are positive for all ω. Consequently, (10) implies (9) and the proof of Theorem 1 is complete

Let us now consider the *circle positivity* of $H(z)$. The polynomial $\tilde{f}(z) \equiv \det H(z)$ is a real self-inversive scalar polynomial

$$\tilde{f}(z) = \sum_{k=0}^{n} a_k(z^k + z^{-k}) \tag{11}$$

where $a_n \neq 0$.

Similarly, as in Theorem 1, we need first to determine positivity of $f(z)$ on C. We recall [3] that

$$\tilde{f}(z) > 0, \qquad \forall z \in \boldsymbol{C} \tag{12}$$

if and only if the self-inversive polynomial

$$f(z) \equiv \sum_{k=0}^{n} a_k(z^{n+k} + z^{n-k}) \tag{13}$$

defined by $f(z) = z^n\tilde{f}(z)$ is such that either $f(z) \not\equiv a_0$ has no zeros on the unit circle $\boldsymbol{C}$ and $\sum_{k=0}^{n} a_k > 0$; or $f(z) \equiv a_0 > 0$.

To eliminate the possibility of zeros on $\boldsymbol{C}$ of the polynomial $f(z)$ we can use the Marden's recursive scheme [12] as proposed in [3]. With an arbitrary polynomial $f(z) \equiv \sum_{k=0}^{n} a_k z^k$, $a_n \neq 0$, we associate the polynomial

$$f^{\square}(z) = z^n f(z^{-1}) = \sum_{k=0}^{n} a_k z^{n-k} \tag{14}$$

and construct the sequence of polynomials

$$f_j(z) = \sum_{k=0}^{n-j} a_k^{(j)} z^k \tag{15}$$

where $f_0(z) \equiv f(z)$ and

$$f_{j+1}(z) = a_0^{(j)} f_j(z) - a_{n-j}^{(j)} f_j^{\square}(z), \qquad j = 0, 1, \cdots, n-1 \tag{16}$$

so that

$$a_k^{(j+1)} = a_0^{(j)} a_k^{(j)} - a_{n-j}^{(j)} a_{n-j-k}^{(j)}. \tag{17}$$

We denote the constant term $a_0^{(j)}$ of each polynomial $f_j(z)$ by δ_j,

$$\delta_{j+1} = [a_0^{(j)}]^2 - [a_{n-j}^{(j)}]^2 = a_0^{(j+1)}, \qquad j = 0, 1, \cdots, n-1 \tag{18}$$

and form the products

$$\prod_k(f) = \delta_1\delta_2 \cdots \delta_k, \qquad k = 1,2,\cdots,n. \tag{19}$$

Then, as proved by Marden [12], if $v = v(f)$ of the products $\prod_k(f)$ are negative and the remaining $\pi = n - v$ are positive, then $f(z)$ has no zeros on the unit circle $\boldsymbol{C} = \{z\colon |z| = 1\}$, v zeros inside $\boldsymbol{C}$, and π zeros outside $\boldsymbol{C}$. As shown by Jury [13], the simple necessary conditions for the case $v = n$ are $f(1) > 0$ and $(-1)^n f(-1) > 0$, or

$$\sum_{k=0}^{n} a_k > 0, \qquad (-1)^n \sum_{k=0}^{n} (-1)^k a_k > 0. \tag{20}$$

When the polynomial to be considered is a self-inversive, as is the case with $f(z)$ in (11), then $f^{\square}(z) \equiv f(z)$ and $f_1(z) \equiv 0$. Thus we have a singular polynomial sequence [12]. On the basis of results obtained in [3], we derive a finite recursive algorithm for counting the number of zeros of $f(z)$ on the unit circle $\boldsymbol{C}$.

Let us construct the polynomial sequence for the polynomial $f(z)$, and then enumerate the polynomials that precede identically zero polynomials in the sequence by f_{j_μ} $(\mu = 1,2,\cdots,\zeta)$. To continue the sequence, each identically zero polynomial f_{j_μ} is replaced by the newly formed polynomial $[f_{j_\mu}'(z)]^{\square} = z^{2n-(j_\mu+1)} f_{j_\mu}'(z^{-1})$ which has the same number of zeros inside the unit circle as proved by Cohn [14], and where $f_{j_\mu}'(z^{-1})$ is obtained by differentiating $f_{j_\mu}(z)$ with respect to z and then substituting z^{-1} for z. Let us denote by v_μ the number of the negative products $\prod_k$ which correspond to the polynomials $f_{j_\mu+1}, f_{j_\mu+2}, \cdots, f_{j_{\mu+1}}$. That is, v_μ is the number of negative products $\prod_k$ between the two consecutive zero rows j_μ and $j_{\mu+1}$. By going through the matter chronologically (similarly as in [2]), we conclude that the number $\nu(f)$ of zeros inside $\boldsymbol{C}$ of the polynomial $f(z)$ is

$$\nu(f) = \sum_{i=1}^{\zeta} \nu_i \tag{21}$$

and the number $\rho(f)$ of zeros on the circle $\boldsymbol{C}$ is

$$\rho(f) = 2[n - \nu(f)]. \tag{22}$$

Now, as in [2], we arrive at the following theorem.

Theorem 2

A Hermitian polynomial $m \times m$ matrix $H(z)$ is circle positive if and only if the polynomial $f(z) \equiv z^n \det H(z)$ satisfies

$$\nu(f) = n, \qquad \sum_{k=0}^{n} a_k > 0 \tag{23}$$

and $H(1) > 0$.

Proof: In order to prove Theorem 2, we notice first that (23) is equivalent to $\det H(z) > 0$, $\forall z \in \boldsymbol{C}$. Then, it remains to show that

$$H(z) > 0, \qquad \forall z \in \boldsymbol{C} \tag{24}$$

implies and is implied by

$$\det H(z) > 0, \qquad \forall z \in \boldsymbol{C} \tag{25a}$$

$$H(1) > 0. \tag{25b}$$

Using the same argument as in the proof of Theorem 1, the equivalence between (24) and (25) is easily established. This proves Theorem 2.

Axis positivity conditions of Theorem 1 were used to test absolute stability of multi-nonlinear systems in [15]. Axis nonnegativity of matrix polynomials can also be verified by the modified Routh test, but the conditions require more computations [2]. Such extended tests can be used effectively to verify optimality and passivity of linear multivariable systems [15]. Similarly, the conditions of Theorem 2 can be used to determine absolute stability properties of discrete nonlinear systems, and can also be extended to provide conditions for circle nonnegativity of polynomial matrices as shown in [3].

To give an estimate of the efficiency of the direct positivity tests when programmed for computer applications, Schaldach [11] calculated the execution time to be approximately equal to $0.025n^2 + 1$ s where n is the degree of the tested polynomial. When standard root-solving programs were used to conclude positivity (e.g., IBM Scientific Subroutine: 1130-CM-02X) the execution time was ten to a hundred times that of the direct test. Needless to say, the root-solving schemes are iterative and cannot be performed in finite number of steps. Furthermore, they are highly sensitive to multiplicity of the zeros of the tested polynomials, the feature that has no effect on the direct tests. Because of these apparent advantages, the direct positivity tests were used in [11] to deliniate regions of positivity with respect to two adjustable coefficients of polynomials up to the tenth degree. Typical computation time was 15 min–1 h. During this time, some 3000 positivity tests were performed, a task that is virtually impossible to complete by using root-solving schemes or the Sturm's test.

In the following section, we will show how the direct positivity tests can be effectively applied to test numerically various stability properties of two-variable polynomials.

III. Stability

Let us consider a real two-variable polynomial

$$h(s,z) \equiv \sum_{p=0}^{l} \sum_{q=0}^{m} b_{pq} s^p z^q \tag{26}$$

where s and z are the two complex variables, and for some p,q, the coefficients b_{pl} and b_{mq} are not both zero. We are interested in determining the conditions under which the polynomial $h(s,z)$ is different from zero simultaneously for $s \in \boldsymbol{P}^c = \{s\colon \operatorname{Re} s \geq 0\}$ and $z \in \boldsymbol{D} = \{z\colon |z| \leq 1\}$, where $\boldsymbol{P}^c$ is the complement of $\boldsymbol{P} = \{s\colon \operatorname{Re} s < 0\}$ and $\overline{\boldsymbol{D}}$ is the closure of $D = \{z\colon |z| < 1\}$. The desired conditions for the stability property

$$h(s,z) \neq 0, \{s \in \boldsymbol{P}^c\} \cap \{z \in \overline{\boldsymbol{D}}\} \tag{27}$$

are obtained in terms of positivity of polynomial Hermitian matrices considered in the previous section.

Following Ansell [5], we can show that (27) is equivalent to

$$h(s,0) \neq 0, \; s \in \boldsymbol{P}^c \tag{28a}$$

$$h(i\omega,z) \neq 0, \; z \in \overline{\boldsymbol{D}}. \tag{28b}$$

Obviously, (27) implies (28). Let us assume that (28) holds. The implicit relationship $h(s,z) = 0$ defines the algebraic function $z = u(s)$, where we assume that $u(s)$ has no singularities for all $s \in \boldsymbol{P}^c$. If $u(s)$ has singularities there, they can be removed by the usual detour similarly as in [7]. Now, from (28b) we have that $|u(i\omega)| > 1$, which by maximum modulus theorem and (28a) implies $|u(s)| > 1$, $s \in \boldsymbol{P}^c$, as shown in [5]. Therefore, (28) implies (27).

All one needs to test the stability property (27) of the two-variable polynomial $h(s,z)$ is the Routh table [2]. It is easy to see that it is true for condition (28a). We use the original Routh table [16] to conclude that the polynomial $g(s) \equiv h(s,0)$ has all its zeros in $\boldsymbol{P}$. To test condition (28b), we can apply the modified Routh array [2]. In analogy to Ansell's use of the Hermite matrix, one can form first the Schur–Cohn matrix [1], [8] for the polynomial $c(z) \equiv z^m h(i\omega, z^{-1})$,

$$c(z) \equiv \sum_{q=0}^{m} c_q z^q \tag{29}$$

where the coefficients c_q are given as

$$c_q = \sum_{p=0}^{l} b_{p,m-q} s^p \tag{30}$$

and $s = i\omega$. The $m \times m$ Schur–Cohn matrix $C = (c_{pq})$ has the elements defined by

$$c_{pq} = \sum_{k=1}^{p} (c_{m-p+k}\bar{c}_{m-q+k} - \bar{c}_{p-k}c_{q-k}), \qquad p \leq q. \tag{31}$$

In (31), $\bar{c}_i$ is the complex conjugate of c_i.

We recall [1] that $C > 0$ if and only if $c(z) = 0$ implies $z \in \boldsymbol{D}$. Since $C = C(i\omega)$ defined by (31) is a Hermitian matrix ($C = C^*$), we have the polynomial $f(\omega^2) \equiv \det C(i\omega)$ as in (5), and can establish $C(s) > 0$, $\forall s \in \boldsymbol{I}$, by applying the modified Routh algorithm to $f(\omega^2)$ and Sylvester inequalities to $C(0) > 0$. Therefore, we have Theorem 3.

Theorem 3

A real two-variable polynomial $h(s,z)$ has the stability property (27) if and only if the polynomials $g(s) \equiv h(s,0)$ and $f(\omega^2) \equiv \det C(i\omega)$ satisfy

$$g(s) = 0 \Rightarrow s \in \boldsymbol{P} \tag{32a}$$

$$v(f) = n, \qquad a_0 > 0 \tag{32b}$$

and $C(0) > 0$.

Theorem 3 follows automatically from (28) and Theorem 1.

Stability of two-variable polynomials arise naturally in the network analysis when positive reality of functions of two complex variables [4], [5] is used. In such cases, we are interested in establishing necessary and sufficient conditions for the polynomial $h(s,z)$ to have the stability property

$$h(s,z) \neq 0, \{s \in \boldsymbol{P}^c\} \cap \{z \in \boldsymbol{P}^c\}. \tag{33}$$

As shown by Ansell [5], the property (33) is equivalent to

$$h(s,1) \neq 0, s \in \boldsymbol{P}^c \tag{34a}$$

$$h(i\omega,z) \neq 0, z \in \boldsymbol{P}^c. \tag{34b}$$

Again, to test (34a) we can use the Routh test. To verify (34b) we follow Ansell's approach and consider the polynomial $d(z) \equiv h(i\omega,z)$,

$$d(z) \equiv \sum_{q=0}^{m} d_q z^q \tag{35}$$

where

$$d_q = \sum_{p=0}^{l} b_{pq} s^p \tag{36}$$

and $s = i\omega$. With $d(z)$ we associate the $m \times m$ Hermite matrix $D = (d_{pq})$ with elements d_{pq} defined by [16]

$$d_{pq} = \begin{cases} (-1)^{1/2(p+q)} \sum_{k=1}^{p} (-1)^k (d_{m-k+1}\bar{d}_{m-p+k} \\ \qquad - \bar{d}_{m-k+1} d_{m-p-q+k}), & p+q \text{ even} \\ (-1)^{1/2(p+q+1)} \sum_{k=1}^{p} (-1)^k (d_{m-k+1}\bar{d}_{m-p+k} \\ \qquad - \bar{d}_{m-k+1} d_{m-p-q+k}), & p+q \text{ odd} \end{cases}$$

and $p \leq q$.

We remember [16] that $D > 0$ if and only if $d(z) = 0$ implies $z \in \boldsymbol{P}$. Since $D = D(i\omega)$ is a Hermitian matrix ($D = D^*$), we have $f(\omega^2) \equiv \det D(i\omega)$ as in (5), and we can state the following theorem.

Theorem 4

A real two-variable polynomial $h(s,z)$ has the stability property (33) if and only if the polynomials $g(s) \equiv h(s,1)$ and $f(\omega^2) \equiv \det D(i\omega)$ satisfy

$$g(s) = 0 \Rightarrow s \in \boldsymbol{P} \tag{37a}$$

$$v(f) = n, \qquad a_0 > 0 \tag{37b}$$

and $D(0) > 0$.

Theorem 4 follows directly from (34) and Theorem 1.

As in the case of Theorem 3, we need only the Routh algorithm to test stability property (33) of the two-variable polynomial $h(s,z)$. If the result of Theorem 4 is compared to that of Ansell [5], we notice that instead of testing positivity of all leading principal minors of the matrix $D(i\omega)$, we need to test only positivity of the determinant of $D(i\omega)$. Furthermore, the Sturm algorithm proposed by Ansell for testing positivity of minors of $D(i\omega)$, is replaced by the systematic Routh test available in the efficient and familiar table form (6) [2].

In the stability analysis of recursive digital filters [6]–[8], it is of interest to establish necessary and sufficient conditions for a polynomial $h(s,z)$ to have the stability property

$$h(s,z) \neq 0, \{s \in \bar{\boldsymbol{D}}\} \cap \{z \in \bar{\boldsymbol{D}}\}. \tag{38}$$

By following Huang [7], one can show that (38) is equivalent to

$$h(s,0) \neq 0, \ s \in \bar{\boldsymbol{D}} \tag{39a}$$

$$h(e^{i\omega},z) \neq 0, \ z \in \bar{\boldsymbol{D}}. \tag{39b}$$

Condition (39a) can be tested by the classical recursive algorithm of Marden [11] described by (14)–(19). That is, the algorithm is used to show that $v(g)$ is equal to the degree of the polynomial $g(s) \equiv s^l h(s^{-1},0)$.

To test condition (39b) in analogy to Ansell's approach [5], we consider the polynomial $c(z) \equiv z^m h(e^{i\omega},z^{-1})$ as in (29) with coefficients c_q defined by (30) and $s = e^{i\omega}$. With the polynomial $c(z)$ we associate again the Schur–Cohn $m \times m$ matrix [8] $C = (c_{pq})$ specified by (31). The matrix $C(e^{i\omega})$ is a Hermitian matrix ($C = C^*$), and we have $f(e^{i\omega}) \equiv \det C(e^{i\omega})$, where $f(z)$ is a self-inversive polynomial as in (13). Then, we arrive at the following theorem.

Theorem 5

A real two-variable polynomial $h(s,z)$ has the stability property (38) if and only if the polynomials $g(s) \equiv s^l h(s^{-1},0)$ and $f(z) = \det C(z)$ satisfy

$$g(s) = 0 \Rightarrow s \in \boldsymbol{D} \tag{40a}$$

$$v(f) = n, \qquad \sum_{k=0}^{n} a_k > 0 \tag{40b}$$

and $C(1) > 0$.

Theorem 5 follows directly from (39) and Theorem 2.

The result of Theorem 5 is considerably simpler than that

of Anderson and Jury [8], and Maria and Fahmy [9], in that only det $C(z)$ need to be tested for positivity on C instead of all the m leading principal minors of the matrix $C(z)$ as proposed in [8], or the same number of the first column entries in the Jury's table as required in [9]. Moreover, positivity of det $C(z)$ on C can be carried out by the recursive scheme of Marden [12] which can be efficiently tabulated as shown by Jury [13]. Recently, the Marden scheme was rewritten by Barnett [17] in terms of the Routh algorithm, which made it possible to determine the stability property (39) by only one algorithm of extreme simplicity, that of Routh. This proposition, however, requires that the Barnett's scheme of Routh be extended to incorporate the self-inversive polynomials and singular sequences in an economical way.

IV. Conclusion

Recursive algorithms were outlined for testing stability of real two-variable polynomials with respect to the open half-plane and the unit disk. Half-plane properties were tested by the Routh table, and the Marden method was used to determine the stability properties relative to the unit circle. When compared with the existing stability criteria [5]–[8], the algorithms offer an outstanding simplicity in that arbitrary nth order polynomials in two variables can be tested by applying only two times the equivalent of the Routh and/or Marden test, and checking positive definitness of the corresponding constant Hermitian nth-order matrix. The simplicity is a consequence of the effective solution to the problem of positivity of polynomial matrices presented in [2], [3], [15] and the suitability of the solution for computer implementation [11].

The outlined algorithms can be extended to multivariable polynomials [18]–[20] and are presently used in the development of positive real tests for multivariable functions and matrices.

References

[1] S. Barnett, *Matrices in Control Theory*. London: Van Nostrand, 1971.
[2] D. D. Šiljak, "New algebraic criteria for positive realness," *J. Franklin Inst.*, vol. 291, pp. 109–120, Feb. 1971.
[3] ——, "Algebraic criteria for positive realness relative to the unit circle," *J. Franklin Inst.*, vol. 295, pp. 469–476, June 1973.
[4] H. Ozaki, and T. Kasami, "Positive real functions of several variables and their applications to variable networks," *IRE Trans. Circuit Theory*, vol. CT-7, pp. 251–260, Sept. 1960.
[5] H. G. Ansell, "On certain two-variable generalizations of circuit theory with applications to networks of transmission lines and lumped reactances," *IEEE Trans. Circuit Theory*, vol. CT-11, pp. 214–223, June 1964.
[6] J. L. Shanks, S. Trietel, and J. H. Justice, "Stability and synthesis of two-dimensional recursive filters," *IEEE Trans. Audio Electroacoust.*, vol. AU-20, pp. 115–128, June 1972.
[7] T. S. Huang, "Stability of two-dimensional recursive filters," *IEEE Trans. Audio Electroacoust.*, vol. AU-20, pp. 158–163, June 1972.
[8] B. D. O. Anderson and E. I. Jury, "Stability test for two-dimensional recursive filters," *IEEE Trans. Audio Electroacoust.*, vol. AU-21, pp. 366–372, Aug. 1973.
[9] G. A. Maria and M. M. Fahmy, "On the stability of two-dimensional digital filters," *IEEE Trans. Audio Electroacoust.*, vol. AU-21, pp. 470–472, Oct. 1973.
[10] E. I. Jury and B. W. Lee, "A stability theory for multinonlinear control systems," in *Proc. 3rd IFAC Cong.*, June 1966, pp. 28A.1–11.
[11] V. H. D. Schaldach, "A computer algorithm to test for positive realness," M.S. thesis, Univ. of Santa Clara, Santa Clara, Calif. Jan. 1973.
[12] M. Marden, *Geometry of Polynomials*. Providence, R. I.: Amer. Math. Soc., 1966.
[13] I. E. Jury, *The Theory and Applications of Z-Transform Methods*. New York: Wiley, 1964.
[14] A. Cohn, "Über die Anzahl der Wurzeln einer algebraischen Gleichung in einem Kreise," *Math. Z.*, vol. 14, pp. 110–148, 1922.
[15] D. D. Šiljak, "A criterion for nonnegativity of polynomial matrices with application to system theory," *Publ. Faculté Elec. Univ. Belgrade, Série: Electron., Telecommun., Automat.*, vol. 96, pp. 163–172, 1973.
[16] S. H. Lehnigh, *Stability Theorems for Linear Motions*. Englewood Cliffs, N.J.: Prentice-Hall, 1966.
[17] S. Barnett, "Application of the Routh array to stability of discrete-time linear systems," *Int. J. Contr.*, vol. 19, pp. 47–55, Jan. 1974.
[18] E. I. Jury, "Stability of multidimensional discrete and continuous filters: Inners approach," in *Proc. 1974 IEEE Int. Symp. Circuits and Systems*, Apr. 1974, pp. 70–75.
[19] ——, *Inners and Stability of Dynamic Systems*. New York: Wiley: 1974 (to be published).
[20] N. K. Bose and E. I. Jury, "Positivity and stability tests for multidimensional filters (discrete-continuous), *IEEE Trans.* ASSP-22, pp. 174–180, June 1974.

9

Reprinted from *IEEE Trans. Circuits Syst.* **CAS-24**(4):201–208 (1977)

Some Stability Properties of Two-Dimensional Linear Shift-Invariant Digital Filters

DENNIS GOODMAN

***Abstract*—This paper presents a detailed discussion of stability of two-dimensional linear digital filters, and the subtle differences between the one-dimensional and two-dimensional cases. In particular, it is shown that the fact that the impulse response trails off to zero, or more stringently is square summable does not guarantee BIBO (bounded-input–bounded-output) stability.**

Necessary conditions for the impulse response to be bounded and sufficient conditions for it to be square summable and to approach zero geometrically along any fixed column (or row) are stated.

I. Introduction

The PURPOSE of this paper is to discuss certain stability properties of two-dimensional linear shift invariant digital filters. Most of these properties have no analogs in the one-dimensional case. The effect of the numerator polynomial on stability will be discussed, and an example in which the necessity of Shank's theorem fails will be presented. Two other examples which exhibit behavior different from the one-dimensional case will also be presented: it will be shown that the fact that the impulse response trails off to zero or, more stringently, is square summable, does not guarantee BIBO stability. Lastly, a necessary condition for the impulse response to be bounded, a sufficient condition for the impulse response to be square summable, and a sufficient condition for the impulse response to approach zero geometrically along any fixed column (or row) will be stated.

Manuscript received February 20, 1976; revised November 23, 1976. This paper was supported by the Joint Services Electronics Program under Contract F44620-71-C-0087.

The author is with the Department of Electrical Engineering and Computer Sciences and the Electronics Research Laboratory, University of California, Berkeley, CA 94720, on leave from the U. S. Naval Research Laboratory, Washington, D. C. 20375.

II. Preliminaries

We will use the symbols $\exists$, $\ni$, and $\forall$ to stand for "there exists," "such that," and "for each," respectively.

We will be concerned with two-dimensional linear shift invariant filters which are causal (i.e., have a first quadrant impulse response) and have a rational transfer function. Hence our transfer functions will be of the form

$$G(z_1,z_2)=\frac{P(z_1,z_2)}{Q(z_1,z_2)}$$

where $P(z_1,z_2)$ and $Q(z_1,z_2)$ are two variable polynomials in z_1 and z_2. We will assume that $Q(0,0)\neq 0$ so that $Q(z_1,z_2)\neq 0$ in some neighborhood $U_\epsilon^2 \triangleq \{(z_1,z_2):|z_1|<\epsilon, |z_2|<\epsilon\}$ of (0,0), hence in U_ϵ^2 the function $G(z_1,z_2)$ is analytic and has power series expansion

$$G(z_1,z_2)=\sum_{m,n=0}^{\infty} g_{mn}z_1^m z_2^n$$

g_{mn} is defined to be the impulse response of $G(z_1,z_2)$, and it is well known that the digital filter represented by $G(z_1,z_2)$ is bounded-input–bounded-output (BIBO) stable iff $\{g_{mn}\}\in l_1$, i.e., $\sum_{m,n=0}^{\infty}|g_{mn}|<\infty$. We will say the impulse response is square summable if $\{g_{mn}\}\in l_2$, i.e., $\sum_{m,n=0}^{\infty}|g_{mn}|^2<\infty$, and we will say the impulse response is bounded if for some finite M we have $|g_{mn}|<M\ \forall m,n$.

We define $U^2 \triangleq \{(z_1,z_2):|z_1|<1, |z_2|<1\}$ to be the open unit bidisc, $\bar{U}^2 \triangleq \{(z_1,z_2):|z_1|\leq 1, |z_2|\leq 1\}$ to be the closed unit bidisc, and $T^2=\{(z_1,z_2):|z_1|=1, |z_2|=1\}$ to be the distinguished boundary of the unit bidisc. Note that T^2 is a proper subset of the boundary of the unit bidisc.

Certain properties of two variable polynomials and rational functions which will be needed later will now be discussed. It is well known that a two variable polynomial is not in general factorable into first-order polynomials; rather, a two variable polynomial can be factored into irreducible factors which are themselves two variable polynomials but which cannot be further factored [1]. (Of course a given polynomial may itself be irreducible.) These irreducible polynomials are unique up to multiplicative constants [1]. Two polynomials which have no irreducible factors in common are said to be mutually prime. Consider the two variable rational function

$$G(z_1,z_2)=\frac{P(z_1,z_2)}{Q(z_1,z_2)}$$

where $P(z_1,z_2)$ and $Q(z_1,z_2)$ are mutually prime. Using the terminology of [2], a point $(z_1,z_2)\ni Q(z_1,z_2)=0$ but $P(z_1,z_2)\neq 0$ will be called a pole or a nonessential singularity of the first kind (such a point is analogous to a pole in the one variable case). A point $(z_1,z_2)\ni Q(z_1,z_2)=P(z_1,z_2)=0$ will be called a nonessential singularity of the second kind (such points have no one variable analog). Clearly, if (z_1,z_2) is a pole, $G(z_1,z_2)=\infty$. If (z_1,z_2) is a nonessential singularity of the second kind, the value of $G(z_1,z_2)$ is undefined and in any open neighborhood Γ of $(z_1,z_2)\exists(z_1^a,z_2^a)\in\Gamma\ni(z_1^a,z_2^a)$ is a pole of $G(z_1,z_2)$. This result may be proved using the argument in [3, pp. 14–15] and follows from the fact that two mutually prime polynomials have at most finitely many common zeros [1].

III. The Effect of the Numerator Polynomial on Stability

Perhaps the most important stability theorem for two-dimensional filters is Shank's theorem which states that $G(z_1,z_2)$ is BIBO stable iff $Q(z_1,z_2)\neq 0\ \forall(z_1,z_2)\in\bar{U}^2$ [4], [5]. Clearly, before applying this theorem, all irreducible factors common to $P(z_1,z_2)$ and $Q(z_1,z_2)$ should first be cancelled to make the numerator and denominator mutually prime; this operation is analogous to cancelling all common poles and zeros in the one-dimensional case. A test for the existence of common factors is given in [6], and an algorithm for extraction of the greatest common factor is given in [7]. Shank's theorem is essentially correct except that cases may arise where $G(z_1,z_2)$ has a nonessential singularity of the second kind on T^2 but $\{g_{mn}\}\in l_1$. Such an example will be given below. For the rest of the paper we will assume that $P(z_1,z_2)$ and $Q(z_1,z_2)$ are mutually prime.

The sufficiency of Shank's theorem can be proved easily. If $Q(z_1,z_2)\neq 0$ in $\bar{U}^2$, then by the continuity of $Q(z_1,z_2)$, there exists a slightly larger bidisc $U_{1+\epsilon}^2 \triangleq \{(z_1,z_2):|z_1|<1+\epsilon, |z_2|<1+\epsilon\}$ such that $Q(z_1,z_2)\neq 0\ \forall(z_1,z_2)\in U_{1+\epsilon}^2\Rightarrow G(z_1,z_2)$ is analytic in $U_{1+\epsilon}^2$ $\Rightarrow\sum_{m,n=0}^{\infty}g_{mn}z_1^m z_2^n$ is absolutely convergent in $U_{1+\epsilon}^2\Rightarrow$ $\sum_{m,n=0}^{\infty}|g_{mn}|<\infty$. The necessity of Shank's theorem may fail to apply under conditions previously mentioned. The following necessary condition does hold, however.

Theorem 1: If $G(z_1,z_2)$ represents a BIBO stable filter, then $G(z_1,z_2)$ has no poles in $\bar{U}^2$ and no nonessential singularities of the second kind on $\bar{U}^2$ except possibly on T^2.

Proof: If $G(z_1,z_2)$ has a pole $(z_1^a,z_2^a)\in\bar{U}^2$, then $G(z_1,z_2)=\sum_{m,n=0}^{\infty}g_{mn}z_1^m z_2^n$ is not absolutely convergent in $\bar{U}^2\Rightarrow\sum_{m,n=0}^{\infty}|g_{mn}|=\infty$, and so the filter is BIBO unstable. To complete the proof we show that if $G(z_1,z_2)$ has a nonessential singularity of the second kind, (z_1^b,z_2^b), in $\bar{U}^2-T^2$ then it has a pole in $\bar{U}^2$. If $(z_1^b,z_2^b)\in U^2$, the result follows from the argument at the end of Section II, and we are left with the case $(z_1^b,z_2^b)\in(\bar{U}^2-T^2)-U^2=\{(z_1,z_2):|z_1|<1, |z_2|=1\}\cup\{(z_1,z_2):|z_1|=1, |z_2|<1\}$. In the following argument we assume $|z_1^b|=1$, the argument for $|z_2^b|=1$ is similar. Either $Q(z_1^b,z_2)$ is identically zero as a function of z_2 or it is not. If $Q(z_1^b,z_2)\equiv 0$, then any z_2^c with $|z_2^c|\leq 1$ and $P(z_1^b,z_2^c)\neq 0$ is such that $(z_1^b,z_2^c)\in\bar{U}^2$ is a pole of $G(z_1,z_2)$. If $Q(z_1^b,z_2)\not\equiv 0$, then setting $\epsilon=\frac{1}{2}(1-|z_2^b|)$, we have by a continuity argument [1] that $\exists\delta>0\ni$ if z_1^c satisfies $|z_1^c-z_1^b|<\delta$ then $\exists z_2^c\ni|z_2^c-z_2^b|<\epsilon$ and $Q(z_1^c,z_2^c)=0$. Choosing z_1^c to satisfy both $|z_1^c|<1$ and $|z_1^c-z_1^b|<\delta$ and then selecting z_2^c as in the previous sentence, we see that (z_1^c,z_2^c) is in U^2 and is either a pole or a nonessential singularity of the second kind of $G(z_1,z_2)$. In either case it follows that $G(z_1,z_2)$ has a pole in $\bar{U}^2$. Q.E.D.

The special case where $Q(z_1,z_2)\neq 0\ \forall (z_1,z_2)\in \bar{U}^2 - T^2$ but $G(z_1,z_2)$ has a nonessential singularity of the second kind on T^2 presents problems: it appears that $G(z_1,z_2)$ may or may not be stable. In particular, both

$$G_1(z_1,z_2) \triangleq \frac{(1-z_1)^8(1-z_2)^8}{2-z_1-z_2} \triangleq \frac{P_1(z_1,z_2)}{Q(z_1,z_2)}$$

and

$$G_2(z_1,z_2) \triangleq \frac{(1-z_1)(1-z_2)}{2-z_1-z_2} \triangleq \frac{P_2(z_1,z_2)}{Q(z_1,z_2)}$$

are transfer functions which have mutually prime numerator and denominator polynomials, and $Q(z_1,z_2)\neq 0$ on $\bar{U}^2$ except at $z_1=z_2=1$. Both $G_1(z_1,z_2)$ and $G_2(z_1,z_2)$ have nonessential singularities of the second kind at $z_1=z_2=1$, but we will show that $G_1(z_1,z_2)$ is BIBO stable and $G_2(z_1,z_2)$ is BIBO unstable. $G_2(z_1,z_2)$ will be considered in Section IV; the stability of $G_1(z_1,z_2)$ is proved below.

Example 1: $G_1(z_1,z_2)$ represents a stable filter. In proving the stability of $G_1(z_1,z_2)$ we will use the notation of [3]; page numbers given below will refer to the appropriate page of [3]. We will give two definitions and prove four lemmas in the course of proving stability. The proof of Lemma 1 is long, tedious, and uninteresting, so it has been relegated to the Appendix.

Definition: $G^*(e^{i\theta_1},e^{i\theta_2}) \triangleq \lim_{r\nearrow 1} G(re^{i\theta_1},re^{i\theta_2})$ (p. 24).

Definition: $G(z_1,z_2)\in H^\infty$ if $G(z_1,z_2)$ is holomorphic in U^2 and $|G(z_1,z_2)|\leqslant M<\infty\ \forall (z_1,z_2)\in U^2$ (p. 51).

Lemma 1: $(\partial^2/\partial\theta_2^2)(\partial^2/\partial\theta_1^2)G_1^*(e^{i\theta_1},e^{i\theta_2})$ exists and is continuous (see Appendix).

Lemma 2: $G_1(z_1,z_2)\in H^\infty$

Proof:

$$|G_1(z_1,z_2)| = \left|\frac{(1-z_1)^8(1-z_2)^8}{2-z_1-z_2}\right|$$

$$= |1-z_1|^7|1-z_2|^7 \frac{1}{\left|\frac{1}{1-z_1}+\frac{1}{1-z_2}\right|}.$$

Since

$$\mathrm{Re}\left\{\frac{1}{1-z}\right\} > \frac{1}{2} \qquad \forall z\in U \triangleq \{z:|z|<1\}$$

$$\mathrm{Re}\left\{\frac{1}{1-z_1}+\frac{1}{1-z_2}\right\} > 1 \qquad \forall (z_1,z_2)\in U^2$$

$$\Rightarrow \frac{1}{\left|\frac{1}{1-z_1}+\frac{1}{1-z_2}\right|} < 1 \qquad \forall (z_1,z_2)\in U^2$$

$$\Rightarrow |G_1(z_1,z_2)| \leqslant 2^7\cdot 2^7\cdot 1 \qquad \forall (z_1,z_2)\in U^2$$

$$= 4^7$$

therefore $G_1(z_1,z_2)\in H^\infty$.

Lemma 3: The Fourier coefficients of $G_1^*(e^{i\theta_1},e^{i\theta_2})$:

$$\hat{g}_{mn} = \frac{1}{(2\pi)^2}\int_0^{2\pi}\int_0^{2\pi} G_1^*(e^{i\theta_1},e^{i\theta_2})e^{-im\theta_1}e^{-in\theta_2}\,d\theta_1\,d\theta_2$$

are equal to the power series coefficients of $G_1(z_1,z_2)$, i.e., $\hat{g}_{mn}=g_{mn}\ \forall m,n$.

Proof: By problem c of p. 53 $G_1(z_1,z_2)\in H^\infty$ (by lemma 2)$\Rightarrow G_1(z_1,z_2)$ is the Poisson integral of $G_1^*(e^{i\theta_1},e^{i\theta_2})$. By Theorem 2.1.4 of p. 21, $G_1(z_1,z_2)$ analytic in $U^2\Rightarrow \hat{g}_{mn}=0$ except for $(m,n)\in z_+ \triangleq \{m,n: m\geqslant 0,\ n\geqslant 0\}$. The series representation of the Poisson integral (p. 17) gives

$$G_1(r_1e^{i\theta_1},r_2e^{i\theta_2}) = \sum_{m,n\in z_+} \hat{g}_{mn} r_1^m e^{im\theta_1} r_2^n e^{in\theta_2}$$

substituting $z_1=r_1e^{i\theta_1}$, $z_2=r_2e^{i\theta_2}$

$$G_1(z_1,z_2) = \sum_{m,n=0}^{\infty} \hat{g}_{mn} z_1^m z_2^n$$

therefore $g_{mn}=\hat{g}_{mn}\ \forall m,n$ by uniqueness of power series.

Lemma 4: $\{\hat{g}_{mn}\}\in l_1$.

Proof:

$$\hat{g}_{mn} = \frac{1}{(2\pi)^2}\int_0^{2\pi}\int_0^{2\pi}\left[G_1^*(e^{i\theta_1},e^{i\theta_2})e^{-im\theta_1}\,d\theta_1\right]e^{-in\theta_2}\,d\theta_2.$$

Integrate the inner integral twice by parts (as in [8, p. 157]) to give

$$\hat{g}_{mn} = \frac{1}{(2\pi)^2}\int_0^{2\pi}\int_0^{2\pi}\left(-\frac{1}{m^2}\right) \cdot \frac{\partial^2}{\partial\theta_1^2} G_1^*(e^{i\theta_1},e^{i\theta_2})e^{-im\theta_1}e^{-in\theta_2}\,d\theta_1\,d\theta_2.$$

Change the order of integration and again integrate the inner integral twice by parts to give

$$\hat{g}_{mn} = \frac{1}{m^2n^2}\frac{1}{(2\pi)^2}\int_0^{2\pi}\int_0^{2\pi}\left[\frac{\partial^2}{\partial\theta_2^2}\frac{\partial^2}{\partial\theta_1^2}G_1^*(e^{i\theta_1},e^{i\theta_2})\right] \cdot e^{-im\theta_1}e^{-in\theta_2}\,d\theta_1\,d\theta_2$$

$$= \frac{1}{m^2n^2}\hat{f}_{mn}$$

where $\hat{f}_{mn}$ are the Fourier coefficients of $(\partial^2/\partial\theta_2^2)(\partial^2/\partial\theta_1^2)G_1^*(e^{i\theta_1},e^{i\theta_2})$. By Lemma 1 this function exists and is continuous; it follows from the Riemann–Lebesgue theorem (in [9, p. 301]) that $\hat{f}_{mn} \underset{m,n\to\infty}{\longrightarrow} 0$. Hence $\exists M<\infty \ni$

$$|\hat{f}_{mn}|\leqslant M \qquad \forall m,n \Rightarrow |\hat{g}_{mn}| \leqslant \frac{M}{m^2n^2} \qquad \forall m,n.$$

$\Rightarrow \{\hat{g}_{mn}\}\in l_1$. This completes the proof of Lemma 4 and also the proof that $G_1(z_1,z_2)$ is BIBO stable as $\{\hat{g}_{mn}\}\in l_1 \Rightarrow \{g_{mn}\}\in l_1$ by Lemma 3.

IV. Two Counterexamples

The two filters

$$G_2(z_1,z_2)=\frac{(1-z_1)(1-z_2)}{2-z_1-z_2}$$

$$G_3(z_1,z_2)=\frac{2}{2-z_1-z_2}$$

exhibit behavior not found in the one-dimensional case. In particular, it is well known that a one-dimensional filter with rational transfer function $H(z)=\Sigma_{m=0}^{\infty}h_m z^m$ is BIBO stable iff $\{h_m\}\in l_2$ or $\lim_{m\to\infty}h_m=0$. $G_2(z_1,z_2)$ is of interest for three reasons: first, because although it is similar to G_1 in that its only singularity on $\bar{U}^2$ is a nonessential singularity of the second kind at $z_1=z_2=1$, it is BIBO unstable; second, because it serves as a warning to anyone who might mistakenly interpret a nonessential singularity of the second kind on T^2 as a "pole-zero cancellation;" third, because although $\{g_{mn}\}\notin l_1$, $\{g_{mn}\}\in l_2$ (hence also $\lim_{m,n\to\infty}g_{mn}=0$). G_3 is interesting because it shows that impulse response behavior different from the one-dimensional case is not confined to filters having nonessential singularities of the second kind; we will show that the impulse response of $G_3(z_1,z_2)$ trails off to zero.

Example 2:

$$G_2(z_1,z_2)\triangleq\frac{(1-z_1)(1-z_2)}{2-z_1-z_2}$$

is BIBO unstable but has square summable impulse response.

Proof: Writing

$$G_2(z_1,z_2)=\frac{1}{\dfrac{1}{(1-z_1)}+\dfrac{1}{(1-z_2)}}$$

and arguing as in Lemma 2; we see that $|G_2(z_1,z_2)|\leqslant 1$ $\forall(z_1,z_2)\in U^2\Rightarrow G_2\in H^\infty$. Noting that $G_2(z_1,z_2)$ is analytic in U^2 and arguing as in Lemma 3 we see that the power series coefficients g_{mn} of $G_2(z_1,z_2)$ equal the Fourier series coefficients $\hat{g}_{mn}$ of $G_1^*(e^{i\theta_1},e^{i\theta_2})$ for all m,n. Since

$$G_2^*(1,1)=\lim_{r\to1}\frac{(1-r)(1-r)}{2-2r}=\lim_{r\to1}\frac{(1-r)}{2}=0$$

and $G_2(z_1,z_2)$ is continuous everywhere else on T^2

$$G_2^*(e^{i\theta_1},e^{i\theta_2})=\begin{cases}G_2(e^{i\theta_1},e^{i\theta_2}), & (\theta_1,\theta_2)\neq(0,0)\\ 0, & (\theta_1,\theta_2)=0.\end{cases}$$

But taking the limit as $(\theta_1,\theta_2)\to0$ from two different directions

$$\lim_{\theta\to0}G_2^*(e^{i\theta},e^{-i\theta})=\lim_{\theta\to0}\frac{(1-e^{i\theta})(1-e^{-i\theta})}{2-e^{i\theta}-e^{-i\theta}}$$

$$=\lim_{\theta\to0}\frac{2-2\cos\theta}{2-2\cos\theta}=1$$

$$\lim_{\theta\to0}G_2^*(e^{i\theta},e^{i\theta})=\lim_{\theta\to0}\frac{(1-e^{i\theta})(1-e^{i\theta})}{2-e^{i\theta}-e^{i\theta}}$$

$$=\lim_{\theta\to0}\frac{1-e^{i\theta}}{2}=0$$

we conclude $\{\hat{g}_{mn}\}\notin l_1$ and so $G_2(z_1,z_2)$ is not BIBO stable.

We now show $\{g_{mn}\}\in l_2$. For $r<1$

$$G_2(re^{i\theta_1},re^{i\theta_2})=\sum_{m,n=0}^{\infty}g_{mn}(re^{i\theta_1})^m(re^{i\theta_2})^n$$

$$=\sum_{m,n=0}^{\infty}\left[g_{mn}r^m r^n\right]e^{im\theta_1}e^{in\theta_2}.$$

Using Parseval's relation (in [9, p. 301]) and the fact that $|G_2(z_1,z_2)|\leqslant1$ $\forall(z_1,z_2)\in U^2$ gives

$$\sum_{m,n=0}^{\infty}|g_{mn}r^m r^n|^2=\frac{1}{(2\pi)^2}\int_0^{2\pi}\int_0^{2\pi}|G_2(re^{i\theta_1},re^{i\theta_2})|^2\,d\theta_1\,d\theta_2$$

$$\leqslant1,\qquad r\in[0,1)$$

hence

$$\sum_{m,n=0}^{\infty}|g_{mn}|^2r^{2(m+n)}\leqslant1,\qquad r\in[0,1).$$

Since

$$\sum_{m,n=0}^{M}|g_{mn}|^2r^{2(m+n)}$$

is continuous in r for any M,

$$\sum_{m,n=0}^{M}|g_{mn}|^2r^{2(m+n)}\leqslant1$$

for $r\in[0,1],\forall M$

$$\Rightarrow\sum_{m,n=0}^{M}|g_{mn}|^2\leqslant1\ \forall M\Rightarrow\sum_{m,n=0}^{\infty}|g_{mn}|^2$$

$$\leqslant1\Rightarrow\{g_{mn}\}\in l_2.$$

Example 3:

$$G_3(z_1,z_2)\triangleq\frac{2}{2-z_1-z_2}=\frac{1}{1-\frac{1}{2}z_1-\frac{1}{2}z_2}$$

is BIBO unstable but has an impulse response $\{g_{mn}\}$ such that $\lim_{m,n\to\infty}g_{mn}=0$.

Proof: $G_3(z_1,z_2)$ is BIBO unstable by Theorem 1. The impulse response of such a filter was calculated in [10] using Cauchy's theorem and was shown to be $g_{mn}=\left(\frac{1}{2}\right)^{m+n}((m+n)!/m!n!)$. Writing $m+n=k$, $g_{mn}=\left(\frac{1}{2}\right)^k\binom{k}{n}$. Since the binomial coefficients have the properties:

$$\text{for } k \text{ even: }\binom{k}{l}\leqslant\binom{k}{k/2},\qquad 0\leqslant l\leqslant k$$

and

$$\text{for k odd:}\quad \binom{k}{l} < \binom{k}{(k+1)/2} < \binom{k+1}{(k+1)/2}, \qquad 0 \le l \le k.$$

We may conclude

$$0 < g_{mn} \le \left(\frac{1}{2}\right)^k \frac{k!}{\left(\frac{k}{2}!\right)^2}, \qquad m+n=k,\ k \text{ even}$$

$$0 < g_{mn} \le \left(\frac{1}{2}\right)^k \frac{(k+1)!}{\left[\left(\frac{k+1}{2}\right)!\right]^2}, \qquad m+n=k,\ k \text{ odd}.$$

Therefore, to show $\lim_{m,n\to\infty} g_{mn}=0$, it suffices to show $\lim_{k\to\infty}\left(\frac{1}{2}\right)^{2k}[(2k)!/(k!)^2]=0$. Applying [11, eq. (6), p. 211]

$$n! = \sqrt{2\pi n}\, n^n e^{-n} e^{[\Omega(n)/12(n+1)]}$$

where $0<\Omega(n)<1\ \forall n$, we see that

$$\lim_{k\to\infty}\left(\frac{1}{2}\right)^{2k}\frac{(2k)!}{(k!)^2}$$

$$= \frac{1}{\sqrt{\pi}} \lim_{k\to\infty} \frac{1}{\sqrt{k}} \exp\left\{\frac{\Omega(2k)}{24k-12} - \frac{\Omega(k)}{6k-6}\right\}.$$

Since $\lim_{k\to\infty} 1/\sqrt{k} = 0$, and $\lim_{k\to\infty} \exp\{(\Omega(2k)/24k-12) - (\Omega(k)/6k-6)\} = 1$; $\lim_{k\to\infty}\left(\frac{1}{2}\right)^{2k}(2k)!/(k!)^2 = 0$, therefore, we conclude $\lim_{m,n\to\infty} g_{mn}=0$. It is also interesting to note that on the diagonal $g_{nn}=\left(\frac{1}{2}\right)^{2n} n!/(n!)^2$, hence $g_{nn}\to 0$ as $1/\sqrt{n}$. This is behavior contrasting with that in the one-dimensional case where if the impulse response approaches zero, it must do so geometrically. It is also interesting to note $\{g_{mn}\}\notin l_2$.

V. Some Stability Theorems

Theorem 2: If $G(z_1,z_2)$ has a bounded impulse response, then $G(z_1,z_2)$ is analytic in U^2 (i.e., $Q(z_i,z_2)\neq 0$ in U^2).

Proof: $|g_{mn}|<M\ \forall m,n \Rightarrow \sum g_{mn} z_1^m z_2^n$ is absolutely convergent $\forall (z_1,z_2)\in U^2 \Rightarrow G(z_1,z_2)$ has no poles in $U^2 \Rightarrow G(z_1,z_2)$ is analytic in U^2 (recall that in Theorem 1 we argued that if $\exists$ a nonessential singularity of the second kind $\in U^2$, $\exists$ a pole $\in U^2$).

Remark: From Example 3 one might be tempted to conclude the converse of Theorem 2; however, the converse is false, e.g.,

$$G_4(z_1,z_2) = \frac{1}{(1-z_1z_2)^2}$$

is analytic in U^2 but has impulse response

$$g_{mn} = \begin{cases} 0, & m \neq n \\ m, & m = n. \end{cases}$$

Theorem 3: If $G(z_1,z_2)$ is bounded in U^2, then $\{g_{mn}\}$ is square summable.

Proof: $G(z_1,z_2) \le M < \infty$ in $U^2 \Rightarrow G(z_1,z_2)$ has no poles in $U^2 \Rightarrow G(z_1,z_2)$ has no nonessential singularities of the second kind in $U^2 \Rightarrow G(z_1,z_2)$ is analytic in $U^2 \Rightarrow G(z_1,z_2)$ has power series expansion $\sum_{m,n=0}^{\infty} g_{mn} z_1^m z_2^n$ which is absolutely convergent in U^2. Using the same argument as in Example 2:

$$\sum_{m,n=0}^{\infty} |g_{mn}|^2 r^{2(m+n)} = \frac{1}{(2\pi)^2}$$

$$\cdot \int_0^{2\pi}\int_0^{2\pi} |G(re^{i\theta_1}, re^{i\theta_2})|^2\, d\theta_1\, d\theta_2 \le M^2, \qquad \forall r \in [0,1).$$

It follows that $\sum |g_{mn}|^2 < \infty$ or $\{g_{mn}\} \in l_2$.

Theorem 4: Let $G(z_1,z_2) = P(z_1,z_2)/Q(z_1,z_2)$. If $Q(z_1,0)\neq 0$ for $z_1 \in \bar{U} \triangleq \{z_1 \ni |z_1| \le 1\}$, then for any fixed n, $g_{mn}\to 0$ geometrically in m and $\sum_{m=0}^{\infty}|g_{mn}|<\infty$.

Proof: Since $Q(0,0)\neq 0$ is assumed, $\exists \epsilon > 0 \ni$

$$\sum_{m,n=0}^{\infty} |g_{mn} z_1^m z_2^n| < \infty, \qquad \forall (z_1,z_2) \in U_\epsilon^2$$

$$U_\epsilon^2 \triangleq \{(z_1,z_2) : |z_1| < \epsilon,\ |z_2| < \epsilon\}$$

so for $(z_1,z_2)\in U^2$ we may iterate the sum to give

$$G(z_1,z_2) = \sum_{n=0}^{\infty} z_2^n \sum_{m=0}^{\infty} g_{mn} z_1^m$$

$$G(z_1,z_2) = \sum_{n=0}^{\infty} z_2^n g_n(z_1)$$

where $g_n(z_1)=\sum_{m=0}^{\infty} g_{mn} z_1^m$. For $(z_1,z_2)\in U_\epsilon^2$ we may differentiate term-wise k times to give

$$\frac{\partial^k}{\partial z_2^k}[G(z_1,z_2)] = \sum_{n=0}^{\infty} \frac{\partial^k}{\partial z_2^k} z_2^n g_n(z_1)$$

setting $z_2=0$ gives

$$\frac{\partial^k}{\partial z_2^k}[G(z_1,z_2)]\Big|_{z_2=0} = k!\, g_k(z_1).$$

The left-hand side is a rational function of the form

$$\frac{P_k(z_1)}{[Q(z_1,0)]^k}$$

so we conclude

$$\sum_{m=0}^{\infty} g_{mn} z_1^m = g_n(z_1) = \frac{1}{n!}\frac{P_n(z_1)}{[Q(z_1,0)]^n}$$

and since $g_n(z_1)$ is a rational function of one variable with denominator nonzero on $\bar{U}$, we conclude $g_{mn} \underset{m\to\infty}{\to} 0$ geometrically for any fixed n, and $\sum_{m=0}^{\infty}|g_{mn}|<\infty$.

Remark: A filter satisfying $Q(z_1,0)\neq 0\ \forall z_1 \in \overline{U}$ and $Q(0,z_2)\neq 0\ \forall z_2 \in \overline{U}$ is not necessarily stable. For example,

$$G(z)=\frac{1}{1-z_1 z_2}$$

has impulse response

$$g_{mn}=\begin{cases}1, & m=n\geqslant 0\\ 0, & \text{otherwise}\end{cases}$$

and hence is unstable, but

$$\sum_{m=0}^{\infty} g_{mn}=1, \quad \forall n \text{ and } \sum_{n=0}^{\infty} g_{mn}=1, \quad \forall m.$$

VI. Conclusion

In this section we summarize the results discussed in this paper. Let

$$G(z_1,z_2)=\frac{P(z_1,z_2)}{Q(z_1,z_2)}=\sum_{m,n=0}^{\infty} g_{mn} z_1^m z_2^n$$

where $P(z_1,z_2)$ and $Q(z_1,z_2)$ are mutually prime and $Q(0,0)\neq 0$. Then the following relationships hold:

a) BIBO stability	$\Leftrightarrow$	$\{g_{mn}\}\in l_1$
b) $Q(z_1,z_2)\neq 0$ in $\overline{U}^2$	$\not\Leftarrow$ / $\Rightarrow$	BIBO stability
c) $Q(z_1,z_2)\neq 0$ in $\overline{U}^2-T^2$	$\Leftarrow$ / $\not\Rightarrow$	BIBO stability
d) $\{g_{mn}\}\in l_2$	$\Leftarrow$ / $\not\Rightarrow$	BIBO stability
e) $\lim g_{mn}=0$	$\Leftarrow$ / $\not\Rightarrow$	$\{g_{mn}\}\in l_1$ or $\{g_{mn}\}\in l_2$
f) $Q(z_1,z_2)\neq 0$ in U^2	$\Leftarrow$ / $\not\Rightarrow$	$\lvert g_{mn}\rvert \leqslant M<\infty, \quad \forall m,n$
g) $\lvert G(z_1,z_2)\rvert \leqslant N<\infty$ in U^2	$\Rightarrow$	$\{g_{mn}\}\in l_2$
h) $Q(z_1,0)\neq 0$ in $\overline{U}$	$\Rightarrow$	$\sum_{m=0}^{\infty} \lvert g_{mn}\rvert<\infty, \quad \forall n.$

Appendix

Proof of Lemma 1: We first determine the values of $G_1^*(e^{i\theta_1},e^{i\theta_2})$. Since $G_1(z_1,z_2)$ is continuous on T_1^2 except at $z_1=z_2=1$, $G_1^*(e^{i\theta_1},e^{i\theta_2})=G_1(e^{i\theta_1},e^{i\theta_2})$ except at $(\theta_1,\theta_2)=(0,0)$. At $(0,0)$, $G_1^*(e^{i0},e^{i0})=\lim_{r\nearrow 1}G_1(r,r)=(1-r)^8(1-r)^8/2(1-r)=0$. Hence

$$G_1^*(e^{i\theta_1},e^{i\theta_2})=\begin{cases}\dfrac{(1-e^{i\theta_1})^8(1-e^{i\theta_2})^8}{2-e^{i\theta_1}-e^{i\theta_2}}, & (\theta_1,\theta_2)\neq(0,0)\\ 0, & (\theta_1,\theta_2)=(0,0).\end{cases}$$

Since $P_1(e^{i\theta_1},e^{i\theta_2})\triangleq(1-e^{i\theta_1})^8(1-e^{i\theta_2})^8$ and $Q(e^{i\theta_1},e^{i\theta_2})\triangleq 2-e^{i\theta_1}-e^{i\theta_2}$ are both differentiable in θ_1 and θ_2 and $Q(e^{i\theta_1},e^{i\theta_2})\neq 0$ except at $(\theta_1,\theta_2)=(0,0)$, applying the formula $(x/y^n)'=(x'y-ny'x)/y^{n+1}$) ($y\neq 0$, n any integer $\geqslant 1$) four times we conclude that $(\partial^2/\partial\theta_2^2)\cdot(\partial^2/\partial\theta_1^2)G_1^*(e^{i\theta_1},e^{i\theta_2})$ exists and is continuous except possibly at $(0,0)$.

Because the notation gets messy in proving existence and continuity at $(0,0)$, we let $\theta\triangleq\theta_1$, $\phi\triangleq\theta_2$ and $G(\theta,\phi)\triangleq G_1^*(e^{i\theta},e^{i\phi})$. The proof proceeds as follows: show, in order

1) $G_\theta(0,0)=0$
2) $G_{\theta\theta}(0,0)=0$
3) $G_{\theta\theta\phi}(0,0)=0$
4) $G_{\theta\theta\phi\phi}(0,0)=0$
5) $G_{\theta\theta\phi\phi}(\theta,\phi)$ is continuous at $(0,0)$.

The proofs of 1), 2), 3), are similar to the proof of 4) and will be omitted. Assuming 3) has been proved

$$G_{\theta\theta\phi}(\theta,\phi)=\begin{cases}\dfrac{\partial}{\partial\phi}\dfrac{\partial^2}{\partial\theta^2}\dfrac{(1-e^{i\theta})^8(1-e^{i\phi})^8}{2-e^{i\theta}-e^{i\phi}}, & (\theta,\phi)\neq(0,0)\\ 0, & (\theta,\phi)=(0,0).\end{cases}$$

Now

$$\frac{\partial}{\partial\phi}\frac{\partial^2}{\partial\theta^2}\frac{(1-e^{i\theta})^8(1-e^{i\phi})^8}{2-e^{i\theta}-e^{i\phi}}=\frac{(1-e^{i\theta})^6(1-e^{i\phi})^7A(e^{i\theta}e^{i\phi})}{(2-e^{i\theta}-e^{i\phi})^3}$$

where $A(e^{i\theta},e^{i\phi})$ is some polynomial in $e^{i\theta}$ and $e^{i\phi}$, and

$$\begin{aligned}G_{\theta\theta\phi\phi}(0,0)&=\lim_{\epsilon\to 0}\frac{G_{\theta\theta\phi}(0,\epsilon)-G_{\theta\theta\phi}(0,0)}{\epsilon}\\ &=\lim_{\epsilon\to 0}\frac{\dfrac{(1-1)^6(1-e^{i\epsilon})^7A(1,e^{i\epsilon})}{(1-e^{i\epsilon})^3}-0}{\epsilon}\\ &=\lim_{\epsilon\to 0}\frac{0-0}{\epsilon}\\ &=0.\end{aligned}$$

Hence

$$G_{\theta\theta\phi\phi}(\theta,\phi)=\begin{cases}\dfrac{\partial^2}{\partial\theta^2}\dfrac{\partial^2}{\partial\theta^2}\dfrac{(1-e^{i\theta})^8(1-e^{i\theta})^8}{2-e^{i\theta}-e^{i\theta}}, & (\theta,\phi)\neq(0,0)\\ 0, & (\theta,\phi)=0\end{cases}$$

and it remains to show the continuity of $G_{\theta\theta\phi\phi}$ at $(0,0)$.

Since for $(\theta,\phi)\neq(0,0)$

$$G_{\theta\theta\phi\phi}(\theta,\phi)=\frac{(1-e^{i\theta})^6(1-e^{i\phi})^6B(e^{i\theta},e^{i\phi})}{(2-e^{i\theta}-e^{i\phi})^4}$$

where $B(e^{i\theta},e^{i\phi})$ is some polynomial in $e^{i\theta}$ and $e^{i\phi}$, it is sufficient to show that

$$\lim_{\|(\theta,\phi)\|\to 0}\frac{|1-e^{i\theta}|^6|1-e^{i\phi}|^6}{|2-e^{i\theta}-e^{i\phi}|^4}=0$$

where $\|(\theta,\phi)\|\triangleq\max\{|\theta|,|\phi|\}$. Simplifying,

$$\frac{|1-ei^{\theta}|^6|1-e^{i\phi}|^6}{|2-e^{i\theta}-e^{i\phi}|^4}=4\frac{[1-\cos\theta]^3[1-\cos\phi]^3}{[3-2\cos\theta-2\cos\phi+\cos(\theta-\phi)]^2}$$

Claim: Let $M(\theta,\phi)\triangleq 3-2\cos\theta+2\cos\phi+\cos(\theta-\phi)$. Then $\exists\epsilon>0\ni 0<\|(\theta,\phi)\|<\epsilon\Rightarrow M(\theta,\phi)>\frac{1}{4}\|(\theta,\phi)\|^4$.

Proof of Claim: In the proof $0(\cdot)$ will denote a function which satisfies $\lim_{x\to 0}0(x)/x=0$. We will show $\exists\epsilon>0\ni$ for (θ,ϕ) with $0<\|\theta,\phi\|<\epsilon$ and $|\theta|>0$, $M(\theta,\phi)>\frac{1}{4}\theta^4$. Since $M(\theta,\phi)=M(\phi,\theta)$ this suffices to prove the claim as reversing the roles of θ and ϕ we get for the same ϵ.

$$0<\|(\theta,\phi)\|<\epsilon,\ |\phi|>0\Rightarrow M(\theta,\phi)>\tfrac{1}{4}\phi^4$$

and so,

$$0<\|(\theta,\phi)\|<\epsilon\Rightarrow M(\theta,\phi)>\max\left\{\tfrac{1}{4}\theta^4,\tfrac{1}{4}\phi^4\right\}=\tfrac{1}{4}\|(\theta,\phi)\|^4.$$

Taking partials with respect to ϕ

$$M_\phi(\theta,\phi)=2\sin\phi-\sin(\phi-\theta)$$
$$M_{\phi\phi}(\theta,\phi)=2\cos\phi-\cos(\phi-\theta).$$

We note that $\exists\epsilon_1>0\ni$ if θ,ϕ satisfy $0\leqslant\|(\theta,\phi)\|<\epsilon_1$, then $M(\theta,\phi)\geqslant 0$, $M_{\phi\phi}(\theta,\phi)>0$ with $M(\theta,\phi)=0$ only at $(0,0)$. Next consider $M_\phi(\theta,\phi)$: since $M_\phi(0,0)=0, M_{\phi\phi}(0,0)=1, M_{\phi\theta}(0,0)=1$, and $M_\phi(\theta,\phi)$ is analytic in θ and ϕ, it follows by the implicit function theorem (pp. 270–273 of [12], the analyticity follows from 10.2.4 on p. 272) that $\exists\epsilon_2>0$ and a unique analytic $\Phi(\theta)\ni M_\phi(\theta,\Phi(\theta))=0$ for $|\theta|<\epsilon_2$. Furthermore, $\Phi(0)=0$ and $d\Phi(\theta)/d\theta|_{\theta=0}=-M_{\phi\theta}(0,0)/M_{\phi\phi}(0,0)=-1$ so that $\Phi(\theta)=-\theta+0(\theta)$. We next select $\epsilon_3>0$, $\epsilon_3\leqslant\min\{\epsilon_1,\epsilon_2\}$ such that $0<|\theta|<\epsilon_3\Rightarrow$

a) $|\Phi(\theta)|<\epsilon_1$

b) $-1.1\leqslant\dfrac{\Phi(\theta)}{\theta}\leqslant-0.9$.

It follows that for $\phi,\theta\ni 0<\|(\theta,\phi)\|<\epsilon_3$ with $|\theta|>0$ we have

$$M(\theta,\phi)\geqslant M(\theta,\Phi(\theta))>0$$

and $(\theta,\Phi(\theta))$ is in the sector bounded by lines of slope -1.1 and -0.9, i.e., $\Phi(\theta)=k\theta$ for some $k\ni-1.1\leqslant k\leqslant-0.9$ (see Fig. 1). We now take the Taylor expansion of $M(\theta,\phi)$ to give

$$M(\theta,\phi)=\left[\theta^2+\phi^2-\tfrac{1}{2}(\theta-\phi)^2\right]+\tfrac{1}{24}\left[(\theta-\phi)^4-2\theta^4-2\phi^4\right]$$
$$+\left\{0_1(\theta^5)+0_2(\phi^5)+0_3(\theta,\phi)^5\right\}$$

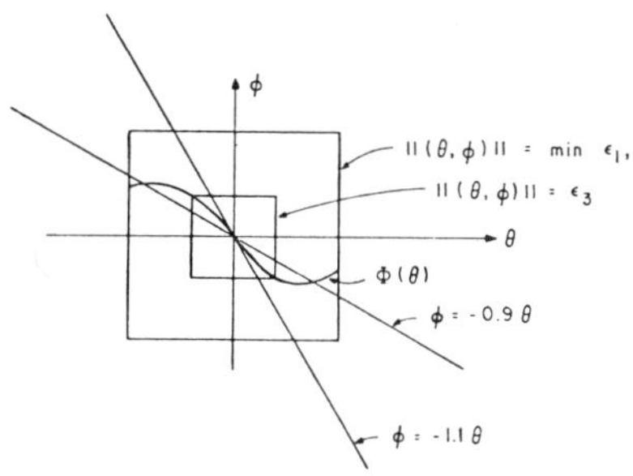

Fig. 1. Behavior of $\Phi(\theta)$ in neighborhood of $(\theta,\phi)=(0,0)$.

set $\phi=k\theta$ to give

$$M(\theta,k\theta)=(1-k)^2\theta^2+\tfrac{1}{24}\left[(1-k)^4-2(1-k^4)\right]\theta^4$$
$$+\left\{0_1(\theta^5)+0_2\left[(k\theta)^5\right]+0_3\left[(1-k)^5\theta^5\right]\right\}$$
$$\geqslant\tfrac{1}{24}\left[(1-k)^4-2(1-k^4)\right]\theta^4$$
$$+\left\{0_1(\theta^5)+0_2\left[(k\theta)^5\right]+0_3\left[(1-k)^5\theta^5\right]\right\}$$

and for $-1.1\leqslant k\leqslant-0.9$

$$M(\theta,k\theta)\geqslant\tfrac{1}{2}\theta^4+\left\{0_1(\theta^5)+0_2(k^5\theta^5)+0_3\left[(1-k)^5\theta^5\right]\right\}$$

therefore $\exists\epsilon_4>0\ni$

$$M(\theta,k\theta)\geqslant\tfrac{1}{4}\theta^4,\qquad\forall\theta:|\theta|<\epsilon_4\quad k:-1.1\leqslant k\leqslant 0.9.$$

Finally, let $\epsilon=\min\{\epsilon_3,\epsilon_4\}$ from which we conclude: for $(\theta,\phi)\ni 0<\|(\theta,\phi)\|<\epsilon_1$, $|\theta|>0: M(\theta,\phi)\geqslant M(\theta,\Phi(\theta))\geqslant\theta^4/4$. As discussed previously, this completes the proof of our claim that for $(\theta,\phi)\ni 0<\|(\theta,\phi)\|<\epsilon$, $M(\theta,\phi)>\frac{1}{4}\|(\theta,\phi)\|^4$. Next, consider the function $1-\cos\theta=+\theta^2/2+0(\theta^3)$. It follows that $\exists\epsilon_5>0\ni[1-\cos\theta]<\theta^2$ for $|\theta|<\epsilon_5$. Hence, for $(\theta,\phi)\ni 0<\|(\theta,\phi)\|<\epsilon_0=\min\{\epsilon,\epsilon_5\}$

$$0\leqslant\frac{4[1-\cos\theta]^3[1-\cos\phi]^3}{[3-2\cos\theta-\cos\phi+\cos(\theta-\phi)]^2}\leqslant\frac{4[\theta^2]^3[\phi^2]^3}{\left[\frac{\|(\theta,\phi)\|^4}{4}\right]^2}$$
$$\leqslant 64\frac{\|\theta,\epsilon\|^{12}}{\|\theta,\phi\|^8}$$
$$\leqslant 64\|\theta,\phi\|^4$$

$$\Rightarrow\lim_{\|\theta,\phi\|\to 0}\frac{4[1-\cos\theta]^3[1-\cos\phi]^3}{[3-2\cos\theta-2\cos\phi+\cos(\theta-\phi)]^2}=0$$

$\Rightarrow G_{\theta\theta\phi\phi}(\theta,\phi)$ is continuous at $(0,0)$.

Q.E.D.

Acknowledgment

The author wishes to thank Profs. C. A. Desoer and E. I. Jury of the Department of Electrical Engineering and

Computer Sciences, and Professors H. Helson, D. Sarason, and C. Stanton of the Department of Mathematics for many helpful discussions. He also wishes to thank Prof. M. Strintzis of the University of Pittsburgh for pointing out an error in the original proof of Theorem 1.

References

[1] G. A. Bliss, *Algebraic Functions*. New York: Amer. Math. Soc., 1933.
[2] W. F. Osgood, *Topics in the Theory of Functions of Several Complex Variables*. New York: Dover, 1966.
[3] W. Rudin, *Function Theory in Polydiscs*. New York: Benjamin, 1969.
[4] T. S. Huang, "Stability of two-dimensional recursive filters," *IEEE Trans. Audio and Electroacoustics*, vol. AI-20, pp. 158–163, June 1972.
[5] J. H. Justice and J. L. Shanks, "Stability criterion for N-dimensional digital filters," *IEEE Trans. on Automatic Control*, vol. AC-18, pp. 284–286, June 1973.
[6] N. K. Bose, "A criterion to determine if two multivariable polynomials are relatively prime," *Proc. IEEE*, vol. 60, pp. 134–135, Jan. 1972.
[7] ——, "An algorithm for GCF extraction from two multivariable polynomials," *Proc. IEEE*, vol. 64, pp. 185–186, Jan. 1976.
[8] L. Schwartz, *Mathematics for the Physical Sciences*. Reading, MA: Addison-Wesley, 1966.
[9] A. Zygmund, *Trigonometric Series*, vol. II. London, England: Cambridge University Press, 1959.
[10] R. M. Mersereau and D. E. Dudgeon, "Two-dimensional digital filtering," *Proc. IEEE*, vol. 63, pp. 510–623, Apr. 1975.
[11] M. R. Spiegel, *Schaum's Outline of Theory and Problems of Advanced Mathematics for Engineers and Scientists*. New York: McGraw-Hill, 1971.
[12] J. Dieudonné, *Foundations of Modern Analysis*. New York: Academic Press, 1969.
[13] C. A. Desoer and M. Vidyasagar, *Feedback Systems: Input-Output Properties*. New York: Academic Press, 1975.
[14] E. I. Jury, *Inners and Stability of Dynamic Systems*. New York: Wiley, 1975.
[15] W. Kaplan, *Operational Methods for Linear Systems*. Reading, MA: Addison-Wesley, 1962.
[16] A. V. Oppenheim and R. W. Schafer, *Digital Signal Processing*. Englewood Cliffs, NJ: Prentice-Hall, 1975.
[17] L. R. Rabiner and B. Gold, *Theory and Application of Digital Signal Processing*. Englewood Cliffs, NJ: Prentice-Hall, 1975.
[18] M. Strintzis, private communication, May 1976.

10

Reprinted from pp. 535–538 of the *1977 IEEE Internat. Conf. on Acoust., Speech, Signal Process. Rec.*, 879 pp.

A STABILITY TEST FOR 2-D RECURSIVE DIGITAL FILTERS USING THE COMPLEX CEPSTRUM*

M. P. Ekstrom and R. E. Twogood
Lawrence Livermore Laboratory
University of California
Livermore, California 94550

ABSTRACT

A procedure is presented for testing the stability of 2-D recursive filters. This test is based on the mapping properties of the complex cepstrum. A numerical algorithm for computing the 2-D cepstrum is described, and subsequently used in calculating numerical examples of the stability test.

I. INTRODUCTION

A principal interest in the design and application of two-dimensional (2-D) recursive or IIR digital filters is in determining their stability. Considerable effort has been expended in developing stability theorems for 2-D recursive filters [1,2] and in formulating algorithmic tests based on these theorems [2,3,4,5].

There are two limitations of this work. First, these results are applicable only to the class of 2-D recursive filters which have unit sample responses $h(m,n)$ which are first quadrant or quarter plane arrays (i.e., for which $h(m,n)$ is non-zero only for $m \geq 0$, $n \geq 0$). In addition, the stability tests are quite complex to implement, computationally.

Recently, a procedure was presented for the spectral factorization of 2-D polynominals [6]. This factorization was used to develop a stability test for the general class of 2-D recursive filters (i.e., half-plane). The basis of this test was the ability of the factorization to separate a 2-D function into multiplicative components, each having a prescribed region of analyticity.

The purpose of this paper is to introduce an alternate and computationally simplified stability test for 2-D recursive digital filters. This test is similar in spirit to the previous factorization approach, although strikingly simpler in structure, and makes use of a subtle mapping property of the 2-D complex cepstrum. The advantages of this approach are its computational reductions (≈50% of the previous test), ease of programming, and applicability to the complete, general class of 2-D recursive filters. The main computational step in the stability test is that of calculating the 2-D complex cepstrum. A numerical algorithm is presented for performing this calculation. Details of an FFT-based computer algorithm for implementing the stability test are also presented along with numerical examples of its application.

*This work was performed under the auspices of the U.S. Energy Research and Development Administration under contract number W-7405-ENG-48 and partially supported by the National Science Foundation under Grant No. ENG-75-15027.

II. STABILITY OF HALF-PLANE RECURSIVE FILTERS

The class of 2-D recursive filter to be considered here is the so-called half-plane recursive filter. These are filters defined over a half-plane, yet recursively computable. An example of such a filter would be described by the recursive form

$$y(k,\ell) = x(k,\ell) - \sum_{\substack{m=0 \\ m+n\neq 0}}^{M} \sum_{n=0}^{N} b(m,n)y(k-m,\ell-n) - \sum_{m=-1}^{-M} \sum_{n=1}^{N} b(m,n)y(k-m,\ell-n) \, . \qquad (1)$$

For convenience and consistency of notation, we define a half-plane region $\mathcal{R}$ to be of the form $\{m \geq 0, n \geq 0\} \cup \{m < 0, n > 0\}$ and $\mathcal{R}_f$ a finite dimensional portion of this half-plane, $\{M \geq m \geq 0, N \geq n \geq 0\} \cup \{-M \leq m < 0, N \geq n > 0\}$. Using this notation we have

$$y(k,\ell) = x(k,\ell) - \sum\sum_{\mathcal{R}_f-(0,0)} b(m,n)y(k-m,\ell-n) \qquad (2)$$

and the filter transfer function

$$H(z_1,z_2) = \frac{1}{B(z_1,z_2)} = \frac{1}{\sum\sum_{\mathcal{R}_f} b(m,n)z_1^{-m}z_2^{-n}} \, . \qquad (3)$$

The filter stability is of course determined by the

location of the essential singularities of $H(z_1,z_2)$ (i.e., the zeros of $B(z_1,z_2)$) in the complex z_1,z_2 space.

Theorem 1 [6]

The half-plane recursive filter $H(z_1,z_2)$ is stable iff

1) $B(z_1,z_2) \neq 0$ on $\{|z_1| = 1, |z_2| \geq 1\}$ and (4)

2) $B(z_1,\infty) \neq 0$ on $\{|z_1| \geq 1\}$ (5)

In this case, we have the filter transfer function

$$H(z_1,z_2) = \frac{1}{B(z_1,z_2)} = \sum\sum_{\mathcal{R}} h(m,n) z_1^{-m} z_2^{-n} \tag{6}$$

Thus, the unit sample response of a stable, half-plane recursive filter is nonzero only over the half-plane, $\mathcal{R}$. Apparently, to implement this stability theorem requires the location of the zeros of $B(z_1,z_2)$. While this cannot be accomplished directly, it does form the basis for practical testing procedures.

In [6] a computationally attractive approach to stability testing was presented. It was based on a 2-D spectral factorization operation involving the autocorrelation function of the filter (with suitable generalization to cover both the half-plane and quarter-plane filters). A main step in this procedure was the use of the cepstrum [7] in performing the desired factorization. The 2-D Z-transform of the cepstrum $\hat{H}(z_1,z_2)$ is defined to be the logarithm of the 2-D Z-transform of an array $h(m,n) \in \ell'$:

$$\begin{aligned}\hat{H}(z_1,z_2) &= \log\,[Z\{h(m,n)\}]\\ &= \log\,[H(z_1,z_2)]\end{aligned} \tag{7}$$

By using the autocorrelation of the filter, the above test involved calculating the logarithm of a real array. While this introduced substantial complexity, it did avoid the problems associated with defining the complex logarithm. Recently [8], the existence of a 2-D complex cepstrum has been proven. Using the existence of this cepstrum, we formulate an alternate stability test which removes the excess complexity of our previous approach.

III. CEPSTRAL STABILITY TEST

For the purposes of our stability discussion, the important property of the cepstral transformation has to do with the analyticity of $\hat{H}(z_1,z_2)$. This property is that the essential singularities and zeros of $H(z_1,z_2)$ map into the essential singularities of $\hat{H}(z_1,z_2)$. Recall that if $H(z_1,z_2)$ is stable, it can be written in a power series for $m,n \in \mathcal{R}$. Because the regions of analyticity of $H(z_1,z_2)$ and $\hat{H}(z_1,z_2)$ are identical, it must also follow that $\hat{H}(z_1,z_2)$ can be similarly expanded:

$$\hat{H}(z_1 z_2) = \sum\sum_{\mathcal{R}} \hat{h}(m,n) z_1^{-m} z_2^{-n} \tag{8}$$

This leads to the following stability theorem on which our test is based.

Theorem 2

The half-plane recursive filter $H(z_1,z_2)$ is stable iff its cepstrum $\hat{h}(m,n)$ has support on $\mathcal{R}$.

Because $\hat{H}(z_1,z_2)$ is analytic on $\{|z_1| = 1, |z_2| \geq 1\}$, $\hat{h}(m,n)$ takes support on the entire half-plane. The additional region of analyticity for $\hat{H}(z_1,\infty)$ on $\{|z_1| \geq 1\}$ ensures that $\hat{h}(m,n) = 0$ for $n < 0$. This constitutes the complete region $\mathcal{R}$ of the m,n plane. (It is relatively straightforward to generalize Theorem 2 to other classes of half-plane and quarter-plane filters. See example 2 in Section V.)

The implementation of this theorem into a stability test is easily achieved. We simply form $B(z_1,z_2)$, given $b(m,n)$ of the filter to be tested, calculate $\hat{B}(z_1,z_2)$, and inverse transform to obtain the cepstrum $\hat{b}(m,n)$. If $\hat{b}(m,n) = 0$ for $m,n \notin \mathcal{R}$, then the filter is stable. If however, $\hat{b}(m,n) \neq 0$ for $m,n \notin \mathcal{R}$, then the filter is unstable.

In the numerical realization of this test, we replace the z-transforms with the DFT. A block diagram of this procedure is shown in Fig. 1. Notice that when performing the required calculations using the DFT, we obtain $\hat{b}_a(m,n)$, an aliased version of $\hat{b}(m,n)$. The error in this approximation is given by

$$\hat{b}_a(m,n) - \hat{b}(m,n) = \sum_{\substack{i,j=-\infty\\ i,j\neq 0,0}}^{+\infty}\sum \hat{b}(m + iP, n + jQ) \tag{9}$$

for a DFT of size P x Q. Because of the rapid decay of the cepstrum, $\hat{b}_a(m,n)$ is generally a quite good approximation to $\hat{b}(m,n)$ for modest-sized FFT's.

The simplicity of the overall calculation is readily apparent. It basically involves computing two each 2-D FFT's and the computation associated with obtaining the complex logarithm.

IV. COMPUTATION OF THE 2-D COMPLEX CEPSTRUM

While the existence of the 2-D cepstrum has been proven, no computational algorithms for performing the cepstral calculations have previously been presented. The problem is that of defining the complex logarithm of the array $B(u,v)$. Because

$$\begin{aligned}B(u,v) &= |B(u,v)| \exp\left\{j \tan^{-1}\left[\frac{\text{Im } B(u,v)}{\text{Re } B(u,v)}\right]\right\}\\ &= |B(u,v)| \exp\{j \arg B(u,v)\}\,,\end{aligned} \tag{10}$$

we can write

$$\hat{B}(u,v) = \log B(u,v)$$
$$= \log |B(u,v)| + j \arg B(u,v) \qquad (11)$$

In order to ensure the analyticity of $\hat{B}(u,v)$, the phase term, arg $B(u,v)$, must be periodic and continuous. Unfortunately, the inverse tangent which is used to compute the phase term yields the principal value of the phase, a number between $-\pi$ and π. Two steps are required to construct the continuous and periodic phase from its principal value. To ensure continuity, an algorithm is required which examines the phase at each point in the transform array and removes the "jumps" of 2π which are present in the principal value. This removal is achieved by adding an appropriate multiple of 2π. This process has been called phase unwrapping . Secondly, we must remove all linear phase terms following the unwrapping to ensure periodicity of the phase.

Several approaches to unwrapping the phase in 2-D appear feasible. The one which we have found to be most successful involves the application of two 1-D phase unwrappings on the axes followed by a 2-D unwrapping over the remainder of the transform array $B(u,v)$. The algorithm proceeds as follows. First, arg $B(u,o)$ and arg $B(o,v)$ are calculated by simply indexing u and v and adding or subtracting 2π whenever a phase discontinuity is encountered. If the FFT size is sufficiently large, phase variations between adjacent points will be $\ll 2\pi$, except where a jump of 2π has occurred.

Following these 1-D unwrappings on the axes, we scan the transform array in the manner indicated in Fig. 2, checking for discontinuities between the "present" phase value (circled lattice point) and the four "previously" unwrapped, neighboring phase values (lattice points marked with x). If a discontinuity is encountered, the multiple of 2π required to remove the discontinuity at that point is added to the phase.

The final step involves the linear phase removal. This is easily accomplished by observing the linear portions in arg $B(u,o)$ and arg $B(o,v)$ of the unwrapped phase. The observed amount of linear phase is subtracted from the 2-D unwrapped phase to give arg $B(u,v)$.

V. EXAMPLES OF APPLICATION

The following two examples illustrate the numerical application of our stability test to both quarter-plane and half-plane digital filters.

Example 1. Quarter-Plane Bandpass Filter
We first test the 6th order filter whose denominator array is given by

b(m,n) (n upward, m to the right, origin 0)

0.015626						
0.09375	0.046875					
0.375	0.28125	0.09375				
0.875	0.9375	0.46875	0.109375			
1.5	1.875	1.3125	0.46875	0.09375		
1.5	2.25	1.875	0.9375	0.28125	0.046875	
1.	1.5	1.5	0.875	0.375	0.09375	0.015625

In computing the cepstrum of $b(m,n)$, we obtained the phase (principal value) function of Fig. 3 using a DFT of dimension 64 x 64. The phase unwrapping procedure successfully removed the phase jumps of 2π resulting in the unwrapped phase of Fig. 4. The resulting cepstrum $\hat{b}_a(m,n)$ is shown in Fig. 5. This array is seen to take significant values only in the first quadrant, hence the filter is judged to be stable. The maximum error of (9) outside the first quadrant is $O(10^{-6})$.

Example 2. Half-Plane Fan Filter
This example demonstrates the versatility of the stability test. We test a fan filter with denominator

b(m,n) (n upward, m to the right, origin 0)

0	0	0	-0.001	
0	-0.002	-0.004	0.003	
0	0.005	-0.009	-0.037	
0.001	0.005	0.050	0.040	
0.001	0.007	0.026	0.218	0.355
0.001	0.003	0.004	-0.152	-0.227
0	-0.002	0.025	0.002	0.051
0	-0.002	-0.025	0.004	-0.010
0	0	0	-0.001	0.001

The modification of our stability test required to accommodate this filter form simply involves determining if the cepstrum $\hat{b}(m,n)$ has support over the same half-plane as the support of $b(m,n)$ above.

The log magnitude and phase of this filter were calculated using a 32 x 32 sized FFT. This phase is shown in Fig. 6. The phase unwrapping algorithm detected (correctly) no jumps or linear terms in the phase function. Thus, using this original phase, we inverse transformed $\hat{B}_a(u,v)$ to obtain $\hat{b}_a(m,n)$ in Fig. 7. We note the cepstrum $\hat{b}_a(m,n)$ has the same support as $b(m,n)$ and conclude that it is stable.

The largest error in $\hat{b}_a(m,n)$ outside its support is $O(10^{-4})$. We have found this to be a typical error when using 32 x 32 FFT's. Repeating this test with a 256 x 256 FFT reduced this error to $O(10^{-15})$.

REFERENCES

[1] J. L. Shanks, S. Treitel, and J. H. Justice, "Stability and synthesis of two-dimensional recursive filters," IEEE Trans. Audio Electroacoust., vol. AU-20, pp. 115-128, June 1972.

[2] T. S. Huang, "Stability of two-dimensional recursive filters," IEEE Trans. Audio Electroacoust., vol. AU-20, pp. 158-163, June 1972.

[3] G. A. Maria and M. M. Fahmy, "On the stability of two-dimensional digital filters," IEEE Trans. Audio Electroacoust., vol. AU-21, pp. 470-472, October 1973.

[4] B. D. O. Anderson and E. I. Jury, "Stability of multidimensional digital filters," IEEE Trans. Circuits and Syst., vol. CAS-21, pp. 300-304, March 1974.

[5] D. D. Siljak, "Stability criteria for two-

variable polynomials," IEEE Trans. Circuits and Syst., vol. CAS-22, pp. 185-189, March 1975.

[6] M. P. Ekstrom and J. W. Woods, "Two-dimensional spectral factorization with applications in recursive filtering," IEEE Trans. Acoust. Speech, Signal Porcessing, Vol. ASSP-24, pp. 115-128, April 1976.

[7] A. V. Oppenheim, R. W. Schafer, and T. G. Stockham, Jr., "Nonlinear filtering of multiplied and convolved signals," Proc. IEEE, vol. 56, pp. 1264-1291, August 1968.

[8] D. E. Dudgeon, "The existence of cepstra for two-dimensional rational polynomials," IEEE Trans. Acoust., Speech and Signal Processing (Correspondence), vol. ASSP-23, pp. 242-243, April 1975.

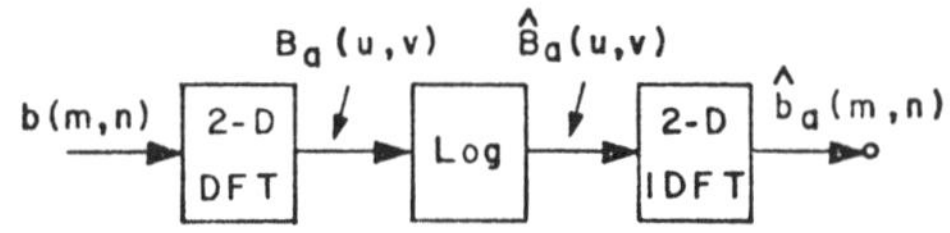

Figure 1. Block diagram of the stability test using 2-D DFT.

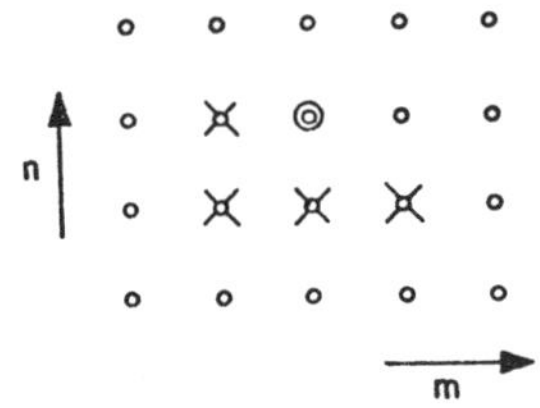

Figure 2. Phase values used in the 2-D phase unwrapping algorithm.

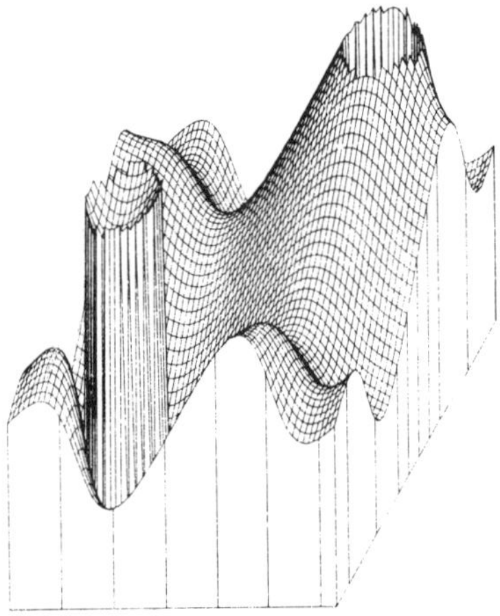

Figure 3. Phase function of Example 1.

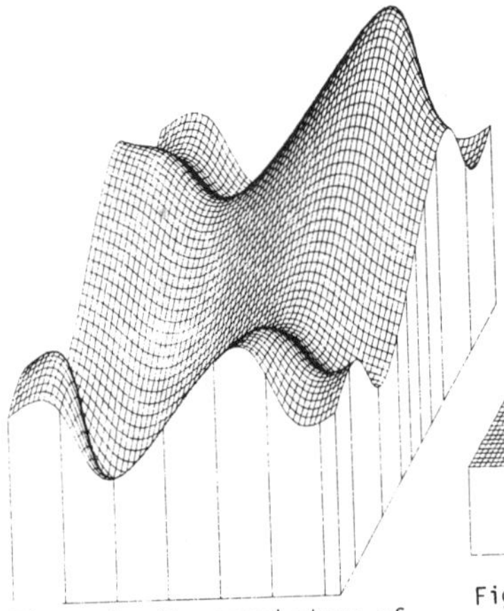

Figure 4. Unwrapped phase of Example 1.

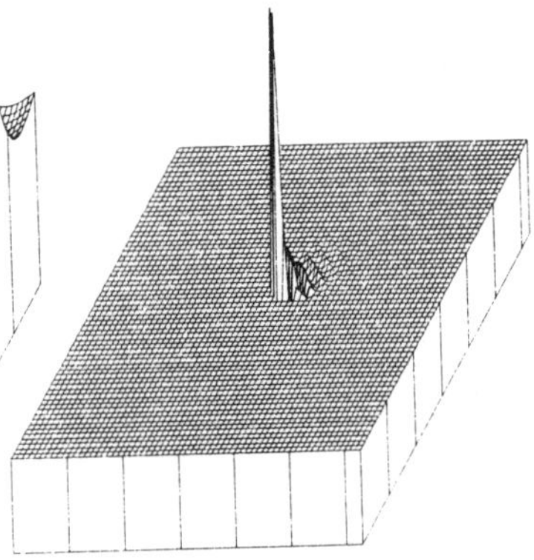

Figure 5. Cepstrum $\hat{b}_a(m,n)$ for the quarter-plane filter of Example 1.

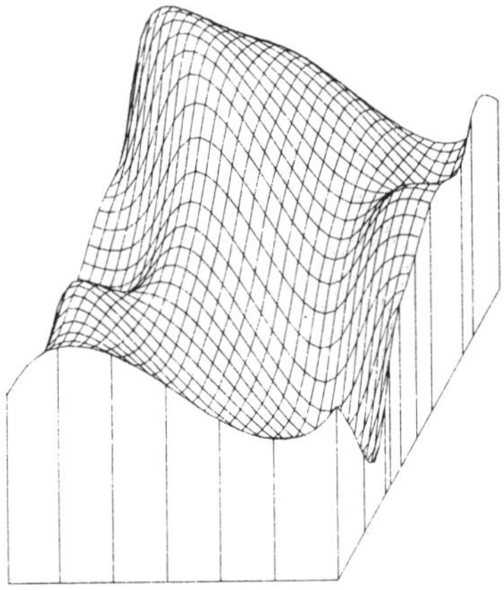

Figure 6. Phase function for Example 2.

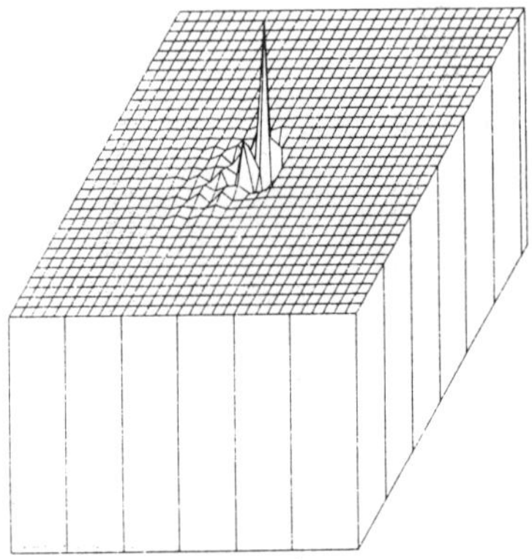

Figure 7. Cepstrum $\hat{b}_a(m,n)$ for the half-plane filter of Example 2.

Editors' Comments on Papers 11 Through 14

DESIGN OF 2-D IIR FILTERS

Almost all work reported on the design of stable, 2-D IIR filters deals with the design of "causal," quarter-plane filters. Both *frequency* domain and *spatial* domain approaches have been presented.

In the *frequency* domain case, design involves determining a stable, IIR transfer function which satisfactorily approximates a frequency domain specification (usually in the form of a 2-D magnitude spectrum). This is a problem of rational approximation. Most of these design efforts can be grouped into two categories, those involving spectral transformations and those involving computer-aided optimization.

In the spectral transformation approach, a 2-D IIR digital filter is designed by using a transformation of complex variables from a 1-D or a 2-D prototype filter with known frequency response characteristics. The prototype filter can be a passive analog filter or another stable digital filter. In the former case, the 2-D digital transfer function is obtained by applying the bilinear transformation on the analog transfer function. However, stability is not always insured by this approach (Goodman 1978). In general, the transformations of variables, more commonly known as spectral transformations, are complex maps which carry a stable rational transfer function into another stable transfer function and can be used to obtain 2-D filters with desirable characteristics.

Historically, the first technique for designing such filters was advanced by Shanks et al. (Paper 1). Here, they consider a one-dimensional recursive filter as a special case of a two-dimensional filter. By rotating the frequency axes of the resultant filter, they arrive at a two-dimensional transfer function. There are two basic problems associated with this technique. First, in general, the rotated 2-D filter may not be stable even though its 1-D prototype is stable. Second, the frequency response of the rotated version is not simply related to that of its parent 1-D filter. This makes it difficult to design 2-D filters with prescribed frequency response characteristics. When using this technique for the design of simple low-order 2-D filters, the stability can be ensured in some cases. Costa and Venetsanopoulos (1974) follow such an approach. They cascade 2-D filter sections obtained by different rotations to obtain stable 2-D low-pass filters with reasonably circularly-symmetric frequency responses. An alternate design procedure also involving the rotation of frequency responses has been presented by Chang and Aggarwal (1976).

In addition to the works mentioned earlier, the spectral transformation approach has also been used by Ahmadi et al. (1976), and Bernabo et al. (1976) to design 2-D IIR filters from 1-D prototypes. Instead of starting with a 1-D prototype, an alternate choice is to start with a 2-D prototype. This type of spectral transformation which converts a known 2-D IIR digital filter prototype to another 2-D IIR digital filter with more desirable characteristic is considered in the paper by Pendergrass et al. (Paper 11). A unified qualitative theory of designing 2-D digital filters via spectral transformations has been advanced very recently by Chakrabarti and Mitra (1977).

In the computer-aided optimization approaches, a nonlinear optimization procedure is used to iteratively adjust the filter coefficients to minimize a specified error criterion. A major difficulty is ensuring the stability of the resultant 2-D digital transfer function. In Paper 12 by Bednar, the differential correction optimization algorithm is used to approximate a given frequency response by a quarter-plane IIR filter. Stability is checked after each iteration of the optimization using a stability testing algorithm similar to that outlined by Anderson and Jury (Paper 7). Because of the computational complexity of the stability test, this technique appears to be most appropriate for the design of low-order filters. Maria and Fahmy (Paper 13) use a ℓ_p optimization technique and avoid this problem by constraining the filters they design to have transfer functions which are products of simple first- and/or second-order terms. This facilitates testing stability of the approximate at each step of the optimization as these low order terms can be tested using a simple set of inequalities associated with the filter coefficients (Paper 6). However, their algorithm may converge to an unstable solution in some cases (Pendergrass 1975).

Alternate optimization approaches involve the design of two-variable passive analog transfer function and use of bilinear transformations to obtain the digital transfer function. Dubois and Blostein (1975) follow such an approach. They first design a two-variable passive (hence guaranteed stable) analog filter via a computer-aided optimization technique and then obtain a 2-D IIR transfer function by applying a double bilinear transformation on the transfer function of the analog filter. The passivity (stability) is ensured by constraining the passive components to always have nonnegative values. A somewhat similar approach has been taken by Ramamoorthy and Bruton (1977), who algebraically express the denominator polynomial of the two-variable, analog transfer function in a suitable form which is always guaranteed to be realizable by a passive network thus ensuring stability.

The 2-D stability testing problem can also be avoided by designing a separable 2-D IIR filter approximating the frequency response characteristic. For a separable filter, the 2-D transfer function $H(z_1,z_2)$ can be expressed as a product of two 1-D transfer functions:

$$H(z_1,z_2) = H_1(z_1) \cdot H_2(z_2). \tag{1.9}$$

In this case, the stability testing reduces to that of checking the stability of 1-D filters, which is considerably simpler. Moreover, a separable filter is also more economical to implement. Twogood and Mitra (Paper 14) describe a computer-aided method of designing separable filters. Their approach makes use of the singular value decomposition (Golub and Reinsch, 1970) of a matrix obtained by sampling the 2-D magnitude response characteristic.

In the *spatial* domain design problem, a filter transfer function is chosen to approximate a finite extent, 2-D impulse response. The earliest work performed on this problem is due to Shanks et al. (Paper 1). They develop a least-squares approach with a spatial error criterion to arrive at a best approximation of the spatial response. Unfortunately, their IIR filter design technique, in general, does not lead to stable filters. To circumvent this problem, the authors suggest using a planar-least-square-inverse (PLSI) approach to arrive at a stable approximation. However, it also appears that this approach may not lead to a stable approximation (Genin and Kamp, 1975). Furthermore, such an approach may lead to a filter whose impulse response does not adequately approximate the prescribed specifications. Another notable contribution in spatial domain designs is due to Parker and Souchon (1977).

The apparent constraint of the above design techniques to the quarter-plane case is an important consideration. In Paper 4, it was shown that a quarter-plane filter was not capable of realizing a *general* magnitude spectrum, and that a half-plane filter was required for this task. Thus, these design approaches will not yield approximates which are arbitrarily close to a prescribed specification by simply increasing the order of the filter. There is, in fact, a finite and irreducible error which is associated with both the quarter-plane form and particular specification. As a consequence, the general performance of these design algorithms should be viewed within this context.

Although many of the above design approaches appear capable of being generalized to the design of half-plane filters, only a limited amount of half-plane filter design work has been performed. Most of this work has been based on the spectral factorization in Part I. This factorization can be adapted to the design of both quarter-plane (multiple) and half-plane filters having either zero or nonzero phase. For designing zero-phase filters (according to a frequency domain specification), the prescribed magnitude-squared frequency response is factored into products of stable 2-D IIR transfer functions which are recursible in specified directions. Several design examples based on this approach are to be found in Papers 3 and 4. A related approach involving a 2-D discrete Hilbert transform has been studied by Dudgeon (1974).

Because, in general, spectral factorization leads to factors which are polynomials of infinite degree, for practical usefulness the process has to be approximated by factors of finite degree. This approximation must be such that the filter meets the prescribed magnitude characteristic within an acceptable tolerance. Both a window procedure (Paper 4) and an optimization procedure (Ekstrom and Woods, 1976) have been studied with encouraging results. The advantage of these techniques is that they are suitable for designing quite high order filters incorporating both the conditions of finite order and satisfactory response characteristics. Because of the importance of this class of IIR filters, more concentrated study into their design is anticipated.

REFERENCES

Ahmadi, M., A. G. Constantinides, and R. A. King. 1976. "Design Technique for a Class of Stable Two-Dimensional Recursive Digital Filters," *Proc. 1976 IEEE Internat. Conf. Acoust., Speech, Signal Process.*, 145–147.

Bernabo, F., P. L. Emiliani, and V. Cappellini. 1976. "Design of 2-Dimensional Recursive Digital Filters," *Electronics Letters* **12**(11):288–289.

Chakrabarti, S., and S. K. Mitra. 1977. "Design of Two-Dimensional Digital Filters via Spectral Transformation," *IEEE Proc.* **65**(6):905–914.

Chang, H., and J. K. Aggarwal. 1976. "Design of Two-Dimensional Recursive Filters by Interpolation," *Proc. 1976 IEEE Internat. Symp. Circuits Syst.*, 369–371.

Costa, J. M., and A. N. Venetsanopoulos. 1974. "Design of Circularly Symmetric Two-Dimensional Recursive Filters," *IEEE Trans. Acoust., Speech, Signal Process.* **ASSP-22**:432–443.

Dubois, E., and M. L. Blostein. 1975. "A Circuit Analogy Method for the Design of Recursive Two-Dimensional Digital Filters," *Proc. 1975 IEEE Internat. Symp. Circuits Syst.*, 451-454.

Dudgeon, D. E. 1974. "Two-Dimensional Recursive Filtering," Sc.D. dissertation, MIT, Department of Electrical Engineering.

Ekstrom, M. P., and J. W. Woods. 1976. "Some Results on the Design of Two-Dimensional Half-Plane Recursive Filters," *Proc. 1976 IEEE Internat. Symp. Circuits Syst.*, 373.

Genin, Y., and Y. Kamp. 1975. "Counter-Example in the Least-Square Inverse Stabilization of 2-D Recursive Filters," *Electronics Letters* **11**:330-331.

Golub, G. H., and C. Reinsch. 1970. "Singular-Value Decomposition and Least Squares Solutions," *Numer. Math.* **14**:403-420.

Goodman, D. 1978. "Some Difficulties with the Double Bilinear Transformation in 2-D Recursive Filter Design," *IEEE Proc. (Letters)*, in press.

Parker, S. R., and L. Souchon. 1977. "Synthesis of N-Dimensional Recursive Digital Filters by Taylor Series Expansion," *IEEE Trans. Circuits Syst.* **CAS-24**(1): 28-34.

Pendergrass, N. A. 1975. "Comments on 'An ℓ_p Design Technique for Two-Dimensional Digital Filters'," *IEEE Trans. Acoust., Speech, Signal Process.* **ASSP-23**:389-390.

Ramamoorthy, P. A., and L. T. Bruton. 1977. "Frequency Domain Approximation of Stable Multi-Dimensional Discrete Recursive Filters," *Proc. 1977 IEEE Internat. Symp. Circuits Syst.*, 654-657.

11

Reprinted from *IEEE Trans. Circuits Syst.* **CAS-23**(1):26–35 (1976)

Spectral Transformations for Two-Dimensional Digital Filters

N. A. PENDERGRASS, STUDENT MEMBER, IEEE, SANJIT K. MITRA, FELLOW, IEEE, E. I. JURY, FELLOW, IEEE

***Abstract*—A theory is developed for spectral transformations for two-dimensional digital filters. It is proved that these transformations take the form of stable two-dimensional all-pass functions and that the result of spectral transformation is stable if the original transfer function is stable. Emphasis is placed on formulating the transformations in a way which makes them easy to use for design, and it is shown that two-dimensional spectral transformations offer a means of quickly obtaining new designs from previous ones. Included is a discussion of applications such as the design and realization of two-dimensional digital filters with tunable characteristics.**

Introduction

OVER THE LAST DECADE, digital technology has become more advanced and reasonable in cost. As a result, applications of digital signal processing are becoming widespread and considerable theory has been developed for the one-dimensional problem. Often, however, data is in the form of numbers regularly or irregularly spaced in a plane. One-dimensional techniques cannot, in most cases, effectively deal with two-dimensional relationships, and for this reason attention is being directed toward the development of a theory for digital manipulation of two-dimensional data. This theory is generally referred to as digital image processing and can be separated into three areas: image encoding, two-dimensional digital filtering, and machine pattern recognition.

This paper is concerned with two-dimensional digital filtering. We define a two-dimensional digital filter as any computational process which transforms a two-dimensional input sequence of numbers into a two-dimensional output sequence of numbers. Furthermore, these two-dimensional sequences are commonly referred to as images even though they may not have been generated as images in the usual sense. For example, the sequence might have a nonspatial coordinate system such as frequency time, or distance time. Usually, however, the results of two-dimensional filtering are displayed as a spatial image for human evaluation.

Manuscript received May 2, 1975; revised August 8, 1975. This work was supported in part by NSF Grant GK-38205.

N. A. Pendergrass was with the Department of Electrical Engineering and Computer Sciences, and the Electronics Research Laboratory, University of California, Berkeley, Calif. 97420. He is now with the Loveland Instruments Division, Hewlett-Packard, Loveland, CO. 80537.

S. K. Mitra is with the Department of Electrical Engineering, University of California, Davis, CA 95616.

E. I. Jury is with the Department of Electrical Engineering and Computer Sciences and the Electronics Research Laboratory, University of California, Berkeley, CA 94720.

There are many applications of two-dimensional digital filters. Usually these applications have one or more of the following objectives:

1) enhancement of the image to make it more acceptable to the human eye,
2) removal of the effects of some degradation mechanism, or
3) separation of features for easier identification or measurement by machine or human.

Perhaps the most widely known application of two-dimensional digital filtering has been in the area of space imagery. Digital processing of satellite images has been used in monitoring environmental effects, earth resources, and urban land use [1]–[4]. In these applications two-dimensional digital filters have been used to enhance or reduce boundaries, remove low-frequency shading effects, reduce noise, and correct for distortion in the imaging system [5], [6].

There are also applications in medicine and biology. For example, using two-dimensional digital processing to reduce spatial low-frequency components in an X-ray image makes features with large high-frequency components, such as a fracture, easier to identify. Contrast enhancement and noise reduction are also used on X-ray images. Additional biomedical uses include removal of scan lines in radioisotope scanning, low-frequency background reduction in photomicrographs, and digital processing of acoustical holograms as a replacement for X-ray images [7]–[11].

Many of the imaging systems mentioned above have counterparts in other sciences and similar benefits are obtained by two-dimensional digital filtering. For example, digital filtering of radiographs is used in nondestructive testing, such as for the measurement of the internal dimensions of nuclear reactor fuel rods [12], [13].

To find an example of processing arrays which are not images in the usual sense, one has only to look into seismic prospecting. In order to gather data about subsurface structure, a number of seismic detectors are placed at stations along a line passing through a shallow bore hole. The digitized outputs of the detectors after an explosion form a two-dimensional array with time along one axis and distance along the other.

Reflected energy arriving at the detectors provides information about the depth and nature of subsurface features. Unfortunately there is a considerable noise problem from wind effects, multiple reflections, and so on.

Two-dimensional processing provides a powerful means of reducing noise and separating signals from different sources for evaluation [14]–[17].

Geophysical data provides additional applications for two-dimensional digital filters. Atmospheric temperature and pressure data have been smoothed by digital means before plotting on weather maps [18]. Similarly, magnetic and gravity measurements can be processed to reduce the effects of surface anomalies in order to identify large subsurface features.

Although this discussion of applications is limited, one can see that they are found in a very wide range of sciences and that virtually any imaging system is a candidate for the use of two-dimensional digital filtering techniques.

Some of the operations performed by two-dimensional digital filters, such as contrast enhancement described above, are nonlinear, although many important filter operations, such as spatial frequency filtering, are linear.

In this paper we are particularly concerned with a method for frequency domain design of linear shift invariant two-dimensional digital filters in which the input and output satisfy a linear constant coefficient difference equation of the form [19]

$$\sum_{k=0}^{M_b}\sum_{l=0}^{N_b} b_{kl}y_{m-k,n-l} = \sum_{k=0}^{M_a}\sum_{l=0}^{N_a} a_{kl}x_{m-k,n-l}$$

where x_{ij} is the input sequence and y_{ij} is the output.

As in one-dimension, these two-dimensional digital filters can be separated into two types: those which have infinite impulse response (IIR) and those which have finite impulse response (FIR). The transfer function of a two-dimensional FIR digital filter is

$$H(Z_1,Z_2) = \sum_{k=0}^{M}\sum_{l=0}^{N} h_{kl}Z_1{}^kZ_2{}^l \qquad (1)^1$$

where the h_{kl}'s are the impulse response coefficient of the filter. The transfer function of a two-dimensional IIR digital filter is

$$H(Z_1,Z_2) = \frac{\sum_{i=1}^{M_a}\sum_{j=1}^{N_a} a_{ij}Z_1^{i-1}Z_2^{j-1}}{\sum_{i=1}^{M_b}\sum_{j=1}^{N_b} b_{ij}Z_1^{i-1}Z_2^{j-1}} = \sum_{k=0}^{\infty}\sum_{l=0}^{\infty} h_{kl}Z_1{}^kZ_2{}^l. \qquad (2)$$

We have assumed that the transfer functions are first quadrant functions [20]; that is, that the impulse response h_{kl} is zero for k or l negative. This condition has been sometimes termed "causal" in the literature; however, when dealing with two-dimensional data, causality does not have the implications for realizability that it does for one-dimensional real time systems. A two-dimensional linear digital filter can be realized with right to left as well as left to right dependence. In addition there may be top to bottom or bottom to top dependence. For this reason the terms first quadrant function, second quadrant function, etc., are to be preferred as a description of a two-dimensional transfer function when referring to the location of its nonzero impulse response.

[1] Note that the two-dimensional z-transform is defined with positive exponents.

Purpose of This Work

The purpose of this work is to develop and investigate a theory for spectral transformation of two-dimensional digital filters. By a spectral transformation we mean a transformation that maps one transfer function into another transfer function with a different spatial frequency response but with some desirable characteristics maintained.

While these transformations are generally known and widely used for one-dimensional digital [19], [21] and continuous [22], [23] filters, no one has yet published an investigation of them for the two-dimensional digital case. The desire for spectral transformations, however, is increased in the two-dimensional problem by the difficulty in obtaining stable IIR filters. This is a result of the greater complexity and the difficulty involved with stability checking [24]. Using spectral transformations in one-dimension, a stable filter design can be altered to produce other stable designs in a quick and straightforward way. A natural question then is whether similar procedures exist for the two-dimensional case.

The Spectral Transformation Problem

Assume that we are given a transfer function of a linear shift invariant two-dimensional digital filter in the form of a rational function as in (1) or (2). We assume that the transfer function is a first quadrant function and stable. This is equivalent to the assumption that the denominator polynomial in (2) satisfies [25], [26]

$$\sum_{i=1}^{M_b}\sum_{j=1}^{N_b} b_{ij}Z_1^{i-1}Z_2^{j-1} \neq 0, \qquad \text{for } |Z_1| \leq 1 \text{ and } |Z_2| \leq 1.$$

We further assume that the a_{ij} and b_{ij} in (1) and (2) are real. This is desirable since it assures that the filter can be realized using real arithmetic.

The problem is to find transformations $G: C^2 \to C^2$ with

$$G: Z_1 = G_1(z_1,z_2)$$
$$Z_2 = G_2(z_1,z_2)$$

which

1) produce stable first quadrant transfer functions from stable first quadrant transfer functions,
2) transform real rational functions into real rational functions, and
3) preserve some important basic characteristics of the amplitude response (such as ripple magnitude in the pass and stop regions) while altering other characteristics (such as cutoff frequencies or the number and shape of stop and pass regions).

These requirements severely restrict the class of transformations that we can consider. First, the condition that transformation of real rational functions result in real rational functions implies that G_1 and G_2 must be real and rational.

Requirement 3) still further restricts G. Let us define a region

$$D = \{(z_1,z_2) \in C^2 \mid |z_1| \leq 1, |z_2| \leq 1\}$$

and denote ∂D as its boundary and $\mathring{D}$ as its interior. The frequency response of a two-dimensional digital filter can be written as the transfer function evaluated on ∂D [19]. Thus the frequency response is

$$H(Z_1,Z_2)\Big|_{\substack{Z_1=e^{-j\omega_1}\\ Z_2=e^{-j\omega_2}}}. \tag{3}$$

Therefore we must have $G: \partial D \to \partial D$ so that

$$|G_1(e^{-j\omega_1},e^{-j\omega_2})| = |G_2(e^{-j\omega_1},e^{-j\omega_2})| = 1$$

for any $\omega_1, \omega_2 \in R$. In other words G_1 and G_2 must be two-dimensional all-pass transfer functions. The general form of the spectral transformation is therefore

$$Z_i = G_i(z_1,z_2)$$

$$= (\pm 1)_i \prod_{l=1}^{K_i} \left\{ \frac{z_1^{N_l^{(i)}} z_2^{M_l^{(i)}} \left(\sum_{n=0}^{N_l^{(i)}} \sum_{m=0}^{M_l^{(i)}} a_{mn}^{(i)} z_1^{-n} z_2^{-m} \right)}{\left(\sum_{n=0}^{N_l^{(i)}} \sum_{m=0}^{M_l^{(i)}} a_{nm} z_1^{n} z_2^{m} \right)} \right\}, \qquad i = 1,2 \tag{4}$$

with $a_{00} = 1$.

The requirements of stability can now be assessed. If G_1 or G_2 is not stable then one of the stable transfer functions $H(Z_1,Z_2) = Z_i$, $i = 1$ or 2, will be transformed to an unstable transfer function. Conversely, if G_1 and G_2 are stable then, by the maximum modulus theorem, $G: \mathring{D} \to \mathring{D}$ [27]. Since $H(Z_1,Z_2)$ is stable by assumption, $H(G_1(z_1,z_2), G_2(z_1,z_2))$ is stable. Thus we have proven that the spectral transformation in (4) takes any stable transfer function to a stable transfer function if and only if G_1 and G_2 are stable.

Notice that the above result concerning stability still holds if $|G_i(e^{-j\omega_1},e^{-j\omega_2})| \leq 1$, $i = 1$ and 2. In this case, however, G does not map $\partial D \to \partial D$ so the amplitude response will be changed in some unknown way. For example, peak magnitude and ripple would not be maintained. This situation is of little interest so we will not treat it further.

General Effects of Transformation

From (3) the frequency response of a two-dimensional digital filter is doubly periodic. The periodicity in ω_1 and ω_2 can be illustrated by drawing the contour plot of the amplitude response of a hypothetical circularly symmetric filter as shown in Fig. 1. Drawing the amplitude response in this manner provides a means of studying the effects of transformations having the form of (4).

Fig. 1. Periodic spatial frequency response.

First, consider the effect of (4) on the origin $(\omega_1,\omega_2) = (0,0)$. The $(\pm 1)_i$ in G_i can be represented as $e^{j\theta_i}$ where θ_i is 0 or $\pm\pi$ if the sign is positive or negative, respectively. The origin is mapped to the point (θ_1,θ_2). Thus the transformation can be used with proper choice of signs to shift the origin of the resulting amplitude response to the points $(0,0)$, $(0,\pm\pi)$, $(\pm\pi,0)$, or $(\pm\pi,\pm\pi)$ on the original response.

Now define N_i as the highest degree of z_1 in G_i after multiplying the terms together. Similarly let M_i be the highest degree of z_2 in G_i. Notice that because of the form of (4) there will be a term $z_1^{N_i} z_2^{M_i}$ in G_i.

The parameters N_i and M_i determine the fundamental nature of the transformation. This can be seen by using the Cauchy mapping theorem [28] to evaluate the effect of transformation on lines in the region $-\pi \leq \omega_1, \omega_2 \leq \pi$. For example, take the ω_1 axis from $-\pi$ to $+\pi$ in Fig. 1. This corresponds to the closed contour $(e^{-j\omega_1},1)$ in C^2. The function $G_1(z_1,1)$ has N_1 zeros and N_1 poles. Because of the form of G_1 the zeros are reciprocals of the poles. But G_1 is stable so all the poles must be outside the z_1 unit circle and therefore all the zeros must be inside. By the Cauchy mapping theorem the contour $(e^{-j\omega_1},1)$ is mapped N_1 times around the unit circle in the $G_1(z_1,1)$ plane. Similarly, that same contour is mapped N_2 times around the unit circle in the $G_2(z_1,1)$ plane.

Ignoring the shift of the origin which was described earlier, the line in Fig. 1 from $(-\pi,0)$ to $(\pi,0)$ is mapped to a locus from $(-N_1\pi,-N_2\pi)$ to $(N_1\pi,N_2\pi)$. In the same way we can determine that the line from $(0,-\pi)$ to $(0,\pi)$ is mapped to a locus from $(-M_1\pi,-M_2\pi)$ to $(M_1\pi,M_2\pi)$ while the line $(-\pi,-\pi)$ to (π,π) is mapped to a locus from $(-(N_1 + M_1)\pi,-(N_2 + M_2)\pi)$ to $((N_1 + M_1)\pi,(N_2 + M_2)\pi)$.

From this discussion and Fig. 1 it is apparent that the values $N_i + M_i$ for $i = 1,2$ determine the number of stop and pass regions included in $-\pi \leq \omega_1$, $\omega_2 \leq \pi$ after transformation of say a low-pass filter. Discussion will be limited to $N_i + M_i \leq 2$, $i = 1,2$. This will cover the transformations of greatest importance. If a larger number of pass or stop regions are needed, a larger $N_i + M_i$ could be chosen and the analysis that follows could be extended.

In order to simplify the analysis and present transformations in an easily used format, the general form of (4) will be treated in cases.

Single Variable Transformations

First we will treat transformations of the form

$$Z_1 = G_1(z_1) = \pm z_1{}^N \frac{\left(1 + \sum_{j=1}^{N} a_j{}^{(1)} z_1{}^{-j}\right)}{\left(1 + \sum_{j=1}^{N} a_j{}^{(1)} z_1{}^{j}\right)}, \qquad N = 1,2$$

$$Z_2 = G_1(z_2) = \pm z_2{}^M \frac{\left(1 + \sum_{j=1}^{M} a_j{}^{(2)} z_2{}^{-j}\right)}{\left(1 + \sum_{j=1}^{M} a_j{}^{(2)} z_2{}^{j}\right)}, \qquad M = 1,2.$$

Of course the variables z_1 and z_2 could be interchanged but that would simply interchange the axis labels after transformation.

These transformations work on the variables Z_1 and Z_2 independently. Therefore relationships for the parameters can be obtained from the one-dimensional transformations developed by Constantinides [21] with reference to Fig. 1 and the previous discussion.

The transformations have been cataloged for convenience in Table I. The drawings illustrate the effect of the given transformations on a hypothetical circularly symmetric filter which could be a high pass, low pass, or bandpass depending on the contour labels.

Notice that each transformation is completely determined by the mapping of one or two points. These points are shown in Table I. Because of the limited number of parameters it is apparent that there are not sufficient degrees of freedom to specify a transformation to produce any arbitrary frequency response from a given one. For example, it is not possible to obtain a circularly symmetric high pass from a circularly symmetric low pass. However, many desirable responses can be obtained as illustrated in Table I and the following examples.

Example 1

Consider the low-pass transfer function obtained by Maria and Fahmy [29]:

$$H(Z_1,Z_2) = (0.00895)\frac{[1 \;\; Z_1 \;\; Z_1{}^2]\begin{bmatrix} 1 & -1.62151 & 0.99994 \\ -1.62151 & 2.63704 & -1.62129 \\ 0.99994 & -1.62129 & 1.00203 \end{bmatrix}\begin{bmatrix} 1 \\ Z_2 \\ Z_2{}^2 \end{bmatrix}}{[1 \;\; Z_1 \;\; Z_1{}^2]\begin{bmatrix} 1 & -1.78813 & 0.82930 \\ -1.78813 & 3.20640 & -1.49271 \\ 0.82930 & -1.49271 & 0.69823 \end{bmatrix}\begin{bmatrix} 1 \\ Z_2 \\ Z_2{}^2 \end{bmatrix}}. \tag{5}$$

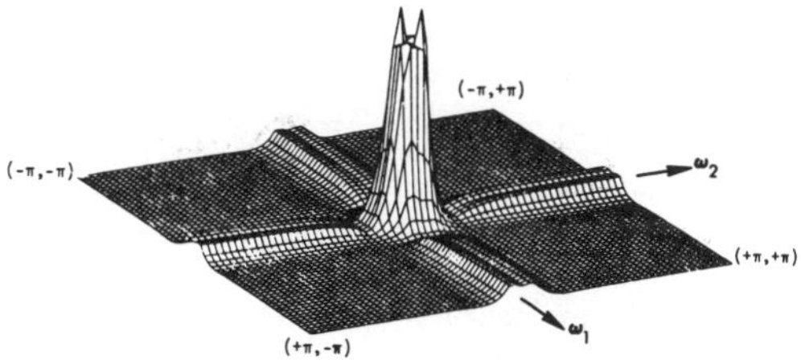

Fig. 2. Original low-pass amplitude response of (5).

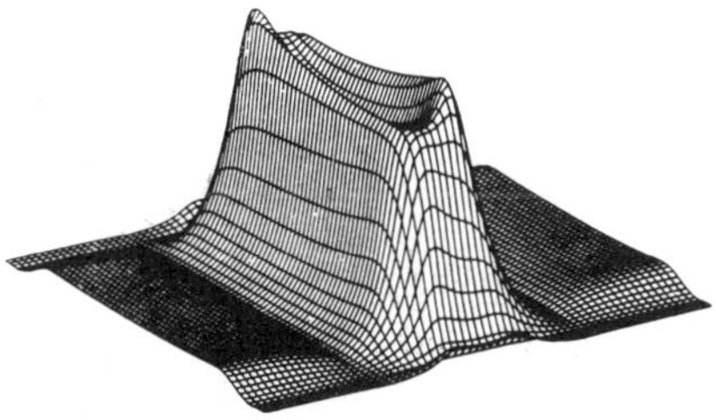

Fig. 3. Result of type I transformation of (5).

Its magnitude response is shown in Fig. 2. It has a value of 0.6 at $\pm 0.08\pi$ on each axis and the peaks are approximately 1.27 in magnitude.

A low pass is desired with magnitude 0.6 at $\pm 0.75\pi$ on the ω_1 axis and $\pm 0.2\pi$ on the ω_2 axis. Using the type I transformation in Table I the transformation is

$$Z_1 = \frac{0.90055 + z_1}{1 + 0.90055 z_1}$$

$$Z_2 = \frac{0.44009 + z_2}{1 + 0.44009 z_2}.$$

This results in a new transfer function

$$H(Z_1,Z_2) = (0.00895)\frac{[1 \;\; Z_1 \;\; Z_1{}^2]\begin{bmatrix} 0.171824 & -0.0513838 & 0.173302 \\ 0.326567 & -0.0960307 & 0.329836 \\ 0.171908 & -0.0509964 & 0.1737713 \end{bmatrix}\begin{bmatrix} 1 \\ Z_2 \\ Z_2{}^2 \end{bmatrix}}{[1 \;\; Z_1 \;\; Z_1{}^2]\begin{bmatrix} 0.0232600 & -0.0327672 & 0.0144259 \\ 0.0207049 & -0.0302811 & 0.0138389 \\ 0.0107713 & -0.0163187 & 0.00797696 \end{bmatrix}\begin{bmatrix} 1 \\ Z_2 \\ Z_2{}^2 \end{bmatrix}}.$$

The new frequency response is shown in Fig. 3.

Example 2

In this example we start with the transfer function of (5). The desired filter is to have two bandpass regions on the ω_1 axis each centered at $\pm 0.6\pi$ with a bandwidth of 0.6π between points on the ω_1 axis with magnitude 0.6. The width of the passband in the ω_2 direction between points of amplitude 0.6 should be 0.08π. A type VIII transformation will meet these requirements and is given by

$$Z_1 = -\frac{-0.832 + 0.052z_1 + z_1^2}{1 + 0.052z_1 - 0.832z_1^2}$$

$$Z_2 = z_2.$$

Let us specify that the transformation maps the ω_1 axis and ω_2 axis onto the 45° line $\omega_1 = \omega_2$. A third point of correspondence will be left arbitrary. Thus we have

$$\begin{aligned} G\colon (\omega,0) &\mapsto (\omega,\omega) \\ (0,\omega) &\mapsto (\omega,\omega) \\ (\omega_1,\omega_2) &\mapsto (\theta_1,\theta_2). \end{aligned} \tag{7}$$

Placing these values into (6) produces

$$\alpha_i = \beta_i = -\gamma_i \tag{8}$$

and

$$\alpha_i = \frac{e^{-j(\omega_1+\omega_2)} - e^{-j\theta_i}}{e^{-j(\omega_1+\theta_i)} - e^{-j(\omega_2+\theta_i)} - e^{-j(\omega_1+\omega_2+\theta_i)} - e^{-j\omega_2} - e^{-j\omega_1} + 1}, \qquad i = 1,2. \tag{9}$$

The transformation produces the new transfer function:

$$H(Z_1,Z_2) = (0.00895)\frac{[1\ \ Z_1\ \ Z_1^2\ \ Z_1^3\ \ Z_1^4]\begin{bmatrix} 0.343086 & -0.549789 & 0.344656 \\ 0.0316427 & -0.0513887 & 0.0314537 \\ -0.574150 & 0.917658 & -0.577895 \\ 0.0316313 & -0.0513453 & 0.0318519 \\ 0.343068 & -0.549271 & 0.345297 \end{bmatrix}\begin{bmatrix} 1 \\ Z_2 \\ Z_2^2 \end{bmatrix}}{[1\ \ Z_1\ \ Z_1^2\ \ Z_1^3\ \ Z_1^4]\begin{bmatrix} 0.0863372 & -0.153695 & 0.0706968 \\ 0.0478634 & -0.0\ 48154 & 0.0388711 \\ -0.008257 & 0.015829 & -0.0076435 \\ 0.0153403 & -0.0285396 & 0.0138986 \\ 0.0337998 & -0.0627717 & 0.0303566 \end{bmatrix}\begin{bmatrix} 1 \\ Z_2 \\ Z_2^2 \end{bmatrix}}.$$

The new amplitude response is shown in Fig. 4.

Two Variable Transformations

Next consider transformations of the form

$$\begin{aligned} Z_i &= G_i(z_1,z_2) \\ &= (\pm)_i \frac{\gamma_i + \beta_i z_1 + \alpha_i z_2 + z_1 z_2}{1 + \alpha_i z_1 + \beta_i z_2 + \gamma_i z_1 z_2}, \qquad i = 1,2. \end{aligned} \tag{6}$$

The parameters are determined by the mapping of three appropriately chosen points. The points must be chosen so that the solution for $\alpha_i,\beta_i,\gamma_i$ is unique and produces stable G_1 and G_2.

Unfortunately, in this form (6) is not suitable for use in design. The specification of the parameters should be easily related to the effect of transformation; however, in two dimensions the cutoff frequencies and other things of interest are not simple points but are continuous lines or regions. Thus the action of the transformation on just three points does not give all of the needed information. This can be remedied by relating parameter choices to the effect of transformation on certain lines in the response plane.

These comments are also true for the one variable transformations already discussed. However, in that case the independence of the transformations on each variable made the effect of transformation easy to predict.

In what follows the signs in (6) will be assumed positive. It is a simple matter to account for negative signs by shifting the origin before transforming.

Huang [26] has shown that the G_i in (6) are stable if and only if

$$|\alpha_i| < 1$$

$$\left|\frac{1 + \alpha_i}{\beta_i + \gamma_i}\right| > 1$$

$$\left|\frac{1 - \alpha_i}{\beta_i - \gamma_i}\right| > 1, \qquad i = 1,2.$$

Thus from (8) the transformation obtained is stable if and only if $-1 < |\alpha_i| < 1/3$ for $i = 1,2$.

The transformation above also maps the −45° line $\omega_1 = -\omega_2$ to the point (0,0). In addition if the point $(\pi/2,\pi/2)$ is chosen to be mapped to the point (θ_1,θ_2), then θ_1 and θ_2 will produce $-1 < |\alpha_i| < 1/3$, $i = 1,2$ if both are less than 2π but greater than approximately 0.7048π. For $(\pi/2,\pi/2)$ a simpler version of (9) can be obtained from (6),

$$\alpha_i = -\frac{\tan\left(\dfrac{\theta_i - \pi}{2}\right)}{2 + \tan\left(\dfrac{\theta_i - \pi}{2}\right)}. \tag{10}$$

Fig. 5 shows the mapping by G of the various lines discussed. The shaded area indicates the region into which the point $(\pi/2,\pi/2)$ can be mapped for a stable result. The dashed line shows a possible mapping of the 45° line $\omega_1 = \omega_2$.

TABLE I
ONE VARIABLE SPECTRAL TRANSFORMATIONS

	Transformation equations		Transformation effect on 
I	$Z_1=\frac{a_1+z_1}{1+a_1z_1}$ $Z_2=\frac{a_2+z_2}{1+a_2z_2}$	$a_i=\frac{-\sin\frac{1}{2}(\beta_i-\omega_{i_c})}{\sin\frac{1}{2}(\beta_i+\omega_{i_c})}$ $i=1,2$	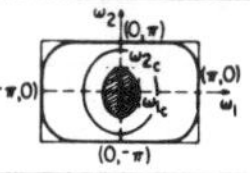
II	$Z_1=-\frac{a_1+z_1}{1+a_1z_1}$ $Z_2=-\frac{a_2+z_2}{1+a_2z_2}$	$a_i=-\frac{\cos\frac{1}{2}(\beta_i-\omega_{i_c})}{\cos\frac{1}{2}(\beta_i+\omega_{i_c})}$	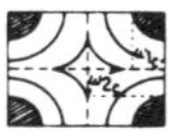
III	$Z_1=-\frac{a_1+z_1}{1+a_1z_1}$ $Z_2=\frac{a_2+z_2}{1+a_2z_2}$	$a_1=-\frac{\cos\frac{1}{2}(\beta_1-\omega_{1_c})}{\cos\frac{1}{2}(\beta_1+\omega_{1_c})}$ $a_2=\frac{-\sin\frac{1}{2}(\beta_2-\omega_{2_c})}{\sin\frac{1}{2}(\beta_2+\omega_{2_c})}$	
IV	$Z_1=\frac{\frac{1-K_1}{1+K_1}-\frac{2a_1}{1+K_1}z_1+z_1^2}{1-\frac{2a_1}{1+K_1}z_1+\frac{1-K_1}{1+K_1}z_1^2}$ $Z_2=\frac{\frac{1-K_2}{1+K_2}-\frac{2a_2}{1+K_2}z_2+z_2^2}{1-\frac{2a_2}{1-K_2}z_2+\frac{1-K_2}{1+K_2}z_2^2}$	$a_i=\frac{\cos(\frac{\omega_i''+\omega_i'}{2})}{\cos(\frac{\omega_i''-\omega_i'}{2})}=\cos\omega_{i_0}$ $K_i=\tan(\frac{\omega_i''+\omega_i'}{2})\tan(\frac{\beta_i}{2})$	
V	$Z_1=\frac{\frac{1-K_1}{1+K_1}-\frac{2a_1}{1+K_1}z_1+z_1^2}{1-\frac{2a_1}{1+K_1}z_1+\frac{1-K_1}{1+K_1}z_1^2}$ $Z_2=\frac{a_2+z_2}{1+a_2z_2}$	a_1, K_1 chosen as in IV a_2 chosen as in I	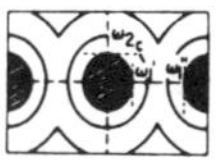
VI	$Z_1=\frac{\frac{1-K_1}{1+K_1}-\frac{2a_1}{1+K_1}z_1+z_1^2}{1-\frac{2a_1}{1+K_1}z_1+\frac{1-K_1}{1+K_1}z_1^2}$ $Z_2=-\frac{a_2+z_2}{1+a_2z_2}$	a_1, K_1 chosen as in IV a_2 chosen as in II	
VII	$Z_1=-\left(\frac{\frac{1-K_1}{1+K_1}-\frac{2a_1K_1}{1+K_1}z_1+z_1^2}{1-\frac{2a_1K_1}{1+K_1}z_1-\frac{1-K_1}{1+K_1}z_1^2}\right)$ $Z_2=-\left(\frac{-\frac{1-K_2}{1+K_2}-\frac{2a_2K_2}{1+K_2}z_2+z_2^2}{1-\frac{2a_2K_2}{1+K_2}z_2-\frac{1-K_2}{1+K_2}z_2^2}\right)$	$a_i=\frac{\cos(\frac{\omega_i''-\omega_i'}{2})}{\cos(\frac{\omega_i''+\omega_i'}{2})}=\cos\omega_{i_0}$ $K_i=\cot(\frac{\omega_i''-\omega_i'}{2})\tan(\frac{\beta_i}{2})$	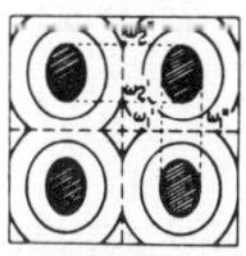
VIII	$Z_1=-\left(\frac{-\frac{1-K_1}{1+K_1}-\frac{2a_1K_1}{1+K_1}z_1+z_1^2}{1-\frac{2a_1K_1}{1+K_1}z_1-\frac{1-K_1}{1+K_1}z_1^2}\right)$ $Z_2=\frac{a_2+z_2}{1+a_2z_2}$	a_1, K_1 chosen as in VII a_2 chosen as in I	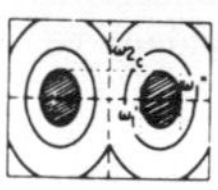
IX	$Z_1=-\left(\frac{-\frac{1-K_1}{1+K_1}-\frac{2a_1K_1}{1+K_1}z_1+z_1^2}{1-\frac{2a_1K_1}{1+K_1}z_1-\frac{1-K_1}{1+K_1}z_1^2}\right)$ $Z_2=-\left(\frac{a_2+z_2}{1+a_2z_2}\right)$	a_1, K_1 chosen as in VII a_2 chosen as in II	
X	$Z_1=-\left(\frac{-\frac{1-K_1}{1+K_1}-\frac{2a_1K_1}{1+K_1}z_1+z_1^2}{1-\frac{2a_1K_1}{1+K_1}z_1-\frac{1-K_1}{1+K_1}z_1^2}\right)$ $Z_2=\frac{\frac{1-K_2}{1+K_2}-\frac{2a_2}{1+K_2}z_2+z_2^2}{1-\frac{2a_2}{1+K_2}z_2+\frac{1-K_2}{1+K_2}z_2^2}$	a_1, K_1 chosen as in VII a_2, K_2 chosen as in IV	

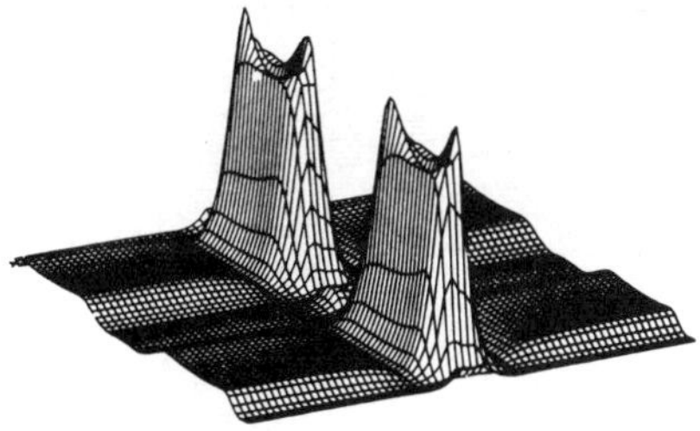

Fig. 4. Result of type VIII transformation of (5).

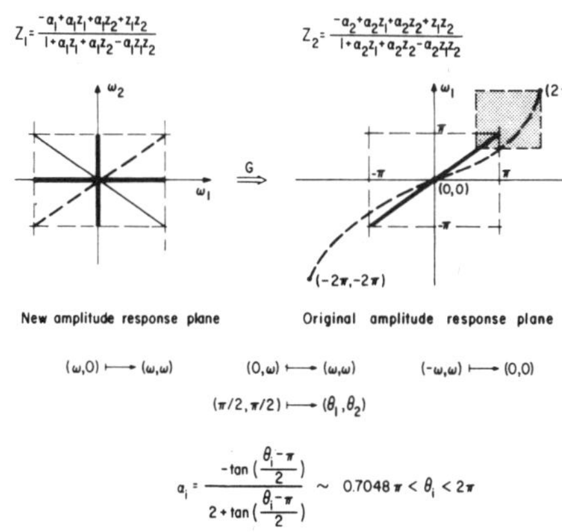

Fig. 5. Two variable transformation.

The following example will demonstrate the above by showing the effect of transformation of a high-pass filter.

Example 3

Consider the transfer function of a high-pass filter [29]:

$$H(Z_1,Z_2) = 0.59944\,\frac{[1\ \ Z_1]\begin{bmatrix}1 & -0.04668\\ -0.04668 & -0.46761\end{bmatrix}\begin{bmatrix}1\\ Z_2\end{bmatrix}}{[1\ \ Z_1]\begin{bmatrix}1 & 0.42602\\ 0.42602 & -0.10692\end{bmatrix}\begin{bmatrix}1\\ Z_2\end{bmatrix}}.$$

A contour plot of the amplitude response is shown in Fig. 6. The point $(\pi/2,\pi/2)$ is to be mapped to $(3\pi/4,15\pi/8)$. Therefore from (6), (8), and (10) the transformation becomes

$$Z_1 = \frac{-0.2612 + 0.2612z_1 + 0.2612z_2 + z_1z_2}{1 + 0.2612z_1 + 0.2612z_2 - 0.2612z_1z_2}$$

$$Z_2 = \frac{0.7154 - 0.7154z_1 - 0.7154z_2 + z_1z_2}{1 - 0.7154z_1 - 0.7154z_2 + 0.7154z_1z_2}.$$

The resulting transfer function is

$$H(Z_1,Z_2) = (0.59944)\,\frac{[1\ \ Z_1\ \ Z_1^2]\begin{bmatrix}1.06618 & -0.625201 & -0.0820382\\ -0.625201 & 0.0018205 & 0.58987\\ -0.0820382 & 0.58987 & -0.675675\end{bmatrix}\begin{bmatrix}1\\ Z_2\\ Z_2^2\end{bmatrix}}{[1\ \ Z_1\ \ Z_1^2]\begin{bmatrix}1.21348 & -0.528443 & -0.326097\\ -0.528443 & 0.446268 & 0.388004\\ -0.326097 & 0.388004 & -0.100284\end{bmatrix}\begin{bmatrix}1\\ Z_2\\ Z_2^2\end{bmatrix}}.$$

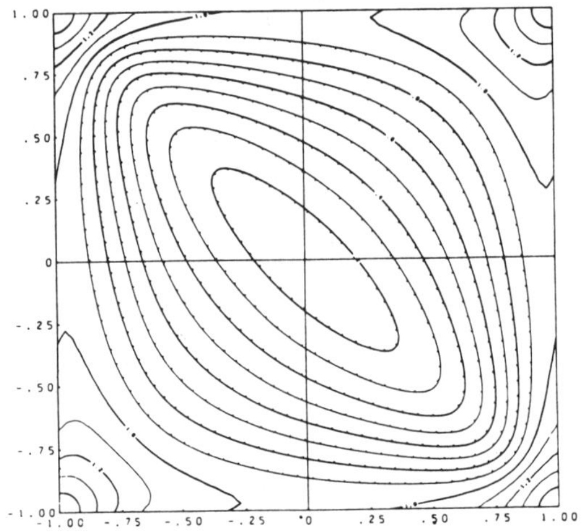

Fig. 6. Contour plot of original high-pass amplitude response of example 3.

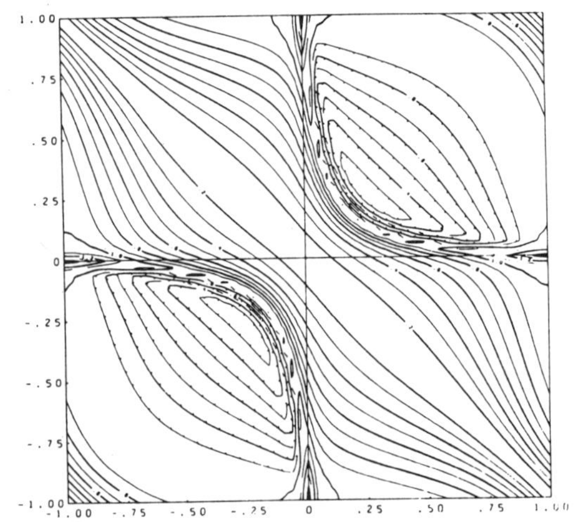

Fig. 7. Result of two variable transformation of example 3.

The contour plot of the resulting amplitude response is shown in Fig. 7.

This example was chosen because of the spacing of the contour lines. Comparison should be made of Figs. 6 and 7 with reference to Fig. 5 to develop an understanding of this transformation.

The two variable transformation described above can be made more general by using one variable transformation both before and after the two variable transformation. Transforming first by a one-variable form allows modification of the response in the neighborhood of the 45° line which will dominate the response after the above two variable form. Following by another one variable transformation makes possible further adjustment of the final response.

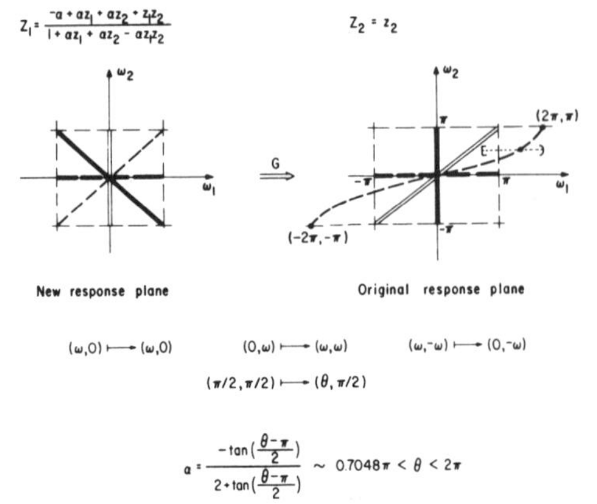

Fig. 8. Hybrid two variable and one variable transformations.

Cascading one variable and two variable transformations increases the control one has over the resulting filter response. However, there are still only a limited number of parameters for specification. Therefore, as with the one variable transformations, it is not possible to control the exact shape of a particular contour line. Yet it is possible (within the stability constraint) to precisely determine one or more points of a contour line so that it approximates the desired shape.

Another transformation can be found which is a hybrid of those so far considered. Take

$$Z_1 = G_1(z_1,z_2) = \frac{-\alpha + \alpha z_1 + \alpha z_2 + z_1 z_2}{1 + \alpha z_1 + z_2 - \alpha z_1 z_2}$$

$$Z_2 = G_2(z_2) = z_2. \tag{11}$$

As before, a specific case was chosen in order to develop a predictable transformation. In this instance the effect on the ω_1 and ω_2 axis and the $-45°$ line is given by

$$\begin{aligned} G: (\omega,0) &\mapsto (\omega,0) \\ (0,\omega) &\mapsto (\omega,\omega) \\ (\omega_1,\omega_2) &\mapsto (\theta,\omega_2) \\ (\omega,-\omega) &\mapsto (0,-\omega). \end{aligned} \tag{12}$$

These relations are illustrated in Fig. 8 and α is calculated according to (9) or (10) as before. The point $(\pi/2,\pi/2)$ is mapped to $(0,\pi/2)$ with θ between 2π and approximately 0.7048π for stability. The dashed line in Fig. 8 shows a possible mapping of the 45° line and the interval over which the $(\pi/2,\pi/2)$ point may be mapped.

Again the transformation may be extended to produce additional response shapes by doing a one variable transformation before and after the above transformation.

An example will be offered next to clarify the previous comments.

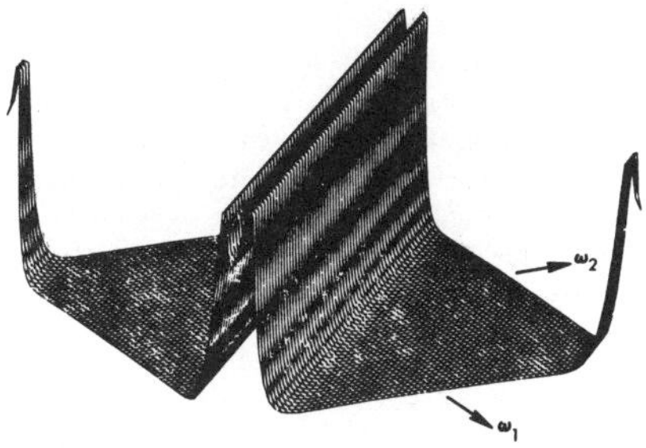

Fig. 9. Result of transformation of (5) using one variable and two variable transformations.

Example 4

Take the low-pass transfer function of (5). The passband is first widened to have the 0.6 magnitude points at $\pm 0.16\pi$ on each axis. This is done using the one variable transformation

$$Z_1 = \frac{0.3405 + z_1}{1 + 0.3405 z_1}$$

$$Z_2 = \frac{0.3405 + z_2}{1 + 0.3405 z_2}.$$

Then a simple two variable transformation is used to place the passband along the $-45°$ line.

$$Z_1 = z_1 z_2$$

$$Z_2 = z_1 z_2.$$

The new tranofer function is

$$H(Z_1,Z_2) = (0.00895) \frac{[1 \;\; Z_1 \;\; Z_1^2 \;\; Z_1^3 \;\; Z_1^4] \begin{bmatrix} 0.318817 & 0 & 0 & 0 & 0 \\ 0 & -0.498350 & 0 & 0 & 0 \\ 0 & 0 & 0.849219 & 0 & 0 \\ 0 & 0 & 0 & -0.495472 & 0 \\ 0 & 0 & 0 & 0 & 0.320952 \end{bmatrix} \begin{bmatrix} 1 \\ Z_2 \\ Z_2^2 \\ Z_2^3 \\ Z_2^4 \end{bmatrix}}{[1 \;\; Z_1 \;\; Z_1^2 \;\; Z_1^3 \;\; Z_1^4] \begin{bmatrix} 0.237861 & 0 & 0 & 0 & 0 \\ 0 & -0.729238 & 0 & 0 & 0 \\ 0 & 0 & 0.888077 & 0 & 0 \\ 0 & 0 & 0 & -0.509699 & 0 \\ 0 & 0 & 0 & 0 & 0.118004 \end{bmatrix} \begin{bmatrix} 1 \\ Z_2 \\ Z_2^2 \\ Z_2^3 \\ Z_2^4 \end{bmatrix}}.$$

The new amplitude response is shown in Fig. 9.

Two-Dimensional Digital Filters with Tunable Characteristics

Besides designing new filters from those previously obtained, there are additional uses for two-dimensional digital spectral transformations.

One important application is in the design of two-dimensional digital filters with tunable characteristics. The type I transformation for example can be used with a given circularly symmetric high-pass (or low-pass) transfer function to produce an approximately elliptically or circularly symmetric high pass (or low pass) with passband edges tunable to any values on the ω_1 or ω_2 axis.

A very flexible filter can be obtained from a cascade of a type I tunable circularly symmetric high-pass and low-pass filter. The resulting bandpass filter would have tunable center frequency and variable bandwidth along each axis. Such a filter would be capable of a very wide range of passband shapes. Other transformations could likewise be used as desired to produce other interesting types of tunable filters.

The most straightforward method of realization of a tunable two-dimensional digital filter is to make each coefficient in the realization adjustable and provide a means of computing each from the tuning parameters. It is simple to write a computer program to do the polynomial multiplication involved in making the spectral transformation. The coefficients can then be recomputed easily as desired. If the original filter has already been realized and if a type I spectral transformation was used, then few changes beyond adjusting the coefficients would be needed to tune the filter. Second-order or two variable transformations would require more extensive revision of the realization because of the increased order.

An alternative approach for changing a realization to introduce a spectral transformation is to replace all Z_1 and Z_2 blocks by realizations of the all-pass functions representing the transformation. This technique makes the tuning parameters directly adjustable parts of the realization. This is very often done in one-dimension and in that case is straightforward. In two dimensions some additional complication results, however.

To illustrate this point, consider a transfer function of the form

$$H(Z_1,Z_2) = \frac{a_{11} + a_{12}Z_2 + a_{21}Z_1 + a_{22}Z_1Z_2}{1 + b_{12}Z_2 + b_{13}Z_2^2 + b_{21}Z_1 + b_{22}Z_1Z_2 + b_{23}Z_1Z_2^2}. \tag{13}$$

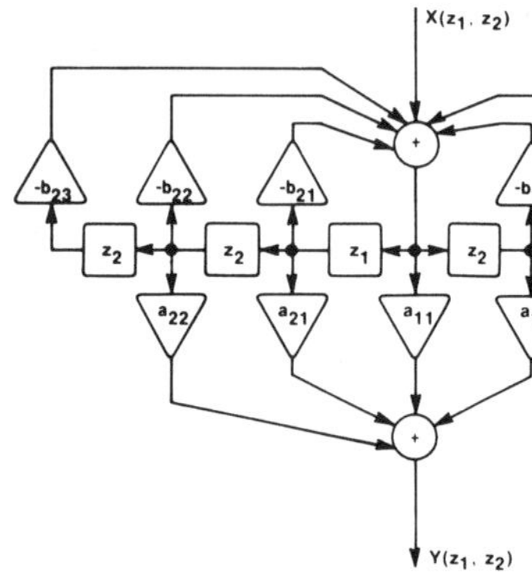

Fig. 10. Cannonic realization of (13).

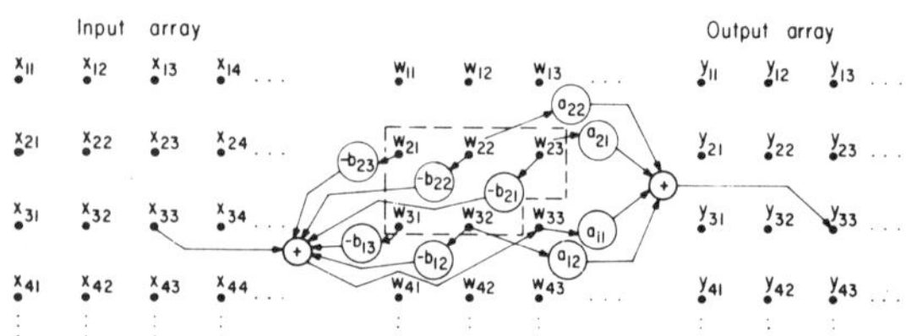

Fig. 11. Another representation of Fig. 8.

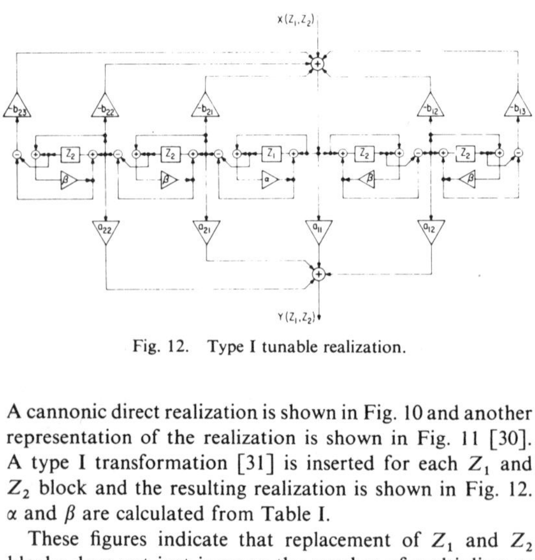

Fig. 12. Type I tunable realization.

A cannonic direct realization is shown in Fig. 10 and another representation of the realization is shown in Fig. 11 [30]. A type I transformation [31] is inserted for each Z_1 and Z_2 block and the resulting realization is shown in Fig. 12. α and β are calculated from Table I.

These figures indicate that replacement of Z_1 and Z_2 blocks does not just increase the number of multipliers as might have been expected. Additional memory is required because the replacement of Z_1 and Z_2 blocks no longer allows the sharing of storage arrays in Fig. 12 as could be done in Fig. 10 and Fig. 11.

Another important application of two-dimensional digital spectral transformations lies in obtaining starting points for an optimization algorithm such as that proposed by Maria and Fahmy [29]. Because of the nonlinear nature of the approximation problem for two-dimensional IIR digital filters, the result from an iterative routine is very sensitive to the initial value chosen. The starting point can radically affect the number of iterations required and whether or not the convergence will be to an optimal or good suboptimal value. Recent examples by Brophy and Salazar have demonstrated this sensitivity in several one-dimensional examples [32].

Conclusions

The development and examples show that filters with many different amplitude responses can be obtained by transformations of a single stable filter. The examples demonstrate that spectral transformations can be used quickly to design a filter approximating a new response if an appropriate previous design is available. It has also been shown that spectral transformations can be used for design and realization of two-dimensional digital filters with tunable characteristics. In addition, it has been pointed out that spectral transformations can provide a useful means of finding a starting point for an iterative design algorithm.

References

[1] A. S. Samulon, "Separation of man-made and natural patterns in high altitude imagery of agricultural areas," *IEEE Trans. Circuits Syst.*, Mar. 1975.
[2] G. J. McMurtry, "Interdisciplinary analysis and interpretations of earth resources technology satellite (ERTS) data," *IEEE Int. Conf. Commun. Rec.*, June 1974, pp. 3881–3883.
[3] R. M. Hoffer, "Techniques for computer-aided analysis of ERTS-1 data, useful in geologic, forest and water resource surveys," Purdue University, Layfayette, IN, LARS Inf. Note 121073, 1973.
[4] R. Elefsen *et al.*, "Urban land-use mapping by machine processing of ERTS-1 multispectral data: A San Francisco Bay area example," Purdue University, Layfayette, IN, LARS Inf. Note 101573, 1973.
[5] G. W. Wecksung and K. Campbell, "Digital image processing at EG & G," *Computer*, vol. 7, no. 5, pp. 63–71, May 1974.
[6] F. C. Billingsly, "Applications of digital image processing," *Appl. Opt.*, vol. 9, Feb. 1970.
[7] R. W. Selzer, "Improving biomedical image quality with computers," NASA Tech. Rep. 32-1336, 1968.
[8] B. H. Mayall, "Digital image processing at Lawrence Livermore Laboratory, Part II: Biomedical applications," *Computer*, vol. 7, no. 5, pp. 81–87, May 1974.
[9] B. P. Hildabrand, *An Introduction to Acoustical Holography*. New York: Plenum, 1972.
[10] A. L. Boyer *et al.*, "Computer reconstruction of images from ultrasonic holograms," in *Acoustical Holography*, vol. 2. New York: Plenum Press, 1970, chap. 15.
[11] C. S. Clark and A. F. Metherell, "Digital image processing of acoustical holograms," *Acoustical Holography*, vol. 5. New York: Plenum Press, 1970, pp. 471–480.
[12] B. R. Hunt and D. H. Janney, "Digital image processing at Los Alamos Scientific Laboratory," *Computer*, vol. 7, no. 5, pp. 57–62, May 1974.
[13] M. P. Ekstrom, "Digital image processing at Lawrence Livermore Laboratory, Part I: Diagnostic radiography applications," *Computer*, vol. 7, no. 5, pp. 72–78, May 1974.
[14] S. Treitel, J. L. Shanks, and C. W. Frasier, "Some aspects of fan filtering," *Geophysics*, vol. 32, pp. 789–800, Oct. 1967.
[15] E. G. Zurflueh, "Applications of two-dimensional linear wavelength filtering," *Geophysics*, vol. 32, pp. 1015–1035, Dec. 1967.
[16] D. Silverman, "The digital processing of seismic data," *Geophysics*, vol. 32, pp. 988–1002, Dec. 1967.
[17] M. B. Dobrin, *Introduction to Geophysical Prospecting*. New York: McGraw-Hill, 1960.
[18] J. L. Holloway, Jr., "Smoothing and filtering of time series and space fields," in *Advances in Geophysics*, vol. 4. New York: Academic Press, 1958, pp. 351–389.
[19] A. V. Oppenheim, R. W. Schafer, *Digital Signal Processing*. Englewood Cliffs, NJ: Prentice-Hall, 1975.
[20] P. Pistor, "Stability criterion for recursive filters," *IBM J. Res. Develop.*, vol. 18, pp. 59–71, Jan. 1974.

[21] A. G. Constantinides, "Spectral transformations for digital filters, *Proc. IEE*, vol. 117, pp. 1585–1590, Aug. 1970.
[22] W. C. Yengst, *Procedures of Modern Network Synthesis*. New York: Macmillan, 1964.
[23] L. Weinberg, *Network Analysis and Synthesis*. New York: McGraw-Hill, 1962.
[24] E. I. Jury, *Inners and Stability of Dynamic Systems*. New York: Wiley, 1974, chap. 6.
[25] J. L. Shanks, S. Treitel, and J. H. Justice, "Stability and synthesis of two-dimensional recursive filters," *IEEE Trans. Audio Electroacoust.* vol. AU-20, pp. 115–128, June 1972.
[26] T. S. Huang, "Stability of two-dimensional recursive filters," *IEEE Trans. Audio Electroacoust.*, vol. AU-20, pp. 158-163, June 1972.
[27] R. C. Gunning and H. Rossi, *Analytic Functions of Several Complex Variables*. Englewood Cliffs, NJ: Prentice-Hall, 1965, chap. 1.
[28] E. I. Jury, *Sampled-Data Control Systems*. New York: Wiley, 1958, chap. 1.
[29] G. A. Maria and M. M. Fahmy, "An *lp* design technique for two-dimensional digital recursive filters," *IEEE Trans. Acoust., Speech, Signal Processing*, vol. ASSP-22, pp. 15–21, Feb. 1974.
[30] S. K. Mitra, A. D. Sagar, and N. A. Pendergrass, "Realizations of two-dimensional recursive digital filters." *IEEE Trans. Circuits Syst.*, vol CAS-22, pp. 177–184, Mar. 1975.
[31] S. K. Mitra and K. Hirano, "Digital all-pass networks," *IEEE Trans. Circuits Syst.*, vol. CAS-21, pp. 688–700, Sept. 1974.
[32] F. Brophy and A. C. Salazar, "Recursive digital filter synthesis in the time domain," *IEEE Trans. Acoust., Speech, Signal Processing*, vol. ASSP-22, Feb. 1974.

12

Reprinted from *IEEE Trans. Circuits Syst.* CAS-22(6):572–574 (1975)

SPATIAL RECURSIVE FILTER DESIGN VIA RATIONAL CHEBYSHEV APPROXIMATION

J. B. Bednar

***Abstract*—The computer implementation of a numerical algorithm for obtaining rational approximations is discussed. This method offers the promise of a comprehensive solution to the problems of two-dimensional recursive digital filter design. Combination of this technique with an easily implemented stability test avoids current stability problems in present methods and places no restrictions as to type on filters that may be designed. Both one- and two-dimensional filters may be designed with this technique.**

I. Introduction

The area of digital filtering is becoming an important topic of research and study. This is due to the widespread applications of such filters. While one-dimensional digital filters have been extensively researched, it is only recently that such investigation is being extended to two-dimensional digital filters [1]–[5].

An important problem in the design of two-dimensional digital filters is the problem of stability of the filter. This problem, which is more complicated than in one dimension, has been the subject of several investigations [4], [5]. The general criterion was obtained by Shanks *et al.* [13] (see also [4] and [5]) and states that if $B(z_1,z_2)$ is the double (z_1,z_2) transform of a filter B, then $1/B(z_1,z_2)$ is stable provided that B does not vanish on the unit bicylinder $D = \{(z_1,z_2): |z_1| \leq 1 \text{ and } |z_2| \leq 1\}$.

Recursive filter synthesis is essentially the ability to calculate a stable rational approximation to a given function of two complex variables. Precisely, we must find two-dimensional polynomials $A(z_1,z_2)$, $B(z_1,z_2)$ such that $1/B(z_1,z_2)$ is stable and the difference between the desired filter function $F(z_1,z_2)$ and the rational function $A(z_1,z_2)/B(z_1,z_2)$ is minimized in some sense. The sense we will use here is that of Chebyshev, so we will be attempting to minimize an expression of the form

$$\Delta = \max_{(z_1,z_2) \text{ in } D} |F(z_1,z_2) - A(z_1,z_2)/B(z_1,z_2)|$$

over the coefficients of the filter A/B. This then becomes a classical problem [6] in approximation theory and part of what we do in this letter is extend the results of Cheney and Loeb [6] to the two-dimensional case.

The Cheney–Loeb algorithm [6] we have in mind, comes in two forms: the one we present and the one called differential

Manuscript received January 23, 1974; revised July 1, 1974. This work was supported in part by the NSF under Grant GK34763.
The author is with the University of Tulsa, Tulsa, Okla. 74104.

correction presented by Barrodale *et al.* [11]. Both versions will produce similar results, but we have as yet been unable to prove stability over the entire bicylinder for the one in [11] although we know of no situation in which it has failed to work properly.

The nature of both of these algorithms is such that an occurrence of instability at any iteration means that the algorithm has converged to an optimum solution. The algorithms are best implemented on the surface $D_s = \{(z_1,z_2): |z_1| = 1 = |z_2|\}$ of the unit bicylinder and so a convenient stability test can serve as an automatic shutdown condition. A stability checking algorithm is presented in a form that makes it highly compatible with these design techniques.

II. Properties of the Design Algorithm

The proof of the algorithm we present has been given in [9], so we summarize only its salient features. The same situation holds for the stability check algorithm [10] which is presented in Section III.

To establish the notation, let $\{a_{ij}\}_{i=0,j=0}^{m_a n_a}$ and $\{b_{ij}\}_{i=0,j=0}^{m_b n_b}$ be the numerator and denominator coefficients of a given stable filter, and $\{c_i\}_{i=0}^{m_c}$ the one dimensional array of length m_c formed by storing $\{a_{ij}\}$ columnwise into $\{c_i\}$ first and then storing $\{b_{ij}\}$ into $\{c_i\}$ beginning at $c_{(m_a n_a+1)}$ in like manner. The c array is then equivalent to the storage allocation that could be used for numerical implementation. With $z = (z_1,z_2)$ in D, we then put

$$P(c,z) = \sum_{i,j=0}^{m_a,n_a} a_{ij} z_1{}^i z_2{}^j,$$

$$Q(c,z) = \sum_{i,j=0}^{m_b,n_b} b_{ij} z_1{}^i z_2{}^j$$

and finally $R(c,z) = P(c,z)/Q(c,z)$. If F is a desired filter function defined on the bicylinder, we set

$$\Delta(c) = \max_{z \text{ in } D} |F(z) - R(c,z)|.$$

If Δ^* denotes the minimum $\Delta(c)$ over all possible $\{a_{ij}\}$ and $\{b_{ij}\}$, then by an extension of a result in [7, p. 53] this minimum is actually achieved by an appropriate set of coefficients. The following algorithm generates a sequence $c^0,c^1,c^2,\cdots$ of filter coefficient arrays in such a way that $\Delta(c^k) \geq \Delta(c^{k+1})$ and $\Delta(c^k) \to \Delta^*$ as $k \to \infty$.

Chebyshev Design Algorithm

Let c^0 be an arbitrarily chosen coefficient array subject to the constraints

$$|c_i{}^0| < 1$$

$$Q(c^0,z) \neq 0, \qquad \text{for all } z \text{ in } D.$$

For each $k \geq 1$ define the auxiliary function

$$\delta_k(c) = \max_{z \text{ in } D} \{|F(z)Q(c,z) - P(c,z)| - \Delta(c^k)|Q(c,z)|\}$$

and select the vector c^{k+1} so as to minimize this function in the standard cube. If $\delta_k(c^{k+1}) = 0$, then c^k is a solution; i.e., $\Delta(c^k) = \Delta^*$ and $Q(c^k,z) \neq 0$ for all z in D. Otherwise, $\Delta(c^k)$ decreases monotonically to Δ^* as $k \to \infty$.

Three shutdown conditions can be obtained. One results from the proof and states that, if $1/Q(c^{k+1},z)$ is unstable, then $R(c^k,z)$ is the optimum solution. Another is that $\delta(c^{k+1}) = 0$ implies that $R(c^k,z)$ is the solution. Finally, $\Delta(c^k)/\Delta(c^{k+1})$ within some preset tolerance of one provides the last.

Since most desired frequency-specified transfer functions are known only on the surface of the unit bicylinder ($|z_1| = 1 = |z_2|$), the above algorithm will have to be implemented in this manner to assure its eventual usefulness. To effectively use the first of the two convergence criteria some way must be determined to decide whether or not $Q(c^k,z) = 0$ in D. This is the subject of the next section.

III. Checking for Stability

It seems reasonable to attempt to find some sort of algorithm for checking stability which would require a modest amount of storage and would take as much advantage of the design code as possible. This can be done by basing such an algorithm on a recent result of Duffin [8]. While not providing a proof of its validity we note that a simple proof can be based on Huang's [4] modification of the stability criterion of the introduction. We do not restate Duffin's algorithm but it can easily be obtained by setting $N = 0$ in the following algorithm.

Recalling that a Schur polynomial $g(w)$ is a polynomial which does not vanish when $|w| \geq 1$, we can immediately observe that Duffin's algorithm can be modified to obtain the following algorithm for testing the stability of digital filters.

Algorithm

For $0 \leq j \leq M$, let

$$D_j(z_1) = \sum_{i=0}^{N} b_{N-i,M-j} z_1{}^i.$$

Assuming that $NM \geq 1$, $b_{00} = 1$ and at least one $b_{i,M} \neq 0$ the polynomial

$$B(z_1,z_2) = \sum_{i,j=0}^{N,M} b_{ij} z_1{}^i z_2{}^j$$

represents a stable filter if and only if $D_M(z_1)$ is a Schur polynomial and

1) $\qquad |D_0(z_1)| < |D_M(z_1)|, \qquad$ when $|z_1| = 1$;

2)

$$\tilde{D}(z_1,z_2) = \sum_{j=0}^{M-1} (\overline{D_M(z_1)} D_{j+1}(z_1) - D_0(z_1)\overline{D_{M-j-1}(z_1)}) z_2{}^j$$

is a Schur polynomial of z_2 for each z_1 with $|z_1| = 1$.

This final algorithm can be implemented digitally as follows:

a) Use Duffin's algorithm to determine if $D_M(z_1)$ has all its roots inside the unit circle. If it does, go to step b); otherwise, terminate because the filter is unstable.

b) Using nonlinear programming techniques, compute

$$\delta = \min_{|z_1|=1} |D_M(z_1)| - |D_0(z_1)|.$$

If $\delta > 0$, go to step c); otherwise, terminate because the filter is unstable.

c) If $M - 1 = 0$, terminate because stability is assured; otherwise, compute

$$D_{j-1}(z) = \bar{D}_M(z_1)D_j(z_1) - D_0(z_1)\bar{D}_{M-j}(z_1)$$

for $j = 1,2,\cdots,M$ and set M to $M - 1$. Return to step b).

Remark: The algorithm can be programmed to calculate the coefficients $D_j(z_1)$ in step c) with a modest increase in storage requirements. In Fortran a function subprogram can be used to calculate $D_j(z_1)$ as a function of the D array, M and j.

IV. Computer Implementation of the Design Technique

Our problem in implementing the design algorithm is to find, at the kth iteration, a coefficient array c so that the function

$$\delta(c) = \max_{z \in D} \{|F(z)Q(c,z) - P(c,z)| - \Delta(c^k)|Q(c,z)|\}$$

is minimized over all possible c with

$$|c_i| \leq 1$$

for all i.

Clearly, $\delta(c)$ is a nonlinear function of c and hence nonlinear programming techniques must be used to calculate the optimum array. Many such techniques are available in the literature and will not be further discussed here. We use the version explained in [12].

Now, finding a properly constrained c to minimize $\delta(c)$ over D is equivalent to finding c to minimize

$$\delta'(c) = \max_{Ds} \{|F(z)Q(c,z) - P(c,z)| - \Delta(c^k)|Q(c,z)|\}$$

in such a manner that $Q(c,z) \neq 0$ over z in D. That this is true follows from the maximum modulus theorem applied to the quantity

$$\beta(z) = |F(z)Q(c,z) - P(c,z)| - \Delta(c^k)|Q(c,z)|$$

whenever c is such that $\delta'(c)$ is minimum. It is necessary to note that in this case $\beta(z) \leq 0$.

The previous remarks now make it possible to suggest a preliminary implementation procedure.

a) Start with some initial c vector, c^0 say, so that

$$Q(c^0,z) \neq 0.$$

b) Using nonlinear programming techniques, find c^k to minimize

$$\delta(c) = \max_{|z|=1} \{|F(z)Q(c,z) - P(c,z)| - \Delta(c^k)|Q(c,z)|\}$$

subject to $|c_i^k| \leq 1$.

c) Use one or all of the shut down conditions to determine if convergence has been attained. If not, repeat step b).

The implementation procedure outlined above is essentially that currently in use by the author. The only problem with it as described is the calculation of $\delta(c)$. Instead of performing the calculation as a maximum or optimization problem, we discretize the domain $|z| = 1$ so that for any c the numerical evaluation of $\delta(c)$ is a finite operation. The value $\Delta(c)$ is then calculated over this same discrete set, thereby avoiding a nonlinear optimization.

The discretization of $|z| = 1$ adds at least one new problem to the design procedure. This of course, is the problem of choosing the mesh over which to do the optimization. Should we choose a mesh at the beginning and keep it constant throughout or should we vary the grid as we iterate? If we choose the latter, then the criteria $\Delta(c^k) \downarrow \Delta^*$ is destroyed and oscillation can result. On the other hand, this latter procedure would seem to offer a better solution to the design problem because it forces the techniques to use more z's in the evaluation of $\delta(c)$. Moreover, it can be shown that the minimum set of points necessary in the mesh is determined by the sum of the degrees of numerator and denominator plus 1 and that convergence is somewhat independent of the choice of these points. We generate our mesh randomly at each iteration.

Example: In this case, we let

$$F(z_1,z_2) = \frac{0.2z_1 + 0.2z_2}{1.0 + 0.2z_1 + 0.2z_2}$$

with $|z| = 1$ being discretized randomly at each iteration step with 20 values (twice as many as needed). After 2.53 minutes and 12 iterations,

$$\frac{P(z)}{Q(z)} = \frac{0.000008827 + 0.1998z_1 + 0.1998z_2}{0.9998 + 0.1999z_1 + 0.1999z_2}$$

$\Delta(c^{12}) = 0.0003077$.

Example: Here we let

$$F(z_1,z_2) = \frac{0.2z_1 + 0.2z_2}{0.2z_1 + 0.2z_2 + 1.0z_1,z_2}$$

with $|z| = 1$ being discretized randomly at each iteration step with 10 values. In this case the stability check in step c) is dropped. After 22 iterations and 3.2 minutes of central processing unit time we have

$$\frac{P(z)}{Q(z)} = \frac{0.01397 - 0.1868z_1 - 0.928z_2 + 0.006963z_1z_2}{-0.004018 - 0.1972z_1 - 0.1678z_2 - 0.9794z_1z_2}$$

$\Delta(c^{18}) = 0.02899$. With the stability check employed,

$$\frac{P(z)}{Q(z)} = \frac{-0.1703 + 0.01417z_1 + 0.06439z_2 + 0.03406z_1z_2}{0.9925 - 0.1473z_1 - 0.03144z_2 - 0.08051z_1z_2}$$

after 5 iterations with $\Delta(c^5) = 0.4702$.

The first example begins with a stable filter function and demonstrates how well the algorithm works. In the second example, an unstable filter specified the desired transfer function and the algorithm converges to it when no stability check is employed. With the stability check engaged convergence is attained earlier.

References

[1] S. Treitel and J. Shanks, "The design of multistage separable planar filters," *IEEE Trans. Geosci. Electron.*, vol. GE-19, June 1971.

[2] E. Darby and C. Davis, "The analysis and design of two-dimensional filters for two-dimensional data," *Geophys. Prospecting*, vol. XV, Sept. 1967.

[3] J. Shanks, "Two-dimensional recursive filters," *SWIECO Rec.*, pp. 1351–1369, 1967.

[4] T. Huang, "Stability of two-dimensional recursive filters," *IEEE Trans. Audio Electroacoust.*, vol. AU-20, pp. 158–163, June 1972.

[5] C. Farmer and B. Bednar, "Stability of spatial digital filter," *Math. Biosc.*, vol. 14, pp. 113–119, 1972.

[6] E. Cheney and H. Loeb, "On rational Chebyshev approximation," *Numer. Math.*, vol. 4, pp. 124–127, 1962.

[7] N. Achieser, *Theory of Approximation.* New York: Ungar, 1956.

[8] R. J. Duffin, "Algorithms for classical stability problems," *SIAM Rev.*, vol. VII, no. 2, Apr. 1969.

[9] C. Farmer and J. B. Bednar, "Chebyshev design of spatial digital filters," submitted to *J. Math. Anal. Appl.*

[10] J. B. Bednar, "An algorithm for checking the stability of linear partial difference equations and two-dimensional digital filters," submitted to *SIAM J. Appl. Math.*

[11] I. Barrodale, M. J. D. Powell, and F. D. K. Roberts, "The differential correction algorithm for rational l_∞ approximation," *SIAM J. Numer. Analysis*, vol. 9, pp. 493–504, Sept. 1972.

[12] R. Fletcher and M. J. D. Powell, "A rapidly convergent descent method for minimization," *Comput. J.*, vol. 6, pp. 163–168, 1963.

[13] J. L. Shanks, S. Trietel, and J. Justice, "Stability and synthesis of two-dimensional recursive filters," *IEEE Trans. Audio Electroacoust.*, vol. AU-20, pp. 115–128, June 1972.

13

Reprinted from *IEEE Trans. Acoust., Speech, Signal Process.*
ASSP-22(1):15-21 (1974)

An l_p Design Technique for Two-Dimensional Digital Recursive Filters

GAMAL A. MARIA, Student Member, IEEE, and
MOUSTAFA M. FAHMY, Member, IEEE

Abstract—In this paper a design technique for the two-dimensional filters is proposed. An optimization algorithm is developed to minimize the p-error criterion under the constraint that the resulting filter be stable. Design of one-dimensional filter may be considered as a special case to which the proposed algorithm is applicable. Several examples are solved to illustrate the technique.

I. Introduction

Recently there has been increasing interest in using the two-dimensional digital filters in processing a variety of two-dimensional signals. Some of the design techniques available for the one-dimensional filters have been extended to the two-dimensional case [1], [2]. In this paper an approximation technique is proposed for the design of two-dimensional recursive filters consisting of first and second order sections connected in cascade. This cascade form has the following advantages over the direct form.[1]

1) The word length that guarantees the stability of the filter after the rounding of its coefficients is shorter. This is due to the fact that the sensitivity of the mapping of the unit circle of the Z_2^{-1}-plane into the Z_1^{-1}-plane under the denominator of the transfer function is lower [3].

2) The error caused by quantization effect and finite coefficient size is lower [4].

3) In two-dimensional filters, the number of multiplications needed for each output using the cascade form is less than that needed when using the direct form.

4) It is easier to check the stability of the cascade filter during the optimization process and make any necessary changes in the coefficients of the unstable section to restore the overall stability. This guarantees that the resulting filter will be stable.

Unlike the one-dimensional filters, it is not always possible, in the two-dimensional case, to factorize the transfer function

$$F(Z_1^{-1}, Z_2^{-1}) = \frac{\sum_{j=0}^{K_1^2} \sum_{i=0}^{K_1^1} a_{ij} Z_1^{-(i-1)} Z_2^{-(j-1)}}{1 + \sum_{\substack{j=0 \\ i,j \neq 0}}^{K_2^2} \sum_{i=0}^{K_2^1} d_{ij} Z_1^{-(i-1)} Z_2^{-(j-1)}}$$

as a product of first and second order sections. Therefore, in this paper the approximating function is chosen, from the start, to be in the following form:

Manuscript received January 31, 1973; revised May 1, 1973 and July 27, 1973.

The authors are with the Department of Electrical Engineering, Queen's University, Kingston, Ont., Canada.

[1] Note that these advantages, except (3), are valid for both one-dimensional and two-dimensional filters.

$$F(Z_1^{-1}, Z_2^{-1}) = A \frac{\prod_{l=1}^{K_1} N^{(l)}(Z_1^{-1}, Z_2^{-1})}{\prod_{l=1}^{K_2} D^{(l)}(Z_1^{-1}, Z_2^{-1})} \quad (1)$$

where

$$N^{(l)}(Z_1^{-1}, Z_2^{-1}) = [1\ Z_1^{-1} \cdots Z_1^{-(n_l^1-1)}] \cdot \begin{bmatrix} 1 & a_{21}^l \cdots & a^l_{1,m_l^1} \\ a_{21}^l & & \\ \vdots & & \\ a^l_{n_l^1,1} & & a^l_{n_l^1,m_l^1} \end{bmatrix} \begin{bmatrix} 1 \\ Z_2^{-1} \\ \vdots \\ Z_2^{-(m_l^1-1)} \end{bmatrix},$$

$$D^{(l)}(Z_1^{-1}, Z_2^{-1}) = [1\ Z_1^{-1} \cdots Z_1^{-(n_l^2-1)}] \cdot \begin{bmatrix} 1 & d_{12}^l \cdots & d^l_{1,m_l^2} \\ d_{21}^l & & \\ \vdots & & \\ d^l_{n_l^2,1} & & d^l_{n_l^2,m_l^2} \end{bmatrix} \begin{bmatrix} 1 \\ Z_2^{-1} \\ \vdots \\ Z_2^{-(m_l^2-1)} \end{bmatrix}$$

and n_l^1, n_l^2, m_l^1, m_l^2 can only assume the numerical value two and three.

In this paper, the problem is first mathematically formulated, and then the optimization technique proposed is developed and illustrated by several examples.

II. Statement of the Problem

The problem considered here can be stated as follows. Let the magnitude characteristic $Y = [Y_{mn}]$ defined on the discrete set of frequency pairs $(\omega_m^{(1)}, \omega_n^{(2)})$, $m = 1, \cdots, M$; $n = 1, \cdots, N$, be given. Let

$$\Phi = [\Phi_1^T, \Phi_2^T, A]^T$$

where

$$\Phi_1^T = [a_{21}^1, a_{31}^1, \cdots, a^1_{n_1^1,1}, a_{12}^1, a_{22}^1, \cdots, a^1_{n_1^1,2}, \cdots, a^1_{n_1^1,m_1^1}, \cdots, a^{K_1}_{n^1_{K_1}, m^1_{K_1}}]$$

$$\Phi_2^T = [d_{21}^1, d_{31}^1, \cdots, d^1_{n_1^2,1}, d_{12}^1, d_{22}^1, \cdots, d^1_{n_1^2,2}, \cdots, d^1_{n_1^2,m_1^2}, \cdots, d^{K_2}_{n^2_{K_2}, m^2_{K_2}}]$$

and the a's, d's, and A are the parameters of the transfer function

$$F(Z_1^{-1}, Z_2^{-1}) = A \frac{\prod_{l=1}^{K_1} N^{(l)}(Z_1^{-1}, Z_2^{-1})}{\prod_{l=1}^{K_2} D^{(l)}(Z_1^{-1}, Z_2^{-1})}$$

defined in Section I. Here $(\omega^{(1)}, \omega^{(2)})$ are related to the variables (Z_1, Z_2) by

$$(Z_{1m}, Z_{2m}) = (e^{j\omega_m^{(1)}\Pi}, e^{j\omega_n^{(2)}\Pi}).$$

Let the performance index $J(\Phi)$ be defined as

$$J(\Phi) = \sum_{m=1}^{M} \sum_{n=1}^{N} [|F_{mn}| - Y_{mn}]^p \quad (2)$$

where p is an even positive integer.[2] It is required to select the elements of the vector Φ that minimize the performance index $J(\Phi)$ under the constraint that the resulting filter $F(Z_1^{-1}, Z_2^{-1})$ be stable.

III. The Optimization Procedure

Here we will use the same optimization procedure used in [5]. The algorithm starts by assuming initial values for the elements of the parameter vector Φ. After each iteration the value of these elements are changed by the increment $\Delta\Phi$ to be chosen such that

$$\frac{dJ(\Phi + \Delta\Phi)}{d\Delta\Phi} \cong \frac{d(J(\Phi) + \nabla J^T \cdot \Delta\Phi + \frac{1}{2}(\Delta\Phi)^T \cdot [H] \cdot \Delta\Phi)}{d\Delta\Phi} = 0.$$

From the above, we get the set of linear equations

$$H\Delta\Phi = -\nabla J \quad (3)$$

where H is the Hessian matrix with the element

$$h(f, s) = \frac{\partial^2 J}{\partial\phi_f\, \partial\phi_s},$$

and ∇J is the gradient vector with the element

$$\nabla J_i = \frac{\partial J}{\partial \phi_i}.$$

This optimization procedure calls for the computation of the gradient vector ∇J and the Hessian matrix H and the solution of the system of linear equations (3) in the unknown $\Delta\Phi$.

The elements of the gradient vector can be calculated as follows:

$$\nabla_k J = \frac{\partial J}{\partial \phi_k} = \sum_{m=1}^{M} \sum_{n=1}^{N} p[|F_{mn}| - Y_{mn}]^{p-1} \frac{\partial |F_{mn}|}{\partial \phi_k}. \quad (4)$$

[2] Here $|F_{mn}| = |F(Z_{1m}^{-1}, Z_{2n}^{-1})|$.

It can be easily shown that

$$\frac{\partial |F_{mn}|}{\partial \phi_k} = \frac{1}{|F_{mn}|} \operatorname{Re}\left(\overline{F}_{mn} \cdot \frac{\partial F_{mn}}{\partial \phi_k}\right) = |F_{mn}| \operatorname{Re}(X_k(m,n)) \tag{5}$$

where $X_k(m, n)$ are the components of $\overline{X}$ defined by

$$\overline{X}^T = (X_1(m,n), \cdots, X_k(m,n), \cdots) = \left(\alpha_{21}^1, \alpha_{31}^1, \cdots, \alpha_{ij}^l, \cdots, \alpha^{K_1}_{n^1_{K_1}, m^1_{K_1}}, \beta_{21}^1, \beta_{31}^1, \cdots, \beta_{ij}^l, \cdots, \beta^{K_2}_{n^2_{K_2}, m^2_{K_2}}, \frac{1}{A}\right)$$

and

$$\alpha_{ij}^l = \frac{Z_{1m}^{-(i-1)} Z_{2m}^{-(j-1)}}{N^{(l)}(Z_{1m}^{-1} Z_{2n}^{-1})}, \quad l = 1, \cdots, K_1, \quad i = 1, \cdots, n_l^1, \quad j = 1, \cdots, m_l^1, \quad i, j \neq 1$$

$$\beta_{ij}^l = -\frac{Z_{1m}^{-(i-1)} Z_{2n}^{-(j-1)}}{D^{(l)}(Z_{1m}^{-1} Z_{2n}^{-1})}, \quad l = 1, \cdots, K_2, \quad i = 1, \cdots, n_l^2, \quad j = 1, \cdots, m_l^2, \quad i, j \neq 1.$$

Substituting (5) in (4), we get

$$\nabla_k J = \sum_{m=1}^{M} \sum_{n=1}^{N} p\, \epsilon_{m,n}^{p-1} |F_{mn}| \operatorname{Re}[X_k(m,n)] \tag{6}$$

where $\epsilon_{m,n} = (|F_{mn}| - Y_{mn})$. Equation (6) is used to compute the gradient vector.

Turning now to the problem of computing the elements of the Hessian matrix defined in equation (3), we proceed as follows:

$$\begin{aligned} h(f,s) &= \frac{\partial^2 J}{\partial \phi_f \partial \phi_s} \\ &= \sum_{m=1}^{M} \sum_{n=1}^{N} p\, \epsilon_{m,n}^{p-1} \frac{\partial^2 |F_{mn}|}{\partial \phi_f \partial \phi_s} + p(p-1)\, \epsilon_{m,n}^{p-2} \frac{\partial |F_{mn}|}{\partial \phi_f} \frac{\partial |F_{mn}|}{\partial \phi_s} \\ &= \sum_{m=1}^{M} \sum_{n=1}^{N} p\, \epsilon_{m,n}^{p-1} \frac{\partial^2 |F_{mn}|}{\partial \phi_f \partial \phi_s} + p(p-1)\, \epsilon_{m,n}^{p-2} \cdot |F_{mn}|^2 \operatorname{Re}(X_f(m,n)) \operatorname{Re}(X_s(m,n)). \end{aligned} \tag{7}$$

To evaluate the first term, we must get the value of $(\partial^2 |F_{mn}|)/(\partial \phi_f \partial \phi_s)$. It can be easily shown that

$$\frac{\partial^2 |F_{mn}|}{\partial \phi_f \partial \phi_s} = \frac{1}{|F_{mn}|} \operatorname{Re}\left(\overline{F}_{mn} \cdot \frac{\partial^2 F_{mn}}{\partial \phi_f \partial \phi_s}\right) + |F_{mn}| \operatorname{Im}(X_f(m,n)) \operatorname{Im}(X_s(m,n)). \tag{8}$$

Noting that

$$\frac{\partial^2 F_{mn}}{\partial a_{ij}^l \partial a_{ce}^l} = 0 \tag{9a}$$

$$\frac{\partial^2 F_{mn}}{\partial A^2} = 0 \tag{9b}$$

$$\frac{\partial^2 F_{mn}}{\partial d_{ij}^l \partial d_{ce}^l} = 2F_{mn} \beta_{ij}^l \beta_{ce}^l \tag{9c}$$

and

$$\frac{\partial^2 F_{mn}}{\partial \phi_f \partial \phi_s} = F_{mn} X_f(m,n) \cdot X_s(m,n) \tag{9d}$$

for all other combination of f, s, it follows that $(\partial^2 |F_{mn}|)/(\partial \phi_f \partial \phi_s)$ takes the following form:

$$\frac{\partial^2 |F_{mn}|}{\partial a_{ij}^l \partial a_{ce}^l} = |F_{mn}| \operatorname{Im}(\alpha_{ij}^l) \operatorname{Im}(\alpha_{ce}^l) \tag{10a}$$

$$\frac{\partial^2 |F_{mn}|}{\partial A^2} = 0 \tag{10b}$$

$$\frac{\partial^2 |F_{mn}|}{\partial d_{ij}^l \partial d_{ce}^l} = |F_{mn}| [2 \operatorname{Re}(\beta_{ij}^l) \operatorname{Re}(\beta_{ce}^l) - \operatorname{Im}(\beta_{ij}^l) \operatorname{Im}(\beta_{ce}^l) \tag{10c}$$

and

$$\frac{\partial^2 |F_{mn}|}{\partial \phi_f \partial \phi_s} = |F_{mn}| \operatorname{Re}(X_f(m,n)) \operatorname{Re}(X_s(m,n)) \tag{10d}$$

for all other combinations of f, s.

Substituting these values in (7) we can compute all the elements of the Hessian matrix.

IV. Steps of the Computer Program Used for Optimization

1) Assume initial values for the elements of the parameter vector Φ.

2) Compute $(X_k(m, n))$ for all k. Start by computing

$$\gamma_{1_l}(m,n) = \frac{1}{N^{(l)}(Z_{1m}^{-1}, Z_{2n}^{-1})}, \quad l = 1, \cdots, K_1$$

and

$$\gamma_{2_l}(m,n) = \frac{1}{D^{(l)}(Z_{1m}^{-1}, Z_{2n}^{-1})}, \quad l = 1, \cdots, K_2.$$

Store these values to be used in computing $|F_{mn}|$.

3) Compute $|F_{mn}|$ using the formula

$$|F_{mn}| = A \frac{\prod_{l=1}^{K_2} |\gamma_{2_l}(m,n)|}{\prod_{l=1}^{K_1} |\gamma_{1_l}(m,n)|}.$$

4) Compute the gradient vector ∇J using (6).

5) Compute the Hessian matrix H using (7), (10).

6) Solve the linear equations (3).

7) Put $\Phi = \Phi + \Delta\Phi$.

8) Examine the stability of the filter [8]. If unstable, successively reduce by half the incremental change in the coefficients of the unstable sections until the filter becomes stable.

9) Repeat these steps (2-8) until the approximation is satisfactory (i.e., change in the performance index is less than a specified number) or a certain number of iteration has been completed.

V. Concluding Remarks

1) If the function Y is symmetrical with respect to $\omega^{(1)}$ and $\omega^{(2)}$, we can put a constraint on the coefficients of $F(Z_1^{-1}, Z_2^{-1})$ such that $d_{ij}^l = d_{ji}^l$, $a_{ij}^l = a_{ji}^l$ for all i, j, l. This reduces the number of parameters to be optimized. The elements of the vector $\overline{X}$ defined in (5) will be redefined as follows:

$$a_{ij}^l = \frac{Z_{1m}^{-(i-1)} Z_{2n}^{-(j-1)} + S Z_{1m}^{-(j-1)} Z_{2n}^{-(i-1)}}{N^{(l)}(Z_{1m}^{-1}, Z_{2n}^{-1})}, \quad j \leq i, \quad i, j \neq 1$$

and

$$\beta_{ij}^l = \frac{-(Z_{1m}^{-(i-1)} Z_{2n}^{-(j-1)} + S Z_{1m}^{-(j-1)} Z_{2n}^{-(i-1)})}{D^{(l)}(Z_{1m}^{-1}, Z_{2n}^{-1})}, \quad j \leq i, \quad i, j \neq 1$$

with

$$S = 1 \text{ for } i \neq j$$

$$S = 0 \text{ for } i = j.$$

Also it is sufficient to compute $|F_{mn}|$ only for $m = 1, \cdots, M$ and n such that $|\omega_n^{(2)}| \leq |\omega_m^{(1)}|$, thus saving considerable amount of computation.

2) This procedure can be used for the design of one-dimensional filters by putting $N = 1$ and $m_l^1 = m_l^2 = 1$ for all l. Here, the proposed algorithm has the advantage of not restricting the search to the class of filters with complex poles and zeros, as is the case with the algorithm suggested by Deczky [9]. Deczky's algorithm excludes the class of filters with real poles and/or zeros within which the optimal filter may lie (as, for example, in the case of wide-band differentiators [6]).

As an illustration of using the proposed algorithm to design a one-dimensional filter, the first example in [6] is solved here using the proposed method. Starting from the same initial conditions, exactly the same results have been obtained after 32 iterations. For this example, it was found that the computation time needed for each iteration of the proposed algorithm is almost twice the computation time needed for each function evaluation in Fletcher-Powell algorithm. So these 32 iterations are equivalent to almost 64 function evaluations in Fletcher-Powell algorithm compared to 155 reported in [6].

3) It was shown by Temes and Zai [7] that J will decrease if $[H]$ is positive definite and $\Delta\phi$ small. To ensure that $[H]$ be so, they suggested a positive definite approximation of the Hessian matrix to be used. An alternative method to ensure the positive definiteness of $[H]$, and which converges faster in the neighborhood of the minimum, is to add a sufficiently large positive constant to the diagonal elements of the Hessian matrix. It was shown [7] that the addition of a positive constant to the diagonal elements of $[H]$ will keep $\Delta\phi$ small and thus J will decrease. This positive constant is decreased in steps and is put to zero near the extremal point. If the extremal point reached is a minimum, $[H]$ will be positive definite and J will continue on decreasing until the minimum is reached. If the extremal point is not a minimum, J will start to increase. In such a case, the search for the minimum should be reinitiated by selecting another set of initial conditions. Of course, the proposed method, as any other nonlinear optimization technique (including Fletcher-Powell technique), does not guarantee that the minimum point reached is the global one.

It is to be noted that, since J goes to infinity at the stability boundary, there exists, in all directions where J decreases, a certain magnitude of $\Delta\phi$ below which the filter will remain in the stability region. This has been discussed in detail in [9]. Also, the fact that a stable minimum does exist has been shown by Deczky [9].

4) For $p = 2$, the coefficient A can be optimized separately using the equation

$$A = \frac{\sum_{m=1}^{M} \sum_{n=1}^{N} Y_{mn} H_{mn}}{\sum_{m=1}^{M} \sum_{n=1}^{N} |H_{mn}|^2}$$

where

$$H_{mn} = \frac{\prod_{l=1}^{K_1} N^{(l)}(Z_{1m}^{-1}, Z_{2n}^{-1})}{\prod_{l=1}^{K_2} D^{(l)}(Z_{1m}^{-1}, Z_{2n}^{-1})}.$$

The above equation can also be used to generate the initial value of A.

VI. Examples

Example 1: In this example we will consider one- and two-dimensional low-pass filters with cutoff frequency equal to one tenth of the Nyquist frequency. The specifications of this filter are as follows:

A) The one-dimensional filter:

$$Y_m = 1 \qquad \omega_m = 0, 0.01, \cdots, 0.09$$

$$= 0.5 \qquad \omega_m = 0.1$$

$$= 0.0 \qquad \omega_m = 0.1, \cdots, 0.2, 0.3, \cdots, 1.0.$$

B) The two-dimensional filter:

$$Y_{mn} = 1 \qquad [(\omega_m^{(2)})^2 + (\omega_n^{(2)})^2]^{1/2} \leq 0.08$$
$$= 0.5 \qquad 0.08 < [(\omega_m^{(1)})^2 + (\omega_n^{(2)})^2]^{1/2} < 0.12$$
$$= 0 \qquad [(\omega_m^{(1)})^2 + (\omega_n^{(2)})^2]^{1/2} \geq 0.12$$
$$\omega_m^{(1)} = 0, 0.02, \cdots, 0.2, 0.4, \cdots, 1.0$$

and

$$\omega_n^{(2)} = 0, 0.02, \cdots, 0.2, 0.4, \cdots, 1.$$

This filter was approximated by one second order section in both the one- and two-dimensional case. The optimization starts with the one-dimensional filter with $p = 2$. After 32 iterations the algorithm converges to the filter

$$F(Z^{-1}) = 0.11024 \frac{1 - 1.64382 Z^{-1} + Z^{-2}}{1 - 1.79353 Z^{-1} + 0.84098 Z^{-2}}$$

with $J(\Phi) = 0.56732$.

Initial conditions for the two-dimensional low-pass filter is taken to be the coefficients of the filter

$$F_0(Z_1^{-1}, Z_2^{-1}) = F(Z_1^{-1}) F(Z_2^{-1}).$$

After 20 more iterations the algorithm converges to

$$F(Z_1^{-1}, Z_2^{-1}) = 0.00895$$

$$\frac{(1\ Z_1^{-1}\ Z_1^{-2}) \begin{bmatrix} 1 & -1.62151 & 0.99994 \\ -1.62151 & 2.63704 & -1.62129 \\ 0.99994 & -1.62129 & 1.00203 \end{bmatrix} \begin{bmatrix} 1 \\ Z_2^{-1} \\ Z_2^{-2} \end{bmatrix}}{(1\ Z_1^{-1}\ Z_1^{-2}) \begin{bmatrix} 1 & -1.78813 & 0.82930 \\ -1.78813 & 3.20640 & -1.149271 \\ 0.82930 & -1.149271 & 0.69823 \end{bmatrix} \begin{bmatrix} 1 \\ Z_2^{-1} \\ Z_2^{-2} \end{bmatrix}}$$

with $J(\Phi) = 1.6361$.

The resulting one- and two-dimensional filters are shown in Fig. 1 and Fig. 2. Three sections of the two-dimensional low-pass filter obtained here connected in series will give peak attenuation of -45 dB and in-band ripple of 0.2 dB (better attenuation performance can still be obtained by optimizing the coefficients of the additional two sections). In [1], the same attenuation has been obtained by 25 × 25 nonrecursive filter with the same passband and twice the transient band. This means that for each output of the recursive filter designed using the proposed method, we need 31 multiplications compared to 625 multiplications when using the nonrecursive filter designed in [1].

Example 2: Here we will consider the wide-band differentiator. This filter will be approximated by: 1) one first order section and 2) two first order sections. To show the effect of p in the resulting filter, the example was solved with $p = 2$, $p = 6$, $p = 10$. The specifications of this filter are as follows:

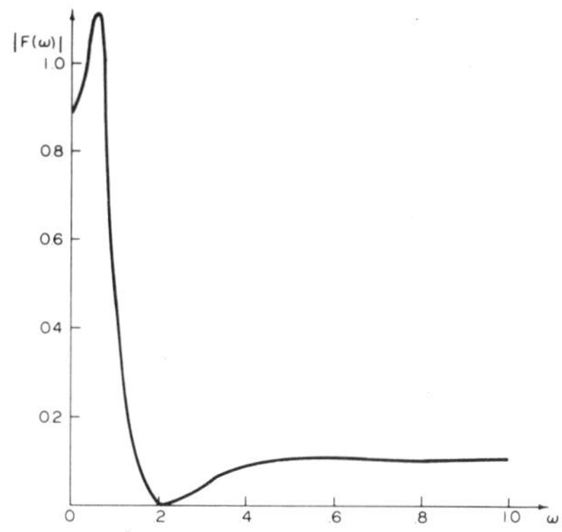

Fig. 1. Example 1: amplitude response of the one-dimensional optimal low-pass filter.

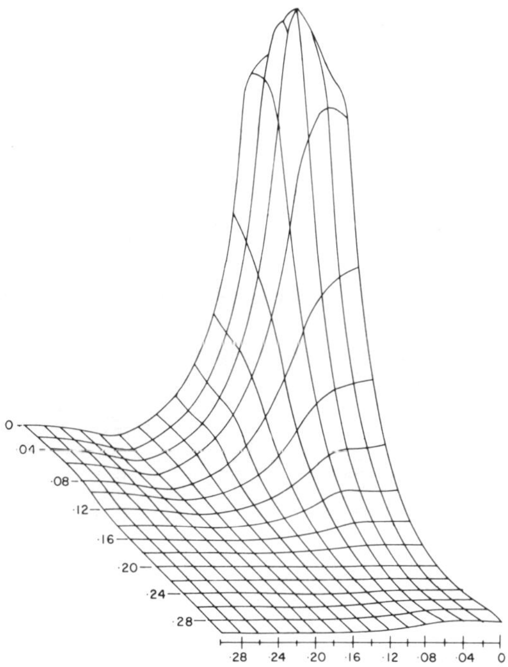

Fig. 2. Example 1: amplitude response of the two-dimensional optimal low-pass filter.

$$Y_{mn} = |\omega_m^{(1)} \omega_n^{(2)}|$$
$$\omega_m^{(1)} = 0, 0.1, \cdots, 1$$
$$\omega_n^{(1)} = 0, 0.1, \cdots, 1.$$

The results of the approximation by one first order section are shown in Table I.

The results of the approximation by two first order sections are as follows.

TABLE I

	Initial Conditions	$p = 2$	$p = 6$	$p = 10$
A	0.11	0.12785	0.12188	0.11580
a_{21}	-0.95	-1.00050	-0.96536	-1.00475
a_{22}	0.95	1.00094	0.95619	1.01491
d_{21}	0.4	0.24616	0.28993	0.29738
d_{22}	0.16	0.06059	0.08637	0.09064
$J(\Phi)$	—	4.7312×10^{-2}	2.8324×10^{-7}	2.4619×10^{-12}
Bounds for the deviation	—	-10 percent	-6.5 percent	-6 percent
From the specifications	—	+5.9 percent	+5.7 percent	+5.4 percent
Number of iterations	—	15	20	20

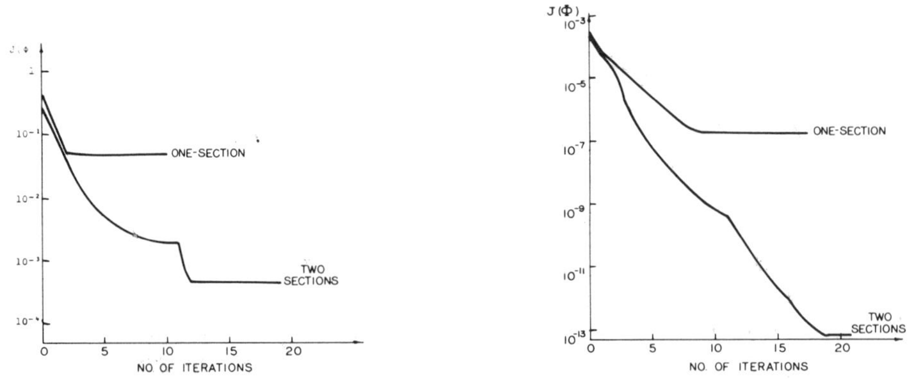

Fig. 3. Example 2. (a) Error versus iteration for $p = 2$. (b) Plot of error versus iteration for $p = 6$.

For $p = 2$, the algorithm converges after 14 iterations to

$$F_2(Z_1^{-1}, Z_2^{-1}) = 0.13452 \frac{(1\ Z_1^{-1})\begin{bmatrix}1 & -1\\ -1 & 1\end{bmatrix}\begin{bmatrix}1\\ Z_2^{-1}\end{bmatrix}(1\ Z_1^{-1})\begin{bmatrix}1 & 0.66411\\ 0.66411 & 0.44104\end{bmatrix}\begin{bmatrix}1\\ Z_2^{-1}\end{bmatrix}}{(1\ Z_1^{-1})\begin{bmatrix}1 & 0.13842\\ 0.13842 & 0.01916\end{bmatrix}\begin{bmatrix}1\\ Z_2^{-1}\end{bmatrix}(1\ Z_1^{-1})\begin{bmatrix}1 & 0.71390\\ 0.71390 & 0.50966\end{bmatrix}\begin{bmatrix}1\\ Z_2^{-1}\end{bmatrix}}$$

with $J(\Phi) = 4.1608 \times 10^{-4}$. The deviation from the specifications is bounded by -0.7 percent and +0.4 percent.

For $p = 6$, the algorithm converges after 21 iterations to

$$F_6(Z_1^{-1}, Z_2^{-1}) = 0.13327 \frac{(1\ Z_1^{-1})\begin{bmatrix}1 & -1.00316\\ -1.00316 & 1.01024\end{bmatrix}\begin{bmatrix}1\\ Z_2^{-1}\end{bmatrix}(1\ Z_1^{-1})\begin{bmatrix}1 & 0.67876\\ 0.67876 & 0.46209\end{bmatrix}\begin{bmatrix}1\\ Z_2^{-1}\end{bmatrix}}{(1\ Z_1^{-1})\begin{bmatrix}1 & 0.14762\\ 0.14762 & 0.02314\end{bmatrix}\begin{bmatrix}1\\ Z_2^{-1}\end{bmatrix}(1\ Z_1^{-1})\begin{bmatrix}1 & 0.72434\\ 0.72434 & 0.52571\end{bmatrix}\begin{bmatrix}1\\ Z_2^{-1}\end{bmatrix}}$$

with $J(\Phi) = 0.6395 \times 10^{-13}$. The deviation from the specifications is bounded by -0.5 percent and +0.4 percent.

The rates of convergence in both cases for $p = 2$ and $p = 6$ are shown in Fig. 3.

Example 3: In this example we will consider the linear high-pass filter with the following specifications:

$$Y_{mn} = [(\omega_m^{(1)})^2 + (\omega_n^{(2)})^2]^{1/2}$$

$$\omega_m^{(1)} = 0, \frac{1}{14}, \cdots, 1.0$$

and

$$\omega_n^{(2)} = 0, \frac{1}{14}, \cdots, 1.0.$$

This filter was approximated by one first order section. The optimization starts with the initial conditions

$$F_0(Z_1^{-1}, Z_2^{-1}) = 0.7 \frac{(1\; Z_1^{-1}) \begin{bmatrix} 1 & -0.67 \\ -0.67 & 0.35 \end{bmatrix} \begin{bmatrix} 1 \\ Z_2^{-1} \end{bmatrix}}{(1\; Z_1^{-1}) \begin{bmatrix} 1 & -0.35 \\ -0.35 & 0.1 \end{bmatrix} \begin{bmatrix} 1 \\ Z_2^{-1} \end{bmatrix}}$$

and $p = 2$. After 27 iterations the algorithm converges to

$$F(Z_1^{-1}, Z_2^{-1}) = 0.59944 \cdot \frac{(1\; Z_1^{-1}) \begin{bmatrix} 1 & -0.04668 \\ -0.04668 & -0.46761 \end{bmatrix} \begin{bmatrix} 1 \\ Z_2^{-1} \end{bmatrix}}{(1\; Z_1^{-1}) \begin{bmatrix} 1 & 0.42602 \\ 0.42602 & 0.10692 \end{bmatrix} \begin{bmatrix} 1 \\ Z_2^{-1} \end{bmatrix}}$$

with $J(\Phi) = 0.28110$.

The amplitude response of the resulting filter is shown in Fig. 4.

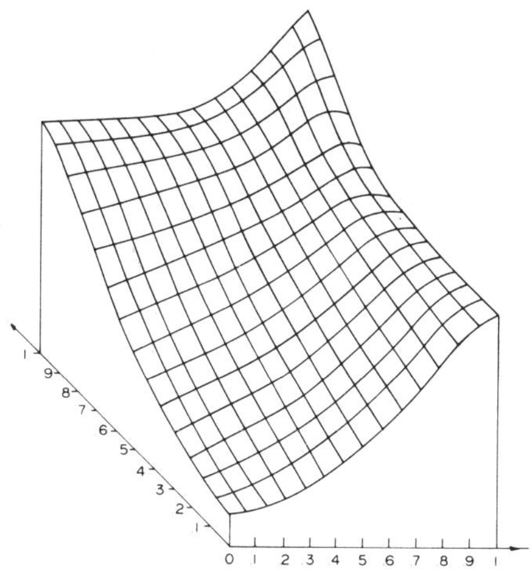

Fig. 4. Example 3: amplitude response of the optimal two-dimensional linear high-pass filter.

References

[1] J. V. Hu and L. R. Rabiner, "Design techniques for two-dimensional digital filters," *IEEE Trans. Audio Electroacoust.*, vol. AU-20, pp. 249-257, Oct. 1972.

[2] E. L. Hall, "Digital filtering of images," Ph.D. dissertation, Univ. of Missouri, Columbia, 1971.

[3] G. A. Maria and M. M. Fahmy, "Effect of rounding on the stability of the two-dimensional digital filters," to be published in *IEEE Trans. Circuits Syst.*, July 1974.

[4] J. B. Knowles and E. M. Olcayto, "Coefficient accuracy and digital filters response," *IEEE Trans. Circuit Theory*, vol. CT-14, pp. 31-41, Mar. 1968.

[5] G. A. Maria and M. M. Fahmy, "l_p^m-approximation by exponentials," *IEEE Trans. Circuit Theory*, vol. CT-20, pp. 71-74, Jan. 1973.

[6] K. Steiglitz, "Computer-aided design of recursive digital filters," *IEEE Trans. Audio Electroacoust.*, vol. AU-18, pp. 123-129, June 1970.

[7] G. C. Temes and D. Y. Zai, "Least *P*th approximation," *IEEE Trans. Circuit Theory*, vol. CT-16, pp. 235-237, May 1969.

[8] T. S. Huang, "Stability of two dimensional recursive filters," *IEEE Trans. Audio Electroacoust.*, vol. AU-20, pp. 158-163, June 1972.

[9] A. G. Deczky, "Synthesis of recursive digital filters using the minimum *p*-error criterion," *IEEE Trans. Audio Electroacoust.*, vol. AU-20, pp. 257-263, Oct. 1972.

14

Reprinted from *IEEE Trans. Acoust., Speech, Signal Process.*
ASSP-25(2):165–169 (1977)

Computer-Aided Design of Separable Two-Dimensional Digital Filters

RICHARD E. TWOGOOD, STUDENT MEMBER, IEEE, AND SANJIT K. MITRA, FELLOW, IEEE

***Abstract*—A procedure for the design of separable two-dimensional digital filters is presented. The computation involved in both the filter design and implementation is shown to be efficient. Several examples are presented which illustrate the application of the technique.**

I. Introduction

THERE HAS BEEN considerable interest in recent years in the design of two-dimensional digital filters [1]-[3]. Typical difficulties encountered in the design techniques have been computational complexity, convergence, and stability analysis. Treitel and Shanks [4] have proposed a computationally straightforward design technique which uses the singular value decomposition (SVD) of a matrix to approximate a spatial nonrecursive filter by a sum of separable two-dimensional filters. The computational gains in filter implementation were shown to be significant for low-order approximations. In this paper we extend the use of the SVD method to the frequency-domain design of separable approximations to desired magnitude specifications $H(u, v)$ for $0 \leqslant u \leqslant 1$ and $0 \leqslant v \leqslant 1$ (F_N = Nyquist frequency = 1 assumed in both directions). One-dimensional design techniques are then used to realize the separable filters. In this manner, one can choose the desired filter support, error norm to minimize, and form of implementation (FIR or IIR) and then find the optimal one-dimensional filters whose outer product forms the separable approximation to the desired response.

Manuscript received April 26, 1976; revised November 29, 1976. This work was performed under the auspices of the U.S. Energy Research and Development Administration under Contract W-7504-ENG-48 and supported in part by a NSF Grant ENG 75-15027.

R. E. Twogood is with the Lawrence Livermore Laboratory, University of California, Livermore, CA 94550.

S. K. Mitra is with the Department of Electrical Engineering, University of California, Davis, CA 95616.

II. Separable Filter Design Algorithm

Assume we have an m by n desired magnitude response sampled for $0 \leqslant u \leqslant 1$ and $0 \leqslant v \leqslant 1$,

$$H_{ij} = H(u_i, v_j), \quad i = 1, 2, \cdots, m \quad j = 1, 2, \cdots, n. \tag{1}$$

We can represent the specified magnitude response as an $m \times n$ matrix $\boldsymbol{H}$ whose i, jth element is H_{ij}. Without loss of generality we can further assume that $n \leqslant m$. As shown in [4], the singular value decomposition of H is given by

$$\boldsymbol{H} = \sum_{i=1}^{n} (\lambda_i)^{1/2} \boldsymbol{\kappa}_i \boldsymbol{\rho}_i \tag{2}$$

where the λ_i are the real, positive eigenvalues of the positive definite matrix $\boldsymbol{H}^T\boldsymbol{H}$, the $\boldsymbol{\rho}_i$ are the normalized eigenrows of $\boldsymbol{H}^T\boldsymbol{H}$, and the $\boldsymbol{\kappa}_i$ are the normalized eigencolumns associated with the n nonvanishing eigenvalues of the positive definite matrix $\boldsymbol{H}\boldsymbol{H}^T$. The retention of only the $t \leqslant n$ largest eigenvalues in the expansion gives a normalized error energy of

$$E_t = 1 - \frac{\sum_{i=1}^{t} \lambda_i}{\sum_{i=1}^{n} \lambda_i}. \tag{3}$$

We show next that the minimal mean square error (MMSE) separable approximation to $\boldsymbol{H}$ is given by

$$H \cong (\lambda_1)^{1/2} \boldsymbol{\kappa}_1 \boldsymbol{\rho}_1 \tag{4}$$

where λ_1 is the dominant eigenvalue, and $\boldsymbol{\rho}_1$ and $\boldsymbol{\kappa}_1$ are the corresponding normalized eigenrow and eigencolumn vectors, respectively. Let a separable approximation to H be given as

$$H \cong c\, r^T \tag{5}$$

where $c = [c_1, c_2, \cdots, c_m]^T$ and $r^T = [r_1, r_2, \cdots, r_n]$. The MMSE separable approximation is then obtained by choosing $\{c_i\}$ and $\{r_i\}$ such that the error

$$J_1 = \sum_{i=1}^{m} \sum_{j=1}^{n} (H_{ij} - c_i r_j)^2 \tag{6}$$

is a minimum. To this end, we differentiate J_1 with respect to each element of c and r, and set the result to zero:

$$\frac{\partial J_1}{\partial c_p} = -2 \sum_{j=1}^{n} H_{pj} r_j + 2c_p \sum_{j=1}^{n} r_j^2 = 0, \quad p = 1, 2, \cdots, m \tag{7a}$$

and

$$\frac{\partial J_1}{\partial r_q} = -2 \sum_{i=1}^{m} H_{iq} c_i + 2r_q \sum_{i=1}^{m} c_i^2 = 0, \quad q = 1, 2, \cdots, n. \tag{7b}$$

The solutions of the above equations are given by

$$c = \frac{H\, r}{r^T r} \tag{8a}$$

and

$$r = \frac{H^T c}{c^T c}. \tag{8b}$$

Substituting (8a) into (8b), we obtain after some algebra

$$H^T H\, r = (r^T r)(c^T c)\, r = \lambda_1\, r \tag{9}$$

indicating that r^T is an eigenrow vector of $H^T H$. Similar arguments show that c is an eigencolumn vector of the matrix $H H^T$. In addition, if r^T is a normalized eigenrow, we have $c^T c = \lambda_1$, or $|c| = (\lambda_1)^{1/2}$. The MMSE separable approximation to H is therefore $(\lambda_1)^{1/2} \boldsymbol{\rho}_1 \boldsymbol{\kappa}_1$ with $\boldsymbol{\rho}_1$ and $\boldsymbol{\kappa}_1$ as defined above.

In order that c and r be realizable as the magnitude response of one-dimensional digital filters, each element of these vectors must be positive. We prove this by contradiction next. Let $c\, r^T$ be the MMSE separable approximation to H and let an element of r^T, r_k be negative. By assumption then the error J_1 as given by (6) is a minimum. We rewrite it as

$$J_1 = \sum_{i=1}^{m} (H_{ik} - c_i r_k)^2 + \sum_{i=1}^{m} \sum_{\substack{j=1 \\ j \neq k}}^{n} (H_{ij} - c_i r_j)^2. \tag{10}$$

In the above, if we replace r_k by $-r_k$, then the new error is lower than J_1. Hence the assertion that $c\, r^T$ is the MMSE separable approximation to H with an element r_k of r^T as negative is false.

The error energy resulting from this separable approximation is therefore

$$E_1 = 1 - \frac{\lambda_1}{\sum_{i=1}^{n} \lambda_i}. \tag{11}$$

Letting $r^T = (\lambda_1)^{1/2} \boldsymbol{\rho}_1$ and $c = \boldsymbol{\kappa}_1$, the problem now is to find the two filters $R(z_1)$ and $C(z_2)$, whose sample magnitude responses are given by

$$|R(e^{ju_i})| = r_i, \qquad i = 1, 2, \cdots, n$$
$$|C(e^{jv_k})| = c_k, \qquad k = 1, 2, \cdots, m. \tag{12}$$

There are many optimization algorithms available for such a task, including the Fletcher-Powell algorithm, least-p (l_p) techniques, and others. The optimization routine used by the authors to perform this step incorporates a Marquardt minimization algorithm with an l_p error criterion. This program can be used to design either FIR or IIR filters. In the FIR case, the user specifies p and the order L of the desired one-dimensional filter, and the optimization program finds a filter

$$G(z) = K\left(1 + \sum_{i=1}^{L} a_i z^{-i}\right) \tag{13}$$

whose magnitude response minimizes

$$\sum_i \{|G(e^{jw_i})| - G_{di}\}^p \quad (p \text{ an even integer}) \tag{14}$$

where G_{di} is the desired sampled response of the one-dimensional filter. For IIR implementation of the filters, the program finds a filter

$$G(z) = K \prod_{i=1}^{L} \frac{(1 + a_{i1} z^{-1} + a_{i2} z^{-2})}{(1 + b_{i1} z^{-1} + b_{i2} z^{-2})} \tag{15}$$

whose magnitude response minimizes the same norm of (14) above. Recursive filters of this form are easily checked for stability because of the quadratic form of the denominator terms. Poles outside the unit circle are reflected to the inside without changing the shape of the magnitude response but necessitating a scaling of the constant K by the magnitude of the reflected pole, as shown by Steiglitz [5].

The steps described above are all easily implemented on a digital computer as subroutines for singular value decomposition and one-dimensional optimization can be readily obtained. The one-dimensional nature of the optimization offers substantial savings over the two-dimensional schemes, which require many more dependent variables and parameters to be optimized. The technique as described above has the obvious limitation that the separable approximation is a good one only for specifications with one dominant eigenvalue, although a possible scheme for using more than one stage in the singular value decomposition is under investigation. A second limitation is imposed by the one-dimensional nature of the digital filters realized in the optimization step. In order to ensure that the impulse responses are real, we must require that the desired two-dimensional specifications satisfy a symmetry across both frequency axes, $H(u, v) = H(-u, v)$ and $H(u, v) =$

$H(u, -v)$, in addition to the normal centrosymmetry condition $H(u, v) = H(-u, -v)$. This requirement is somewhat less restrictive than other standard assumptions made in two-dimensional digital filter design, such as circular symmetry or symmetry across a 45° line, $H(u, v) = H(v, u)$.

III. Computational Considerations

As indicated earlier, the technique for the design of separable approximations to two-dimensional digital filters described above is straightforward and computationally efficient. The primary reason for designing separable (or sum of separable) filters, however, is the resulting computational savings in the filter implementation [4]. For an M by N picture and an m by n planar filter, the normal convolution requires $MNmn$ "multiply and add" operations. On the other hand, if the filter is separable and $mn << MN$, the implementation would require approximately $MN(m+n)$ operations. The reduction is possible because a separable filter can be implemented as a sequence of one-dimensional convolutions [4], [6]. For $m = n = 20$ (a typical filter support) the separable approximation gains a factor of 10 in computation over the nonseparable 20 by 20 implementation. Recursive forms of separable filters are similarly more efficient than nonseparable recursive filters, although the gains are not as dramatic because the size of the filters decreases.

A third possible implementation of linear filtering in two dimensions involves the use of the two-dimensional fast Fourier transform (FFT). The FFT method, which requires computation proportional to $MN \log_2 (MN)$, is often more efficient than the convolution method for fairly small filter supports m by n, depending on the computer used and the memory available [7]. The separable approximation allows much larger values for m and n before the FFT becomes competitive, and for practical applications the spatial implementation of the separable filter is much more efficient than the FFT method.

IV. Examples

Three examples are now presented which illustrate the application of the design technique described above. First, consider the ideal low-pass rectangular filter shown in Fig. 1, where

$$H(u_i, v_j) = \begin{cases} 1, & 0 \leqslant u_i \leqslant 0.4, \quad 0 \leqslant v_j \leqslant 0.4 \\ 0, & 0.4 < u_i \leqslant 1 \quad \text{or} \quad 0.4 < v_j \leqslant 1. \end{cases} \tag{16}$$

Note that only one quadrant of the (u, v) plane is specified because of the required symmetry across both axes mentioned above. The grid size chosen was 50 by 50. The singular value decomposition gave only one nonzero eigenvalue, $\lambda_1 = 400$. The error in the separable approximation is therefore zero, as it should be since $H(u_i, v_j) = r(u_i)c(v_j)$. The weighted eigenrow associated with λ_1 is

$$r^T = (1, 1, 1, 1, 1, 1, 1, 1, 1, 1, 1, 1, 1, 1, 1, 1, 1, 1, 1, 1, 1, 0, \cdots, 0) \tag{17}$$

and the associated eigencolumn is the same, $c = r$. This means that only one one-dimensional filter need be realized (a result

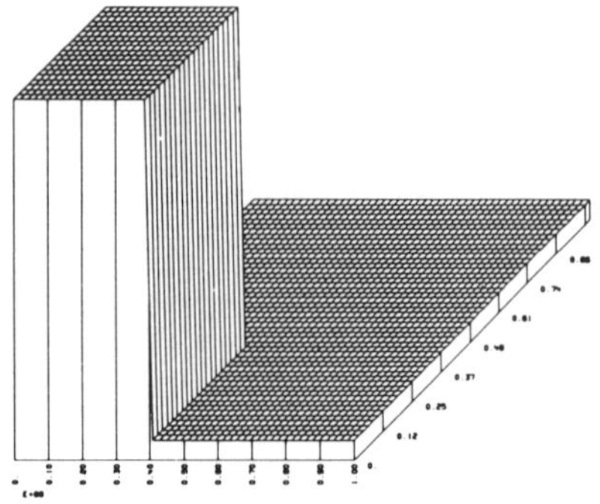

Fig. 1. Ideal rectangular filter.

of symmetry across the 45° line, $H(u, v) = H(v, u)$ in the specification). The optimization routine found the filter

$$R(z) = C(z) = 0.1138 \left(\frac{1 - 0.4426z^{-1} + 0.9815z^{-2}}{1 - 0.8834z^{-1} + 0.365z^{-2}} \right) \times \left(\frac{1 + 1.191z^{-1} + 0.95z^{-2}}{1 - 0.7897z^{-1} + 0.9063z^{-2}} \right) \tag{18}$$

which minimized the error norm of (14) with $p = 2$. The separable filter $H_s(z_1, z_2) = R(z_1)\,C(z_2)$ has the magnitude response shown in Fig. 2. Defining the error

$$J_e = \sum_{i=1}^{m} \sum_{j=1}^{n} (H(u_i, v_j) - |H_s(u_i, v_j)|)^2 \tag{19}$$

the filter shown has $J_e = 9.159$. In this example, the entire error in the final separable realization is due to the inability to realize the one-dimensional filters exactly. In most cases, however, the major cause of error is the approximation of the desired response H with only the dominant stage in its singular value decomposition.

The second example is the ideal circular low-pass filter similar to the one used by Hu and Rabiner [3],

$$H(u, v) = \begin{cases} 1, & r \leqslant R \\ 0, & r > R \end{cases} \tag{20}$$

where $r = (u^2 + v^2)^{1/2}$. Fig. 3 shows the ideal filter with $R = 0.6$, sampled on a 25 by 25 grid. The only nonzero eigenvalues of H were found to be $\lambda_1 = 165.15$, $\lambda_2 = 9.805$, $\lambda_3 = 1.807$, $\lambda_4 = 0.939$, $\lambda_5 = 0.716$, $\lambda_6 = 0.579$, $\lambda_7 = 0.378$, $\lambda_8 = 0.277$. Again we need to realize only one one-dimensional filter since $r = c$; the optimization routine found

$$R(z) = C(z) = 0.2813 \frac{1 + 1.591z^{-1} + 0.9591z^{-2}}{1 - 0.3161z^{-1} + 0.2788z^{-2}}. \tag{21}$$

The response of the resulting separable approximation $H_s(z_1, z_2) = R(z_1)C(z_2)$ is shown in Fig. 4. The error energy from the SVD approximation is

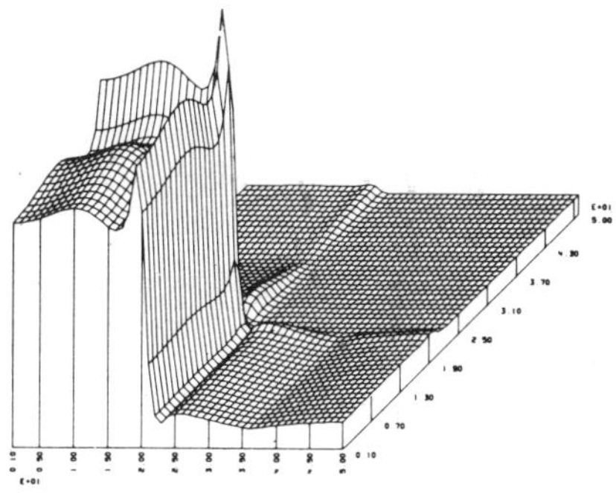

Fig. 2. Response of the separable approximation.

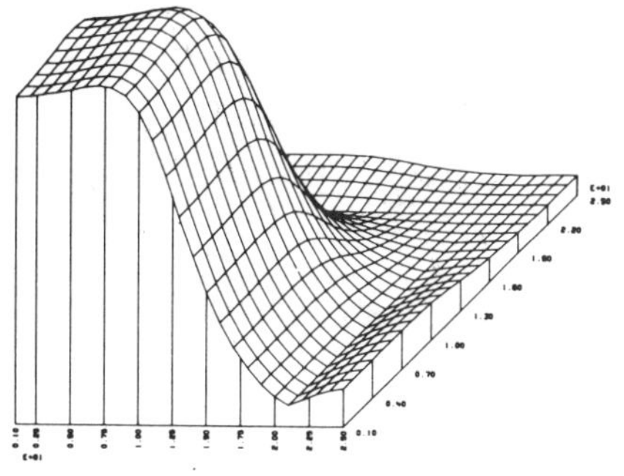

Fig. 4. Response of the separable approximation to Fig. 3.

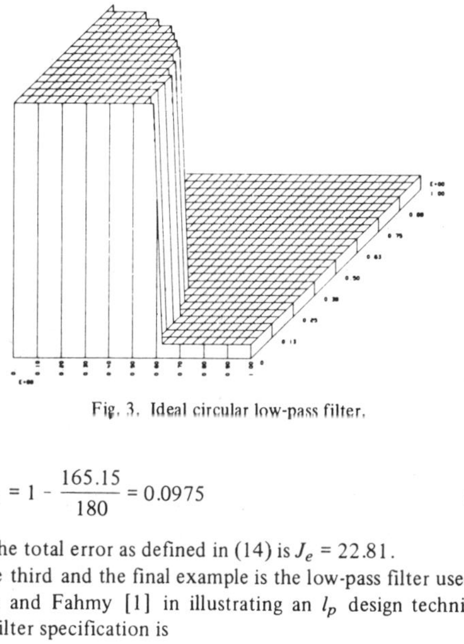

Fig. 3. Ideal circular low-pass filter.

$$E_1 = 1 - \frac{165.15}{180} = 0.0975 \tag{22}$$

and the total error as defined in (14) is $J_e = 22.81$.

The third and the final example is the low-pass filter used by Maria and Fahmy [1] in illustrating an l_p design technique. The filter specification is

$$H(u_i, v_j) = \begin{cases} 1, & 0 \leqslant r \leqslant 0.08 \\ 0.5, & 0.08 < r < 0.12 \\ 0, & 0.12 \leqslant r \leqslant 1 \end{cases}$$

$$u_i = 0, 0.02, \cdots, 0.2, 0.4, 0.6, 0.8, 1$$

$$v_j = 0, 0.02, \cdots, 0.2, 0.4, 0.6, 0.8, 1 \tag{23}$$

where $r = (u_i^2 + v_j^2)^{1/2}$, as shown in Fig. 5. The nonzero eigenvalues are $\lambda_1 = 19.23$, $\lambda_2 = 0.250$, $\lambda_3 = 0.182$, $\lambda_4 = 0.162$, $\lambda_5 = 0.113$, $\lambda_6 = 0.061$. The separable approximation $(\lambda_1)^{1/2}$ $\boldsymbol{\rho}_1$ $\boldsymbol{\kappa}_1$ thus introduces a normalized error energy of only

$$E_1 = 1 - \frac{19.23}{20} = 0.038. \tag{24}$$

The associated eigenrow and eigencolumn are again equivalent because of the 45° symmetry of the specification, with

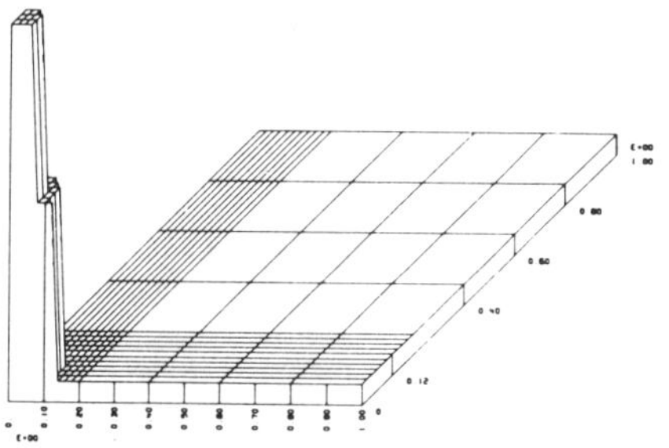

Fig. 5. Ideal low-pass filter specification (from Maria and Fahmy [1]).

$$\boldsymbol{r}^T = \boldsymbol{c}^T = \{1.203, 1.010, 1.010, 0.907, 0.508, 0.450, 0.117, 0, 0, 0, 0, 0, 0, 0, 0\}. \tag{25}$$

An optimal one-dimensional filter was found to be

$$R(z) = C(z) = 0.06582 \frac{1 - 1.514z^{-1} + 0.9962z^{-2}}{1 - 1.784z^{-1} + 0.8166z^{-2}}. \tag{26}$$

The total error in the realization of the second-order separable filter $R(z_1)C(z_2)$ is $J_e = 1.353$. Note that this is better than the "optimal" filter found for the same specification by Maria and Fahmy in [1], where $J_e = 1.6361$ for the second-order, nonseparable, two-dimensional recursive filter realized with their l_p technique. Thus we have found a separable filter which would require about one half the computation to implement as the result in [1], but which has approximately the same error. The separable result is shown in Fig. 6.

V. Discussion

The purpose of this paper has been to propose a computationally efficient technique for separable filter design. As shown in the examples, the approximation can be quite good in some cases. The ease of the design and savings in computation in filter implementation indicate that the separable approximation should be considered before utilizing more sophisticated and costly design techniques.

It should be noted that in the procedure suggested in this paper, two levels of approximation are involved. The first

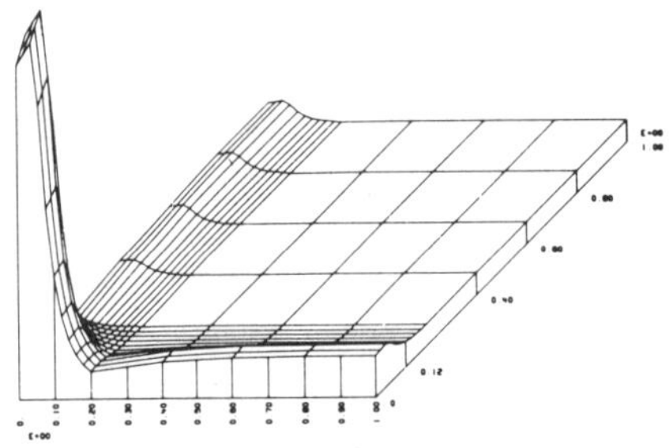

Fig. 6. Response of the separable approximation to Fig. 5.

level of approximation always leads to a globally optimum solution with respect to a MMSE separable approximation of the two-dimensional magnitude specification. In the second level of approximation, computer-aided optimization techniques are used to obtain one-dimensional transfer functions which are "close" to the one-dimensional magnitude specifications obtained in the previous step. Since in general the use of an iterative optimization technique may lead to a local optimum, the separable approximation obtained using the two-level procedure of this paper may or may not be a globally optimum solution. In the case of the third example considered here, the total error in the realization of the separable approximation is less than that obtained using a direct nonseparable approximation [1] which is most likely due to the latter result being a locally optimum solution.

For highly nonseparable filters, other methods such as the l_p-techniques of Maria and Fahmy [1] might be more appropriate. The extension of our technique described above to sum-of-separable filters involves the realization of separable terms with a specified phase, as the eigenvectors of the higher-order terms in the SVD decomposition include sign changes. The authors are considering techniques for the realization and the implementation of these filters, in addition to other decompositions or transformations to aid in the separable approximation.

References

[1] G. A. Maria and M. A. Fahmy, "An l_p-design technique for two-dimensional digital recursive filters," *IEEE Trans. Acoust., Speech, Signal Processing*, vol. ASSP-22, pp. 15–21, Feb. 1974.

[2] J. L. Shanks, S. Treitel, and J. H. Justice, "Stability and synthesis of two-dimensional recursive filters," *IEEE Trans. Audio Electroacoust.*, vol. AU-20, pp. 115–128, Oct. 1972.

[3] J. V. Hu and L. R. Rabiner, "Design techniques for two-dimensional digital filters," *IEEE Trans. Audio Electroacoust.*, vol. AU-20, pp. 249–257, Oct. 1972.

[4] S. Treitel and J. L. Shanks, "The design of multistage separable planar filters," *IEEE Trans. Geosci. Electron.*, vol GE-9, pp. 10–27, Jan. 1971.

[5] K. Steiglitz, "Computer-aided design of recursive digital filter," *IEEE Trans. Audio Electroacoust.*, vol. AU-18, pp. 123–129, June 1970.

[6] L. R. Rabiner and B. Gold, *Theory and Application of Digital Signal Processing*. Englewood Cliffs, NJ: Prentice-Hall, 1975, pp. 441–442.

[7] M. P. Ekstrom and R. E. Twogood, "Recursive and nonrecursive digital filtering of images: Some computational considerations," Lawrence Livermore Laboratory, University of California, Livermore, Rep. UCRL-79479, Nov. 1973.

Editors' Comments on Papers 15 Through 19

DESIGN OF 2-D FIR FILTERS

Design of FIR or nonrecursive filters involves determining a finite set of impulse response coefficients such that the filter characteristics satisfy a prescribed specification. This is a problem of polynomial approximation. In contrast to the IIR design problem, filter stability is not an issue. As a result, the design problem here is, in some sense, less complex than that of the 2-D IIR filter.

Two well-known techniques in the 1-D case are easily extendable to the 2-D case. The first, a windowing approach, involves performing an inverse 2-D Fourier transform to the given frequency response $H(e^{j\omega_1}, e^{j\omega_2})$ to obtain the ideal impulse response $h_{m,n}$. This response is, in general, of infinite extent. An FIR approximation is obtained by multiplying the $h_{m,n}$ array by a finite extent window function $w_{m,n}$. The resultant FIR filter $\tilde{h}_{m,n}$ then has a frequency response $\tilde{H}(e^{j\omega_1}, e^{j\omega_2})$ which is given by the 2-D convolution of the frequency responses of $h_{m,n}$ and $w_{m,n}$. Paper 15 by Huang develops several circularly symmetric 2-D windows which can be used for this purpose. Examples of 2-D FIR filters designed using Huang's windows are to be found in Rabiner and Gold (1975).

The second approach to the 2-D FIR filter design is also a straightforward extension of that developed for the 1-D case. This is the frequency sampling approach. Here by sampling the frequency response function $H(e^{j\omega_1},e^{j\omega_2})$ at equal intervals on ω_1 and ω_2 from 0 to 2π, the Discrete Fourier Transform (DFT) is determined. A simple inverse DFT leads to the impulse response coefficients of the FIR filter. By making some of the frequency samples adjustable, the error between the frequency response of the FIR filter and the specified frequency response at various other frequency values can be minimized. Hu and Rabiner (Paper 16) have successfully applied linear programming techniques to minimize such errors in the design of linear phase FIR filters. This second approach unfortunately requires a very large amount of computer time even for filters of moderate complexity. To reduce this computational burden, Hu and Rabiner employed symmetry of the filter array about both axes (so-called quadrilateral symmetry) and about the 45 degree line in the first quadrant (so-called octogonal symmetry). This reduces the number of independent terms in the filter array by approximately a factor of eight. Fiasconaro (1973) subsequently developed a more basic modification of the approach which is also slightly more efficient, computationally.

One of the most powerful 1-D FIR digital filter design techniques is that due to Parks and McClellan (1972), which is based on the Remez exchange algorithm. Kamp and Thiran (Paper 17) have extended the Remez algorithm to the 2-D case for designing FIR filters approximating circularly symmetric characteristics. These designs are performed to optimize a weighted Chebyshev error norm. A similar approach has also been independently advanced by Mersereau et al. (1975). As with the linear programming techniques, these methods require a substantial amount of computer time for their implementation and appear to be useful only for the design of small filter arrays having at least quadrilateral symmetry.

A design procedure originally proposed by McClellan (Paper 18) avoids these computational problems. It is used for the design of linear phase 2-D FIR filters and is perhaps the most powerful design algorithm presently available. McClellan's method involves converting a linear phase 1-D filter to a 2-D filter by a simple change of frequency variables. With an appropriate choice of this spectral transformation, the resulting 2-D filter may realize certain optimal (Chebyshev) characteristics. With this technique, the more difficult part of the design problem is in obtaining the 1-D filter parameters. Importantly, there are a number of effective algorithms available for this purpose (Rabiner and Gold, 1975; Oppenheim and Schafer, 1975), most of which are computationally quite manageable. Mersereau et al. (Paper 19) have generalized McClellan's scheme with regard to alternate choices for the transformation. They have designed moderately large filter arrays (on the order of 60 X 60) with excellent response characteristics. This procedure is constrained to the design of filters having quadrilateral symmetry.

A number of other direct 2-D FIR filter design techniques have also been proposed. In this direction, notable contributions have been made by Kamp and Thiran (1974) and Manry and Aggarwal (1976).

REFERENCES

Fiasconaro, J. G. 1973. "The Design of Two-Dimensional Nonrecursive Digital Filters," *Proc. 1973 Nat'l. Telecommun. Conf.*, I, 12A/1-8.

Kamp, Y., and J. P. Thiran. 1974. "Maximally-Flat Nonrecursive Two-Dimensional Digital Filters," *IEEE Trans. Circuits Syst.* **CAS-21**:437-449.

Manry, M. T., and J. K. Aggarwal. 1976. "The Design of Multidimensional FIR Digital Filters by Phase Correction," *IEEE Trans. Circuits Syst.* **CAS-23**: 185-199.

Mersereau, R. M., D. B. Harris, and H. S. Hersey. 1975. "An Efficient Algorithm for the Design of Equiripple Two-Dimensional FIR Digital Filters," *Proc. 1975 IEEE Internat. Symp. on Circuits Syst.*, 443-446.

Oppenheim, A. V., and R. W. Schafer. 1975. *Digital Signal Processing.* Englewood Cliffs, N.J.: Prentice-Hall.

Parks, T. W., and J. H. McClellan. 1972. "Chebyshev Approximation for Nonrecursive Digital Filters with Linear Phase," *IEEE Trans. Circuit Theory* **CT-19**: 189-194.

Rabiner, L. R., and B. Gold. 1975. *Theory and Application of Digital Signal Processing.* Englewood Cliffs, N.J.: Prentice-Hall.

15

Reprinted from *IEEE Trans. Audio Electroacoust.* **AU-20**(1):88–89 (1972)

TWO-DIMENSIONAL WINDOWS

T. S. Huang[1]

Department of Electrical Engineering,
M.I.T. Lincoln Laboratory[2]

I. Introduction

Two-dimensional windows find applications in many diverse fields, such as the spectral estimation of random fields, the design of two-dimensional digital filters, optical apodization, and antenna array design.

Many good one-dimensional windows have been devised [1]–[4]; however, relatively few two-dimensional windows have been investigated [5], [6]. In this correspondence, we establish a result which enables us to get good two-dimensional windows from good one-dimensional windows.

II. The Problem

We first review briefly the one-dimensional problem. Let the Fourier transform of a function $f(x)$ be $F(u)$. For some reason, we want to truncate $f(x)$:

$$g(x) = f(x)w(x) \tag{1}$$

$$w(x) = 0, \qquad \text{for } |x| > A \tag{2}$$

where A is a constant.

The Fourier transform of $g(x)$ is

$$G(u) = F(u) \otimes W(u) \tag{3}$$

Manuscript received June 22, 1971.
This work was performed at M.I.T. Lincoln Laboratory, Lexington, Mass., under Air Force Contract F19628-70-C-0230.

where $W(u)$ is the Fourier transform of $w(x)$, and $\otimes$ denotes convolution. Our problem is to choose an appropriate shape for the window function $w(x)$ such that $G(u)$ is close to $F(u)$ and in any region surrounding a discontinuity of $F(u)$, $G(u)$ will not contain excessive ripples. It is well known that the Fourier transform $W(u)$ of a good window $w(x)$ should have a big central peak and small sidelobes.

In two dimensions, the problem is entirely similar. Let the two-dimensional Fourier transform of a function $f_2(x, y)$ be $F_2(u, v)$, and let

$$g_2(x, y) = f_2(x, y)w_2(x, y) \tag{4}$$

where

$$w_2(x, y) = 0, \qquad \text{for } |x^2 + y^2| > A^2. \tag{5}$$

The Fourier transform of $g_2(x, y)$ is

$$G_2(u, v) = F_2(u, v) \otimes W_2(u, v) \tag{6}$$

where $W_2(u, v)$ is the Fourier transform of $w_2(x, y)$. The problem is to choose an appropriate shape for the two-dimensional window function $w_2(x, y)$ such that $G_2(u, v)$ is close to $F_2(u, v)$ and in the neighborhood of a discontinuity of $F_2(u, v)$, $G_2(u, v)$ does not contain excessive ripples.

III. The Result

Intuitively, we feel that if $w(x)$ is a good symmetrical one-dimensional window, then

$$w_2(x, y) = w(\sqrt{x^2 + y^2}) \tag{7}$$

will be a good two-dimensional window. This is indeed partially verified by the following two examples. The first example is

$$w(x) = \begin{cases} 1, & \text{for } |x| \le 1 \\ 0, & \text{for } |x| > 1. \end{cases} \tag{8}$$

Then the Fourier transform is

$$W(u) = \frac{2 \sin u}{u} \tag{9}$$

whose first sidelobe peak is about 20 percent of the peak at $u = 0$. The corresponding two-dimensional window

$$w_2(x, y) = \begin{cases} 1, & \text{for } |x^2 + y^2| \le 1 \\ 0, & \text{for } |x^2 + y^2| > 1 \end{cases} \tag{10}$$

has the two-dimensional Fourier transform

$$W_2(u, v) = \frac{2\pi J_1(\sqrt{u^2 + v^2})}{\sqrt{u^2 + v^2}} \tag{11}$$

whose first sidelobe peak is only about 12 percent of the peak at $u = 0 = v$.

The second example is

$$w(x) = \begin{cases} 1 - x, & \text{for } |x| \le 1 \\ 0, & \text{for } |x| > 1. \end{cases} \tag{12}$$

The Fourier transform is

$$W(u) = \left[\frac{\sin \frac{u}{2}}{\frac{u}{2}} \right]^2 \tag{13}$$

whose sidelobe peak is about 4 percent of the peak at $u = 0$. The corresponding two-dimensional window

[1] Now with the Department of Electrical Engineering, Massachusetts Institute of Technology, Cambridge, Mass.
[2] Operated with support from the U.S. Air Force.

$$w_2(x, y) = \begin{cases} 1-\sqrt{x^2+y^2}, & \text{for } |x^2+y^2| \leq 1 \\ 0, & \text{for } |x^2+y^2| > 1 \end{cases} \quad (14)$$

has a two-dimensional Fourier transform

$$W_2(u, v) = 2\pi \left[\rho^{-3} \int_0^{\rho} J_0(t)\, dt - \rho^{-2} J_0(\rho) \right] \quad (15)$$

where $\rho = \sqrt{u^2+v^2}$, whose first sidelobe peak is only about 2 percent of the peak at $u=0=v$.

The above comparisons are, however, unfair. Because when we convolve a window with a discontinuity, what count in the one-dimensional case are the areas under the sidelobes, while in the two-dimensional case what count are the volumes under the sidelobes. A fair comparison would be to look at the result of the convolution. One thing we can say along this line is contained in the following theorem which is the main result of this correspondence.

Theorem: If a symmetrical one-dimensional window $w(x)$ and a two-dimensional window $w_2(x, y)$ are related by

$$w_2(x, y) = w(\sqrt{x^2 + y^2}) \quad (16)$$

then their Fourier transforms $W(u)$ and $W_2(u, v)$ satisfy the relation

$$\frac{1}{2\pi} W_2(u, v) \otimes H_2(u, v) = W(u) \otimes H(u) \quad (17)$$

where

$$H(u) = \begin{cases} 1, & \text{for } u \geq 0 \\ 0, & \text{for } u < 0 \end{cases} \quad (18)$$

$$H_2(u, v) = \begin{cases} 1, & \text{for } u \geq 0 \text{ and all } v \\ 0, & \text{for } u < 0 \text{ and all } v \end{cases} \quad (19)$$

and $\otimes$ denotes convolution.

Proof: We first show that

$$W(u) = \frac{1}{2\pi} \int_{-\infty}^{\infty} dv W_2(u, v). \quad (20)$$

By definition,

$$w_2(x, y) = \frac{1}{4\pi^2} \int_{-\infty}^{\infty} du\, dv W_2(u, v) e^{j(xu+yv)}.$$

Whence

$$w_2(x, 0) = \frac{1}{2\pi} \int_{-\infty}^{\infty} du\, e^{jxu} \cdot \left[\frac{1}{2\pi} \int_{-\infty}^{\infty} dv W_2(u, v) \right]. \quad (21)$$

But from (16),

$$w_2(x, 0) = w(x).$$

Therefore from (21), $1/2\pi \int_{-\infty}^{\infty} dv W_2(u, v)$ is the Fourier transform of $w(x)$. This established (20). Now,

$$W(u) \otimes H(u) = \int_{-\infty}^{\infty} dt W(t) H(u-t) = \int_{-\infty}^{u} dt W(t) \quad (22)$$

and

$$W_2(u, v) \otimes H_2(u, v) = \iint_{-\infty}^{\infty} dt\, ds W(t, s) H_2(u-t, v-s) = \int_{-\infty}^{u} dt \int_{-\infty}^{\infty} ds W(t, s). \quad (23)$$

From (22) and (20), we have

$$W(u) \otimes H(u) = \frac{1}{2\pi} \int_{-\infty}^{u} dt \int_{-\infty}^{\infty} ds W(t, s) = \frac{1}{2\pi} W_2(u, v) \otimes H_2(u, v)$$

by virtue of (23).

IV. The Design of Two-Dimensional Nonrecursive Ideal Low-Pass Filters

One way of designing one-dimensional nonrecursive digital filters is the so-called window method [7]. To fix ideas, let us consider the design of an ideal low-pass filter. The ideal frequency response is

$$F(u) = \begin{cases} 1, & \text{for } |u| \leq B \\ 0, & \text{for } |u| > B \end{cases} \quad (24)$$

where B is a constant. (This frequency response is actually repeated periodically because the impulse response is sampled.) The inverse Fourier transform $f(x)$ of $F(u)$, which is the impulse response, has infinite duration, but in reality we have to use a finite-duration impulse response. So we truncate $f(x)$ by using a window:

$$g(x) = f(x) w(x) \quad (25)$$

$$w(x) = 0, \quad \text{for } |x| > A \quad (26)$$

where A is a constant. The actual frequency response we are getting is then

$$G(u) = F(u) \otimes W(u). \quad (27)$$

Suppose now we wish to design a two-dimensional ideal low-pass filter with an ideal frequency response

$$F_2(u, v) = F(\sqrt{u^2+v^2}) = \begin{cases} 1, & \text{for } |u^2+v^2| \leq B^2 \\ 0, & \text{for } |u^2+v^2| > B^2. \end{cases} \quad (28)$$

And we truncate the two-dimensional impulse response $f_2(x, y)$ by a two-dimensional window $w_2(x, y)$. Then the actual frequency response we are getting is

$$G_2(u, v) = F_2(u, v) \otimes W_2(u, v). \quad (29)$$

Let us assume that the widths of $W(u)$ and $W_2(u, v)$ are much smaller than B, then near the discontinuities of $F(u)$, viz., $u = \pm B$, $G(u)$ is essentially equal to the convolution of $W(u)$ and a one-dimenional step function, and similarly, nears the discontinuities of $F_2(u, v)$, viz., $u^2+v^2=B^2$, $G_2(u, v)$ is essentially equal to the convolution of $W_2(u, v)$ and a two-dimensional step function. It therefore follows from our theorem that if

$$w_2(x, y) = w(\sqrt{x^2 + y^2}) \quad (30)$$

then

$$\frac{1}{2\pi} G_2(u, v) \approx G(\sqrt{u^2 + v^2}). \quad (31)$$

This means that we can design a good two-dimensional low-pass filter by using the window $w_2(x, y)$ as given by (30), if $w(x)$ is a good window to use in designing a good one-dimensional low-pass filter.

V. Summary

We have shown that if $w(x)$ is a good symmetrical one-dimensional window, then $w_2(x, y) = w(\sqrt{x^2+y^2})$ is a good circularly symmetrical two-dimensional window.

References

[1] R. B. Blackman and J. W. Tukey, *The Measurement of Power Spectra.* New York: Dover, 1959.

[2] R. B. Blackman, *Linear Data-Smoothing and Prediction in Theory and Practice.* Reading, Mass.: Addison-Wesley, 1965.

[3] J. F. Kaiser, "Digital filters," in *System Analysis by Digital Computer*, F. F. Kuo and J. F. Kaiser, Eds. New York: Wiley, 1966, ch. 7.

[4] H. D. Helms, "Nonrecursive digital filters: Design methods for achieving specifications on frequency response," *IEEE Trans. Audio Electroacoust.*, vol. AU-16, pp. 336–342, Sept. 1968.

[5] A. Papoulis, *Systems and Transforms with Applications in Optics.* New York: McGraw-Hill, 1969.

[6] J. L. Allen, "The theory of array antennas (with emphasis on radar applications)," M.I.T. Lincoln Lab., Lexington, Mass., Tech. Rep. 323, Memo DDC AD-422945, July 1963.

[7] L. R. Rabiner, B. Gold, and C. A. McGonegal, "An approach to the approximation problem for nonrecursive digital filters," *IEEE Trans. Audio Electroacoust.*, vol. AU-18, pp. 83–106, June 1970.

16

Reprinted from *IEEE Trans. Audio Electroacoust.* **AU-20**(4):249–257 (1972)

Design Techniques for Two-Dimensional Digital Filters

JAMES V. HU and LAWRENCE R. RABINER

***Abstract*—The theory for designing finite-duration impulse response (FIR) digital filters can readily be extended to two or more dimensions. Using linear programming techniques, both frequency sampling and optimal (in the sense of Chebyshev approximation over closed compact sets) two-dimensional filters have been successfully designed. Computational considerations have limited the filter impulse response durations (in samples) to 25 by 25 in the frequency sampling case, and to 9 by 9 in the optimal design case. However, within these restrictions, a large number of filters have been investigated. Several of the issues involved in designing two-dimensional digital filters are discussed.**

Introduction

Techniques for designing one-dimensional digital filters have been investigated for several years, and are fairly well established in the literature [1]–[7]. Very few of these techniques, however, have been extended to two or more dimensions. Since there is no lack of interest in two-dimensional signal processing, it is worth investigating those design procedures that are readily extendable to two dimensions. In this paper, the frequency sampling technique and optimal filter design procedures are discussed in the case of the design of circularly symmetric two-dimensional low-pass filters.

Before discussing the results obtained, it is worth reviewing the state of the art in designing two-dimensional digital filters. Two of the most common classes of one-dimensional digital filters are usually called recursive and nonrecursive filters. (The alternate terminology infinite impulse response (IIR) and finite impulse response (FIR) has recently been proposed [8] and will be used throughout this paper.) IIR filters can be efficiently realized by factoring their z-transforms into a cascade of real or complex pole-zero networks and realizing the filter using these simple sections. In two dimensions, however, the two-dimensional z-transforms cannot generally be factored into lower order systems [9], [10]. This leads to many design problems, such as the difficulty in determining the stability of the filter, as well as in finding an efficient realization where the coefficients may be truncated to a reasonable number of bits. Because of these difficulties, very little work has been done on designing IIR two-dimensional digital filters.

For FIR digital filters, the problems of stability do not exist in one dimension as the z-transform is a finite polynomial. Thus design techniques in one dimension are often directly extendable to two or more dimensions by appropriate modifications to the design procedure. Huang [11] has already shown how the windowing technique may be extended to two dimensions by forming two-dimensional circularly symmetric windows from one-dimensional windows. The remainder of this paper discusses the design of circularly symmetric frequency sampling filters and optimal (equiripple) filters.

Manuscript received May 2, 1972. This paper was presented at the IEEE Arden House Workshop on Digital Filtering, Harriman, N. Y., January 1972.

The authors are with Bell Telephone Laboratories, Inc., Murray Hill, N. J. 07974.

Theory of Two-Dimensional FIR Filters

Let $h(n_1, n_2)$ be the impulse response of a two-dimensional filter where $-\infty < n_1, n_2 < \infty$. The two-dimensional z transform of the filter is defined as

$$H(z_1, z_2) = \sum_{n_1=-\infty}^{\infty} \sum_{n_2=-\infty}^{\infty} h(n_1, n_2) z_1^{-n_1} z_2^{-n_2}. \quad (1)$$

If the resulting filter is to be stable, the impulse response must satisfy the condition

$$\sum_{n_1=-\infty}^{\infty} \sum_{n_2=-\infty}^{\infty} |h(n_1, n_2)| < \infty. \quad (2)$$

For IIR filters (when any one of the four limits on the summation is infinite), it is generally quite difficult to determine whether or not a particular filter is stable, i.e., whether (2) is satisfied. For FIR filters, on the other hand, stability is always guaranteed since all four limits on the summation are finite and $|h(n_1, n_2)| < \infty$.

If we restrict ourselves to the case when $h(n_1, n_2)$ is a finite sequence defined from $0 \le n_1 \le N_1 - 1$, $0 \le n_2 \le N_2 - 1$, then (1) becomes

$$H(z_1, z_2) = \sum_{n_1=0}^{N_1-1} \sum_{n_2=0}^{N_2-1} h(n_1, n_2) z_1^{-n_1} z_2^{-n_2}. \quad (3)$$

The "frequency response" of the filter is obtained by evaluating (3) for $z_1 = e^{j\omega_1}$ and $z_2 = e^{j\omega_2}$, giving

$$H(e^{j\omega_1}, e^{j\omega_2}) = \sum_{n_1=0}^{N_1-1} \sum_{n_2=0}^{N_2-1} h(n_1, n_2) \exp\left[-j(\omega_1 n_1 + \omega_2 n_2)\right]. \quad (4)$$

The two-dimensional FIR filter may be realized in several ways. The simplest technique is a direct-convolution realization using the relation

$$y(n_1, n_2) = \sum_{m_1=0}^{N_1-1} \sum_{m_2=0}^{N_2-1} h(m_1, m_2) x(n_1 - m_1, n_2 - m_2) \quad (5)$$

where $x(n_1, n_2)$ is the input to the filter and $y(n_1, n_2)$ is the filter output.

The discrete Fourier transform (DFT) relations for the filter may be obtained by evaluating (4) at the discrete set of values

$$\omega_{k_1} = k_1\left(\frac{2\pi}{N_1}\right), \qquad k_1 = 0, 1, \cdots, N_1 - 1 \tag{6}$$

$$\omega_{k_2} = k_2\left(\frac{2\pi}{N_2}\right), \qquad k_2 = 0, 1, \cdots, N_2 - 1 \tag{7}$$

giving[1]

$$H(k_1, k_2) = \sum_{n_1=0}^{N_1-1}\sum_{n_2=0}^{N_2-1} h(n_1, n_2) \cdot \exp\left[-j2\pi\left(\frac{k_1 n_1}{N_1} + \frac{k_2 n_2}{N_1}\right)\right]. \tag{8}$$

The inverse DFT is readily obtained as

$$h(n_1, n_2) = \frac{1}{N_1 N_2}\sum_{k_1=0}^{N_1-1}\sum_{k_2=0}^{N_2-1} H(k_1, k_2) \cdot \exp\left[j2\pi\left(\frac{k_1 n_1}{N_1} + \frac{k_2 n_2}{N_2}\right)\right]. \tag{9}$$

It is interesting to note that two-dimensional DFT's, in (8), may be evaluated using one-dimensional DFT's by partitioning (8) as

$$H(k_1, k_2) = \sum_{n_1=0}^{N_1-1} \underbrace{\left[\sum_{n_2=0}^{N_2-1} h(n_1, n_2) e^{-j2\pi(n_2 k_2/N_2)}\right]}_{g(n_1, k_2)} e^{-j2\pi(n_1 k_1/N_1)} \tag{10}$$

and realizing that the inner summation is a one-dimensional DFT where n_1 is held fixed; once all the inner summation DFT's have been calculated (calling the results $g(n_1, k_2)$, the outer summation is a DFT where k_2 is held fixed.

A frequency interpolation formula from the DFT coefficients may be derived by inserting (9) into (4), giving

$$H(e^{j\omega_1}, e^{j\omega_2}) = \sum_{n_1=0}^{N_1-1}\sum_{n_2=0}^{N_2-1}\left[\frac{1}{N_1 N_2}\sum_{k_1=0}^{N_1-1}\sum_{k_2=0}^{N_2-1} H(k_1, k_2) \cdot \exp\left[j2\pi\left(\frac{k_1 n_1}{N_1} + \frac{k_2 n_2}{N_2}\right)\right]\right] \cdot \exp\left[-j(\omega_1 n_1 + \omega_2 n_2)\right]. \tag{11}$$

Interchanging orders of summation, and summing over the n_1 and n_2 indices, gives

$$H(e^{j\omega_1}, e^{j\omega_2}) = \sum_{k_1=0}^{N_1-1}\sum_{k_2=0}^{N_2-1} H(k_1, k_2) A(k_1, k_2, \omega_1, \omega_2) \tag{12}$$

where

$$A(k_1, k_2, \omega_1, \omega_2) = \frac{1}{N_1 N_2}\left[\frac{1-\exp(-jN_1\omega_1)}{1-\exp\left[j\left(2\pi\frac{k_1}{N_1} - \omega_1\right)\right]}\right] \cdot \left[\frac{1-\exp(-jN_2\omega_2)}{1-\exp\left[j\left(2\pi\frac{k_2}{N_2} - \omega_2\right)\right]}\right]. \tag{13}$$

Equation (12) serves as the basis for designing frequency sampling two-dimensional filters. As seen from (12), the continuous frequency response of the filter is a linear combination of shifted interpolating functions $(A(k_1, k_2, \omega_1, \omega_2))$ weighted by the DFT coefficients $H(k_1, k_2)$. The DFT coefficients are called frequency samples as they exactly specify the value of the frequency response at uniformly spaced frequencies. For designing frequency sampling filters, the majority of the frequency samples are given specific values depending on the frequency response being approximated. The remaining unspecified frequency samples are left as free variables to be optimized according to some minimization criterion. The details of frequency sampling design of two-dimensional low-pass filters will be left for the next section.

For designing optimal (equiripple) filters whose peak error of approximation, over several closed areas in the ω_1, ω_2 plane, is minimized, (4) can be applied directly.[2] In this case, the filter coefficients are the impulse response samples that are chosen to minimize the approximation error using an optimization procedure. Since the filter frequency response of (4) is still linear in the filter coefficients $(h(n_1, n_2))$, optimization procedures similar to those used to design frequency sampling filters may be applied. The details of this method are discussed in a later section.

Finally, it can be observed from (4) that if certain symmetries are maintained in the filter impulse response coefficients (similar statements can be made about the DFT coefficients), the filter's frequency response becomes purely real, to within a linear phase shift in each of the two dimensions. The necessary symmetries are

$$h(n_1, n_2) = h(N_1 - 1 - n_1, n_2) = h(n_1, N_2 - 1 - n_2). \tag{14}$$

The application of (14) to (4) gives (assuming N_1, N_2 even)

[1] The terminology $H(k_1, k_2)$ rather than $H(e^{j\omega_{k_1}}, e^{j\omega_{k_2}})$ is used for notational convenience.

[2] Equivalently, by making all the DFT coefficients variable, (12) can be applied directly for designing optimal (equiripple) filters.

$$H(e^{j\omega_1}, e^{j\omega_2}) = e^{-j\omega_1((N_1-1)/2)}e^{-j\omega_2((N_2-1)/2)} \cdot \sum_{n_1=0}^{(N_1/2)-1} \sum_{n_2=0}^{(N_2/2)-1} h(n_1, n_2) \cdot \cos\left[\left(\frac{N_1-1}{2}\right) - n_1\right] \cdot \cos\left[\left(\frac{N_2-1}{2}\right) - n_2\right] \tag{15}$$

which is purely real, except for the linear phase terms outside the summation.

Design of Frequency Sampling Filters

For the specific case of linear phase frequency sampling filters, (12) and (13) can be modified through use of the symmetry relations on the DFT coefficients which are of the form

$$H(k_1, k_2) = |H(k_1, k_2)| e^{j\theta(k_1,k_2)} \tag{16}$$

where

$$|H(k_1, k_2)| = |H(k_1, N_2 - k_2)| = |H(N_1 - k_1, k_2)| \tag{17}$$

$$\theta(k_1, k_2) = \theta(k_1) + \theta(k_2) \tag{18}$$

$$\theta(k_1) = \begin{cases} -\dfrac{2\pi}{N_1} k_1\left(\dfrac{N_1-1}{2}\right), & k_1 = 0, 1, \cdots, NU \\ \dfrac{2\pi}{N_1}(N_1 - k_1)\left(\dfrac{N_1-1}{2}\right), & k_1 = NU+1, \cdots, N_1 - 1 \end{cases} \tag{19}$$

$$\theta(k_2) = \begin{cases} -\dfrac{2\pi}{N_2} k_2\left(\dfrac{N_2-1}{2}\right), & k_2 = 0, 1, \cdots, NV \\ \dfrac{2\pi}{N_2}(N_2 - k_2)\left(\dfrac{N_2-1}{2}\right), & k_2 = NV+1, \cdots, N_2 - 1 \end{cases} \tag{20}$$

and

$$NU = \begin{cases} N_1/2, & N_1 \text{ even} \\ \dfrac{N_1-1}{2}, & N_1 \text{ odd} \end{cases} \tag{21}$$

$$NV = \begin{cases} N_2/2, & N_2 \text{ even} \\ \dfrac{N_2-1}{2}, & N_2 \text{ odd.} \end{cases} \tag{22}$$

Whenever N_1 or N_2 is even, the additional constraints

$$\theta(k_1, N_2/2) = \theta(N_1/2, k_2) = 0 \tag{23}$$

$$H(k_1, N_2/2) = H(N_1/2, k_2) = 0 \tag{24}$$

must be maintained.

Applying these constaints into (12) and (13) yields the following equations (after considerable arithmetic manipulations):

$$H(e^{j\omega_1}, e^{j\omega_2}) = \exp\left[-j\left[\omega_1\left(\frac{N_1-1}{2}\right) + \omega_2\left(\frac{N_2-1}{2}\right)\right]\right] \cdot \frac{1}{N_1N_2} \cdot \left[H(0,0)\alpha(\omega_1, N_1)\alpha(\omega_2, N_2) + \sum_{k_1=1}^{NU} |H(k_1, 0)| \alpha(\omega_2, N_2)\beta(\omega_1, k_1, N_1) + \sum_{k_2=1}^{NV} |H(0, k_2)| \alpha(\omega_1, N_1)\beta(\omega_2, k_2, N_2) + \sum_{k_1=1}^{NU}\sum_{k_2=1}^{NV} |H(k_1, k_2)| \beta(\omega_1, k_1, N_1)\beta(\omega_2, k_2, N_2)\right] \tag{25}$$

where

$$\alpha(\omega, N) = \frac{\sin \dfrac{\omega N}{2}}{\sin \dfrac{\omega}{2}} \tag{26}$$

$$\beta(\omega, k, N) = \frac{\sin\left[\left(\dfrac{\omega}{2} - \dfrac{\pi k}{N}\right)N\right]}{\sin\left(\dfrac{\omega}{2} - \dfrac{\pi k}{N}\right)} + \frac{\sin\left[\left(\dfrac{\omega}{2} + \dfrac{\pi k}{N}\right)N\right]}{\sin\left(\dfrac{\omega}{2} + \dfrac{\pi k}{N}\right)} \tag{27}$$

$$NU = \begin{cases} N_1/2 - 1, & N_1 \text{ even} \\ (N_1 - 1)/2, & N_1 \text{ odd} \end{cases} \tag{28}$$

$$NV = \begin{cases} N_2/2 - 1, & N_2 \text{ even} \\ (N_2 - 1)/2, & N_2 \text{ odd.} \end{cases} \tag{29}$$

Although cumbersome in appearance, (25) is seen to consist basically of a sum of simple interpolating functions that are shifted (up or down) in frequency depending on k_1 and k_2. As such, (25)—without the leading linear phase terms—is the basic design equation for frequency sampling filters.

To apply (25) to the design of a circularly symmetric low-pass filter, consider Fig. 1 in which the filter response regions are plotted in the ω_1, ω_2 plane. The passband is the region where

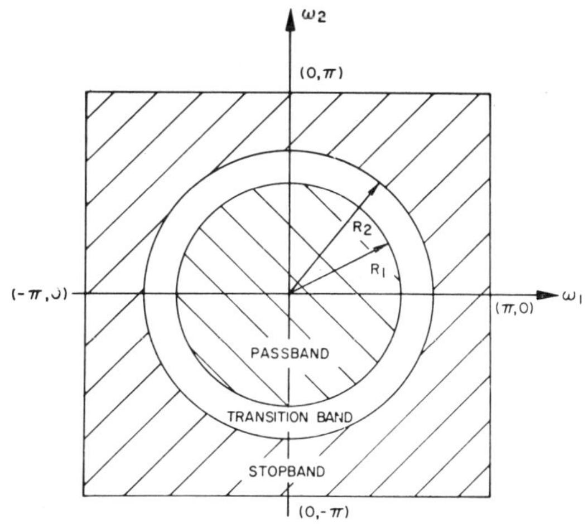

Fig. 1. Regions in the (ω_1, ω_2) plane over which the filter response is specified for a low-pass filter.

$$\rho(\omega_1, \omega_2) = (\omega_1^2 + \omega_2^2)^{1/2} \le R_1. \tag{30}$$

The transition band is the region where

$$R_1 < \rho(\omega_1, \omega_2) < R_2 \tag{31}$$

and the stopband is the region where

$$\rho(\omega_1, \omega_2) \ge R_2. \tag{32}$$

With respect to Fig. 1, the DFT points, or frequency samples, are located on an N_1 by N_2 grid of points in the ω_1, ω_2 plane as shown in Fig. 2 for $N_1=9$, $N_2=9$. To design a frequency sampling approximation to the ideal circularly symmetric response

$$\hat{H}(e^{j\omega_1}, e^{j\omega_2}) = \begin{cases} 1, & \rho(\omega_1, \omega_2) \le R_1 \\ 0, & \rho(\omega_1, \omega_2) \ge R_2 \end{cases} \tag{33}$$

the DFT values that occur inside the passband are given a value of 1.0; those that occur in the stopband are given a value 0.0; those that lie in the transition band are free variables whose value will be chosen by a minimization criterion.

To simplify both the equations and the computation involved in performing the desired minimizations, the following assumptions will be used in the remainder of this paper. First, we assume $N_1 = N_2 =$ odd integer. Next, we assume

$$H(k_1, k_2) = H(k_2, k_1) \tag{34}$$

i.e., there is symmetry across the 45° diagonals in the ω_1, ω_2 plane. This is a reasonable assumption for designing circularly symmetric filters as (34) is valid in the ideal case of zero approximation error. Finally, we assume the linear phase term in (25) can be disregarded in terms of the design procedure.

With these assumptions in mind, (25) can be simplified to the form

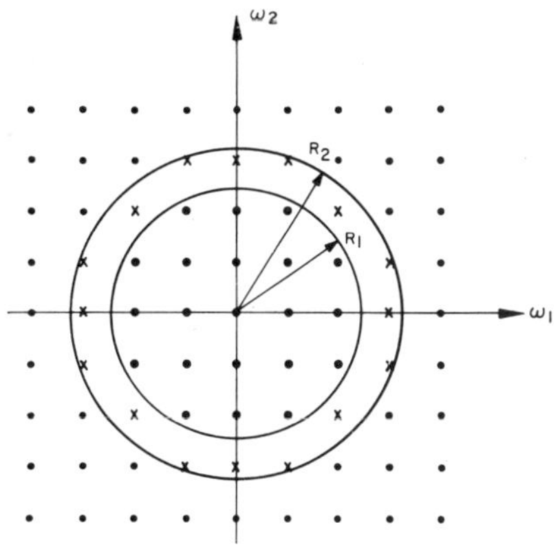

Fig. 2. Locations of frequency samples (DFT points) for a 9 by 9 grid.

$$H(e^{j\omega_1}, e^{j\omega_2}) = H_F(\omega_1, \omega_2) + \sum_{m=1}^{M} T_m H_m(\omega_1, \omega_2) \tag{35}$$

where $H_F(\omega_1, \omega_2)$ represents the contribution of the fixed DFT coefficients (the 1.0's), T_m represents the amplitude of the mth DFT coefficient in the transition band, $H_m(\omega_1, \omega_2)$ represents the interpolation function appropriate to the mth transition coefficient, and M is the total number of transition coefficients. The pertinent design equations can now be put in the form

$$1 - \alpha\delta \le H(e^{j\omega_1}, e^{j\omega_2}) \le 1 + \alpha\delta, \qquad \text{for } \rho(\omega_1, \omega_2) \le R_1 \tag{36}$$

$$-\delta \le H(e^{j\omega_1}, e^{j\omega_2}) \le \delta, \qquad \text{for } \rho(\omega_1, \omega_2) \ge R_2 \tag{37}$$

where δ represents the peak approximation error in the stopband, and $\alpha\delta$ (α any constant) represents the peak approximation error in the passband. The design goal is to choose the transition samples so as to minimize δ.

The above problem is readily seen to be a linear programming [7] problem by evaluating (36) and (37) at a dense set of points in the passband and stopband and by writing the equations explicitly as linear inequalities in the form

$$\left.\begin{aligned} \sum_{m=1}^{M} T_m H_m(\omega_1, \omega_2) - \alpha\delta &\le 1 - H_F(\omega_1, \omega_2) \\ -\sum_{m=1}^{M} T_m H_m(\omega_1, \omega_2) - \alpha\delta &\le -1 + H_F(\omega_1, \omega_2) \end{aligned}\right\} \text{for all points with } \rho(\omega_1, \omega_2) \le R_1 \tag{38}$$

$$\left.\begin{aligned} \sum_{m=1}^{M} T_m H_m(\omega_1, \omega_2) - \delta &\le -H_F(\omega_1, \omega_2) \\ -\sum_{m=1}^{M} T_m H_m(\omega_1, \omega_2) - \delta &\le H_F(\omega_1, \omega_2) \end{aligned}\right\} \text{for all points with } \rho(\omega_1, \omega_2) \ge R_2 \tag{39}$$

choosing $(T_1, T_2, \cdots, T_M)$ to minimize δ.

TABLE I

Stopband Attenuation for Low-Pass Filters (Filter Attenuation in Decibels)

	$(R_2-R_1)\cdot(12.5/\pi)$				
$R_1\cdot(12.5/\pi)$	1	1.5	2	2.5	3
0			−42.8		−69.28
1			−37.37		−65.08
2			−42.06		
3			AF		−67.92
4			−38.56		
5			−41.83		NC
6			AF		
7			AF		
8			NC		
8.5				NC	
9			−44.68		
9.5		−31.85		NC	
10	−14.42	NC	−48.35		
10.5	−21.87	−33.72			
11	−11.25				

AF = algorithm failed for undetermined reasons.
NC = convergence was not attained within a specified number of iterations.

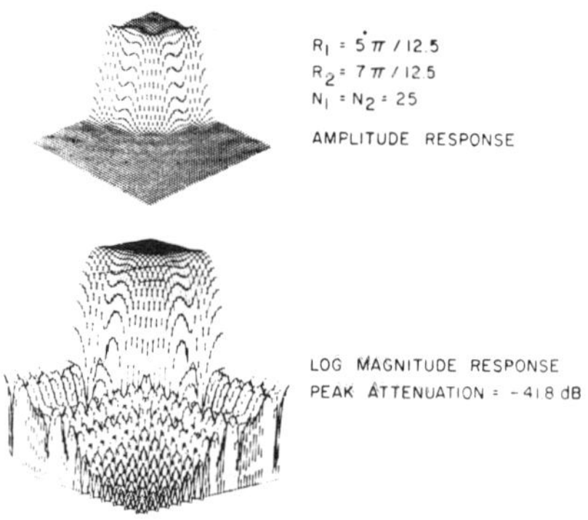

Fig. 3. Amplitude and log magnitude response of a typical low-pass filter for a transition width of $R_2-R_1=2\pi/12.5$.

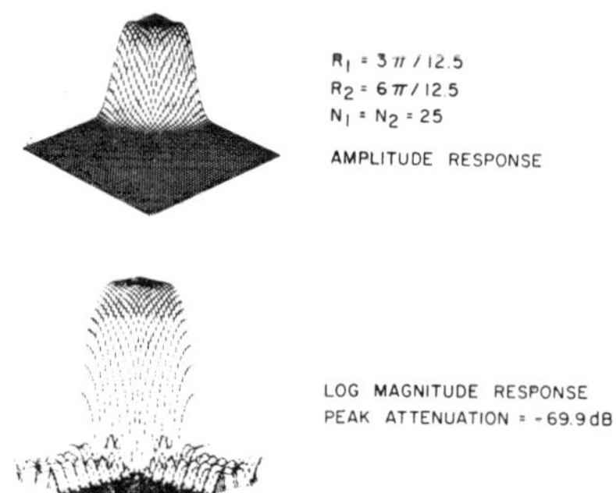

Fig. 4. Amplitude and log magnitude response of a typical low-pass filter for a transition width of $R_2-R_1=3\pi/12.5$.

Results of Low-Pass Filter Design

The linear programming problem of (38) and (39) was solved on a large central computer (GE-635) using the scientific subroutine APMM [12]. A 64–1 interpolation was used throughout the stopband and passband to evaluate $H(e^{j\omega_1}, e^{j\omega_2})$ over a fine grid. Table I presents a summary of results[3] obtained for the case where $N_1=N_2=N=25$ and $\alpha \gg 1$, i.e., in-band ripple was not constrained. The entries in this table give values for 20 $\log_{10}\delta$ as a function of R_1, the in-band radius, and R_2-R_1, the width of the transition band. For many of the entries in the table, the linear program subroutine did not converge within a specified number of iterations (on the order of 200) and the run was terminated. Each filter that was designed required on the order of 6-min computation time to determine the transition coefficients. This was due to the large number of constraint equations (on the order of several thousand), even though the number of variable coefficients was small (on the order of 10).

[3] The exact filter coefficients corresponding to the entries in Table I are available in [13].

If the passband ripple is unconstrained, the results of Table I indicate that a transition width of $2(\pi/12.5)$ gives about 40-dB stopband attenuation, whereas a width of $3(\pi/12.5)$ gives about 60-dB attenuation. Figs. 3 and 4 show two-dimensional perspectives [14] of the frequency response of two of the filters in Table I. Fig. 3 shows linear and log magnitude responses for a filter with $R_1=5\pi/12.5$, $R_2-R_1=2\pi/12.5$. Fig. 4 shows linear and log magnitude responses for a filter with $R_1=3\pi/12.5$, $R_2-R_1=3\pi/12.5$. These figures clearly show that the peak in-band ripple occurs at the edges of the passband, and the peak out-of-band ripple occurs at the edges of the stopband. (The filter response only in the region $0\le\omega_1,\ \omega_2\le\pi$ is shown in these figures as the response in the rest of the $\omega_1,\ \omega_2$ plane is determined from symmetry considerations.)

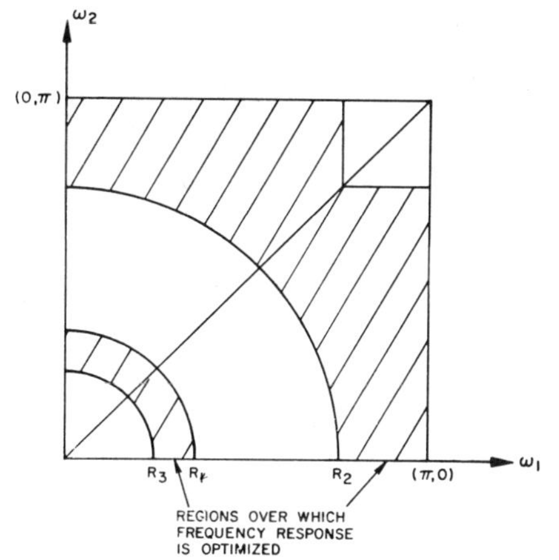

Fig. 5. Modified regions in the (ω_1, ω_2) plane over which the filter response is optimized for a circularly symmetric low-pass filter.

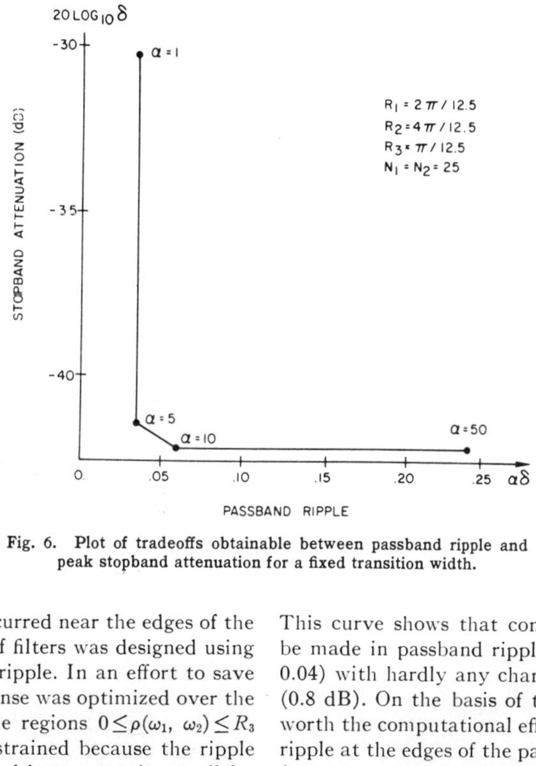

Fig. 6. Plot of tradeoffs obtainable between passband ripple and peak stopband attenuation for a fixed transition width.

Since the worst ripples occurred near the edges of the transition band, a new set of filters was designed using the constraint on passband ripple. In an effort to save computation, the filter response was optimized over the region shown in Fig. 5. The regions $0 \leq \rho(\omega_1, \omega_2) \leq R_3$ and $\omega_1 > \omega_2 > R_2$ were unconstrained because the ripple in the passband and stopband becomes quite small far from the band edges. A filter was designed with $R_3 = \pi/12.5$, $R_1 = 2\pi/12.5$, $R_2 = 4\pi/12.5$, and with $N_1 = N_2 = 25$. Values of α of 50, 10, 5, and 1 were used. Fig. 6 gives a plot of the measured tradeoffs between the stopband ripple $20 \log_{10} \delta$ and the passband ripple $\alpha\delta$. This curve shows that considerable improvement can be made in passband ripple (a reduction from 0.24 to 0.04) with hardly any change in stopband attenuation (0.8 dB). On the basis of these results, it would seem worth the computational effort to constrain the in-band ripple at the edges of the passband. Figs. 7–10 show the frequency responses for the cases $\alpha = 50$ and $\alpha = 1$, as well as magnitude contour plots (magnitude ≥ 0.75) for these two cases. In the contour plots, a dotted contour is drawn each 0.05. The passband and the transition region for the $\alpha = 1$ case are significantly more circularly symmetric than for the $\alpha = 50$ case, because the $\alpha = 50$

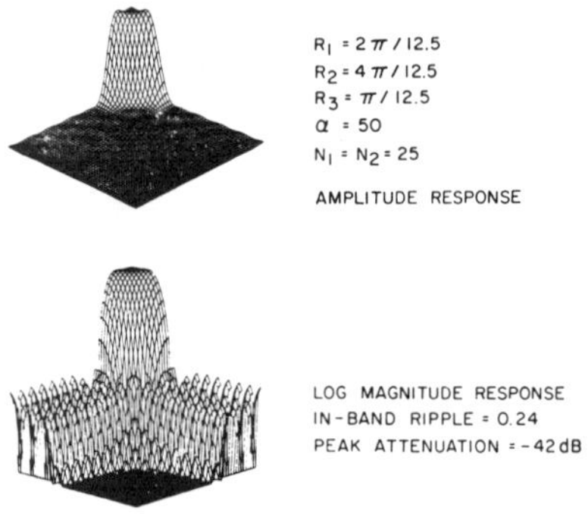

Fig. 7. Amplitude and log magnitude response of a low-pass filter with $\alpha = 50$.

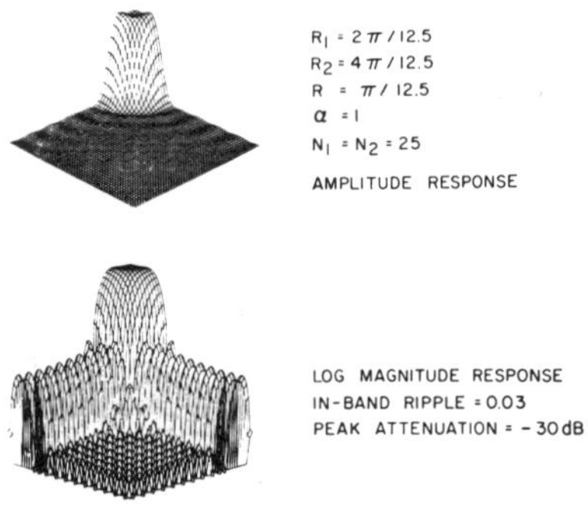

Fig. 8. Amplitude and log magnitude response of a low-pass filter with $\alpha = 1$.

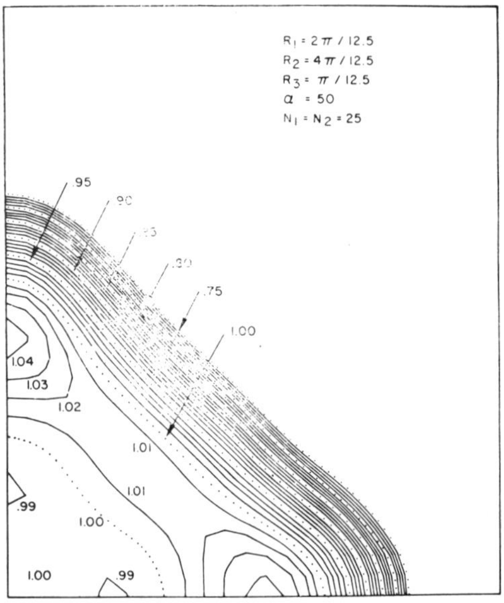

Fig. 9. Contour plot of passband and transition band response for the filter of Fig. 7.

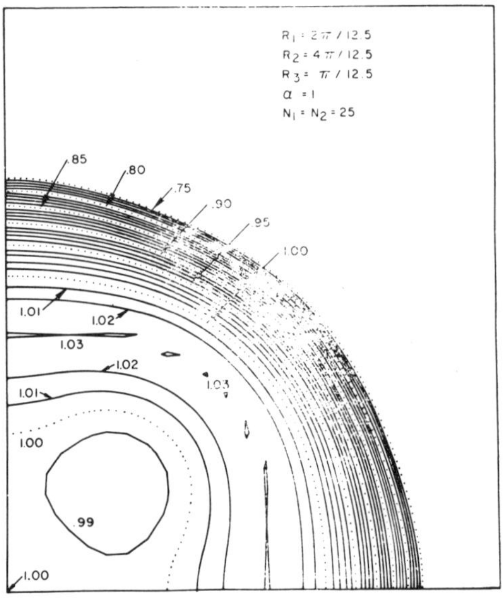

Fig. 10. Contour plot of passband and transition band response for the filters of Fig. 8.

case essentially has an unconstrained passband and thus the requirements for maintaining a good approximation to circular symmetry in the passband are not met.

Design of Optimal (Equiripple) Filters

By allowing all the impulse response coefficients to vary, or equivalently all the DFT coefficients, filters can be designed which are optimal in the sense of Chebyshev approximation over closed regions. For the case when all the DFT coefficients are varied, as was done here, the filter frequency response is evaluated using (25) and treating all the filter coefficients $H(k_1, k_2)$ as unknowns. Optimal approximations to circularly symmetric filters may be designed (using linear programming techniques) by evaluating (25) at a dense set of points in both the passband and the stopband and minimizing both the stopband ripple δ and the passband ripple $\alpha\delta$.

Because of the increased amount of computation, necessitated both by the need to evaluate $H(e^{j\omega_1}, e^{j\omega_2})$ over the entire ω_1, ω_2 plane (except, of course, the transition band) and by the increased number of variables, the size of the impulse responses considered was reduced to $N_1 = N_2 = 9$. Within these limitations, four optimal (equiripple) filters were designed. A summary of the results for these four cases is given in Table II. Figs. 11 and 12 show the frequency response of the

TABLE II
Optimal Filters

Filter 1	$R_1 = 2\pi/4.5$, $R_2 = 3\pi/4.5$, $\alpha = 1$ in-band ripple = 0.0867 out-of-band attenuation = −21.24 dB
Filter 2	$R_1 = \pi/4.5$, $R_2 = 2\pi/4.5$, $\alpha = 10$ in-band ripple = 0.287 out-of-band attenuation = −30.84 dB
Filter 3	$R_1 = 1.5\pi/4.5$, $R_2 = 3\pi/4.5$, $\alpha = 10$ in-band ripple = 0.079 out-of-band attenuation = −41.97 dB
Filter 4	$R_1 = 2\pi/4.5$, $R_2 = 3\pi/4.5$, $\alpha = 10$ in-band ripple = 0.235 out-of-band attenuation = −32.51 dB

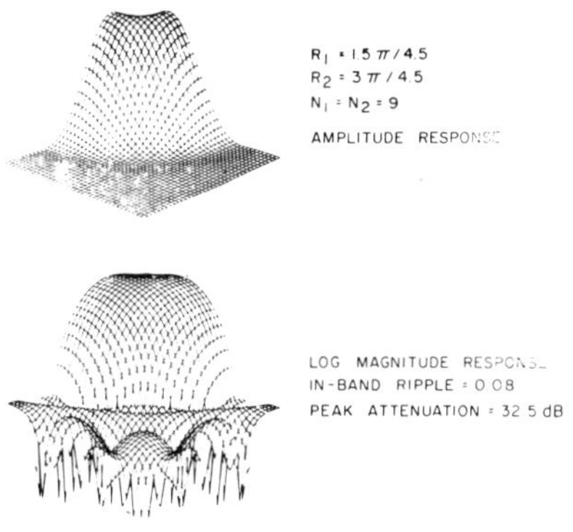

Fig. 11. Amplitude and log magnitude response of an optimal low-pass filter.

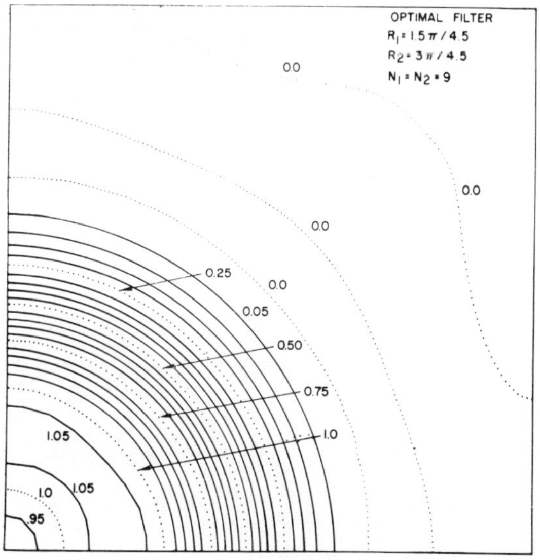

Fig. 12. Contour plot of the response of the filter of Fig. 11.

third filter in Table II and a contour plot of the log magnitude response, respectively. As the contour lines in the passband and transition region show, the passband response is a good approximation to a circularly symmetric response. Fig. 11 also shows the interesting phenomenon that there are equiripple ridges, as well as peaks. Thus the ripple of the frequency response may attain a maximum or minimum in a continuous region of the ω_1, ω_2 plane (corresponding to the equiripple ridge), as well as at isolated unconnected values. This is an added complexity to the one-dimensional design case where, because of the well-ordered frequency scale, the ripple peaks alternate between maxima and minima. This phenomenon suggests that simply counting points at which $H(e^{j\omega_1}, e^{j\omega_2})$ attains a maximum value can lead to significant errors in designing equiripple filters. For this reason several of the one-dimensional design algorithms [3], [5], [6] are not readily extended to two dimensions.

Conclusions

This paper represents preliminary work on techniques for designing two-dimensional digital filters. It shows that the frequency sampling and optimal (equiripple) design methods can be carried over to the two-dimensional design problem directly, although at considerable computational cost. Future efforts along these lines should be aimed at lowering these costs.

Acknowledgment

The authors wish to thank Mrs. N. Graham for her work in devising a novel algorithm for making the perspective plots shown.

References

[1] J. F. Kaiser, "Digital filters," in *Systems Analysis by Digital Computer*, F. Kuo and J. F. Kaiser, Eds. New York: Wiley, 1966, ch. 7.
[2] B. Gold and C. M. Rader, *Digital Processing of Signals*. New York: McGraw-Hill, 1969.

[3] H. W. Schuessler, "On the approximation problem in the design of digital filters," in *Proc. 5th Annual Princeton Conf. Information Sciences and Systems*, Mar. 1971, pp. 54–63.

[4] L. R. Rabiner, B. Gold, and C. A. McGonegal, "An approach to the approximation problem for nonrecursive digital filters," *IEEE Trans. Audio Electroacoust.*, vol. AU-18, pp. 83–106, June 1970.

[5] E. Hofstetter, A. V. Oppenheim, and J. Siegel, "A new technique for the design of nonrecursive digital filters," in *Proc. 5th Annual Princeton Conf. Information Sciences and Systems*, Mar. 1971, pp. 64–72.

[6] T. W. Parks and J. H. McClellan, "Chebyshev approximation for nonrecursive digital filters with linear phase," *IEEE Trans. Circuit Theory*, vol. CT-19, pp. 189–194, Mar. 1972.

[7] L. R. Rabiner, "The design of finite impulse response digital filters using linear programming techniques," *Bell Syst. Tech. J.*, Aug. 1972.

[8] ——, "Techniques for designing finite-duration impulse-response digital filters," *IEEE Trans. Commun. Technol.*, vol. COM-19, pp. 188–195, Apr. 1971.

[9] J. L. Shanks, "The design of stable two-dimensional recursive filters," in *Proc. M. J. Kelly Communications Conf.*, Univ. of Missouri, Columbia, 1970.

[10] ——, "Two-dimensional recursive filters," in *SWIEECO Rec.*, 1969, pp. 19E1–19E8.

[11] T. S. Huang, "Two-dimensional windows," *IEEE Trans. Audio Electroacoust.* (Corresp.), vol. AU-20, pp. 88–89, Mar. 1972.

[12] *IBM System/360 Programming Manual*, Scientific Subroutine Package (360 A-CM-03 X), Fortran Program APMM, pp. 283–288.

[13] J. Hu, "Frequency sampling design of two-dimensional finite impulse response digital filters," M.S. thesis, Mass. Inst. Tech., Cambridge, June 1972.

[14] N. Graham, "Perspective drawing of surfaces with hidden line elimination," *Bell Syst. Tech. J.*, vol. 51, pp. 843–861, Apr. 1972.

17

Reprinted from *IEEE Trans. Circuits Syst.* **CAS-22**(3):208–218 (1975)

Chebyshev Approximation for Two-Dimensional Nonrecursive Digital Filters

Y. KAMP AND J. P. THIRAN, MEMBER, IEEE

***Abstract*—The Remez exchange algorithm is extended for the design of two-dimensional nonrecursive digital filters approximating circularly symmetrical low-pass specifications according to a weighted Chebyshev error norm. Since the approximating function does not satisfy the Haar condition, the optimal solution is not necessarily unique and a straightforward extension of the one-dimensional exchange method may fail to converge. It is shown how the algorithm has to be complemented with a perturbation technique in order to force convergence under all circumstances. In the case of nonuniqueness the solution provided by the algorithm is a vertex of the polyhedron containing all optimal solutions, and a method is given which allows one to compute an adjacent vertex located on the same edge, thereby allowing the successive determination of all the vertices defining the polyhedron. For this case also, a procedure is described which selects, among all optimal solutions, the best one according to some additional criterion. Finally, the efficiency and accuracy aspects of the algorithm are considered and practical conclusions are drawn as an aid for the designer.**

I. Introduction

Work in the domain of two-dimensional digital filters has been up to now mainly concerned with nonrecursive designs because they escape the stability problem which is inherent in recursive filters [1] and because they can achieve a strictly linear phase which is a desirable feature for the distortionless processing of images.

Two-dimensional nonrecursive digital filters can be obtained by extending some of the methods already available for the design of one-dimensional filters. This was done for the windowing technique [2], linear programming [3], and for designs according to a maximally flat criterion [4]. Recently [5], an ingenious solution was proposed to transform a one-dimensional Chebyshev design into a two-dimensional design through a change of variables. The method is limited by the fact that the optimality of the result is tied to special classes of cutoff boundaries and, in particular, optimal approximations of specifications with exact circular symmetry cannot be achieved. This demonstrates the need for a method in which the two-dimensional nature of the problem is effectively taken into account. In this line the present paper investigates the problems connected with the Chebyshev approximation by functions of two variables. As a result, an extension of the Remez exchange algorithm [6] is presented for the design of two-dimensional nonrecursive filters which approximate low-pass specifications having exactly circular pass- and stop-band limits. Moreover, most of the results obtained here also apply to the design of a much wider class of filters.

Manuscript received March 11, 1974; revised July 5, 1974.

The authors are with the MBLE Research Laboratory, Brussells, Belgium.

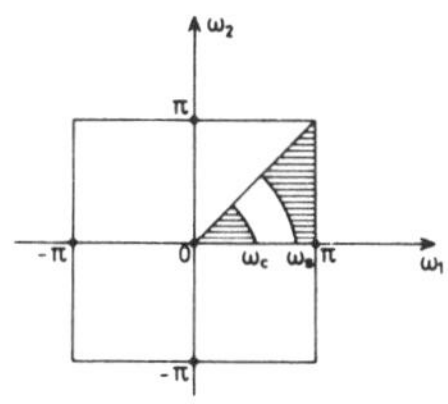

Fig. 1. Approximation domain.

II. The Approximation Problem and its Solution

A. Statement of the Problem

It has been shown in [4] that the requirement of circular symmetry imposes the following form for the frequency response of a nonrecursive two-dimensional digital filter

$$G(e^{j\omega_1},e^{j\omega_2}) = \sum_{I=0}^{N}\sum_{J=0}^{I} B_{IJ}(\cos I\omega_1 \cos J\omega_2 + \cos J\omega_1 \cos I\omega_2). \quad (1)$$

For the discussion of the approximation problem, it will prove more convenient to rewrite it as

$$f(x) = \sum_{i=1}^{m} a_i f_i(x) \quad (2)$$

where

$$\begin{aligned} f_i(x) &= \cos I\omega_1 \cos J\omega_2 + \cos J\omega_1 \cos I\omega_2, \\ x &= (\omega_1,\omega_2), \\ a_i &= B_{IJ}, \\ i &= I(I+1)/2 + J + 1, \\ m &= (N+1)(N+2)/2. \end{aligned} \quad (3)$$

The purpose is to design a circularly symmetric low-pass filter and the function to be approximated is thus

$$F(x) = \begin{cases} 1 & \text{if } \omega_1^2 + \omega_2^2 \le \omega_c^2 \\ 0 & \text{if } \omega_1^2 + \omega_2^2 \ge \omega_s^2 \end{cases}$$

where ω_c and ω_s are, respectively, the pass- and stopband limits. Since the frequency response (1) is already symmetrical with respect to both axes and both diagonals of the frequency plane, it will be sufficient to control its behavior in the shaded region of Fig. 1. In fact, this continuous domain will be replaced by a discrete finite point set $X = \{x_1,x_2,\cdots,x_n\}$ with $n > m$.

If V denotes the linear space of approximating functions (2) when $\{a_i\} \in \boldsymbol{R}^m$, the following linear Chebyshev approximation problem has to be solved: find $f^* \in V$ which minimizes the norm

$$\|f - F\| = \max_{x \in X} w(x)|f(x) - F(x)| \quad (4)$$

where the weight function $w(x)$ is strictly positive on X. The minimum of this norm is the Chebyshev deviation

$$H = \min_{f \in V} \|f - F\|. \quad (5)$$

The solution lies in a generalization of the Remez algorithm which will be summarized in the following two sections by using the approach of Stiefel [6].

B. Reference and Reference Function

The theory of the exchange algorithm relies on the following definitions. For the approximating function (2) the *characteristic vector* ϕ_j relative to a point x_j is defined by $\phi_j = [f_1(x_j), f_2(x_j), \cdots, f_m(x_j)]^T$.

A *reference* $R \subset X$ is a set of $(m+1)$ distinct points $R = \{x_1, x_2, \cdots, x_{m+1}\}$ such that the corresponding matrix $[\phi_1, \phi_2, \cdots, \phi_{m+1}]$ has rank m. The existence of at least one reference in X is equivalent to the claim that the base functions $f_i(x)$ are linearly independent on X.

A *basis* is a set of m linearly independent characteristic vectors. By definition, a reference contains at least one basis. Let then $B = [\phi_1, \phi_2, \cdots, \phi_m]$ be a basis for R. The *characteristic relation of a reference* expresses the characteristic vector ϕ_{m+1} of the reference as a linear combination of the other vectors

$$\phi_{m+1} = \sum_{j=1}^{m} \lambda_j \phi_j \quad (6)$$

where

$$\Lambda = [\lambda_1, \lambda_2, \cdots, \lambda_m]^T = B^{-1}\phi_{m+1}. \quad (7)$$

By putting

$$\lambda_{m+1} = -1 \quad (8)$$

one derives from (6)

$$\sum_{j=1}^{m+1} \lambda_j f(x_j) = 0, \qquad \forall f \in V \quad (9)$$

and $[{}^{\Lambda}_{-1}]$ is then called the *vector of characteristic coefficients.*

The *error* $h(x)$ associated with an approximating function is defined by

$$h(x) = w(x)[f(x) - F(x)]. \quad (10)$$

A function $f(x)$ is a *reference function for R* if

$$\operatorname{sgn} h(x_j) = \eta \operatorname{sgn} \lambda_j, \qquad 1 \le j \le m+1 \quad (11)$$

where $\eta = \pm 1$ and where sgn x is arbitrary for $x = 0$. By (8) the latter definition is equivalent to

$$h(x_j)h(x_{m+1})\lambda_j \le 0, \qquad 1 \le j \le m. \quad (12)$$

In the one-dimensional problem the characteristic coefficients have alternating signs and (11) becomes the well-known alternation property of the reference function. A *leveled reference function for R* satisfies (11) and in addition $|h(x_j)| = |h|$ for $1 \le j \le m+1$. Owing to (9), (10), and (11), the *reference deviation* $|h|$ of a reference can be computed as

$$|h| = \left|\sum_{j=1}^{m+1} \lambda_j F(x_j)\right| \Big/ \left|\sum_{j=1}^{m+1} |\lambda_j| w^{-1}(x_j)\right| \quad (13)$$

and the *coefficient vector* $A = [a_1, a_2, \cdots, a_m]^T$ of the leveled reference function by

$$A^T = \left[F(x_1) + \frac{h(x_1)}{w(x_1)}, \cdots, F(x_m) + \frac{h(x_m)}{w(x_m)}\right] B^{-1}. \quad (14)$$

From (9), (10), (11), and (13), Theorem 1 follows.

Theorem 1 [6]*:* If $f(x)$ is a reference function for R with associated error $h(x)$, then the reference deviation of R is given by

$$|h| = \sum_{j=1}^{m+1} |\lambda_j||h(x_j)|w^{-1}(x_j) \Big/ \sum_{j=1}^{m+1} |\lambda_j| w^{-1}(x_j).$$

C. Exchange Algorithm

Theorem 2 [6]*:* Let $f(x)$ be a reference function for R, $h(x)$ the corresponding error and $x_k \notin R$ an arbitrary point in X. There exists a point $x_l \in R$ such that by exchanging x_l with x_k one obtains a new reference R' for which $f(x)$ is still a reference function. The point to be exchanged is found according to the following rule. For the reference $\{x_1, x_2, \cdots, x_m, x_k\}$, let

$$\phi_k = \sum_{j=1}^{m} \mu_j \phi_j \quad (15)$$

be the characteristic relation with

$$M = [\mu_1, \mu_2, \cdots, \mu_m]^T = B^{-1}\phi_k. \quad (16)$$

Case 1: If

$$h(x_j)h(x_k)\mu_j \le 0, \qquad 1 \le j \le m \quad (17)$$

then $x_l = x_{m+1}$.

Case 2: If

$$E = \{x_j, 1 \le j \le m \mid h(x_j)h(x_k)\mu_j > 0\} \ne - \quad (18)$$

the index l of the point to be exchanged is such that

$$|\lambda_l/\mu_l| = \min_{j \in E} |\lambda_j/\mu_j|. \quad (19)$$

It is indeed easily shown that for the new reference R' a condition of the type (12) is satisfied. The new characteristic relation is (15) in Case 1 and results from the elimination of ϕ_l between (6) and (15) in Case 2, giving

$$\phi_{m+1} = \frac{\lambda_l}{\mu_l}\phi_k + \sum_{\substack{j=1 \\ \ne l}}^{m} \lambda_j' \phi_j \quad (20)$$

with

$$\lambda_j' = \lambda_j - \lambda_l \mu_j / \mu_l, \qquad 1 \le j \ne l \le m. \quad (21)$$

Let it be noted also that for Case 2 it is not necessary to compute the inverse of the new basis B' by matrix inversion.

Since B and B' only differ by one characteristic vector, $(B')^{-1}$ can directly be derived from B^{-1} [7]. This may, however, lead to important error accumulation, and it is therefore suggested to periodically restart the algorithm with a fresh inverse generated by matrix inversion in order to keep the reference well behaved.

Method of Ascent: Using the preceding theorems a monotonically increasing sequence of reference deviations is constructed as follows.

Let $f(x)$ be the leveled reference function for R and $h(x)$ the associated error. If

$$\max_X |h(x)| = |h| \tag{22}$$

$f(x)$ is an *optimal solution* for the Chebyshev approximation $F(x)$ on X [6]. Else let $x_k \notin R$ be a point of X such that

$$|h(x_k)| > |h|. \tag{23}$$

By the above exchange Theorem 2, a new reference R' is constructed whose characteristic relation is either (15) or (20). The new reference deviation $|h'|$ results then from Theorem 2.

Case 1:

$$|h'| = \frac{|h(x_k)|w^{-1}(x_k) + |h| \sum_{j=1}^{m} |\mu_j| w^{-1}(x_j)}{w^{-1}(x_k) + \sum_{j=1}^{m} |\mu_j| w^{-1}(x_j)} \tag{24}$$

and by (23), $|h'| > |h|$.

Case 2:

$$|h'| = \frac{|h(x_k)\lambda_l \mu_l^{-1}| w^{-1}(x_k) + |h| \left[w^{-1}(x_{m+1}) + \sum_{\substack{j=1\\ \neq l}}^{m} |\lambda_j'| w^{-1}(x_j) \right]}{|\lambda_l \mu_l^{-1}| w^{-1}(x_k) + w^{-1}(x_{m+1}) + \sum_{\substack{j=1\\ \neq l}}^{m} |\lambda_j'| w^{-1}(x_j)} \tag{25}$$

and (23) again implies $|h'| > |h|$ except in the case $\lambda_l = 0$ which will be considered in the next section. The fact that the reference deviation is strictly increasing is of prime importance for the convergence proof of the algorithm because in a finite number of iterations a final reference is obtained for which (22) holds.

Multiple Exchange: In order to reduce the number of iterations, it is advantageous to increase as much as possible the new reference deviation $|h'|$. To this end, it would be preferable to introduce in the new reference not only a single point but several points where the absolute value of the error exceeds the reference deviation $|h|$. For the one-dimensional case [8], this is easy to achieve because the sign alternation of the λ_j allows the simultaneous exchange of all the reference points. For two-variable approximations there is no such simple rule for the sign of the characteristic coefficients but one can try to perform a multiple exchange in the following way.

In the neighborhood of each reference point, the local maximum of the absolute value of the error will be retained if its value is larger than the reference deviation. Let then $(x_k^{(1)}, x_k^{(2)}, \cdots, x_k^{(r)})(1 \le r \le m+1)$ be the set of the distinct local maxima. One calculates the vector $M_1 = B^{-1}\phi_k^{(1)}$ where $\phi_k^{(1)}$ denotes the characteristic vector of $x_k^{(1)}$ and the exchange Theorem 2 gives then the point x_l of the reference R which has to be exchanged. Let R_1 be the new reference, B_1 the corresponding basis. The same sequence of operations is repeated for the second maximum $x_k^{(2)}$ by computing $M_2 = B_1^{-1}\phi_k^{(2)}$ but this second exchange is effectively performed only if it does not eliminate $x_k^{(1)}$. One proceeds in this way for all local maxima with the rule that an exchange will not be carried out if it would eliminate one of the previously introduced maxima.

D. Perturbation Technique for Degenerate Problems

An approximation problem becomes *degenerate for a reference R* if the corresponding characteristic relation contains at least one vanishing λ_j. The occurrence of degeneracies is a direct consequence of the fact that practically no two-variable functions satisfy the Haar condition [9]. According to the methods of ascent described in the preceding section, one has Theorem 3.

Theorem 3: A *static exchange* $|h'| = |h|$ occurs if and only if the problem is degenerate with respect to R and if among the points x_j of R for which $\lambda_j = 0$, one at least satisfies the inequality $h(x_j)h(x_k)\,\mu_j > 0$ where μ_j is defined by (16).

The possibility of static exchanges for degenerate problems entails that the same reference can be repeated periodically. In these cases of cycling, the above exchange algorithm does not converge and one has to resort to a special perturbation technique. This technique was originally developed in linear programming [7] to cope with similar convergence problems and was later on adapted to suit the exchange algorithm. The adaptation presented here is similar to but more efficient than that of Bittner [10] and Descloux [11]. Like in linear programming, the perturbation technique is only of theoretical importance because it ensures convergence without exception, but in practice no degenerate cases have been known to cycle for this problem.

The perturbation technique consists in replacing the degenerate problem P by a perturbed problem $P(\varepsilon)$ satisfying the twofold condition that it never becomes degenerate and that it coincides with the original problem when ε tends to zero. In this way, the convergence of the perturbed problem also insures that of the degenerate problem. For convenience it will be assumed that the points of R have been ordered such that $\lambda_j = 0\ (1 \le j \le p)$ and $\lambda_j \ne 0\ (p+1 \le j \le m)$. According to Theorem 3, it is also assumed that at least one of the points $x_j\ (1 \le j \le p)$ is an element of the set E defined in (18). To construct the perturbed problem, the vector ϕ_{m+1} of the degenerate problem is replaced by $\phi_{m+1}(\varepsilon) = \phi_{m+1} + \sum_{j=1}^{p} \varepsilon^j \phi_j$

where ε is a small positive quantity. The characteristic relation of the perturbed problem is then

$$\phi_{m+1}(\varepsilon) = \sum_{j=1}^{m} \lambda_j(\varepsilon)\phi_j \tag{26}$$

where, by (7),

$$\begin{cases} \lambda_j(\varepsilon) = \varepsilon^j, & 1 \leq j \leq p \\ \lambda_j(\varepsilon) = \lambda_j \neq 0, & p+1 \leq j \leq m \end{cases} \tag{27}$$

thus showing that the perturbed problem is nondegenerate. Since by hypothesis some indices j $(1 \leq j \leq p)$ belong to E, the exchange Theorem 2 will give a point x_l with $1 \leq l \leq p$, whence

$$\lambda_l(\varepsilon) = \varepsilon^l. \tag{28}$$

The characteristic relation (20) of the new reference R' becomes here

$$\phi_{m+1} = \lambda_k'(\varepsilon)\phi_k + \sum_{\substack{j=1 \\ \neq l}}^{m} \lambda_j'(\varepsilon)\phi_j$$

with

$$\begin{cases} \lambda_j'(\varepsilon) = \lambda_j(\varepsilon) - \lambda_l(\varepsilon)\mu_j/\mu_l, & (1 \leq j \neq l \leq m) \\ \lambda_k'(\varepsilon) = \lambda_l(\varepsilon)/\mu_l. \end{cases} \tag{29}$$

The new reference deviation $|h'(\varepsilon)|$ is still given by (25) provided that the characteristic coefficients are replaced by (29). Owing to (28),

$$\lim_{\varepsilon \to 0} |h'(\varepsilon)| = |h|$$

and the exchange of the perturbed problem corresponds to a static exchange for the original problem. In addition, it is seen that the $\lambda_j'(\varepsilon)$ $(j = k,\ 1 \leq j \neq l \leq m)$ are polynomials without constant term and which are linearly independent because $\lambda_j'(\varepsilon)$ is the only polynomial having a term in ε^j. In particular, $\lambda_j'(\varepsilon) \not\equiv 0$ and the perturbed problem is still nondegenerate.

In order to compute the leveled reference function for R' at the next iteration, one should take

$$\operatorname{sgn} h'(x_j) = \eta \lim_{\varepsilon \to 0} [\operatorname{sgn} \lambda_j'(\varepsilon)], \quad j = k, \quad 1 \leq j \neq l \leq p$$

which means that the sign is determined by the coefficient of the lowest power of ε in $\lambda_j'(\varepsilon)$. For this iteration, a new set E' has to be defined analogously to (18) and the perturbed problem continues if among the $\lambda_j'(\varepsilon)$ $(j = k, 1 \leq j \neq l \leq p)$ some have still indices belonging to E'. During the subsequent iterations, the perturbed problem remains nondegenerate because linear combinations of the linearly independent polynomials $\lambda_j'(\varepsilon)$ $(j = k, 1 \leq j \neq l \leq p)$ are again linearly independent. All the exchanges of the perturbed problem correspond to static exchanges for the original one.

If among the $\lambda_j'(\varepsilon)$ $(j = k, 1 \leq j \neq l \leq p)$ no index belongs to E', then the perturbed problem stops. Indeed, a point x_q $(k \neq q \geq p + 1)$ is then exchanged such that by (29) and (27) $\lambda_q'(0) = \lambda_q \neq 0$. A relation similar to (25) for the new reference R'' shows then that

$$\lim_{\varepsilon \to 0} |h''(\varepsilon)| > |h|$$

which corresponds to a nonstatic exchange for the original problem.

Bittner [10] makes the perturbed problem unnecessarily complicated by defining $\phi_{m+1}(\varepsilon) = \phi_{m+1} + \sum_{j=1}^{n} \varepsilon^j \phi_j$ instead of (26). The resulting computations are much more time-consuming since normally $p < m \ll n$. Descloux [11], on the other hand, oversimplifies the perturbed problem by taking $\phi_{m+1}(\varepsilon) = \phi_{m+1} + \varepsilon \sum_{j=1}^{p} \alpha_j\phi_j$ where the α_j are arbitrary constants. Indeed, (27) becomes then $\lambda_j(\varepsilon) = \varepsilon\alpha_j$ $(1 \leq j \leq p)$ and it is no longer possible to guarantee that the coefficients $\lambda_j'(\varepsilon)$ $(1 \leq j \neq l \leq p)$ in (29) are nonzero. This means that the perturbed problem can be degenerate and its convergence is thus no longer insured.

For the problem defined in Section II-A several cases of degeneracy can theoretically exist. However, experience has shown that in practice only one type of degeneracy actually occurs which is characterized by the fact that the reference contains exactly $(N + 2)$ points on the frequency axis ω_1, in short, degeneracy with respect to the ω_1-axis. At these points the base functions $f_i(x)$ in (3) reduce to linear combinations of $\cos I\omega_1$ $(0 \leq I \leq N)$ and therefore, like in the one-variable case, the characteristic vector of an arbitrary point on ω_1 is a linear combination of the characteristic vectors corresponding to the remaining $(N + 1)$ points on ω_1. By the definitions of Section II-B, a basis for such a reference contains exactly $(N + 1)$ points on ω_1 while the $(N + 2)$th point plays the role of x_{m+1} in R. Consequently, in the vector Λ only the characteristic coefficients corresponding to the points on ω_1 are nonzero.

III. Nonuniqueness

Since the approximating function (2) does not satisfy the Haar condition on X, the Chebyshev approximation problem stated in Section II-A has not necessarily a unique solution.

A. The Set of Best Chebyshev Approximations

In order to compute all best approximations, the approach used in [10] will now be extended. It is based on the concept of *extreme optimal solution* which is an optimal solution such that 1) the error has at least $m + 1$ *extremal points*, i.e., points where it attains the Chebyshev deviation, and 2) the approximating function is a reference function for $m + 1$ extremal points.

By the assumption of independent base functions on X, the set of best Chebyshev approximations are characterized by the following theorem [11].

Theorem 4: The set C of best Chebyshev approximations is bounded. In the m-dimensional space of the coefficient vector A, it is a convex polyhedron whose vertices are the extreme optimal solutions.

Since a convex polyhedron is the convex hull of its vertices, the determination of C amounts to computing all extreme optimal solutions. Starting from the extreme optimal solution obtained by the exchange algorithm, we will first determine an *adjacent extreme optimal solution* which belongs to the same *edge* of C, i.e., the set of optimal solutions corresponding to the same $m - 1$ extremal points with independent characteristic vectors.

In order to express the dependence of $h(x)$ on the coefficient vector A, one rewrites the error function (10) at the point x_i as

$$h_i(A) = w_i(\phi_i^T A - F_i) \quad (30)$$

with $w_i = w(x_i)$ and $F_i = F(x_i)$. One can then define C as the admissible solutions A of the set of inequalities

$$\begin{cases} h_i(A) \leq H \\ -h_i(A) \leq H, \quad (1 \leq i \leq n) \end{cases} \quad (31)$$

where H is the Chebyshev deviation.

Let A be the extreme optimal solution obtained by the exchange algorithm with no more than $m + 1$ extremal points $\{x_j; 1 \leq j \leq m + 1\}$ such that one has (12) with (6), and suppose we want to compute the adjacent extreme optimal solution A^* located on the edge $h_j(A^*) = h_j(A)$ $(1 \leq j \leq m - 1)$. From (30), this edge is still defined by

$$\phi_j^T(A^* - A) = 0, \quad (1 \leq j \leq m - 1). \quad (32)$$

The extreme optimal solution A^* will be different from A if $\operatorname{sgn} h_m(A)h_m(A^*) < H$ or, by (30), if

$$h_m(A)\phi_m^T(A^* - A) < 0. \quad (33)$$

The infinite number of A^* satisfying (32) together with (33) must now be restricted by the remaining constraints of (31). First, one has $\operatorname{sgn} h_{m+1}(A)h_{m+1}(A^*) \leq H$ which, by means of (30), (6), (32), and (33), can be rewritten as $h_{m+1}(A) \cdot h_m(A)\lambda_m \geq 0$. Obviously, this inequality is compatible with (12) only if $\lambda_m = 0$ which yields

$$h_{m+1}(A^*) = h_{m+1}(A). \quad (34)$$

Therefore, the number of extreme optimal solutions which are adjacent to the solution obtained by the exchange method is equal to the number of zero characteristic coefficients corresponding to the final reference of the exchange algorithm. In particular, if every characteristic coefficient is different from zero, the Chebyshev approximation is unique. On the other hand, all best approximations have the same error at the extremal points relative to the nonzero characteristic coefficients: according to the terminology used in [12], they form a *critical point set*.

In addition to (32)–(34), the admissible values of A^* are finally restricted by the set of inequalities

$$\begin{cases} -\operatorname{sgn} h_m(A)h_m(A^*) \leq H \\ \pm h_i(A^*) \leq H, \quad (m + 1 < i \leq n) \end{cases}$$

which, by means of (30), (32), (33), and the relations $\phi_i = \sum_{j=1}^{m} \mu_{ij}\phi_j$ $(m + 1 < i \leq n)$, are equivalent to

$$\begin{cases} -\operatorname{sgn} h_m(A)\phi_m^T(A^* - A) \leq \dfrac{2H}{w_m} \\ -\operatorname{sgn} h_m(A)\phi_m^T(A^* - A) \leq \dfrac{H \mp h_i(A)}{\mp w_i\mu_{im} \operatorname{sgn} h_m(A)}, \\ \qquad \text{for } \{m + 1 < i \leq n \mid \mp\mu_{im}h_m(A) > 0\} \end{cases}$$

where the three upper and lower signs correspond to each other. Consequently, one has Theorem 5.

Theorem 5: The extreme optimal solution A^* adjacent to A and located on the edge (32) with $\lambda_m = 0$, is characterized by the extremal point set $\{x_1, \cdots, x_{m-1}, x_m^*, x_{m+1}\}$ where x_m^* corresponds to the minimum of the following expressions

$$\begin{cases} \dfrac{2H}{w_m} \\ \dfrac{H \mp h_i(A)}{w_i|\mu_{im}|}, \quad \text{for } \{m + 1 < i \leq n \mid \mp\mu_{im}h_m(A) > 0\}. \end{cases}$$

The errors at the points $\{x_1, \cdots, x_{m-1}, x_{m+1}\}$ of the old reference are unchanged whereas the error sign at x_m^* is given by $\operatorname{sgn}[-h_m(A)]$ if $x_m^* = x_m$ and by $\operatorname{sgn}[-\mu_{im}h_m(A)]$ if $x_m^* = x_i$ $(m + 1 < i \leq n)$.

When there are t zero characteristic coefficients in (6), the above theorem allows one to compute t adjacent extreme optimal solutions. If the t points x_m^* together with the error sign are identical, it is readily seen that the number of vertices of C is equal to $t + 1$: the set of best approximations is thus completely defined. Otherwise, the process must be repeated for each new extreme optimal solution and the number of vertices may be very high.

Finally, it should be pointed out that, if the optimal solution obtained by the exchange algorithm has more than $m + 1$ extremal points, the number of adjacent extreme optimal solutions can be smaller than t: in fact, the best approximation may be unique even if some coefficients of the characteristic relation (6) are zero.

B. Strict Approximation

In case of nonuniqueness of the best approximation, one is faced with the problem of choosing one of the infinite number of solutions belonging to C. One way to overcome this difficulty is the strict approximation as proposed by Rice [12]. Suppose that in the characteristic relation corresponding to the extremal points of the optimal solution obtained by the exchange method, one has $\lambda_i \neq 0$ $(1 \leq i \leq t)$ and $\lambda_i = 0$ $(t < i \leq m)$. As mentioned above, all elements of C have the same error at the critical point set $\{x_1, x_2, \cdots, x_t, x_{m+1}\}$ and, consequently, at all points x_k depending on this set, i.e., such that $\phi_k = \sum_{i=1}^{t} \mu_{ki}\phi_i$. If X^* denotes this set of points where all best approximations have identical errors, the strict approximation is the element of C whose error has the minimum deviation in $X - X^*$.

Since the matrix of the characteristic vectors corresponding to the points $x^* \in X^*$ has rank t, each base function $f_i(x^*)$ can be expressed in terms of t independent functions $\{g_1(x^*), \cdots, g_t(x^*)\}$ as

$$f_i(x^*) = \sum_{j=1}^{t} b_{ij}g_j(x^*), \quad (1 \leq i \leq m) \quad (35)$$

where the matrix $[b_{ij}]$ has rank t. By suitable ordering of the base functions f_i, it is always possible to make the first t rows of $[b_{ij}]$ linearly independent. Let $\{g_1, \cdots, g_m\}$ be a set of new base functions such that the first t functions are solution of the nonsingular system of equations

$$f_i(x) = \sum_{j=1}^{t} b_{ij}g_j(x), \quad (1 \leq i \leq t)$$

whereas the $m - t$ last ones are given by

$$g_k(x) = f_k(x) - \sum_{j=1}^{t} b_{kj} g_j(x), \qquad (t < k \leq m). \quad (36)$$

The approximating function (2) then becomes

$$f(x) = G_1(x) + G_2(x) \quad (37)$$

in which

$$\begin{cases} G_1(x) = \sum_{i=1}^{t} \alpha_i g_i(x) \\ G_2(x) = \sum_{i=t+1}^{m} a_i g_i(x) \end{cases} \quad (38)$$

with $\alpha_i = \sum_{j=1}^{m} a_j b_{ij}$. For $x = x^*$, by (36) and (35), $G_2(x^*) = 0$ and $f(x^*) = G_1(x^*)$. Therefore, since the functions $\{g_1, \cdots, g_t\}$ are independent on X^*, all optimal approximating functions have the same coefficients α_i. The strict approximation will then be obtained by approximating $F(x) - G_1(x)$ by $G_2(x)$ on the set $X - X^*$. It may be possible that this new problem has no unique solution either in which case a further decomposition of the approximating function must be carried out.

As mentioned in Section II-D, exhaustive computations have yielded only degeneracies with respect to the ω_1-axis, in which case $t = N + 1$ and $X^* = \{x^* = (\omega_1, 0)\}$. Applying the above procedure to (1), one gets

$$G_1(x) = \sum_{I=0}^{N} \alpha_I \cos I\omega_1 \cos I\omega_2$$

with

$$\alpha_I = 2B_{II} + \sum_{J=0}^{I-1} B_{IJ} + \sum_{J=I+1}^{N} B_{JI} \quad (39)$$

and

$$G_2(x) = -\sum_{I=1}^{N} \sum_{J=0}^{I-1} B_{IJ}(\cos I\omega_1 - \cos J\omega_1) \cdot (\cos I\omega_2 - \cos J\omega_2). \quad (40)$$

In fact, $G_1(x^*)$ is the best approximation on the ω_1-axis and, as is well known [8], approximation by one-dimensional filters is very fast and accurate. In order to take advantage of this property, one can directly compute the strict approximation by first performing the approximation on the ω_1-axis which yields the optimal solution $\{\alpha_I; 0 \leq I \leq N\}$ with Chebyshev deviation H^*. Once the coefficients B_{IJ} $(1 \leq I \leq N; 0 \leq J \leq I - 1)$ have been determined by the approximation outside the ω_1-axis, it remains to deduce B_{II} $(0 \leq I \leq N)$ from relation (39). Obviously, one does not know from the outset whether the problem is degenerate with respect to the final reference of the exchange algorithm but, thanks to the following theorem, the method turns out to be advantageous even in case of nondegeneracy.

Theorem 6: Let $R^* = \{x_i^*; 1 \leq i \leq t + 1\}$ be the final reference of the exchange algorithm on X^* with Chebyshev deviation H^*. If $R' = \{x_i; 1 \leq i \leq m - t + 1\}$ denotes a reference on $X - X^*$ with reference deviation $|h|$ greater than H^*, it is possible to build from R^* and R' a reference R on the whole region X, whose reference deviation is greater than H^*.

The two references R^* and R' correspond to the characteristic relations

$$G_1(x_{t+1}^*) = \sum_{i=1}^{t} \lambda_i^* G_1(x_i^*), \qquad (\lambda_i^* \neq 0)$$

and

$$G_2(x_{m-t+1}) = \sum_{i=1}^{m-t} \mu_i G_2(x_i) \quad (41)$$

with

$$h(x_{m-t+1}) h(x_i) \mu_i \leq 0, \qquad (1 \leq i \leq m - t). \quad (42)$$

As $G_1(x^*) = f(x^*)$, $f(x)$ is also a reference function for $\bar{R} = \{x_1^*, \cdots, x_t^*, x_1, \cdots, x_{m-t}, x_{t+1}^*\}$ with reference deviation H^* and characteristic relation

$$f(x_{t+1}^*) = \sum_{i=1}^{t} \lambda_i^* f(x_i^*) + \sum_{i=1}^{m-t} \lambda_i f(x_i) \quad (43)$$

in which all λ_i are zero. By (37), (41) becomes

$$f(x_{m-t+1}) = \sum_{i=1}^{m-t} \mu_i f(x_i) + G_1(x_{m-t+1}) - \sum_{i=1}^{m-t} \mu_i G_1(x_i). \quad (44)$$

Since $G_1(x)$ is given in (38) in terms of t independent functions, any $G_1(x_j)$ for $1 \leq j \leq m - t + 1$ can be expressed as a linear combination of t independent values $G_1(x_i^*) = f(x_i^*)$ $(1 \leq i \leq t)$ and (44) becomes

$$f(x_{m-t+1}) = \sum_{i=1}^{t} \mu_i^* f(x_i^*) + \sum_{i=1}^{m-t} \mu_i f(x_i). \quad (45)$$

By applying the exchange theorem to the characteristic relations (43) and (45), it is now possible to introduce the point x_{m-t+1} with deviation $|h| > H^*$ in $\bar{R}$ in order to form R. This exchange cannot be a static one because condition (42) holds for all zero λ_i in (43) and the reference deviation of R will thus be larger than H^*.

In consequence of this theorem, as soon as the reference deviation $|h|$ of the exchange method for $X - X^*$ becomes greater than H^*, one can restart the exchange algorithm for the whole region X with a reference deviation greater than H^*, thus avoiding any further degeneracy with respect to the ω_1-axis.

IV. Numerical Results

In order to save computation time, it is advantageous to use a Cartesian grid for the discrete point set X which is to replace the continuous approximation domain of Fig. 1. Indeed, evaluating (1) along a vertical or horizontal line reduces to univariable computation. Nevertheless, an accurate representation of the continuous domain requires to include the intersections of the grid lines with the passband and stopband boundary circles. If one characterizes the quality of approximation by the precision factor $(h_{max} - H)/H$ where h_{max} is the maximum of the absolute value of the error function on the continuous approximation domain, experience shows that a Cartesian grid of about 25 N by 25 N points is sufficient for getting a precision factor on the order of 10^{-3}, which is quite acceptable for practical purposes.

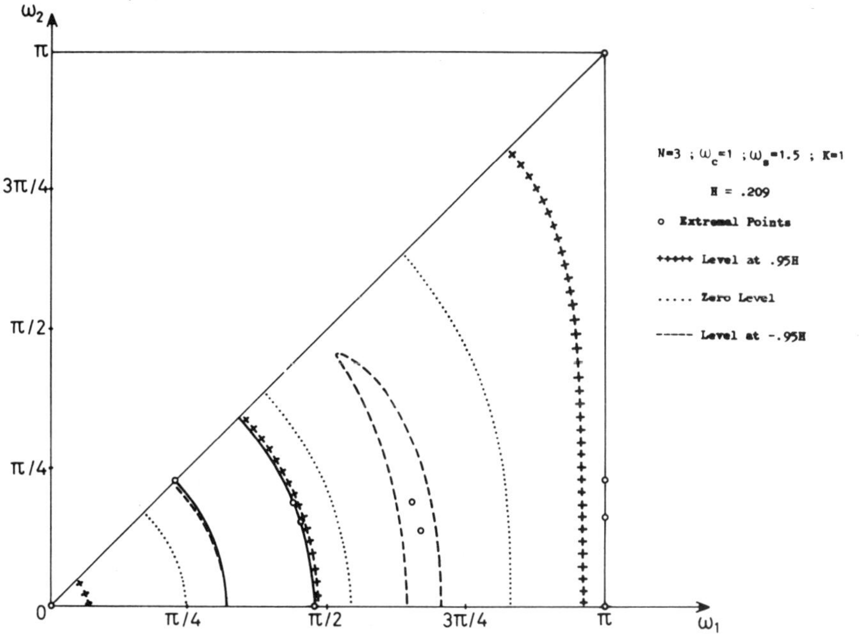

Fig. 2. Typical shape of the optimal error function.

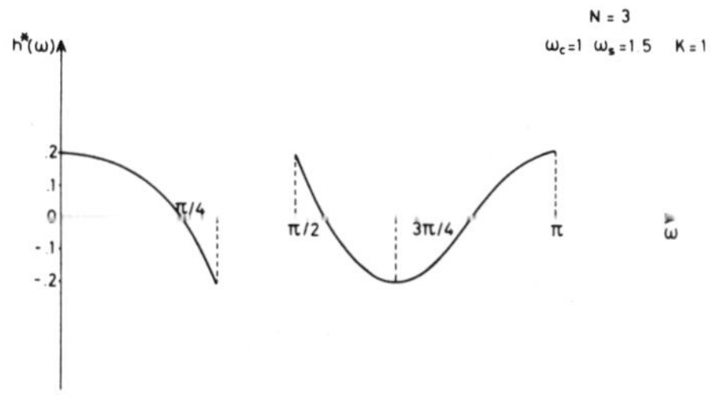

Fig. 3. Optimal one-dimensional error curve.

The weight function $w(x)$ introduced in the norm (4) is chosen equal to K in the passband and to unity in the stopband to get different tolerances in both bands.

A typical example of the optimal error function is presented in Fig. 2 for the third-degree filter with $\omega_c = 1$, $\omega_s = 1.5$, and $K = 1$. One recognizes the error shape of the optimal one-dimensional filter drawn in Fig. 3, rotated in all directions with the final reference points located approximately on circular ridges whose radii correspond to the abscissas of the extremal points of Fig. 3. In fact, on a circle ($\omega_1 = R\cos t$, $\omega_2 = R\sin t$), each base function (3) can be expanded in terms of Bessel functions of the first kind as

$$\cos I\omega_1 \cos J\omega_2 + \cos J\omega_1 \cos I\omega_2 = 2J_0[R(\sqrt{I^2 + J^2})] + 4\sum_{k=1}^{\infty} J_{4k}[R(\sqrt{I^2 + J^2})] \cos 4k\alpha_{IJ} \cos 4kt$$

where $\alpha_{IJ} = \sin^{-1}[I/(I^2 + J^2)^{1/2}]$. Since the Bessel function $J_n(x)$ has a zero of multiplicity n at the origin, it has negligible values when its argument x is small with respect to the order n. For the above-mentioned third-degree filter, on the boundary circle $\omega_s = 1.5$, the approximating function can be represented with an error on the order of 10^{-3}, as a second-degree trigonometric polynomial in $4t$. Hence, since there are three extremal points on ω_1, the error function attains approximately the Chebyshev deviation at all points of the circle. The circular symmetry of the error function is less pronounced for circles with larger radii and, in particular, the ridge corresponding to the last extremal point of Fig. 3, i.e., the Nyquist frequency, is the straight line $\omega_1 = \pi$. As, for $\omega_1 = \pi$, the approximating function is a third-degree polynomial in ω_2, the four extremal points located on $\omega_1 = \pi$ produce a ridge of strictly constant height for all ω_2. In order to take advantage of this property during the exchange algorithm, one can split the approximating function into two parts according to (37) such that

$$G_1(x) = \sum_{I=0}^{N} \alpha_I (-1)^I \cos I\omega_1 \cos I\omega_2$$

$$G_2(x) = -\sum_{I=1}^{N} \sum_{J=0}^{I-1} B_{IJ}[\cos I\omega_1 - (-1)^{I+J}\cos J\omega_1] \cdot [(-1)^{I+J}\cos I\omega_2 - \cos J\omega_2]$$

with

$$\alpha_I = (-1)^I 2B_{II} + \sum_{J=0}^{I-1} (-1)^J B_{IJ} + \sum_{J=I+1}^{N} (-1)^J B_{JI}.$$

TABLE I

$\omega_c=1$; $\omega_s=2$; K=1

Precision factor = 10^{-3}

N	Single Exchange		Multiple Exchange	
	Iteration Number	Computation Time (Sec)	Iteration Number	Computation Time (Sec)
1	1	9	1	9
2	3	9	3	9
3	16	46	18	54
4	25	114	15	118
5	85	564	30	401
6	184	2276	51	1276

TABLE II

$\omega_c=1$; $\omega_s=2$; K=1

Precision factor = 10^{-2}

N	Single Exchange		Multiple Exchange	
	Iteration Number	Computation Time (Sec)	Iteration Number	Computation Time (Sec)
1	1	9	1	9
2	2	8	2	8
3	9	29	9	41
4	15	85	9	77
5	68	534	21	360
6	128	1714	16	463

TABLE III

N=1 ; $\omega_c=1.5$; $\omega_s=2$; K=10

Filter Coefficients	Extreme Optimal Solution A_1	Extreme Optimal Solution A_2	Strict Approximation
A_{00}	.793917	.790117	.790699
$A_{10} = A_{01}$	.0552223	.0571223	.0568314
A_{11}	.017049	.016099	.0162445

Since $G_2(x) = 0$ for $\omega_1 = \pi$, imposing a constant value of $f(x)$ on $\omega_1 = \pi$ amounts to canceling all coefficients α_l except α_0. It turns out that the exchange algorithm modified this way yields very good results, at least within the precision factor of 10^{-3}, if π belongs to the extremal point set of the best univariable approximation. On the other hand, since nonuniqueness generally occurs only when π is not an extremal point of the optimal unidimensional error, the following rule of thumb was found to be very efficient for solving the approximation problem. It consists in performing the approximation in two steps:

1) approximation along the ω_1-axis by a univariable approximating function which yields the extremal point set $R^* = \{x_k^*; 1 \leq k \leq N + 2\}$;

2) if $x_{N+2}^* = \pi$, approximation in the whole region by maintaining a constant value of the error on $\omega_1 = \pi$; and

2′) if $x_{N+2}^* \neq \pi$, approximation by means of the procedure of Section III-B which allows a direct computation of the strict approximation.

In both cases, the points of the initial reference for starting the exchange algorithm are equally spaced on circles of radius x_k^* and their number is proportional to the index k.

As regards the rate of convergence of the iterative process, a necessary condition [13] for the Remez algorithm, applied to univariable functions consisting of m base functions, to converge quadratically, is that the optimal error function has exactly $m + 1$ extremal points: on the contrary, the convergence is slower if their number exceeds $m + 1$. In particular, this property has been verified for one-dimensional filters [14]. Although such a theorem has not been proved for bivariable functions, it is reasonable to expect a similar phenomenon for two-dimensional filters. Since the optimal error exhibits ridges of extrema, the convergence is slow as evidenced by Table I which gives the iteration number and the computation time (on the Electrologica X8 computer: cycle time = 1 μs) of the single and multiple exchange algorithms in the case of $\omega_c = 1$, $\omega_s = 2$, $K = 1$ and for various degrees. If both algorithms are comparable for N ranging from 1 to 4, the multiple exchange method becomes quite advantageous when the degree is higher. In order to stress the importance of the precision factor on the computation time, Table II indicates how the figures of Table I vary when the precision factor is relaxed from 10^{-3} to 10^{-2}: from these results, it appears that the difficulty for designing the filters, especially the highest degree filter, strongly depends on the quality of approximation which is required by the practical problem.

To serve as an illustration for nonuniqueness, Table III presents several optimal solutions relative to the parameters $N = 1$, $\omega_c = 1.5$, $\omega_s = 2$, and $K = 10$. When degeneracy with respect to the ω_1-axis occurs for a first-degree polynomial, there is only one vanishing characteristic coefficient, and the number of extreme optimal solutions is exactly equal to 2. Therefore, the two extreme optimal solutions A_1 and A_2 given in the first two columns of Table III completely define the set of best approximations and one can verify that the strict approximation of the third column is the convex combination $\delta A_1 + (1 - \delta)A_2$ with $\delta = 0.153124$.

Another example of nonuniqueness results from the data set $N = 3$, $\omega_c = 1.5$, $\omega_s = 2$, and $K = 1$. In that case, there are tens of different extreme optimal solutions and it is no longer possible to mention all of them. As regards the strict approximation, it is worth pointing out that performing the approximation outside the ω_1-axis with the function (40) yields four extremal points located on the parallel of the Cartesian grid closest to the ω_1-axis, which corresponds to a new degenerate problem. To remove this difficulty, it is thus necessary to apply the procedure of Section III-B to the reduced approximation problem. Fig. 4 shows the optimal error function of the strict approximation together with the three critical point sets.

If the exchange algorithm is used to compute an extreme optimal solution, since the problem is degenerate with respect to the final reference, the last iterations consist of static exchanges. In order to ensure unconditional convergence, one has to resort to the perturbation technique of Section I-D. For the example at hand, the number of iterations of the unperturbed and perturbed exchange algorithms is equal to 36 and 41, respectively, while the corresponding computation times are 92 s and 137 s. Therefore, although both iteration numbers are comparable,

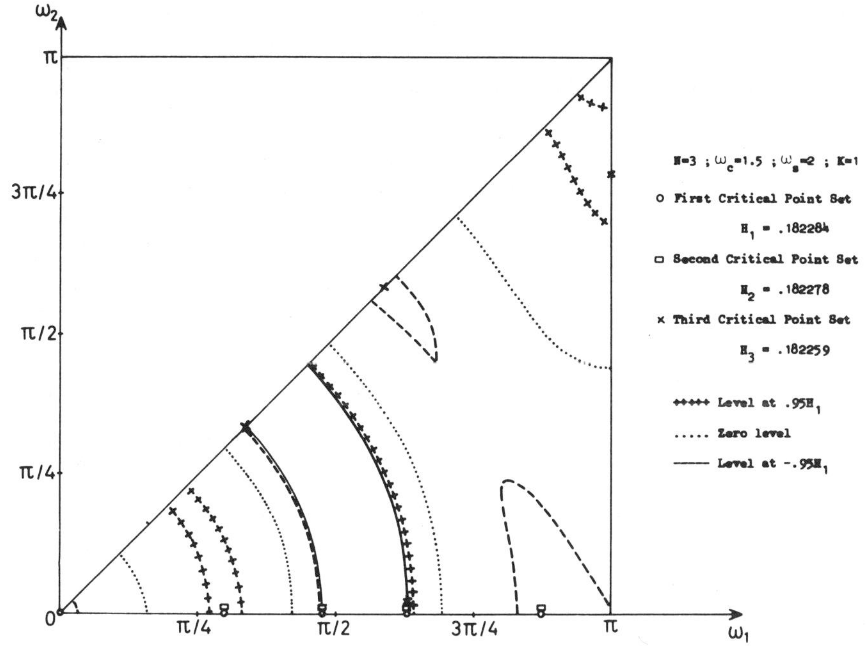

Fig. 4. Optimal error function and critical point sets of a strict approximation.

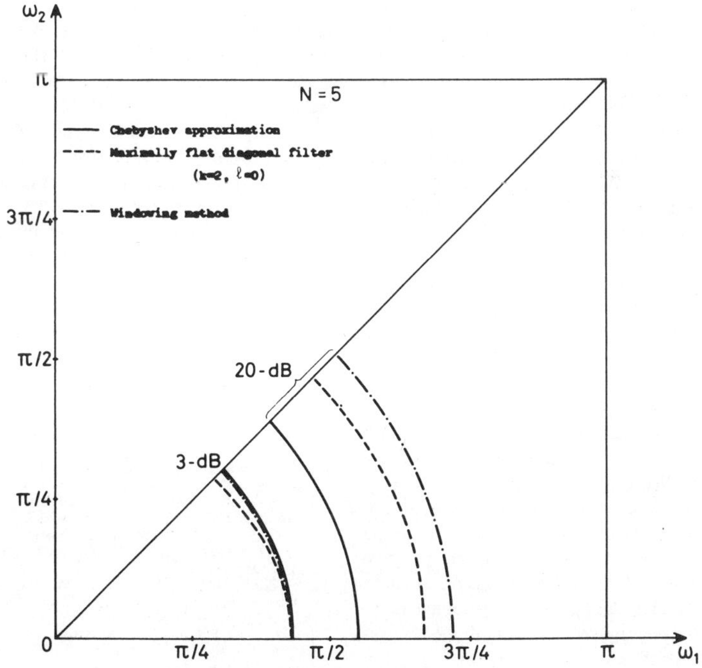

Fig. 5. Comparison with other design methods.

the computation time of the perturbed algorithm is higher owing to the more involved computations at each iteration step. On the other hand, experience has proven that the unperturbed exchange algorithm never failed to converge. Hence, as pointed out in Section I-D, the perturbation method is more important from the theoretical standpoint than from the practical one.

Hu and Rabiner [3] have used linear programming to design frequency sampling and optimal (in the sense of Chebyshev approximation) two-dimensional filters. Results for the latter design have been reported only for a few filters of the fourth degree and were found to be comparable with those obtained by the exchange algorithm of this paper. By lack of a detailed description of their methods, it seems likely that the small discrepancies are due to different discretization schemes of the continuous approximation domain. As regards the frequency sampling design, the comparison was made on a few samples of the filters of degree 12 published in [3]. For these examples, design by Chebyshev approximation showed that a degree 7 was already sufficient to achieve the same filtering performance.

To visualize the superiority of the Chebyshev approximation on other design methods, a fifth-degree filter was computed having a 3-dB passband and a 20-dB stopband (Fig. 5). For comparison, the 3- and 20-dB contour lines have been plotted for a maximally flat diagonal filter [4] and for a filter designed with a windowing technique [2], both having the same degree and the same passband limit along the ω_1-axis. The circular symmetry of these designs is comparable but the selectivity given by the Chebyshev approximation is significantly better, i.e., smaller transition bandwidth for substantially the same passband boundary.

V. Conclusion

The preceding analysis has clearly shown that the Chebyshev approximation of two-dimensional nonrecursive filters is fundamentally more difficult and inherently slower than in the one-dimensional case. The reasons behind this deserve some further comments.

First, two-variable approximating functions do generally not satisfy the Haar condition and this is reflected by the existence of degeneracies. If degeneracy occurs before the end of the algorithm it may fail to converge because the same reference is repeated periodically. To force convergence even in this case, it was shown how to complement the algorithm with a suitable perturbation technique. This technique is only of theoretical interest, however, because it is computationally very costly and because in practice cycling of the algorithm has never been observed.

A second difference is that the optimal solution is not characterized by a simple alternation of the error like in the one-dimensional problem and, as a consequence, the rule for exchanging one point of the reference becomes more complicated. This fact makes the one-by-one exchange more time-consuming and even prevents a simultaneous exchange, a procedure which precisely makes the algorithm so efficient in the single variable case. An improvement in this direction was made, however, by proposing a method which allows to exchange several points simultaneously, and the experiments have demonstrated that it leads to important reductions of the run time when the degree is high.

Further, the maximum detection over a two-dimensional region is much longer than the corresponding work over an interval and is made even slower by the ridges in the error function whose existence was already mentioned in [3]. Finally, the computation time is also increased by the fact that in two dimensions there exists no counterpart of the Lagrange interpolation formula for the leveled reference function and by the fact that the number of coefficients is proportional to the square of the degree.

These considerations show that the nonfulfilled Haar condition is not the only reason for the slowness of the two-dimensional approximation. Several measures were taken to improve this situation. Besides the multiple exchange already mentioned, a heuristic procedure was proposed where the approximation is first performed on the ω_1-axis or where the value of the approximating function is kept constant over $\omega_1 = \pi$. Its advantage lies in a significant saving of computation time and although it does not always give the optimal solution, the accuracy of the result is in most cases satisfactory. It is clear that future efforts should be aimed at further reducing the run time.

Comparison with other design methods (frequency sampling, maximally flat criterion, and windowing technique) has made evident the superiority of the Chebyshev design for what regards selectivity. For this norm, the one-by-one exchange is known to be more economical than linear programming [15] and this speed advantage is increased by the possibility of performing a multiple exchange.

This paper has solved the typical problem of designing a two-dimensional nonrecursive filter which approximates circularly symmetrical low-pass specifications. Most of the results, however, are not restricted to circular symmetry or low-pass specifications but apply to the design of any filter whose frequency response can be put under the form (2) within a linear phase factor.

Acknowledgment

The authors wish to thank J. Praet for his programming of the exchange algorithm.

References

[1] T. S. Huang, W. F. Schreiber, and O. J. Tretiak, "Image processing," *Proc. IEEE*, vol. 59, pp. 1586–1609, Nov. 1971.

[2] T. S. Huang, "Two-dimensional windows," *IEEE Trans. Audio Electroacoust.*, vol. AU-20, pp. 88–89, Mar. 1972.

[3] J. V. Hu and L. R. Rabiner, "Design techniques for two-dimensional digital filters," *IEEE Trans. Audio Electroacoust.*, vol. AU-20, pp. 249–257, Oct. 1972.

[4] Y. Kamp and J. P. Thiran, "Maximally flat nonrecursive two-dimensional digital filters," *IEEE Trans. Circuits Syst.*, vol. CAS-21, pp. 437–449, May 1974.

[5] J. H. McClellan, "On the design of one-dimensional and two-dimensional FIR digital filters," Ph.D. dissertation, Rice University, Houston, Tex., Apr. 1973.

[6] E. Stiefel, "Über diskrete und lineare Tschebyscheff-Approx-

imationen," *Numer. Math.*, vol. 1, pp. 1–28, 1959.

[7] S. I. Gass, *Linear Programming.* New York: McGraw-Hill, 1958.

[8] T. W. Parks and J. H. McClellan, "Chebyshev approximation for nonrecursive digital filters with linear phase," *IEEE Trans. Circuit Theory*, vol. CT-19, pp. 189–194, Mar. 1972.

[9] L. Collatz, "Approximation von Functionen bei einer und bei mehreren unabhängigen Veränderlichen," *Z. Angew. Math. Mech.*, vol. 36, pp. 198–211, May/June 1956.

[10] L. Bittner, "Das Austauschverfahren der linearen Tschebyscheff-Approximation bei nicht erfüllter Haarscher Bedingung," *Z. Angew. Math. Mech.*, vol. 41, pp. 238–256, June 1961.

[11] J. Descloux, "Dégénérescence dans les approximations de Tschebyscheff linéaires et discrètes," *Numer. Math.*, vol. 3, pp. 180–187, 1961.

[12] J. R. Rice, *The Approximation of Functions*, vol. 2, *Nonlinear and Multivariate Theory.* Reading, Mass.: Addison-Wesley, 1969.

[13] G. Meinardus, *Approximation von Funktionen und ihre numerische Behandlung.* Berlin: Springer, 1964.

[14] A. N. Godard, "Design of optimum FIR bandpass filters through frequency transformations," *Philips Res. Rep.*, vol. 29, pp. 243–252, June 1974.

[15] E. Stiefel, "Note on Jordan elimination, linear programming and Tchebycheff approximation," *Numer. Math.*, vol. 2, pp. 1–17, 1960.

18

Reprinted from pp. 247–251 of the *Seventh Annu. Princeton Conf. Inf. Sci. and Syst., Proc.* (1973), 584 pp.

THE DESIGN OF TWO-DIMENSIONAL DIGITAL FILTERS BY TRANSFORMATIONS

J. H. McClellan
Rice University

Abstract

This paper describes a technique for designing two-dimensional (2-D) Finite Impulse Response (FIR) digital filters by application of a change of variables to an optimal min-max one-dimensional (1-D) filter. The advantages of such an approach are that very large filter arrays can be designed in an amount of time comparable to that required to do the 1-D design, and in many cases the filters which are obtained are optimal.

Among the types of filters which can be treated by such an approach are circularly symmetric lowpass and bandpass filters. Several examples are presented to illustrate the method and to compare it with existing designs.

I. Introduction

Two-dimensional (2-D) digital signal processing is very important in such areas as geophysical seismic signal processing and image enhancement for picture processing. However, many of the techniques for designing one-dimensional (1-D) digital filters have not been extended to 2-D digital filter design. In particular, in the area of finite impulse response (FIR) linear phase digital filter design the iterative design methods based on the Remez exchange algorithm have not been extended to the 2-D case, and so at this time there is no efficient design procedure for optimum Chebychev filters available in two dimensions. Since the problem of 2-D design in the FIR linear phase case is a linear approximation problem, linear programming can be applied[1], but this approach has met with only limited success. Linear programming is slow and has been limited to the design of 9x9 filters. This is because of the size of the problem. For example, the design of an optimum 31x31 FIR filter involves optimization over 16x16=256 parameters. Therefore, it is important to consider methods of design which may yield suboptimum filters but which are computationally more efficient. Within such a class of methods are the windowing technique[2] and the frequency sampling technique[1]. This paper describes a new approach based on transformations of 1-D filters, and shows in addition that many of the new designs are optimal min-max 2-D filters.

The technique consists of transforming a 1-D filter into a 2-D filter by a change of variables. The class of filters which can be approximated by this method is limited, but included in the class are many important filters, such as those filters whose magnitude is constant over regions of the 2-D frequency plane (e.g., lowpass and bandpass circularly symmetric filters). The main advantage of this technique is that it is very fast because the computing time is devoted almost entirely to the 1-D design which is known to be efficient[3]. Furthermore, it is possible to prove that many filters designed by this method are optimal in the Chebychev sense.

II. Problem Formulation

A two-dimensional digital filter with impulse response matrix $h(k,p)$ $k=0,1,\dots N_1-1$; $p=0,1,\dots N_2-1$ has a frequency response defined by the 2-D Discrete Fourier transform

$$H(F_1,F_2) = \sum_{k=0}^{N_1-1}\sum_{p=0}^{N_2-1} h(k,p)\exp(-j2\pi(F_1K+F_2p))$$

If the impulse response is constrained so that it is $N_1 \times N_2$, where N_1 and N_2 are odd integers, and the symmetry conditions

$$h(N_1-m-1,p) = h(m,p) \quad m = 0,1,\dots n_1 = \tfrac{1}{2}(N_1-1)$$

$$h(k,N_2-m-1) = h(k,m) \quad m = 0,1,\dots n_2 = \tfrac{1}{2}(N_2-1)$$

are imposed, then the frequency response can be rewritten

$$H(F_1,F_2)=\exp(-j2\pi(n_1F_1 + n_2F_2)) \sum_{k=0}^{n_1}\sum_{p=0}^{n_2} a(k,p)\cos 2\pi kF_1 \cos 2\pi pF_2 \qquad (1)$$

where

$$a(0,0) = h(n_1,n_2)$$

$$a(0,p) = 2h(n_1,n_2-p) \quad p = 1,\dots n_2$$

$$a(k,0) = 2h(n_1-k,n_2) \quad k = 1,\dots n_1$$

$$a(k,p) = 4h(n_1-k,n_2-p) \quad k = 1,\dots n_1;\ p = 1,\dots n_2$$

Let

$$\hat{H}(F_1,F_2) = \sum_{k=0}^{n_1}\sum_{p=0}^{n_2} a(k,p)\cos 2\pi kF_1 \cos 2\pi pF_2$$

Since $\exp(-j2\pi(n_1F_1 + n_2F_2))$ has magnitude one, $|\hat{H}(F_1,F_2)|$ is the magnitude of the frequency response.

A one dimensional filter with impulse response $h(k)$, $k = 0,1,\dots N-1$ has a frequency response defined by the Fourier transform

$$G(F) = \sum_{k=0}^{N-1} h(k)\exp(-j2\pi kF)$$

If the length of the impulse response is odd $N = 2n+1$ and symmetric, $h(k) = h(N-1-k)$, $k=0,1,\dots n$ then

$$G(F) = \exp(-j2\pi nF) \sum_{k=0}^{n} b(k)\cos 2\pi kF \qquad (2)$$

where $b(0) = h(n)$ and $b(k) = 2h(n-k), k=1,\dots n$

Letting

$$\hat{G}(F) = \sum_{k=0}^{n} b(k)\cos 2\pi kF$$

the magnitude of the frequency response is $|\hat{G}(F)|$.

The change of variables to be described depends on the fact that both $\hat{G}(F)$ and $\hat{H}(F_1,F_2)$ are sums of cosine functions. $\hat{G}(F)$ can be rewritten as

$$\hat{G}(F) = \sum_{k=0}^{n} \hat{b}(k) (\cos 2\pi F)^k$$

by the choice of suitable coefficients $\hat{b}(k)$. This can be seen by letting $x = \cos 2\pi F$. Then $\cos 2\pi k F = \cos k(\arccos x) = T_k(x)$, where T_k is the k-th Chebychev polynomial. Thus,

$$\sum_{k=0}^{n} b(k)T_k(x) = \sum_{k=0}^{n} \hat{b}(k) x^k$$

where both sides of the equation are now polynomials in x.

The following change of variables is proposed

$$\cos 2\pi F = A \cos 2\pi F_1 + B \cos 2\pi F_2 + C \cos 2\pi F_1 \cos 2\pi F_2 + D \qquad (3)$$

Substitution of (3) into G(F) yields a bivariate polynomial in $\cos 2\pi F_1$ and $\cos 2\pi F_2$

$$\sum_{k=0}^{n} \sum_{p=0}^{n} \hat{a}(k,p)(\cos 2\pi F_1)^k (\cos 2\pi F_2)^p$$

which can be rewritten uniquely in the same form as $H(F_1,F_2)$ namely,

$$\sum_{k=0}^{n} \sum_{p=0}^{n} a(k,p) \cos 2\pi k F_1 \cos 2\pi p F_2$$

Thus, it has been shown that the proposed change of variables will preserve the proper form for a 2-D filter and the resulting filter will be a square array (i.e., $N_1 = N_2$).

III. Change of Variables

The change of variables (3) defines a mapping from the interval [0,0.5] of the 1-D frequency axis to the square [0,0.5] × [0,0.5] in the 2-D frequency plane. In order to be able to use the change of variables to produce a desired frequency response (i.e., pick A,B,C, and D), it is necessary to describe the mechanics of the transformation (3).

Solving (3) for F_2 as a function of F_1,

$$F_2 = \frac{1}{2\pi} \text{Arccos} \left\{ \frac{\cos 2\pi F - D - A\cos 2\pi F_1}{B + C\cos 2\pi F_1} \right\} \qquad (4)$$

it is easy to see that for a fixed F we get a curve in the (F_1,F_2) plane, and along this curve the transformed 2-D frequency response is a constant equal to the value of the 1-D frequency response at F. As F varies we get a family of

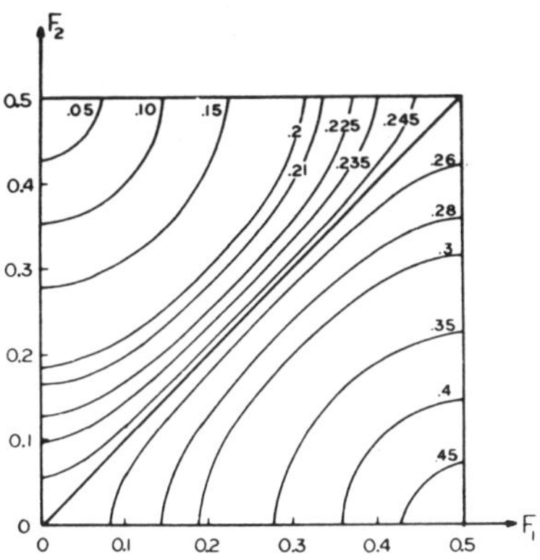

Figure 1. Contour lines for equiripple fan filter as F varies from 0 to 0.5 in steps of 0.05.

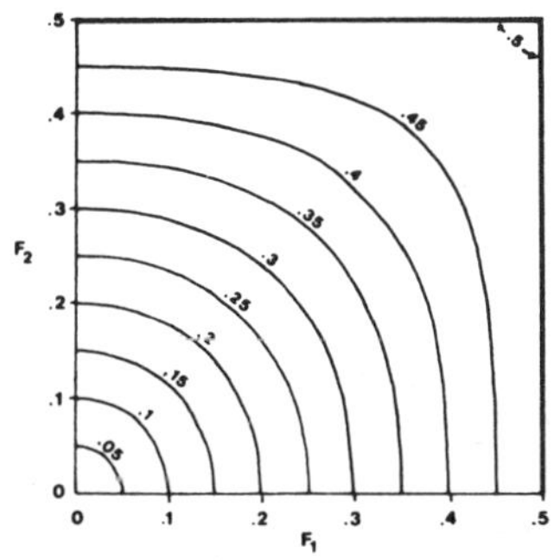

Figure 2. Contour lines for circularly symmetric filter as F varies from 0 to 0.5 in steps of 0.05.

contours which completely describe the transformed 2-D frequency response. For example, A = -B = 0.5, C = D = 0 gives the contours of Figure 1; whereas, A = B = C = -D = 0.5 gives the contours of Figure 2. The former has been used in the design of equiripple fan filters[4], and the latter can be used to design circularly symmetric filters[1].

Letting $u = \cos 2\pi F_1$, $v = \cos 2\pi F_2$, and $x = \cos 2\pi F$, (4) is rewritten as

$$v = \frac{x - D - Au}{B + Cu} \qquad (5)$$

As x varies from -1 to +1, v as a function of u is a family of curves in the square [-1,1] × [-1,1]. Fixing x and calculating the derivative,

$$\frac{dv}{du} = (-AB + CD - Cx)(B + Cu)^{-2}$$

shows that v is either a monotone increasing or decreasing function of u, because the sign of the derivative doesn't change as u varies from -1 to +1. Thus, the same thing is true for F_2 vs. F_1; namely, F_2 is monotone increasing or decreasing as a function of F_1. If the derivative is zero, it is identically zero, meaning that v is a constant. In order to avoid such horizontal contours, it is sufficient to assume that $|C| \leq |A|$. Similarly, to exclude vertical contours one considers u as a function of v and concludes that $|C| \leq |B|$ is a sufficient condition. Note that the equal sign will permit horizontal or vertical contours at the boundaries of square $[0,0.5] \times [0,0.5]$ as in the contour line for $F = 0.5$ in Figure 2.

In order to insure that the mapping from F to (F_1,F_2) is well defined, it is necessary that A,B,C, and D be chosen so that the right hand side of (3) is between +1 and -1 for all $F \in [0,0.5]$. With this constraint there are two cases: all contours increasing or all contours decreasing.

If all the contours are increasing, as in Figure 1, then the point $F = 0$ is mapped to the point $(F_1 = 0, F_2 = 0.5)$ and the point $F = 0.5$ to $(0.5,0)$, or vice-versa. Thus,

$$\pm 1 = A - B - C + D$$

$$\mp 1 = -A + B - C + D$$

which simplifies to $A = \pm 1 - B$ and $C = D$. If all the contours are decreasing, as in Figure 2, $F = 0$ is mapped to $(0,0)$ and $F = 0.5$ to $(0.5, 0.5)$, or vice-versa. Thus,

$$\pm 1 = A + B + C + D$$

$$\mp 1 = -A - B + C + D$$

which simplifies to $A = \pm 1 - B$ and $C = -D$. In both cases, the number of parameters to choose has been reduced from four to two.

IV. Design Procedure

Figure 3 shows the desired specification of a 2-D circularly symmetric lowpass filter. Between the pass region and the stop region we have allowed for a transition region, which is bounded by the arcs of two different circles. Since the arcs are strictly decreasing functions, $C = -D$ and $A = 1 - B$. The object of the design procedure is to choose A and C such that for some $F = F_p$ the resulting contour line closely approximates the inner circular arc and for some $F = F_s$ the resulting contour line approximates the outer arc. Then one designs an optimal Chebychev 1-D lowpass filter with passband cutoff frequency F_p and stopband cutoff frequency F_s, using the Remez exchange algorithm and applies the change of variables to get the desired 2-D filter.

It is well known[5] that the transition width of a 1-D lowpass filter affects the attenuation

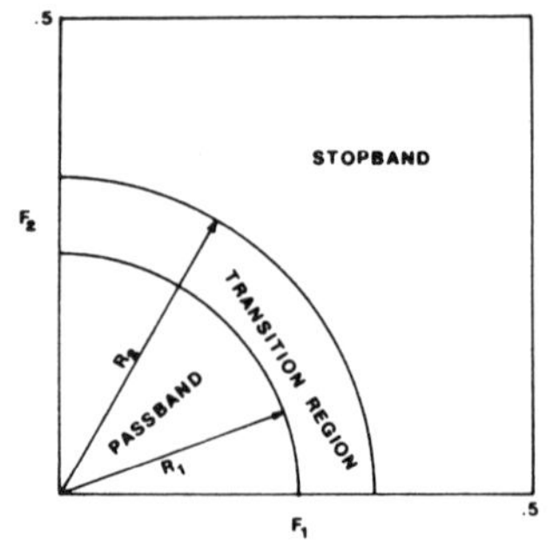

Figure 3. Desired specifications for a circularly symmetric lowpass filter.

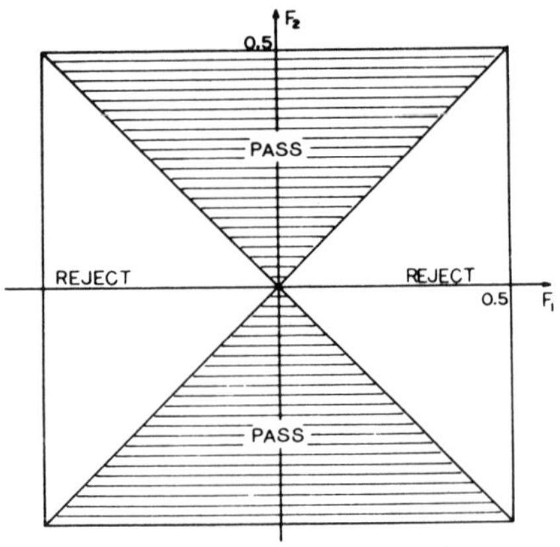

Figure 4. Desired specifications for a 2-D fan filter.

which can be obtained from such a filter. Therefore, in the above procedure, one would like to be able to design a 1-D filter whose transition width is as large as possible. This can be done by picking A and C such that the contours are bunched together closely in the transition region. The best choice for A and C is 0.5 in order to approximate circular symmetry (Fig. 2).

The design procedure for fan filters is similar. Figure 4 shows the desired response. Since the cutoff boundary is an increasing function, $C = D$ and $A = 1 + B$. The best choices for A,B,C and D are $A = -B = 0.5$ and $C = D = 0$ (Fig. 1). The details of fan filter design have been described elsewhere[4].

V. Optimality of the 2-D Filters

When using the change of variables described above it is often possible to show that the 2-D filter which is obtained is an optimal Chebychev approximation to the desired frequency response. The argument used to prove this assertion is to recognize that along a line in the 2-D frequency square we are trying to approximate a desired 1-D frequency response. Hence, the best 1-D approximation to this desired 1-D frequency response yields a lower bound for the error in approximating the desired 2-D frequency response over the entire frequency square. If this lower bound can be achieved then the resulting 2-D filter is a best Chebychev approximation. Let's illustrate the argument for the case of circularly symmetric filters.

Circularly symmetric 2-D filters are obtained from 1-D filters by using the change of variables (3) with $A = B = C = -D = 0.5$. Restricted to the line $F_2 = 0$, the 2-D response is a 1-D response in the variable F_1. Since $\cos 2\pi F =$
$0.5 \cos 2\pi F_1 + 0.5 + 0.5 \cos 2\pi F_1 - 0.5 = \cos 2\pi F_1$, the 2-D response along $F_2 = 0$ is exactly the same as the optimal 1-D response used to generate the 2-D filter. Thus the deviation of the 1-D filter is a lower bound for the deviation of the 2-D filter. However, since the change of variables preserves the equiripple nature of the error, the deviation of the 2-D response is equal to this lower bound and the filter is optimal in the Chebychev sense.

A similar argument can be made to prove that fan filters designed with the change of variables $A = 0.5$, $B = -0.5$, and $C = D = 0$ are optimal Chebychev filters. In this case the line to be used is $F_2 = 0.5 - F_1$. We remark that in an earlier paper[4] on fan filter design it was felt that this technique might not give an optimum Chebychev filter, but now we have proved the optimality of those designs.

VI. Results

The above methods have been applied to several different types of filters to illustrate the flexibility of the method considered here. Figure 5 shows a lowpass circularly symmetric 9 × 9 filter (with $A = B = C = -D = 0.5$) for comparison with an optimal filter designed by linear programming (see filter 3 in Table II of reference 1). Using the above argument, it is possible to show that both filters have exactly the same error in the passband ($\delta_p = 0.079$) and the stopband ($\delta_s = 0.0079$), since along the line $F_2 = 0$ they are exactly the same filter. Both filters are optimal min-max filters, the difference being that the passband edges and stopband edges are slightly different. Figure 6 shows the difference in cutoff boundaries for the two filters. The dashed lines are circular arcs which represent the linear programming design, the solid lines are the edges for the design by the change of variables.

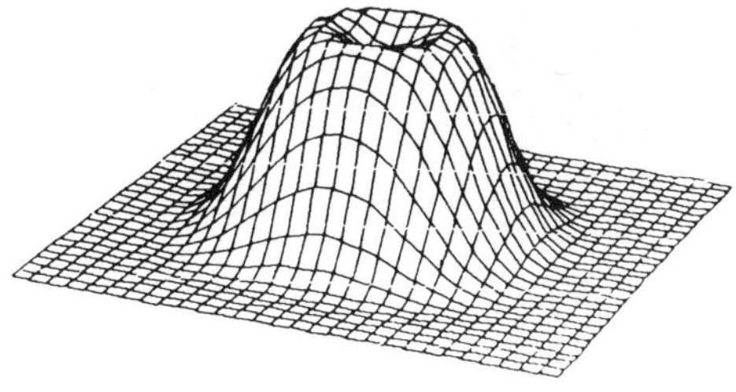

Figure 5. 9 × 9 circularly symmetric lowpass filter $R_1 = 0.1666\ldots$, $R_2 = 0.333\ldots$, $\delta_p = 0.079$, and $\delta_s = 0.0079$.

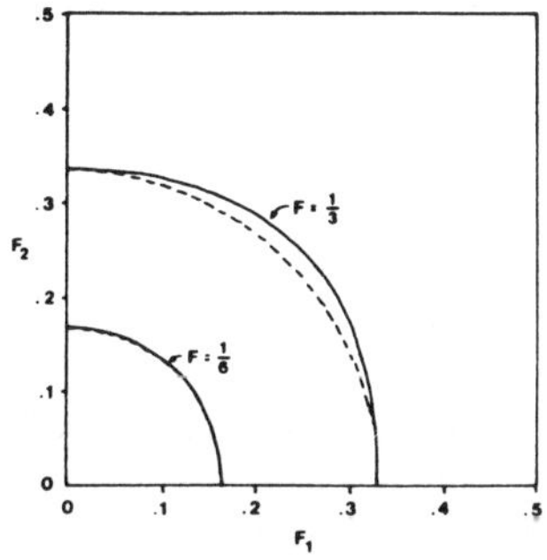

Figure 6. Comparison of cutoff contours for 9 × 9 circularly symmetric lowpass filter. Solid lines denote change of variables, dashed lines denote linear programming.

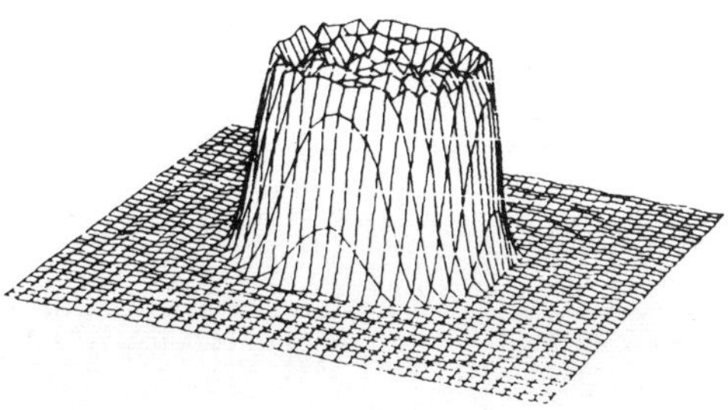

Figure 7. 31 × 31 circularly symmetric lowpass filter. $R_1 = 0.2$, $R_2 = 0.25$, $\delta_p = 0.0755$, and $\delta_s = -42.45$db.

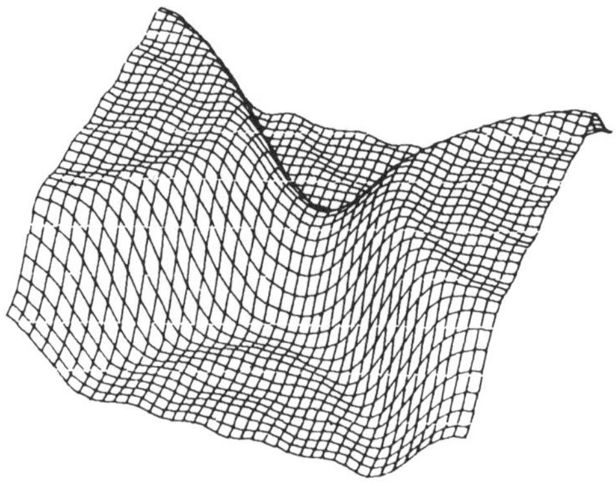

Figure 8. 21 x 21 fan filter. $\delta_p = \delta_s = 0.0229$.

The important feature of this new method is that it is not restricted to the design of such small filters as the length 9 x 9 filter above. With a larger filter it is possible to obtain smaller deviations and/or a sharper cutoff. Figure 7 shows the magnitude of the frequency response for a 31 x 31 circularly symmetric lowpass filter with a transition region of width 0.05. The design time for this filter was approximately 5 seconds on an IBM 370/155 computer.

Figure 8 shows the magnitude response of a 21 x 21 fan filter. The 1-D lowpass filter used to generate the fan filter has a passband cutoff at $F_p = 0.21$ and a stopband cutoff at $F_s = 0.29$. The resulting deviation is $\delta_p = \delta_s = 0.0229$.

VII. Summary

A method for transforming 1-D digital filters into 2-D digital filters has been described. This method has the advantage that it is fast, because only a 1-D filter must be designed, and that often the resulting 2-D filter is optimal in the Chebychev sense. Thus, several important types of 2-D filters can be designed using this technique.

References

1. J. V. Hu and L. R. Rabiner, "Design Techniques for Two-Dimensional Digital Filters", IEEE Trans. Audio Electroacoustics, vol. AU-20, pp. 249-257, Oct. 1972.

2. T. S. Huang, "Two-Dimensional Windows", IEEE Trans. Audio Electroacoustics (Corresp.), vol. AU-20, pp. 88-89, March 1972.

3. T. W. Parks and J. H. McClellan, "Chebyshev Approximation for Non-Recursive Digital Filters with Linear Phase", IEEE Trans. Circuit Theory, vol. CT-19,pp. 189-194, March 1972.

4. J. H. McClellan and T. W. Parks, "Equiripple Approximation of Fan Filters", Geophysics, vol. 37, pp. 573-583, August 1972.

5. O. Herrmann, L. R. Rabiner, and D. S. K. Chan, "Practical Design Rules for Optimum Finite Impulse Response Lowpass Digital Filters", submitted for publication.

19

Reprinted from *IEEE Trans. Circuits Syst.* **CAS-23**(7):405–414 (1976)

McClellan Transformations for Two-Dimensional Digital Filtering: I — Design

RUSSELL M. MERSEREAU, MEMBER, IEEE, WOLFGANG F. G. MECKLENBRÄUKER, AND THOMAS F. QUATIERI, JR., MEMBER, IEEE

***Abstract*—This paper discusses the design of two-dimensional (2-D) linear-phase FIR digital filters by transformations of one-demensional (1-D) filters, using a technique first presented by McClellan. His original transformations are generalized and several algorithms are presented for the design of the generalized transformations. Examples are included to demonstrate the versatility of the design method.**

I. Introduction

TWO-DIMENSIONAL (2-D) digital filters have found numerous applications within the larger context of 2-D digital signal processing. In the area of computer image processing, for example, they are used for the enhancement of low-quality images such as X-ray photographs and for the compensation of linear optical degradations. They are also used for preprocessing data in 2-D pattern recognition systems. They have found uses in nonlinear filtering strategies for dynamic range reduction and contrast enhancement and as prefilters in various schemes for picture encoding. 2-D digital filters are also becoming more useful in the processing of sequential data derived from multidimensional arrays as encountered in geophysical exploration, earthquake/nuclear test detection, sonar, radar, and radio astronomy. Such filtering has always been expensive, however, in terms of the computational effort and memory that was required to implement the filter. Those advances in technology which have produced smaller, faster, and cheaper computing hardware have made 2-D digital filtering increasingly practical.

In the design of one-dimensional (1-D) digital filters it is frequently sufficient, particularly if the filtered signal is speech, to design the filter to approximate a given magnitude characteristic with little regard to the phase response. In all of the aforementioned 2-D examples, however, it is necessary to control both the magnitude and phase responses of the desired digital filter. Usually, the phase is required to be either linear or zero. Huang *et al.* [1] have recently demonstrated, for example, that linear phase is important in the processing of images, a result also known from television.

In this paper a design technique first presented by McClellan [2], [3] for the design of certain 2-D finite-duration impulse response (FIR) filters with linear phase is analyzed and generalized, and a design procedure for these filters is presented which is both fast and efficient.

Other techniques exist for the design of 2-D linear-phase FIR filters. Some of these have been known for several years and most are generalizations of familiar 1-D design algorithms. These include windowing [4], frequency sampling [5], linear programming, and multiple-exchange Chebyshev techniques [6], [7]. Of these the windowing technique is the easiest to apply and it is the most general, for it can be used to find filters of any order with any magnitude and phase characteristic. If linear phase is desired, the other three techniques will generally produce a better approximation to an ideal response for a filter of given order, but the design time required by these algorithms can be considerable.

Filters designed by any of these procedures are normally implemented in one of two ways [8]—by performing the indicated convolution directly, which is efficient for low-order filters, or by using transform techniques which become more efficient for larger filters. If the fast Fourier transform is used to implement a filter with a real impulse response, the crossover point for the two implementations corresponds to an impulse response of size 10×10, roughly, for large input arrays. Number theoretic transforms can be used in place of Fourier transforms with even greater efficiency in some cases [9].

With the McClellan transformation the time required to design a filter is quite modest, and as a result filters of very high order can be readily designed. These filters, being FIR filters, can either be implemented directly or by using transform techniques. In addition, there exists a third implementation for these filters which involves fewer multiplications than the direct convolution. For a direct convolution, a filter whose impulse response is $(N \times N)$ points requires on the order of N^2 multiplications per output value computed. A filter designed using a generalized McClellan transformation can be implemented with

Manuscript received June 12, 1975; revised March 15, 1976. This work was supported in part by the Advanced Research Projects Agency monitored by ONR under Contract N00014-75-C-0951-NR049-308 and in part by the National Science Foundation under Grant ENG71-02319-A02.

R. M. Mersereau and T. F. Quatieri are with the Department of Electrical Engineering and Computer Science and the Research Laboratory of Electronics, Massachusetts Institute of Technology, Cambridge, MA 02139.

W. F. G. Mecklenbräuker was a Visiting Scientist with the Department of Electrical Engineering and Computer Science, Massachusetts Institute of Technology, Cambridge, MA 02139. He is now with the Philips Research Laboratories, Eindhoven, The Netherlands.

a number of multiplications per output point which is proportional to N. If the number of necessary multiplications is used as a measure of implementation efficiency, this implementation is more efficient than either the direct convolution or the FFT approach for filters of size up to roughly 45×45 points. The details of this implementation scheme are presented in a companion paper [10].

In the present paper we shall present a detailed discussion of the generalized McClellan transformation in Section II, and in Section III we shall present some algorithms for determining the parameters which it requires. In Section IV we shall present some examples.

II. The Transformation

The McClellan transformation is a transformation of a 1-D zero-phase FIR filter into a 2-D zero-phase FIR filter by a substitution of variables. It can be applied to 1-D filters of odd length and in one special case also to filters of even length. The even-length case is considered in Appendix I.

For a 1-D filter of length $2N+1$ to be zero phase its impulse response $h(n)$ must have Hermitian symmetric coefficients. Thus if $h(n)$ is real it must also be even. Denoting the frequency response by $H(e^{j\omega})$, we can thus write

$$H(e^{j\omega})=h(0)+\sum_{n=1}^{N} h(n)\left[e^{-j\omega n}+e^{j\omega n}\right] = h(0)+\sum_{n=1}^{N} 2h(n)\cos \omega n. \quad (1)$$

2-D digital filters which have frequency responses of the form

$$H(e^{j\omega_1},e^{j\omega_2})=\sum_{m=0}^{M_1}\sum_{n=0}^{M_2} a(m,n)\cos m\omega_1 \cos n\omega_2 \quad (2)$$

are examples of 2-D zero-phase filters. Equation (2) does not represent the most general class of such filters, but many useful filters are in this class and it is the only class with which we shall be concerned here. The impulse response of such a system $h(m,n)$ is a real sequence which is an even function of each of its arguments.

The generalized McClellan transformation converts 1-D filters of the form (1) into 2-D filters of the form (2) by means of the substitution

$$\cos\omega=\sum_{p=0}^{P}\sum_{q=0}^{Q} t(p,q)\cos p\omega_1 \cos q\omega_2. \quad (3)$$

The relation between $h(n)$ and $h(m,n)$ can be seen by rewriting (1) in the form

$$H(e^{j\omega})=\sum_{n=0}^{N} b(n)\left[\cos\omega\right]^n \quad (4)$$

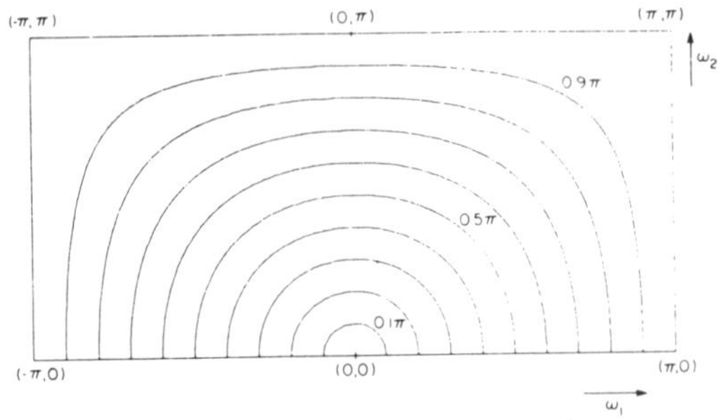

Fig. 1. Contours of the original McClellan transformation. Each contour is labeled with its corresponding 1-D frequency ω.

and performing the substitution indicated in (3)

$$H(e^{j\omega_1},e^{j\omega_2}) = \sum_{n=0}^{N} b(n)\left[\sum_{p=0}^{P}\sum_{q=0}^{Q} t(p,q)\cos p\omega_1 \cos q\omega_2\right]^n. \quad (5)$$

By exploiting the recurrence formula for Chebyshev polynomials, this can be rewritten in the form of (2). The 2-D filter which results is of size $(2NP+1)\times(2NQ+1)$. McClellan considered only the case $P=Q=1$.

It follows implicitly from (3) that points in the frequency response of the 1-D filter are mapped to contours in the (ω_1,ω_2) plane. Furthermore, the shape of these contours is determined only by the parameters $\{t(p,q), 0\leqslant p\leqslant P, 0\leqslant q\leqslant Q\}$. The variation of the frequency response from contour to contour, on the other hand, is controlled by the impulse response of the 1-D filter $h(n)$.

The problem of designing filters by the McClellan transformation thus splits into two parts—the design of the contour parameters $\{t(p,q)\}$ and the design of the 1-D prototype filter $h(n)$. In this paper we shall only be concerned with the first of these problems; several well-known efficient algorithms are available for the design of 1-D FIR filters. (See, for example, the texts by Rabiner and Gold [11] or Oppenheim and Schafer [12].)

Once a transformation has been developed that, for example, produces nearly circular contours, it can be used with several different 1-D prototypes to produce different 2-D designs. As an example, consider the original first-order transformation of McClellan for circular contours. This has $P=Q=1$ and

$$-t(0,0)=t(1,0)=t(0,1)=t(1,1)=\frac{1}{2}. \quad (6)$$

As seen in Fig. 1, the contours which result are nearly circular for low values of ω and become increasingly squared for larger values of ω. Thus, if this transformation is used with a low-pass prototype filter, the passband of the 2-D filter is a disk centered at the origin, whereas if a bandpass prototype is used, the 2-D passband is an annulus.

III. The Contour Approximation Problem

In this section we shall consider the problem of choosing the transformation parameters $t(p,q)$ so that the contours (ω = constant) produced by the transformation

$$\cos\omega = F(\omega_1,\omega_2) \tag{7}$$

$$F(\omega_1,\omega_2)=\sum_{p=0}^{P}\sum_{q=0}^{Q} t(p,q)\cos p\omega_1 \cos q\omega_2 \tag{8}$$

in the (ω_1,ω_2) plane have some desired shape. For some problems it is sufficient to control the shape of a single contour. An example might be in the design of a circular low-pass filter approximating a constant value in the passband and in the stopband. Here the only important contour for approximation is the band edge; whether other contours in the passband or stopband correspond to true circles or not is not important. In other problems it is necessary to constrain the shape of all contours in the frequency plane. This is typically the case for filters where the ideal response is not piecewise constant. These two problems are quite different and the contour approximation problem must be formulated differently in the two cases. We shall consider both problems here. The design algorithms will be developed gradually by beginning with some intuitive techniques and developing them into fairly powerful algorithms.

A. The Constrained Approximation Approach

Let us use as a working example the following: we want to design a 2-D low-pass filter whose passband is in the shape of an ellipse whose minor axis is ω_a and whose major axis is $\omega_b, \omega_b>\omega_a$. The contour in the frequency plane to which the passband edge of the 1-D prototype should map is then described by the relation

$$\left(\frac{\omega_1}{\omega_a}\right)^2+\left(\frac{\omega_2}{\omega_b}\right)^2=1. \tag{9}$$

Suppose that we wish to use a first-order transformation of the form

$$\cos\omega = F(\omega_1,\omega_2)=t(0,0)+t(1,0)\cos\omega_1+t(0,1)\cos\omega_2 + t(1,1)\cos\omega_1\cos\omega_2. \tag{10}$$

If the prototype filter is low pass and the 2-D filter is to be low pass, we might first require that the 1-D origin map to the 2-D origin. This gives the constraint equation

$$t(0,0)+t(1,0)+t(0,1)+t(1,1)=1. \tag{11}$$

One variable, say $t(1,1)$, is thus constrained to be a function of the other three, while those three are still unconstrained. To find values for these free variables we can first solve the equation

$$\cos\omega = F(\omega_1,\omega_2) \tag{12}$$

for ω_2 in terms of ω, ω_1 and the free mapping parameters. This yields the equation

$$\omega_2 = G(\omega,\omega_1,t(0,0),t(1,0),t(0,1)) = \cos^{-1}\left[\frac{\cos\omega - t(0,0)-t(1,0)\cos\omega_1}{t(0,1)+t(1,1)\cos\omega_1}\right]. \tag{13}$$

We can then define an error function

$$E_1(\omega_1)=G(\omega_0,\omega_1,\boldsymbol{t})-\omega_b\sqrt{1-\left(\frac{\omega_1}{\omega_a}\right)^2} \tag{14}$$

where ω_0 is the cutoff frequency of the prototype. The parameters $\boldsymbol{t}$ can then be chosen to minimize some function of $E_1(\omega_1)$, such as its integral square value (L_2 approximation) or maximum absolute value (L_∞ or Chebyshev approximation). There are two difficulties with this formulation. First, the error function is not a linear function of the unknown parameters; thus nonlinear optimization routines must be used for the minimization. Second, for transformations other than those of first order we cannot find an explicit relation for G—the relation is only defined implicitly.

A suboptimum approach reformulates the problem as a linear approximation problem. If the mapping were exact then as the elliptical contour was traversed, the value of $F(\omega_1,\omega_2)$ would be constant. In particular, we would then have

$$\cos\omega_0 = t(0,0)+t(1,0)\cos\omega_1+t(0,1)\cos\left[\omega_b\sqrt{1-\left(\frac{\omega_1}{\omega_a}\right)^2}\right] + t(1,1)\cos\omega_1\cos\left[\omega_b\sqrt{1-\left(\frac{\omega_1}{\omega_a}\right)^2}\right] \tag{15}$$

where ω_0 is the passband cutoff of the prototype. If the mapping is not exact, however, the equality in (15) will only be approximate and we can define an error function as

$$E_2(\omega_1)=\cos\omega_0 - t(0,0)-t(1,0)\cos\omega_1 - t(0,1)\cos\left[\omega_b\sqrt{1-\left(\frac{\omega_1}{\omega_a}\right)^2}\right] - (1-t(0,0) - t(0,1)-t(1,0))\cos\omega_1\cos\left[\omega_b\sqrt{1-\left(\frac{\omega_1}{\omega_a}\right)^2}\right]. \tag{16}$$

This error is now a linear function of $t(0,0)$, $t(0,1)$, and $t(1,0)$, and thus linear optimization routines can be used to minimize the integral squared error or the maximum absolute error. The former problem can be solved trivially using a classical least squares formulation, and the parameters for minimizing the maximum absolute error can be found by linear programming.

It has been observed from several examples that the

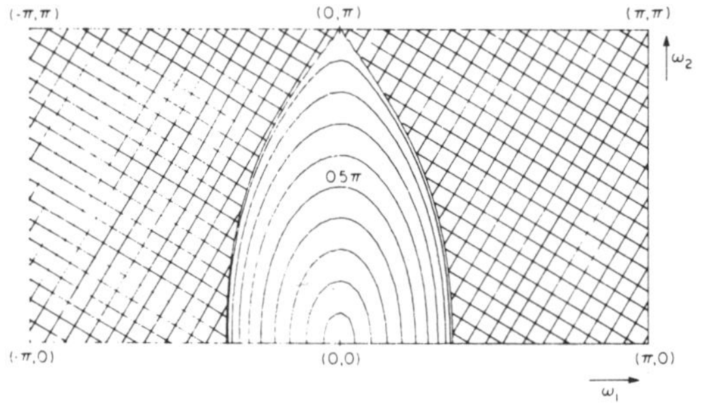

Fig. 2. First-order transformation approximating an elliptical contour. In the crosshatched area $|f(\omega_1,\omega_2)|>1$.

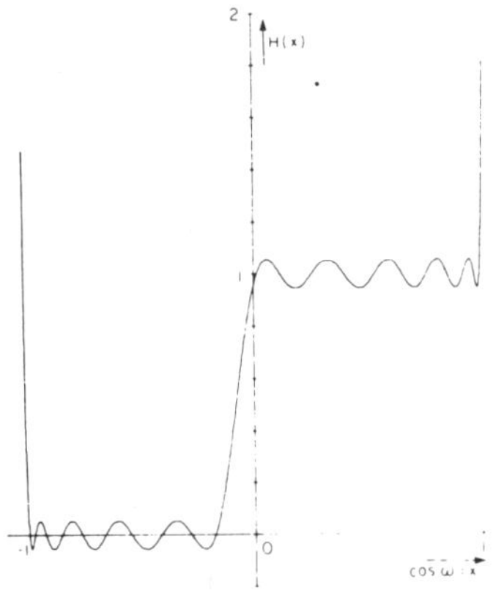

Fig. 3. Frequency response of a 41-point low-pass filter as a function of $x=\cos\omega$.

contours which result from a least squares minimization of $E_2(\omega_1)$ are essentially the same as those which result from a Chebyshev minimization. Furthermore, for a first-order transformation, at least, the minimization of $E_2(\omega_1)$ effectively minimizes $E_1(\omega_1)$. Thus, due to computational considerations, we have preferred to determine the transformation parameters by means of a least squares minimization of $E_2(\omega_1)$.

Performing a least squares fit to the elliptical contour of (9) with $\omega_a=0.25\pi$, $\omega_b=0.5\pi$, and $\omega_0=0.5\pi$, the following parameters result:

$$t(0,0)=-2.58366$$
$$t(1,0)=2.58426$$
$$t(0,1)=0.16411.$$

The contours consistent with this transformation are shown in Fig. 2. The crosshatched areas in the frequency plane denote those values of ω_1 and ω_2 for which $|F(\omega_1,\omega_2)|>1$. Since the substitution performed by the transformation is $\cos\omega=F(\omega_1,\omega_2)$, the frequency response in the crosshatched areas corresponds to complex values of ω. In Fig. 3 we show the frequency response of a possible prototype filter with an impulse response of length 41. The frequency response is plotted, not as a function of ω but as a function of $x=\cos\omega$. This function is a polynomial of twentieth degree, and for $|x|>1$ the magnitude of the frequency response grows as x^{20}. It is these large values of the frequency response which are mapped into the crosshatched regions of the frequency plane.

As a practical matter a transformation will only be useful if the coefficients of the mapping $t(p,q)$ satisfy the relation

$$\left|\sum_{p=0}^{1}\sum_{q=0}^{1} t(p,q)\cos p\omega_1 \cos q\omega_2\right| \leqslant 1,\qquad 0\leqslant\omega_1\leqslant\pi,\ 0\leqslant\omega_2\leqslant\pi. \tag{17}$$

Fortunately, it is possible to design the transformation parameters without regard to this constraint and to then perform a simple linear scaling on the final parameters so that (17) is satisfied. The contours defined by $F(\omega_1,\omega_2)$ = constant will not be changed by using

$$F'(\omega_1,\omega_2)=c_1F(\omega_1,\omega_2)-c_2 \tag{18}$$

as a transformation for any c_2 and nonzero c_1. If $F_{\max}$ denotes the maximum value of $F(\omega_1,\omega_2)$ and $F_{\min}$ denotes its minimum value, by choosing

$$c_1=2/(F_{\max}-F_{\min})$$
$$c_2=c_1\cdot F_{\max}-1 \tag{19}$$

the condition

$$|F'(\omega_1,\omega_2)|=\left|\sum_{p=0}^{1}\sum_{q=0}^{1} t'(p,q)\cos p\omega_1\cos q\omega_2\right|\leqslant 1$$

must always be true. The shape of the contours is unchanged; all that changes is the 1-D frequency ω which is associated with each contour. In particular, if $F(\omega_1,\omega_2)$ was originally associated with the value ω, $F'(\omega_1,\omega_2)$ defined by (18) will be associated with the value ω' where

$$\omega'=\cos^{-1}(c_1\cos\omega-c_2). \tag{20}$$

In most cases this implies that the 1-D prototype must be designed after the transformation has been designed.

By shifting and scaling the elliptical transformation shown in Fig. 2, the contours in Fig. 4 result. The cutoff frequency of the prototype is now $\omega_0'=0.25\pi$. The ellipse with minor axis 0.25π is the same as the one in Fig. 2, but now the frequency response will be well behaved over the whole frequency plane. In Fig. 5 we show the final 41×41 filter that results from the shifted and scaled transformation.

We have experimented with two other schemes for guaranteeing that the inequalities of (17) are satisfied. One is to perform the minimization of $E_2(\omega_1)$ using linear

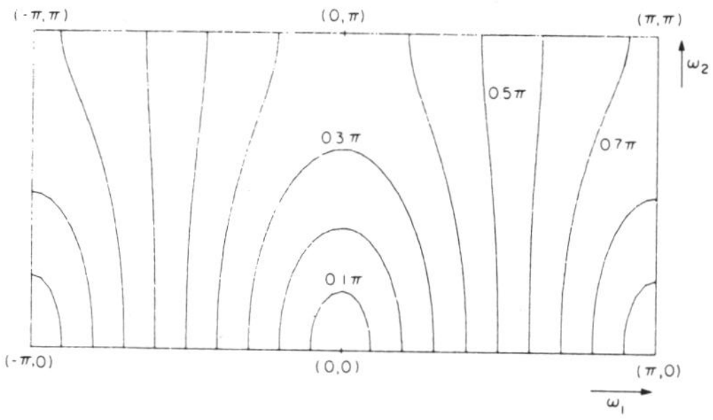

Fig. 4. Contours of the scaled and shifted version of the transformation of Fig. 2 ($c_1 = 0.29244$, $c_2 = -0.70756$).

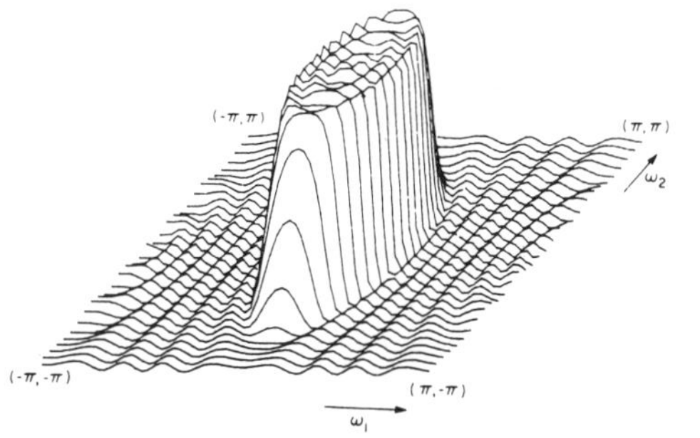

Fig. 5. Frequency response of a 41×41-point low-pass filter designed using the transformation shown in Fig. 4. The 1-D prototype filter has the specifications $\omega_p = 0.25\pi$, $\omega_s = 0.325\pi$, and is optimal in the Chebyshev sense with $\delta_p = \delta_s = 0.0238$.

programming which can incorporate the inequalities of (17) directly as part of its input specification. This procedure is computationally expensive, however, whereas shifting and scaling a least squares design is computationally modest and the transformations which result from linear programming are no better than those which result from the simpler scheme.

An alternate procedure is to consider cos ω_0 as a free variable. We still have a linear approximation, and generally a useful transformation will result although it is not guaranteed. When a solution results which satisfies the inequalities of (17), it has been observed to be identical to the solution obtained by fixing cos ω_0 and then shifting and scaling the resulting parameters. Since shifting and scaling guarantees that the inequalities of (17) will be satisfied, we regard it as the superior approach.

This procedure can now be stated in a more general form. This algorithm can design contours of any shape. It can design more than one contour at once and it can be used with mappings of any degree. Let us state the procedure for a two-contour mapping. Assume that we wish to map the point $\omega = \omega_a$ to the contour $\omega_1 = f_a(\phi)$, $\omega_2 = g_a(\phi)$ and $\omega = \omega_b$ to the contour $\omega_1 = f_b(\psi)$, $\omega_2 = g_b(\psi)$. We can then form the set of error equations

$$k_a e_a(\phi_i) = \cos \omega_a - \sum_{p=0}^{P} \sum_{q=0}^{Q} t(p,q) \cos f_a(\phi_i) \cos g_a(\phi_i)$$
$$k_b e_b(\psi_i) = \cos \omega_b - \sum_{p=0}^{P} \sum_{q=0}^{Q} t(p,q) \cos f_b(\psi_i) \cos g_b(\psi_i) \quad (21)$$

on some set of discrete samples of ϕ and ψ. k_a and k_b are weighting constants that allow a differential weighting of the error function on the different contours. We can then minimize, for example, the quantity

$$E_1 = \max_{i,j} |e_a(\phi_i), e_b(\psi_j)| \quad (22)$$

or

$$E_2 = \sum_i e_a^2(\phi_i) + \sum_j e_b^2(\psi_j) \quad (23)$$

subject to equality constraints of the form

$$\sum_{p=0}^{P} \sum_{q=0}^{Q} r_{pq} t(p,q) = s \quad (24)$$

and inequality constraints of the form

$$\left| \sum_{p=0}^{P} \sum_{q=0}^{Q} t(p,q) \cos p\omega_{1i} \cos q\omega_{2i} \right| \leq 1. \quad (25)$$

We can minimize E_1 by linear programming, or if the inequality constraints (25) are not present we can minimize E_2 using least squares techniques. After the $t(p,q)$ have been determined, the inequalities can be satisfied by shifting and scaling the resulting transformation at the expense of the equality constraints. (Some equality constraints such as (11) are generally necessary in a one-contour design to prevent a trivial solution. Once a nontrivial solution has been found, however, they are generally not important.)

If it is desired to constrain the shapes of all contours simultaneously, this approach can still be followed, but a bit of trial and error may be necessary. Quatieri [13] has presented some approaches that should be useful in this case. One approach is to apply the aforementioned technique constraining more than one contour. Another technique is to apply additional equality or inequality constraints. For example, if all contours are to have the same general shape, constraining two contours and requiring that the 1-D frequency be mapped linearly to both axes in the 2-D frequency plane:

$$\omega \rightarrow (\omega, 0)$$
$$\omega \rightarrow (0, \omega) \quad (26)$$

will result in a set of fairly concentric contours.

B. Another Formulation of the Contour Design Problem

The preceding design algorithm is simple to implement, and for first- or second-order transformations or for designing several contours simultaneously, it is preferable to the algorithms which follow. That algorithm used an optimization routine to get a best fit to a given contour shape. When transformations of degree higher than 1 are used, however, it is possible for the 2-D transition region to be narrower than the transition region of the 1-D prototype. Thus, if a 2-D piecewise constant design is desired, it may be advantageous to simultaneously optimize the transition width of the 2-D filter. The following algorithm accomplishes this.

The design problem for 2-D equiripple filters is normally stated in the following terms [6]: find the best $(2N+1)\times(2N+1)$ approximation in the Chebyshev sense to an ideal function which has value 1 in its passband $\{(\omega_1,\omega_2)\in I_p\}$ and value 0 in its stopband $\{(\omega_1,\omega_2)\in I_s\}$. A best approximation is one (not necessarily unique) for which the maximum deviation of the (weighted) error function on $\{(\omega_1,\omega_2)\in I_p \cup I_s\}$ is minimum.

We can also state the McClellan transformation design problem in this form. In particular, we can ask: given regions I_p and I_s, what is the best McClellan transformation of degree $P\times Q$ in the sense that the deviation between the 2-D filter which ultimately results and the ideal filter function is minimized over the union of the passband and stopband? In our case the deviation will effectively be minimized then the transition bandwidth of the 1-D prototype is maximized. We can thus set the problem up in the following fashion:

$$\cos\omega_s \geqslant \sum_p \sum_q t(p,q)\cos p\omega_1 \cos q\omega_2 \geqslant -1, \qquad \omega_1,\omega_2\in I_s$$

$$\cos\omega_p \leqslant \sum_p \sum_q t(p,q)\cos p\omega_1 \cos q\omega_2 \leqslant 1, \qquad \omega_1,\omega_2\in I_p. \tag{27}$$

The coefficients $t(p,q)$ should then be chosen to maximize $\omega_s-\omega_p$. This is a nonlinear optimization problem. We can replace it by a linear problem which can be solved as a linear program by maximizing instead the quantity $\cos\omega_p-\cos\omega_s$. The values $\cos\omega_p$ and $\cos\omega_s$ are the cosines of the 1-D passband and stopband cutoff frequencies. They are then used to design the 1-D prototype.

This formulation of the problem tends to give better transformations than the contour matching formulation in the sense of a wider 1-D transition width for single contour problems. For $P=Q=1$ the two techniques are comparable. This technique, on the other hand, does require a linear program for its solution and will not work well on other than piecewise constant ideal functions.

The form of (27) suggests still another algorithm which can be implemented without linear programming. The inequalities (27) are reminiscent of those used to define a 2-D equiripple filter if the $t(p,q)$ are considered as the impulse response of the filter.

If we thus design a $(2P+1)\times(2Q+1)$ 2-D filter with passband region I_p and stopband region I_s using, say, the equiripple algorithm of Kamp and Thiran [6] or Mersereau *et al.* (each of which is considerably more efficient than linear programming), we can convert its impulse response coefficients into the transformation coefficients $t(p,q)$. If δ_p and δ_s are the passband and stopband ripples, respectively, of the resulting $(2P+1)\times(2Q+1)$ filter, then

$$\omega_p=\cos^{-1}\left\{\frac{1+\delta_s-3\delta_p}{1+\delta_p+\delta_s}\right\} \tag{28}$$

$$\omega_s=\cos^{-1}\left\{\frac{3\delta_s-\delta_p-1}{1+\delta_p+\delta_s}\right\}. \tag{29}$$

The transformation coefficients can be obtained from the impulse response coefficients by

$$\begin{aligned} t(0,0)&=c_1h(0,0)-c_2 \\ t(p,0)&=2c_1h(p,0), && p\neq 0 \\ t(0,q)&=2c_1h(0,q), && q\neq 0 \\ t(p,q)&=4c_1h(p,q), && p\neq 0 \quad q\neq 0 \end{aligned} \tag{30}$$

where

$$c_1=\frac{2}{1+\delta_p+\delta_s} \tag{31}$$

$$c_2=\frac{-\delta_s+\delta_p+1}{1+\delta_p+\delta_s}. \tag{32}$$

The transition band of the 1-D prototype $\omega_s-\omega_p$ which results is slightly wider than if linear programming is used to solve (27) directly, but the sacrifice is well worth the reduced computational burden.

IV. Examples

In this last section we present a few additional examples of filters designed using the McClellan transformation. The contours were designed using the preceding algorithms. In Figs. 6 and 7 we show a circularly symmetric low-pass filter of degree 31×31 designed using a third-order transformation. This transformation was derived from a 7×7 2-D circularly symmetric equiripple FIR low-pass filter. The coefficients of the transformation are given in Appendix II.

In Fig. 8 an example of a 27×27 fan filter is shown. The first-order transformation has the coefficients $t(0,0)=t(1,1)=0$, $t(0,1)=-t(1,0)=0.5$, which are due to McClellan and Parks [2]. Such filters are useful for velocity discrimination in seismic processing.

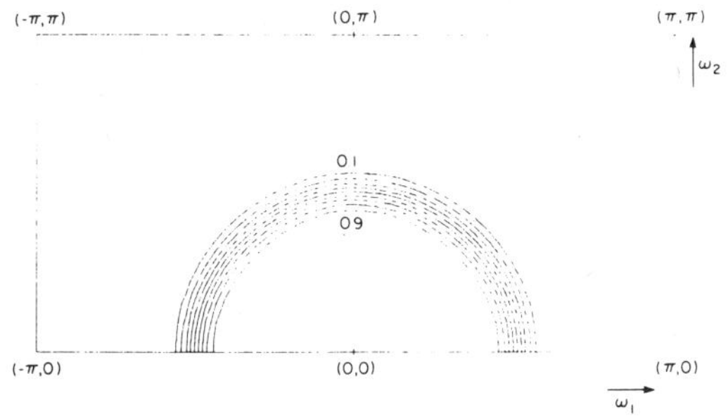

Fig. 6. Contours of the frequency response of a 31×31-point circularly symmetric low-pass filter showing the variation in the transition band. This filter is designed by a third-order transformation which is derived from a 7×7-point 2-D circularly symmetric equiripple FIR low-pass filter with $\omega_p = 0.4\pi$, $\omega_s = 0.6\pi$, and $\delta_p = \delta_s = 0.12696$.

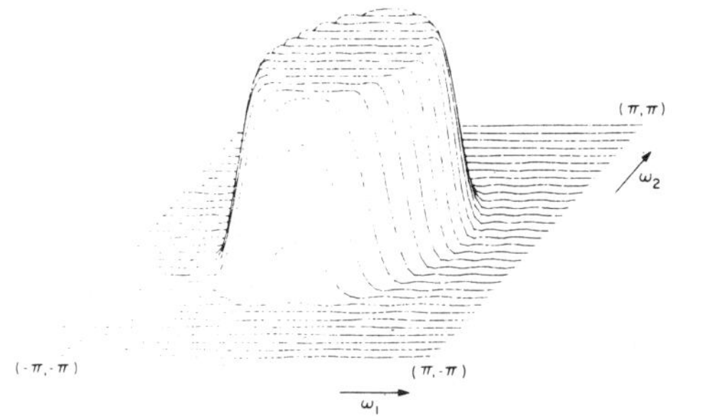

Fig. 7. Frequency response of the low-pass filter whose contours are shown in Fig. 6. The 1-D prototype filter has length 11 and is optimal in the Chebyshev sense with $\omega_p = 0.29716\pi$, $\omega_s = 0.70284\pi$, and $\delta_p = \delta_s = 0.00523$.

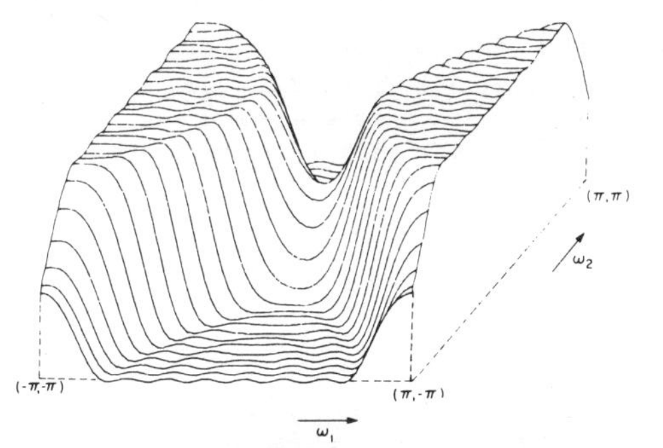

Fig. 8. Frequency response of a 27×27-point fan filter derived from a first-order transformation. The 1-D prototype filter is optimal in the Chebyshev sense with $\omega_p = 0.43\pi$, $\omega_s = 0.57\pi$, and $\delta_p = \delta_s = 0.01157$.

An example of a filter which is not piecewise constant is depicted in Figs. 9 and 10. The transformation was designed by the contour matching approach. The contours were to be nearly circular at radius 0.85π and radius

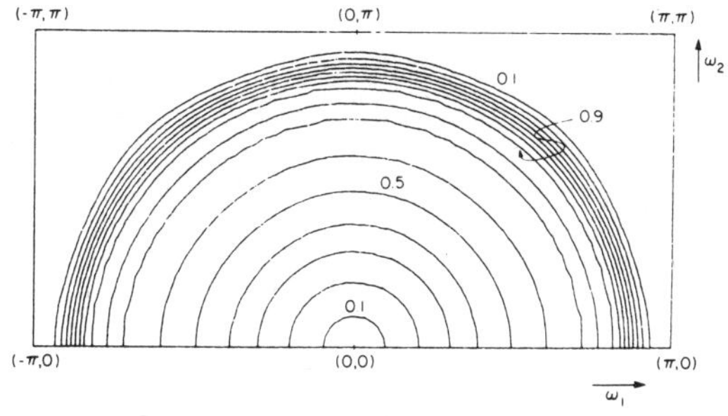

Fig. 9. Contours of the linearly increasing frequency response of a 61×61-point filter derived from a second-order transformation.

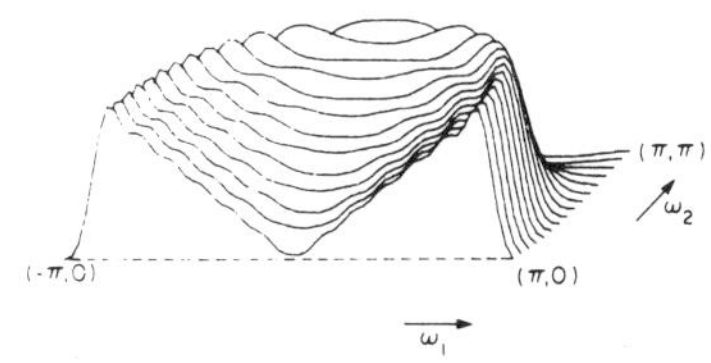

Fig. 10. Frequency response for $\omega_2 > 0$ of the filter whose contours are shown in Fig. 9.

0.25π, and the 1-D ω axis was mapped linearly to both the ω_1 and ω_2 axes. The resulting transformation function $F(\omega_1, \omega_2)$ then required slight shifting and scaling. The shift and scale were compensated for in the design of the 1-D prototype which was designed using a 31-point Kaiser window. The transformation is of second order and the final 2-D filter is 61×61. Filters of this form are useful in the reconstruction of objects from their projections and transverse axial tomography [14].

To show the versatility of the design methods for transformations as described in this paper, Fig. 11 shows three examples of possible 2-D FIR filters designed using various transformations.

V. Summary

The purpose of this paper is to show that the generalized McClellan transformation is a viable technique for the design of 2-D filters with quadrilateral symmetry. Two basic algorithms for estimating the transformation parameters are described. Of these the algorithm in Section III-A is to be preferred for first-order transformations or when several contours must be matched simultaneously. The second algorithm is to be preferred when a higher order transformation is desired to design a piecewise constant transfer function.

To illustrate the efficiency of these techniques, every example in this paper was programmed on a desk calculator and the longest design took less than $\frac{1}{2}$ h. This is equivalent to a few seconds on a large systems computer.

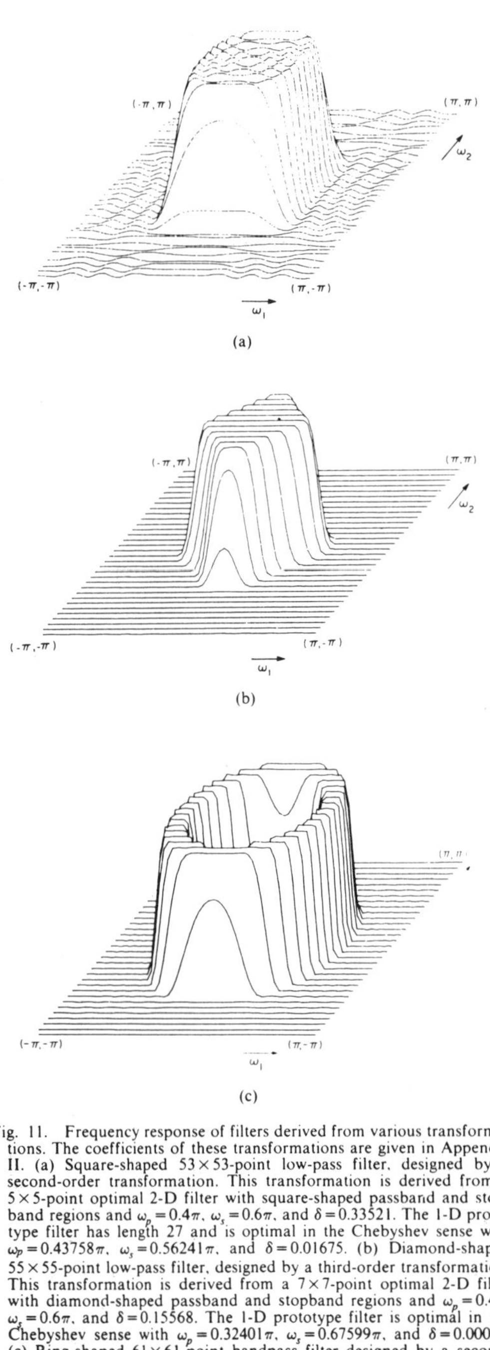

Fig. 11. Frequency response of filters derived from various transformations. The coefficients of these transformations are given in Appendix II. (a) Square-shaped 53×53-point low-pass filter, designed by a second-order transformation. This transformation is derived from a 5×5-point optimal 2-D filter with square-shaped passband and stopband regions and $\omega_p=0.4\pi$, $\omega_s=0.6\pi$, and $\delta=0.33521$. The 1-D prototype filter has length 27 and is optimal in the Chebyshev sense with $\omega_p=0.43758\pi$, $\omega_s=0.56241\pi$, and $\delta=0.01675$. (b) Diamond-shaped 55×55-point low-pass filter, designed by a third-order transformation. This transformation is derived from a 7×7-point optimal 2-D filter with diamond-shaped passband and stopband regions and $\omega_p=0.4\pi$, $\omega_s=0.6\pi$, and $\delta=0.15568$. The 1-D prototype filter is optimal in the Chebyshev sense with $\omega_p=0.32401\pi$, $\omega_s=0.67599\pi$, and $\delta=0.00075$. (c) Ring-shaped 61×61-point bandpass filter designed by a second-order transformation. The transformation is designed to approximate a circularly symmetric passband at $\omega_0=0.5\pi$. The 1-D prototype filter is optimal in the Chebyshev sense with $\omega_p=0.12\pi$, $\omega_s=0.3\pi$, and $\delta=0.00345$.

Not only are the filters easy to design, but they can be economically implemented as well. They represent, in our opinion, the most viable technique available at this time for designing and implementing 2-D FIR zero-phase filters.

Appendix I

McClellan Transformations for Filters of Size $2N\times 2N$

Linear-phase 2-D FIR filters whose impulse responses are nonzero only over a square whose sides are $2N$ points long can also be designed using the McClellan transformation in one special case. Because the technique is only valid for one particular transformation, it is of less value than the transformations for filters with odd-length sides, but this special case does include certain nearly circularly symmetric filters.

Consider a 1-D linear-phase digital filter of length $2N$. Assume that the impulse response of the filter $h(n)$ is nonzero for n in the range $-N+1\leqslant n\leqslant N$. Since the filter has linear phase, $h(n)$ obeys the symmetry constraint

$$h(n)=h(1-n), \qquad -N+1\leqslant n\leqslant N. \tag{A1}$$

Its frequency response can then be written as

$$H(e^{j\omega})=e^{-j\frac{\omega}{2}}\sum_{n=1}^{N}h(n)2\cos\omega\left(n-\frac{1}{2}\right). \tag{A2}$$

From (A2) it is seen that $H(e^{j\pi})=0$, and thus we can write

$$H(z)=(1+z^{-1})\hat{H}(z). \tag{A3}$$

Similarly, if $h(m,n)$ is the impulse response of a $2M\times 2M$ linear-phase FIR digital filter which is nonzero over the region $-M+1\leqslant m\leqslant M$, $-M+1\leqslant n\leqslant M$, the frequency response of the filter can be written as

$$H(e^{j\omega_1},e^{j\omega_2})=e^{-j(\omega_1+\omega_2)/2}\sum_{m=1}^{M}\sum_{n=1}^{M}4h(m,n)\cdot\cos\omega_1\left(m-\frac{1}{2}\right)\cos\omega_2\left(n-\frac{1}{2}\right). \tag{A4}$$

From (A4) it is readily seen that $H(e^{j\pi},e^{j\omega_2})=H(e^{j\omega_1},e^{j\pi})=0$. Thus $H(z_1,z_2)$ can be written as

$$H(z_1,z_2)=(1+z_1^{-1})(1+z_2^{-1})\hat{H}(z_1,z_2). \tag{A5}$$

$\hat{H}(z)$ is of length $2N-1$, and similarly $\hat{H}(z_1,z_2)$ is of size $(2M-1)\times(2M-1)$. It is thus possible to find filters $\hat{H}(z_1,z_2)$ from the filters $\hat{H}(z)$ by using a McClellan transformation for odd-length sequences as discussed in Section II. We can thus develop a four-step procedure for generating a 2-D filter from a 1-D design.

1) Determine the appropriate $2N$ point prototype filter $h(n)$.

2) From $h(n)$ determine the $2N-1$ point sequence $\hat{h}(n)$, where $\hat{h}(n)$ is the inverse z transform of $\hat{H}(z)$ defined in (A3). This is a simple numerical operation.

3) Using a McClellan transformation, transform $\hat{h}(n)$ to $\hat{h}(m,n)$. The resulting filter will be of length $(2M-1)\times(2M-1)$.

4) Convolve $\hat{h}(m,n)$ with $p(m,n)$ to get $h(m,n)$ where $p(m,n)$ is the inverse 2-D z transform of $P(z_1,z_2)$

$$P(z_1,z_2)=(1+z_1^{-1})(1+z_2^{-1}). \tag{A6}$$

The filter which results from this procedure is linear phase and of size $2N\times 2N$. The contours of the mapping in step 3) are altered by the convolution of step 4). This procedure thus yields the desired contours only in the special case where the contours of the transformation used in 3) correspond to the contours

$$|P(e^{j\omega_1},e^{j\omega_2})|=\text{constant}. \tag{A7}$$

Condition (A7) is true only for scaled and shifted versions of the transformation

$$-t(0,0)=t(0,1)=t(1,0)=t(1,1)=\frac{1}{2} \tag{A8}$$

which is McClellan's original transformation for nearly circular contours. Thus the only useful transformation for 2-D filters whose sides are even integers is the transformation depicted in Fig. 1. It is nonetheless fairly useful for designing approximately circularly symmetric transfer functions. The filters designed must also be of the low-pass type in that $H(e^{j\pi},e^{j\omega_2})=H(e^{j\omega_1},e^{j\pi})=0$ must hold. This, however, is a property of linear-phase $2N\times 2N$ filters designed by any method and is not a constraint introduced by the transformation.

Appendix II

In this Appendix the coefficients of the transformations are given, which are used in Figs. 6 and 11. Since these transformations are all symmetric in ω_1 and ω_2, it follows that $t(p,q)=t(q,p)$.

Transformation for Fig. 6:

	$t(p,q)$	p = 0	1	2	3
q	0	−0.46485	0.42187	0.15087	−0.10937
	1		0.69151	0.083514	−0.17052
	2			−0.18001	−0.09853
	3				−0.0073362

Transformation for Fig. 11(a):

	$t(p,q)$	p = 0	1	2
q	0	−0.43121	0.30365	0.005325
	1		0.68633	0.008787
	2			0.10836

Transformation for Fig. 11(b):

	$t(p,q)$	p = 0	1	2	3
q	0	−0.55078	0.31577	0.14760	0.04362
	1		0.50364	0.20770	0.0016933
	2			0.097659	−0.18861
	3				−0.12855

Transformation for Fig. 11(c):

	$t(p,q)$	p = 0	1	2
q	0	−0.12403	0.49995	−0.37602
	1		0	−0.49995
	2			−0.12393

References

[1] T. S. Huang, J. W. Burnett, and A. G. Deczky, "The importance of phase in image processing filters," *IEEE Trans. on Acoustics, Speech and Signal Processing*, vol. ASSP-23, no. 6, pp. 529–542, December 1975.

[2] J. H. McClellan and T. W. Parks, "Equiripple approximation of fan filters," *Geophysics*, vol. 7, pp. 573–583, 1972.

[3] J. H. McClellan, "The design of two-dimensional digital filters by transformations," *Proc. 7th Annual Princeton Conf. Information Sciences and Systems*, pp. 247–251, 1973.

[4] T. S. Huang, "Two-dimensional windows," *IEEE Trans. Audio and Electroacoustics*, vol. AU-20, pp. 88–90, March 1972.

[5] J. V. Hu and L. R. Rabiner, "Design techniques for two-Dimensional digital filters," *IEEE Trans. Audio and Electroacoustics*, vol. AU-20, pp. 249–257, October 1972.

[6] Y. Kamp and J. P. Thiran, "Chebyshev approximation for two-dimensional nonrecursive digital filters," *IEEE Trans. Circuits and Systems*, vol. CAS-22, no. 3, pp. 208–218, March 1975.

[7] R. M. Mersereau, D. B. Harris, and H. S. Hersey, "An efficient algorithm for the design of equiripple two-dimensional FIR digital filters," *Proc. 1975 IEEE Int. Symp. on Circuits and Systems*, pp. 443–446, April 1975.

[8] R. M. Mersereau and D. E. Dudgeon, "Two-dimensional digital filtering," *Proc. IEEE*, vol. 63, no. 4, pp. 610–623, April 1975.

[9] C. M. Rader, "On the application of the number theoretic transforms of high speed convolution to two-dimensional filtering," *IEEE Trans. Circuits and Systems*, vol. CAS-22, no. 6, p. 575, June 1975.

[10] W. F. G. Mecklenbräuker and R. M. Mersereau, "McClellan transformations for two-dimensional digital filtering: II. Implementation," to be published.

[11] L. R. Rabiner and B. Gold, "Theory and implementation of FIR digital filters," ch. 3 in *Theory and Application of Digital Signal Processing*, Prentice-Hall, Englewood Cliffs, NJ, 1975

[12] A. V. Oppenheim and R. W. Schafer, "Digital filter design techniques," ch. 5 in *Digital Signal Processing*, Prentice-Hall, Englewood Cliffs, NJ, 1975.

[13] T. F. Quatieri, Jr., "The design of two-dimensional digital filters by generalized McClellan transformations," M.Sc. Thesis, Department of Electrical Engineering and Computer Science, Massachusetts Institute of Technology, Cambridge, MA., June 1975.

[14] R. M. Mersereau and A. V. Oppenheim, "Digital reconstruction of multi-dimensional signals from their projections," *Proc. IEEE*, vol. 62, pp. 1319–1338, October 1974.

Part II

STATISTICAL, TWO-DIMENSIONAL DIGITAL SIGNAL PROCESSING

Editors' Comments on Papers 20 Through 23

20 HELSTROM
Image Restoration by the Method of Least Squares

21 PRATT
Generalized Wiener Filtering Computation Techniques

22 EKSTROM and ALGAZI
Optimum Design of Two-Dimensional Nonrecursive Digital Filters

23 LEBEDEV and MIRKIN
Smoothing of Two-Dimensional Images Using the "Composite" Model of a Fragment

DESIGN OF NONRECURSIVE PROCESSORS

The preceeding papers in this volume deal with 2-D digital signal processing of a deterministic nature. The design of these processors or digital filters was based on achieving a deterministic design criterion, usually in the form of a frequency domain specification. Such design problems generally arise in filtering applications requiring the enhancement and/or suppression of specific components (e.g., noise) of the input sequence. The translation of this requirement into a filter specification must preceed the design stage considered in Papers 1–19.

In this section, we consider an alternate and perhaps more general approach to processor design. This approach is statistically based and involves developing the processor from the classical viewpoint of extracting a signal sequence in the presence of additive noise. Probabilistic models are assumed for both the signal and noise sequences, i.e., they are both assumed to be generated by random mechanisms. The design problem is that of finding the digital processor or filter which gives the best signal estimate (in a statistical sense) from the noise corrupted sequence. Statistical error criteria are used in these designs. Here, rather than design a processor to filter (suppress) a specific signal or noise sequence as is generally the case in deterministic designs, the processor is chosen to be optimal over an ensemble of signal and noise sequences. While this offers a somewhat simplistic comparison between the deterministic and probabilistic approaches, it does point out the contrasting viewpoints associated with each design approach.

As in the deterministic case, statistical processors can also be grouped into nonrecursive and recursive categories. Historically, the first work involving the design of 2-D nonrecursive digital processors was the true benchmark paper of Helstrom (Paper 20). Although presented in the context of removing instrument distortion from optical images (so-called image restoration), this paper extends the minimum mean-square error estimation technique of Wiener to the problem of processing 2-D random fields. Continuous and discrete formalisms are described for the design of optical and digital processors, respectively.

In developing the digital processor, the observable data $f(m,n)$ is represented as a linear transformation (known and space invariant) of the desired signal $s(m,n)$, combined with additive noise. Both the signals and noise are modeled as spatial stochastic processes, gaussian and zero-mean with known covariances. The estimate of the desired signal, $\hat{s}(m,n)$, is taken to be a weighted linear combination of the data:

$$\hat{s}_{m,n} = \sum_{m'} \sum_{n'} h_{m,m';n,n'} f_{m',n'}. \tag{2.1}$$

Because the observed data array is of finite extent, this estimate is said to be based on a finite sample.

Design of the estimator involves picking the weighting array $h_{m,m';n,n'}$ to minimize the mean-square error

$$\epsilon = E\left\{[\hat{s}_{m,n} - s_{m,n}]^2\right\} \tag{2.2}$$

between the estimate and true value of the signal. In this error expression, $E\{\cdot\}$ is an expectation taken over the ensemble of signal and noise arrays. Thus, the error is a measure of the statistical concentration of the point difference.

Helstrom employs a matric formalism to characterize this extremal problem and to solve for the optimal weighting array. His formalism is discussed in the context of its application to a variety of estimation problems and its relation to a similar (deterministic) least-squares design. The major computations in implementing this design are involved with the inversion of potentially large covariance matrices. This may be a severe problem in many applications.

Pratt (Paper 21) considers the use of unitary transforms in reducing the computational requirements associated with this Wiener filtering. In filter design and application, he formulates the required matrix operations in terms of the unitary transforms of the data and covariance arrays. The transforms are chosen to yield sparse matrices which can be manipulated with reduced computational complexity. A discussion of selective computations to further reduce the requirements is presented along with its effect on filter performance.

In Papers 20 and 21, the mean-square estimates were obtained using an optimal, linear shift-variant filtering of the data. An alternate approach of Ekstrom and Algazi (Paper 22) involves constraining the Wiener filter to be shift-invariant. Thus, in contrast to Eq. (2.1), the signal estimate is taken to be

$$\hat{s}_{m,n} = \sum_{m'} \sum_{n'} h_{m-m',n-n'} f_{m',n'}. \tag{2.3}$$

As before, the filter array (in this case $h_{m,n}$) is chosen to minimize the mean-square error ϵ in Eq. (2.2).

Ekstrom and Algazi's design procedure is presented for the case of filtering 2-D homogeneous random fields. Its computational implementation requires the inversion of large-scale Toeplitz matrices. A number of numerically efficient algorithms are available for this purpose (Bareiss, 1969; Akaike, 1973; and Ekstrom, 1974). The apparent advantages of the shift-invariant filtering are its reduced computational complexity in design, and particularly in application, and the ease with which the order of the filter may be controlled. The disadvantage is, of course, the reduced filter performance which derives from the constraint.

In the three previous contributions, the available data are modeled as the realization of a homogeneous, 2-D random field. Nonrecursive Wiener processors were derived to filter the data based on a mean-square error criterion. In contrast, Lebedev and Mirkin (Paper 23) adopt a composite model for the fields which leads to a multicategory Wiener filtering problem. Here, the model involves additive noise and multiple classes or categories of random fields in the signal, with each category distinguished by its covariance. A gaussian probability density is associated with the occurence of each category in the data. With this model, the optimal mean-square estimate is obtained by Wiener filtering the data for each category (using the approaches of Papers 20–22, for example), and forming the estimate as the weighted sum of all the filter outputs. The weights are the a posteriori probabilities that the point is a member of the respective category. While this work has not appeared in the open literature and consequently not received much attention, it does appear to have significant potential for 2-D applications where the traditional approaches do not adequately model the locally nonhomogeneous information in the signal.

REFERENCES

Akaike, H. 1973. "Block Toeplitz Matrix Inversion," *SIAM J. Appl. Math.* **23**:234–241.

Bareiss, E. H. 1969. "Numerical Solution of Linear Equations with Toeplitz and Vector Toeplitz Matrices," *Numer. Math.* **13**:404–424.

Ekstrom, M. P. 1974. "An Iterative Improvement Approach to the Numerical Solution of Vector Toeplitz Systems," *IEEE Trans. Computers* **C-23**:320–325.

20

Reprinted from *Opt. Soc. Amer. J.* **57**(3):297–303 (1967)

Image Restoration by the Method of Least Squares

CARL W. HELSTROM*
Westinghouse Research Laboratories, Pittsburgh, Pa. 15235
(Received 15 August 1966)

The restoration of optical images, as well as the unfolding of spectroscopic and other data that have been convolved with a window function or an instrumental impulse response, can be viewed as the solution of an integral equation. Solution of such an integral equation when the data are corrupted by noise or experimental error is treated as the problem of finding an estimate that is a linear functional of the data and minimizes the mean squared error between the true solution and itself. The estimate depends on assumptions about the spectral densities of the images and the noise, the choice of which is discussed. Coherent optical processing and digital processing are described.

INDEX HEADINGS: Image restoration; Fourier transform.

1. IMAGE RESTORATION AS THE SOLUTION OF AN INTEGRAL EQUATION

THE degradation suffered by images in optical systems can often be described as a convolution of the true or geometrical image with a spread function.[1] Let $J_0(\mathbf{r})$ be the illuminance at a point $\mathbf{r}=(x,y)$ of the image plane if there were no degradation, and $J(\mathbf{r})$ be the actual illuminance. Then at least over a limited region, or "isoplanatism-patch,"[2]

$$J(\mathbf{r})=\int S(\mathbf{r}')J_0(\mathbf{r}-\mathbf{r}')d^2\mathbf{r}', \tag{1.1}$$

where $S(\mathbf{r})$ is the spread function and $d^2\mathbf{r}'=dx'dy'$ the differential of area. (All integrals can be taken over the infinite plane if the integrands are properly defined.)

* Present address: Department of Applied Electrophysics, University of California, San Diego, La Jolla, California 92037.

[1] E. L. O'Neill, *Introduction to Statistical Optics* (Addison-Wesley Publishing Co., Reading, Mass., 1963), p. 20.

[2] P. B. Fellgett and E. H. Linfoot, Phil. Trans. Roy. Soc. A247, 369 (1955).

It is convenient to normalize the spread function so that

$$\int S(\mathbf{r})d^2\mathbf{r}=1. \tag{1.2}$$

If the system were perfect, the spread function would be a two-dimensional delta-function, $S(\mathbf{r})=\delta(\mathbf{r})$, and $J(\mathbf{r})$ would equal $J_0(\mathbf{r})$.

Other impairments of an image can also be expressed as convolutions. If an image moving with velocity v in the x direction is recorded photographically in an exposure of duration T, the transmittance of the developed film is a function of an integrated illuminance given in Eq. (1.1), where the spread function can be taken as

$$\begin{aligned} S(\mathbf{r})=S(x,y)&=(vT)^{-1}\delta(y), \quad -\tfrac{1}{2}vT\leq x\leq \tfrac{1}{2}vT, \\ &=0, \quad |x|>\tfrac{1}{2}vT. \end{aligned} \tag{1.3}$$

The restoration of the true image $J_0(\mathbf{r})$ can be considered as the solution of the integral Eq. (1.1), whose kernel is the spread function $S(\mathbf{r})$, assumed known. The method that first presents itself is to apply the two-dimensional Fourier transform. Putting[3]

$$\begin{aligned} j(\boldsymbol{\omega})&=\int J(\mathbf{r})\exp(-i\boldsymbol{\omega}\cdot\mathbf{r})d^2\mathbf{r}, \\ s(\boldsymbol{\omega})&=\int S(\mathbf{r})\exp(-i\boldsymbol{\omega}\cdot\mathbf{r})d^2\mathbf{r}, \end{aligned} \tag{1.4}$$

we can write the solution of the integral equation as

$$\begin{aligned} J_0(\mathbf{r})&=\int j_0(\boldsymbol{\omega})\exp(i\boldsymbol{\omega}\cdot\mathbf{r})d^2\boldsymbol{\omega}/(2\pi)^2, \\ j_0(\boldsymbol{\omega})&=j(\boldsymbol{\omega})/s(\boldsymbol{\omega}). \end{aligned} \tag{1.5}$$

This method often does not work because the illuminance $J(\mathbf{r})$ cannot be measured precisely enough. Indeed, the illuminance $J(\mathbf{r})$ actually observed is given not by Eq. (1.1), but by

$$J(\mathbf{r})=\int S(\mathbf{r}')J_0(\mathbf{r}-\mathbf{r}')d^2\mathbf{r}'+N(\mathbf{r}), \tag{1.6}$$

where $N(\mathbf{r})$ represents experimental error, background radiation, and, in photography, the effect of granularity of the emulsion. We can encompass all these in the term spatial noise. The noise $N(\mathbf{r})$ cannot be known in advance and differs from experiment to experiment.

Most optical spread functions attenuate high spatial frequencies; $|s(\boldsymbol{\omega})|\rightarrow 0$ as $|\boldsymbol{\omega}|\rightarrow\infty$. An instructive analogy is the passage of a signal through a low-pass filter. To restore the original signal, the output must be passed through the inverse filter, which amplifies the high frequencies. By this time, however, noise has inevitably been added, and it contains frequencies far beyond those of the signal. These noise frequencies are so strengthened by the inverse filter as to blot out the desired restoration.

An attempt to treat the optical problem by Fourier transforms as in Eq. (1.5) is frustrated in the same way. A further difficulty arises when, as with the kernel in Eq. (1.3), the Fourier transform $s(\boldsymbol{\omega})$ of the spread function contains zeros at finite values of $\boldsymbol{\omega}$. Because of the noise, the Fourier transform $j(\boldsymbol{\omega})$ does not vanish at the same values of $\boldsymbol{\omega}$, and the quotient $j(\boldsymbol{\omega})/s(\boldsymbol{\omega})$ acquires troublesome infinities.

The solution of such integral equations involving experimental data is a common problem in physics. It arises in nuclear and optical spectroscopy, in photometry, in astronomy, and indeed wherever a finite instrumental response or entrance window convolves with the quantity to be measured. A method of avoiding certain of the difficulties in optics has recently been demonstrated by Harris.[4] Other approaches have been given by Trumpler and Weaver,[5] Kahn,[6] Phillips,[7] and Twomey.[8,9] Here we propose a method that makes use of the statistical properties of the noise. It is based on the theory of minimum-mean-square prediction and estimation.

2. AN ESTIMATED SOLUTION OF THE INTEGRAL EQUATION

With the spatial noise $N(\mathbf{r})$ unknown, Eq. (1.6) cannot be solved directly; the most that can be done is to produce an estimate $\hat{J}_0(\mathbf{r})$ of the solution. If both the image illuminance $J_0(\mathbf{r})$ and the noise $N(\mathbf{r})$ are viewed as spatial stochastic processes, the best estimate is naturally defined as the one that maximizes the posterior probability density of $J_0(\mathbf{r})$, given $J(\mathbf{r})$, as determined by Bayes's rule.[10]

The processes $J_0(\mathbf{r})$ and $N(\mathbf{r})$ are independent. If they were also gaussian, with zero means and known covariances, the posterior probability density would be greatest for the linear estimate

$$\hat{J}_0(\mathbf{r})=\int M(\mathbf{r}')J(\mathbf{r}-\mathbf{r}')d^2\mathbf{r}', \tag{2.1}$$

in which the estimating kernel $M(\mathbf{r})$ is chosen so as to

[3] In general we shall use a lower-case letter to denote the Fourier transform of a spatial function designated by the corresponding upper-case letter.

[4] J. L. Harris, Sr., J. Opt. Soc. Am. **56**, 569 (1966).

[5] R. J. Trumpler and H. F. Weaver, *Statistical Astronomy* (University of California Press, Berkeley, Calif., 1953), Ch. 1.4, p. 95 ff.

[6] F. D. Kahn, Proc. Cambridge Phil. Soc. **51**, 519 (1955).

[7] D. L. Phillips, J. Assoc. Comp. Mach. **9**, 84 (1962).

[8] S. Twomey, J. Assoc. Comp. Mach. **10**, 97 (1963).

[9] S. Twomey, J. Franklin Inst. **279**, 95 (1965).

[10] For continuous processes such as these, a probability density must be treated by a limiting procedure as samples of the process are taken closer and closer together in the plane. The assertions of the text remain valid.

minimize the mean squared error

$$E=\mathbf{E}[\hat{J}_0(\mathbf{r})-J_0(\mathbf{r})]^2 \tag{2.2}$$

between the estimate and the true illuminance, **E** denoting an expected value.[11]

Although the noise may often be at least approximately treated as a gaussian process, the images to be restored rarely fall into that category. Nevertheless, if an estimator is optimum for a class of gaussian processes whose structure resembles that of the finest images to be encountered, it can be expected to be effective when applied to these images as well. It is admitted that a detailed statistical specification of the class of images anticipated would lead, if the optimization problem could be solved, to a superior estimator; but it would almost certainly be more complicated than the linear estimator of Eq. (2.1), and of restricted applicability besides. The linear estimator that minimizes the mean-squared error, Eq. (2.2), therefore recommends itself as an initial step in the search for good image restoration.

The integral in Eq. (2.1) may be said to represent the action of a restoring filter, and such a linear estimate may be indeed be realizable by coherent optical processing.[12,13] If the illuminances are sampled at discrete points, a discrete counterpart obtained by approximating the integrals in Eqs. (1.6) and (2.1) by summation formulas can be executed by a digital computer. The estimating kernel $M(\mathbf{r})$ depends on the spread function $S(\mathbf{r})$ and on the properties of the images and the noise. We turn to the problem of calculating it.

Of the limited region in which the degradation of the image can be described by Eq. (1.1) there is no reason to favor one part over another, and the stochastic processes representing the images $J_0(\mathbf{r})$ and the noise $N(\mathbf{r})$ may as well be taken as spatially stationary. Since it is chiefly the deviation

$$\Delta J_0(\mathbf{r})=J_0(\mathbf{r})-\mathbf{E}J_0(\mathbf{r})$$

from an over-all mean illuminance that is of interest, the representative processes can without loss be taken to have zero mean values.

The mean value $\mathbf{E}J_0(\mathbf{r})$, which for a spatially ergodic process is equal to the limit as A goes to infinity of the spatial average $A^{-1}\int_A J_0(\mathbf{r})d^2\mathbf{r}$, over a region of area A, does not usually contribute to the identification of an image, and we need not be concerned with its estimation. With the mean value of the noise $N(\mathbf{r})$ taken as 0, however, a good estimate of the mean of $J_0(\mathbf{r})$, in view of Eq. (1.2), is simply the integral of the observed illuminance $J(\mathbf{r})$ over a large region of the image plane, divided by the area of that region.

The independent, zero-mean gaussian processes $\Delta J_0(\mathbf{r})$ and $N(\mathbf{r})$ are completely characterized by their spatial covariance functions

$$\begin{aligned}\Phi_0(\mathbf{r})&=\mathbf{E}\Delta J_0(\mathbf{r}')\Delta J_0(\mathbf{r}'+\mathbf{r})\\ \Phi_n(\mathbf{r})&=\mathbf{E}N(\mathbf{r}')N(\mathbf{r}'+\mathbf{r})\end{aligned} \tag{2.3}$$

or by the Fourier transforms of these, $\varphi_0(\boldsymbol{\omega})$ and $\varphi_n(\boldsymbol{\omega})$, which are the spatial spectral densities of the image and noise processes, respectively. When Eqs. (1.6) and (2.1) are substituted into the error expression of Eq. (2.2) and expected values are formed, the mean-squared error becomes

$$\begin{aligned}E=\Phi_0(0)-2\int\!\!\int M(\boldsymbol{\tau})S(\mathbf{r})\Phi_0(\boldsymbol{\tau}+\mathbf{r})d^2\boldsymbol{\tau}d^2\mathbf{r}+\int\!\!\int M(\boldsymbol{\tau}_1)M(\boldsymbol{\tau}_2)\\ \left\{\int\!\!\int S(\mathbf{r}_1)S(\mathbf{r}_2)\Phi_0(\boldsymbol{\tau}_1+\mathbf{r}_1-\boldsymbol{\tau}_2-\mathbf{r}_2)d^2\mathbf{r}_1d^2\mathbf{r}_2+\Phi_n(\boldsymbol{\tau}_1-\boldsymbol{\tau}_2)\right\}d^2\boldsymbol{\tau}_1d^2\boldsymbol{\tau}_2\\ =\int\{\varphi_0(\boldsymbol{\omega})-[m(\boldsymbol{\omega})s(\boldsymbol{\omega})+m^*(\boldsymbol{\omega})s^*(\boldsymbol{\omega})]\varphi_0(\boldsymbol{\omega})+|m(\boldsymbol{\omega})|^2[|s(\boldsymbol{\omega})|^2\varphi_0(\boldsymbol{\omega})+\varphi_n(\boldsymbol{\omega})]\}d^2\boldsymbol{\omega}/(2\pi)^2.\end{aligned} \tag{2.4}$$

This can be written, as in the usual treatment of Wiener filtering,[14,15]

$$E=\int\left\{\left|(m\boldsymbol{\omega})-\frac{s^*(\boldsymbol{\omega})\varphi_0(\boldsymbol{\omega})}{\gamma(\boldsymbol{\omega})}\right|^2\gamma(\boldsymbol{\omega})+\varphi_0(\boldsymbol{\omega})\varphi_n(\boldsymbol{\omega})/\gamma(\boldsymbol{\omega})\right\}d^2\boldsymbol{\omega}/(2\pi)^2,$$

with

$$\gamma(\boldsymbol{\omega})=|s(\boldsymbol{\omega})|^2\varphi_0(\boldsymbol{\omega})+\varphi_n(\boldsymbol{\omega}),$$

from which it is evident that the Fourier transform $m(\boldsymbol{\omega})$ of the estimating kernel that minimizes the mean-squared error E is

$$m(\boldsymbol{\omega})=[|s(\boldsymbol{\omega})|^2\varphi_0(\boldsymbol{\omega})+\varphi_n(\boldsymbol{\omega})]^{-1}s^*(\boldsymbol{\omega})\varphi_0(\boldsymbol{\omega}). \tag{2.5}$$

The minimum mean-squared error attained by this estimator $M(\mathbf{r})$ is then

$$E_{\min}=\int[|s(\boldsymbol{\omega})|^2\varphi_0(\boldsymbol{\omega})+\varphi_n(\boldsymbol{\omega})]^{-1}\times\varphi_0(\boldsymbol{\omega})\varphi_n(\boldsymbol{\omega})d^2\boldsymbol{\omega}/(2\pi)^2. \tag{2.6}$$

[11] D. Middleton, *An Introduction to Statistical Communication Theory* (McGraw-Hill Book Co., New York, N. Y., 1960), Section 21.4, p. 994 ff.

[12] L. J. Cutrona, E. N. Leith, C. J. Palermo, and L. J. Porcello, Trans. IRE **IT-6**, 386 (1960).

[13] A. Vander Lugt, Trans. IEEE **IT-10**, 139 (1964).

[14] N. Wiener, *The Extrapolation, Interpolation and Smoothing of Stationary Time Series* (John Wiley & Sons, Inc., New York, N. Y., 1949), p. 84.

[15] H. W. Bode and C. E. Shannon, Proc. IRE **38**, 417 (1950).

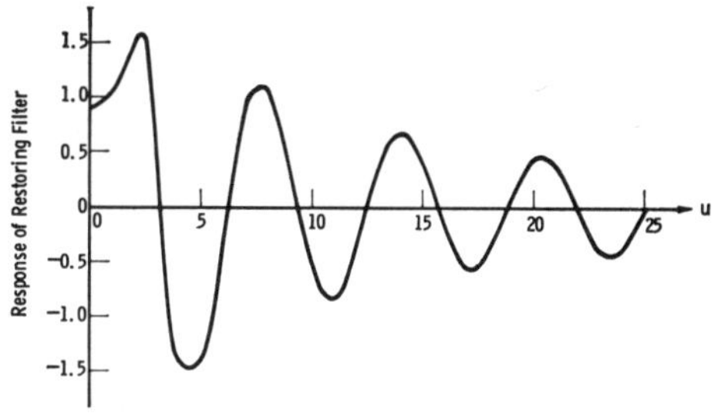

FIG. 1. Restoring filter for a moving image. $u=\frac{1}{2}vT\omega_x$.

There remains the question of what covariance functions or spectral densities to adopt for the images $J_0(\mathbf{r})$ and the spatial noise $N(\mathbf{r})$. For the images there is usually a certain distance δ representing the size of the finest details that should be resolved. The smaller this dimension δ, the larger the minimum error $E_{\min}$. Nothing is to be gained by adopting a value of δ smaller than the distances over which the features of the images $J_0(\mathbf{r})$ change significantly. Thus δ corresponds to a correlation distance for the stochastic processes representing the images. Their covariance function $\Phi_0(\mathbf{r})$ has widths in x and y of the order of δ, and their spectral density $\varphi_0(\boldsymbol{\omega})$ widths in each direction of the order of $2\pi/\delta$. Since little more than this can usually be said about the images, a plausible choice for the spectral density $\varphi_0(\boldsymbol{\omega})$ is simply

$$\begin{aligned}\varphi_0(\boldsymbol{\omega})&=4\pi\sigma^2/W^2, \quad |\boldsymbol{\omega}|\leq W=2\pi/\delta\\&=0, \quad |\boldsymbol{\omega}|>W,\end{aligned} \tag{2.7}$$

where $\sigma^2=\mathrm{Var}\, J_0(\mathbf{r})$ is the mean-squared deviation of the image illuminances from their average value.

The noise $N(\mathbf{r})$ usually has finer structure than the images to be restored. Its correlation distance is much smaller than δ and its spectral density $\varphi_n(\boldsymbol{\omega})$ much broader than $\varphi_0(\boldsymbol{\omega})$. In most cases there is little reason to take $\varphi_n(\boldsymbol{\omega})$ as anything but constant, $\varphi_n(\boldsymbol{\omega})\equiv\varphi_n$. This choice corresponds to assuming that the variations $N(\mathbf{r})$ are like what the communications engineer calls white noise. Measurements of the illuminance $J(\mathbf{r})$ however close together have components $N(\mathbf{r})$ that are statistically independent, as is to be expected when experimental error and stray incoherent background radiation are taken into account.

The spatial frequency response $m(\boldsymbol{\omega})$ of the restoring filter depends on the ratio φ_n/φ_0, which is now a constant, but a rough estimate of this ratio usually suffices. The spectral density φ_n of the noise can be estimated after substituting for the images $J_0(\mathbf{r})$ a completely blank source producing the same average illuminance. A large number of measurements are then made of the integrated illuminance

$$J_T=\int_A J(\mathbf{r})d^2\mathbf{r}$$

over separate regions of area A whose diameters are much greater than the correlation distance of the spatial noise. The sample variance of the values of J_T so obtained provides an estimate of the product $A\varphi_n$.

As an example we show in Fig. 1 the spatial frequency response $m(\boldsymbol{\omega})$ of the restoring filter for an image moving with velocity v in the x direction. The kernel of the integral equation is given in Eq. (1.3); its Fourier transform is

$$s(\boldsymbol{\omega})=(\sin u)/u, \quad \boldsymbol{\omega}=(\omega_x,\omega_y), \quad u=\tfrac{1}{2}vT\omega_x. \tag{2.8}$$

Since here the spreading occurs only in the x direction, it is reasonable to cut off the spectral density $\varphi_0(\boldsymbol{\omega})$ of the representative image processes only in the ω_x direction and to put

$$\begin{aligned}\varphi_0(\boldsymbol{\omega})&=\pi\sigma^2/W, \quad -W\leq\omega_x\leq W,\\&=0, \quad |\omega_x|>W.\end{aligned} \tag{2.9}$$

Any distortion in the y direction due to the finite aperture of the imaging system is left unamended. The restoring filter now has the spatial-frequency response

$$\begin{aligned}m(\boldsymbol{\omega})&=u(\sin u)/(\sin^2 u+Bu^2), \quad |\omega_x|\leq W,\\&=0, \quad \omega_x>W,\\u&=\tfrac{1}{2}vT\omega_x, \quad B=W\varphi_n/\pi\sigma^2,\end{aligned} \tag{2.10}$$

which is plotted vs u in Fig. 1 for a ratio $B=0.1$. It is

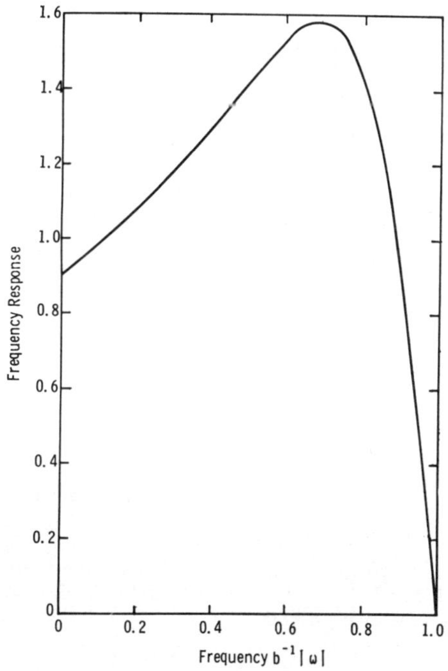

FIG. 2. Restoring filter for a slit-diffracted image.

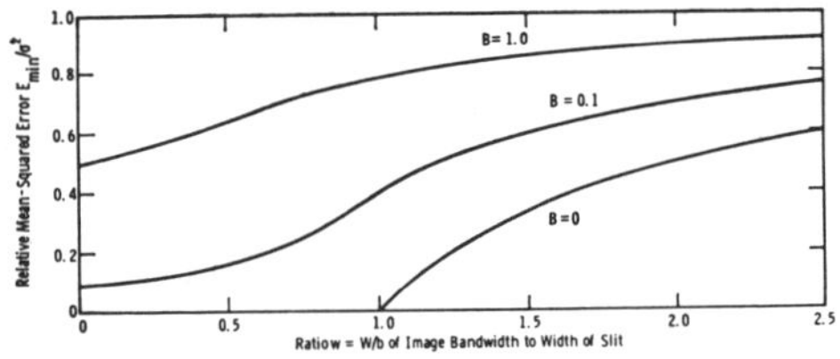

FIG. 3. Minimum mean squared error in restoration: slit diffraction.

symmetrical about $u=0$ and is to be cut off at a value of u equal to $\frac{1}{2}vTW$.

The restoration of an image that has suffered diffraction at a slit can also be viewed from the present standpoint. Let incoherent light be passing through an infinite slit of width a from an object that can be supposed at infinity. The image is then focused on a plane at a distance F from a cylindrical lens. Then if $b=2\pi a/\lambda F$, where λ is the wavelength of the light, the spread function is

$$S(\mathbf{r})=2(\pi bx^2)^{-1}\sin^2(bx/2)\delta(y). \quad (2.11)$$

Its Fourier transform is

$$\begin{aligned} s(\boldsymbol{\omega})&=1-b^{-1}|\omega_x|, & |\omega_x|&\leqslant b,\\ &=0, & |\omega_x|&>b, \end{aligned} \quad (2.12)$$

which is the convolution with itself of a rectangular function of width b. Thus b can be identified with the optical bandwidth of the system.

If we take the noise as having a uniform spectral density φ_n and attribute to the geometrical image the same spectral density as in Eq. (2.10), the frequency response of the restoring filter is, by Eq. (2.5),

$$\begin{aligned} m(\boldsymbol{\omega})&=[(1-b^{-1}|\omega_x|)^2+B]^{-1}(1-b^{-1}|\omega_x|),\\ &\qquad\qquad |\omega_x|<\min(W,b),\\ &=0, \qquad |\omega_x|>\min(W,b)\\ B&=W\varphi_n/\pi\sigma^2. \end{aligned} \quad (2.13)$$

Figure 2 shows this response function for a value of $B=0.1$.

The minimum mean-squared error $E_{\min}$ is obtained by substituting $s(\boldsymbol{\omega})$ into the one-dimensional form of Eq. (2.6), which after integration yields

$$\begin{aligned} E_{\min}/\sigma^2&=(\beta/w)\{\tan^{-1}(\beta^{-1})-\tan^{-1}[\beta^{-1}(1-w)]\},\\ &\qquad\qquad 0<w\leqslant 1,\\ &=(\beta/w)[\tan^{-1}(\beta^{-1})+\beta^{-1}(w-1)], \quad w>1 \quad (2.14)\\ w&=W/b, \quad \beta=B^{\frac{1}{2}}. \end{aligned}$$

This is plotted in Fig. 3 vs $w=W/b$ for $B=\beta^2=0$, 0.1, 1.0. For $W/b=0$, $E_{\min}/\sigma^2=B/(1+B)$. In the limit of no noise at all, $B\rightarrow 0$,

$$\begin{aligned} E_{\min}/\sigma^2&\rightarrow 0, \quad 0<W<b,\\ E_{\min}/\sigma^2&\rightarrow 1-b/W, \quad W>b. \end{aligned}$$

Thus if the width a of the slit is greater than $\lambda FW/2\pi$, the representative image processes can be restored perfectly in the absence of noise, but not if $a<\lambda FW/2\pi$.

The spatial response $m(\boldsymbol{\omega})$ of the restoring filter for circular diffraction looks much like that shown in Fig. 2, but it appears difficult to evaluate the minimum mean-squared error analytically.

3. THE DATA PROCESSING

Optical Filtering

The image might be restored optically by use of coherent light, as described by Cutrona *et al.*[12] and Vander Lugt.[13] A transparency must be prepared whose amplitude transmittance deviates from a reference value by an amount proportional to the illuminance $J(\mathbf{r})$. Plane parallel coherent light passes through the transparency and thence to a convergent lens. In the focal plane beyond the lens is a second transparency whose amplitude transmittance is proportional to the function $m(\boldsymbol{\omega})$ of Eq. (2.5). The coordinates $\mathbf{x}$ in the focal plane are related to $\boldsymbol{\omega}$ by

$$\mathbf{x}=\lambda F\boldsymbol{\omega}/2\pi,$$

where λ is the wavelength of the coherent light and F is the focal length of the lens.

The field amplitude of the light falling on this transparency is the Fourier transform $j(\boldsymbol{\omega})$ of $J(\mathbf{r})$, and the transmitted light has a field amplitude $m(\boldsymbol{\omega})j(\boldsymbol{\omega})$. This light then passes through a second lens whose focal plane coincides with that of the first. The emergent light has a field amplitude proportional to the estimate $\hat{J}_0(\mathbf{r})$, which can be obtained by allowing the light to fall on a photographic plate of known characteristics. The limitation of the spectral density $\varphi_0(\boldsymbol{\omega})$ of the images to a spatial band of radius $|\boldsymbol{\omega}|=W$, as in Eq. (2.7), corresponds to placing an aperture stop in the common focal plane of the two lenses.

Unless $m(\boldsymbol{\omega})$ is real and positive, preparation of the filter transparency may be difficult. Holographic techniques[16] might be applied, however, by making the amplitude transmittance proportional to

$$\mu+m(\boldsymbol{\omega})\exp i\boldsymbol{\omega}\cdot\mathbf{a}+m^*(\boldsymbol{\omega})\exp(-i\boldsymbol{\omega}\cdot\mathbf{a}),$$

where μ is a constant large enough to prevent this expression from becoming negative, and where $\mathbf{a}$ is a suitable vector. The field amplitude of the transmitted light is then proportional to

$$j(\boldsymbol{\omega})[\mu+m(\boldsymbol{\omega})\exp i\boldsymbol{\omega}\cdot\mathbf{a}+m^*(\boldsymbol{\omega})\exp(-i\boldsymbol{\omega}\cdot\mathbf{a})];$$

after Fourier transformation by the second lens, the restored image is displaced by the vector $\mathbf{a}$ from a replica of the original image.

[16] G. W. Stroke, *An Introduction to Coherent Optics and Holography* (Academic Press, New York, London, 1966), p. 79ff.

Digital Processing

If the values of the illuminance $J(\mathbf{r})$ are sampled at a rectangular grid of points, and if the illuminance $J_0(\mathbf{r})$ of the restored image is desired only at the intersections of a similar rectangular grid, the integrations in Eqs. (1.6) and (2.1) are replaced by finite summations obtained by applying a quadrature formula such as Simpson's or the trapezoidal rule. Taking the points in each grid in a certain fixed order we can write the resulting equations in matrix form,

$$\mathbf{J}=\mathbf{S}\mathbf{J}_0+\mathbf{N}, \quad \hat{\mathbf{J}}_0=\mathbf{M}\mathbf{J}, \tag{3.1}$$

where $\mathbf{J}$, $\mathbf{J}_0$, and $\mathbf{N}$ are column vectors of sample values, of the functions $J(\mathbf{r})$, $J_0(\mathbf{r})$, and $N(\mathbf{r})$, respectively. $\mathbf{S}$ and $\mathbf{M}$ are matrices and $\hat{\mathbf{J}}_0$ is a column vector of estimated illuminances at the grid points on the restored image. The number n of points at which the illuminance $J(\mathbf{r})$ is measured is generally larger than the number m of points at which the illuminance $J_0(\mathbf{r})$ is to be estimated. The matrix $\mathbf{S}$ is then $n\times m$, and the matrix $\mathbf{M}$ is $m\times n$.

The elements of the estimating matrix $\mathbf{M}$ are to be chosen to minimize an average error that can be written

$$E=\mathrm{Tr}\mathbf{E}\mathbf{G}(\hat{\mathbf{J}}_0-\mathbf{J}_0)(\tilde{\hat{\mathbf{J}}}_0-\tilde{\mathbf{J}}_0), \tag{3.2}$$

where again $\mathbf{E}$ denotes the expected value. Here $\tilde{}$ indicates the transpose of a matrix, and $\mathbf{G}$ is a positive-definite, symmetric error matrix. The simplest kind of error matrix is the identity matrix $\mathbf{I}$, which weights errors at all points equally.

The matrix $\mathbf{M}$ that minimizes the error E turns out to be independent of the precise form of $\mathbf{G}$. Under assumptions akin to those made in Sec. 2, it is given by[17]

$$\mathbf{M}=\boldsymbol{\varphi}_0\tilde{\mathbf{S}}(\mathbf{S}\boldsymbol{\varphi}_0\tilde{\mathbf{S}}+\boldsymbol{\varphi}_n)^{-1}, \tag{3.3}$$

and the minimum error is

$$E_{\min}=\mathrm{Tr}\mathbf{G}(\boldsymbol{\varphi}_0-\mathbf{M}\mathbf{S}\boldsymbol{\varphi}_0). \tag{3.4}$$

Here

$$\boldsymbol{\varphi}_0=\mathbf{E}\Delta\mathbf{J}_0\Delta\tilde{\mathbf{J}}_0, \quad \Delta\mathbf{J}_0=\mathbf{J}_0-\mathbf{E}\mathbf{J}_0, \quad \boldsymbol{\varphi}_n=\mathbf{E}\mathbf{N}\tilde{\mathbf{N}}, \tag{3.5}$$

are covariance matrices of the sampled image and spatial noise processes. The derivation of Eq. (3.3) is a straightforward exercise in differential calculus.

Equivalent to taking the spatial noise $N(\mathbf{r})$ to be white is the assumption that the covariance matrix $\boldsymbol{\varphi}_n$ is diagonal: $\boldsymbol{\varphi}_n=\bar{\varphi}_n\mathbf{I}$, where $\mathbf{I}$ is the identity matrix. If the spacings of the grid of points at which the restored image $\hat{J}_0(\mathbf{r})$ are evaluated are of the order of the desired resolution distance δ, the matrix $\boldsymbol{\varphi}_0$ can also be taken as diagonal: $\boldsymbol{\varphi}_0=\bar{\varphi}_0\mathbf{I}$; for if structural variations in the image $J_0(\mathbf{r})$ take place in distances of the order of δ, values of the representative stochastic processes at points separated by δ must be nearly uncorrelated.

Under these assumptions the column vector $\hat{\mathbf{J}}_0$ of estimated illuminances and the minimum error E are given by

$$\hat{\mathbf{J}}_0=\tilde{\mathbf{S}}(\mathbf{S}\tilde{\mathbf{S}}+\alpha\mathbf{I})^{-1}\mathbf{J}=(\tilde{\mathbf{S}}\mathbf{S}+\alpha\mathbf{I})^{-1}\tilde{\mathbf{S}}\mathbf{J}, \quad \alpha=\varphi_n/\varphi_0, \tag{3.6}$$

$$E_{\min}=\bar{\varphi}_n\mathrm{Tr}\mathbf{G}(\tilde{\mathbf{S}}\mathbf{S}+\alpha\mathbf{I})^{-1}. \tag{3.7}$$

In the limit $\bar{\varphi}_0\gg\bar{\varphi}_n$ the *a priori* density of the image illuminance is so broad as to be nearly uniform, and in effect no prior knowledge of the strength of the image relative to that of the noise is being assumed. We then find that the best estimate of the illuminances is given by the column vector

$$\tilde{\mathbf{J}}_0=(\tilde{\mathbf{S}}\mathbf{S})^{-1}\tilde{\mathbf{S}}\mathbf{J}, \quad \alpha\ll 1, \tag{3.8}$$

and the minimum error is

$$E_{\min}=\bar{\boldsymbol{\varphi}}_n\mathrm{Tr}\mathbf{G}(\tilde{\mathbf{S}}\mathbf{S})^{-1}. \tag{3.9}$$

This estimate represents a solution of the linear equations $\mathbf{J}=\mathbf{S}\mathbf{J}_0$ by the method of least squares.[18] The continuous counterpart of this procedure is to take the response $m(\boldsymbol{\omega})$ of the restoring filter equal to $1/s(\boldsymbol{\omega})$ for $|\boldsymbol{\omega}|\leq W$, and to 0 for $|\boldsymbol{\omega}|>W$. It may be unsafe if the Fourier transform $s(\boldsymbol{\omega})$ of the spread function has any zeros in the region $|\boldsymbol{\omega}|\leq W$.

The $m\times n$ matrix $(\tilde{\mathbf{S}}\mathbf{S}+\alpha\mathbf{I})^{-1}\tilde{\mathbf{S}}$ yielding the estimate $\tilde{\mathbf{J}}_0$ as in Eq. (3.6) has elements significantly different from zero only in places that connect points in the image plane separated by distances of the order of the width of the spread function $S(\mathbf{r})$. Hence although the images contain such a large number of grid points that they tax the storage capacity of a digital computer, the images to be processed can be broken down into sections of a size governed by the dimensions of the spread function, which is usually much smaller.

If it is desired to smooth the solution, as when it is being evaluated at points spaced by somewhat less than the "correlation distance" δ of the true image, a covariance matrix $\boldsymbol{\varphi}_0$ that is nondiagonal can be used without greatly increasing the amount of computation. A possible form might be

$$\boldsymbol{\varphi}_0=||\varphi_{0,rs}||=||\bar{\varphi}_0\exp(-|\mathbf{r}_r-\mathbf{r}_s|/\delta)||.$$

The noise is still assumed white, $\boldsymbol{\varphi}_n=\bar{\varphi}_n\mathbf{I}$. The estimates are then given by vector

$$\hat{\mathbf{J}}_0=(\tilde{\mathbf{S}}\mathbf{S}+\bar{\varphi}_n\boldsymbol{\varphi}_0^{-1})^{-1}\tilde{\mathbf{S}}\mathbf{J},$$

and the minimum-mean-squared error by

$$E_{\min}=\bar{\varphi}_n\mathrm{Tr}\mathbf{G}(\tilde{\mathbf{S}}\mathbf{S}+\bar{\varphi}_n\boldsymbol{\varphi}_0^{-1})^{-1}.$$

[17] R. Deutsch, *Estimation Theory* (Prentice-Hall, Inc., Englewood Cliffs, N. J., 1965), p. 66.

[18] A. Reiz, Arkiv Mat. Astron. Fysik **29A**, #29, 1 (1943).

An estimate of this form was proposed by Twomey on somewhat different grounds.[8,9]

One-dimensional convolution-integral equations, such as arise in spectroscopy and photometry, can be treated in the same digital fashion. The linear equations that replace them after sampling are of the same form as those in Eq. (3.1), and what we have said about the nature of the covariance functions can usually be applied to those other domains as well.

As a one-dimensional example, we present the convolution of a gaussian function

$$J_0(x)=(2\pi)^{-\frac{1}{2}}\exp(-x^2/2) \tag{3.10}$$

with a rectangular kernel of width $S=1.5$ and height 1. What corresponds to the geometrical image is now the function

$$\begin{aligned} J_g(x)&=\tfrac{1}{2}[I(x+\tfrac{1}{2}S)-I(x-\tfrac{1}{2}S)],\\ I(x)&=(2/\pi)^{\frac{1}{2}}\int_0^x \exp(-t^2/2)dt, \end{aligned} \tag{3.11}$$

which was calculated at the 71 points $x_k=k\eta$, $-35<k\leq 35$, $\eta=0.125$, to yield a set of pristine data $z_k=J_g(x_k)$. The data vector $\mathbf{J}$ consisted of the 71 numbers

$$J_k=[(z_k+e_k')^2+e_k''^2]^{\frac{1}{2}},\quad -35\leq k\leq 35, \tag{3.12}$$

where the e_k', e_k'' were 142 pseudorandom numbers generated by a computer in such a way as to be approximately normally distributed with mean 0 and standard deviation 0.001. The data J_k were formed as in Eq. (3.12) to ensure their being positive, as are the data in any experiments in which an intensity of some kind is being measured.

The values of the estimate $\hat{J}_0(x)$ were sought at 29 points $x_j=jh$, $-14\leq j\leq 14$, $h=0.25$. The solution was assumed equal to 0 outside the interval $-3.5<x<3.5$; the true solution, Eq. (3.10), is less than 10^{-3} for $|x|>3.5$. To put the integral equation into the form of Eq. (3.1), the integral was approximated by Simpson's rule, which involves replacing the integrand over each interval of length $2h$ by a parabola passing through the values of the integrand at the two end-points and at the midpoint. The results of applying the method of Eq. (3.6) to these data are plotted in Fig. 4 along the curve of the true solution $J_0(x)$. Two values of α, 0 and $6.25\cdot 10^{-3}$, were used, with only slightly different results.

The results obtained above may be compared with those from the direct method of solving an integral equation with such a rectangular kernel. It employs the relation

$$dJ_g/dx=J_0(x+\tfrac{1}{2}S)-J_0(x-\tfrac{1}{2}S), \tag{3.13}$$

which is obtained by differentiating the integral equation. If we denote by J_k' the numerical derivative of the data at the kth point, we can write Eq. (3.13) in the form

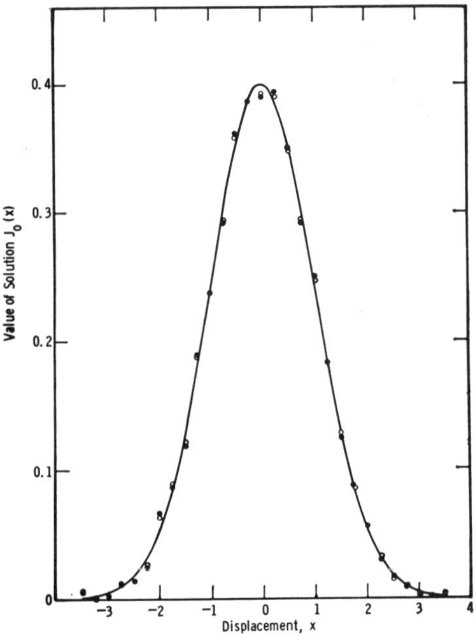

FIG. 4. Restoration of a gaussian function. ○: $\alpha=0$, ●: $\alpha=6.25\cdot 10^{-3}$.

$$\begin{aligned} J_{0k}&=J_{k-6}'+J_{0,k-12}, && -29\leq k\leq 0,\\ J_{0k}&=J_{0,k+12}-J_{k+6}', && 0\leq k\leq 29, \end{aligned} \tag{3.14}$$

in which the spacing of the data is $\eta=0.125=S/12$. By assuming $J_{0k}=0$, $|k|>35$, we should be able to derive values of the solution for $|k|\leq 29$ by numerically differentiating the data J_k, starting from each end of the set, and substituting into Eq. (3.14). The numerical differentiation formula was obtained by differentiating Newton's forward interpolation formula, ten terms of which were used.[19] Applied to the pristine data $z_k=J_g(x_k)$, this method produced the correct solution with three-figure accuracy over most of the range. Applied to the very same data J_k as were used in calculating the points on Fig. 4, it yielded a set of positive and negative numbers varying over a range from −1.4 to +1.5 and bearing no resemblance whatever to the true solution. It is necessary to smooth such fluctuations in order to obtain a useful solution, although how best to do so in this case is not apparent.

ACKNOWLEDGMENT

I wish to thank Robert Fardo for programming and executing the numerical example on the digital computer.

[19] J. R. Scarborough, *Numerical Mathematical Analysis* (The Johns Hopkins Press, Baltimore, Md., 1930), pp. 48, 114.

21

Reprinted from *IEEE Trans. Computers* **C-21**(7):636–641 (1972)

Generalized Wiener Filtering Computation Techniques

WILLIAM K. PRATT, MEMBER, IEEE

***Abstract*—The classical signal processing technique known as Wiener filtering has been extended to the processing of one- and two-dimensional discrete data by digital operations with emphasis on reduction of the computational requirements. In the generalized Wiener filtering process a unitary transformation, such as the discrete Fourier, Hadamard, or Karhunen–Loéve transform is performed on the data that is assumed to be composed of additive signal and noise components. The transformed data is then modified by a filter function, and the inverse transformation is performed to obtain the discrete system output. The filter function is chosen to provide the best mean square estimate of the signal portion of the input data.**

It is shown that all unitary data transformations provide the same minimum mean square error performance. However, by using fast transform algorithms, the computational requirements for some of the transforms, e.g., Fourier and Hadamard, can be reduced significantly. Furthermore, with certain data transforms, a selective computational procedure will result in a significant reduction in computation load, with only a minimal increase in estimation error. An example of the application of the generalized Wiener filtering process to image enhancement is presented.

***Index Terms*—Data transforms, filtering, Fourier transform, Hadamard transform, image enhancement, Karhunen–Loéve transform, two-dimensional signal processing, Wiener filter.**

I. INTRODUCTION

WIENER filtering is a classical technique of signal estimation that has been applied, primarily, to one-dimensional continuous signals, with analysis and implementation based upon continuous Fourier signal theory [1]. It is possible, of course, to perform Wiener filtering operations on time sampled signals and extend the technique to two dimensions. Furthermore, the filtering operation can be implemented by unitary transformations other than the Fourier transform [2]–[4]. Finally, it is possible to significantly reduce the computational requirements without severely affecting performance by a technique of selective computation.

II. ONE-DIMENSIONAL FILTERING

Fig. 1 is a block diagram of a generalized one-dimensional Wiener filtering system. A zero-mean M-element data-column vector f composed of additive zero-mean signal s, and noise n components is the input to the system. The signal and noise are assumed uncorrelated with each other. In the system a unitary transformation operation, utilizing an $M \times M$ transformation matrix A is performed on the input data vector yielding

$$F = Af = As + An \equiv S + N. \tag{1}$$

Table I defines several unitary transforms that have been used for digital signal processing applications. Next, the transform input vector is multiplied by the $M \times M$ filter matrix G. Note that, in general, G is not a diagonal matrix. If G contains off-diagonal terms, the filtering operation, denoted as *vector* filtering, combines the spectral components of the input data vector. On

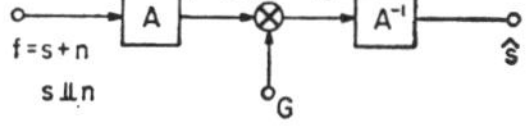

Fig. 1. Generalized Wiener filtering.

TABLE I
UNITARY DATA TRANSFORMS

Fourier: Discrete sinusoidal basis vectors

$$A = \frac{1}{\sqrt{N}}\left[\exp\left(-\frac{2\pi i}{N}ux\right)\right] \qquad A^{-1} = A^*$$

where

N Data vector length.
x Data vector element index.
u Transformed data vector element index.

Computation requires about $N \log_2 N$ complex multiply and add operations.

Hadamard: Sequency ordered orthogonal rectangular waveform basis vectors

$$A = H_N P \qquad A^{-1} = A$$

where

P Row permutation index.

$$H_N = \frac{1}{\sqrt{2}}\begin{bmatrix} H_{N/2} & H_{N/2} \\ H_{N/2} & -H_{N/2} \end{bmatrix}$$

$$H_2 = \frac{1}{\sqrt{2}}\begin{bmatrix} 1 & 1 \\ 1 & -1 \end{bmatrix}.$$

Computation requires about $N \log_2 N$ add operations.
Note: Many other Hadamard matrices of dimension N mod 4 are known to exist.

Karhunen-Loéve:

$$A = K \qquad A^{-1} = K^T$$

where

$KC_f K^{-1} = \mathrm{diag}\,[\lambda(i)]$.
C_f Covariance matrix of data vector.
K Eigenvectors of C_f.
$\lambda(i)$ Eigenvalues of C_f.

Computation requires N^2 multiply and add operations.
Note: If C_f is a *Markov process covariance matrix, i.e.*, $C_f = [\rho^{|x_i - x_j|}]$ where ρ is the adjacent element correlation factor, the eigenvalues and eigenvectors can be found recursively.

Manuscript received November 16, 1971; revised March 3, 1972. This work was supported in part by the Air Force Office of Scientific Research under Contract F44620-70-C-0113 and in part by the Advanced Research Projects Agency monitored by the Air Force Eastern Test Range under Contract F08606-72-C-0008. A preliminary version of this paper was presented at the IEEE, UMC, Two-Dimensional Digital Signal Processing Conference, Columbia, Mo., October 6–8, 1971.

The author is with the Department of Electrical Engineering, University of Southern California, Los Angeles, Calif. 90007.

the other hand, if G contains no off-diagonal terms, the filtering operation called *scalar* filtering simply weights each spectral component of the input vector individually. In the final step an inverse unitary transform operation is performed to obtain the resultant estimate

$$\hat{s} = A^{-1}GF = A^{-1}GAf. \tag{2}$$

The filter matrix G is chosen in conjunction with the transformation matrix A so that the mean square error (mse) between the signal and its estimate is minimized.

Statistical Representation

For many classes of data, a data vector f can be considered a sample of a random process (stationary or nonstationary) of known mean $\bar{f}$ and with a covariance matrix

$$C_f = \overline{(f - \bar{f})(f^* - \bar{f}^*)^T}. \tag{3}$$

The data covariance matrix of a unitary transform of data vector $F = Af$, likewise is given by

$$C_F = \overline{(F - \bar{F})(F^* - \bar{F}^*)^T} \tag{4}$$

which reduces to [5]

$$C_F = AC_fA^{*T} = AC_fA^{-1}. \tag{5}$$

It should be recognized that (5) can be regarded as a separable two-dimensional unitary transform of C_f for purposes of computation.

Generalized Wiener Filtering

In the design of the generalized Wiener filtering system of Fig. 1 a filter matrix G is chosen to minimize the mse

$$e = \text{Tr}\left\{\overline{(s - \hat{s})(s - \hat{s})^T}\right\} \tag{6}$$

between the signal s and its estimate $\hat{s}$. The expression for the mse can be rewritten in terms of transform domain quantities as

$$e = \text{Tr} \left\{\overline{(A^{-1}S - A^{-1}G(S + N))(A^{-1}S - A^{-1}G(S + N))^T}\right\}. \tag{7}$$

Making use of the fact that

$$\overline{SS^T} = C_S = AC_sA^{*T}$$
$$\overline{NN^T} = C_N = AC_nA^{*T}$$
$$\overline{SN^T} = \overline{NS^T} = 0 \tag{8}$$

one obtains

$$e = \text{Tr}\left\{C_S - 2GC_S + GG(C_S + C_N)\right\}. \tag{9}$$

Now, a straightforward minimization of (9) by Lagrangian techniques yields the optimum filter matrix

$$G_O = C_S(C_S + C_N)^{-1}. \tag{10}$$

Thus, the optimum filter matrix may be determined from the transform spectral densities of the signal and noise vectors. Alternatively, the optimum filter can be found by a two-dimensional transformation

$$G_O = Ag_OA^{*T} \tag{11}$$

of a matrix

$$g_O = C_s(C_s + C_n)^{-1} \tag{12}$$

based upon the covariance matrices of the signal and noise. The matrix g_O will be called the response matrix, in analogy with the impulse response of a linear system.

Substitution of the optimum filter matrix into (9) gives the minimum mse

$$e_{\min} = \text{Tr}\left\{C_SC_N(C_S + C_N)^{-1}\right\}. \tag{13}$$

Alternatively, the mse can be written in terms of data domain covariance matrices as

$$e_{\min} = \text{Tr}\left\{C_sC_n(C_s + C_n)^{-1}\right\}. \tag{14}$$

Note that from (14), the minimum mse is independent of the type of unitary transform employed. However, the character of the filter matrix G is dependent upon the unitary transform. The conclusion is that one is free to choose the transform that will minimize the computational processes entailed in filter generation and filter operation.

Table II lists the number of multiplications required to perform the Wiener filtering operations of (2). For long vector lengths the ratio of the multiplication requirements of the Fourier or Hadamard transforms to the identity operator approaches unity and the Karhunen–Loéve transfer requires more than twice as many multiplications as the identity operator. The computational requirements for filter generation are considered in the next section.

TABLE II
OPTIMAL WIENER FILTERING COMPUTATION REQUIREMENTS

Transform	Approximate Number of Multiplication/Addition Operations
Identity	M^2
Karhunen–Loéve	$2M^2+M$
Fourier/Hadamard	$M^2+2M \log_2 M$

Note: $M=2^m$ where m is an integer.

Wiener Filtering Example

Consider the mean square estimation of a Markov process signal in the presence of white noise. The signal and noise covariance matrices are given by

$$C_s = \begin{bmatrix} 1 & \rho & \rho^2 & \cdots \rho^m \\ \rho & 1 & \rho & \cdots \rho^{m-1} \\ \rho^2 & \rho & 1 & \cdots \rho^{m-2} \\ \rho^m & & & \cdots 1 \end{bmatrix} \tag{15}$$

where ρ is the adjacent element correlation and by

$$C_n = \begin{bmatrix} k_o & & & & 0 \\ & k_o & & & \\ & & \cdot & & \\ & & & \cdot & \\ 0 & & & & k_o \end{bmatrix}. \qquad (16)$$

The optimum filter could be computed by the techniques of (10) or (11), but in either case it would be necessary to perform a matrix inverse operation. This task can be avoided, for the special case of this example, by a matrix diagonalization procedure. First, the signal plus noise covariance matrix is diagonalized by the matrix K_s of eigenvectors of the signal covariance matrix. Thus,

$$K_s(C_s + C_n)K_s^{-1} = \text{diag}\,[k_o + \lambda_s(i)], \qquad i = 1, 2, \cdots, N \qquad (17)$$

where $\lambda_s(i)$ are the eigenvalues of C_s. Then, since

$$(C_s + C_n)^{-1} = K_s^{-1}\,\text{diag}\left[\left(\frac{1}{k_o + \lambda_s(i)}\right)\right] K_s \qquad (18)$$

the response matrix may be written directly as

$$g = C_s K_s^{-1}\,\text{diag}\left[\left(\frac{1}{k_o + \lambda_s(i)}\right)\right] K_s. \qquad (19)$$

With C_s expressed in its diagonal form one obtains

$$g = K_s^{-1}\,\text{diag}\left[\frac{\lambda_s(i)}{(k_o + \lambda_s(i))}\right] K_s. \qquad (20)$$

The general term of the filter function g is then

$$g\{i,j\} = \sum_{t=1}^{N} K_s\{t, i\} K_s\{t, j\} \frac{\lambda_s(t)}{k_o + \lambda_s(t)}. \qquad (21)$$

As shown in the Appendix the eigenvectors and eigenvalues of the Markov process covariance matrix C_s can be found recursively for any correlation factor ρ. Thus, the response matrix g can be obtained for large size matrices relatively easily.

The filter matrix may now be determined by performing the two-dimensional transformation of the response matrix. Several cases of interest are listed below:

$$\text{Fourier: } G_O = FgF^* \qquad (22)$$

$$\text{Hadamard: } G_O = HgH \qquad (23)$$

Karhunen–Loéve:

$$G_O = K_s g K_s^{-1} = \text{diag}\left[\frac{\lambda_s(i)}{k_o + \lambda_s(i)}\right] \qquad (24)$$

identity:

$$G_O = IgI = K_s^{-1}\left(\text{diag}\left[\frac{\lambda_s(i)}{k_o + \lambda_s(i)}\right]\right) K_s. \qquad (25)$$

Fig. 2 contains magnitude displays of the Fourier,

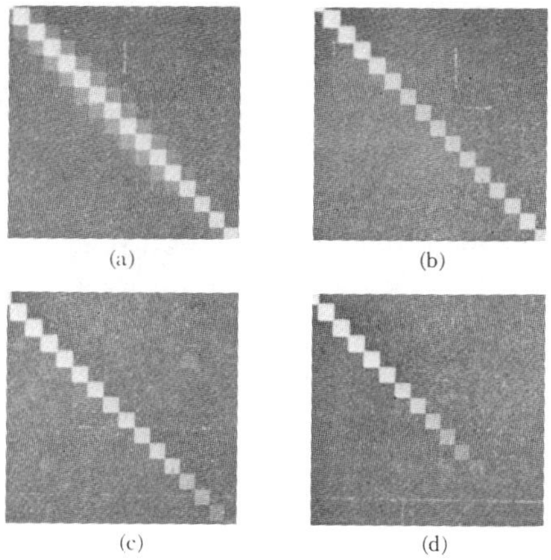

Fig. 2. Magnitude displays of Wiener filter matrices. (a) Identity. (b) Fourier. (c) Hadamard. (d) Karhunen–Loéve.

Hadamard, Karhunen–Loéve, and identity Wiener filter matrices for a vector length of 16 elements, $\rho = 0.9$ and $k_o = 0.1$ (a signal-to-noise ratio of 10). Comparing the filter planes, it is apparent that the filter characteristic changes drastically for different unitary transforms. For the Karhunen–Loéve transform, the filtering operation is a scalar multiplication, while for the identity transform most of the elements of the filter matrix are of relatively large magnitude. The Fourier and Hadamard filter matrices contain large magnitude terms along the diagonal and terms of decreasing magnitude away from the diagonal.

Suboptimal Wiener Filtering

When considering only the amount of computation required to perform the Wiener filtering operation, as opposed to the computation requirement for the filter generation, it is obvious that the bulk of the computation for the Fourier, Hadamard, and identity transform systems can be attributed to the filter matrix multiplication operation. The Karhunen–Loéve transform system only requires a scalar filter multiplication, but entails two matrix multiplication operations for the transformations. As a compromise, consideration has been given to the development of filter matrices that contain a relatively large number of zero entries. With such matrices the filter operation could be performed with a reduced number of computations. The goal in the filter design, of course, is to maintain the mse performance as close to the optimum filter level as possible.

The suboptimal Wiener filtering design problem can be formulated as an optimization problem whereby the filter matrix G is chosen to minimize the mse

$$e = \text{Tr}\,\{C_S - 2GC_S + GG(C_S + C_N)\} \qquad (26)$$

under the constraint that certain selected elements of G are zero. The zero elements of G are so placed that fast

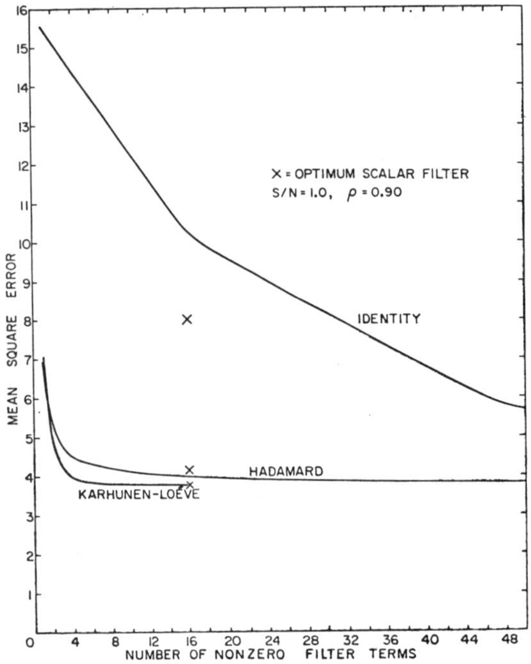

Fig. 3. Mean square error performance of Wiener filters.

computational algorithms can be employed in the filter matrix multiplication.

As a special case of this technique consider the filter design when the filter matrix is constrained to be a diagonal matrix. The optimum filter is found to be [3]

$$G = \text{diag}\left[\frac{C_S(i, i)}{C_S(i, i) + C_N(i, i)}\right] \tag{27}$$

and the resultant mse is

$$e = N - \sum_{i=1}^{N} \frac{C_S^2(i, i)}{C_S(i, i) + C_N(i, i)}. \tag{28}$$

Conceivably, one could determine the filter design when the filter matrix contains two elements per row, and continue with additional matrix elements until the desired mse is achieved. Unfortunately, the optimization problem rapidly becomes quite complex.

One alternative to the optimization is to use selected elements of the optimum vector Wiener filter matrix as the suboptimal filter matrix. For example, only the diagonal and near diagonal terms of the optimum filter matrix could be retained. Another technique is to select only those terms of large magnitude. Fig. 3 illustrates the mse performance of the identity, Hadamard, and Karhunen–Loéve vector and scalar Wiener filters as a function of the number of largest magnitude filter terms not set to zero in the filter matrix. The filters are 16×16 element matrices based upon a Markov process signal

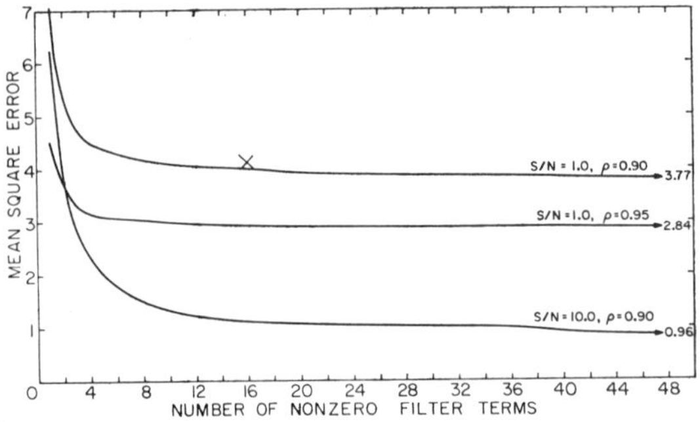

Fig. 4. Mean square error performance of the Hadamard–Wiener filter as a function of signal-to-noise ratio and signal correlation.

plus additive white noise. The crosses indicate the optimum scalar filter performance. It should be noted from the figure that the identity filter performs quite poorly compared to the other transforms because there is no energy compaction in the transform domain. Furthermore, the performance of the Hadamard transform for as few as about ten nonzero terms is quite close to the optimum performance when all 256 terms are employed. Fig. 4 shows the mse performance of the Hadamard suboptimal Wiener filter as a function of signal-to-noise ratio and signal correlation.

III. Two-Dimensional Filtering

The Wiener filtering concept can be extended to two-dimensional filtering quite readily. Let $f(x, y)$ represent an $M \times M$ data matrix. Then, the two-dimensional unitary transform of the data $F(u, v)$, itself an $M \times M$ matrix, may be defined as

$$F(u, v) = \sum_{x=0}^{M-1} \sum_{y=0}^{M-1} f(x, y) a(x, y, u, v) \tag{29}$$

where $a(x, y, u, v)$ is the unitary transform kernel. The Wiener filtering operation is then described by

$$\hat{S}(u', v') = \sum_{u=0}^{M-1} \sum_{v=0}^{M-1} F(u, v) G(u, v, u', v') \tag{30}$$

where $G(u, v, u', v')$ is the Wiener filter function. It should be noted that (30) indicates that to obtain each element of the filtered data in the transform domain $\hat{S}(u', v')$ at coordinate (u', v'), it is necessary to multiply each element of $F(u, v)$ by a weighting function $G(u, v, u', v')$, that could be different for each (u, v, u', v'), and then sum over all values (u, v). This general operation is denoted as *tensor* filtering in analogy with the vector operation of (2). Finally, an inverse transformation is performed to obtain the estimate

$$\hat{s}(x, y) = \sum_{u'=0}^{M-1} \sum_{v'=0}^{M-1} \hat{S}(u', v') b(x, y, u', v') \tag{31}$$

where $b(x, y, u', v')$ is the kernel of the inverse trans-

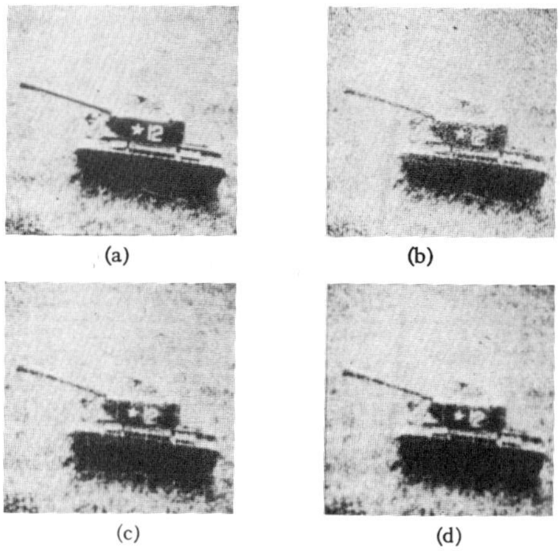

Fig. 5. Image enhancement examples. (a) Original. (b) Original plus noise. (c) Optimal Hadamard filtering. (d) Suboptimal filtering.

formation defined by the kernel $a(x, y, u, v)$. If the two-dimensional unitary transform kernels are separable so that computations may be sequentially performed upon the rows and columns of the arrays, the transformations of (29) and (31) simplify to

$$[F(u, v)] = [a(y, v)][f(x, y)][a(x, u)] \tag{32}$$

and

$$[\hat{s}(x, y)] = [b(y, v')][\hat{S}(u', v')][b(x, u')] \tag{33}$$

where the brackets indicate matrix representations of the appropriate arrays. Furthermore, if the Wiener filtering operation is limited to sequential operations on the rows and columns of $F(u, v)$, then

$$[\hat{S}(u', v')] = [G_y(v, v')][F(u, v)][G_x(u, u')] \tag{34}$$

where $[G_x(u, u')]$ and $[G_y(v, v')]$ are the Wiener filter matrices for the rows and columns, respectively, of the original data array. Under these conditions two-dimensional Wiener filtering simply becomes a sequential application of one-dimensional filtering to the rows and columns of the data array.

Image Enhancement Applications

The generalized Wiener filtering technique has been applied to the enhancement of images as a test of its validity. Fig. 5(a) shows an original image containing a toy tank. The image contains 256×256 elements and 64 gray scales. In Fig. 5(b) uniformly distributed noise has been added to the original to produce an image with a signal-to-noise ration of unity. Fig. 5(c) illustrates the effect of optimal Hadamard transform Wiener filtering in 16×16 blocks over the image surface. There is an obvious improvement in image quality, however, the grid pattern caused by the small block size is noticeable. The grid pattern in the filtered images can obviously be eliminated by filtering over the entire image, but at the expense of computational complexity. Alternatively, the grid pattern can be minimized by overlapping the filtering blocks or by spatial averaging at the interface between blocks. In Fig. 5(d) the filtering operation was performed with 75 percent of the filter elements set to zero. The 25 percent of the filter elements retained are the largest magnitude elements of the filter matrix. Since the locations of the zero elements of the filter matrix are known prior to the filtering operation, a computational algorithm can be employed to avoid multiplication by the zeros, and therefore, reduce the filtering computation requirements by the 75 percent factor. There is no apparent difference in the visual results for the optimal and suboptimal filtering examples.

IV. Summary

It has been shown that classical Wiener filtering operations can be performed through the use of various unitary transforms. The amount of computation required for the filter implementation and filter operations depends upon the particular unitary transform employed. In many applications it is possible to choose the unitary transform to minimize the computational load. A technique of selective computation has been introduced to reduce the Wiener filtering computational requirements. With this technique the computational load can be reduced significantly with only a moderate degradation in mse performance. Finally, Hadamard transform Wiener filtering has been successfully applied to enhance the quality of a noisy image.

Appendix

A Recursive Technique for Computation of Eigenvalues and Eigenvectors of a Markov Process Covariance Matrix

Consider a Markov process covariance matrix of the form

$$[C] = \begin{bmatrix} 1 & R & R^2 & R^3 & \cdots & R^n \\ R & 1 & R & R^2 & \cdots & R^{n-1} \\ R^2 & R & 1 & R & \cdots & R^{n-2} \\ \vdots & & & & & \vdots \\ R^{n-1} & & \cdots & R & 1 & R \\ R^n & & \cdots & R^2 & R & 1 \end{bmatrix} \tag{35}$$

where $R \leq 1$. This is a matrix of finite Toeplitz form, for which an associated real valued function

$$f(X) = \sum_{n=-\infty}^{\infty} R^{|n|} e^{inX} = \frac{1 - R^2}{1 - 2R \cos X + R^2} \tag{36}$$

may be defined [6]. Toeplitz [7] has shown that the eigenvalues λ_k of $[C]$ are related to the roots of $f(X)$ by

$$\lambda_k = \frac{1 - R^2}{1 - 2R \cos X_k + R^2}. \tag{37}$$

The roots of $f(X)$ may be determined indirectly by solution of the characteristic equation

$$D_n(\lambda_k) \equiv |C - I\lambda_k| = 0 \qquad (38)$$

in terms of the X_k, where $[I]$ is the identity matrix. Grenander and Szego have shown the characteristic equation can be written as a recurrence relation

$$D_n(\lambda_k) = [1 - \lambda_k + R^2(1 + \lambda_k)]D_{n-1}(\lambda_k) - R^2\lambda_k^2 D_{n-2}(\lambda_k), \qquad n = 2, 3, 4, \cdots \qquad (39)$$

with

$$D_0(\lambda_k) = 1 - \lambda_k \qquad (40)$$

and

$$D_1(\lambda_k) = (1 - \lambda_k)^2 - R^2. \qquad (41)$$

Substitution for λ_k from (3) yields

$$D_n(\lambda_k) = \frac{(-\lambda_k R)^{n+1}}{1 - R^2}\left\{\frac{\sin(n+2)X_k}{\sin X_k} - \frac{2R\sin(n+1)X_k}{\sin X_k} + \frac{R^2 \sin nX_k}{\sin X_k}\right\} = 0. \qquad (42)$$

Then, by solving for the X_k in order, such that

$$\pi > X_1 > X_2 > \cdots > X_{n+1} > 0 \qquad (43)$$

the eigenvalues are obtained in increasing order from (37).

Let

$$[H_k] = [h_{1,k}, h_{2,k}, \cdots, h_{n+1,k}]^T$$

be the eigenvector corresponding to the eigenvalue λ_k. Then

$$[C][H_k] = \lambda_k[H_k] \qquad (44)$$

or

$$[H_k] = \lambda_k[C]^{-1}[H_k]. \qquad (45)$$

For a covariance matrix of the form (35) the matrix inverse is [8]

$$[C]^{-1} = \frac{1}{1-R^2}\begin{bmatrix} 1 & -R & 0 & 0 & \cdot & 0 \\ -R & (1+R^2) & -R & 0 & \cdot & 0 \\ 0 & -R & 1+R^2 & -R & \cdot & 0 \\ \cdot & \cdot & \cdot & \cdot & \cdot & \cdot \\ \cdot & \cdot & \cdot & \cdot & \cdot & \cdot \\ 0 & \cdot & \cdot & \cdot & 1+R^2 & -R \\ 0 & \cdot & \cdot & \cdot & -R & 1 \end{bmatrix}. \qquad (46)$$

A simple recursion relation can now be formulated.

$$h_{1,k} = \frac{\lambda_k}{1-R^2}[h_{1,k} - Rh_{2,k}] \qquad (47)$$

$$h_{i,k} = \frac{\lambda_k}{1-R^2}[-Rh_{i-1,k} + (1+R^2)h_{i,k} - Rh_{i+1,k}], \qquad 1 < i < n+1 \qquad (48)$$

$$h_{n+1,k} = \frac{\lambda_k}{1-R^2}[-Rh_{n,k} + h_{n+1,k}]. \qquad (49)$$

Then arbitrarily setting $h_{1,k}=1$, each component of the eigenvector may be successively computed for each eigenvalue. Note that $h_{1,k}\neq 0$ since $h_{1,k}=0$ implies that $h_{1,k}=0$ for all i. The eigenvectors may be normalized, as desired, after their computation.

Acknowledgment

The author wishes to thank W. Chen and D. Messing of the Image Processing Laboratory, University of Southern California, for providing valuable assistance in the experimental portions of this project.

References

[1] W. B. Davenport, Jr., and W. L. Root, *Random Signals and Noise*. New York: McGraw-Hill, 1968.
[2] J. Pearl, "Basis-restricted transformations and performance measures for spectral representation," *IEEE Trans. Inform. Theory* (Corresp.), vol. IT-17, pp. 751–752, Nov. 1971.
[3] ——, "Walsh processing of random signals," presented at the Symp. Appl. Walsh Functions, Washington D. C., Apr. 1971.
[4] W. K. Pratt, "Linear and nonlinear filtering in the Walsh domain," presented at the Symp. Appl. Walsh Functions, Washington, D. C., Apr. 1971.
[5] ——, "A comparison of digital image transforms," presented at the Mervin J. Kelley Commun. Conf., Univ. Missouri, Rolla, Sept. 1970.
[6] U. Grenander and G. Szego, *Toeplitz Forms and Their Applications*. Berkeley: Univ. California Press, 1958.
[7] O. Toeplitz, "Zur Theorie der Quadratischen und Bilinearen Formen von Unendlichvielen Veranderlichen, I. Teil: Theorie der L-Formen," *Math. Ann.*, vol. 70, 1911.
[8] A. Arcese, P. H. Mengert, and E. W. Trombini, "Image detection through bipolar correlation," *IEEE Trans. Inform. Theory*, vol. IT-16, pp. 534–541, Sept. 1970.

22

Reprinted from pp. 263–268 of the *Fourth Asilomar Conf. on Circuits and Syst., Proc.* (1970), 707 pp.

OPTIMUM DESIGN OF TWO–DIMENSIONAL NONRECURSIVE DIGITAL FILTERS

M. P. Ekstrom

Lawrence Radiation Laboratory, University of California, Livermore

V. R. Algazi

Department of Electrical Engineering, University of California, Davis

Abstract

An algorithm is presented for the design of two-dimensional nonrecursive digital filters. The filter input and desired output are random fields with known covariance arrays. The design is based on minimizing a mean-square error criterion subject to constraints on the orders of the filter. The numerical aspects of the design algorithm and implementation of the filter are treated extensively.

1. INTRODUCTION

Numerous applications in the areas of image and geological data processing involve the discrete filtering of two-dimensional signals.[1,2,3] One approach to the processing of such two-dimensional data is line-by-line filtering of the data arrays using one-dimensional digital filters. This is convenient because the filter can be implemented easily and some fairly classical design methods, set in the frequency domain, are available.

However, this approach ignores the two-dimensional coherence of the data and may result in an unsatisfactory performance. Although some techniques have been proposed recently for the design of two-dimensional digital filters,[4] these are direct extensions of the one-dimensional frequency domain methods and subject to similar performance limitations. In the design applications we consider, the filter performance cannot be meaningfully specified in the frequency domain. Also, the difficulties in implementing two-dimensional filters lead us naturally to limit the complexity of the filter.

With these considerations in mind, we present in this paper an algorithm for the optimum design of two-dimensional nonrecursive digital filters of constrained dimension. The filter input and desired output are assumed known in a statistical sense and are characterized by their respective covariance arrays. The output is estimated by a weighted, linear combination of input samples. These weights comprise the impulse response array of the digital filter. With this model, mean-square estimation techniques are applied to derive the optimum filter array. This approach leads to the solution of a matric equation for the two-dimensional filter.

The most substantial numerical problem in the design routine is the inversion of a dense matrix. This problem is avoided in two special-case design problems where the covariance arrays are taken to be decomposable. The implementation of the designed filter using a two-dimensional FFT also is discussed.

2. PROBLEM STATEMENT

The digital filter design problem is shown schematically in Figure 1. We consider the desired

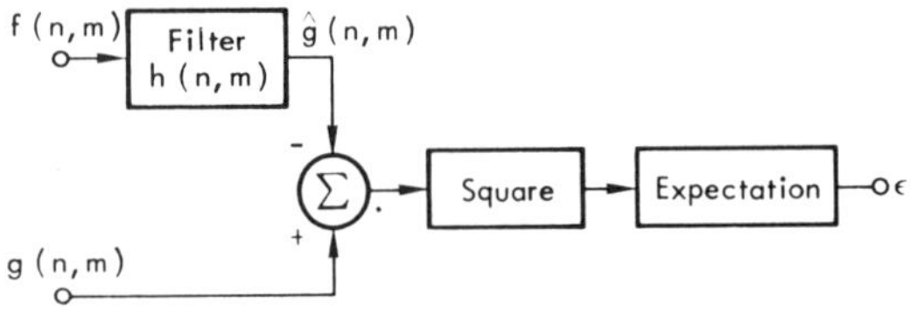

FIGURE 1. DIGITAL FILTER DESIGN PROBLEM

output array $\{g(n,m)\}$ to be defined on a sampling grid in the entire n,m plane.* The estimate of each member of this array, $\hat{g}(n,m)$, is restricted to be a shift-invariant linear combination of a finite number of past and future values of the input array $\{f(n,m)\}$. Specifically, we write

$$\hat{g}(n,m) = \sum_{i=A}^{B} \sum_{j=C}^{D} f(n-i, m-j)h(i,j) \tag{1}$$

for all n,m.** The array $\{h(n,m)\}$ is the filter impulse response or point spread function[5] and has been constrained to be of finite memory and anticipation. It is confined to the region

$$n = -A, \ldots, 0, \ldots, B$$
$$m = -C, \ldots, 0, \ldots, D. \tag{2}$$

The input, $\{f(n,m)\}$, is taken to be a sample array from a zero-mean, homogeneous random field with the two-dimensional autocovariance array $\{R_{ff}(n,m)\}$. Its crosscovariance with the input is given by $\{K_{gf}(n,m)\}$.

*For simplicity in notation, the sampling grid has been normalized; consequently, n,m are integers.

**A, B, C, and D are taken to be positive integers.

3. DESIGN ALGORITHM

Given the statistical nature of the input array the mean-square error criterion* of Figure 1 is a natural measure of the digital filter performance. In the filter design, we determine an estimate $\{\hat{g}(n,m)\}$ of $\{g(n,m)\}$ which minimizes the mean-square error

$$\epsilon = E\{[g(n,m) - \hat{g}(n,m)]^2\} \ \forall n,m \tag{3}$$

in which E{ } is an expectation taken over the ensemble of possible input arrays. To solve for the weighting array $\{h(i,j)\}$ which minimizes (3) for all i, j as in (1), we make use of the orthogonality principle:[7]

> For each n,m, the filter array $\{h(i,j)\}$ which minimizes the mean-square error ϵ is such that the difference
>
> $$g(n,m) - \sum_{i=-A}^{B} \sum_{j=-C}^{D} f(n-i,m-j)h(i,j)$$
>
> is orthogonal to each of the elements in the input array used in the estimate. That is,

$$E\{[g(n,m) - \hat{g}(n,m)]\, f(n-\lambda, m-\tau)\} = 0 \tag{4}$$

$$\lambda = -A, \ldots, 0, \ldots, B$$
$$\tau = -C, \ldots, 0, \ldots, D.$$

We substitute (1) into (4) to give

$$E\left\{\left[g(n,m) - \sum_i \sum_j f(n-i,m-j)\, h(i,j)\right] \times f(n-\lambda, m-\tau)\right\} = 0. \tag{5}$$

Taking the expectation operation inside the double summation, we have

$$E\{g(n,m)\, f(n-\lambda, m-\tau)\}$$
$$= \sum_i \sum_j E\{f(n-i, m-j)\, f(n-k, m-l)\}\, h(i,j)\,. \tag{6}$$

*A weighted mean-square distortion measure[6] may be more meaningful for some types of data. It can be shown that the optimum digital filter minimizes both the weighted and unweighted mean-square distortion measures.

However, these expectations are simply values of the covariance functions:

$$E\{g(n,m)\, f(n-\lambda, m-\tau)\} = K_{gf}(\lambda, \tau) \tag{7a}$$

$$E\{f(n-i, m-j)\, f(n-\lambda, m-\tau)\} = R_{ff}(\lambda-i, \tau-j). \tag{7b}$$

Thus we have the following set of equations to solve for $\{h(i, j)\}$:

$$K_{gf}(\lambda, \tau) = \sum_{i=-A}^{B} \sum_{j=-C}^{D} R_{ff}(\lambda-i, \tau-j)\, h(i, j) \tag{8}$$

$$\lambda = -A, \ldots, 0, \ldots, B$$
$$\tau = -C, \ldots, 0, \ldots, D\,.$$

To complete the design algorithm, we write the system of equations given in (8) in matric form* as

$$\underline{k}_{gf} = \underline{R}_{ff}\, \underline{h} \tag{9a}$$

$$\begin{bmatrix} \underline{k}_{gf}^{-A} \\ \underline{k}_{gf}^{-A+1} \\ \vdots \\ \underline{k}_{gf}^{B} \end{bmatrix} = \begin{bmatrix} \underline{R}_{ff}^{0} & \underline{R}_{ff}^{1} & & \underline{R}_{ff}^{A+B} \\ \underline{R}_{ff}^{1} & \underline{R}_{ff}^{0} & & \\ \vdots & & \ddots & \\ \underline{R}_{ff}^{A+B} & & & \underline{R}_{ff}^{0} \end{bmatrix} \begin{bmatrix} \underline{h}^{-A} \\ \underline{h}^{-A+1} \\ \vdots \\ \underline{h}^{B} \end{bmatrix} \tag{9b}$$

where the general form of each of the subvectors and submatrices is

$$\left(\underline{k}_{gf}^{W}\right)^T = [K_{gf}(W, -C) \ldots K_{gf}(W, 0) \ldots K_{gf}(W, D)] \tag{10a}$$

$$(\underline{h}^W)^T = [h(W, -C) \ldots h(W, 0) \ldots h(W, D)] \tag{10b}$$

$$\underline{R}_{ff}^{W} = \begin{bmatrix} R_f(W, 0) & R_f(W, 1) \ldots & R_f(W, C+D) \\ R_f(W, 1) & R_f(W, 0) & \\ \vdots & & \ddots \\ R_f(W, C+D) & & R_f(W, 0) \end{bmatrix}. \tag{10c}$$

*Underlined lower and upper case letters represent column vectors and matrices, respectively.

The elements of these subvectors and submatrices can be identified directly from (8) and (10); $\underline{k}_{gf}^{W}$ is a column vector made up of the W<u>th</u> row of the crosscovariance array $\{K_{gf}(n, m)\}$. Similarly, $\underline{h}^W$ corresponds to the W<u>th</u> row of the filter impulse response array. Each of the submatrices of $\underline{R}_{ff}$ is a covariance matrix constructed from the appropriate row of the input covariance array.

The optimum digital filter response or point spread array $\{h(n, m)\}$ for all n, m follows formally from (9) by inverting $\underline{R}_{ff}$:

$$\underline{h} = \underline{R}_{ff}^{-1}\, \underline{k}_{gf}\,. \tag{11}$$

The design algorithm as given in (11) is capable of treating a very broad class of problems requiring only <u>a priori</u> knowledge of the covariance arrays. In an actual design, the dimensions assigned to the filter represent a balance between conflicting design objectives that deal with filter implementation and performance. Although it is clear that a digital filter of smaller dimensions will be easier to implement, it should be clear also that the filter performance generally will improve with increasing dimensions.

Unfortunately, a dimensional constraint is also imposed by the design algorithm. The difficulty arises in the numerical realization of (11). $\underline{R}_{ff}$, the matrix to be inverted, is of dimension (A+B+1) (C+D+1), which for moderate to large filters may be quite sizeable. For example, the design of a 50 × 50 filter requires the inversion of a dense matrix of dimension 2500. Even for low-order filters (< 20), the storage and inversion of $\underline{R}_{ff}$ may lead to long computations and numerical instabilities. In some useful special cases, however, filters of larger dimensions can be designed with this algorithm.

4. REALIZATIONS FOR DECOMPOSABLE COVARIANCE ARRAYS

The numerical realization of the design algorithm is substantially simpler when the various covariance arrays are decomposable; i.e., at each point of definition

$$R(n, m) = R_x(n)\, R_y(m)\,. \tag{12}$$

With the grid of sample points a Cartesian product, the two-dimensional function space of this class of covariance arrays is a direct product space.[8] It follows that the block covariance vectors factor into direct products of smaller dimensioned vectors.

The decomposibility condition is of great practical importance in that it eliminates most of the numerical problems mentioned above.* Before considering these cases, we introduce the direct or tensor product of matrices and list a few of its properties which will be useful in the development to follow.

4.1 MATRIX DIRECT PRODUCT

We consider the square matrices A and B of dimension u, v, respectively. The direct product of $\underline{A}$ and $\underline{B}$, written $\underline{A} \otimes \underline{B}$, is defined by the partitioned matrix

$$\underline{A} \otimes \underline{B} = \begin{bmatrix} a_{11}\underline{B} & a_{12}\underline{B} & \cdots & a_{1u}\underline{B} \\ a_{21}\underline{B} & a_{22}\underline{B} & & \\ \vdots & & & \\ a_{u1}\underline{B} & \cdots & & a_{uu}\underline{B} \end{bmatrix} \tag{13}$$

of dimension u · v where

$$\underline{A} = \begin{bmatrix} a_{11} & a_{12} & \cdots & a_{1u} \\ a_{21} & a_{22} & & \\ \vdots & & & \\ a_{u1} & \cdots & & a_{uu} \end{bmatrix} . \tag{14}$$

As the direct product is a particular kind of general matrix product, it satisfies all the algebraic properties of ordinary matrix multiplication. This follows directly from the definition (13). The two identities involving the direct product which we will use in the sequel are

$$(\underline{A} \otimes \underline{B})^{-1} = \underline{A}^{-1} \otimes \underline{B}^{-1} \tag{15a}$$

$$(\underline{A} \otimes \underline{B})\,(\underline{C} \otimes \underline{D}) = \underline{A}\,\underline{C} \otimes \underline{B}\,\underline{D} \tag{15b}$$

where the dimensions of the matrices are assumed to be such that the operations are defined.

4.2 AUTOCOVARIANCE DECOMPOSABLE

For the autocovariance array $\{R_{ff}(n, m)\}$ decomposable, we have

$$R_{ff}(n, m) = R_x(n)\, R_y(m) \tag{16}$$

$$n = 0, 1, \ldots, A + B$$
$$m = 0, 1, \ldots, C + D .$$

To determine the effect of (16) on the numerical realization of (11), we first note that each of the submatrices of $\underline{R}_{ff}$ can be written as

$$\underline{R}_{ff}^{W} = R_x(W) \begin{bmatrix} R_y(0) & R_y(1) & \cdots & R_y(C{+}D) \\ R_y(1) & R_y(0) & & \\ \vdots & & \ddots & \\ R_y(C{+}D) & & & R_y(0) \end{bmatrix} \tag{17}$$

$$= R_x(W)\, \underline{R}_y .$$

Consequently the fully matrix to be inverted, $\underline{R}_{ff}$, can be written in partitioned form as

$$\underline{R}_{ff} = \begin{bmatrix} R_x(0)\,\underline{R}_y & R_x(1)\,\underline{R}_y & R_x(A{+}B)\,\underline{R}_y \\ R_x(1)\,\underline{R}_y & R_x(0)\,\underline{R}_y & \\ \vdots & & \ddots \\ R_x(A{+}B)\,\underline{R}_y & & R_x(0)\,\underline{R}_y \end{bmatrix} . \tag{18}$$

Letting

$$\underline{R}_x = \begin{bmatrix} R_x(0) & R_x(1) & \cdots & R_x(A{+}B) \\ R_x(1) & R_x(0) & & \\ \vdots & & \ddots & \\ R_x(A{+}B) & & & R_x(0) \end{bmatrix} , \tag{19}$$

the block covariance array $\underline{R}_{ff}$ is clearly a direct product form (in keeping with our previous vector space interpretation of the decomposability), and

$$\underline{R}_{ff} = \underline{R}_x \otimes \underline{R}_y \tag{20}$$

*It also occurs frequently in applications. For instance, the covariance of images are decomposable to a good approximation.[9]

where $\underline{R}_x$ and $\underline{R}_y$ are covariance matrices.

Using the direct product identity (15a), we determine the filter array given in (11):

$$\begin{aligned} h &= (\underline{R}_{ff})^{-1}\,\underline{k}_{gf} \\ &= (\underline{R}_x \otimes \underline{R}_y)^{-1}\,\underline{k}_{gf} \\ &= (\underline{R}_x^{-1} \otimes \underline{R}_y^{-1})\,\underline{k}_{gf}\,. \end{aligned} \tag{21}$$

The computational differences in calculating (21) and (11) are quite significant. Rather than invert a matrix of dimension (A+B+1) (C+D+1), we need only invert two smaller matrices, $\underline{R}_x$ and $\underline{R}_y$, of dimension A+B+1 and C+D+1, respectively. This is, of course, a simpler task (i.e. numerically feasible). Further, the matrices to invert not only are smaller but are of a form more easily invertible. Both $\underline{R}_x$ and $\underline{R}_y$ are covariance matrices and are members of a general class of matrices, finite-dimensional Toeplitz forms. An efficient algorithm, due to Trench,[(10)] exists for the machine inversion of such matrices.

As a concluding comment here, we note that while the matrix inversion is now much simpler, we still must store the full inverse, $\underline{R}_{ff}^{1}$, in calculating $\underline{h}$. Fortunately, (21) can be written and calculated in a partitioned form. This allows the row-by-row computation of the filter array with relatively small amounts of rapid-access storage and efficient use of backing storage (for block transfers).

4.3 AUTO- AND CROSSCOVARIANCE DECOMPOSABLE

In addition to the decomposibility of $\{R_{ff}(n, m)\}$, let us assume that the array $\{K_{gf}(n, m)\}$ takes the form

$$K_{gf}(n, m) = K_x(n)\, K_y(m)$$

$$n = -A, \ldots, 0, \ldots, B$$

$$m = -C, \ldots, 0, \ldots, D. \tag{22}$$

Proceeding as before, we now note that the cross-covariance vector $\underline{k}_{gf}$ can be written as the direct product of two vectors. Letting

$$\underline{k}_x^T = [K_x(-A), \ldots, K_x(0), \ldots, K_x(B)] \tag{23a}$$

$$\underline{k}_y^T = [K_y(-C), \ldots, K_y(0), \ldots, K_y(D)], \tag{23b}$$

it follows that

$$\underline{k}_{gf} = \underline{k}_x^T \otimes \underline{k}_y^T\,. \tag{24}$$

When we use (15b) and (24), (21) becomes

$$\begin{aligned} \underline{h} &= (\underline{R}_{ff})^{-1}\,\underline{k}_{gf} \\ &= (\underline{R}_x^{-1} \otimes \underline{R}_y^{-1})\,(\underline{k}_x \otimes \underline{k}_y) \\ &= (\underline{R}_x^{-1}\,\underline{k}_x) \otimes (\underline{R}_y^{-1}\,\underline{k}_y)\,. \end{aligned} \tag{25}$$

This may also be written as

$$\underline{h} = \underline{h}_x \otimes \underline{h}_y \tag{26}$$

with

$$\underline{h}_x = \underline{R}_x^{-1}\,\underline{k}_x \tag{27a}$$

$$\underline{h}_y = \underline{R}_y^{-1}\,\underline{k}_y\,. \tag{27b}$$

As might be expected, the general two-dimensional problem has degenerated to two each one-dimensional problems. In addition to the computational advantages due to decomposability of the autocovariance, the storage requirements are now drastically reduced, since (25) can be computed by one-dimensional vector storage. It is interesting to note that (27) represents the specialization of our synthesis techniques to problems of one independent variable.

5. FILTER IMPLEMENTATION

The usual method of implementing one-dimensional nonrecursive digital filters involves the use of the Fast Fourier Transform (FFT).[(11)] The filtering operation, a discrete aperiodic convolution, is formulated first as a periodic convolution of appropriately defined sequences. The convolution is then calculated in the transform domain (as a product) by means of the FFT.

Our implementation of the filtering operation defined in (2) was developed by extending this method to two dimensions. While the numerical details are somewhat more involved, conceptually the two-dimensional case follows directly from the one-dimensional as introduced by Stockham.[12] A periodic equivalent to (2) is developed by augmenting the input and filter-impulse-response arrays with "grace" borders to prevent spatial aliasing. With this extended region of definition, each of the augmented arrays is transformed via the FFT, the transformed arrays are multiplied together, and the product array is inverse-transformed to give the filter output, $\{g(n, m)\}$, for all n, m, as in (1).

The primary operation, the two-dimensional FFT, is performed by repeated application of a one-dimensional FFT routine. By means of a base-2 FFT algorithm, the machine calculation of the above transform requires 2NM rapid-access memory locations and a computing time proportional to $NM \log_2 NM$. The filtering operation may be timed accordingly.

With reference specifically to the processing of pictures by computer, the results presented here apply to noise removal or filtering, to edge enhancement in the presence of noise, to the reduction of quantization effects, and to the restoration of images degraded by the modulation transfer function of optical systems and by noise. Our experience in applying the design algorithm to the reduction of quantization noise may serve as an example. An image was degraded by quantization of the grey scale to 3 and 4 bits. An 11×19 filter was designed and implemented on the CDC 7600 computer by means of the FFT. The filtering resulted in a notable improvement of image quality, principally in the high detail regions of the image. The computation time was on the order of one minute for a 512×512 image array.

ACKNOWLEDGMENT

This research was supported partly by the U.S. Atomic Energy Commission, and partly by NASA under grant NGL 05-003-404.

REFERENCES

(1) R. Nathan, "Digital Video-Data Handling," Jet Propulsion Laboratory Technical Report 32-877, January 1966.

(2) C. W. Helstrom, "Image Restoration by the Method of Least Squares," J. Opt. Soc. Am. **57**, 297-303 (1967).

(3) K. S. Fu, D. A. Landgrebe, and T. L. Phillips, "Information Processing of Remotely Sensed Agricultural Data," Proc. IEEE **57**, 639-653 (1969).

(4) E. C. Mahen, E. L. Hall, and S. J. Dwyer III, "General Isotropic Spatial Design Techniques," Proc. 3rd Asilomar Conf. on Circuits and Systems, December 1969.

(5) A. Papoulis, Systems and Transforms with Applications in Optics. New York: McGraw-Hill, 1968.

(6) D. J. Sakrison and V. R. Algazi, "Comparison of Line by Line and Two Dimensional Encoding of Random Images" (to appear in IEEE Tran. Info. Theory).

(7) A. Papoulis, Probability, Random Variables and Stochastic Processes. New York: McGraw-Hill, 1965.

(8) P. R. Halmos, Finite-Dimensional Vector Spaces. Princeton, N.J.: D. Van Nostrand, 1958.

(9) E. R. Kretzmer, "Statistics of Television Signals," Bell Syst. Tech. J. **31**, 751-763 (1952).

(10) S. Zohar, "Toeplitz Matrix Inversion: The Algorithm of W. F. Trench," J. Ass. Comput. Mach. **16**, 592-601 (1969).

(11) B. Gold and C. M. Rader, Digital Processing of Signals. New York: McGraw-Hill, 1969.

(12) T. G. Stockham, Jr., "Highspeed Convolution and Correlation," Joint Computer Conference, AFIPS Proc. **28**, 229-233 (1966).

23

SMOOTHING OF TWO-DIMENSIONAL IMAGES USING THE "COMPOSITE" MODEL OF A FRAGMENT

D. S. Lebedev and L. I. Mirkin

This article was translated by Addis Translations International, Portola Valley, California, from the original Russian, pages 57–62 in Iconics–Digital Holography–Image Processing, *Institute for Problems in Information Transmission, Academy of Sciences, USSR, 1975, 150pp.*

The smoothing of images, that is, the attenuation of distortion caused by noise in the image-reproduction system (photographic, television, etc.), is a task of practical importance.

Smoothing is possible because the image and the noise have different properties. In the statistical approach to smoothing, the effectiveness of the algorithm used depends on how complete a statistical description it gives of the image and the noise. Ordinarily, noise is marked by a simple statistical structure; the noise properties either are known or can be obtained through simple measurements. But for images, it is difficult to make a sufficiently complete measurement of the statistical characteristics. This job can be made easier through the use of a constructive model of the image.

In the smoothing of one-dimensional signals, the gaussian model of signals often finds use. This model leads to the Wiener algorithm[1] of linear smoothing, which gives the minimum root-mean-square (RMS) deviation of the smoothed signal from the original one. This kind of smoothing takes place in a filter with the frequency response

$$K(\omega) = S(\omega)/(S(\omega) + N(\omega)), \tag{1}$$

where $S(\omega)$ and $N(\omega)$ are the Wiener (energy) spectra of the signal and the noise and ω is the frequency. The theory of Wiener smoothing can readily be generalized to the case of n-dimensional signals; the optimum frequency response of the smoothing filter will be, as before, described by Eq. (1), in which ω is now the n-dimensional frequency.

The application of Wiener linear smoothing to an image does not always improve the image quality as visually rated. This is true because noise, as a rule, has high-intensity components up to the highest spatial frequencies, while the signal is mainly concentrated at low spatial frequencies, with the $S(\omega)$ values falling off rapidly at high frequencies. Therefore, in view of Eq. (1), the smoothing filter will suppress the high-frequency components of the image, which convey important information about the edges and fine details of the imaged object. The smoothed image becomes blurred.

As D. Gabor[2] showed, a spectral description of the statistical properties of an image is insufficient, for it does not reflect the locally anisotropic structure (even though the image as a whole is isotropic).

To describe such a structure, the present article proposes a "composite" model of an image fragment. This model is used for synthesizing a smoothing algorithm that estimates the image elements with the minimum rms deviation. The algorithm was executed on a digital computer.

"Composite" model of an image fragment. A monochromatic, stationary image can be described as a brightness distribution $u(t)$, where t (a pair of numbers) represents the coordinates of a point in the image plane. Below we shall consider only discretized images with integral coordinates.

An ensemble of images represents a random field. The probabilistic description of such a field gives the value of $P(u)$, the n-dimensional combined probability density of the image fragment $u = u(t_1), \ldots, u(t_n)$ consisting of n elements.

We assume that there are M classes of fragments, distinguished by the types of correlational links between the elements defining the structure.

Some classes are formed of fragments with isotropic structure (without domination by links in some direction on the image plane); others, by fragments with some anisotropy. Let $P_\theta(u)$ be the n-dimensional probability density of a fragment u, on the condition that the fragment belongs to the class θ ($\theta = 1, \ldots, M$) and $\pi(\theta)$ is the probability distribution of the classes

$$\left(\sum_{\theta=1}^{M} \pi(\theta) = 1\right).$$

Then

$$P(u) = \sum_{\theta=1}^{M} \pi(\theta) P_\theta(u). \tag{2}$$

Expression (2) is the decomposition of the density $P(u)$ in the system of densities $P_\theta(u)$, $\theta = 1, \ldots, M$. This representation is especially useful when $\pi(\theta)$ is approximated closely by a small collection of gaussian distributions:

$$P_\theta(u) = |R_\theta^{-1}| \exp\left\{-\frac{1}{2} u' R_\theta^{-1} u\right\}', \theta = 1, \ldots, M, \tag{3}$$

where R_θ^{-1} is the inverse of the covariance matrix R_θ corresponding to class θ, and the prime denotes transposition.

The results of statistical measurements on fragments of real images show that a good description of the fragments can be obtained with a model having as few as five classes. Four of the classes correspond to the predominant correlational links in the four directions making angles of 0°, 45°, 90°, and 135° to the horizontal (these directions being chosen because the image is shown on a square grid). The fifth class describes fragments with "isotropic" structure.* For several images we have measured the matrices R_θ and the probability distribution $\pi(\theta)$, $\theta = 0, 1, 2, 3, 4$.

Smoothing algorithm. Let there be observed an image with noise, additively superimposed, independent of the image, and with a known probability density. It is required to find the optimum (from the standpoint of minimum rms deviation) estimate $\hat{u}(t)$ of an image element $u(t)$ from an n-element fragment of the observed image $v_t = (v(t+\tau_0), \ldots, v(t+\tau_{n-1}))$, where $\tau_1, \ldots, \tau_{n-1}$ are given points lying in a neighborhood of the point $\tau_0 = (0,0)$.

It is known that the optimum estimate of an element $u(t)$ is the *a posteriori* tentative mean value

$$\hat{u}(t) = \int u(t)P(u_t|v_t)du_t, \tag{4}$$

where

$$P(u_t|v_t) = \frac{P(u_t)q(v_t|u_t)}{\int P(u_t)q(v_t|u_t)du_t}. \tag{5}$$

In these expressions du_t denotes $du(t+\tau_0), \ldots, du(t+\tau_{n-1})$; $u_t = (u(t+\tau_0), \ldots, u(t+\tau_{n-1}))$; and $q(v_t|u_t)$ is the tentative density of the observed fragment v_t for a given fragment u_t. Using Eq. (2), we rewrite Eq. (5) in the following form:

$$P(u_t|v_t) = \sum_{\theta=1}^{M} P_\theta(u_t|v_t)\pi(\theta|v_t), \tag{6}$$

where

$$P_\theta(u_t|v_t) = \frac{P_\theta(u_t)q(v_t|u_t)}{\int P_\theta(u_t)q(v_t|u_t)du_t} \tag{7}$$

and

$$\pi(\theta|v_t) = \frac{\pi(\theta)\int P_\theta(u_t)q(v_t|u_t)dv_t}{\sum_{\theta=1}^{M} \pi(\theta)\int P_\theta(u_t)q(v_t|u_t)du_t}. \tag{8}$$

*"Isotropic" denotes a structure in which there are two pairs of directions with correlational links equal in magnitude: the vertical-horizontal pair and the pair of diagonal directions. The correlations of the two pairs may differ in magnitude.

Substituting Eq. (6) into (4), we obtain

$$\hat{u}(t) = \sum_{\theta=1}^{M} \pi(\theta|\nu_t)\hat{u}_\theta(t), \tag{9}$$

where

$$\hat{u}_\theta(t) = \int u(t)P_\theta(u_t|\nu_t)du_t. \tag{10}$$

The quantity $\hat{u}_\theta(t)$ is a tentative estimate of $u(t)$ for the given class of the fragment. The estimate $\hat{u}(t)$ is a weighted sum of the tentative estimates. The weight of each tentative estimate is the *a posteriori* probability (8) that the given fragment ν_t belongs to class θ.

Let the noise have a gaussian distribution with a known covariance matrix N. Then

$$q(\nu_t|u_t) = |N^{-1}|\exp\left\{-\frac{1}{2}(\nu_t - u_t)'N^{-1}(\nu_t - u_t)\right\}. \tag{11}$$

Substituting Eqs. (11) and (3) into (7), we then obtain from Eq. (10) the well-known formula of Wiener:[1]

$$\hat{u}_\theta(t) = \sum_{\ell=1}^{n} a_{1\ell}\nu(t + \tau_{\ell-1}), \tag{12}$$

where $a_{1\ell}$ is an element of the matrix $A = R_\theta(R_\theta + N)^{-1}$.

In this case, the *a posteriori* probability of state (8) becomes

$$\pi(\theta|\nu_t) = \frac{\pi(\theta)\exp\left\{-\frac{1}{2}\nu_t'(R_\theta + N)^{-1}\nu_t\right\}}{\sum_{\theta=1}^{M} \pi(\theta)\exp\left\{-\frac{1}{2}\nu_t'(R_\theta + N)^{-1}\nu_t\right\}}. \tag{13}$$

As expression (12) shows, the tentative estimate $\hat{u}_\theta(t)$ can be found with the help of a linear filter. If ν_t is replaced by $\nu_s = (\nu(s + \tau_0), \ldots, \nu(s + \tau_{n-1}))$, that is, if the observed fragment is shifted at the filter output, then we obtain the estimate $\hat{u}_\theta(s)$.

The implementation of algorithm (9) requires the construction of M linear filters and M devices for computing $\pi(\theta|\nu_t)$. It turns out that the $\hat{u}_\theta$ from each filter must be multiplied by the corresponding $\pi(\theta|\nu_t)$ value and all the products must be added.

Algorithm (9) can be interpreted in the following way. For each class of fragments a specific mode of smoothing is used, implemented by the corresponding linear filter. If, for example, for some θ the matrix R_θ describes only the horizontal correlational links, then the estimate $\hat{u}_\theta(t)$ should be obtained by smoothing only in the horizontal direction.

Realization of the algorithm on a digital computer. The smoothing algorithm (9) was implemented on a Minsk-22 digital computer. The image, 128 X 128 elements quantized at 64 levels (6 bits), was input and output through scanning and printing optical-mechanical hardware built around phototelegraphic equipment.

For the simulation of a noisy image, guassian white noise was superimposed on the original. The computer generated the noise by the method used by BPF[3].

The fragment observed contained 5 X 5 elements, and the element to be estimated lay in the center of the fragment. We assumed that there were five classes of fragments; thus we used five R_θ matrices, of which one corresponded to "isotropic" correlational links and four were anisotropic. These matrices corresponded to links along only one of four lines passing through the center (vertical, horizontal, and two diagonals at ± 45°). Thus there were five smoothing modes:* over the whole fragment of twenty-five elements, and along line segments containing five elements each. The five estimates obtained from the five smoothing procedures were summed with weights equal to the computed *a posteriori* probabilities.

Figure 1 shows the original image with, superimposed on it, noise whose rms value was 5 percent of the brightness range of the image. Figure 2 shows the image obtained through smoothing of the noisy image by algorithm (9). Figure 3 is the result of Wiener smoothing. A comparison of Figures 2 and 3 shows that algorithm (9), described above, gives an image with less blurriness than does the Wiener algorithm. Consequently, the model used gives a better fit to the image structure than does the Gaussian model.

Figures 1, 2, and 3.

*Algorithms with several smoothing modes have appeared earlier.[4,5] In these algorithms, however, one of the particular estimates was used as the resultant estimate, rather than the weighted sum as in algorithm (9).

REFERENCES

1. *N. Wiener.* The Interpolation, Extrapolation and Smoothing of Stationary Time Series, N.Y., 1949.
2. *D. Gabor.* The Smoothing and Filtering of Two-Dimensional Images.–Progress in Biocybernetics, v. 2 Amsterdam, 1965.
3. L. P. Yaroslavskii, M. A. Rabinovich, and L. I. Mirkin, "Metod generirovaniya gaussovskikh korrelirovannykh psevdo-sluchainykh chisel na EVM" [Method of Generating Gaussian Correlated Pseudorandom Numbers on the Computer], *Zh. Vychisl. Mat. Mat. Fiz.*, 1972 (5).
4. *R. E. Graham.* "Snow Removal–A Noise–Stripping Process for picture Signals." IRE Transaction on Information Theory, 1962. **IT-8,** No. 2.
5. G. G. Vainshtein, "Prostranstvennaya fil'tratsiya izobrazhenii sredstvami analogovoi vychislitel'noi tekhniki" [Spatial Filtration of Images Using Analog Computing Hardware], in the collection *Ikonika. Prostranstvennaya fil'tratsiya izobrazhenii. Fotograficheskie sistemy* [Iconics. Spatial Filtration of Images. Photographic Systems]. Moscow: Izdatel'stvo "Nauka," 1970.

Editors' Comments on Papers 24 Through 28

24 NAHI
Role of Recursive Estimation in Statistical Image Enhancement

25 POWELL and SILVERMAN
Modeling of Two-Dimensional Covariance Functions with Application to Image Restoration

26 ABOUTALIB and SILVERMAN
Restoration of Motion Degraded Images

27 WOODS and RADEWAN
Kalman Filtering in Two Dimensions

28 FEICK and STRINTZIS
Constrained Optimal Linear Recursive Estimation of Two-Dimensional Processes

DESIGN OF RECURSIVE PROCESSORS

As in the 1-D case, recursive filtering techniques are of interest in 2-D because of their potential for both on-line or real-time operation and reduced computational complexity. Statistical recursive filtering based on mean-square estimation has found extensive application in 1-D signal processing since its introduction by Kalman. Although the extensions of Wiener filtering techniques to 2-D are relatively direct (on reflection), the 2-D extension of Kalman's recursive formalism suffers some severe difficulties. These difficulties derive from two sources. The first is related to the signal modeling and the second to computational requirements of the estimator formalism.

The *a priori* statistical information on which the nonrecursive estimator designs are based are the covariance arrays. In developing recursive estimators, recursive models are employed. The classical 2-D autoregressive, moving average model is characterized by the difference equation

$$s_{m,n} = \sum_{m'} \sum_{n'} a_{m-m',n-n'} w_{m',n'} - \sum_{m'} \sum_{n'} b_{m-m',n-n'} s_{m',n'} \qquad (2.4)$$

where $w_{m,n}$ is a white noise driving process. The limits for m',n' are such that values in a compact neighborhood of (m,n) are incorporated in the sum. The specific form is essentially determined by the structure of the recursion, i.e., the order of computing the output points. In contrast to the 1-D case, there are a countable number of ways this may occur.

With proper choice of coefficients, $\{a\}$ and $\{b\}$, the output array $\{s_{m,n}\}$ may have a prescribed covariance. In the equivalent 1-D modeling problem, the $\{a\}$ and $\{b\}$ sequences are obtained from the Z-transform of the covariance by a spectral factorization. Unfortunately, as indicated in Papers 3, 4, and 5, this spectral factorization is more complicated in 2-D and the exact factorization generally results in infinite dimensional models. Some recent work of Ekstrom and Twogood (1977) based on the factorization algorithm in Paper 4 and numerical optimization techniques suggests that finite-dimensional models may be obtained in approximation. However, most recursive processors have been developed using the separable, exponential covariance model of Kretzmer (1952). To a certain extent, the widespread utilization of Kretzmer's model is due to the fact that it can be spectrally factored using 1-D factorizations. While there exists some experimental evidence that this model adequately describes the gross correlation features of so-called natural images, its use in general 2-D filtering applications is clearly restrictive.

The computational difficulties in the direct 2-D Kalman filter are due to the excessive dimension of the state vector. Woods and Radewan (1976) have found that the size of the state vector is not totally determined by the order of the model (as in 1-D) but varies in proportion to the dimension of the data array. Thus, the storage and arithmetic operations required to implement the direct filter are prohibitively large for data arrays of practical interest.

As a consequence of this situation and the close connection between modeling and structure of recursive filters, a number of recursive estimation schemes have been developed by approximating the statistics of the 2-D signals using 1-D recursive models. These models are chosen to match various versions of traditional 1-D Kalman estimators. Nahi

(Paper 24) was one of the first to study recursive estimation of 2-D random fields. In Paper 24 he addresses the additive signal and noise problem. Both signal and noise fields were assumed to be homogeneous, and the data array scanned in a row sequential format (line-by-line). Because the resulting 1-D sequence is nonstationary, Nahi approximates its statistics using a stationary model based on the Kretzmer exponential covariance. Standard Kalman techniques are then used to process the scanned data. For additional information on this approach, see Nahi and Assefi (1972).

A vector scanning procedure is also discussed in Paper 24 and a related paper (Nahi and Franco, 1973). Here, the data array is partitioned into strips and each strip vector-scanned a column at a time. Because the vector-scan process is homogeneous, the Kalman filter can be readily derived. It is expected, however, that estimator performance may be severely degraded in regions of the data array near strip boundaries. Overlapping strips may resolve this question at the cost of increased computational complexity.

Powell and Silverman (Paper 25) studied the modeling of these scanning processes in somewhat greater depth. They consider the filtering problem of estimating a signal (2-D) distorted by a linear transformation in the presence of additive noise (the restoration problem in Paper 21), and present Kalman estimators based on the vector-scanning models. Aboutalib and Silverman (Paper 26) apply these techniques to the specific problems of restoring images degraded by motion blur, space invariant, and space variant. As in the earlier papers, the models in Papers 25 and 26 are based on the Kretzmer exponential covariance.

A departure from the previous approaches to recursive Kalman filtering was taken by Woods and Radewan (Paper 27). Using the general class of 2-D autoregressive moving average models of Eq. (2.4), they derive two direct 2-D Kalman estimators, a point processor and a vector processor. As mentioned above, the computational complexity involved with implementing these processors exactly is prohibitive. In the case of both types of filters, they present approximate realizations which require substantially reduced computations and allow practical implementations. The point processor is approximated by restricting the estimate update calculations to a subset of the state vector. That is, only elements which are within a prescribed region near the point are updated. The approach taken in the vector filter is to reduce the length of the state vector by reducing the size of the data array. This is accomplished by dividing the original data into overlapping strips, and processing each strip independently. The strip width is chosen to exceed the correlation distances of the desired signal. Because the state vector varies in proportion to the strip width, it is desirable to limit the width to achieve the desired reduction in computations. Depending on the size of the data array and correlation models involved, these reductions may be quite sizable—the authors quote 2–3 orders of magnitude, for example. They conjecture that the performance loss associated with these approximations is not of practical significance.

Feick and Strintzis (Paper 28) derive a class of recursive estimators which also allows direct control of the estimator order (i.e., computational requirement). Their method is similar in spirit to a 1-D form of Papoulis (1965) and assumes the availability of no autoregressive, moving average models, only the fundamental covariance arrays. Some potential questions regarding existence/uniqueness in their design algorithm apparently remain to be resolved.

Some additional work regarding 2-D recursive formalisms which may be of interest is contained in the papers of Angel and Jain (1973), Habibi (1972), Nahi and Habibi (1975), and Strintzis (1976).

REFERENCES

Angel, E. and A. K. Jain. 1973. "A Dimensionality Reducing Model for Distributed Filtering," *IEEE Trans. Autom. Control* **AC-18**:59–62.

Ekstrom, M. P., and R. E. Twogood. 1977. "Finite Order Recursive Models for Two-Dimensional Random Fields," *Proc. 20th Midwest Symp. on Circuits and Syst.,* 188–189.

Habibi, A. 1972. "Two-Dimensional Bayesian Estimate of Images," *IEEE Proc.* **60**: 878–883.

Kretzmer, E. R. 1952. "Statistics of Television Signals," *Bell Syst. Tech. J.* **31**: 751–763.

Nahi, N. E., and T. Assefi. 1972. "Bayesian Recursive Image Estimation," *IEEE Trans. Comput.* **C-21**:734–738.

Nahi, N. E., and C. A. Franco. 1973. "Recursive Image Enhancement-Vector Processing," *IEEE Trans. Commun.* **COM-21**:305–311.

Nahi, N. E., and A. Habibi. 1975. "Decision-Directed Recursive Image Enhancement," *IEEE Trans. Circuits and Syst.* **CAS-22**:286–293.

Papoulis, A. 1965. *Probability, Random Variables and Stochastic Processes.* New York: McGraw-Hill.

Strintzis, M. G. 1976. "Comments on 'Two-Dimensional Bayesian Estimate of Images'," *IEEE Proc.* **64**:1255–1257.

Woods, J. W., and C. H. Radewan. 1976. "Reduced Update Kalman Filter—A Two-Dimensional Recursive Processor," *Proc. 1976 Johns Hopkins Conf. Info. Sci. and Syst.,* 21–26.

24

Reprinted from *IEEE Proc.* **60**(7):872-877 (1972)

Role of Recursive Estimation in Statistical Image Enhancement

NASSER E. NAHI, MEMBER, IEEE

***Abstract*—Recursive linear and nonlinear estimation concepts are applied to the problem of image enhancement, whereby procedures for the derivation of the estimators are developed. These procedures are applicable to enhancement of images characterized statistically by their mean and correlation function. Design of an approximate time invariant and a time varying estimator is discussed and the implications of vector scanning are investigated. Feasibility of the approach is illustrated by means of examples, and the computational advantages are presented.**

I. Introduction

IMAGE ENHANCEMENT is the function of restoring an image which has been distorted due to reflections from spurious objects, inaccuracies in the sensing mechanism, or corruption introduced during transmission. In many cases the distortion can be characterized by a random process (noise) with known statistics. Furthermore, in certain applications, the *a priori* knowledge of the image itself is only of statistical nature. In other words, the image does not contain characteristic features which would then have to be utilized in the enhancement process.

The image enhancement, subject to the above restrictions, describes the classical problem of statistical estimation and filtering, where one basically attempts to filter out the noise from an observation. The most general and computationally efficient procedures available at present to perform this filtering operation are the recursive estimation methods of Kalman filtering and their nonlinear extensions [1]. However, here it is necessary that the observation be a function of one independent variable, such as time, in contrast to an image which is defined on a plane. The purpose of this paper is to introduce procedures for converting the planar statistical information of the image and the noise into a form suitable for application of recursive estimation procedures. The effectiveness and computational simplicity of this method of image enhancement will be clearly demonstrated.

Manuscript received November 8, 1971; revised March 6, 1972. This work was supported in part by the Advanced Research Projects Agency of the Department of Defense and was monitored by the Air Force Eastern Test Range under Contract F08606-72-C-0008.

The author is with the Department of Electrical Engineering, University of Southern California, Los Angeles, Calif.

II. Outline of the Method

The statistical information of the image and the noise is assumed to be available in terms of the mean and the two-dimensional autocorrelation functions of the signal and noise. The random processes representing the image and the noise are assumed wide-sense stationary. Furthermore, it is assumed that the noise is additive. The observation is scanned horizontally with uniform speed, producing the scanner output $s(t)$, where t represents the time variable. Due to the assumed additive property of the noise, the scanner output $s(t)$ can be decomposed into functions $s_I(t)$ and $s_n(t)$ where the subscripts I and n represent the image and noise, respectively, and

$$s(t) = s_I(t) + s_n(t). \quad (1)$$

The signal $s(t)$ may be vector valued representing vector scanning, i.e., the process of scanning more than one picture line at a time. The following are the required steps for the design of the image enhancer.

1) Determine the autocorrelation function of $s_I(t)$ and $s_n(t)$ utilizing the spatial autocorrelation functions of the image and noise.

2) Determine dynamic (difference-equation) models of $s_I(t)$ and $s_n(t)$.

3) Design appropriate recursive estimator utilizing the above dynamic models.

Due to the periodic nature of the scanner operation, the random processes $s_I(t)$ and $s_n(t)$ are in general nonstationary. Determination of dynamic models for such processes is very complex and procedures exist only for some special cases. Often such dynamic systems may not exist at all. Consequently, procedures are developed in Sections IV and V to determine dynamic models representing approximately the statistics of $s_I(t)$ and $s_n(t)$.

III. Derivation of Autocorrelation of Scanner Output

Let us first consider the scalar (one-line) scanning of the image, and denote the correlation function of $s_I(t)$ by $\phi_I(t, t+\tau)$. The derivation of noise correlation function $\phi_n(t, t+\tau)$ is identical. Denote the horizontal position variable (a continuous variable) by z, $0 \leq z \leq Z$, and the vertical variable by an integer $n=1, 2, \cdots, N$ representing the nth scanned line. The brightness function is then defined by $b(z, n)$. Without any loss of generality, the reference value of brightness level is chosen such that $b(z, n)$ has zero mean. The random process $b(z, n)$ is assumed wide-sense stationary with autocorrelation function $R_I(z, n)$ where

$$Eb(z_1, n_1)b(z_2, n_2) \overset{\circ}{=} R_I(z_2 - z_1, n_2 - n_1) \overset{\circ}{=} R_I(z, n) \quad (2)$$

where E is the mathematical expectation.

The scanner output is $s_I(t)$, its horizontal speed is v and, without any loss of generality, the vertical movement is assumed to take zero time. We wish to determine the autocorrelation function $\phi_I(t, t+\tau)$, i.e., $Es_I(t)s_I(t+\tau)$ in terms of $R_I(z, n)$, v, and Z, where E is the mathematical expectation.

The variables t and τ can be equivalently expressed by

$$\begin{aligned} t &= jT + \sigma, \qquad j = 0, 1, \cdots, N-1, \quad 0 \leq \sigma \leq T \\ \tau &= iT + \gamma, \qquad i = \cdots, -1, 0, 1 \cdots, \\ &\qquad 0 \leq t + \tau \leq NT, \quad 0 \leq \gamma \leq T \end{aligned} \quad (3)$$

where $T = Z/v$ is the time required to traverse one horizontal line. The scanner output can now be written as

$$\begin{aligned} s_I(t) &= b[v\sigma, j+1] \\ s_I(t+\tau) &= b[v(\sigma+\gamma), i+j+1], \qquad \sigma+\gamma \leq T \quad (4) \\ &= b[v(\sigma+\gamma-T), i+j+2], \qquad \sigma+\gamma > T. \quad (5) \end{aligned}$$

Hence

$$\begin{aligned} \phi_I(t, t+\tau) &= R_I[v\gamma, i], \qquad \sigma+\gamma \leq T \\ &= R_I[v(\gamma - T), i+1], \qquad \sigma + \gamma > T. \end{aligned} \quad (6)$$

Clearly $\phi_I(t, t+\tau)$ is a function of σ and γ (or equivalently t and τ) and hence the random process $s_I(t)$ is nonstationary.

IV. A Stationary Enhancement Process

A stationary correlation function approximation to $\phi_I(t, t+\tau)$ can be obtained by randomizing [2] the variable t over the intervals $jT \leq t \leq (j+1)T$, $j=1, 2, \cdots$. Equivalently from (3), we may randomize σ over the range $0 \leq \sigma \leq T$. While the range of σ is fixed, its density function can be chosen arbitrarily. A uniform density has the intuitive appeal that all points of the picture are weighed equally, and consequently is chosen here. Finally, an average correlation function is denoted by $r_I(\tau)$ and defined by

$$r_I(\tau) = E_\sigma \phi_I[jT + \sigma, (j+i)T + \sigma + \gamma]. \quad (7)$$

For σ distributed as described, the probability of the event $[\sigma+\gamma \leq T]$ can easily be determined in terms of the correlation variable γ

$$P_r[(\sigma+\gamma) \leq T] = \frac{T-\gamma}{T}. \quad (8)$$

Performing the expectation in (7), utilizing (6) and (8), yields

$$\begin{aligned} r_I(\tau) &= \frac{T-\gamma}{T} R_I[v\gamma, i] + \frac{\gamma}{T} R_I[v(\gamma - T), i+1] \\ \tau &= iT + \gamma. \end{aligned} \quad (9)$$

The autocorrelation function $r_I(\tau)$ represents a stationary random process denoted in the following by $y_I(t)$. Realization of this random process by a first-order vector differential equation with white noise input described by (10)–(13) is necessary for the application of recursive estimation of Kalman type,

$$\mathring{x}(t) = A_I x(t) + B_I u(t) \quad (10)$$

$$y_I(t) = C_I x(t) \quad (11)$$

$$Eu(t)u'(t+\tau) = Q_I \delta(\tau) \quad (12)$$

$$Ey_I^2(0) = r_I(0) \quad (13)$$

where $x(t)$ is an n-dimensional vector, $\delta(\tau)$ is the Dirac delta function, prime denotes transpose, and

$$Ey_I(t)y_I(t+\tau) = r_I(\tau). \quad (14)$$

General procedures for determination of the matrix parameters A_I, B_I, C_I, and Q_I are given in [3] and [1]. A special procedure was utilized in [5], where $r_I(\tau)$ was approximated by retaining a few terms of the Fourier series expansion of the periodic portion of $r_I(\tau)$. This latter procedure will be utilized in this paper.

A similar model can be derived for the noise term. However, for sake of simplicity of presentation, the noise term is assumed here to be white noise with zero mean.[1] Hence the noise corrupted observation, i.e., the scanner output, is given by

$$s(t) = C_I x(t) + v(t) \quad (15)$$

$$Ev(t) = 0$$

$$Ev(t)v'(t+\tau) = R\delta(\tau) \quad (16)$$

where R is a given quantity. Since we intend to utilize a digital image enhancement procedure, the above model is discretized yielding the following difference equation model.

$$x(k+1) = Ax(k) + Bu(k), \qquad k = 1, 2, \cdots \quad (17)$$

$$s(k) = Cx(k) + v(k) \quad (18)$$

$$Eu(k) = Ev(k) = 0$$

$$Eu(k)u'(j) = K\Delta(i-k) \quad (19)$$

$$Ev(k)v'(j) = \sigma^2 \Delta(j-k) \quad (20)$$

where $s(k)$ represents the sampled value of the scanner output and $\Delta(\cdot)$ is the Kronecker delta function. The sampling inter-

[1] The mean of the noise is not, in general, zero. However, its value can be simply subtracted from the observation to yield $y(t)$ used above.

val is chosen as T/N. This will produce N observations for each horizontal scan. Since there are N (horizontal) scan lines, the final discrete observation is an $N \times N$ grid. The matrix parameters A, B, C, K, and σ are given [1] in terms of the parameters of the continuous model:

$$A = e^{A_I(T/N)} \tag{21}$$

$$BKB' = \int_0^{T/N} e^{A_I(T/N)} e^{-A_I\xi} B_I Q_I B_I' e^{-A_I'\xi} e^{A_I'(T/N)} d\xi \tag{22}$$

$$C = C_I. \tag{23}$$

On the basis of this discrete model, with parameters as specified by (21)–(23), a recursive estimator may now be designed [1]. The equations are given here for the sake of completeness.[2]

$$\hat{x}(k) = [I - F(k)C]A\hat{x}(k-1) + F(k)s(k) \tag{24}$$

$$F(k) = P(k)C'[CP(k)C' + \sigma^2]^{-1} \tag{25}$$

$$P(k+1) = A[I - F(k)C]P(k)A' + BKB' \tag{26}$$

$$\hat{x}(0) = 0$$

$$CP(0)C' = r_I(0).$$

The estimate of the image is therefore $C\hat{x}(k) \triangleq \hat{y}_I(k)$, i.e., $\hat{y}_I(k)$ is the best estimate of $y_I(k)$, obtained recursively in real time, where $s(k)$ is the observation associated with the grid point which is being estimated.

It is clear that image enhancement, from the point of view of scanner output, represents an interpolation problem, i.e., it is desired to determine the best estimate of $y_I(k)$, $0 \leq k \leq N^2$, given $s(0), \cdots, s(N^2)$. In general, the interpolation requires a far more complicated procedure [1] than standard Kalman filtering. However, since for the image processing considered here the length of data is fixed (N^2) and, furthermore, the observation is usually available for additional repeated processing, it is possible to obtain two estimate values of $y_I(k)$, denoted $\hat{y}_I(k)$ and $\bar{y}_I(k)$, one by running the scanner in one direction starting, for example, at top left corner of the picture and the other by running the scanner in reverse direction starting at bottom right corner. Associated with these estimates are estimation error variances denoted by $\hat{\sigma}^2(k) = C\hat{P}(k)C'$ and $\bar{\sigma}^2(k) = C\bar{P}(k)'C'$, respectively. The two estimates $\hat{y}_I(k)$ and $\bar{y}_I(k)$ can then be combined to yield the optimal interpolated (smoothed) value $y_I^*(k)$ where

$$y_I^*(k) = \frac{\hat{\sigma}^2(k)}{\hat{\sigma}^2(k) + \bar{\sigma}^2(k)} \bar{y}_I(k) + \frac{\bar{\sigma}^2(k)}{\bar{\sigma}^2(k) + \hat{\sigma}^2(k)} \hat{y}_I(k). \tag{27}$$

Example

Using a 32×32 grid, Figs. 1 and 2 represent two images, where the gray level is chosen as 6.1 for the image and -1 for the portion outside the image. The area of the image is 144 grid units. These images are sample functions of a two-dimensional random process with autocorrelation function

$$R_I(z, i) = 6.1e^{-4.35|z| - 0.136|i|}.$$

The gray levels 6.1 and -1 were chosen to yield zero mean for these sample functions. The parameters of $R_I(z, i)$ were obtained by experimentally (using a digital computer) determining an average of sample correlation functions for a variety of shapes with the area 144 and gray levels 6.1, -1 occupying different portions of the picture and at different orientations. A plot of $r_I(\tau)$ using (9) and $R_I(z, i)$ above for $v = T = 1$ is presented in Fig. 3. Using the fact that $r_I(\tau)$ is a positive even function having a pseudoperiodic nature, an approximation to the function is obtained and denoted by $r_a(\tau)$ where

$$r_a(\tau) = 6.1e^{-0.136|\tau|}[0.333 + 0.405 \cos 2\pi\tau + 0.101 \cos 4\pi\tau].$$

A plot of $r_a(\tau)$ also appears in Fig. 3.

The usefulness of this approximation is in that each of the terms on the right hand side can be modeled by a separate differential equation [1], the first term by a first-order and second and third terms each by a second-order system. This procedure simply leads us to a fifth-order differential equation model, while the direct realization of this size system by

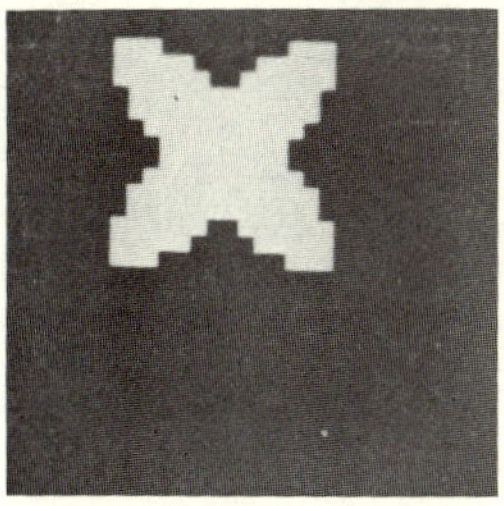

Fig. 1. Image number 1 used in the example.

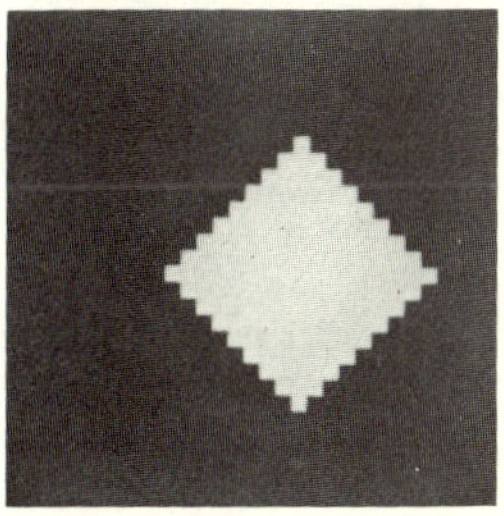

Fig. 2. Image number 2 used in the example.

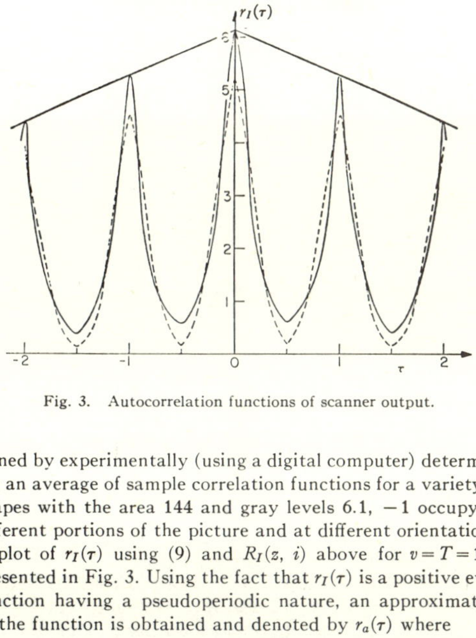

Fig. 3. Autocorrelation functions of scanner output.

[2] In many cases only a slight degradation in performance results if the steady state solution of (26) is utilized in place of $P(k)$. This leads to considerable computational simplification since the estimator reduces to (24) only, with a constant gain $F(k)$.

general methods of [3] is very cumbersome. The numerical results, referring to (10)–(13), are given by

$$A_I = \begin{bmatrix} -0.136 & 0 & 0 & 0 & 0 \\ 0 & 0 & 1 & 0 & 0 \\ 0 & -39.4 & -0.27 & 0 & 0 \\ 0 & 0 & 0 & 0 & 1 \\ 0 & 0 & 0 & -157.7 & -0.27 \end{bmatrix}$$

$$B_I = \begin{bmatrix} 0.743 & 0 & 0 \\ 0 & 0.820 & 0 \\ 0 & 4.92 & 0 \\ 0 & 0 & 0.410 \\ 0 & 0 & 5.04 \end{bmatrix}$$

$$C_I = [1 \quad 1 \quad 0 \quad 1 \quad 0]$$

$$Q_I = \begin{bmatrix} 1 & 0 & 0 \\ 0 & 1 & 0 \\ 0 & 0 & 1 \end{bmatrix}.$$

The corresponding parameter values for the discrete system, (17)–(20), are given by

$$A = \begin{bmatrix} 0.996 & 0 & 0 & 0 & 0 \\ 0 & 0.983 & 0.031 & 0 & 0 \\ 0 & -1.22 & 0.97 & 0 & 0 \\ 0 & 0 & 0 & 0.926 & 0.03 \\ 0 & 0 & 0 & -4.77 & 0.913 \end{bmatrix}$$

$$BKB' = \begin{bmatrix} 0.02 & 0 & 0 & 0 & 0 \\ 0 & 0.02 & 0.12 & 0 & 0 \\ 0 & 0.12 & 0.60 & 0 & 0 \\ 0 & 0 & 0 & 0.01 & 0.07 \\ 0 & 0 & 0 & 0.07 & 0.49 \end{bmatrix}$$

$$C = C_I$$

$$K = Q_I$$

and the noise variance σ^2 is chosen as 9. The signal-to-noise ratio is defined by ρ, where

$$\rho \triangleq \frac{\text{peak-to-peak variation of signal}}{\sigma}.$$

Two values of ρ, namely 7.1/3 and 7.1/10, were chosen. The signal $s(k)$ was generated by using the image as described above and adding white noise with variance σ^2. The estimator equations of (24)–(26) were implemented with results given in Figs. 4 and 5 for the two images. Figs. 4(a) and 5(a) represent the scanner output, i.e., the unprocessed image. Figs. 4(b) and 5(b) represent the estimated values. The interpolated values are represented in Figs. 4(c) and 5(c) where for simplicity $\hat{P}(k)$ and $\tilde{P}(k)$ in (27) were assumed equal. This is suspected to yield slightly inferior results only at the top and bottom few lines since corresponding estimators have not yet reached a steady-state operation indicated by $\hat{P}(k) = \tilde{P}(k) =$ constant. Finally, the average of the operation on the picture and its transpose is given in Figs. 4(d) and 5(d), where the latter represents a 14-dB improvement.

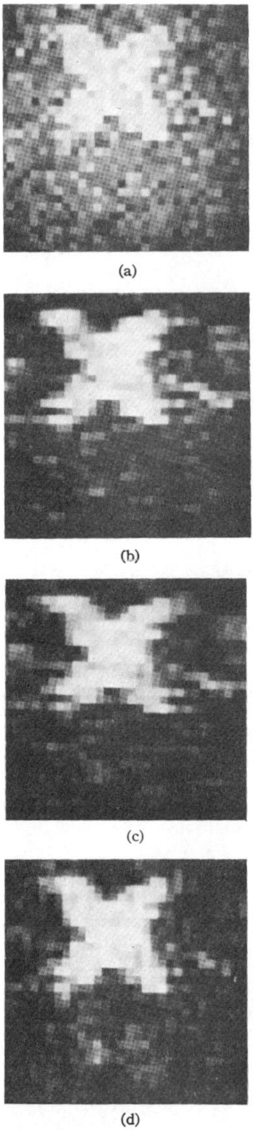

Fig. 4. Picture number 1 and its estimates.

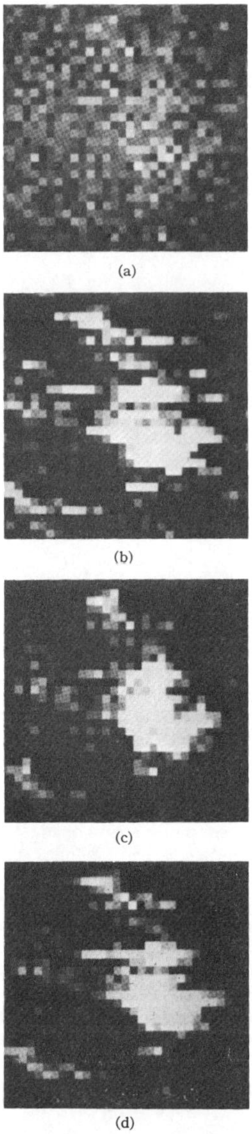

Fig. 5. Picture number 2 and its estimates.

V. A Nonstationary Enhancement Process

The random process representing the scanner output is a nonstationary process having the autocorrelation function $\varphi_I(t, t+\tau)$ given by (6). Determination of a dynamic model representing this correlation function is not, at present, feasible, since it also does not have the basic "separability" property [3]. In the preceding section this autocorrelation was approximated by randomizing the variable σ, a time variable representing the position of the scanner on a line, over the interval $[0, T]$. This resulted in a stationary correlation function.

In order to achieve a better approximation to the original nonstationary correlation function, the interval $[0, T]$ is subdivided into M mutually exclusive intervals, $T_1, \cdots, T_M$ where

$$\sum_{m=1}^{m} T_m = T.$$

The variable σ in (6) is now randomized over each of the subintervals, again using a uniform density function. Consequently a stationary correlation function will be obtained for each subinterval. Denoting this by $r_I(\tau, m)$, (9) is now replaced by

$$r_I(\tau, m) = \Gamma_m R_I[v\gamma, j] + (1-\Gamma_m) R_I[v(\gamma - T), j+1] \quad (28)$$
$$\tau = iT + \gamma$$

where Γ_m is defined as follows:

$$\Gamma_m = \begin{cases} 1, & \gamma < \sum_{m+1}^{m} T_i \\ \dfrac{\sum_m^m T_i - \gamma}{T_m}, & \sum_m^m T_i \le \gamma \le \sum_{m+1}^m T_i \\ 0, & \gamma > \sum_m^m T_i. \end{cases}$$

Each autocorrelation $r_I(\tau, m)$ is now modeled, as in the previous section, resulting in a discrete model described by (17)–(20). However, the model parameters are functions of index m. Similarly, the parameters in the estimator equations (24)–(26), i.e., the parameters A, BQB', are replaced by $A(m)$, $B(m)Q(m)B'(m)$, where the estimator uses the mth value of parameters when the observation from the mth segment is processed. Consequently, the estimator is nonstationary (time varying). It may be noted that the only additional complexity in the implementation of the time varying estimator is the simple task of changing parameter values of the recursive operation NM times during the image enhancement process.

VI. Consideration of Certain Nonlinearities

In the preceding analysis only linear systems have been considered. Although there are a variety of causes of nonlinearities, an important source is the input–output relation of the sensor producing the scanner output. This is due to the inability to produce a signal exactly in proportion to the gray level, over a wide range. Sensor saturation may be cited as one cause of this nonlinearity. This type of nonlinearity, if significant, can be easily handled by recursive estimators similar to those described above.

Let us represent the operation of the scanner by

$$s(k) = g[Cx(k)] + v(k) \quad (29)$$

where g is a nonlinear function of its argument and (29) reduces to (18) when g is linear. A recursive estimator for $y_I(k) = Cx(k)$ given (17), (19), (20), and (29) is derived in [1]. The results are given in the following for the one-step predictor:

$$\hat{x}(k+1) = A\hat{x}(k) + F(k)\{s(k) - g[C\hat{x}(k)]\}$$

where $F(k)$ is obtained from a modified form of (25), (26) by replacing for the parameters C' and σ^2 the quantities

$$\left.\frac{\partial g[Cx(k)]}{\partial x(k)}\right|_{x(k)\to\hat{x}(k)}$$

and

$$\frac{1}{\sigma^2}\left[\frac{\partial g[Cx(k)]}{\partial x(k)}\right]'_{x(k)\to\hat{x}(k)} \left[\frac{\partial g[Cx(k)]}{\partial x(k)}\right]_{x(k)\to\hat{x}(k)} \quad (30)$$

respectively.

VII. Vector Scanning

Vector scanning, i.e., scanning more than one picture line at a time, may be utilized as a means of reducing the total required time for scanning. Once the picture has been scanned, however, the computer may be programmed to read the data as a scalar scanner. On the other hand, whether physically a scalar or a vector scanner has been utilized, the data may be read as in a vector scanning. This latter case will be considered in this section and its implication to the model of the scanner output will be discussed. Specifically, utilizing an example, it will be shown that, from the point of view of recursive image enhancement, a vector scanning may result in a lower order (hence computationally simpler) estimator than the scalar scanning, both achieving the same degree of approximation to the given statistical information.

The procedure for design of an estimator for vector scanning is very similar to that of scalar scanning. Let $s(t)$ be the scanner output of an M-dimensional scanner, hence an M vector. The correlation function $r_I(\tau)$, (9), now represents an $M\times M$ matrix with its l, mth element given by

$$\frac{T-\gamma}{T} R_I[v\gamma, jM + m - l] + \frac{\gamma}{T} R_I[v(\gamma - T), (j+1)M + m - l] \quad (31)$$
$$\tau = jT + \gamma.$$

The problem now is the realization of the matrix correlation function by a system of differential equations such as (10)–(13). Procedures for this are given in [3], [6]. It is clear that to the extent that we realize the scalar correlation function (9) or the matrix correlation function (31) *exactly*, the resultant estimation performances will be identical [6]. This is because the statistical information in the two representations are identical. However, as shown in the example of Section IV, in practice it is necessary to realize an approximation to the correlation function to keep the order of the estimator acceptably low. It is in this regard that the vector scanning may result in a lower order estimator. In part, it can be shown that, from the point of view of computational complexity, there exists an optimum size for the scanning vector depending on the expected size of the image [6]. Without a detailed treatment, these points are illustrated by means of an example.

Example

Let us consider an image with two-dimensional correlation function

$$R(z, n) = e^{-16|z| - 0.5|n|}.$$

A typical sample of this random process on a 32×32 grid is an object with horizontal and vertical dimension of 4 grid points. Let $T = v = 1$. Considering scalar scanning first, the correlation function (9) yields

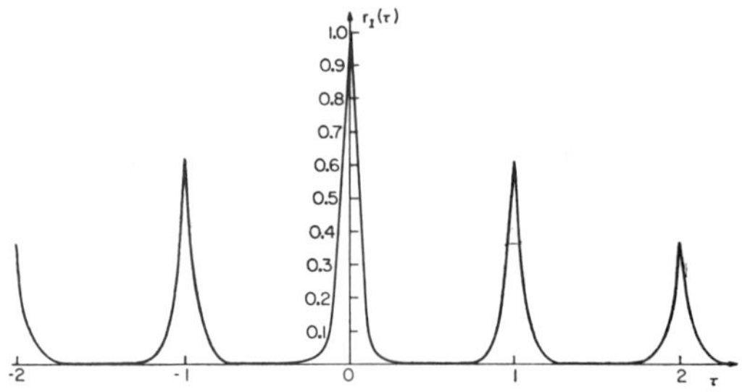

Fig. 6. Autocorrelation function of scalar scanner output.

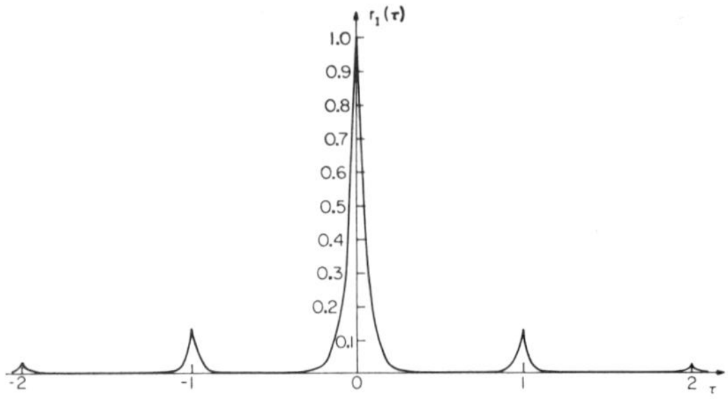

Fig. 7. Autocorrelation function of vector scanner output.

$$r_I(\tau) = (1-\gamma)e^{-16|\gamma|-0.5|i|} + \gamma e^{-16|1-\gamma|-0.5|i+1|}$$

$$\tau = i + \gamma.$$

This correlation function is approximated by

$$r_a(\tau) = a_0 e^{-0.5|\tau|} + a_1 e^{-0.5|\tau|}\cos 2\pi\tau + a_2 e^{-0.5|\tau|}\cos 4\pi\tau$$

which leads to a fifth-order estimator (see example, Section IV). The coefficient $a_0 = 0.125$, $a_1 = 0.22$, and $a_2 = 0.163$ will provide a least-squares fit to $r_I(\tau)$. A plot of $r_I(\tau)$ appears in Fig. 6.

Next, let us consider a four-dimensional vector scanning. The l, mth element of the correlation matrix (31) is now given by

$$r_I(\tau, l, m) = (1-\gamma)e^{-16|\gamma|-0.5|4i+m-l|} + \gamma e^{-16|1-\gamma|-0.5[4(i+1)+m-l]}$$

$$\tau = i + \gamma.$$

This autocorrelation function, for any l, m, is approximated by one exponential function. The term $r_I(\tau, 1, 1)$ is then approximately

$$r_a(\tau, 1, 1) = e^{-16|\tau|}.$$

Consequently the associated estimator with be a fourth-order equation (in contrast with fifth-order for the scalar case). A plot of $r_I(\tau, 1, 1)$ appears in Fig. 7. The case $l = m = 1$ is chosen for illustration since it represents the worst effect of the approximation. Comparison of Figs. 6 and 7 reveals that it is possible to achieve a much better approximation to the original data in case of a vector scanning, while the resultant estimator is also of a lower order. In other words, to achieve the same quality of approximation many terms in $r_a(\tau)$ are needed, in contrast with $r_I(\tau, 1, 1)$, which can be well approximated by one term. Additional advantages of the vector model such as elimination of the required averaging operation leading to (31) are discussed in [6].

VIII. Conclusions

The role of recursive linear (Kalman) and nonlinear (extended Kalman) filtering in image enhancement process has been established. The procedure is applicable to those images characterized statistically by mean and correlation function. It was shown that both scalar and vector scanning can be considered with possible computational reduction for the latter. This approach to image enhancement enables us to apply the extensive results obtained during the past decade in the area of recursive estimation to the problem of image enhancement. To cite a few examples, the problem of lack of precise knowledge of the image and/or noise correlation functions, $R_I(z, i)$ and $R_n(z, i)$, can be handled by means of adaptive estimation [7], [8]. If the additional complexity of adaptive estimation is not desirable, a bounding filter [9] can be designed to cope with the problem of lack of precise *a priori* data, whereby an estimate is obtained with a known minimum upper bound for the estimation error variance. The problem of nonuniform noise over the picture can be handled by simply allowing the term σ^2 in the estimator equations (24)–(26) to vary with k. The problem of missing data [1] (i.e., from not having observed the gray level associated with certain grid points, say during the transmission–reception of an image) can be trivially handled by setting the corresponding $F(k)$ in (24), (26) equal to zero.

Finally, the recursive estimation approach is very attractive in light of computational considerations. This is specially true if steady-state filter gains are utilized [$F(k)$ = constant in (24)]. The recursive nature of the operation and the fact that the operation is directly on the scanner output (i.e., in time domain) make the approach well suited to computer implementation. Furthermore, the order of the estimator (which is a direct indication of computational complexity) is invariant with the fineness of the grid. For example, for the problem in Section IV, the same fifth-order estimator equation would have been sufficient if in place of 32 scan lines one had to deal with 10 000 lines.

Acknowledgment

The author wishes to thank Professor W. K. Pratt of the Department of Electrical Engineering, University of Southern California, Los Angeles, for valuable discussions of this problem.

References

[1] N. E. Nahi, *Estimation Theory and Application*. New York: Wiley, 1969.
[2] L. E. Franks, "A model for random video process," *Bell Syst. Tech. J.*, vol. 45, pp. 609–630, Apr. 1966.
[3] H. L. VanTrees, *Detection, Estimation and Modulation Theory*. New York: Wiley, 1971, Pt. II.
[4] A. Papoulis, *Probability, Random Variables and Stochastic Processes*. New York: McGraw-Hill, 1965.
[5] N. E. Nahi and T. Assefi, "Bayesian recursive image estimation," *Proc. Two Dimensional Digital Image Processing* (Univ. Missouri, Columbia, Oct. 1971).
[6] N. E. Nahi and C. Franco, "Recursive image enhancement: Vector scanning," to be published.
[7] N. E. Nahi and B. M. Schaefer, "Decision-directed adaptive recursive estimators: Divergence prevention," *IEEE Trans. Automat. Contr.*, vol. AC-17, pp. 61–68, Feb. 1972.
[8] I. M. Weiss, "A survey of discrete Kalman–Bucy filtering with unknown noise covariances," presented at AIAA Guidance and Control Conf., Santa Barbara, Calif., 1970.
[9] N. E. Nahi and I. M. Weiss, "Bounding filters in the presence of inexactly known parameters," presented at IEEE Decision and Control Conf., Miami, Fla., Dec. 1971.

25

Reprinted from *IEEE Trans. Autom. Control* AC-19(1):8–12 (1974)

Modeling of Two-Dimensional Covariance Functions with Application to Image Restoration

STEVEN R. POWELL, MEMBER, IEEE, AND LEONARD M. SILVERMAN, MEMBER, IEEE

***Abstract*—Exact models are derived for the horizontal line scan of monochromatic pictorial images whose light intensity at each point can be represented by a class of two-dimensional homogeneous random fields. Recursive restoration schemes are presented for images of this type whose degradation can be represented by additive white noise.**

I. Introduction

A GREAT deal of recent work in the processing of two-dimensional monochromatic images has focused on recursive digital techniques for restoration and enhancement. Recursive techniques are appealing since they can be used in on-line processing and require less computer memory when processing off-line. In this paper we shall describe how modeling and estimation procedures familiar to the control engineer can be utilized to recursively restore a large class of images.

It has been known for some time that the light intensity at each point of an ensemble of uncorrupted images can be represented by a two-dimensional homogeneous random field. For a large class of images the covariance function of this field can be approximated by a factorable decaying exponential function of its arguments [16]–[19] and the image degradation can be represented by an uncorrelated additive white-noise field [1],[24]. A Kalman–Bucy filtering approach which has had successful results for degraded images that can be represented in this way is due to Nahi and Assefi [1].

By scanning the uncorrupted image one line at a time along N equispaced horizontal lines, the image field can be transformed into a scalar-valued stochastic process $s(t)$. The degraded image and noise fields can be similarly transformed into scalar-valued stochastic processes, $z(t)$ and $s(t)$, respectively, so that $z(t) = s(t) + n(t)$. Kalman–Bucy filtering theory cannot be directly applied to find the linear least-squares (lls) filtered estimate of $s(t)$, however, since $s(t)$ is a nonstationary process with a nonseparable covariance function and will therefore not admit a finite-dimensional linear Markov realization (MR) [3]. This problem is resolved by approximating $s(t)$ by a wide-sense stationary process $s^*(t)$. An MR of $s^*(t)$ then follows easily by factorization of the spectral density function. Improved results have recently been obtained with a better approximation to $s(t)$ [25].

Under the same assumptions Nahi and Franco [4] have found that somewhat better image restoration can be achieved by partitioning the image into N/M horizontal strips of equal size and then scanning each strip with one sweep of its M equispaced horizontal lines. The process $\mathbf{s}_i(t)$ which results from scanning the ith strip, $i = 1,2,\cdots, N/M$, is then an M-dimensional wide-sense stationary vector process whose MR can also be found by spectral factorization. The Kalman–Bucy filter derived from this model of $\mathbf{s}_i(t)$ can then be used to obtain the lls filtered estimate of $\mathbf{s}_i(t)$.

In the remainder of this chapter we shall use the scanning approach to model a large class of two-dimensional random fields. After some preliminary definitions and a review of modeling theory we shall exhibit an *exact* realization of the scalar scan process $s(t)$. This realization will not be Markov since continuous changing of the initial conditions and delays in the output will be required. Nevertheless, because only one linear time-invariant system is needed for state generation, the model is of relatively simple form. We shall then show how an MR of the vector scan process $\mathbf{s}_i(t)$ can be derived from the realization of $s(t)$ and how Kalman–Bucy and recent whitening filter techniques can be utilized to achieve lls estimation of $s(t)$. Finally, various techniques of suboptimal estimation will be discussed.

II. Relation of the Scan Process to the Random Field

Let $f(x,y)$ be a two-dimensional scalar-valued random field defined on $[0,T] \times [0,N-1]$ where $[0,T]$ is a continuous interval and $[0,N-1]$ represents the set of nonnegative integers from 0 to $N-1$. Suppose $f(x,y)$ is homogeneous with zero mean and factorable covariance function given by

$$k(x_1 - x_2, y_1 - y_2) = \alpha(x_1 - x_2)\beta(y_1 - y_2) \quad (1)$$

where $\alpha(x_1 - x_2)$ and $\beta(y_1 - y_2)$ are continuous on $[0,T]$ and $[0,N-1]$, respectively. Since $k(x_1 - x_2, y_1 - y_2)$ is a covariance function on $[0,T] \times [0,N-1]$, $\alpha(x_1 - x_2)$ and $\beta(y_1 - y_2)$ are also covariance functions on $[0,T]$ and $[0,N-1]$, respectively.

In the image restoration literature [1] $f(x,y)$ has been called the brightness field of an ensemble of images and represents the light intensity at the point (x,y) of that ensemble. It will become obvious that factorability of the

Manuscript received June 12, 1972. Paper recommended by P. Frost, Past Chairman of the IEEE S-CS Estimation and Identification Committee. This work was supported in part by U.S. Army Research Office Grant DA-ARO-D-31-124-71-637 and in part by National Science Foundation Grant GK-33416.

S. R. Powell was with the Department of Electrical Engineering–Systems, University of Southern California, Los Angeles, Calif. He is now with Bell Laboratories, Murray Hill, N. J. 07974.

L. M. Silverman is with the Department of Electrical Engineering–Systems, University of Southern California, Los Angeles, Calif.

covariance function is necessary for our derivation. In the context of image restoration this restriction is not unreasonable since a large body of experimental evidence indicates that a variety of pictorial data can be adequately modeled in this way [16]–[19]. Otherwise, (1) can be considered as a first approximation to the actual covariance function.

Let the process $s(t)$ be defined on $[0,(N-1)T]$ by

$$s(t) = \begin{cases} f(t,0), & 0 \le t \le T \\ f(t-kT,k), & kT < t \le (k+1)T, \\ & k = 1,2,\cdots,N-1. \end{cases} \quad (2)$$

$s(t)$ defined in this manner has been referred to as the output of the scalar scanner [4] and is a scalar-valued cyclostationary process with zero mean and covariance function

$$R_s(t,\tau) = \begin{cases} \alpha(|t-\tau|)\beta(0), & i=j \\ \alpha(|t-\tau-(i-j)T|)\beta(i-j), & i>j \\ \alpha(|\tau-t-(j-i)T|)\beta(j-i), & i<j, \end{cases} \quad (3)$$

for $iT < t \le (i+1)T$ and $jT < \tau \le (j+1)T$.[1]

III. Modeling of Wide-Sense Stationary Processes

Let $\omega(x)$, defined on the continuous interval $(-\infty,\infty)$, be a white noise-free zero-mean scalar-valued wide-sense stationary stochastic process with covariance function $r(x_1 - x_2)$. Then it is well-known that a time-invariant Markov realization (MR) for $\omega(x)$ on $(-\infty,\infty)$ (i.e., a finite-dimensional linear dynamical time-invariant system defined on $(-\infty,\infty)$ which generates a zero-mean output process $y(x)$ with covariance function $r(x_1 - x_2)$ when driven by white noise) will exist only if the Laplace transform $\Phi(s)$ of $r(x_1 - x_2)$ is proper rational. The transfer function $h(s)$ of that MR will satisfy

$$\Phi(s) = h(s)h(-s) \quad (4)$$

and will have its poles in the left half-plane. Alternatively, an MR for $\omega(x)$ on $(-\infty,\infty)$ will exist only if there exist an integer n, a constant $n \times n$ matrix A, and constant n-vectors b and c such that

$$r(x_1 - x_2) = c'e^{A|x_1-x_2|}b. \quad (5)$$

Assumption 1:

$$\alpha(x_1 - x_2) = c_\alpha' e^{A_\alpha|x_1-x_2|}b_\alpha \quad (6)$$

where A_α is asymptotically stable and, without loss of generality, has minimal dimension n.

Lemma 1: Let $\omega_\alpha(x)$ be a zero-mean scalar process with covariance function $\alpha(x_1 - x_2)$. Suppose $\omega_\alpha(x)$ satisfied Assumption 1. Then a time-invariant MR of $\omega_\alpha(x)$ on $[0,T]$ is given by

$$\begin{aligned} \dot{z} &= Fx + gu, & Eu(x) &= 0, Eu(x_1)u(x_2) = \delta(x_1 - x_2), \\ y &= h'x, & Ez(0) &= 0,\ Ez(0)z'(0) = P_0, \end{aligned} \quad (7)$$

$$Ez(0)u(x) = 0, \qquad x \in [0,T],$$

where (F, g, h') is a state-variable realization of $h(s)$ obtained by spectrally factoring $\Phi(s)$ as in (4) and P_0 is the unique nonnegative definite symmetric solution of

$$FP_0 + P_0F' + gg' = 0. \quad (8)$$

Proof: The proof is given in [8].

Remark 1: A time-invariant MR can also be obtained by specializing the recursive procedures given in [15] for realizing nonstationary processes with covariance functions of the form

$$r(x_1,x_2) = a'(x_1)b(x_2). \quad (9)$$

Kailath and Geesey [26] have shown that a causally invertible MR of $\omega(x)$ will necessarily be time-varying and will be in some sense unique. It has also been shown [27] that of all Markov realizations of $\omega(x)$ the causally invertible MR will have minimum state covariance.

Now let $\omega(k)$, defined on the set of integers $(-\infty,\infty)$, be a white noise-free zero-mean scalar process with covariance function $r(i-j)$. Then an MR of $\omega(k)$ on $(-\infty,\infty)$ will exist only if $r(i-j)$ has a proper rational z-transform $\Phi(z)$. The transfer function $h(z)$ of a time-invariant MR then satisfies

$$\Phi(z) = h(z)h(-z) \quad (10)$$

and has its poles within the unit circle [9]. Alternatively, a MR for $\omega(k)$ on $(-\infty,\infty)$ will exist only if there exist an integer n, a constant $n \times n$ matrix A, and constant n-vectors b and c such that

$$r(i-j) = c'e^{A|i-j|}b. \quad (11)$$

Assumption 2:

$$\beta(i-j) = c_\beta' e^{A_\beta|i-j|}b_\beta, \quad (12)$$

where, without loss of generality, A_β is of minimal dimension.

Lemma 2: Let $\omega_\beta(k)$, defined on $[0,N-1]$, be a zero-mean scalar process whose covariance function satisfies Assumption 2. Then the impulse response function $h_\beta(m,n)$ of a zero initial state MR of $\omega_\beta(k)$ on $[0,N-1]$ satisfies

$$\boldsymbol{B} = \boldsymbol{H}_\beta \boldsymbol{H}_\beta', \quad (13)$$

where

$$\boldsymbol{B} = \begin{bmatrix} \beta(0) & \beta(1) \cdots \beta(N-1) \\ \beta(1) & \beta(0) \cdots \beta(N-2) \\ \vdots & \vdots \quad \ddots \quad \vdots \\ \beta(N-1) & \beta(N-2) \cdots \beta(0) \end{bmatrix} \quad (14)$$

and

$$\begin{bmatrix} h_\beta(0,0) & & O \\ h_\beta(1,0) & h_\beta(1,1) & \\ \vdots & \vdots & \ddots \\ h_\beta(N-1,0) & h_\beta(N-1,1) \cdots & h_\beta(N-1,N-1) \end{bmatrix}. \quad (15)$$

[1] Here, and in the sequel, the interval is closed if the left endpoint is zero.

Proof: It is well known [10] that $h_\beta(m,n)$ satisfies

$$\beta(i-j) = \sum_{k=0}^{N-1} h_\beta(i,k)h_\beta(j,k) \tag{16}$$

with $h_\beta(m,n) = 0$, $m < n$. The result follows by direct calculation.

IV. Model of the Scan Process

Using Lemmas 1 and 2 we can exhibit an exact model of the scan process $s(t)$.

Theorem 1: Let $\omega_\alpha(x)$ and $\omega_\beta(k)$ be the zero-mean wide-sense stationary processes defined in Lemmas 1 and 2, respectively. Let (F,g,h',P_0) and $h_\beta(m,n)$ be the MR of $\omega_\alpha(x)$ on $[0,T]$ defined in (7) and the impulse response of the zero-initial state MR of $\omega_\beta(k)$ on $[0,N-1]$ defined in (13), respectively. Then the output process on the continuous interval $[0,NT]$ of the following system has zero mean and covariance function given by (3):

$$\dot{x}(t) = Fx(t) + gu(t),\ x(iT) = \gamma_i, \qquad i = 0,1,\cdots,N-1$$

$$y(t) = \sum_{k=0}^{i} h_\beta(i,k)h'x(t-(i-k)T), \qquad iT < t \leq (i+1)T \tag{17}$$

where

$$Eu(t) = 0,\ Eu(t)u(\tau) = \delta(t-\tau),\ E\gamma_i u(t) = 0 \forall i, \forall t,$$

and

$$E\gamma_i\gamma_j' = \begin{cases} 0, & i \neq j \\ P_0, & i = j. \end{cases} \tag{18}$$

Proof: The proof relies on the fact that

$$h'Ex(t_1)x'(t_2)h = \begin{cases} c_\alpha' e^{A\alpha|t_1-t_2|}b_\alpha, & iT < t_1,t_2 \leq (i+1)T, \\ & i = 0,1,\cdots,N-1, \\ 0, & \text{elsewhere} \end{cases}$$

since the initial states and the input processes on different intervals are uncorrelated (i.e., $E\gamma_m\gamma_n' = 0$ and $Eu(t-mT)u(t-nT) = 0$ for $m \neq n$). Only the kth term, $k = 1,2,\cdots,i\wedge j$, of each output combination is thus correlated. The covariance function of the output process then becomes

$$Ey(t)y(\tau) = \sum_{k=0}^{i\wedge j} h_\beta(i,k)h'E[x(t-(i-k)T)x'(\tau-(j-k)T)]hh_\beta(j,k)$$

$$= \begin{cases} c_\alpha' e^{A\alpha|t-\tau|}b_\alpha\beta(0), & i = j \\ c_\alpha' e^{A\alpha|t-\tau-(i-j)T|}b_\alpha\beta(i-j), & i > j \\ c_\alpha' e^{A\alpha|\tau-t-(j-i)T|}b_\alpha\beta(i-j), & i < j, \end{cases}$$

where the last equation follows from Lemmas 1 and 2.

Remark 2: It is clear that (17) is not an MR of $s(t)$ on $[0,NT]$ since delays and uncorrelated initial conditions are required. It should also be noted that Assumption 2 is not necessary to derive (17). The same derivation follows for any set of constants $h_\beta(m,n)$, $0 \leq m, n \leq N-1$, satisfying (13). However, $h_\beta(m,n)$ can be identified as the impulse response function of a zero-initial state MR of $\omega_\beta(k)$ on $[0,N-1]$ only if Assumption 2 is in force.

Let the M-dimensional, $M \leq N$, vector process

$$\mathbf{s}'(t) = [f(t,0) \,\vdots\, f(t,1) \,\vdots\, \cdots \,\vdots\, f(t,M-1)] \tag{19}$$

be defined for $0 \leq t \leq T$. In image restoration $\mathbf{s}(t)$ will represent the light intensity of an ensemble of images at points with horizontal coordinate t located on the first M lines [4] and has been called the output of the vector scanner. The vector scan process $\mathbf{s}(t)$ has the property that it will admit a time-invariant MR on $[0,T]$ of order nM. This MR is found directly from Theorem 1 by state augmentation.

Theorem 2: Let $\mathbf{s}(t)$ be the zero-mean wide-sense stationary process defined in (19). Let (F,g,h',P_0) and $\mathbf{H}_\beta$ be defined in (7) and (13), respectively. Then the following system is an MR of $\mathbf{s}(t)$ on $[0,T]$:

$$\begin{aligned} \dot{\mathbf{x}} &= \mathbf{F}\mathbf{x} + \mathbf{G}\mathbf{u}, & E\mathbf{x}(0) &= 0,\ E\mathbf{x}(0)\mathbf{x}'(0) = \mathbf{P}_0, \\ \mathbf{y} &= \mathbf{H}'\mathbf{x}, & E\mathbf{u}(t) &= 0,\ E\mathbf{u}(t)\mathbf{u}'(\tau) = I_{nM}\delta(t-\tau), \\ & & E\mathbf{x}(0)\mathbf{u}'(t) &= 0 \forall t \geq 0, \end{aligned} \tag{20}$$

where

$$\mathbf{F} = \begin{bmatrix} F & & O \\ & \ddots & \\ O & & F \end{bmatrix} \qquad \mathbf{G} = \begin{bmatrix} g & & O \\ & \ddots & \\ O & & g \end{bmatrix}$$

$$\mathbf{H}' = \mathbf{H}_\beta \begin{bmatrix} h' & & O \\ & \ddots & \\ O & & h' \end{bmatrix}, \text{ and } \mathbf{P}_0 = \begin{bmatrix} P_0 & & O \\ & \ddots & \\ O & & P_0 \end{bmatrix}.$$

Proof: The proof follows directly from Theorem 1 by letting $\mathbf{x}'(t) = [x'(t-(M-1)T) \,\vdots\, x'(t-(M-2)T) \,\vdots\, \cdots \,\vdots\, x'(t)]$, $\mathbf{u}'(t) = [u(t-(M-1)T) \,\vdots\, u(t-(M-2)T) \,\vdots\, \cdots \,\vdots\, u(t)]$, and $\mathbf{y}'(t) = [y(t-(M-1)T) \,\vdots\, y(t-(M-2)T) \,\vdots\, \cdots \,\vdots\, y(t)]$, $(M-1)T \leq t \leq MT$. Since (20) is time-invariant the interval of definition $0 \leq t \leq T$ can be chosen.

Remark 4: In [4] a zero-mean random field $f(x,y)$ with covariance function

$$k(x_1-x_2, i-j) = e^{-\mu_h|x_1-x_2|}\beta(|i-j|) \tag{21}$$

is considered. An MR on $[0,T]$ for this random field can be given by (20) where

$$F = -\mu_h,\ g = \sqrt{2\mu_h},\ h = 1, \text{ and } P_0 = 1. \tag{22}$$

V. Linear Least-Squares Estimation of $f(x,y)$ in Additive Noise

In this section we shall consider recursive linear least-squares (lls) schemes for estimating $f(x,y)$ in the presence of additive noise. We shall assume that the zero-mean homogeneous random field

$$z(x,y) = f(x,y) + n(x,y) \tag{23}$$

where

$$En(x,y) = 0,\ En(x_1,y_1)n(x_2,y_2) = r\delta(x_1-x_2, y_1-y_2), \quad r > 0$$

and

$$Ef(x_1,y_1)n(x_2,y_2) = 0 \forall x_1,x_2,y_1,y_2$$

is defined on $[0,T] \times [0,N-1]$. Let $\boldsymbol{n}(t)$ be the M-dimensional vector process defined by

$$\boldsymbol{n}'(t) = [n(t,0) \vdots n(t,1) \vdots \cdots \vdots n(t,M-1)]. \quad (24)$$

Then $\boldsymbol{n}(t)$ is a zero-mean white noise process with covariance function $R_n(t-\tau)$ given by

$$R_n(t-\tau) = \boldsymbol{R}\delta(t-\tau) \quad (25)$$

where $\boldsymbol{R} = rI_M$ is a positive definite symmetric matrix. Furthermore,

$$Es(t)n'(\tau) = 0, \qquad 0 \le t, \tau \le T. \quad (26)$$

Define the M-dimensional vector process $\boldsymbol{z}(t)$ by

$$\boldsymbol{z}(t) = \boldsymbol{s}(t) + \boldsymbol{n}(t). \quad (27)$$

Then $\hat{\boldsymbol{s}}(t|t)$, the lls estimate of $\boldsymbol{s}(t)$ given $\{\boldsymbol{z}(\sigma), 0 \le \sigma \le t\}$, can be found recursively as the output of a Kalman–Bucy filter [2] and its kth element, $k = 0,1,\cdots,M-1$, will be the lls estimate of $f(t,k)$ given $\{z(\xi,\eta), 0 \le \xi \le t, 0 \le \eta \le M-1\}$. In this case, $\hat{\boldsymbol{s}}(t|t)$, $0 \le t \le T$, is given by

$$\begin{aligned}\dot{\hat{\boldsymbol{x}}}(t|t) &= [\boldsymbol{F} - \pi(t)\boldsymbol{H}\boldsymbol{R}^{-1}\boldsymbol{H}']\hat{\boldsymbol{x}}(t|t) \\ &\quad + \pi(t)\boldsymbol{H}\boldsymbol{R}^{-1}\boldsymbol{z}(t), \hat{\boldsymbol{x}}(0|0) = 0 \\ \hat{\boldsymbol{s}}(t|t) &= \boldsymbol{H}'\hat{\boldsymbol{x}}(t|t), \end{aligned} \quad (28)$$

where $\hat{\boldsymbol{x}}(t|t)$ is the lls filtered estimate of the state of (20) and

$$\pi(t) = E[(\boldsymbol{x}(t) - \hat{\boldsymbol{x}}(t|t))(\boldsymbol{x}(t) - \hat{\boldsymbol{x}}(t|t))'] \quad (29)$$

satisfies the Riccati equation

$$\begin{aligned}\dot{\pi}(t) &= \boldsymbol{F}\pi(t) + \pi(t)\boldsymbol{F}' - \pi(t)\boldsymbol{H}\boldsymbol{R}^{-1}\boldsymbol{H}'\pi(t) \\ &\quad + \boldsymbol{G}\boldsymbol{G}', \pi(0) = \boldsymbol{P}_0. \end{aligned} \quad (30)$$

In many applications $\{z(\xi,\eta), 0 \le \xi \le T, 0 \le \eta \le M-1\}$ is given and a recursive lls estimate of $f(t,k)$, $k = 0,1,\cdots,M-1$ is desired. This smoothed estimate is given by the kth element of $\hat{\boldsymbol{s}}(t|T)$, the lls estimate of $\boldsymbol{s}(t)$ given $\{\boldsymbol{z}(\sigma), 0 \le \sigma \le T\}$. $\hat{\boldsymbol{s}}(t|T)$ can be found as the output of the recursive fixed interval smoother [13]

$$\begin{aligned}\boldsymbol{s}(t|T) &= \boldsymbol{H}'[\pi^{-1}(t) + \pi_B^{-1}(T-t)]^{-1}[\pi^{-1}(t)\hat{\boldsymbol{x}}(t|t) \\ &\quad + \pi_B^{-1}(T-t)\hat{\boldsymbol{x}}_B(T-t|T-t)] \end{aligned} \quad (31)$$

where the backward time filtered estimate $\hat{\boldsymbol{x}}_B(t|t)$ and error covariance $\pi_B(t)$ satisfy

$$\begin{aligned}\dot{\hat{\boldsymbol{x}}}_B(T-t|T-t) &= [\boldsymbol{F} - \pi_B(t)\boldsymbol{H}\boldsymbol{R}^{-1}\boldsymbol{H}']\hat{\boldsymbol{x}}_B(t|t) \\ &\quad - \pi_B(t)\boldsymbol{H}\boldsymbol{R}^{-1}\boldsymbol{z}(t), \hat{\boldsymbol{x}}_B(T|T) = 0 \end{aligned} \quad (32)$$

and

$$\begin{aligned}\dot{\pi}_B(T-t) &= -\boldsymbol{F}\pi_B(T-t) - \pi_B(T-t)\boldsymbol{F}' \\ &\quad - \pi_B(T-t)\boldsymbol{H}\boldsymbol{R}^{-1}\boldsymbol{H}'\pi_B(T-t) + \boldsymbol{G}\boldsymbol{G}', \quad \pi_B(T) = \boldsymbol{P}_0. \end{aligned} \quad (33)$$

Remark 5: Nahi and Franco [4] have had considerable success in image restoration by replacing $\pi(t)$ and $\pi_B(t)$ with their steady-state values. In this case (31) becomes

$$\hat{\boldsymbol{s}}(t|T) = \tfrac{1}{2}\boldsymbol{H}'[\hat{\boldsymbol{x}}(t|t) + \hat{\boldsymbol{x}}_B(T-t|T-t)]. \quad (34)$$

An alternate procedure for lls estimation of nonstationary signal processes in additive nonsingular white noise has been described by Kailath and Geesey [20] and Anderson and Moore [21] and involves obtaining a whitening filter for the observation process. This procedure requires the solution to only one Riccati equation, which occurs in obtaining an invertible MR of the observation process. Since the previous approach requires the solution to one Riccati equation (30) in the estimation stage and may require an additional one in modeling the scan process, the whitening filter procedure may, in many cases, prove to be more efficient. Another desirable feature of this procedure is that it can be readily extended to colored and correlated noise problems.

First, we note that the observation process $\boldsymbol{z}(t)$ defined in (27) is an M-dimensional zero-mean vector process with covariance function given by

$$\begin{aligned}Ez(t)z'(\tau) &= \boldsymbol{C}'e^{\boldsymbol{A}t}e^{-\boldsymbol{A}\tau}\boldsymbol{B}1(t-\tau) \\ &\quad + \boldsymbol{B}'e^{-\boldsymbol{A}'t}e^{\boldsymbol{A}'\tau}\boldsymbol{C}1(\tau-t) + \boldsymbol{R}\delta(t-\tau), \end{aligned} \quad (35)$$

where

$$\boldsymbol{A} = \begin{bmatrix} A_\alpha & & \boldsymbol{0} \\ & \ddots & \\ \boldsymbol{0} & & A_\alpha \end{bmatrix}, \quad \boldsymbol{B} = \begin{bmatrix} b_\alpha & & \boldsymbol{0} \\ & \ddots & \\ \boldsymbol{0} & & b_\alpha \end{bmatrix},$$

and

$$\boldsymbol{C}' = \begin{bmatrix} c_\alpha' & & \boldsymbol{0} \\ & \ddots & \\ \boldsymbol{0} & & c_\alpha' \end{bmatrix}.$$

Then [21] $\hat{\boldsymbol{s}}(t|t)$ is given by

$$\hat{\boldsymbol{s}}(t|t) = \boldsymbol{C}'\boldsymbol{x}_w(t), \qquad 0 \le t \le T, \quad (36)$$

where $\boldsymbol{x}_w(t)$, the state of the whitening filter for $\boldsymbol{z}(t)$, is generated by

$$\begin{aligned}\dot{\boldsymbol{x}}_w(t) &= [\boldsymbol{A} - (\boldsymbol{B} - \boldsymbol{P}(t)\boldsymbol{C})\boldsymbol{R}^{-1}\boldsymbol{C}']\boldsymbol{x}_w(t) \\ &\quad + (\boldsymbol{B} - \boldsymbol{P}(t)\boldsymbol{C})\boldsymbol{R}^{-1/2}\boldsymbol{z}(t), \qquad \boldsymbol{x}_w(0) = 0 \\ \boldsymbol{\nu}(t) &= -\boldsymbol{R}^{-1/2}\boldsymbol{C}'\boldsymbol{x}_w(t) + \boldsymbol{R}^{-1/2}(t)\boldsymbol{z}(t), \end{aligned} \quad (37)$$

$\boldsymbol{P}(t)$ satisfies the Riccati equation

$$\begin{aligned}\dot{\boldsymbol{P}}(t) &= \boldsymbol{P}(t)\boldsymbol{A}' + \boldsymbol{A}\boldsymbol{P}(t) + [\boldsymbol{P}(t)\boldsymbol{C} - \boldsymbol{B}]\boldsymbol{R}^{-1}[\boldsymbol{P}(t)\boldsymbol{C} - \boldsymbol{B}]', \\ &\qquad \boldsymbol{P}(0) = 0, \end{aligned} \quad (38)$$

and $\boldsymbol{R}^{1/2}$ is the unique symmetric positive definite square root of $\boldsymbol{R}$. Furthermore, $\hat{\boldsymbol{s}}(t|T)$ is given by

$$\boldsymbol{s}(t|T) = \boldsymbol{C}'\boldsymbol{x}_w(t) + [\boldsymbol{B} - \boldsymbol{C}'\boldsymbol{P}(t)]\boldsymbol{\lambda}(t), \quad (39)$$

where

$$\begin{aligned}\dot{\boldsymbol{\lambda}}(t) &= -[\boldsymbol{A}' - \boldsymbol{C}\boldsymbol{R}^{-1}(\boldsymbol{B}' - \boldsymbol{C}'\boldsymbol{P}(t))]\boldsymbol{\lambda}(t) - \boldsymbol{C}\boldsymbol{R}^{-1}\boldsymbol{\nu}(t), \\ &\qquad \boldsymbol{\lambda}(T) = 0. \end{aligned} \quad (40)$$

Existence of the filter and smoother on $[0,T]$ is guaranteed by minimality of A_α and hence $\boldsymbol{A}$ [8]. Similar whitening filter procedures also exist for lls estimation of signals in additive singular correlated white noise or colored noise that can itself be modeled as the output of a linear dynamical system [22], [23].

VI. Suboptimal Estimation Procedures

Because of their simplicity, the lls estimation procedures of the previous section might seem quite appealing. In many applications, however, storage capacity will place limitations on their digital computer implementation. For example, in the image restoration problem considered by Nahi and Franco, simultaneous estimation of all $M = N = 32$ lines of the image would require a filter of dimension $nN = 32$. Good suboptimal estimation schemes are therefore necessary.

One such scheme, which has had considerable success in image restoration [4], is to partition $[0,T] \times [0,N-1]$ into $[0,T] \times [k,k+M]$, $k = 0,M,2M,\cdots,(N+1)-M$. Then the appropriate estimator described in the previous section can be utilized for recursive lls estimation in each strip $[0,T] \times [k,k+M]$ separately. This approach will work well if $k(x_1 - x_2, y_1 - y_2)$ is relatively small when (x_1,y_1) and (x_2,y_2) are located in different strips.

Another technique is suggested by the model (17) of the scalar scan $s(t)$. An estimate $\hat{f}(x_1,j_1)$ of $f(x_1,j_1)$ can be obtained as the sum of two independent estimates:

$$\hat{f}(x_1,j_1) = \hat{f}_1(x_1,j_1) + \hat{f}_2(x_1,j_1), \tag{41}$$

where

$$\hat{f}_1(x_1,j_1) = \sum_{i=0}^{j_1-1} \epsilon_i \hat{f}(x_1,i) \tag{42}$$

represents a linear combination of the estimates at points with $x = x_1$ on the previous j_1 lines and $\hat{f}_2(x_1,j_1)$ represents a contribution from line j_1 only. A great reduction in computation can then be achieved by the reasonable assumption that only a subset of the j_1 lines "close" to line j_1 will actually affect the estimate $\hat{f}_1(x_1,j_1)$. Such an estimation scheme is currently being investigated.

Conclusions

In this paper we have considered horizontal line scanning methods for pictorial images whose brightness functions form a class of two-dimensional homogeneous random fields. These fields are characterized by covariance functions which are factorable functions of the two orthogonal spatial coordinates. Each of these functions is itself a covariance function which admits a linear time-invariant finite-dimensional Markov realization.

An exact realization of the single line and multiple line scanning processes of such images was exhibited and recursive image restoration schemes based on these scanning models were explored. It was shown that a Kalman–Bucy or whitening filter approach can be utilized to restore a corrupted image provided the dimension of the filter is sufficiently large. If implementation of such a filter on a digital computer is prohibitive, filtering and smoothing schemes suggested by the single line scan model may be acceptable approximations. A comparison of this method with other available methods is currently under investigation.

References

[1] N. E. Nahi and T. Assefi, "Bayesian recursive image estimation," *IEEE Trans. Comput.* (Short Notes), vol. C-21, pp. 734–738, July 1972.

[2] R. E. Kalman and R. S. Bucy, "New results in linear filtering and prediction theory," *Trans. ASME* (J. Basic Eng.), ser. D, vol. 83, pp. 95–107, Dec. 1961.

[3] R. E. Kalman, "Linear stochastic filtering theory—Reappraisal and outlook," in *Proc. Symp. System Theory* (Poly. Inst. of Brooklyn, N. Y.), pp. 197–205, 1965.

[4] N. E. Nahi and C. A. Franco, "Application of Kalman filtering to image enhancement," in *Proc. IEEE Conf. Decision and Control*, New Orleans, La., Dec. 1972, pp. 63–65.

[5] D. C. Youla, "On the factorization of rational matrices," *IRE Trans. Inform. Theory*, vol. IT-7, pp. 172–189, July 1961.

[6] M. C. Davis, "Factoring the spectral matrix," *IEEE Trans. Automat. Contr.*, vol. AC-8, pp. 296–305, Oct. 1963.

[7] M. G. Strintzis, "A solution to the matrix factorization problem," *IEEE Trans. Inform. Theory*, vol. IT-18, pp. 225–232, Mar. 1972.

[8] J. B. Moore and B. D. O. Anderson, "Simulation of stationary stochastic processes," *Proc. Inst. Elec. Eng.* (London), vol. 115, Feb. 1968.

[9] N. E. Nahi, *Estimation Theory and Applications.* New York: Wiley, 1969.

[10] P. Whittle, *Prediction and Regulation by Linear Least-Squares Methods.* London: English Universities Press, 1954, pp. 48–49.

[11] M. R. Gevers, "Structural properties of realizations of discrete time Markov processes," Stanford University, Stanford, Calif., Tech. Rep. 7050-19, ISL, May 1972.

[12] T. Kailath and P. Frost, "An innovations approach to least-squares estimation—Part II: Linear smoothing in additive white noise," *IEEE Trans. Automat. Contr.*, vol. AC-13, pp. 655–660, Dec. 1968.

[13] D. Q. Mayne, "A solution of the smoothing problem for linear dynamic systems," *Automatica*, vol. 4, pp. 73–92, 1966.

[14] D. C. Fraser and J. E. Potter, "The optimum linear smoother as a combination of two optimum linear filters," *IEEE Trans. Automat. Contr.* (Short Papers), vol. AC-14, pp. 387–390, Aug. 1969.

[15] B. D. O. Anderson, J. B. Moore, and S. G. Loo, "Spectral factorization of time-varying covariance functions," *IEEE Trans. Inform. Theory*, vol. IT-15, pp. 550–557, Sept. 1969.

[16] E. R. Kretzmer, "Statistics of television signals," *Bell Syst. Tech. J.*, pp. 751–763, July 1952.

[17] L. E. Franks, "A model for the random video process," *Bell Syst. Tech. J.*, Apr. 1966.

[18] T. S. Huang, "Subjective effect of two-dimensional pictorial noise," *IEEE Trans. Inform. Theory*, vol. IT-11, pp. 43–53, Jan. 1965.

[19] A. Habibi and P. A. Wintz, "Image coding by linear transformation and block quantization," *IEEE Trans. Commun. Technol.*, vol. COM-19, pp. 50–62, Feb. 1971.

[20] T. Kailath and R. Geesey, "An innovations approach to least squares estimation—Part IV: Recursive estimation given lumped covariance functions," *IEEE Trans. Automat. Contr.*, vol. AC-16, pp. 720–727, Dec. 1971.

[21] B. D. O. Anderson and J. B. Moore, "The Kalman–Bucy filter as a true time-varying Wiener filter," *IEEE Trans. Syst., Man, Cybern.*, vol. SMC-1, pp. 119–128, Apr. 1971.

[22] T. Kailath and R. Geesey, "An innovations approach to least-squares estimation—Part V: Innovations representations and recursive estimation in colored noise," *IEEE Trans. Automat. Contr.*, vol. AC-18, pp. 435–453, Oct. 1973.

[23] S. R. Powell, "Modeling of nonstationary stochastic processes with application to image restoration," Ph.D. dissertation, Univ. Southern California, Los Angeles, Sept. 1973.

[24] C. W. Helstrom, "Image restoration by the method of least squares," *J. Opt. Soc. Amer.*, vol. 57, pp. 297–303.

[25] T. Assefi, "Bayesian recursive image restoration," Ph.D. dissertation, Univ. Southern California, Los Angeles, Sept. 1973.

[26] T. Kailath and R. Geesey, "Covariance factorization—An explication via examples," in *Proc. 2nd Asilomar Conf. Circuits and Systems*, Calif., 1968.

[27] B. D. O. Anderson and T. Kailath, "The choice of signal-process models in Kalman–Bucy filtering," *J. Math. Anal. Appl.*, vol. 35, pp. 659–668, Sept. 1971.

26

Reprinted from *IEEE Trans. Circuits Syst.* **CAS-22**(3):278–286 (1975)

Restoration of Motion Degraded Images

A. O. ABOUTALIB AND L. M. SILVERMAN

***Abstract*—A new recursive procedure, based on state-space techniques and the use of a Kalman–Bucy filter is presented to solve the restoration problem of images degraded by motion in one direction. Both the noise-free and the noisy cases are considered.**

I. Introduction

WHENEVER there is relative motion between an object and the imaging system during exposure, the recorded image is degraded. Recently, there has been a great deal of interest in *a posteriori* technique for restoring motion-degraded images. A common method utilized to solve this problem is based on first characterizing the motion by means of a function, termed the point spread function (PSF), and considerable work has been devoted to finding such functions [1]–[5]. Once the motion degradation has been described mathematically via the PSF a more difficult task is to process the recorded image in the hope of restoring the image to the original object intensity function.

When the motion degradation is space invariant, with no noise present during recording, and when the recorded image extends over an infinite or semi-infinite interval in space, then the inverse filter in the Fourier domain is applicable. The infinite interval condition holds when a small object is photographed on a background of known intensity or when the recorded area is much larger than the extent of the blur. This technique, known as inverse filtering, has been described in detail in several papers [3], [4], [6]–[9]. If the Fourier transform of the PSF has zeros at some spatial frequencies then the inverse filter becomes a poor restoration technique, especially in the presence of noise. Slepian [6] related the difficulties with zeros to the constraint that the inverse filter must reproduce the background over an infinite region. A modified inverse filter to take care of this problem has been suggested in [2], [7], [8]. In the presence of noise, linear least square error filters are more effective [9], [10].

Manuscript received April 7, 1974; revised September 25, 1974. This work was supported by NSF under Grant GK-33416A # 1.

The authors are with the Department of Electrical Engineering, University of Southern California, Los Angeles, Calif. 90007.

When the motion degradation is space variant, the Fourier transform approach can not be used and the restoration problem becomes more complex. However, for some class of moving systems, especially aerial systems, it is possible to decompose the overall space-variant operation into an equivalent cascade of a space-invariant operation and a time-independent geometrical distortion, such decomposition reveals immediately the solution to the restoration problem by inverting the geometrical distortion and implementing the inverse filter [4].

In this paper, we introduce a new recursive approach to the restoration problem for images degraded by motion in one direction. This approach is based on modeling the motion by a linear discrete-time dynamical system. This leads to a simple recursive scheme for restoration when the motion PSF is known exactly and the image is noise free by use of a similar model of an exact inverse for the blurring system. The inverse then provides a computationally effective algorithm for restoring the image. The aforementioned problem of zeros does not arise since an exact model operating over the finite picture is utilized. In addition, boundary conditions enter naturally as initial states in the inverse model. The algorithm is generally faster than approaches which utilize fast Fourier transforms. In the presence of noise the blurring model provides a convenient way to obtain the best linear mean square estimate of the image by incorporating the modeling ideas developed in [22]–[24] and the use of a Kalman–Bucy filter.

In Section II, we review the characterization of motion degradation and the PSF. The restoration of noise free motion degraded images is then presented in Section III. The noisy problem is considered in Section IV where the motion degraded image is corrupted by additive white noise, and the best linear mean square estimate of the image is obtained. In Section V, we consider the restoration of images degraded by space-variant motion. Conclusions and extensions are discussed in Section VI.

II. Review of Motion Degradation and the Point Spread Function

Given the physical description of the motion, a general linear model for describing the degradation from object to image in the absence of noise is the superposition integral

$$b_b(x_1,x_2) = \iint_{-\infty}^{\infty} h(x_1,x_2;u_1,u_2)b_0(u_1,u_2)\,du_1\,du_2 \quad (2\text{-}1)$$

where $b_0(u_1,u_2)$ is the object intensity at the spatial point (u_1,u_2) in the object coordinate frame, $b_b(x_1,x_2)$ is the image intensity recorded by the imaging system at the spatial point (x_1,x_2) in the image coordinate frame, and $h(x_1,x_2;u_1,u_2)$ is the PSF, which is the response at (x_1,x_2) to a unit impulse at (u_1,u_2).

If $h(x_1,x_2;u_1,u_2)$ depends on each of the four spatial variables $(x_1,x_2;u_1,u_2)$ then the blurring system is said to be linear space variant [LSV]; if $h(x_1,x_2;u_1,u_2)$ depends only on the spatial differences $(x_1 - u_1)$; $(x_2 - u_2)$ then the system is linear space-invariant [LSI] and (2-1) reduces to a convolution

$$b_b(x_1,x_2) = \iint_{-\infty}^{\infty} h(x_1 - u_1, x_2 - u_2)b_0(u_1,u_2)\,du_1\,du_2. \quad (2\text{-}2)$$

Given the mathematical description of the relative movement, the corresponding PSF can be derived [1]–[5] and we have

$$b_b(x_1,x_2) = \iint_R h(x_1,x_2;u_1,u_2)b_0(u_1,u_2)\,ds_R + \beta(x_1,x_2) \quad (2\text{-}3)$$

where R is the motion trace and $\beta(x_1,x_2)$ is the response due to the background; $\beta(x_1,x_2) = 0$ if the background is of zero intensity.

For images in which the motion blur occurs in a straight line along one spatial coordinate, say along u_1 (the horizontal coordinate) the degraded image may be viewed as a family of degraded line images and may be restored line by line. In this case the dependence on the vertical coordinate can be dropped, and if the background is of zero intensity then (2-3) reduces to

$$b_b(x_1) = \int_{R_1} h(x_1,u_1)b_0(u_1)\,du_1 \quad (2\text{-}4)$$

where R_1 is the extent of the motion along u_1. For space-invariant blurring systems this becomes

$$b_b(x_1) = \int_{R_1} h(x_1 - u_1)b_0(u_1)\,du_1. \quad (2\text{-}5)$$

One discrete version of (2-5) is

$$b_b(k) = \sum_{l \in L} h(k - l)b_0(l) \quad (2\text{-}6)$$

where k,l take on integer values and L is to be determined from R_1. We will be concerned throughout this paper with digitized objects whose motion degradation can be described in a discrete form. The PSF $h(i)$ can be written as

$$h(i) = c_0\Delta(i) + c_1\Delta(i - 1) + \cdots + c_m\Delta(i - m) \quad (2\text{-}7)$$

where $\Delta(\cdot)$ is the Kronecker delta function, m is extent of the motion blur, and c_j is the value of the PSF $h(j)$ at j (the horizontal variable) for $j = 0,1,2,\cdots,m$. Consequently (2-6) can be viewed as

$$\begin{aligned} b_b(k) &= [c_0 + c_1D + c_2D^2 + \cdots + c_mD^m]b_0(k) \\ &= H(D)b_0(k) \end{aligned} \quad (2\text{-}8)$$

where D is a delay operator such that

$$b_0(k - i) = D^ib_0(k), \qquad \text{for } i = 0,1,\cdots,m$$

$H(D)$ can be viewed as the z transform $H(z)$ of the PSF $h(i)$ after replacing D by $1/z$. $H(z)$ is termed the transfer function of the blurring system; it has a finite number of terms and is a proper rational function if the corresponding PSF is causal.

III. Restoration of Noise-Free Motion-Degraded Images [LSI]

We will consider the case where the motion degradation can be described by (2-6).

The method we propose is to obtain a state-space model (A,f,g,d) that realizes the motion transfer function $H(z)$ as follows.

$$x(k+1) = Ax(k) + fb_0(k) \quad (3\text{-}1)$$

$$b_b(k) = gx(k) + db_0(k), \qquad k = 0, 1, \cdots, N-1 \quad (3\text{-}2)$$

where $x(k) \in R^m$ is the state vector, m is the extent of the motion blur, $b_0(k) \in R^1$ is the input to the system which is the object intensity at the spatial point k, $b_b(k) \in R^1$ is the output of the system which is the blurred image intensity at the spatial point k as recorded by the imagining system, A,f,g are constant matrices of order $(m \times m)$, $(m \times 1)$, $(1 \times m)$, respectively, d is a scalar, k is a horizontal variable takes on integer values $(0, 1,\cdots, N-1)$, and N is the number of pixels contained in the line image.

(A,f,g,d) represents a single-input single-output linear discrete system that realizes the blurring transfer function $H(z)$ in the sense that

$$H(z) = g(zI - A)^{-1}f + d \quad (3\text{-}3)$$

where

$$H(z) = c_0 + \frac{c_1}{z} + \frac{c_2}{z^2} + \cdots + \frac{c_m}{z^m}.$$

In general there is no unique solution to the realization problem and different realizations of the same transfer function can have distinct characteristics [12]–[15]. It should be noted that a realization of a transfer function exists if and only if it is a proper rational function.

The realization we will utilize in this paper has a canonical structure which will be seen to lead to efficient computation and is described by the following matrices

$$A = \begin{bmatrix} 0 & \frac{-c_2}{c_0} & \frac{-c_3}{c_0} & \cdots & \frac{-c_m}{c_0} \\ 0 & 0 & 0 & \cdots & 0 \\ 0 & 1 & 0 & \cdots & 0 \\ \vdots & \vdots & \vdots & \cdots & \vdots \\ 0 & 0 & 0 & \cdots & 0 \end{bmatrix}$$

$$f = \begin{bmatrix} c_1 \\ -c_0 \\ 0 \\ \vdots \\ 0 \end{bmatrix}$$

$$g = [1 \quad 0 \quad 0 \quad \cdots \quad 0]$$

$$d = c_0$$

with the initial state $x(0) = 0$.

For moving objects with background of nonzero intensity the initial state vector $x(0)$ is nonzero and can be determined from the knowledge of the motion PSF (in terms of the coefficients c_i for $i = 0,1,\cdots,m$) and the background intensity $b_0(-j)$ for $j = 1,2,\cdots,m$ as follows

$$x(0) = [x_1(0)x_2(0)\cdots x_m(0)]'$$

where

$$x_1(0) = c_1b_0(-1) + c_2b_0(-2) + \cdots + c_mb_0(-m)$$
$$x_2(0) = -c_0b_0(-1)$$
$$x_3(0) = -c_0b_0(-2)$$
$$\vdots$$
$$x_m(0) = -c_0b_0(-m+1).$$

The response $b_b^{ic}(k)$ due to the nonzero initial condition is given by

$$b_b^{ic}(k) = gA^kx(0).$$

For $k \geq m$, $A^k = 0$ and consequently $b_b^{ic}(k) = 0$ which shows as is to be expected that the background has no effect beyond the first m pixels.

Since (A,f,g,d) represents a linear dynamical system, the zero state response and the response due to the initial condition combine to give the system total response (decomposition property) and, due to the recursive nature of the system representation, (A,f,g,d) represents the blurring system with any background of known intensity where the initial condition vector $x(0)$ represents the background intensity.

Our approach to the noise free restoration problem is to obtain another linear system which is an exact inverse to the blurring system (A,f,g,d). It can be verified [16]–[18] that the system $(\hat{A},\hat{f},\hat{g},\hat{d})$ given in the following is an exact inverse.

$$z(k+1) = \hat{A}z(k) + \hat{f}b_b(k) \quad (3\text{-}4)$$

$$b_0(k) = \hat{g}z(k) + \hat{d}b_b(k), \qquad k = 0, 1,\cdots, N-1 \quad (3\text{-}5)$$

where

$$\hat{A} = A - fd^{-1}g = \begin{bmatrix} \frac{-c_1}{c_0} & \frac{-c_2}{c_0} & \cdots & \frac{-c_{m-1}}{c_0} & \frac{-c_m}{c_0} \\ 1 & 0 & \cdots & 0 & 0 \\ 0 & 1 & \cdots & 0 & 0 \\ \vdots & & & \vdots & \\ 0 & 0 & \cdots & 1 & 0 \end{bmatrix}$$

$$\hat{f}' = (fd^{-1})' = \begin{bmatrix} \frac{c_1}{c_0} & -1 & \cdots & 0 & 0 \end{bmatrix}$$

$$\hat{g} = \begin{bmatrix} -\frac{1}{c_0} & 0 & 0 & \cdots & 0 \end{bmatrix}$$

$$\hat{d} = \frac{1}{c_0}$$

with $z(0) = x(0)$.

Note that these equations provide a straightforward recursive scheme for exact restoration if the motion PSF is known exactly in terms of the coefficients c_i for $i = 0,1,2,\cdots,m$.

In fact, the inverse system $(\hat{A},\hat{f},\hat{g},\hat{d})$ can be obtained directly as a realization of the transfer function $1/H(z)$ without going through the realization of the blurring system and the initial state $z(0)$ can be determined from the knowledge of the coefficients c_i for $i = 0,1,2,\cdots,m$ and the background intensity $b_0(-j)$ for $j = 1,2,\cdots,m$ as before.

As an example, suppose the object translates at a constant velocity v along the horizontal direction during the exposure interval $[0,T]$, the extent of the motion blur m is then equal to vT, let $m = vT = 10$ pixel elements, and without any loss of generality we let $T = 1$, then

$$h(i) = \frac{1}{v}\,\Delta(i) + \frac{1}{v}\,\Delta(i-1) + \cdots + \frac{1}{v}\,\Delta(i-10).$$

As a result the inverse system $(\hat{A},\hat{f},\hat{g},\hat{d})$ has dimension equal to 10 and has the following structure

$$\begin{bmatrix} x_1 \\ x_2 \\ x_3 \\ \vdots \\ x_{10} \end{bmatrix}(k+1) = \begin{bmatrix} -1 & -1 & -1 & \cdots & -1 \\ 1 & 0 & 0 & \cdots & 0 \\ 0 & 1 & 0 & \cdots & 0 \\ \vdots & & & & \\ 0 & 0 & 0 & \cdots & 0 \end{bmatrix} \begin{bmatrix} x_1 \\ x_2 \\ x_3 \\ \vdots \\ x_{10} \end{bmatrix}(k) + \begin{bmatrix} 1 \\ -1 \\ 0 \\ \vdots \\ 0 \end{bmatrix} b_b(k)$$

$$b_0(k) = -vx_1(k) + vb_0(k), \qquad k = 0, 1, \cdots, N-1$$

where $x(0)$ represents the background intensity.

Fig. 1 shows a computer simulation of (256 by 256) images. The motion degradation in Fig. 1(b) is severe and much of the information in the high spatial frequencies is lost. The restored image in Fig. 1(c) has all the details of the original image in Fig. 1(a).

The Structure of the Inverse System and the Computation Aspects

The computational aspects of the inverse system $(\hat{A},\hat{f},\hat{g},\hat{d})$ will now be discussed. Observe first that its dimension is equal to the extent of the motion blur, and the number of multiplications required to restore one pixel is $m + 3$. To restore a line image of N pixels requires $N(m + 3)$ multiplications. On the other hand, Fourier transform of a line image of N pixels using the fast Fourier algorithm requires $2N \log_2 N$ multiplications and as a result, the number of multiplications required to restore a line image using the inverse Fourier filter technique is $4N \log_2 N + N$. When compared with $N(m + 3)$ this shows that the inverse system requires less number of multiplications that the Fourier inverse filter if the motion blur extends up to $\frac{1}{8}$ the width of the image for 256 by 256 images and up to $\frac{1}{2}$ the width of the image for 32 by 32 images.

The stability of the inverse system can be determined from the c_i's since they are the coefficients of the characteristic equation of $\hat{A}$. Standard tests for discrete time systems can be utilized [28]. Note that since $c_i \geq 0$ for all i then a sufficient condition for stability [30] is that

$$\sum_{i=1}^{m} c_i < c_0.$$

This condition holds for sharply peaked motion PSF

(a)

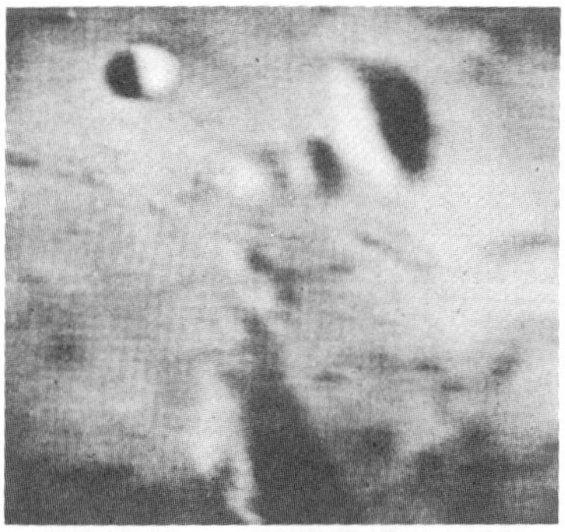

(b)

(c)

Fig. 1. (a) Lunar object. (b) Motion degraded lunar object (10 pixels). (c) Restored image of (b).

(motion with constant acceleration). A necessary and sufficient condition for stability is that the spectral radius of the matrix $\hat{A}$, $\rho(\hat{A}) < 1$ [30].

Since the inverse system is an exact model operating on a finite image, errors will not build up due to inverse system instability if the motion PSF is known exactly (in terms of exact knowledge of the coefficients c_i's). Problems of error accumulation may arise, however, due to inexact c_i's and the instability of the inverse system.

If the degraded image is not noise free then errors may accumulate if the inverse system is unstable [19]–[21]. A general smoothing scheme is developed in the next section for the general noisy case which is stable whether or not the inverse system is stable.

IV. Restoration of Motion Degraded Images Corrupted by Additive White Noise

With the assumption that the noise degradation in the observation is a result of white noise added to the motion degraded image, a convenient way of obtaining the best linear mean square estimate of the image is to apply Kalman–Bucy filtering theory and use the previous work done by Nahi and Franco [22], Nahi and Assefi [23], and Powell and Silverman [24].

As we presented in the previous section, a difference equation model (A,f,g,d) that realizes the motion transfer function $H(z)$ is described by (3-1), (3-2). With white noise of zero mean and known variance added to the motion degraded image, a model that describes both motion and noise degradation is given by

$$x(k+1) = Ax(k) + fb_0(k) \tag{4-1}$$

$$\tilde{b}_b(k) = gx(k) + db_0(k) + v(k) \tag{4-2}$$

where the white noise process $v(\cdot)$ is such that

$$Ev(k) = 0 \tag{4-3}$$

$$Ev(k_1)v(k_2) = \sigma^2 \Delta(k_1 - k_2), \qquad k = 0, 1, 2, \cdots, N-1 \tag{4-4}$$

and where the variance σ^2 is assumed known, $\Delta(\cdot)$ is Kronecker delta function and $\tilde{b}_b(k)$ is the noise corrupted motion degraded image. We select as a measure of the signal-to-noise ratio

$$\rho = \frac{\text{peak-to-peak variation of the signal}}{2\sigma}.$$

Given the correlation function of the output of the line scanner of the object intensities (scanning the object horizontally one line at a time in a sequential order) a difference equation model driven by white noise can be developed which possess a response characteristic that matches that of the scanner output in a statistical sense [23]. Such a model is given by another dynamical system of the form

$$r(k+1) = \alpha r(k) + \beta u(k) \tag{4-5}$$

$$b_0(k) = \gamma r(k), \qquad k = 0, 1, 2, \cdots, N-1 \tag{4-6}$$

where

$$Eu(k) = 0 \tag{4-7}$$

$$Eu(k_1)u'(k_2) = K\,\Delta(k_2 - k_1). \tag{4-8}$$

Procedures for determination of the matrices α, β, γ, and K under various constraints are given in [23], [25], [26].

The two models represented by (4-1)–(4-4) and (4-5)–(4-8) can now be augmented in one model as

$$\omega(k+1) = \begin{bmatrix} A & f\gamma \\ 0 & \alpha \end{bmatrix} \omega(k) + \begin{bmatrix} 0 \\ \beta \end{bmatrix} u(k) \tag{4-9}$$

$$\tilde{b}_b(k) = [g \quad d\gamma]\,\omega(k) + v(k), \qquad k = 0, 1, 2, \cdots, N-1 \tag{4-10}$$

where $\omega'(k) = [x'(k)\ r'(k)]$.

On the basis of this discrete model a recursive estimator may be designed [25] using a standard Kalman–Bucy approach yielding the best linear mean square estimate $\hat{\omega}(k)$ of $\omega(k)$. The filter equations are given below for the sake of completeness

$$\hat{\omega}(k) = [I - F(k)\bar{g}]\bar{A}\hat{\omega}(k-1) + F(k)\tilde{b}_h(k) \tag{4-11}$$

$$F(k) = P(k)\bar{g}'[\bar{g}P(k)\bar{g}' + \sigma^2]^{-1} \tag{4-12}$$

$$P(k+1) = \bar{A}[I - F(k)\bar{g}]P(k)\bar{A}' + \bar{B}K\bar{B}' \tag{4-13}$$

where

$$\bar{A} = \begin{bmatrix} A & f\gamma \\ 0 & \alpha \end{bmatrix} \qquad \bar{B} = \begin{bmatrix} 0 \\ \beta \end{bmatrix} \qquad \bar{g} = [g \quad d\gamma]$$

and $P(k)$ is the estimation error covariance matrix. Equations (4-11)–(4-13) together with the initial conditions $P(0)$ and $\hat{\omega}(0)$ which should be specified as a part of available data, describe the operation of the filter and the best linear mean square estimate of the image becomes

$$\hat{b}_0(k) = \gamma\hat{r}(k) = \tilde{g}_1\hat{\omega}(k) \tag{4-14}$$

where $\tilde{g}_1 = [0 \ \gamma]$.

Stability of the filtering scheme is guaranteed by the controllability and observability of the augmented system. As an example, consider the motion degraded image in Fig. 1(b). Additive white noise of zero mean and variances ${\sigma_1}^2 = 10$ and ${\sigma_2}^2 = 100$ are added to the motion degraded image ($\rho_1 \simeq 40$ and $\rho_2 = 10$, respectively) and the noise corrupted motion degraded images are shown in Figs. 2(a) and (c). The random process that represents the brightness level of the object is assumed wide sense stationary with exponential autocorrelation function $R(z,i)$

$$R(z,i) = R(0,0) \exp(-\mu_h|z| - \mu_v|i|)$$

where $R(0,0)$, μ_h, and μ_v are to be determined; in this example $R(z,i)$ is given by

$$R(z,i) = 763.606 \exp(-0.103|z| - 0.062|i|).$$

As a result the difference equation model that possesses response characteristic that matches that of the line scanner output of the object in a statistical sense can be developed [23], [27] and is given by

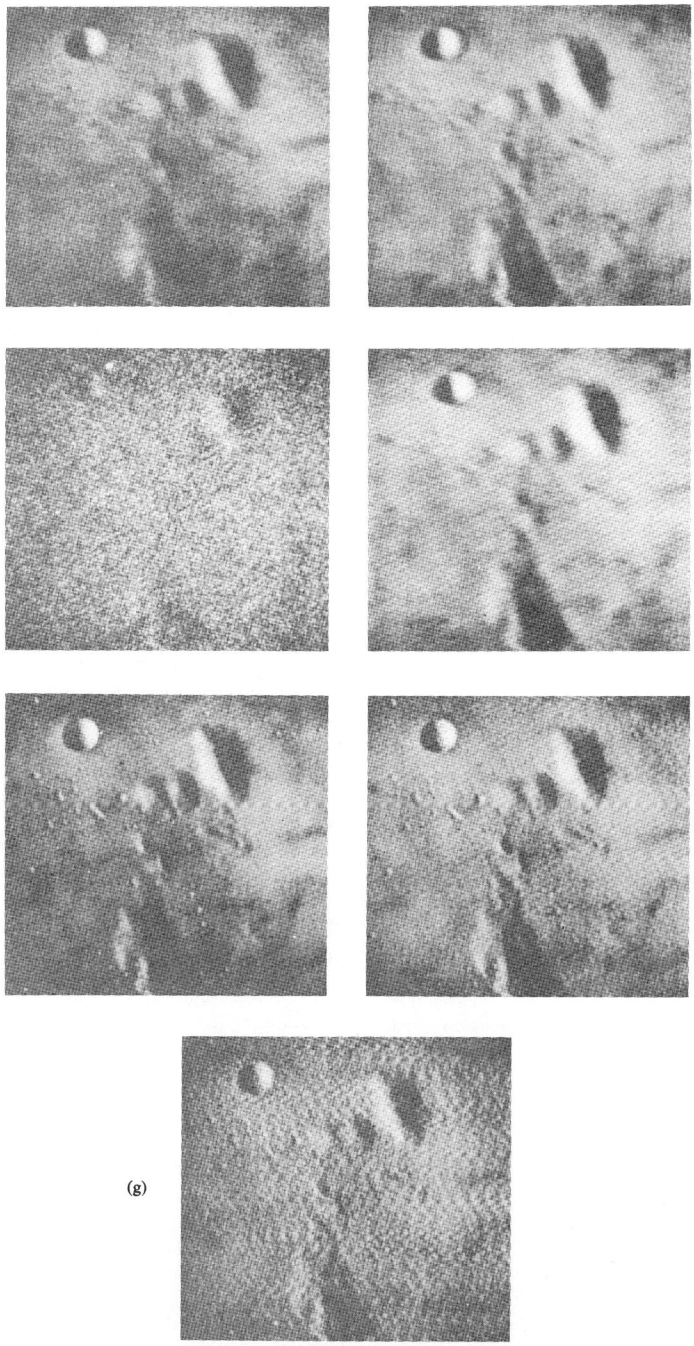

(a) (b) (c) (d) (e) (f) (g)

Fig. 2. (a) Corrupted motion degraded image ($\sigma_1^2 = 10$). (b) Restored image of (a) (scalar scan). (c) Noise corrupted motion degraded image ($\sigma_2^2 = 100$). (d) Restored image of (c) (scalar scan). (e) Restored image of (b) using Kalman filter ($\sigma^2 = 0$). (f) Restored image of (a) (vector scan). (g) Restored image of (c) (vector scan).

$$r(k+1) = \alpha r(k) + \beta u(k)$$
$$b_0(k) = \gamma r(k)$$

where

$$\alpha = \begin{bmatrix} 0.9997 & 0.0 & 0.0 & 0.0 & 0.0 \\ 0.0 & 0.9996 & 0.0039 & 0.0 & 0.0 \\ 0.0 & -0.1541 & 0.9992 & 0.0 & 0.0 \\ 0.0 & 0.0 & 0.0 & 0.9987 & 0.0039 \\ 0.0 & 0.0 & 0.0 & -0.6164 & 0.9983 \end{bmatrix}$$

$$\beta K \beta' = \begin{bmatrix} 0.0270 & 0.0 & 0.0 & 0.0 & 0.0 \\ 0.0 & 0.0536 & 0.3146 & 0.0 & 0.0 \\ 0.0 & 0.3146 & 1.8447 & 0.0 & 0.0 \\ 0.0 & 0.0 & 0.0 & 0.0485 & 0.5474 \\ 0.0 & 0.0 & 0.0 & 0.5474 & 6.1730 \end{bmatrix}$$

$$\gamma = [1 \quad 1 \quad 0 \quad 1 \quad 0 \,].$$

The matrices $\bar{A}$, $\bar{g}$, and $\bar{B}K\bar{B}'$ can be computed as indicated and need not be displayed explicitly here. The covariance $P(k)$ in (4-13) approximately reaches its steady state value after about 100 pixels independently of the initial condition. As a result, we can use the steady state value of $P(k)$ to process the entire image to obtain a good suboptimal estimate. The restored images using this approach are shown in Figs. 2(b) and (d).

It can be readily verified [32] utilizing the filter (4-11)–(4-13), that as $\sigma^2 \to 0$ the estimate obtained using the Kalman–Bucy filter approaches that obtained by the inverse system. In Fig. 2(e) the object is restored from the motion degraded image in 1(b) using Kalman filter by setting $\sigma^2 = 0$.

Nahi and Franco [22] and Powell and Silverman [24] introduced vector scanning, where several lines of the image are scanned at a time. In this case $b_0(k)$ and $b_b(k)$ are sx 1 vectors, with components $b_0{}^i(k)$ and $b_b{}^i(k)$, $i = 1,2,\cdots,s$ where s is the number of lines scanned at a time. It is evident [22] that vector scanning accounts for the vertical correlation, consequently better results in restoration are expected by using vector scanning. The blurring system represented by (3-1) and (3-2) becomes a multi-input multi-output system which is composed of s decoupled single-input single-output systems each having dimension equal to m as follows

$$\begin{bmatrix} x_1 \\ x_2 \\ \vdots \\ x_s \end{bmatrix}(k+1) = \begin{bmatrix} A & 0 & \cdots & 0 \\ 0 & A & \cdots & 0 \\ \vdots & & & \vdots \\ 0 & 0 & \cdots & A \end{bmatrix} \begin{bmatrix} x_1 \\ x_2 \\ \vdots \\ x_s \end{bmatrix}(k) + \begin{bmatrix} f & 0 & \cdots & 0 \\ 0 & f & \cdots & 0 \\ \vdots & & & \vdots \\ 0 & 0 & \cdots & f \end{bmatrix} \begin{bmatrix} b_0{}^1 \\ b_0{}^2 \\ \vdots \\ b_0{}^s \end{bmatrix}(k) \quad (4\text{-}15)$$

$$\begin{bmatrix} b^1 \\ b_b{}^2 \\ \vdots \\ b_b \end{bmatrix}(k) = \begin{bmatrix} g & 0 & \cdots & 0 \\ 0 & g & \cdots & 0 \\ \vdots & & & \vdots \\ 0 & 0 & \cdots & g \end{bmatrix} \begin{bmatrix} x_1 \\ x_2 \\ \vdots \\ x_s \end{bmatrix}(k) + \begin{bmatrix} d & 0 & \cdots & 0 \\ 0 & d & \cdots & 0 \\ \vdots & & & \vdots \\ 0 & 0 & \cdots & d \end{bmatrix} \begin{bmatrix} b_0{}^1 \\ b_0{}^2 \\ \vdots \\ b_0{}^s \end{bmatrix}(k). \quad (4\text{-}16)$$

Where the superscripts refer to the appropriate lines in the scanned band.

To illustrate vector scanning, the noise corrupted motion degraded images in Figs. 2(a) and (c) are restored by scanning the image 2 lines at a time. The difference equation model that possess a response characteristic that matches that of the vector scanner output of the object in a statistical sense can be developed [22] and is given by

$$r(k+1) = \alpha r(k) + \beta u(k)$$
$$b_0(k) = \gamma r(k)$$

where

$$\alpha = 0.902$$

$$\beta K \beta' = \begin{bmatrix} 0.186 & 0 \\ 0 & 0.186 \end{bmatrix} \qquad \gamma = \begin{bmatrix} 22.632 & 15.856 \\ 15.856 & 22.632 \end{bmatrix}.$$

The augmented model can be computed, and the filter (4-11)–(4-13) can be utilized. The restored images are shown in Figs. 2(f) and (g) and are considerably sharper than Figs. 2(b) and (d) obtained using scalar scanning.

V. Restoration of Space-Variant Motion Degraded Images

For space-variant motion degradation, where the PSF is a causal function, the motion degraded image intensity can be written as

$$b_b(k) = c_0(k)b_0(k) + c_1(k)b_0(k-1) + \cdots + c_{m_k}(k)b_0(k-m_k) \quad (5\text{-}1)$$

where m_k is the extent of the motion blur at k. Let

$$M = \max_{0 \le i \le N-1} \{m_i\}. \quad (5\text{-}2)$$

It can be verified that a difference equation model that realizes the space-variant motion blur is given by

$$x(k+1) = A(k)x(k) + f(k)b_0(k) \quad (5\text{-}3)$$

$$b_b(k) = gx(k) + d(k)b_0(k), \qquad k = 0, 1, \cdots, N-1 \quad (5\text{-}4)$$

where

$$A(k) = \begin{bmatrix} 0 & c_2(k+1) & c_3(k+1) & \cdots & c_M(k+1) \\ 0 & 0 & 0 & \cdots & 0 \\ 0 & 1 & 0 & \cdots & 0 \\ 0 & 0 & 1 & \cdots & 0 \\ \vdots & \vdots & & & \vdots \\ 0 & 0 & 0 & \cdots 1 & 0 \end{bmatrix}$$

$$f(k) = \begin{bmatrix} c_1(k+1) \\ 1 \\ 0 \\ 0 \\ \vdots \\ 0 \end{bmatrix}$$

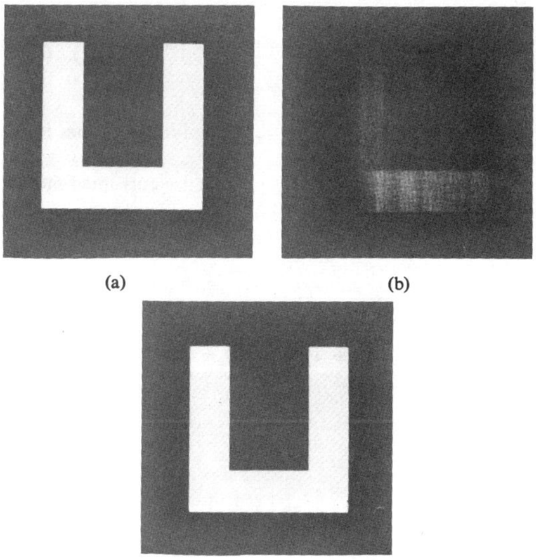

(a) (b) (c)

Fig. 3. (a) The original image. (b) The motion blurred image. (c) Restored image of (b).

$$g = [1 \quad 0 \quad 0 \quad \cdots \quad 0]$$

$$d(k) = c_0(k)$$

and $c_i(k) = 0$ for all $i > m_k$.

The exact inverse of the blurring system model described by (5-3) and (5-4) is given by

$$z(k+1) = \hat{A}(k)z(k) + \hat{f}(k)b_b(k) \tag{5-5}$$

$$b_0(k) = \hat{g}(k)z(k) + \hat{d}(k)b_b(k), \quad k = 0, 1, 2, \cdots, N-1 \tag{5-6}$$

where

$$\hat{A}(k) = A(k) - f(k)d^{-1}(k)g = \begin{bmatrix} \dfrac{-c_1(k+1)}{c_0(k)} & c_2(k+1) & c_3(k+1) & \cdots & c_M(k+1) \\ -\dfrac{1}{c_0(k)} & 0 & 0 & \cdots & 0 \\ 0 & 1 & 0 & \cdots & 0 \\ \vdots & & & & \\ 0 & 0 & 0 & \cdots & 0 \end{bmatrix}$$

$$\hat{f}'(k) = (f(k)d^{-1}(k))' = \left[\frac{c_1(k+1)}{c_0(k)} \quad \frac{1}{c_0(k)} \quad 0 \quad \cdots \quad 0\right]$$

$$\hat{g}(k) = \left[-\frac{1}{c_0(k)} \quad 0 \quad 0 \quad \cdots \quad 0\right]$$

$$\hat{d}(k) = \frac{1}{c_0(k)}$$

with $z(0) = x(0)$ and $c_0(k) \neq 0$ for $k = 0, 1, 2, \cdots, N-1$.

As an example, consider the one-dimensional motion PSF

$$h(k,l) = \begin{cases} \dfrac{2}{l+1}, & \left[\dfrac{2k-1}{3}\right] \leq l \leq k \\ 0, & \text{elsewhere} \end{cases}$$

where $[(2k-1)/3]$ is the integer value of $((2k-1)/3)$; $m_i = k - [(2k-1)/3]$ and $c_i(k) = 2/[1 + (k-i)]$ for $i \leq m_k$ and $c_i(k) = 0$ for $i > m_k$. The original is a (32 by 32) image of the letter U with background of zero intensity as shown in Fig. 3(a). The image is degraded by the PSF $h(k,l)$ and the degraded image is shown in Fig. 3(b), and

$$M = \max_{0 \leq i \leq N-1} \{m_i\} = 11.$$

The degraded image is restored using the inverse system model and the restored image is shown in Fig. 3(c).

VI. Conclusions

A new recursive technique has been presented for the restoration of images degraded by motion in one direction. It has been established that in the absence of noise the original image can be recovered exactly with a speed of computation faster than techniques based on fast Fourier transforms. If noise is also present the scheme can be readily extended and as demonstrated, yields images of high accuracy. A further advantage of the approach presented is that it extends directly to the space variant case without the use of *ad hoc* modeling schemes.

At present we are studying images degraded by motion which trace a curve in the plane and initial work has shown that the techniques given here can be applied in a straight forward manner [31], [32]. The noisy space variant case is also being examined.

The modeling approach adopted here should also be useful when the PSF is not known exactly. It is felt that system identification techniques and adaptive estimation should be applicable.

Acknowledgment

The authors would like to thank Prof. A. Sawchuk of the University of Southern California for many useful discussions.

References

[1] R. V. Shack, "The influence of image motion and shutter operation on the photographic transfer function," *Appl. Opt.*, vol. 3, pp. 1171–1181, 1964.

[2] M. M. Sondi, "Image restoration: the removal of spatially invariant degradations," *Proc. IEEE*, vol. 60, pp. 842–853, July 1972.
[3] T. S. Huang, W. F. Schreiber, and O. J. Tretiak, "Image processing," *Proc. IEEE*, vol. 59, pp. 1586–1609, Nov. 1971.
[4] A. A. Sawchuck, "Space-variant image motion degradation and restoration," *Proc. IEEE*, vol. 60, pp. 854–861, July 1972.
[5] E. B. Brown "V/H image motion in aerial cameras," *Photogrammetric Eng.*, vol. 31, pp. 308–323, 1965.
[6] D. Slepian, "Restoration of photographs blurred by image motion," *Bell Syst. Tech. J.*, vol. 46, pp. 2353–2362, 1967.
[7] J. L. Harris, Sr., "Image evaluation and restoration," *J. Opt. Soc. Amer.*, vol. 56, pp. 569–574, May 1966.
[8] L. J. Cutrona and W. D. Hall, "Some considerations in post-facto blur removal," in Eval. of Mot. Deg. Images, M. Nagel, Ed. NASA Tech. Rep. sp-193, Washington, D.C., pp. 139–148, 1968.
[9] C. W. Helstrom, "Images restoration by the method of least squares," *J. Opt. Soc. Amer.*, vol. 57, pp. 297–303, 1967.
[10] D. Slepian, "Linear least-squares filtering of distorted images," *J. Opt. Soc. Amer.*, vol. 57, no. 7, pp. 918–923, July 1967.
[11] D. P. MacAdam, "Digital image restoration by constrained deconvolution," *J. Opt. Soc. Amer.*, vol. 60, pp. 1617–1627, Dec. 1970.
[12] L. M. Silverman, "Realization of linear dynamical systems," *IEEE Trans. Automat. Contr.*, vol. AC-16, pp. 554–567, Dec. 1971.
[13] ——, "Synthesis of impulse response matrices by internally stable and passive realizations," *IEEE Trans. Circuit Theory*, vol. CT-15, pp. 238–245, Sept. 1968.
[14] D. C. Youla, "The synthesis of linear dynamical systems from prescribed weighting patterns," *SIAM J. Appl. Math.*, vol. 14, pp. 527–549, 1966.
[15] A. J. Tether, "Construction of minimal linear state-variable models from finite input-output data," *IEEE Trans. Automat. Contr.*, vol. AC-15, pp. 427–436, Aug. 1970.
[16] R. W. Brockett, "Poles, zeros and feedback: state space interpretation," *IEEE Trans. Automat. Contr.*, vol. AC-10, pp. 129–135, Apr. 1965.
[17] L. M. Silverman, "Properties and application of inverse systems," *IEEE Trans. Automat. Contr.*, vol. AC-13, pp. 436–437, Aug. 1968.
[18] J. L. Massey and M. K. Sain, "Inverse of linear sequential circuits," *IEEE Trans. Computers*, vol. C-17, pp. 330–337, Apr. 1968.
[19] S. Twomey, "On the numerical solution of Fredholm integral equations of the first kind by the inversion of linear system produced by quadrature," *J. Assoc. Comp. Mach.*, vol. 10, pp. 97–101, Jan. 1963.
[20] D. L. Phillips, "A technique for the numerical solution of certain integral equations of the first kind," *J. Assoc. Comp. Mach.*, vol. 9, pp. 84–97, Jan. 1962.
[21] S. Twomey, "The application of numerical filtering to the solution of integral equations encountered in indirect sensing measurements," *J. Franklin Inst.*, vol. 279, pp. 95–109, Feb. 1965.
[22] N. E. Nahi and C. A. Franco, "Recursive image enhancement-vector processing," *IEEE Trans. Commun.*, vol. COM-21, pp. 305–311, Apr. 1973.
[23] N. E. Nahi and Assefi, "Bayesian recursive image estimation," *IEEE Trans. Computers*, vol. C-21, pp. 734–738, July 1972.
[24] S. R. Powell and L. M. Silverman, "Modeling of two-dimensional random fields with application to image restoration," *IEEE Trans. Automatic Control*, vol. AC-19, pp. 8–13, Feb. 1974.
[25] N. E. Nahi, *Estimation Theory and Application.* New York: Wiley, 1969.
[26] H. L. VanTrees, *Detection, Estimation and Modulation Theory.* New York: Wiley, 1971, pt. II.
[27] A. O. Aboutalib and L. M. Silverman, "State space image motion restoration," in *Proc. 1973 IEEE Conf. Decision and Control*, pp. 731–738, Dec. 1973.
[28] L. A. Zadeh and C. A. Desoer, *Linear System Theory: The State Space Approach.* New York: McGraw-Hill, 1963.
[29] E. K. Blum, *Numerical Analysis and Computations: Theory and Practice.* Addison-Wesley, 1972.
[30] R. S. Varga, *Matrix Iterative Analysis.* Englewood Cliffs, N.J., Prentice-Hall, 1962.
[31] A. O. Aboutalib and L. M. Silverman, "Restoration of images degraded by curvilinear motion," presented at the 2nd Int. Joint Conf. Pattern Recognition, Copenhagen, Denmark, Aug. 1974.
[32] A. O. Aboutalib, "Restoration of images degraded by motion," Ph.D. dissertation, Univ. Southern California, Los Angeles, June 1974.

27

Reprinted from *IEEE Trans. Inf. Theory* **IT-23**(4):473–482 (1977)

Kalman Filtering in Two Dimensions

JOHN W. WOODS, MEMBER, IEEE, AND CLARK H. RADEWAN, MEMBER, IEEE

***Abstract*—The Kalman filtering method is extended to two dimensions. The resulting computational load is found to be excessive. Two new approximations are then introduced. One, called the strip processor, updates a line segment at a time; the other, called the reduced update Kalman filter, is a scalar processor. The reduced update Kalman filter is shown to be optimum in that it minimizes the post update mean-square error (mse) under the constraint of updating only the nearby previously processed neighbors. The resulting filter is a general two-dimensional recursive filter.**

I. INTRODUCTION

IN RECENT YEARS there have been many attempts to extend Kalman filtering to two dimensions [1]–[3]. This is a direct result of the method's great success in one dimension. In the two-dimensional (2-*D*) case, the enormity of the data calls particularly for an efficient recursive processor. Unfortunately, the previous efforts to achieve a truly recursive 2-*D* Kalman filter were of only limited success because of both the difficulty in establishing a suitable 2-*D* recursive model and also the high dimension of the resulting state vector. In fact, a straightforward extension of one-dimensional (1-*D*) Kalman filter techniques would result in a number of state variables proportional to N for the filtering of an $N \times N$ digital image.

In this paper, we propose two schemes which to a large extent overcome the computational problems that have precluded the use of 2-*D* Kalman-like processors. We develop two new approximations: one to the 2-*D* Kalman vector processor (updates a line at a time), called the Kalman strip filter; and another to the 2-*D* Kalman scalar processor, (updates a point at a time) called the reduced update filter. Both of these processors offer computationally effective solutions to the problem of 2-*D* Kalman filtering. These results have application to a wide range of problems involving estimation on 2-*D* data fields, including the image processing problem where the utility of linear filtering is well-established. The theory includes the case of space-variant models, allowing a better match to local source statistics, in turn permitting a greater noise reduction. Further, it should be possible to treat nonlinear space-variant models via techniques similar to those of the 1-*D* extended Kalman filtering [5].

Manuscript received May 3, 1976; revised October 25, 1976. This work was supported in part by the U.S. Energy Research and Development Administration under contract number W-7504-ENG-48 and in part by the Defense Advanced Research Projects Agency of the Department of Defense under ARPA Order No. 2436.

J. W. Woods was with Lawrence Livermore Laboratory, Livermore, CA. He is now with the Electrical and Systems Engineering Department, Rensselaer Polytechnic Institute, Troy, NY 12181.

C. H. Radewan is with the Lawrence Livermore Laboratory, Livermore, CA 94550.

We start with a brief review of the concept of state and its role in 1-*D* Kalman filtering. Then we define the 2-*D* Kalman scalar and vector filters, and we point out their undesirable computational properties in that the state vector grows with the image size. Next, we present the Kalman strip filter and the reduced update Kalman filter. Finally, we present examples of application of the filters in a simulated data environment.

The Concept of State

The concept of state is central to linear system theory in one dimension. The state is defined as the minimum information about the past and present needed to determine all future responses given the future input [4]. For systems governed by differential or difference equations, the state is simply a sufficient set of initial conditions for the one-point boundary value problem. The state concept carries over to the case of stochastic differential and difference equations in Kalman filtering. In fact, the state vector of the dynamical system signal model determines the order of the Kalman calculations. In this random case, the state can be defined as the minimum amount of information about past and present estimates needed to determine an optimal causal estimate of future response given future noisy observations. The dynamical model for this case is given by [5]

$$\boldsymbol{s}(m) = \boldsymbol{F}\boldsymbol{s}(m-1) + \boldsymbol{G}\boldsymbol{u}(m) \tag{1}$$

$$\boldsymbol{r}(m) = \boldsymbol{H}\boldsymbol{s}(m) + \boldsymbol{v}(m), \tag{2}$$

where $\boldsymbol{s}$ is the signal state vector, $\boldsymbol{r}$ is the observation or received signal, $\boldsymbol{v}$ is the observation noise, and $\boldsymbol{u}$ is the random process which generates the signal. The matrices $\boldsymbol{F}$, $\boldsymbol{G}$, and $\boldsymbol{H}$ are, respectively, the system, drive, and observation matrices.

In (1) and (2), $\boldsymbol{u}$ and $\boldsymbol{v}$ are uncorrelated white Gaussian zero mean noise sources with correlation

$$\boldsymbol{E}[\boldsymbol{u}(m)\boldsymbol{u}^T(k)] = \boldsymbol{Q}_u\delta_{mk}$$

and

$$\boldsymbol{E}[\boldsymbol{v}(m)\boldsymbol{v}^T(k)] = \boldsymbol{Q}_v\delta_{mk},$$

where $\boldsymbol{Q}_u$ and $\boldsymbol{Q}_v$ are correlation matrices and δ_{mk} is the Kronecker delta. The Kalman filtering equations for this model are [5]

$$\hat{\boldsymbol{s}}_b(m) = \boldsymbol{F}\hat{\boldsymbol{s}}_a(m-1) \tag{3}$$

$$\boldsymbol{P}_b(m) = \boldsymbol{F}\boldsymbol{P}_a(m-1)\boldsymbol{F}^T + \boldsymbol{G}\boldsymbol{Q}_u\boldsymbol{G}^T \tag{4}$$

$$\boldsymbol{K}(m) = \boldsymbol{P}_b(m)\boldsymbol{H}^T(\boldsymbol{H}\boldsymbol{P}_b(m)\boldsymbol{H}^T + \boldsymbol{Q}_v)^{-1} \tag{5}$$

$$\hat{\boldsymbol{s}}_a(m) = \hat{\boldsymbol{s}}_b(m) + \boldsymbol{K}(m)[\boldsymbol{r}(m) - \boldsymbol{H}\hat{\boldsymbol{s}}_b(m)] \tag{6}$$

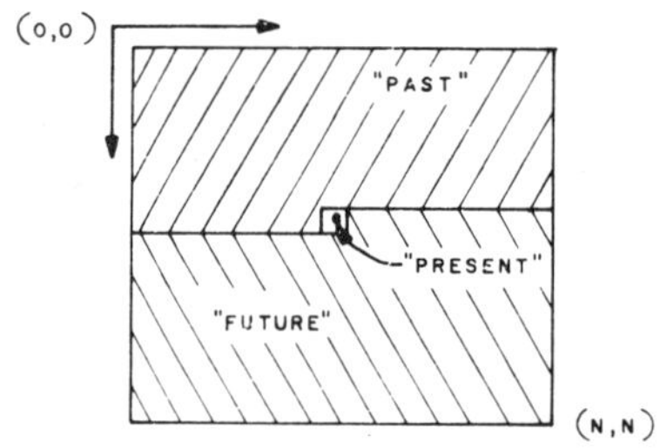

Fig. 1. Scalar scanning rectangular region using raster scan.

$$P_a(m) = [I - K(m)H]P_b(m), \tag{7}$$

where

$$P_i(m) \triangleq E[(s - \hat{s})_i(s - \hat{s})_i^T], \qquad i = a,b$$

is the error covariance matrix. The subscripts a and b indicate "after" and "before" updating, respectively.

These equations have the following simple interpretation. First, in (3), we project the last estimate forward using the dynamics of the system model. Then in (6) we update this estimate using the new observation $r(m)$. Equation (5) is the gain matrix for the updating in (6). The remaining equations (4) and (7) are the error covariances necessary to calculate the new gain matrix.

II. Kalman Filtering in Two Dimensions

A) A Scalar 2-D Kalman Filter

When we try to extend the above formalism to two dimensions, we encounter a problem concerning the 1-D definition of state as we will now show. First, we consider a particular scanning of a discrete 2-D square region consisting of an $N \times N$ regularly spaced lattice. Since the scanning does not qualitatively affect the results, we will assume the raster scan, i.e., left to right, advance one line, then repeat. Continuing with this procedure for the next N-2 lines we complete the scan. Thus at any point in the picture, some points will be the "past," one point will be the "present," and the remaining points will be the "future." These words will thus have their conventional meaning with respect to the order in which the points are processed (see Fig. 1). Next, we generalize our dynamical model (1). We obtain, by application of a recently developed 2-D spectral factorization routine [6], the $(M \times M)$th order (see Appendix) scalar equation,[1]

$$s(m,n) = \sum_{\mathcal{R}_{\oplus+}} c_{kl}s(m-k,n-l) + u(m,n), \tag{8}$$

where $\mathcal{R}_{\oplus+} = \{m \geq 0, n \geq 0\} \cup \{m < 0, n > 0\}$, which is to be compared to the scalar version of (1),

$$s(m) = \sum_{k=1}^{M} c_k s(m-k) + u(m). \tag{9}$$

[1] Inclusion of zeroes in the model leads to additional complications. To simplify the discussion, we consider only all-pole dynamical systems.

The model of (8) is called a nonsymmetric half-plane (NSHP) model [7].

Now we are in a position to appreciate the difficulty with state in two dimensions. For while (8) only uses $O(M^2)$ points for each computation of a new output point, so that the amount of computation is related to the order of the filter as in (9), when (8) moves across the image, values of s will be needed from the previous M lines. Thus the memory requirements of (8) are not of order related to the order of the filter as in (9). Because of this large memory requirement, the state vector for (8) must contain the M previous lines and hence has $O(MN)$ components.

It is convenient to take the part of the state vector on the support of the filter and give it a special name. Thus we define a new term *pseudo-state* as the minimum part of the state vector needed to compute the present output given the present input. This pseudo-state has a number of elements related to the order of the filter and hence its size determines the order of computation of the filter. However, the storage requirements are related to the size of the state vector. In the ordinary 1-D case, the pseudo-state vector is the state vector.

Next we will consider the Kalman filtering equations for this model. We will find that both the per point computation and the memory grow with N^2, the square of the width of the image field. Because of this unsatisfactory situation, we then turn to some approximate filters which greatly reduce the computation while only slightly affecting the accuracy of the estimate.

To obtain the Kalman filtering equations for the signal model of (8) and the observation equation

$$r(m,n) = s(m,n) + v(m,n), \tag{10}$$

where u and v are independent white Gaussian sources, we note that the scanning operation transforms the 2-D problem into an equivalent 1-D problem. Thus (8) can be put into the form (1) by defining the state vector of $M(N+1)$ components

$$\begin{aligned} \mathbf{s}(m,n) = [&s(m,n), s(m-1,n), \\ &\cdots, s(1,n); s(N,n-1), \cdots, s(1,n-1); \\ &\cdots; s(N,n-M), \cdots, s(m-M,n-M)]^T. \end{aligned}$$

With this transformation, we obtain

$$\mathbf{s}(m,n) = F\mathbf{s}(m-1,n) + Gu(m,n) \tag{11}$$

with corresponding observational model

$$r(m,n) = H\mathbf{s}(m,n) + \mathbf{v}(m,n) \tag{12}$$

which is identical in form to (1) and (2) with line by line scanning understood. Thus we could immediately write down the Kalman equations. They would be (3) to (7) with the above interpretation of the $\mathbf{s}$ vector. The difficulty with these equations is the amount of computation and memory requirements associated with them. For example, consider an $(M \times M)$th order system model as in (8) and an observation region consisting of an $N \times N$ square with $N \gg M$. Then the dimension of the matrix equations (3) to (8) is approximately MN. This means that in general the order of the computation is $O(M^3N^3)$. However, taking advantage of the spatial invariant structure of the signal model, this can be reduced to $O(M^2N^2)$ as will be shown below. For $M = 4$ and $N = 100$, we require on the order of 160 000 multiplies and adds per output point. The overall total computation for the 10 000 element picture would be $O(10^9)$. At 1 μs/operation, the computer time for such a calculation would be 10^3 s or approximately 20 min. In addition, the data storage problems are immense. To store $P_{b,a}$ at each stage we need $O(MN)^2$ storage locations. For the above M and N, this is 160 000 words of storage to be accessed at each picture point! These numbers tell us that the exact 2-D Kalman scalar filter is computationally unmanageable, given today's state of the computer art. However, not only that, it is also wasteful of computation as will be pointed out below.

Next we consider the 2-D Kalman vector filter [8]. This is also an exact processor as is the scalar one mentioned above; however, the vector processor observes a line at a time instead of a point at a time.

B) A Kalman Vector Filter

In this formulation, we consider a vector scanning [8] consisting of a (horizontal) line at a time, as is necessary to do optimal Kalman vector filtering. We can write a recursive signal model

$$\boldsymbol{s}(n) = \boldsymbol{F}\boldsymbol{s}(n-1) + \boldsymbol{G}\boldsymbol{u}(n) \tag{13}$$

with vector observations

$$\boldsymbol{r}(n) = \boldsymbol{H}\boldsymbol{s}(n) + \boldsymbol{v}(n). \tag{14}$$

Here

$$\boldsymbol{s}(n) \triangleq [s(1,n), \cdots, s(N,n); s(1,n-1), \cdots, s(N,n-1); \cdots; s(1,n-M+1), \cdots, s(N,n-M+1)]^T$$

and

$$\boldsymbol{r}(n) = [r(1,n), \cdots, r(N,n)]^T.$$

Since (13) and (14) are of the form of (1) and (2), the Kalman equations are (3) thru (7) with the above interpretations for the $\boldsymbol{s}$ and $\boldsymbol{r}$ vectors. We note that these matrix equations are of order $O(MN)$ as in the 2-D Kalman scalar filter. However, each iteration of the vector filter yields N estimates (a whole line), so for the example given for the scalar filter, the vector filter computation times are reduced by a factor of N yielding an $O(M^3N^2)$ computation. The storage requirements are the same and hence this is again a computationally unmanageable solution.[2]

We will see below how splitting the picture up into partially overlapping strips can greatly reduce the computation of this vector filter. Further reductions in the order of the computation can be realized by using the special block form of the matrices $\boldsymbol{F}$, $\boldsymbol{G}$, and $\boldsymbol{H}$ in (13) and (14). These reductions in computation permit the practical implementation of the Kalman vector filter. As in the 2-D scalar case, we can construct the recursive dynamical model (13) by using a 2-D spectral factorization procedure [6].

[2] Note that the term vector processor was used in [8] to refer to a processor less than a line wide. However, we will reserve this term for the full width processor.

III. New Developments

In this section, we will summarize new approximate methods for realizing the 2-D Kalman scalar and vector filters [10], [11]. While both these techniques provide 2 to 3 orders of magnitude reduction in computation and storage, it can be anticipated that the reduction in the optimality of the estimate will be only slight. The reason for this surprising result is that the cause of the excessive computations is "purely theoretical," that is, the very uncommon but possible occurrence of high correlation at great distances. We will first develop an approximate Kalman vector filter called the strip processor. Then we will present a scalar filter called the reduced update Kalman filter.

A) A Kalman Strip Processor

The Kalman strip filter arises from the following considerations. First, although the vector filter has a state vector whose dimension is the width of the picture as is required for theoretical optimality, experience suggests that very few images will have the large correlation at great distances to justify such computation. Rather, most images show significant correlation only over rather compact neighborhoods of a given point. Thus we consider processing strips of the picture independently. The strips are overlapped with only the middle points being used as the final estimate. If the strip width is greater than the correlation distance of the signal and noise field, then close to optimum performance can result for points near the center. Processing the picture in these strips can greatly reduce the order of the computation and storage.

Second, we observe that if the data are stationary in the horizontal direction, we can save computation and storage by the following scheme: take the strips in the vertical direction and note that the filters for the different strips will all be the same. Thus the Kalman covariance equations (4) and (7) and the gain equation (5) need be calculated only for the first strip. Then the corresponding line in each of the remaining strips may be updated via (6) using the same gain matrix. In this way, storage need be provided for the covariance and gain matrices only in the first strip. Thus these two simplifications allow us to construct a good approximation to the Kalman vector filter which is computationally manageable. We will now elaborate on the details of constructing such a Kalman strip filter. This will consist of two parts: first, a discussion of the model, second, a discussion of the filter for that model.

1) The Strip Dynamical Model: The recursive signal model is the same as given in (13) with vector observations given in (14) with the exception that the vectors now are of dimension MW, where W is the strip width ($W \ll N$). We start the model development from a general NSHP recursive model (see Appendix), and develop from it a vector recursive model,

$$\boldsymbol{s}(n) = \boldsymbol{F}\boldsymbol{s}(n-1) + \boldsymbol{G}\boldsymbol{u}(n) \tag{15}$$

$$\boldsymbol{r}(n) = \boldsymbol{H}\boldsymbol{s}(n) + \boldsymbol{v}(n), \tag{16}$$

where

$$\boldsymbol{s}(n) \triangleq [s(1,n), \cdots, s(W,n); s(1,n-1), \cdots, s(W,n-1); \cdots; s(1,n-M+1), \cdots, s(W,n-M+1)]^T$$

and

$$\boldsymbol{r}(n) \triangleq [r(1,n), \cdots, r(W,n)]^T.$$

To obtain $r = s + v$, choose

$$\boldsymbol{H} = [I:0: \cdots :0]^T.$$

Now $\boldsymbol{F}$, $\boldsymbol{G}$, and $\boldsymbol{Q}_u$ must be chosen so that s approximately satisfies (8). The conversion from (8) to (15) is most easily seen in the Z-transform domain. Taking the Z-transform of (8), we see that the system function transforming s into u is

$$\frac{U(z_1,z_2)}{S(z_1,z_2)} = 1 - C(z_1,z_2), \tag{17}$$

where

$$C(z_1,z_2) = \sum_{\mathcal{R}_{\oplus+}} c_{pq} z_1^{-p} z_2^{-q}. \tag{18}$$

Equation (17) can be factored as

$$1 - C(z_1,z_2) = (1 - C(z_1,\infty))(1 - C^{(1)}(z_1,z_2)), \tag{19}$$

where

$$C(z_1,\infty) = \sum_{p>0} c_{p0} z_1^{-p},$$

and where $C^{(1)}(z_1,z_2)$ contains only $q > 0$ terms and is generally of infinite order in z_1^{-1}.

Now $C^{(1)}$ can be approximately represented by $\boldsymbol{F}$ since it represents an operation on the previous rows. The first factor on the right side of (19) can be inverted and represented by $\boldsymbol{G}$. In this way we arrive at the following form for $\boldsymbol{F}$:

$$\boldsymbol{F} = \begin{bmatrix} \boldsymbol{A}_1 & \boldsymbol{A}_2 & \cdots & \boldsymbol{A}_M \\ \boldsymbol{I} & \boldsymbol{O} & \cdots & \boldsymbol{O} \\ \boldsymbol{O} & \boldsymbol{I} & \boldsymbol{O} & \boldsymbol{O} \\ \boldsymbol{O} & \cdots & \boldsymbol{O}\;\boldsymbol{I} & \boldsymbol{O} \end{bmatrix}, \tag{20}$$

which is $MW \times MW$ with $W \times W$ blocks $\boldsymbol{I}$ and $\boldsymbol{A}_m$; $m = 1, \cdots, M$. Here

$$\boldsymbol{A}_m = [\boldsymbol{I} - \boldsymbol{C}_0]^{-1} \boldsymbol{C}_m \tag{21}$$

where the $\boldsymbol{C}_m$ are given by

$$(\boldsymbol{C}_m)_{ij} = c_{m,i-j}. \tag{22}$$

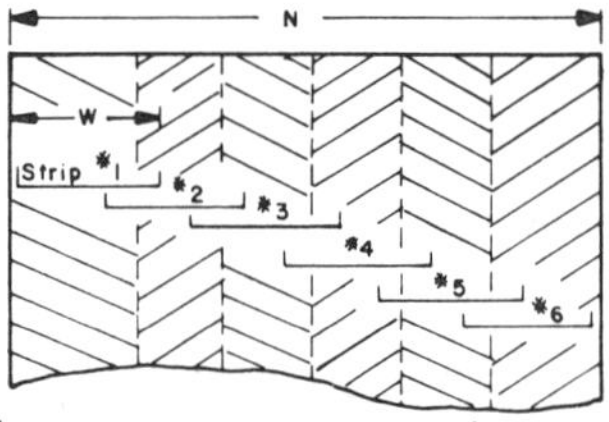

Fig. 2. Kalman strip processing example (dotted lines indicate estimate boundaries).

In like manner, we could set

$$\boldsymbol{G} = [(\boldsymbol{I} - \boldsymbol{C}_0)^{-1} \vdots \boldsymbol{O} \vdots \boldsymbol{O} \vdots \boldsymbol{O} \vdots \cdots \vdots \boldsymbol{O}]^T.$$

However, the edge effects due to the small strip width W would be expected to cause significant errors. To lessen this problem, we choose $\boldsymbol{G}$ to make the correlations of models (15) and (8) agree exactly in the current row. This procedure leaves uncorrected slight errors in the correlations with the $(M - 1)$ previous rows in the model. Since the errors are both small and concentrated near the strip edges, they are expected to be insignificant in effect on the "saved" estimates in the middle section of the strip.

2) *The Kalman Strip Filter:* The Kalman equations for this model are

$$\hat{\boldsymbol{s}}_b^i(n) = \boldsymbol{F}\hat{\boldsymbol{s}}_a^i(n-1) \tag{23}$$

$$\hat{\boldsymbol{s}}_a^i(n) = \hat{\boldsymbol{s}}_b^i(n) + \boldsymbol{K}^1(n)[\boldsymbol{r}^i(n) - \boldsymbol{H}\hat{\boldsymbol{s}}_b^i(n)], \tag{24}$$

where i is the number of the strip. In addition, to calculate the gain matrix for the nth line we must compute

$$\boldsymbol{P}_b^1(n) = \boldsymbol{F}\boldsymbol{P}_a^1(n-1)\boldsymbol{F}^T + \boldsymbol{G}\boldsymbol{Q}_u\boldsymbol{G}^T \tag{25}$$

$$\boldsymbol{K}^1(n) = \boldsymbol{P}_b^1(n)H^T[\boldsymbol{H}\boldsymbol{P}_b^1(n)\boldsymbol{H}^T + \boldsymbol{Q}_v]^{-1} \tag{26}$$

$$\boldsymbol{P}_a^1(n) = [\boldsymbol{I} - \boldsymbol{K}^1(n)\boldsymbol{H}]\boldsymbol{P}_b^1(n). \tag{27}$$

We note that for the horizontally stationary case, (25) through (27) need only be computed once for each line, while the estimates (23) and (24) are computed for all the strips at the nth line before proceeding to line $(n + 1)$. Initial conditions are needed for (23) and (25) at the first line. They are given by

$$\boldsymbol{s}_b^i(1) = \boldsymbol{E}[\boldsymbol{s}^i(1)] = \boldsymbol{o} \tag{28}$$

and

$$\boldsymbol{P}_b^1(1) = \boldsymbol{E}[\boldsymbol{s}^i(1)\boldsymbol{s}^{iT}(1)] = \boldsymbol{R}_s, \tag{29}$$

where $\boldsymbol{R}_s$ is the $MW \times MW$ correlation matrix of the model.

Fig. 2 shows a diagram for strip processing where the middle third of the elements are retained for the final estimate of s. The overlap would require a factor of three in computation. In the scheme used in the example of Section IV, a width of $W = 25$ is used with retention of the middle 15 elements. In this case, the "extra" computation due to the overlap is reduced to a factor of two. The amount of "extra" computation can be traded off for uniformity in the estimation error.

3) Optimization of Strip Width: To minimize the amount of computation for the completely nonstationary case, we can choose the number of retained elements N_R and the strip width W consistent with obtaining near optimum performance. To keep the saved points at least D from the strip edge, where D is the "correlation distance" of the signal s, we get

$$W = 2D + N_R.$$

The number of computations per point is proportional to

$$\frac{(2D+N_R)^3}{N_R}, \qquad \text{for } N \gg 2D.$$

The minimum occurs at

$$N_R^o = D \tag{30}$$

which yields

$$W^o = 3D \tag{31}$$

to minimize computation.

When the data are stationary horizontally, then the number of computations is asymptotically

$$\frac{(2D+N_R)^2}{N_R}, \qquad \text{for } N \gg 2D.$$

The minimum turns out to be

$$N_R^o = 2D$$

which yields

$$W^o = 4D. \tag{32}$$

B) Reduced Update Kalman Filtering

In this section, we will treat an approximation to the 2-D Kalman scalar processor presented in Section II-A. This involves the dynamical model (8) and the scalar observation (10). As indicated earlier, the state vector for this filter is $O(MN)$ dimensional, where M is the order of the recursive model and N is the width of the picture. Thus all these points must be updated in an equation equivalent to (6). The main concept of the reduced update filter is that the Kalman equations are composed of two steps: a prediction part and an update part. Now the prediction part, as set forth in (3), is a computationally straightforward projection of $O(M^2)$ previous estimates. However, the update part involves calculations involving each of the $O(MN)$ random variables in the state vector. Since $N \gg M$, we can reduce the bulk of the computation by reducing the update process. Thus we choose to update only those elements of the state vector within a certain distance of the point currently being processed, (m,n). We can expect this procedure to result in a good approximation because significant updates will be confined to a region around the observation at (m,n). Therefore, omitting the update of points far away should only minimally impact performance. For convenience of notation, we take this region to be the support of the NSHP recursive signal model in (8). Hence, there are only $O(M^2)$ points to update, i.e., we only update the pseudo-state vector for each observation. In the remainder of this section, we will derive the equations for the optimal updating of the pseudo-state vector. We will see that these equations can point the way to further simplifications which are expected again to result in only slightly suboptimum performance.

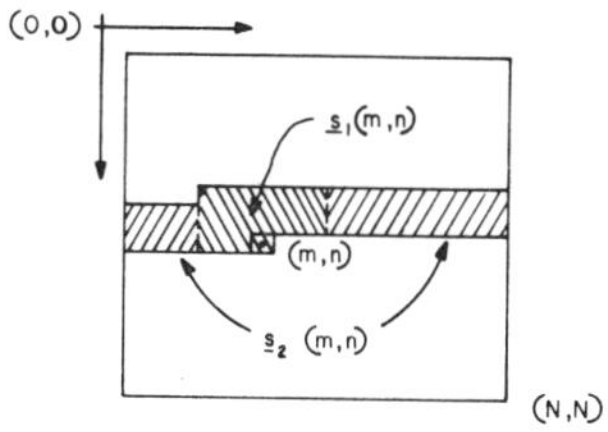

Fig. 3. Assignment of points to partitioned state vector.

Let the signal model be generated by

$$s(m,n) = \sum_{\mathcal{R}_{\oplus+}} c_{kl} s(m-k, n-l) + w(m,n), \tag{33}$$

where $\{w(m,n)\}$ is a zero mean homogeneous white Gaussian source with variance σ_w^2. The received signal is

$$r(m,n) = s(m,n) + v(m,n), \tag{34}$$

where $\{v(m,n)\}$ is zero mean homogeneous white Gaussian noise, independent of $\{w(m,n)\}$, with variance σ_v^2.

Let the field $\{r(m,n)\}$ be observed over an $N \times N$ rectangular region. Next, introduce a vector notation for the pseudo-state vector corresponding to a scalar line by line scan, viz.,

$$\begin{aligned} \mathbf{s}_1(m,n) \triangleq [&s(m,n), \cdots, s(m-M,n); s(m+M,n-1), \\ &\cdots, s(m-M,n-1); \\ &\cdots; s(m+M,n-M), \cdots, s(m-M,n-M)]^T. \end{aligned} \tag{35}$$

Order the remaining points of the state vector onto $\mathbf{s}_2(m,n)$. Thus the resulting assignment of points is as shown in Fig. 3. With this convention the state vector can be written

$$\mathbf{s}(m,n) = (\mathbf{s}_1^T(m,n), \mathbf{s}_2^T(m,n))^T. \tag{36}$$

The state dynamical model can then be written

$$\mathbf{s}(m,n) = \mathbf{C}\mathbf{s}(m-1,n) + \mathbf{w}(m,n), \tag{37}$$

where $\mathbf{C}$ is the system propagation matrix determined by $\{c_{kl}\}$ and by the ordering of the state vector $\mathbf{s}(m,n)$. Note that (37) holds for all the points (m,n), except near the boundaries where boundary conditions must be incorporated. The drive vector is given as

$$\mathbf{w}(m,n) \triangleq (w(m,n), 0, \cdots, 0)^T. \tag{38}$$

The scalar observation equation is

$$r(m,n) = \mathbf{H}\mathbf{s}(m,n) + v(m,n), \tag{39}$$

where

$$\mathbf{H} = (1, 0, \cdots, 0). \tag{40}$$

We can partition the matrix C similarly to $\boldsymbol{s}$ as

$$C = \begin{pmatrix} C_{11} & C_{12} \\ C_{21} & C_{22} \end{pmatrix}. \tag{41}$$

It then turns out that C_{11} and C_{12} contain all the $\{c_{kl}\}$ terms, and the remainder of C constitutes a shift transformation. Thus (37) may be rewritten as

$$\boldsymbol{s}_1(m,n) = C_{11}\boldsymbol{s}_1(m-1,n) + \boldsymbol{w}_1(m,n) + C_{12}\boldsymbol{s}_2(m-1,n), \tag{42}$$

where $\boldsymbol{w}$ has been partitioned similarly to $\boldsymbol{s}$. Equation (42) focuses on the computation to be performed at (m,n). That is, $\boldsymbol{s}_2$ requires only the shifting of previously computed values. We can similarly partition $\boldsymbol{H} = (\boldsymbol{H}_1, \boldsymbol{H}_2)$ with $\boldsymbol{H}_2 = \boldsymbol{O}$ to get a new observation equation

$$r(m,n) = \boldsymbol{H}_1\boldsymbol{s}_1(m,n) + v(m,n). \tag{43}$$

1) Derivation of Reduced Update Filter: Assume that the received array $\{r(m,n)\}$ is scanned in line by line fashion. Then the Kalman filtering equations can be written immediately from (3) through (7) as discussed in Section I.

a) Extrapolation:

$$m \to m+1$$

$$\boldsymbol{P}_b(m,n) = \boldsymbol{C}\boldsymbol{P}_a(m-1,n)\boldsymbol{C}^T + \boldsymbol{G}\boldsymbol{Q}_w\boldsymbol{G}^T \tag{44}$$

$$\hat{\boldsymbol{s}}_b(m,n) = \boldsymbol{C}\hat{\boldsymbol{s}}_a(m-1,n) \tag{45}$$

b) Update:

$$\boldsymbol{K}(m,n) = \boldsymbol{P}_b(m,n)\boldsymbol{H}^T[\boldsymbol{H}\boldsymbol{P}_b(m,n)\boldsymbol{H}^T + \sigma_v^2]^{-1} \tag{46}$$

$$\hat{\boldsymbol{s}}_a(m,n) = \hat{\boldsymbol{s}}_b(m,n) + \boldsymbol{K}(m,n)[r(m,n) - \boldsymbol{H}\hat{\boldsymbol{s}}_b(m,n)] \tag{47}$$

$$\boldsymbol{P}_a(m) = [\boldsymbol{I} - \boldsymbol{K}(m,n)\boldsymbol{H}]\boldsymbol{P}_b(m,n). \tag{48}$$

If an arbitrary gain matrix is used instead of the one prescribed in (46), the covariance $\boldsymbol{P}_a$ given in (48) changes to [5]

$$\boldsymbol{P}_a(m,n) = (\boldsymbol{I} - \boldsymbol{K}(m,n)\boldsymbol{H})\boldsymbol{P}_b(m,n)(\boldsymbol{I} - \boldsymbol{K}(m,n)\boldsymbol{H})^T + \boldsymbol{K}(m,n)\sigma_v^2\boldsymbol{K}^T(m,n). \tag{49}$$

Let $\boldsymbol{K}(m,n)$ have the form

$$\boldsymbol{K}(m,n) = \begin{bmatrix} \boldsymbol{K}_1(m,n) \\ \hline \boldsymbol{O} \end{bmatrix}, \tag{50}$$

where the partitioning corresponds to that of $\boldsymbol{s}(m,n)$. We now choose $\boldsymbol{K}_1(m,n)$ to minimize the trace of $\boldsymbol{P}_a(m,n)$ as given in (49) for $\boldsymbol{K}(m,n)$ as given in (50). The result is the reduced update Kalman filter.

c) Extrapolation:

$$m \to m+1$$

$$\boldsymbol{P}_b(m,n) = \boldsymbol{C}\boldsymbol{P}_a(m-1,n)\boldsymbol{C}^T + \boldsymbol{G}\boldsymbol{Q}_w\boldsymbol{G}^t \tag{51}$$

$$\hat{\boldsymbol{s}}_{1b}(m,n) = \boldsymbol{C}_{11}\hat{\boldsymbol{s}}_{1a}(m-1,n) + \boldsymbol{C}_{12}\hat{\boldsymbol{s}}_{2a}(m-1,n). \tag{52}$$

d) Update:

$$\boldsymbol{K}_1(m,n) = \boldsymbol{P}_{11,b}(m,n)\boldsymbol{H}_1^T(\boldsymbol{H}_1\boldsymbol{P}_{11,b}(m,n)\boldsymbol{H}_1^T + \sigma_v^2)^{-1} \tag{53}$$

$$\hat{\boldsymbol{s}}_{1a}(m,n) = \hat{\boldsymbol{s}}_{1b}(m,n) + \boldsymbol{K}_1(m,n)[r(m,n) - \boldsymbol{H}_1\hat{\boldsymbol{s}}_{1b}(m,n)] \tag{54}$$

$$\boldsymbol{P}_{11,a}(m,n) = [\boldsymbol{I} - \boldsymbol{K}_1(m,n)\boldsymbol{H}_1]\boldsymbol{P}_{11,b}(m,n) \tag{55a}$$

$$\boldsymbol{P}_{12,a}(m,n) = [\boldsymbol{I} - \boldsymbol{K}_1(m,n)\boldsymbol{H}_1]\boldsymbol{P}_{12,b}(m,n), \tag{55b}$$

where $\boldsymbol{P}_a$ and $\boldsymbol{P}_b$ have been partitioned similarly to $\boldsymbol{s}$.

Equations (51)–(55) can provide great computational savings over the standard Kalman filtering equations. To understand these equations better, it is helpful to convert them back to scalar notation. First, we note that (52) will become, in scalar notation,

$$\hat{s}_b^{(m,n)}(m,n) = \sum_{\mathcal{R}_{\oplus+}} c_{kl}\hat{s}_a^{(m-1,n)}(m-k,n-l), \tag{56}$$

since it represents propagation of the previous estimates through the dynamics of the system. In these scalar equations, the superscript indicates the step in the filtering, while the argument represents the position of the data. Equation (51) represents the error in this predicted estimate; thus it becomes

$$R_b^{(m,n)}(m,n;k,l) = \sum_{op} c_{op}R_a^{(m-1,n)}(m-o,n-p;k,l), \qquad (k,l) \in \mathcal{S}_{\oplus+}^{(m,n)} \tag{57}$$

$$R_b^{(m,n)}(m,n;m,n) = \sum_{kl} c_{kl}R_b^{(m,n)}(m,n;m-k,n-l) + \sigma_w^2, \tag{58}$$

where $\mathcal{S}_{\oplus+}^{(m,n)}$ is the support of the state vector $\boldsymbol{s}(m,n)$.

Equation (53) computes a $\boldsymbol{K}_1(m,n)$ which has the same support as $\boldsymbol{s}_1(m,n)$, namely, $\mathcal{R}_{\oplus+}^{(m,n)}$, the support of the pseudo-state vector. The scalar equation identical to (53) is

$$K^{(m,n)}(i,j) = \frac{R_b^{(m,n)}(m,n;i,j)}{R_b^{(m,n)}(m,n;m,n) + \sigma_v^2}, \qquad (i,j) \in \mathcal{R}_{\oplus+}^{(m,n)}. \tag{59}$$

Similarly, (54) becomes

$$\hat{s}_a^{(m,n)}(i,j) = \hat{s}_b^{(m,n)}(i,j) + K^{(m,n)}(m-i,n-j)[r(m,n) - \hat{s}_b^{(m,n)}(m,n)], \qquad (i,j) \in \mathcal{R}_{\oplus+}^{(m,n)}. \tag{60}$$

Finally, (55a) and (55b) both are expressible as the set of scalar equations

$$R_a^{(m,n)}(i,j;k,l) = R_b^{(m,n)}(i,j;k,l) - K^{(m,n)}(m-i,n-j)R_b^{(m,n)}(m,n;k,l), \quad \text{for } (i,j) \in \mathcal{R}_{\oplus+}^{(m,n)};\ (k,l) \in \mathcal{S}_{\oplus+}^{(m,n)}. \tag{61}$$

The reduced update Kalman filter comprising (56) to (61) has been derived as an optimal approximation to the 2-D Kalman scalar filter. The prediction part of the Kalman filter is left unchanged, and the update is optimized under the constraint of updating only the nearest previously processed neighbors. It is thus perhaps not clear that the resulting reduced update filter is overall optimal in the class of 2-D recursive filters of the same order as itself because the prediction part of the Kalman filter was

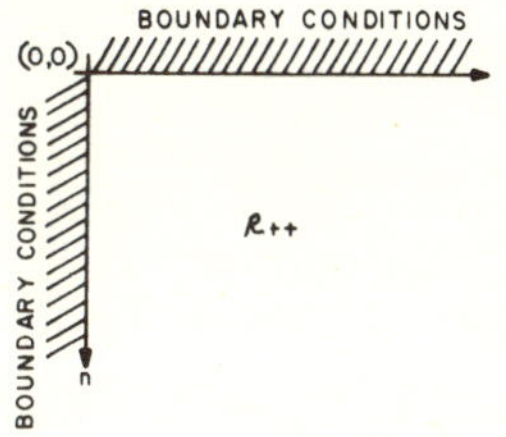

Fig. 4. Data region to be filtered by reduced update equations.

held fixed in the above derivation. However, it turns out that this procedure is indeed overall optimal as will now be seen.

Theorem: The reduced update Kalman filter is optimal under its constraint, i.e., it minimizes the mse over the class of spatially varying, linear, 2-D recursive filters of similar order.

Proof: Consider filtering the region $\mathcal{R}_{\oplus+}$ with boundary conditions (bc) on the upper and left edges as shown in Fig. 4. The method of proof will be induction. The signal model is given by (33) with observations given by (34). For the first point $(m,n) = 0$, obviously the best prediction will be

$$\hat{s}_b^{(0,0)}(0,0) = \sum_{\mathcal{R}_{\oplus+}} c_{kl} s^{(\mathrm{bc})}(-k,-l). \tag{62}$$

If this is followed by the optimal constrained update, we obtain

$$\hat{s}_a^{(0,0)}(0,0) = \hat{s}_{\mathrm{mmse}}(0,0),$$

the best mse estimate subject to constraint. Now assume we have the best constrained mse estimate at (m,n). Then the best prediction based on the pseudo-state is

$$\hat{s}_b^{(m+1,n)}(m+1,n) = \sum c_{kl} \hat{s}_a^{(m,n)}(m+1-k, n-l), \tag{63}$$

as can be seen by applying to (33) the conditional expectation operator $E[\cdot \mid \hat{s}_1(m,n)]$. Finally, the optimal constrained update generates the optimal constrained mse estimate at $(m+1,n)$. This completes the proof.

From this theorem follows the very important fact that for homogeneous data the Kalman reduced update filter will converge to an optimal 2-D NSHP recursive filter for estimating the data. Now, in one dimension, such a filter could be obtained via Wiener's spectral design procedure [9]. However, the same procedure in the 2-D case leads to an infinite-order filter, thus the optimal finite-order filter is not obtained. Hence, the reduced update Kalman filter is also particularly attractive from the standpoint of design of spatially invariant filters for homogeneous random fields.

2) Order of Computation: Next we investigate the order of computation of the reduced update Kalman filter. We will consider each of (56) thru (61) separately. First, (56) will be $O(M^2)$ for an Mth order NSHP model. Equation (56) will thus be $O(M^2)$ for each (k,l). Since there are $O(MN)$ points in $\mathcal{S}_{\oplus+}^{(m,n)}$, we obtain $O(M^3N)$ for (57). Equation (58) is $O(M^2)$, as are (58) and (60). Equation (61) is simple, but has to be computed for each pair (i,j) and (k,l) with $(k,l) \in \mathcal{S}_{\oplus+}^{(m,n)}$, thus giving a computational total of $O(M^3N)$ as in (57). Summing up, we get the overall total computation per point as $O(M^3N)$. This is to be compared with $O(M^3N^3)$ for the general Kalman scalar filter and $O(M^2N^2)$ for the $(M \times M)$th order half-plane filter model.

The overall savings of a factor N^2 results from two simplifications. First, the reduced update has reduced the orders of the matrices from $MN \times MN$ to $M^2 \times MN$. Second, for the $(M \times M)$th order filter model, the scalar equations (56) to (61) write only the newly computed values at point (m,n). Equation (51), for example, contains many more error covariance values than (57) and (58). However, only the first row (and column) of $\boldsymbol{P}_b(m,n)$ actually changes. The other elements simply get shifted. The scalar equations were written with respect to a fixed reference, the origin in the data plane. Thus no shifting appears in that notation. Another way of looking at this is as follows. The $\boldsymbol{C}$ matrix is composed of zeroes and ones except for the first row. These zeroes and ones simply serve to shift the data in the state vector, one place down as each new element is put in. This computation can be avoided by using indirect addressing and simply writing the new element over the oldest element in storage. Then a pointer or indirect address can be used to keep track of the "head" of the vector. This is essentially the reason for not counting the shifting operations, as they can be avoided with minimal computational effort on many machines.

For the reasons outlined in the previous paragraph, (56) thru (61) not only provide a convenient way to perceive the two-dimensional nature of the reduced update filter, but also present its essential computational aspects. The main computation was observed to be in (57) and (61). Equation (56) computes the error covariance between the "predicted point" (m,n) and the previous estimates in the filters state vector. Experience suggests this will be peaked at the point (m,n) with rapid decay with distance from (m,n). Thus it is a reasonable approximation to compute (57) only in a fixed size region including $\mathcal{R}_{\oplus+}^{(m,n)}$. This reasoning can also be applied to (61), where (k,l) would be restricted to a region significantly smaller than $\mathcal{S}_{\oplus+}^{(m,n)}$ and of fixed size for increasing N. Calling such a region $\mathcal{T}_{\oplus+}^{(m,n)}$ we can rewrite (57) and (61) as

$$R_b^{(m,n)}(m,n;k,l) = \sum c_{op} R_a^{(m-1,n)}(m-o, n-p; k,l), \qquad (k,l) \in \mathcal{T}_{\oplus+}^{(m,n)} \tag{64}$$

$$\begin{aligned} R_a^{(m,n)}(i,j;k,l) &= R_b^{(m,n)}(i,j;k,l) \\ &\quad - K^{(m,n)}(m-i, n-j) R_b^{(m,n)}(m,n;k,l), \\ &\quad \text{for } (i,j) \in \mathcal{R}_{\oplus+}^{(m,n)};\ (k,l) \in \mathcal{T}_{\oplus+}^{(m,n)}. \end{aligned} \tag{65}$$

These approximate reduced update equations reduce the order of the computation to $O(M^4)$, a constant with respect to N. Fig. 5 sets forth the envisioned region assignment. As mentioned above, the adverse effects of substituting $\mathcal{T}$

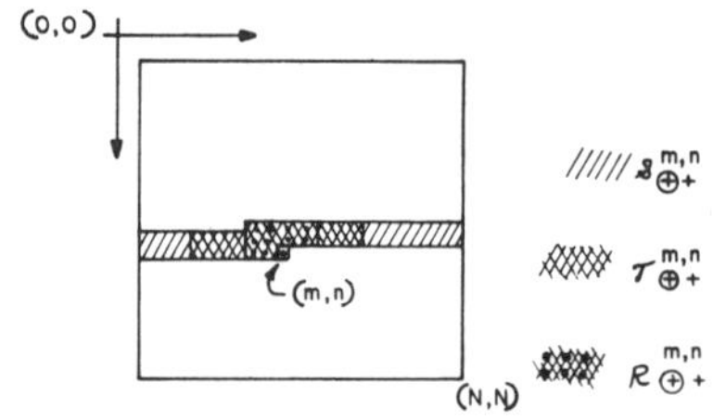

Fig. 5. Region assignment for $\mathcal{T}_{\oplus+}^{(m,n)}$, covariance update region.

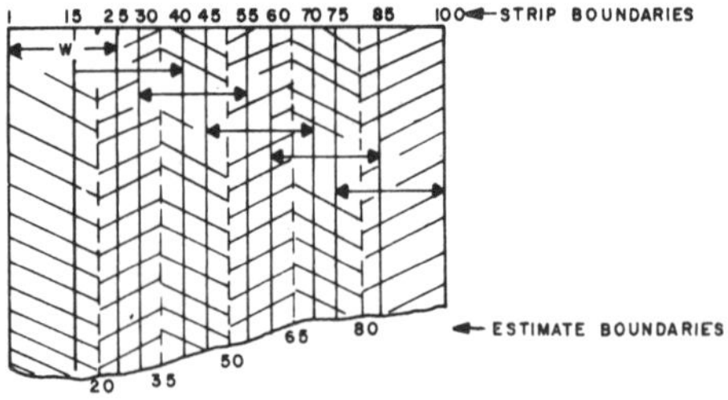

Fig. 6. Strip arrangement detail.

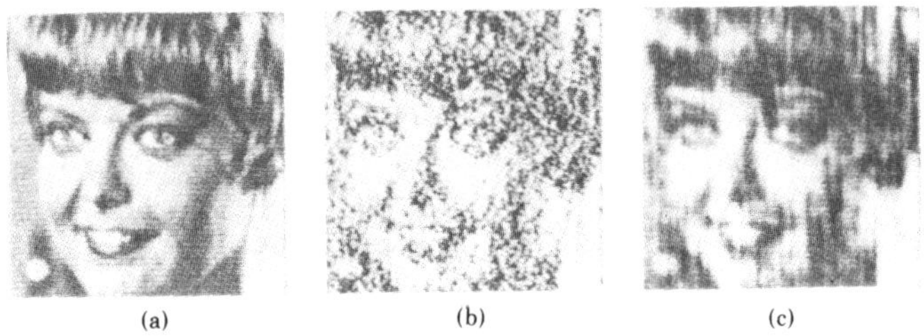

Fig. 7. Strip processing example. (a) Original. (b) Noisy (SNR = 1). (c) Estimate.

for $\mathcal{S}$ are expected to be minimal for most pictures. This is so both for the reasons mentioned previously and also because (59), the Kalman gain calculation does not directly make use of points outside $\mathcal{R}_{\oplus+}^{(m,n)}$. Thus the effect of this "truncation" would seem to be at worst second-order.

If we look at memory requirements, we find that they are dominated by the need to store the $\boldsymbol{R}_a$ error covariances. For the 2-D Kalman scalar filter, this storage will be (M^2N^2). For the reduced update filter we get the same amount of storage, only a portion of which is accessed at every point. For the approximate reduced update filter, we obtain $O(M^3N)$. A little thought reveals that this is the minimal possible dependence on N, i.e., linear, for a spatially varying processor.[3] Thus we cannot hope for further improvements here. However, for homogeneous regions, these storage requirements can be greatly relaxed. Then one only has to run the processor over a much smaller region to obtain near convergence to the steady-state filter. Subsequently, only (56) and (60) need be computed at a substantially lower amount of computational effort.

IV. Experimental Results

A noisy picture was processed with the Kalman strip filter. The signal was a 100×100 element center segment from a standard SMPTE test picture. The 8-bit original had an estimate of its mean removed prior to processing. The picture was then scaled so that its variance equaled one. White Gaussian noise was added at unit variance to produce a SNR of unity. The original with mean removed was also used to generate a (10×4)th order $+\oplus$ NHSP model [6], [7]. (See Appendix). This turns out to be a model for the transposed pictures. Thus the noisy image was transposed and inserted into the Kalman strip filter program with the above NSHP model. The strip arrangement is shown in detail in Fig. 6. There are six strips, each 25-points wide, with five-element overlap on each side for the interior strips. The middle 15 elements of the strips cover the picture and provide a fairly uniform error covariance. The mean value was added back and the estimate images transposed for presentation.

Fig. 7(a) shows the original image. Fig. 7(b) shows the noisy image. Fig. 7(c) shows the strip filter estimate. The measured SNR improvement was 7.7 dB. This compares favorably with previous results. As the model and noise were stationary, convergence was obtained; in this case, in 10 lines. The computer time was under 15 s on a CDC 7600 computer. The error covariance values were within 3 percent of their minimum value over the 15-element middle region of the strip. On the strip edge, it was as much as 60 percent above the minimum error variance value which was near the center.

The noisy picture of Fig. 7(b) was also modeled with a (3×3)th order $\oplus+$ NSHP factor and processed by the approximate reduced update Kalman filter for various sizes of $\mathcal{T}_{\oplus+}$. In particular, region half-widths of 3, 5, 7, and 9 were tried. The results were approximately the same in all cases, indicating both that the error decorrelates fast for our model and that the filter is fairly robust with respect to this type of error. The measured SNR improvement was 8.3 dB with output shown as Fig. 8. Convergence

[3] To see this, note that the processor must minimally have access to the error variances for the entire previous line.

Fig. 8. Reduced update estimate.

was obtained in 10 lines or less with a run time of 40 s. This run time could be greatly shortened if the filter were updated only for the first 10 columns in the first 10 lines instead of the entire first 10 lines. The error covariance values of a sampling of the columns indicate that such a change in the algorithm would have negligible effect on the final result. This is because the chosen covariance model, as is typical of images, does not show substantial correlation over distances on the order of ten or more pixels.

V. Conclusions

The Kalman strip filter was introduced as a new approximation to the Kalman vector filter which processes a line at a time. The new filter is not limited to separable correlation functions and provides nearly uniform estimation error over the image by incorporating strip overlap. An optimum strip width was chosen to minimize computation. The reduced update Kalman filter was introduced and shown to be optimum in that it minimizes the post update mse under the constraint of updating only the nearby previously processed neighbors. It was shown that for all-pole data models, the Kalman reduced update filter coverges to an optimum 2-*D* recursive filter in the homogeneous case.

Appendix

Spectral Factorization in Two Dimensions [6]

In this Appendix, we review some of the properties of two-dimensional (2-*D*) spectral factorization that are pertinent to the development of 2-*D* recursive models. Since our recursive model consists of a recursive filtering of spatially white Gaussian noise, the problem can be thought of as a filter design problem.

In [7] it is shown that under mild conditions, one may factor a 2-*D* spectrum into nonsymmetric half-plane (NSHP) factors. The support of the filter numerator a and denominator b is restricted to a nonsymmetric half-plane, that is, a half-plane with the negative axis removed, as shown in Fig. 9. This removal allows the filter to be recursible. Fig. 10 shows how the NSHP support of b permits the 2-*D* recursion from left to right and top to bottom. This recursion can be accomplished precisely, because only previously computed points are used in the computation of the present output.

By the order of an NSHP filter, we mean the following: let the support of b and a be given by the union of an $(M_1 + 1) \times (M_2 + 1)$ square and an $M_1 \times M_2$ square situated as shown in Fig. 11. We will agree to call such a filter an $(M_1 \times M_2)$th-order NSHP

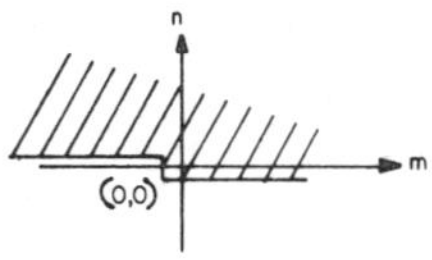

Fig. 9. Nonsymmetric half-plane.

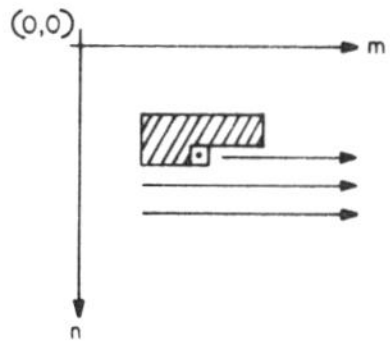

Fig. 10. Recursion of filter with NSHP support.

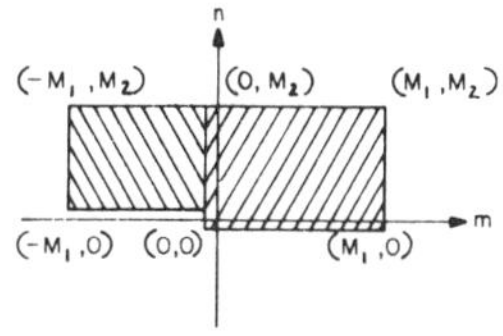

Fig. 11. Diagram illustrating concept of order of NSHP filter.

filter. This notation is then indicative of the highest powers of z_1, z_2, z_1^{-1}, and z_2^{-1} in the filter's 2-*D* *Z*-transform

$$B(z_1,z_2) = \sum_{m=0}^{M_1} \sum_{n=0}^{M_2} b_{mn} z_1^{-m} z_2^{-n} + \sum_{m=-1}^{-M_1} \sum_{n=-1}^{-M_2} b_{mn} z_1^{-m} z_2^{-n}. \qquad \text{(A1)}$$

As mentioned above, these recursive factors arise from a spectral factorization, in this case, that of the signal spectrum S_s. In general, the exact factors of S_s will be infinite-order, however, finite-order approximations can result in a specified small error. We can obtain such an approximation, in the simplest case, by simply truncating the factors to finite support. Thus we can solve the following rational approximation problem.

Given $\epsilon > 0$, choose the NSHP filter order $M_1 \times M_2$ high enough so that

$$\left|\left| S_s(u,v) - \left|\frac{A(e^{ju},e^{jv})}{B(e^{ju},e^{jv})}\right|^2 \right|\right| < \epsilon, \qquad \text{(A2)}$$

where A and B are the 2-*D* *Z*-transforms of the filter numerator a and denominator b, respectively, and where $||\cdot||$ is a suitable functional norm, e.g., L^1, L^2, L^∞, etc., and most importantly where the resulting 2-*D* recursive filter is *stable*. In [6], a window method is advanced as a simple way to design these filters. This method was used to obtain the models in this paper. More elaborate and efficient design methods are presently under investigation.

References

[1] A. Habibi, "Two-dimensional Bayesian estimate of images," *Proc. IEEE*, vol. 60, p. 878–883, July 1972.

[2] N. E. Nahi, "Role of the recursive estimation in statistical image enhancement," *Proc. IEEE*, vol. 60, p. 872–877, July 1972.

[3] S. R. Powell and L. M. Silverman, "Modeling of two-dimensional covariance functions with application to image restoration," *IEEE Trans. Auto. Control*, vol. AC-19, p. 8–13, Feb. 1974.

[4] P. Padulo and M. A. Arbib, *System Theory*. Philadelphia, PA: W. B. Saunders, 1974, p. 21.

[5] A. P. Sage and J. L. Melsa, *Estimation Theory with Applications to Communications and Control*. New York: McGraw-Hill, 1971, p. 89–90.

[6] M. P. Ekstrom and J. W. Woods, "Two-dimensional spectral factorization with applications in recursive digital filtering," *IEEE Trans. Acoust., Speech and Signal Proc.*, vol. ASSP-24, pp. 115–128, Apr. 1976.

[7] J. W. Woods and M. P. Ekstrom, "Nonsymmetric half-plane recursive filters—characterization, stability theory and test," in *Proceedings 1975 Int'l Sympos. on Circuits and Systems*, Newton, MA, Apr. 1975, p. 447–450.

[8] N. E. Nahi and C. A. Franco, "Recursive image enhancement by vector scanning," *IEEE Trans. Commun.*, vol. C-21, pp. 305–311, April 1973.

[9] N. Wiener, *The Extrapolation, Interpolation, and Smoothing of Stationary Time Series*. New York: Wiley, 1949.

[10] J. W. Woods and C. H. Radewan, "The Kalman strip filter—a two-dimensional recursive vector processor," in *Proc. 9th Asilomar Conference Circuits, Sys. & Comp.*, Pacific Grove, CA, Nov. 1975.

[11] J. W. Woods and C. H. Radewan, "Reduced update Kalman filter—a two-dimensional recursive processor," presented at *Johns Hopkins Conf. on Inform. Sci. & Sys.*, Baltimore, MD, Apr. 1976.

28

Reprinted from pp. 11-24–11-26 of the *1975 Internat. Conf. Commun. Rec.*

CONSTRAINED OPTIMAL LINEAR RECURSIVE ESTIMATION OF TWO-DIMENSIONAL PROCESSES

Rodolfo Feick Michael G. Strintzis
University of Pittsburgh

ABSTRACT

We introduce in this paper a class of linear recursive estimation methods which rely on minimal modeling and which are direct and highly efficient under arbitrary computational and storage restrictions. Correlation sequences are assumed to be known with reasonable accuracy and thus, the primary application area envisioned is composed of discrete communication and identification problems. The method is particularly valuable in application to two-dimensional image processing problems, since in this area storage requirements are particularly pronounced and no accurate dynamical models exist which lead to practicable estimation schemes.

INTRODUCTION

The artificial separation of modeling and estimation techniques, and the effects of this separation are strikingly demonstrated in current work on estimation for two-dimensional images. To begin with, the reluctance to integrate modeling and estimation has led to the formation of two distinct classes of models. The first class consists of models (such as "Markov Mesh"[1] models) which are accurate but so unwieldy and complicated that they have not yet led to any practicable image restoration algorithm. Because of this difficulty, an entirely different class of dynamical models [2,3,4,5] has been built more or less to fit known estimation procedures, the Kalman-Bucy procedure in particular. In this second class of models the correlation functions ascribed to images are two-dimensional, stationary, and factorable into the product of the respective point correlations along each of the two principal axes.[2,3,4,5] The "optimum" algorithms that have been developed for these models often require arbitrary truncation to meet the severe storage and computational limitations inherent to on-line processing. Thus a suboptimal algorithm will result even if the underlying model assumptions were realistic. It is easy to see that factorability of the covariance function is not a very realistic assumption. In fact, it would be reasonable to assume that the correlation is a function of the euclidean distance of points on the plane. Such correlations are not factorable.

In this paper we illustrate the development of computationally simple recursive linear filters of a prespecified recursion order. While these filters may be inferior to the unconstrained optimum linear filters, the computational complexity of the latter and the consequent need for truncation may result in a degradation of performance offsetting any advantage of the optimum filter. The proposed filters are general in that they are applicable to any type of covariance function. The orders of the recursion, i.e., the storage requirements, are prespecified as design parameters of the filter.

Finally, before turning to the description of the filters we point out that the fact that simple examples were chosen to illustrate the method, does not mean that the applicability of the algorithm is limited to these simple cases. In fact, the extension from the scalar to the vector case is conceptually straightforward although notationally more cumbersome.

Description of the Filter. The concepts involved in the design of the recursive two-dimensional filter can be illustrated most easily and compactly in the one-dimensional scalar case. Let

$$\left\{x(k)\right\}_{k=-\infty}^{k=\infty}$$

be a scalar valued, zero mean wide sense stationary stochastic process with known but arbitrary covariance sequence $R(n)$; $n=1,2,\ldots$ We wish to obtain the minimum mean square error estimate $\hat{x}(k)$ of $x(k)$ under the constraint that:

$$\hat{x}(k) = a(k)\,\hat{x}(k-1) + b(k)\,y(k) \tag{1}$$

where $y(k)$ is a measurement of $x(k)$, for simplicity assumed corrupted by additive white noise $w(k)$ uncorrelated with the process $x(k)$, i.e.:

$$y(k) = x(k) + w(k) \tag{2}$$

We will assume that the first available measurement is $y(o)$. The values of $a(k)$ and $b(k)$ for which the kth-step error $e_k=E[\hat{x}(k)-x(k)]^2$ is minimized can be determined from the following orthogonality conditions.

$$E[(\hat{x}(k) - x(k))\,\hat{x}(k-1)] = 0 \tag{3}$$

$$E[(\hat{x}(k) - x(k))\,y(k)] = 0 \tag{4}$$

Substitution of (1) into (3) and (4) yields:

$$a(k)E[\hat{x}(k-1)\hat{x}(k-1)]+b(k)E[\hat{x}(k-1)y(k)]=E[\hat{x}(k-1)x(k)] \tag{5}$$

$$a(k)E[\hat{x}(k-1)y(k)]+b(k)E[y(k)y(k)]=E[x(k)y(k)] \tag{6}$$

these equations can be simplified by use of the orthogonality conditions and assumptions:

$$E[\hat{x}(k-1)\,\hat{x}(k-1)] = E[\hat{x}(k-1)\,x(k-1)] \tag{7}$$

$$E[\hat{x}(k-1)y(k)] = E[\hat{x}(k-1)x(k)] \tag{8}$$

$$E[y(k)\,x(k)] = R(0) \tag{9}$$

$$E[y(k)y(k)] = R(0) + V \tag{10}$$

where $V= E[w(k)w(k)]$ is the noise power. (11)

Substituting these relations into (5) and (6) we can write:

$$\begin{bmatrix} E[\hat{x}(k-1)x(k-1)] & E[\hat{x}(k-1)x(k)] \\ E[\hat{x}(k-1)x(k)] & R(0)+V \end{bmatrix}\begin{bmatrix} a(k) \\ b(k) \end{bmatrix}=\begin{bmatrix} E[\hat{x}(k-1)x(k)] \\ R(0) \end{bmatrix} \tag{12}$$

It follows from (1) that:

$$E[\hat{x}(k-1)x(k-1)] = b(k-1)R(0)+a(k-1)E[\hat{x}(k-2)x(k-1)] \tag{13}$$

$$\begin{aligned} E[\hat{x}(k-1)x(k)] = {} & b(k-1)R(1)+a(k-1)b(k-2)R(2) + \\ & a(k-1)a(k-2)b(k-3)R(3)+ \\ & a(k-1)a(k-2)a(k-3)b(k-4)R(4)+\ldots \end{aligned} \tag{14}$$

We note that the calculation of $E[\hat{x}(k-1)x(k-1)]$ requires only the storage of $b(k-1)$, $a(k-1)$ and

$E[\hat{x}(k-2)x(k-1)$ from the previous step. The calculation of $E[\hat{x}(k-1)x(k)]$ will in general require a "growing memory" since all of the previously obtained gains $a(\cdot)$ and $b(\cdot)$ are needed. The algorithm appears to converge reasonably fast irrespective of the initial conditions. A set of initial conditions can be obtained for example by noting that since for k=0 there exists no previous estimate it is reasonable to set $\hat{x}(0)=b(0)y(0)$ and from (6), $b(0) = \frac{R(0)}{R(0)+V}$, $a(0) = 0$. The least squares error at every step is easily seen to be $b(k)V$. The fact that except for the trivial case where $R(n) = R(0)\lambda^n$ the algorithm requires a "growing memory" to store all previously calculated values of $a(\cdot)$ and $b(\cdot)$ until steady state is reached suggests that it may be advantageous to obtain these steady state values directly from (12). Once the steady state gains are known it may be possible to use these values directly if some deterioration in performance during the transient is acceptable. A different and very promising scheme leading to a nongrowing-memory filter is to determine a covariance function of the form $R(0)\lambda^n$ such that the corresponding "Kalman filter" has steady state gains equal to the steady state gains obtained from (12). Such a model could be termed a "task oriented Markov model" since it is specified expressly so as to lead to efficient completion of the specific estimation task.

The steady state equations obtained from (12) by assuming $a(i)=a$ and $b(i)=b$ for every integer i can be shown to be:

$$z(a) = \frac{b-1}{b-1+a^2} \tag{15}$$

$$z(a) = \frac{1}{b}\left(1-b\frac{V}{R(0)}\right) \tag{16}$$

where,

$$z(a) = \sum_{i=0}^{\infty} \frac{R(i)}{R(0)} a^i \tag{17}$$

It must be noted that if the solution to this system of nonlinear equations is not unique it is possible that a solution does not satisfy the orthogonality equations. In fact (12) is a reformulation of the orthogonality conditions arrived at by virtue of relation (7), the validity of which hinges upon the assumption that "previous" gains satisfy the orthogonality conditions. In the steady state case, however, previous gains are assumed equal to the unknown gains being calculated. It is therefore possible that (15) and (16) have a solution such that (7) is not satisfied and consequently such that (15) and (16) are not equivalent to the original orthogonality conditions. It is clear on the other hand that a solution to the orthogonality conditions must satisfy (15) and (16) as well. While for this simple one-dimensional case general conditions for the uniqueness of the solution of (15) and (16) are easily obtained, the problem becomes such more significant in the two-dimensional case. The simplest way out of the difficulty is to use the original equations (5) and (6) instead of (12). Unfortunately, this procedure results in equations which at least for the two-dimensional case are computationally much more involved. For this reason, in the two-dimensional case it was found more convenient to solve the simplified equations and to check a posteriori if the solution obtained satisfies the orthogonality conditions.

It is quite clear that a filter as simple as the one described by equation (1) is of only limited applicability. However, the general idea described above can be extended to the case of vector valued processes and filters making use of more than one previous estimate and more than one measurement.

THE TWO DIMENSIONAL FILTER

Let $\{x(k,s)\}$ be a two-dimensional. zero mean, wide sense stationary, stochastic process with covariance function $R(i,j)$. We assume again additive white observation noise $w(k,s)$ uncorrelated with the process.

$$y(k,s) = x(k,s) + w(k,s) \tag{18}$$

We define the noise power V as $E[w(k,s)]^2$. The form of the proposed filter is:

$$\hat{x}(k,s) = a_1(k,s)\hat{x}(k,s-1) + a_2(k,s)\hat{x}(k-1,s) + a_3(k,s)\hat{x}(k-1,s-1) + b(k,s)y(k,s) \tag{19}$$

where $\hat{x}(k,s)$ is the estimate of $x(k,s)$. (See Figure 1.) The choice of $\hat{x}(k,s-1)$, $\hat{x}(k-1,s)$, $\hat{x}(k-1,s-1)$ as the "past" estimates is arbitrary and is based only on the intuitive appeal of the scheme, although a theoretical connection exists with the first-order Markov Mesh model[1].

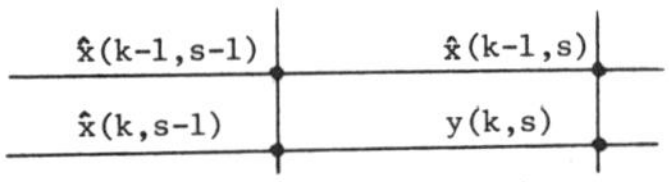

Fig. 1. Estimation Scheme

As in the one-dimensional case, we require that the estimation error be orthogonal to the variables being used by the estimator: $\hat{x}(k,s-1)$, $\hat{x}(k-1,s)$ $\hat{x}(k-1,s-1)$ and $y(k,s)$. The following set of linear equations is the counterpart of (12) in the two-dimensional case. If [A] is the matrix:

$$\begin{bmatrix} E_0(k,s-1) & E_4(k-1,s-1) & E_2(k-1,s-1) & E_3(k,s-1) \\ E_4(k-1,s-1) & E_0(k-1,s) & E_3(k-1,s-1) & E_2(k-1,s) \\ E_2(k-1,s-1) & E_3(k-1,s-1) & E_0(k-1,s-1) & E_1(k-1,s-1) \\ E_3(k,s-1) & E_2(k-1,s) & E_1(k-1,s-1) & R(o,o)+V \end{bmatrix}$$

then:

$$[A] \times \begin{bmatrix} a_1(k,s) \\ a_2(k,s) \\ a_3(k,s) \\ b(k,s) \end{bmatrix} = \begin{bmatrix} E_3(k,s-1) \\ E_2(k-1,s) \\ E_1(k-1,s-1) \\ R(o,o) \end{bmatrix} \tag{20}$$

where:

$$E_0(i,j) = E[\hat{x}(i,j)x(i,j)] \tag{21-a}$$
$$E_1(i-1,j-1) = E[\hat{x}(i-1,j-1)x(i,j)] \tag{21-b}$$
$$E_2(i-1,j) = E[\hat{x}(k-1,j)x(i,j)] \tag{21-c}$$
$$E_3(i,j-1) = E[\hat{x}(i,j-1)x(i,j)] \tag{21-d}$$
$$E_4(i-1,j-1) = E[\hat{x}(i-1,j)\hat{x}(i,j-1)] \tag{21-e}$$

The minimum step-by-step mean square error $E_{min}(k,s)$ is easily seen to be $b(k,s)V$.

At every point (k,s) the corresponding gains can be calculated as the solution of the system of equations (20). The coefficients of the gain vector and the right hand side of (20) are obtained in terms of previous gains. The corresponding, rather complex, relations are omitted here to avoid extending this paper unnecessarily.

As in the one-dimensional case the steady state gains can be determined from the set of nonlinear equations which results from (20) where $a_1(k,s)$, $a_2(k,s)$, $a_3(k,s)$, and $b(k,s)$ are assumed to have evolved to constants. The nature of the equations makes a closed form solution impossible and only iterative procedures will lead to the steady state gains. As has been discussed before, it must be verified that the obtained solution satisfies the orthogonality conditions. It is easy to see that an iterative solution to the original orthogonality equations (i.e., the counterparts of (5) and (6) for the two-dimensional case) requires a much greater computational effort than the solution of (20) under steady state conditions. A computer program to obtain the steady state gains by an iterative procedure has been developed for the two-dimensional filter described by equation (19).

A Solution Algorithm

For the sake of completeness we describe below the simple method that has been employed to solve for the steady state gains a_1, a_2, a_3, and b. Because of its inherent simplicity Picard's algorithm was used: From an initial guess for a_1, a_2, a_3, b we calculate the steady state coefficients E_o, E_1, E_2, E_3 and E_4; a better approximation to the solution is then obtained from (20) and the procedure is repeated until convergence is achieved. The most significant difficulty lies in the calculation of E_o, E_1, E_2, E_3, and E_4 since these quantities are now infinite power series in the variables a_1, a_2, a_3, b.

Convergence of the algorithm was obtained in general by choosing at every step of the iterative procedure the new values for a_1, a_2, a_3, and b as weighted means of the old values used to calculate E_o, E_1, E_2, E_3 and E_4, and the solution of (20). Direct substitution of the solution of (20) as the new set of gains led to divergence in many cases. While it is clear that the iterative solution of (20) requires a considerable computational effort, it must be kept in mind that all the necessary calculations can be carried out off-line, the actual enhancement algorithm being extremely simple once the optimum gains have been determined.

The computer program has been tested for various type of covariance functions and signal-to-noise ratios R(o,o)/V.

If $E = \lim_{k,s\to+\infty} \left\{E_{min}(k,s)\right\}$, the amount of noise reduction V/E that can be obtained from this filter appears to be very close to the best to date reported filter performance. A valid comparison cannot be made since all of the to date reported results are based on tests performed on actual pictures, whereas the results of this work are of purely statistical nature. The table below shows the results obtained for some representative cases. The performance index of the filter N has been expressed in db N = 10 log V/E = 10 log 1/b

PART I

$R(i,j) = R(o,o)\exp(-\sqrt{ci^2 + dj^2})$

R(0,0)/V	c	d	N
0.83	0.25	0.25	4.4
0.83	0.04	0.04	5.8
0.83	0.00766	0.00766	7.4

Part II

$R(i,j) = R(0,0)\exp(-ci-dj)$

R(0,0)/V	c	d	N
0.83	0.5	0.5	4.4
0.83	0.2	0.2	5.9
0.83	0.0875	0.0875	7.7

Table 1 Summary of Results

The performance of a vector scanning recursive filter[4] for an image with covariance R(i,j) = R(0,0) exp (-ci-dj) as indicated in the last row of Table 1 has been reported to be around 7.9 db; more conventional scalar scanning results in slightly poorer performance.

CONCLUSIONS

The feasibility of designing efficient filters subject to computational complexity constraints has been demonstrated. The only information necessary for the design of the filter is the statistical properties of the process to be estimated. No modeling effort is required. The method is of general applicability in that no restrictions on the covariance function are necessary. Several questions concerning these constrained filters are as yet unanswered. Subjects of current investigation in this area include robustness of the algorithm, the specific method for optimally allocating the available storage to previous estimates and previous measurements, clarification of the relation to two-dimensional autoregressive-moving-average models and the identification of fixed-memory suboptimal substitutes for the growing memory transitory state of the algorithm.

REFERENCES

[1] Abend, Harley and Kanal, "Classification of Binary Random Patterns," IEEE Trans. on Info. Theory, Vol. IT-11, pp. 538-544, October, 1965.

[2] N. E. Nahi and T. Assefi, "Bayesian Recursive Image Estimation," IEEE Trans. Comput. (Short Note), Vol. C-21, pp. 734-738, July, 1972.

[3] N. E. Nahi, "Role of Recursive Estimation in Statistical Image Enhancement," Proc. IEEE, Vol. 60, pp. 872-877, July, 1972.

[4] N. E. Nahi and C. A. Franco, "Recursive Image Enhancement Vector Processing," IEEE Trans. Comm. Vol. COM-19, pp. 305-311, April, 1973.

[5] S. R. Powell and L. M. Silverman, "Modeling of Two-Dimensional Covariance Functions with Application to Image Restoration," IEEE Trans. Auto. Cont., Vol. AC-19, pp. 8-12, February, 1974.

Part III

IMPLEMENTATION TECHNIQUES

Editors' Comments on Papers 29, 30, and 31

29 MANRY and AGGARWAL
Picture Processing Using One-Dimensional Implementations of Discrete Planar Filters

30 MERSEREAU and DUDGEON
The Representation of Two-Dimensional Sequences as One-Dimensional Sequences

31 TREITEL and SHANKS
The Design of Multistage Separable Planar Filters

ONE-DIMENSIONAL IMPLEMENTATIONS

The last part of this volume is concerned with implementation of the 2-D digital signal processors discussed in Parts I and II. In both the deterministic and statistical design cases, the implementation of non-recursive or FIR filters involves realizing the convolution sum of Eq. (1.3) or its equivalent. In contrast, the realization of recursive or IIR filters (in the steady state) is based on a direct computation of the 2-D difference equation of Eq. (1.1) or its equivalent.

Consideration of filter implementations in two dimensions appears in many regards to involve similar questions to those arising in 1-D filtering. As can be seen from the two equations referred to above, in direct form, the output sample of the 2-D filter is computed using the two arithmetic operations, multiplication and addition. Analogous to the 1-D case, storage of intermediate output values for future calculations is also required in 2-D recursive filtering. However, in contrast to the 1-D case in which interest in filter implementation has centered around special purpose hardware, most practical impelmentations of 2-D filters have utilized general purpose computing systems. This is principally due to the limited number of on-line signal processing applications in two dimensions, and the extensive computational resources necessary for 2-D filter implementation and data display. As a consequence, the implementation topics addressed in this section are not typical of those of the 1-D case.

The first three papers deal with approaches to reducing the computational complexity associated with 2-D filtering by using 1-D digital filters. The paper by Manry and Aggarwal (Paper 29) treats the problem of realizing a 2-D recursive filter as a 1-D recursive filter. To this end, the authors introduce two integer valued functions over the integers. The purpose of the first function, called the order function, is to describe the sectioning of the 2-D array following a specific scheme, and the purpose of the second function is to describe the scheme of concatenation of these sections to obtain a 1-D array. The 1-D filter derived from a 2-D filter by using the above-mentioned sectioning and concatenation is a time-varying filter. The authors then propose two approximate implementations of the time-varying filter by time-invariant filters and examine the frequency response, stability, error behavior of approximations, and the storage requirements for these approximate time-invariant 1-D implementations.

Mersereau and Dudgeon (Paper 30), on the other hand, limit themselves to the so-called first approximate filter of Manry and Aggarwal (Paper 29), and consider primarily the implementation of 2-D nonrecursive filters. The authors develop a 1-D finite-extent MN point sequence $\{g_k\}$ (called the *projection*) by concatenating the columns of the finite extent 2-D array $\{x_{m,n}\}$ of size MxN in accordance with the relation

$$g_{(Nm+n)} = x_{m,n} \quad \text{for } 0 \leqslant m \leqslant M-1,\ 0 \leqslant n \leqslant N-1. \tag{3.1}$$

If $G(z)$ is the 1-D Z-transform of $\{g_k\}$, and $X(z_1,z_2)$ is the 2-D Z-transform of $\{x_{m,n}\}$, then the two Z-transforms are related as

$$G(z) = X(z^N, z). \tag{3.2}$$

Considering the 1-D DFT as a regular 2-D array of samples, the authors introduce a modified 2-D DFT and use it for computing convolutions. In a similar vein, they apply the modified DFT for processing the output of a drum scanner. Finally, they also use the idea of projections in designing 2-D FIR filters, where the design is performed by determining a 1-D filter corresponding to a "critical slice" and then mapping it to a 2-D function by back projection.

An entirely different approach to the 1-D implementation of a 2-D FIR filter has been advanced by Treitel and Shanks (Paper 31). The authors consider the implementation of a 2-D FIR filter

$$H(z_1,z_2) = \sum_{m=0}^{M-1} \sum_{n=0}^{N-1} h_{m,n} z_1^{-m} z_2^{-n} \tag{3.3}$$

as a parallel connection of several 2-D FIR filters which are described by separable transfer functions. That is, the 2-D transfer function is expressed as:

$$H(z_1,z_2) \cong \sum_{i=1}^{k} H_1^{(i)}(z_1)H_2^{(i)}(z_2) \tag{3.4}$$

where $k \leqslant N \leqslant M$. To achieve this result, the authors apply the singular-value decomposition technique to the impulse response matrix $[h_{m,n}]$. In general, $H_1^{(i)}(z_1)$ is a $(M-1)^{\text{th}}$ order 1-D filter and $H_2^{(i)}(z_2)$ is a $(N-1)^{\text{th}}$ order 1-D filter. If $k < MN/(M+N)$, then an implementation of $H(z_1,z_2)$ via Eq. (3.4) requires fewer multiplications and is more efficient than the original direct implementation of Eq. (3.3). A similar approach has been recently described by Baglai and Smirnov (1975).

Another approach has been proposed by Suresh and Shenoi (1976). This involves connecting in parallel several separable 2-D FIR filters. Here they express the MxN impulse response matrix $[h_{m,n}]$ as a finite sum of separable matrices with the first of these matrices having the dimension MxN and each of the succeeding matrices in the sum having dimensions smaller than the preceding one by unity in both row and column indices.

REFERENCES

Baglai, R. D., and K. K. Smirnov. 1975. "Computer Processing of 2-Dimensional Signals," *Zh. Vychisl. Mat. and Mat. Fiz.* (USSR) **15**(1):241–248.

Suresh, B. R., and B. A. Shenoi. 1976. "Exact Realization of 2-Dimensional Filters by Separable Filters," *Electron. Lett.* **12**(10):242–244.

29

Reprinted from *IEEE Trans. Acoust., Speech, Signal Process.*
ASSP-22(3):164-173 (1974)

Picture Processing Using One-Dimensional Implementations of Discrete Planar Filters

M. T. MANRY, STUDENT MEMBER, IEEE, AND J. K. AGGARWAL, SENIOR MEMBER, IEEE

***Abstract*—The problem of implementing a two-dimensional recursive filter as a one-dimensional recursive filter is examined. The results of the present paper show that the exact one-dimensional implementation of a planar recursive filter is a time-varying filter. However, planar filters may be approximated by one-dimensional time-invariant recursive filters. The frequency response, stability, and storage requirements of the approximate filters are derived and illustrated.**

I. INTRODUCTION

TWO-DIMENSIONAL filters are often used in the processing of discrete two-dimensional images. The techniques used include two-dimensional convolution, the two-dimensional fast-Fourier transform, and two-dimensional recursive filtering. The questions of stability arising in the study of two-dimensional recursive filters [1]–[3] are not completely solved. However, the reductions in storage and computation time often make their use worthwhile. Two descriptions of planar filtering are the causal recursive two-dimensional difference equation

$$g(M,N) = \sum_{m=0}^{r} \sum_{n=0}^{s} b_{mn} f(M-m, N-n) - \sum_{\substack{m=0 \\ m \neq n \text{ if } n=0}}^{p} \sum_{n=0}^{q} c_{mn} g(M-m, N-n), \quad (1)$$

and the two-dimensional z transform transfer function

$$G(z_1,z_2)/F(z_1,z_2) = \sum_{m=0}^{r} \sum_{n=0}^{s} b_{mn} z_1^m z_2^n \Big/ \sum_{m=0}^{p} \sum_{n=0}^{q} c_{mn} z_1^m z_2^n. \quad (2)$$

We assume that $c_{00} = 1$. G and g denote output quantities, while F and f denote input quantities. The integers, r, s, p, and q, are nonnegative. The integer variables M and m denote horizontal position and displacement, respectively, while the integer variables N and n denote vertical position and displacement, respectively. We assume that $g(M,N) = f(M,N) = 0$ for $M < 0$ and/or $N < 0$, and that the output $g(M,N)$ is evaluated for $M \geq 0$, $N \geq 0$. The transfer function of (2) can be expanded as follows:

$$G(z_1,z_2)/F(z_1,z_2) = \sum_{m=0}^{\infty} \sum_{n=0}^{\infty} \tilde{h}_{mn} z_1^m z_2^n. \quad (3)$$

Manuscript received September 8, 1973; revised January 10, 1974. This work was supported in part by the National Science Foundation under Grant GK 32595X and in part by the Joint Services Electronics Program under Contract F44620-71-C-0091.

The authors are with the Department of Electrical Engineering and the Electronics Research Center, University of Texas, Austin, Tex. 78712.

Using the impulse response coefficients of (3), (1) can be written as

$$g(M,N) = \sum_{m=0}^{M} \sum_{n=0}^{N} \tilde{h}_{mn} f(M-m, N-n). \quad (4)$$

Further, we can write

$$\tilde{H}(z_1,z_2) = G(z_1,z_2)/F(z_1,z_2). \quad (5)$$

If the M,N plane is mapped in a one-to-one manner into a set of integers, the input and output intensities, $f(M,N)$ and $g(M,N)$, can be replaced by the one-dimensional input and output intensities, $\bar{f}(K)$ and $\bar{g}(K)$, and (1) can be replaced by a one-dimensional difference equation. In the following, a one-dimensional implementation of (1) is formulated. The error involved, when an exact one-dimensional implementation is approximated by a time-invariant one-dimensional implementation, is expressed for a particular mapping. The storage advantages of the approximate one-dimensional filters are discussed.

II. THE ORDER FUNCTION $Q(M,N)$

Definition: The order function $Q(M,N)$ is defined as an integer function which maps the discrete nonnegative M,N plane onto the set of nonnegative integers in a one-to-one fashion.

The function gives the order in which the points on the M,N plane are filtered using (1). Also, for each nonnegative M and N, $Q(M,N)$ equals the number of intensities which are filtered before the intensity $f(M,N)$ is filtered. If $K = Q(M,N)$, and K is a nonnegative integer variable, there exists a unique inverse function $Q^{-1}(K)$ such that $Q^{-1}(K) = (M,N)$. Let

$$Q(M,N) = N + \sum_{i=0}^{M+N} i, \quad (6)$$

where $M \geq 0$ and $N \geq 0$. This mapping is shown in Fig. 1. From the values of $Q(M,N)$, which are shown below each point, the output intensities are evaluated in the order: $g(0,0)$, $g(1,0)$, $g(0,1)$, $g(2,0)$, $g(1,1)$,$\cdots$. The example mapping is one-to-one.

III. PROPERTIES OF $Q(M,N)$

Order functions have the following properties.

Property 1: $Q(0,0) = 0$.

Property 2: $Q(M,N) \leq -1$ for $M < 0$ and/or $N < 0$.

Property 3: $Q(M,N) < \infty$ for $M < \infty$ and $N < \infty$.

Property 4: $Q(M,N) > Q(M-m, N-n)$ for non-

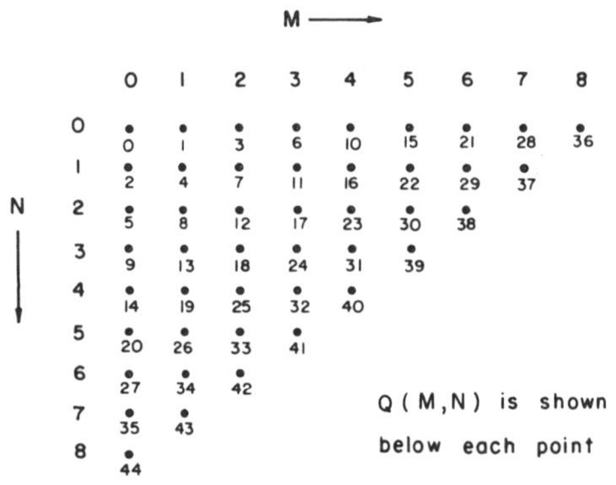

Fig. 1. First order function example.

negative M,N,m, and n, where m and n are not both zero.

Property 5: $Q(M,N) \geq (M+1)(N+1) - 1$.

From Property 1, $g(0,0)$ is the first output intensity to be calculated. If Property 2 did not hold, some of the initial condition intensities would have to be calculated. If M and N are finite and we want to calculate a finite number of intensities before $g(M,N)$, Property 3 is necessary. Property 4 is necessary if a causal equation, such as (1), is used to calculate output intensities. Property 5 follows from Properties 1 and 4.

IV. THE FUNCTION $J(m,n,Q(M,N))$

When (1) is put into one-dimensional form, $\bar{g}(K)$ will be calculated, where $K = Q(M,N)$ and $\bar{g}(K) = g(M,N)$. Since $g(M,N)$ is calculated using terms such as $g(M-m, N-n)$, we need a function $J(m,n,Q(M,N))$ such that $\bar{g}(K - J(m,n,K)) = g(M-m, N-n)$.

Definition: For a given function $Q(M,N)$, $J(m,n,Q(M,N)) = Q(M,N) - Q(M-m,N-n)$ for nonnegative m,n,M,N.

From Property 4, $J(m,n,Q(M,N)) > 0$ for nonnegative M,N,m,n, and where m and n are not both zero. Also $J(0,0,Q(M,N)) = 0$.

Theorem 1:

$$\sum_{i=1}^{A} J(m,n,Q(m\cdot i,n\cdot i)) = Q(m\cdot A,n\cdot A)$$

for nonnegative m,n, and for a positive integer A.

Proof: $J(m,n,Q(m\cdot i,n\cdot i)) = Q(m\cdot i,n\cdot i) - Q(m\cdot i - m,n\cdot i - n)$. Thus

$$\sum_{i=1}^{A} J(m,n,Q(m\cdot i,n\cdot i))$$
$$= (Q(m,n) - Q(0,0)) + (Q(2\cdot m,2\cdot n) - Q(m,n)) + (Q(3\cdot m,3\cdot n) - Q(2\cdot m,2\cdot n)) + \cdots + (Q(A\cdot m,A\cdot n) - Q(A\cdot m - m,A\cdot n - n))$$
$$= Q(A\cdot m,A\cdot n).$$

Q.E.D.

TABLE I
TABULATION OF $J(0,1,K)$ AND $J(1,0,K)$ FOR THE MAPPING OF FIG. 1

M	N	K	$J(0,1,K)$	$J(1,0,K)$
0	0	0	1	1
1	0	1	2	1
0	1	2	2	3
2	0	3	4	2
1	1	4	3	2
0	2	5	3	6
3	0	6	7	3
2	1	7	4	3
1	2	8	4	3
0	3	9	4	10
4	0	10	11	4
3	1	11	5	4
2	2	12	5	4
1	3	13	5	4
0	4	14	5	15
5	0	15	16	5
4	1	16	6	5
3	2	17	6	5
2	3	18	6	5
1	4	19	6	5

V. THE ONE-DIMENSIONAL FILTERING EQUATION

Theorem 2: If $f(M,N) = g(M,N) = 0$ for $M < 0$ and/or $N < 0$, if $\bar{f}(Q(M,N)) = f(M,N)$, and if

$$\bar{g}(K) = \sum_{m=0}^{r}\sum_{n=0}^{s} b_{mn}\bar{f}(K - J(m,n,K)) - \sum_{\substack{m=0\ n=0 \\ m\neq n \text{ if } n=0}}^{p\quad q} c_{mn}\bar{g}(K - J(m,n,K)), \quad (7)$$

then $\bar{g}(Q(M,N)) = g(M,N)$. This follows easily from (1) and (7).

Let $Q(M,N)$ be given by (6) for $M \geq 0$ and $N \geq 0$ and let $Q(M,N) = -1$ for $M < 0$ and/or $N < 0$. This mapping is shown in Fig. 1 for nonnegative M and N. The planar filter, represented by

$$g(M,N) = f(M,N) - 0.5g(M-1,N) - 0.5g(M,N-1), \quad (8)$$

is one-dimensionally implemented as

$$\bar{g}(K) = \bar{f}(K) - 0.5\bar{g}(K - J(1,0,K)) - 0.5\bar{g}(K - J(0,1,K)). \quad (9)$$

The functions $J(0,1,K)$ and $J(1,0,K)$ are listed in Table I.

VI. TIME INVARIANCE

A one-dimensional filter, described by (7), is time-invariant if b_{mn}, c_{mn}, and $J(m,n,K)$ are independent of K for $0 \leq m \leq \max(r,p)$ and $0 \leq n \leq \max(s,q)$. We have assumed, in (7), that b_{mn} and c_{mn} are independent of K and that $J(m,n,K)$ is dependent on K.

Theorem 3: There does not exist a function $Q(M,N)$ such that the function $J(m,n,Q(M,N))$ is independent of the value of $Q(M,N)$, for nonnegative integers M,N, and for positive integers m,n.

Proof: Use Property 5 of $Q(M,N)$ with Theorem 1, assuming that M,N,m, and n are nonnegative integers. Assume that $J(m,n,K) = \bar{J}(m,n)$, a function independent of K, for nonnegative m,n,K.

$$\sum_{i=1}^{A} J(m,n,Q(m\cdot i,n\cdot i)) = Q(m\cdot A,n\cdot A)$$

$$Q(M,N) \geq (M+1)(N+1) - 1$$

$$\sum_{i=1}^{A} J(m,n,Q(m\cdot i,n\cdot i)) = \sum_{i=1}^{A} \bar{J}(m,n) = A\cdot\bar{J}(m,n)$$

$$Q(m\cdot A,n\cdot A) \geq (m\cdot A+1)(n\cdot A+1) - 1$$

$$A\cdot\bar{J}(m,n) \geq (m\cdot A+1)(n\cdot A+1) - 1.$$

Therefore, $\bar{J}(m,n) \geq ((m\cdot A+1)(n\cdot A+1) - 1)/A$ for any positive integer A.

Since $Q(m,n) - Q(0,0) = J(m,n,Q(m,n)) = \bar{J}(m,n)$, it follows that

$$Q(m,n) \geq ((m\cdot A+1)(n\cdot A+1) - 1)/A. \quad (10)$$

Since Theorem 1 holds for any positive integer A, let A increase to infinity. Using (10), $Q(m,n) \rightarrow \infty$ as $A \rightarrow \infty$. This contradicts Property 3 of the function $Q(M,N)$. Therefore, the theorem is proved by contradiction.

Q.E.D.

Corollary: In every region on the nonnegative M,N plane which includes the points $(M,N) = (0,0)$, $(0,1)$, and $(1,0)$, there does not exist a function $Q(M,N)$ such that $J(0,1,Q(M,N))$ and $J(1,0,Q(M,N))$ are independent of the value of $Q(M,N)$.

Proof: Since $Q(0,0) = 0$ and only the points $(0,1)$ and $(1,0)$ satisfy Property 5 of $Q(M,N)$ with equality, for $Q(M,N) = 1$, either $Q(0,1) = 1$ or $Q(1,0) = 1$.

Case 1: Assume $Q(1,0) = 1$. Then $J(1,0,Q(1,0)) = Q(1,0) - Q(0,0) = 1$. Since $Q(0,1) > 1$ and $Q(-1,1) \leq -1$, $J(1,0,Q(0,1)) > 2$. For this case, then, $J(1,0,Q(0,1)) \neq J(1,0,Q(1,0))$ and $J(1,0,Q(M,N))$ is therefore not independent of $Q(M,N)$.

Case 2: Assume $Q(0,1) = 1$. Then $J(0,1,Q(0,1)) = Q(0,1) - Q(0,0) = 1$. $J(0,1,Q(1,0)) = Q(1,0) - Q(1,-1) > 2$ since $Q(1,0) > 1$ and $Q(1,-1) \leq -1$. For this case $J(0,1,Q(M,N))$ is not independent of $Q(M,N)$.

Q.E.D.

Theorem 3 shows that no time-invariant implementation of (7) exists which is valid for the entire M,N plane. The corollary shows that (7) is time-varying for any sequence of input intensities which includes $\bar{f}(Q(0,0))$, $\bar{f}(Q(0,1))$, and $\bar{f}(Q(1,0))$.

There is no saving in computation time or storage when (1) is implemented exactly using (7). Ignoring the computations of $M-m$, $N-n$, and $K - J(m,n,K)$, (1) and (7) require the same storage, for a given order function, and the same number of additions and multiplications. Since $J(m,n,K)$ must be calculated and stored, and since the mappings of $f(M,N)$ onto $f(K)$ and $\bar{g}(K)$ onto $g(M,N)$ are required, the use of (7) requires more computation time and storage than the use of (1).

VII. DERIVATION OF THE FIRST APPROXIMATE IMPLEMENTATION

Let

$$Q(M,N) = N + P(M + [N/P](L-1)) \quad (11)$$

for nonnegative integers M and N, where P and L are positive integers, and $[x]$ is the integer part of x. (i.e., $[4.3] = 4$ and $[-4.3] = -4$). The expression of (11) maps, onto the set of nonnegative integers, an infinite number of vertically stacked subarrays, each of which is L points in horizontal length and P points in vertical height. Because of Property 3 of the function $Q(M,N)$, we let (11) describe the order function for H of these vertically stacked subarrays, where H is a finite positive integer. For $M > L-1$, $N > H\cdot P - 1$, $M < 0$, or $N < 0$, we do not describe $Q(M,N)$ explicitly. For the mapping of Fig. 2, we have two subarrays with horizontal lengths of eight points and vertical heights of ten points. Therefore, for Fig. 2, $L = 8$, $P = 10$, and $H = 2$. For $0 \leq m \leq M \leq L-1$ and $0 \leq n \leq N \leq H\cdot P - 1$, $J(m,n,Q(M,N))$ is calculated as follows:

$$\begin{aligned} J(m,n,Q(M,N)) &= Q(M,N) - Q(M-m,N-n) \\ &= (N + P(M + [N/P](L-1))) \\ &\quad - (N - n + P(M-m \\ &\quad + [(N-n)/P](L-1))) \\ &= n + P\cdot m + P(L-1)([N/P] \\ &\quad - [(N-n)/P]). \end{aligned}$$

For large portions of the region defined by $0 \leq M \leq L-1$ and $0 \leq N \leq H\cdot P - 1$, $[N/P] = [(N-n)/P]$ and

$$J(m,n,Q(M,N)) = n + P\cdot m \quad (12)$$

Let $P = 10$, $L = 8$, $H = 2$, $p = r = 2$, and $q = s = 3$ for a planar filter and the mapping of Fig. 2. In the figure the points for which $J(m,n,Q(M,N)) \neq n + P\cdot m$, for $0 \leq n \leq 3$ and $0 \leq m \leq 2$, are blocked off. If P and L are allowed to increase, a proportionately smaller number of points are blocked off. This means that as P and L increase, (12) is valid for a proportionately greater number of points.

On approximating the indices using (12), we obtain the following approximate filter from (7):

$$\bar{g}(K) = \sum_{m=0}^{r}\sum_{n=0}^{s} b_{mn}\bar{f}(K - (n + P\cdot m))$$

$$- \sum_{\substack{m=0 \\ m\neq n \text{ if } n=0}}^{p}\sum_{n=0}^{q} c_{mn}\bar{g}(K - (n + P\cdot m)). \quad (13)$$

Equation (13) is a time-invariant difference equation which approximates the exact filtering operation of (7) for $Q(M,N)$ specified by (11).

VIII. STABILITY OF THE FIRST APPROXIMATE FILTER

Equation (13) can be z-transformed and the filtering process can be expressed in transfer function form. Where

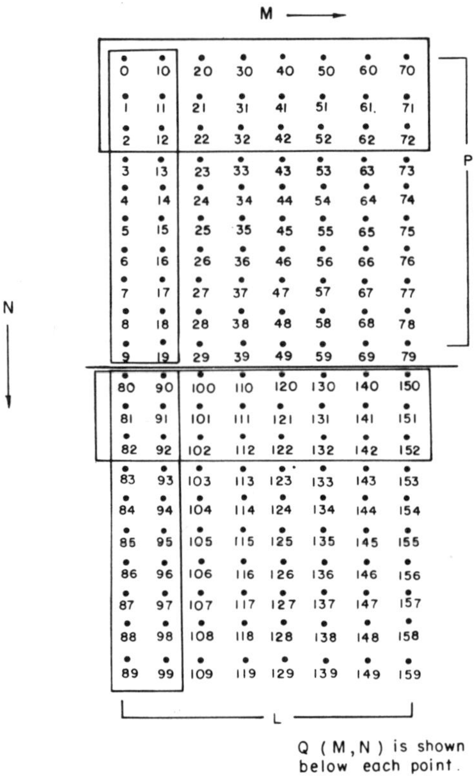

Fig. 2. Second order function example.

$\tilde{G}(z)$ is the z transform of $\tilde{g}(K)$ and $\tilde{F}(z)$ is the z transform of $\tilde{f}(K)$,

$$\tilde{G}(z)/\tilde{F}(z) = \sum_{m=0}^{r} \sum_{n=0}^{s} b_{mn} z^{n+P\cdot m} / \sum_{m=0}^{p} \sum_{n=0}^{q} c_{mn} z^{n+P\cdot m}. \quad (14)$$

The transfer function (14) results from (2) when z_1 is replaced by z^P and z_2 is replaced by z. Assuming that

$$1/ \sum_{m=0}^{p} \sum_{n=0}^{q} c_{mn} z_1^m z_2^n = \sum_{m=0}^{\infty} \sum_{n=0}^{\infty} h_{mn} z_1^m z_2^n, \quad (15)$$

$G(z_1,z_2)/F(z_1,z_2)$ is stable if (as seen in [1])

$$\sum_{m=0}^{\infty} \sum_{n=0}^{\infty} | h_{mn} | < \infty. \quad (16)$$

Theorem 4: The one-dimensional discrete causal filter described by (14) is stable if the two-dimensional discrete causal filter described by (2) is stable.

Proof: Assume the filter described by (2) is stable. If

$$1/ \sum_{m=0}^{p} \sum_{n=0}^{q} c_{mn} z^{n+P\cdot m} = \sum_{i=0}^{\infty} \bar{h}_i z^i, \quad (17)$$

we must prove that $\sum_{i=0}^{\infty} | \bar{h}_i | < \infty$. In (15) replace z_1 by z^P and z_2 by z. This results in

$$1/ \sum_{m=0}^{p} \sum_{n=0}^{q} c_{mn} z^{n+P\cdot m} = \sum_{m=0}^{\infty} \sum_{n=0}^{\infty} h_{mn} z^{n+P\cdot m}.$$

Collect terms for which $n + P\cdot m = i$.

$$\bar{h}_i = \sum_{\substack{m=0 \\ n+P\cdot m = i}}^{\infty} \sum_{n=0}^{\infty} h_{mn}. \quad (18)$$

Since (2) is stable (16) holds. Since

$$\sum_{m=0}^{\infty} \sum_{n=0}^{\infty} | h_{mn} | = \sum_{i=0}^{\infty} \Big(\sum_{\substack{m=0 \\ n+P\cdot m = i}}^{\infty} \sum_{n=0}^{\infty} | h_{mn} | \Big),$$

then

$$\sum_{i=0}^{\infty} \Big(\sum_{\substack{m=0 \\ n+P\cdot m = 1}}^{\infty} \sum_{n=0}^{\infty} | h_{mn} | \Big) \geq \sum_{i=0}^{\infty} \Big| \sum_{\substack{m=0 \\ n+P\cdot m = i}}^{\infty} \sum_{n=0}^{\infty} (h_{mn}) \Big|,$$

by the triangle inequality, and

$$\sum_{i=0}^{\infty} \Big| \sum_{\substack{m=0 \\ n+P\cdot m = i}}^{\infty} \sum_{n=0}^{\infty} (h_{mn}) \Big| = \sum_{i=0}^{\infty} | \bar{h}_i |.$$

$$\therefore \quad \sum_{i=0}^{\infty} | \bar{h}_i | < \infty.$$

Q.E.D.

IX. FREQUENCY RESPONSE OF THE APPROXIMATE FILTERS

The approximate one-dimensional filter $\bar{H}(z^P,z)$ has $\bar{H}[\exp(-jP\theta), \exp(-j\theta)]$ as its frequency response while the original two-dimensional filter $\bar{H}(z_1,z_2)$, has $\bar{H}[\exp(-j\theta_1), \exp(-j\theta_2)]$ as its frequency response. As shown in Theorem 5, the one dimensional filter's response is a sampled version of the original planar filter's response. In this section $\lfloor x \rfloor$ is x rounded to the nearest integer, (i.e., $\lfloor 4.2 \rfloor = 4$ and $\lfloor 4.5 \rfloor = 5$).

Theorem 5: If the relationship between θ_1, θ_2, and θ is described by the parametric equations,

$$\theta_2 = \theta \quad (19)$$

and

$$\theta_1 = P\theta - 2\pi \lfloor P\theta/2\pi \rfloor \quad (20)$$

then

$$\bar{H}[\exp(-jP\theta), \exp(-j\theta)] = \bar{H}[\exp(-j\theta_1), \exp(-j\theta_2)].$$

Proof: Assume that (19) and (20) hold. From (20),

$$| \theta_1/2\pi | = | P\theta/2\pi - \lfloor P\theta/2\pi \rfloor |.$$

Since $| x - \lfloor x \rfloor | \leq 0.5$, it follows that $| \theta_1/2\pi | \leq 0.5$ and $| \theta_1 | \leq \pi$. Assuming that $-\pi \leq \theta_1 \leq \pi$, it is also true that $-\pi \leq \theta_1 \leq \pi$ and $-\pi \leq \theta_2 \leq \pi$. Using (19),

$$\exp(-j\theta_2) = \exp(-j\theta). \quad (21)$$

Using (20),

$$\exp(-j\theta_1) = [\exp(-jP\theta)]\{\exp(j2\pi\lfloor P\theta/2\pi \rfloor)\}$$
$$= \exp(-jP\theta). \quad (22)$$

Using (21) and (22), $\bar{H}(\exp(-j\theta_1), \exp(-j\theta_2)) = H[\exp(-jP\theta), \exp(-j\theta)]$. Q.E.D.

Let $P = 8$. The locus of the parametric equations (19) and (20) consists of diagonal parallel lines, as shown in Fig. 3. Assume that Fig. 3 is the amplitude response of a low-pass planar recursive filter, that the amplitude of the shaded region of the figure is 1, and that the amplitude elsewhere is 0. Using Theorem 5 and the diagonal lines of Fig. 3, the amplitude response of the approximate one-dimensional filter is constructed and shown in Fig. 4.

In general, the approximate filter's frequency response is formed from P diagonal lines of the planar filter's frequency response. However, if P is an even integer as in the example, the first and last of the lines must be counted as one line for this result to hold. If we let P increase, the sampling of the planar frequency response becomes proportionately finer. As we shall see in the next section, the error of the approximate filtering method decreases as P increases.

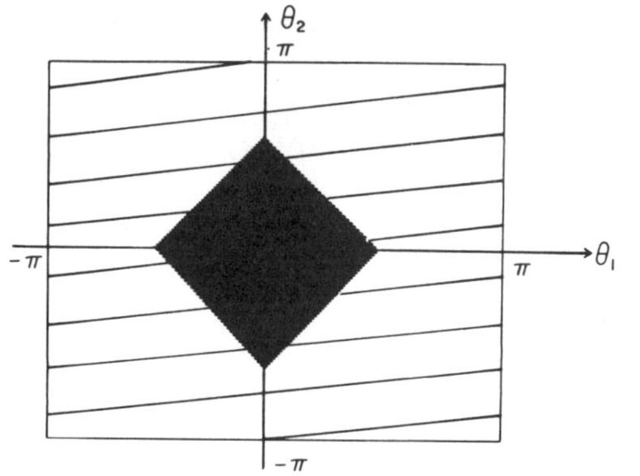

Fig. 3. Planar filter amplitude response.

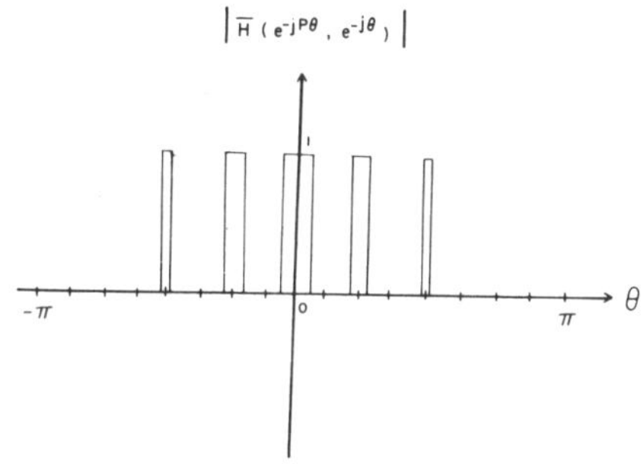

Fig. 4. Approximate filter amplitude response.

X. THE ERROR EQUATIONS

Since we are concerned with implementations of two-dimensional filters, we find it convenient to express the error of the first approximate filtering method in two-dimensional form. To express (13) two-dimensionally, we replace K with $Q(M,N)$, giving

$$\tilde{g}(Q(M,N)) = \sum_{m=0}^{r}\sum_{n=0}^{s} b_{mn}\bar{f}(Q(M,N) - (n + P\cdot m))$$
$$- \sum_{\substack{m=0 \\ m\neq n \text{ if } n=0}}^{p}\sum_{n=0}^{q} c_{mn}\tilde{g}(Q(M,N) - (n + P\cdot m)). \quad (23)$$

We define the error resulting from the use of (13) as follows:

$$\epsilon(M,N) = g(M,N) - \tilde{g}(Q(M,N)). \quad (24)$$

If a planar filter is approximated using (13), $\epsilon(M,N)$ can be calculated for a total of $H\cdot P\cdot L$ points where $0 \leq N \leq H\cdot P - 1$. The average absolute error $\bar{\epsilon}$ is an empirical average of $|\epsilon(M,N)|$, which is defined as follows:

$$\bar{\epsilon} = (1/(H\cdot P\cdot L))\sum_{M=0}^{L-1}\sum_{N=0}^{H\cdot P-1} |\epsilon(M,N)|. \quad (25)$$

In this section, we find expressions for $\epsilon(M,N)$ and upper bounds on $|\epsilon(M,N)|$ and $\bar{\epsilon}$.

Theorem 6: If $g(M,N)$, $Q(M,N)$, $\tilde{g}(Q(M,N))$, and $\epsilon(M,N)$ are described by (1), (11), (23), and (24), respectively, then

$$\epsilon(M,N) = d(M,N) - \sum_{\substack{m=0 \\ m\neq n \text{ if } n=0}}^{p}\sum_{n=0}^{q} c_{mn}\epsilon(M-m, N-n), \quad (26)$$

where

$$d(M,N) = \sum_{m=0}^{r}\sum_{n=0}^{s} b_{mn}(f(M-m,N-n)$$
$$- \bar{f}(Q(M,N) - (n + P\cdot m)))$$
$$- \sum_{\substack{m=0 \\ m\neq n \text{ if } n=0}}^{p}\sum_{n=0}^{q} c_{mn}(\tilde{g}(Q(M-m,N-n))$$
$$- \tilde{g}(Q(M,N) - (n + P\cdot m))). \quad (27)$$

Let $\bar{q} = \max\{q,s\}$ and $\bar{p} = \max\{r,p\}$.

Corollary: Given P, L, H, $\bar{q}$, and $\bar{p}$, there are at least $H(L - \bar{p})(P - \bar{q})$ ordered integer pairs (M,N) such that $d(M,N) = 0$.

Theorem 6 follows when (23) is subtracted from (1). The corollary can be proved using (11) and (27).

Using (15), (26) can be replaced by the nonrecursive equation,

$$\epsilon(M,N) = \sum_{m=0}^{M}\sum_{n=0}^{N} h_{M-m,N-n}d(m,n). \quad (28)$$

Let U be the set of ordered integer pairs (M,N) such that $d(M,N) \neq 0$ if $(M,N) \in U$, where $0 \leq M \leq L - 1$ and $0 \leq N \leq H\cdot P - 1$. From the corollary to Theorem 6, the complement of U has a minimum of $H(L - \bar{p})(P - \bar{q})$

members. Therefore, U has a maximum of $(H\cdot P\cdot L - H(L-\bar{p})(P-\bar{q}))$ or $H(\bar{p}P+\bar{q}L-\bar{p}\bar{q})$ members. This means that

$$\sum_{\substack{m=0\\(m,n)\in U}}^{L-1}\sum_{n=0}^{H\cdot P-1} 1 \le H(\bar{p}P+\bar{q}L-\bar{p}\bar{q}). \tag{29}$$

Let $P = 10$, $L = 8$, $\bar{p} = 2$, and $\bar{q} = 3$. In Fig. 2 every point for which $(M,N) \in U$ is blocked off. The set U has $H(\bar{q}\cdot L + \bar{p}\cdot P - \bar{q}\cdot\bar{p})$, or 76 members for this example.

Equation (28) can be rewritten as follows:

$$\epsilon(M,N) = \sum_{\substack{m=0\\(m,n)\in U}}^{M}\sum_{n=0}^{N} h_{M-m,N-n}d(m,n). \tag{30}$$

Using (30), we can place an upper bound on $|\epsilon(M,N)|$.

Theorem 7: If a planar filter is approximately realized using (13) and $K_x = \max_{m,n}|d(m,n)|$, then

$$|\epsilon(M,N)| \le K_x \sum_{\substack{m=0\\(m,n)\in U}}^{M}\sum_{n=0}^{N} |h_{M-m,N-n}|.$$

Corollary: If

$$f_x = \max\{|f(M,N)|\}$$
$$0 \le M \le L-1$$
$$0 \le N \le H\cdot P-1$$

and

$$\tilde{H}_x = \lim_{\substack{M\to\infty\\N\to\infty}} \sum_{m=0}^{M}\sum_{n=0}^{N} |\tilde{h}_{mn}|$$

then

$$K_x \le f_x\Big(\sum_{m=0}^{r}\sum_{n=0}^{s}|b_{mn}| + \tilde{H}_x\sum_{m=0}^{p}\sum_{n=0}^{q}|c_{mn}|\Big).$$

Theorem 7 is easily derived from (30). The corollary can be proved using (2), (3), (13), (14), and (27). In the following theorem, we find an upper bound on $\bar{\epsilon}$ as a function of P and L.

Theorem 8: If

$$H_x = \lim_{M,N\to\infty}\sum_{m=0}^{M}\sum_{n=0}^{N}|h_{mn}|$$

and (13) is used in the filtering of an image, then

$$\bar{\epsilon} < (\bar{q}/P + \bar{p}/L - (\bar{q}\cdot\bar{p})/(P\cdot L))H_xK_x.$$

Proof: Combining (25) and (30),

$$\bar{\epsilon} = (1/H\cdot P\cdot L))\sum_{M=0}^{L-1}\sum_{N=0}^{H\cdot P-1}\Big|\sum_{\substack{m=0\\(m,n)\in U}}^{M}\sum_{n=0}^{N} h_{M-m,N-n}\,d(m,n)\Big|$$

$$< (1/(H\cdot P\cdot L))\sum_{M=0}^{L-1}\sum_{N=0}^{H\cdot P-1}\sum_{\substack{m=0\\(m,n)\in U}}^{L-1}\sum_{n=0}^{H\cdot P-1}|h_{M-m,N-n}|\cdot|d(m,n)|$$

$$= (1/(H\cdot P\cdot L))\sum_{\substack{m=0\\(m,n)\in U}}^{L-1}\sum_{n=0}^{H\cdot P-1}|d(m,n)| \cdot \sum_{M=0}^{L-1}\sum_{N=0}^{H\cdot P-1}|h_{M-m,N-n}|$$

$$< (1/(H\cdot P\cdot L))H_x\sum_{\substack{m=0\\(m,n)\in U}}^{L-1}\sum_{n=0}^{H\cdot P-1}|d(m,n)|\,|1|$$

$$\le (1/(H\cdot P\cdot L))H_xK_x\sum_{\substack{m=0\\(m,n)\in U}}^{L-1}\sum_{n=0}^{H\cdot P-1}1.$$

Using (29)

$$\bar{\epsilon} < (1/(H\cdot P\cdot L))H_x\cdot K_x\cdot H(\bar{p}\cdot P+\bar{q}\cdot L-\bar{p}\cdot\bar{q})$$
$$= (\bar{q}/P+\bar{p}/L-(\bar{p}\cdot\bar{q})/(P\cdot L))H_xK_x.$$

Q.E.D.

Corollary: Given a stable planar recursive filter and an input array of infinite height and width, assume that a portion of the input array is to be filtered using (13) and that

$$f_{xm} = \max_{\substack{0\le M\le\infty\\0\le N\le\infty}}|f(M,N)|.$$

It follows that

$$\lim_{\substack{L\to\infty\\P\to\infty}}\bar{\epsilon} = 0.$$

Proof: From Theorem 8,

$$\lim_{\substack{P\to\infty\\L\to\infty}}(\bar{q}/P+\bar{p}/L-(\bar{q}\cdot\bar{p})/(P\cdot L)) = 0. \tag{31}$$

Using the corollary to Theorem 7 and the fact that $f_x \le f_{xm}$,

$$\lim_{\substack{P\to\infty\\L\to\infty}}K_x \le f_{xm}\Big(\sum_{m=0}^{r}\sum_{n=0}^{s}|b_{mn}| + \tilde{H}_x\sum_{m=0}^{p}\sum_{n=0}^{q}|c_{mn}|\Big). \tag{32}$$

Since the planar filter is stable, $\tilde{H}_x$ and H_x are finite. From (31) and (32),

$$\lim_{\substack{L\to\infty\\P\to\infty}}\bar{\epsilon} = 0.$$

Q.E.D.

From the corollary to Theorem 8, it is advantageous to increase P and L. Since an increase in P means an increase in the order of the approximate filter, from (14), the size of P is limited by the maximum field length of the machine used as a filter.

XI. THE STORAGE ADVANTAGE OF THE FIRST APPROXIMATE FILTER

For large two-dimensional images, the first approximate filtering method has a storage advantage over exact two-dimensional filtering methods. If S_A is the number of input and output intensities that must be stored for the first approximate filter, S_{ER} is the number of intensities that must be stored when an image is filtered by rows two-dimensionally, and the integer constants p, q, r, s, P, and L are as defined previously, then

$$S_A = P(r + p) + s + q + 1 \tag{33}$$

and

$$S_{ER} = L(s + q) + r + p + 1. \tag{34}$$

Let $P = 100$, $L = 1000$, $p = r = 2$, and $q = s = 3$, as in Figs. 5 and 6. The output intensity $g(L - 1,3)$ is to be calculated using (1) and the output intensity $\tilde{g}(Q(L - 1,199))$ is to be calculated using (13). Fig. 5 represents the input image, and Fig. 6 represents the output image for both filtering methods. The intensities, which are blocked off in both figures, are needed for the calculations of $g(L - 1,3)$ and $\tilde{g}(Q(L - 1,199))$, or must be stored for later calculations. In order to show the storage for the calculations of $g(L - 1,3)$ and $\tilde{g}(Q(L - 1,199))$, on the same two figures, it is convenient to have the points $(L - 1,3)$ and $(L - 1,199)$ some distance apart. From the figures, exact two-dimensional filtering requires the storage of $L(s + q) + r + p + 1$, or 6005 intensities. The first approximate filtering method requires the storage of $P(r + p) + s + q + 1$, or 407 intensities, for this example.

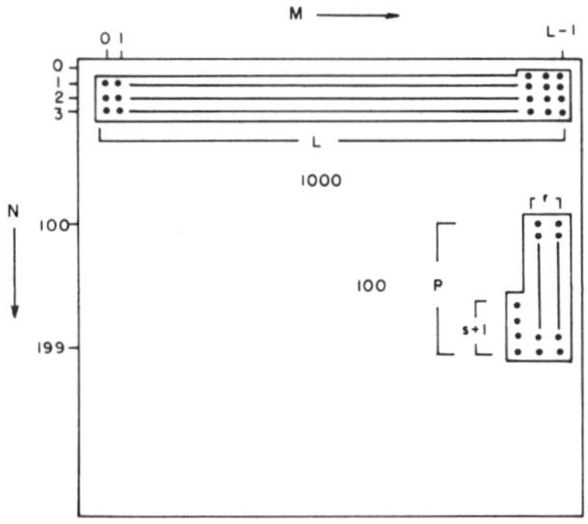

Fig. 5. Input intensities $f(M,N)$ and $\tilde{f}(Q(M,N))$.

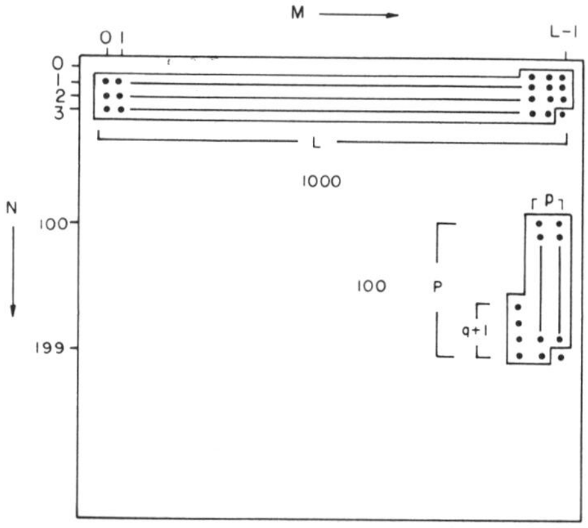

Fig. 6. Output intensities $g(M,N)$ and $\tilde{g}(Q(M,N))$.

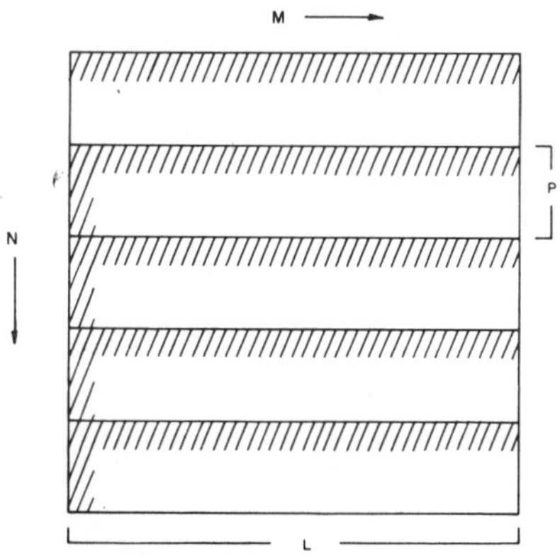

Fig. 7. Distribution of $|\epsilon(M,N)|$.

XII. THE SECOND APPROXIMATE ONE-DIMENSIONAL IMPLEMENTATION

When the first approximate one-dimensional implementation is used, $|\epsilon(M,N)|$ can be very large in the upper portions of each array, as illustrated in Fig. 7. In this section, we propose a second approximate one-dimensional implementation which retains the storage advantage of the first method but lacks the horizontal bands of large $|\epsilon(M,N)|$.

XIII. DEVELOPMENT OF THE SECOND APPROXIMATE METHOD

Assume that a given stable planar filter is to be realized one-dimensionally by the first approximate method and that the values of P and L have been picked. Assume that the upper bounds on $|\epsilon(M,N)|$ have been found, for $0 \leq M \leq L - 1$ and $0 \leq N \leq H \cdot P - 1$, using Theorem 7. Assume that this upper bound is unacceptably large for the first A horizontal lines of each subarray, where A is some positive integer satisfying $1 \leq A < P$. Instead of filtering the image by the first approximate method, divide the input image into arrays with heights of P points such that these arrays overlap, as shown in Fig. 8. Assuming that Fig. 8 shows the partitioned input image, each shaded region contains the same set of intensities as the lowest rectangular region, with an area of $L \cdot A$ points, which is in the array above. Filter the partitioned image by the approximate method. Letting Fig. 8 now represent the output image, $|\epsilon(M,N)|$ may attain large values in the shaded regions. After discarding the shaded regions, we have a partitioned output image, as in Fig. 9, which can be combined into a final output image. The interior of the final image has no regions with large $|\epsilon(M,N)|$. Since we have not discarded the top portion of the first output

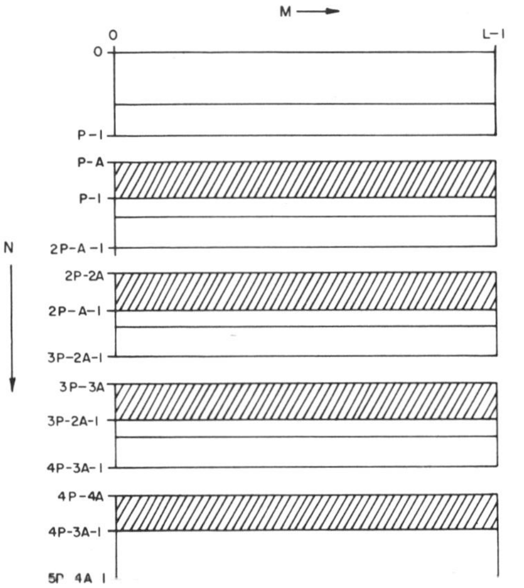

Fig. 8. Divided image.

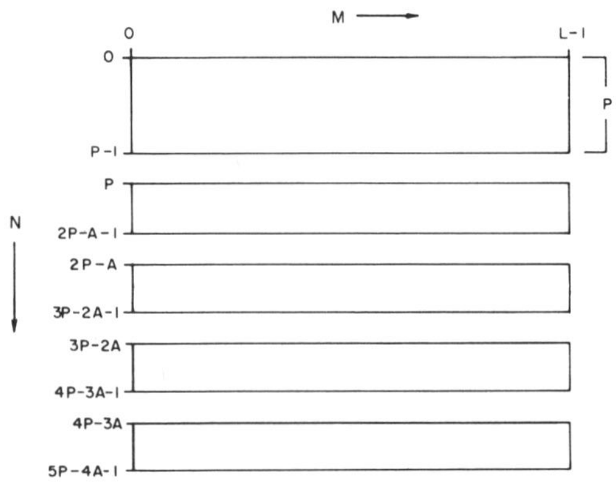

Fig. 9. Final image before recombination.

array or the left edge of any of the output arrays, the top and left edges of the output image have large error. If we assume that initial condition error is acceptable, the edge error can be ignored.

XIV. NUMERICAL RESULTS

In this section, we give a numerical example which compares the exact and approximate filtering methods.

We used the CDC 6600 computer at the University of Texas for the examples in this section. A subroutine, using the CDC 6600 library function RANF(X), was used to generate two-dimensional images, which were approximately white and Gaussian. Assuming that $U(M,N)$ is a white Gaussian image with zero mean and a standard deviation of *std*1, assuming that $V(M,N)$ is a white Gaussian image with zero mean and a standard deviation of *std*2, assuming that $U(M,N)$ and $V(M,N)$ are uncorrelated, and assuming $U(M,N) = V(M,N) = 0$ for $M < 0$ or $N < 0$, we generated input images as follows:

$$f_1(M,N) = \sum_{m=0}^{r} \sum_{n=0}^{s} b_{mn} U(M-m, N-n) - \sum_{\substack{m=0 \\ m \neq n \text{ if } n=0}}^{p} \sum_{n=0}^{q} c_{mn} f_1(M-m, N-n), \quad (35)$$

$$f(M,N) = f_1(M,N) + V(M,N). \quad (36)$$

The filter coefficients [3] used in (35), which are the same as those used in the filtering operations, are expressed in the planar transfer function,

$$G(z_1,z_2)/F(z_1,z_2) = (1 + 0.3z_1 + 0.4z_2 + 0.1z_1z_2)/(1 - 1.5z_1 - 1.2z_2 + 1.8z_1z_2 + 0.6z_1^2 + 0.5z_2^2 - 0.72z_1^2z_2 - 0.75z_1z_2^2 + 0.29z_1^2z_2^2). \quad (37)$$

This is a low-pass spatial filter.

The input image $f(M,N)$ was formed, using (36), by adding a white Gaussian image $V(M,N)$ to a correlated Gaussian image $f_1(M,N)$.

XV. AN EXAMPLE COMPARING THE APPROXIMATE AND EXACT FILTERING METHODS

Using (35)–(37), we generated an input image $f(M,N)$, where $0 \leq M \leq 63$, $0 \leq N \leq 39$, $std1 = 2.5$, and $std2 = 0.5$. Using $P = 20$, $L = 64$, and $A = 10$, we filtered the input image $f(M,N)$ by the exact two-dimensional method, by the first approximate method, and by the second approximate method. The exact two-dimensional method produced an output $g(M,N)$. The first approximate method produced an output $\tilde{g}(Q(M,N))$. The second approximate method produced an output image $g_1(M,N)$. In displaying the images, we used a gray code with ten levels, with darker shades representing the more positive intensities. The input image $f(M,N)$ is shown in Fig. 10, $g(M,N)$ in Fig. 11, $\tilde{g}(Q(M,N))$ in Fig. 12, $g_1(M,N)$ in Fig. 13, $|\epsilon(M,N)|$ in Fig. 14, and $|g(M,N) - g_1(M,N)|$ in Fig. 15. In Figs. 14 and 15, the most negative intensity possible is zero; and the darker shades, therefore, represent larger magnitudes of error. Comparing Fig. 14 with Fig. 15, the second approximate method's output image lacks the band of large error present in the first approximate method's output image.

XVI. CONCLUSION

Since a large number of order functions exist, every planar filter has many one-dimensional implementations. Because there is no saving in storage or reduction in com-

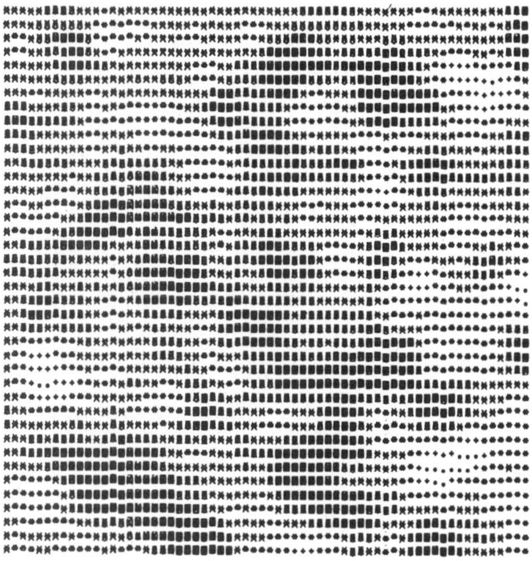

Fig. 10. $f(M,N)$.

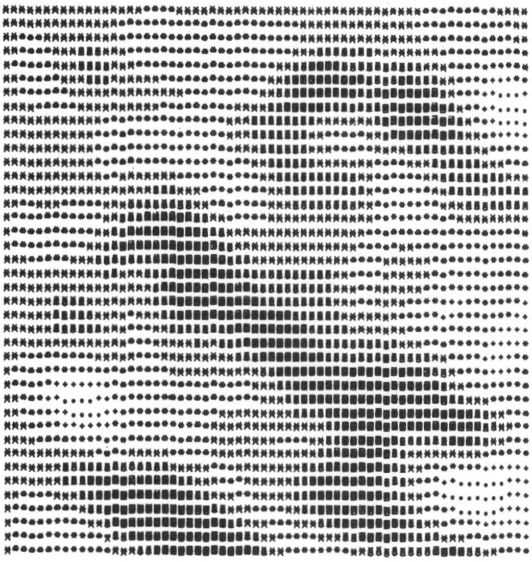

Fig. 11. $g(M,N)$.

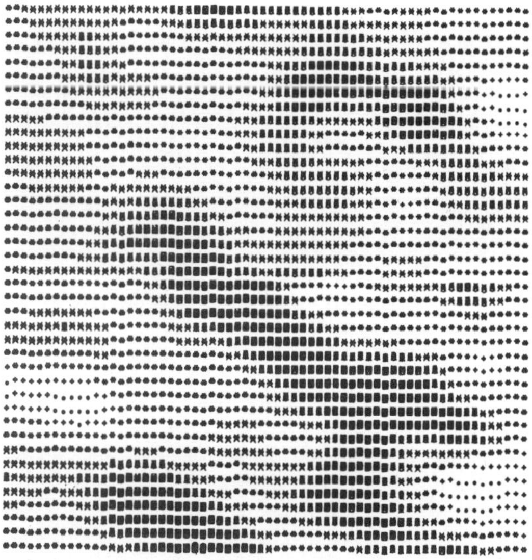

Fig. 12. $\tilde{g}(Q(M,N))$.

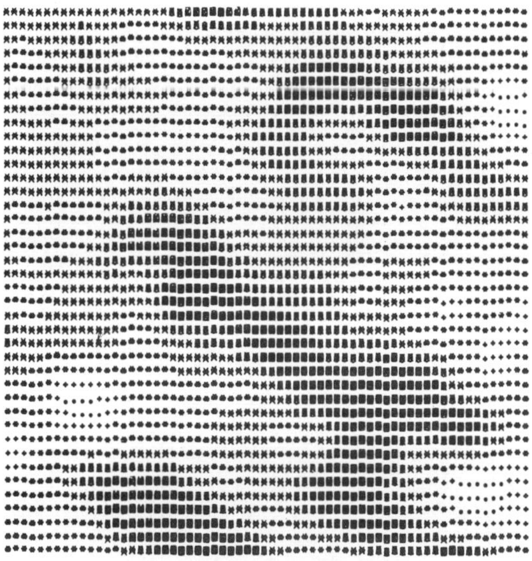

Fig. 13. $g_1(M,N)$.

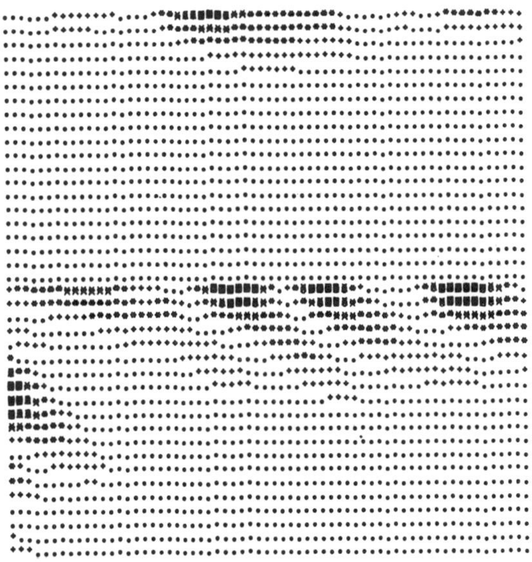

Fig. 14. $|\epsilon(M,N)|$.

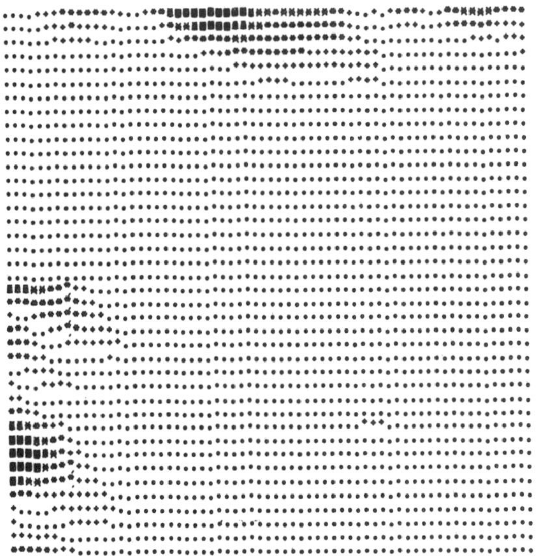

Fig. 15. $|g(M,N) - g_1(M,N)|$.

putation time when a planar filter is exactly one-dimensionally implemented, and since planar filters can't be one-dimensionally implemented with time-invariant filters, there is little reason to implement two-dimensional filters, precisely, in one-dimensional form. However, when large planar images are filtered, approximate time-invariant one-dimensional implementations require much less storage capacity than exact two-dimensional implementations. By filtering images in overlapping segments and increasing the one-dimensional filter's order, the error of approximate one-dimensional filtering can be reduced.

REFERENCES

[1] T. S. Huang, "Stability of two-dimensional recursive filters," *IEEE Trans. Audio and Electroacoust.*, vol. AU-20, pp. 158–163, June 1972.

[2] J. L. Shanks, "Two-dimensional recursive filters," in *SWIEEECO Rec.*, Apr. 1969, pp. 19E1–19E8.

[3] J. L. Shanks, S. Treitel, and J. H. Justice, "Stability and synthesis of two-dimensional recursive filters," *IEEE Trans. Audio and Electroacoust.*, vol. AU-20, pp. 115–128, June 1972.

30

Reprinted from *IEEE Trans. Acoust., Speech, Signal Process.*
ASSP-22(5):320–325 (1974)

The Representation of Two-Dimensional Sequences as One-Dimensional Sequences

RUSSELL M. MERSEREAU, MEMBER, IEEE, AND DAN E. DUDGEON, MEMBER, IEEE

***Abstract*—A number of signal processing techniques which have been developed for processing one-dimensional sequences do not generalize to the processing of two-dimensional signals, largely due to the absence of a two-dimensional factorization theorem. In an attempt to circumvent this problem, a specific representation of two-dimensional sequences as one-dimensional sequences is presented in this paper. Using this mapping several two-dimensional problems can be viewed as one-dimensional problems and approached using one-dimensional techniques. This representation is valid both for signals of finite extent and for the more general class of signals with rational z-transforms. In this paper we consider applications of these techniques for high speed convolution, processing of drum scans, and two-dimensional finite impulse response (FIR) filter design.**

I. INTRODUCTION

IN THIS paper we consider a specific representation of two-dimensional discrete arrays (sequences) as one-dimensional sequences. Aside from providing a more complete understanding of two-dimensional systems, there are a number of possible practical advantages to such a representation. It suggests the possibility of designing and implementing two-dimensional systems as one-dimensional systems. In this way, many of the filter design algorithms and analytical techniques which are well understood in one-dimension can be extended to two-dimensions, and many of the difficulties associated with two-dimensional systems, such as the lack of a factorization theorem, can be partially circumvented. In addition, as will shortly be explained, there are certain image processing devices such as television cameras and drum scanners which can be conveniently understood in terms of such mappings.

Manuscript received January 4, 1974. This work was supported in part by the Advanced Research Projects Agency of the Department of Defense, monitored by the ONR under Contract N0014-67-A-0204-0064 and by the National Science Foundation under Grant GK-31353.

The authors are with the Research Laboratory of Electronics and Department of Electrical Engineering, Massachusetts Institute of Technology, Cambridge, Mass. 02139.

II. REPRESENTATION OF SIGNALS OF FINITE DURATION

A two-dimensional array of finite duration $\{f(m,n), 0 \leq m \leq M-1, 0 \leq n \leq N-1\}$ can be mapped to a one-dimensional sequence $g(p)$ by means of the mapping

$$g(Nm+n) = f(m,n) \qquad 0 \leq m \leq M-1, \quad 0 \leq n \leq N-1. \qquad (1)$$

The samples of the sequence $g(p)$ are formed by concatenating the columns of the array $f(m,n)$. Due to the limited extent of $f(m,n)$, the maping of (1) is readily seen to be invertible. It is not necessary that M be finite. In the case where m is allowed to vary over the range $-\infty < m < \infty$, the sequence $g(p)$ is infinitely long but is still invertible.

If this mapping is considered in the z-plane, then

$$G(z) = \sum_{p=0}^{MN-1} g(p)z^{-p}$$
$$= \sum_{m=0}^{M-1}\sum_{n=0}^{N-1} g(Nm+n)z^{-(Nm+n)}$$
$$= \sum_{m=0}^{M-1}\sum_{n=0}^{N-1} f(m,n)z^{-Nm}z^{-n}$$
$$G(z) = F(z^N,z) \qquad (2)$$

where $F(z_1,z_2)$ is the two-dimensional z-transform of $f(m,n)$.

$$F(z_1,z_2) = \sum_{m=0}^{M-1}\sum_{n=0}^{N-1} f(m,n)z_1^{-m}z_2^{-n}. \qquad (3)$$

Thus the z-transform of $g(p)$ corresponds to the z-transform of $f(m,n)$ evaluated along the contour $z_1 = z_2^N$. The function $G(z) = F(z^N,z)$ shall be referred to as a slice of $F(z_1,z_2)$ and $g(p)$ shall be referred to as a projection of $f(m,n)$. These terms are motivated by definitions for the corresponding functions of continuous variables [1].

If instead of looking in the z-plane we restrict our attention to the Fourier plane $z_1 = \exp[j\omega_1]$, $z_2 = \exp[j\omega_2]$ then (2) becomes

$$G[\exp(j\omega)] = F[\exp(jN\omega), \exp(j\omega)]. \qquad (4)$$

Because $F[\exp(j\omega_1), \exp(j\omega_2)]$ is doubly periodic in ω_1 and ω_2 with period 2π in each, $G[\exp(j\omega)]$ corresponds to the evaluation of $F[\exp(j\omega_1), \exp(j\omega_2)]$ along a series of N parallel lines in the two-dimensional Fourier plane. Each of the N parallel lines forms an angle $\theta = \tan^{-1} 1/N$ with the ω_1-axis, as illustrated in Fig. 1.

The projection of (1) is not the only one we could have chosen. For example

$$g(2Nm+n) = f(m,n) \quad 0 \le m \le M-1, \quad 0 \le n \le N-1$$
$$= 0 \quad \text{otherwise} \qquad (5)$$

would work equally well. This is equivalent to appending N zeros to each column before concatenating the columns. It is also an invertible mapping and the Fourier transform of this $g(p)$ is a slice which consists of $2N$ lines of $F[\exp(j\omega_1), \exp(j\omega_2)]$ which form an angle $\theta = \tan^{-1} 1/2N$ with the ω_1-axis. Other projections exist which also transform to slices of $F[\exp(j\omega_1), \exp(j\omega_2)]$, but not all of them are invertible. A slice (or projection) which can be inverted will be referred to as a critical slice (or critical projection). Some general theorems concerning the existence of critical slices are contained in Mersereau [2],

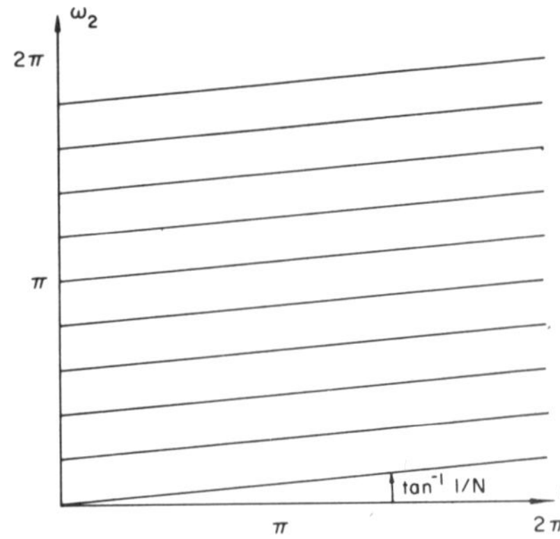

Fig. 1. Lines in the two-dimensional Fourier plane along which $F[\exp(j\omega_1), \exp(j\omega_2)]$ and $G(e^{j\omega})$ are numerically equal.

although for present purposes we shall consider only the slice defined by (4).

Equation (1) demonstrates how $f(m,n)$ can be recovered from $g(Nm+n)$. To see how $F[\exp(j\omega_1, \exp(j\omega_2)]$ can be recovered from $G[\exp(j\omega)]$ we can represent $F[\exp(j\omega_1), \exp(j\omega_2)]$ as

$$F[\exp(j\omega_1), \exp(j\omega_2)] = \sum_{m=0}^{M-1}\sum_{n=0}^{N-1} g(Nm+n) \cdot \exp[-j(\omega_1 m + \omega_2 n)]. \qquad (6)$$

But

$$g(Nm+n) = \frac{1}{2\pi}\int_{-\pi}^{\pi} G[\exp(j\omega)] \cdot \exp[j\omega(Nm+n)]\,d\omega. \qquad (7)$$

Substituting this expression into (6) gives

$$F[\exp(j\omega_1), \exp(j\omega_2)] = \sum_{m=0}^{M-1}\sum_{n=0}^{N-1} \frac{1}{2\pi}\int_{-\pi}^{\pi} G[\exp(j\omega)] \cdot \exp[j\omega(Nm+n)]\,d\omega \exp[-j(\omega_1 m + \omega_2 n)].$$

Rearranging terms and evaluating the sums this becomes

$$F[\exp(j\omega_1), \exp(j\omega_2)] = \exp\left[-j\left(\frac{M-1}{2}\omega_1 + \frac{N-1}{2}\omega_2\right)\right] \cdot \frac{1}{2\pi}\int_{-\pi}^{\pi} G[\exp(j\omega)] \cdot \frac{\sin\frac{1}{2}M(N\omega-\omega_1)\sin\frac{1}{2}N(\omega-\omega_2)}{\sin\frac{1}{2}(N\omega-\omega_1)\sin\frac{1}{2}(\omega-\omega_2)} \cdot \exp\left[j\frac{MN-1}{2}\omega\right]d\omega. \qquad (8)$$

In the general case where $-\infty \le m \le \infty$ (8) becomes

$$F[\exp(j\omega_1), \exp(j\omega_2)] = \sum_{k=0}^{N-1} G\left\{\exp\left[j\left(\frac{\omega_1 + 2\pi k}{N}\right)\right]\right\} \cdot \frac{\sin \frac{1}{2}N[(\omega_1 + 2\pi k)/N - \omega_2]}{\sin \frac{1}{2}[(\omega_1 + 2\pi k)/N - \omega_2]} \cdot \exp\left[-j\frac{N-1}{2}\left(\frac{\omega + 2\pi k}{N} - \omega_2\right)\right] \quad (8a)$$

which can simply be interpreted as an interpolation formula in the ω_2 direction.

III. A MODIFIED TWO-DIMENSIONAL DISCRETE FOURIER TRANSFORM (DFT)

From the previous section it was observed that an MN point sequence $g(p)$ could be formed from an $M \times N$ array $f(m,n)$ by selecting the samples of the array column by column. The sequence $g(p)$ thus has a DFT from which it can be recovered, which consists of MN evenly spaced samples of $G(e^{j\omega})$. Interpreted in the two-dimensional Fourier plane associated with $F[\exp(j\omega_1), \exp(j\omega_2)]$ this says that the one-dimensional DFT of $g(p)$ can be considered as a regular (although not rectangular) two-dimensional array of samples of $F[\exp(j\omega_1), \exp(j\omega_2)]$ which must, of necessity, completely specify the function $f(m,n)$. With this is mind let us define a modified two-dimensional DFT, $F_M(k,l)$, of $f(m,n)$ according to the following:

$$F_M(k,l) = \sum_{m=0}^{M-1}\sum_{n=0}^{N-1} f(m,n) \cdot \exp\left[-j\frac{2\pi}{MN}(Nm+n)(k+Ml)\right] \quad (9)$$

$$0 \le k \le M-1$$
$$0 \le l \le N-1$$

which can be inverted by

$$f(m,n) = \frac{1}{MN}\sum_{k=0}^{M-1}\sum_{l=0}^{N-1} F_M(k,l) \cdot \exp\left[j\frac{2\pi}{MN}(Nm+n)(k+M1)\right]. \quad (10)$$

Equations (9) and (10) can be computed using an MN point one-dimensional fast Fourier transform (FFT) algorithm and inverse FFT, respectively. From (3) it is observed that

$$F_M(k,l) = F(z_1,z_2)\Big|_{\substack{z_1 = \exp[j(2\pi/M)k] \\ z_2 = \exp[j(2\pi/MN)(k+Ml)].}} \quad (11)$$

The samples of $F[\exp(j\omega_1), \exp(j\omega_2)]$ which correspond to $F_M(k,1)$ for $M = 20$, $N = 10$ are shown in Fig. 2. The

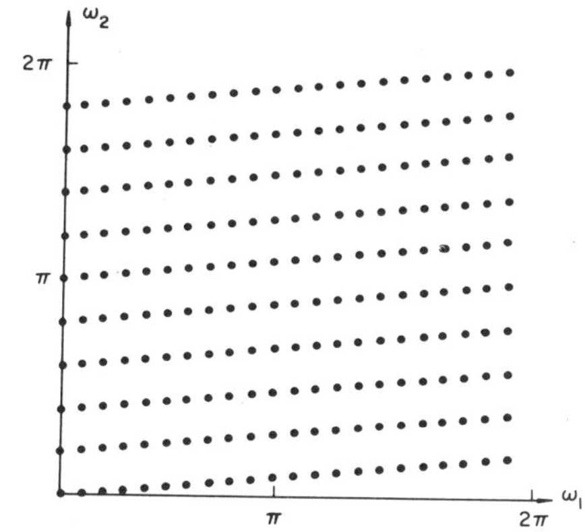

Fig. 2. The raster of samples in the two-dimensional Fourier plane which corresponds to the 20 × 10 point modified two-dimensional DFT, $F_M(k,l)$.

z-transform $F(z_1,z_2)$ can be obtained from $F_M(k,l)$ by

$$F(z_1,z_2) = \frac{1}{MN}\sum_{k=0}^{M-1}\sum_{l=0}^{N-1} F_M(k,l)\frac{1 - z_1^{-M}}{1 - z_1^{-1}\exp[j(2\pi k/M)]} \cdot \frac{1 - z_2^{-N}\exp[j(2\pi k/M)]}{1 - z_2^{-1}\exp[j(2\pi/MN)(k+Ml)]}. \quad (12)$$

By comparison, the standard two-dimensional DFT computes the sum

$$F_S(k,l) = \sum_{m=0}^{M-1}\sum_{n=0}^{N-1} f(m,n)\exp\left[j2\pi\left(\frac{km}{M} + \frac{ln}{N}\right)\right] \quad (13)$$

$$0 \le k \le M-1$$
$$0 \le l \le N-1.$$

Thus,

$$F_S(k,l) = F(z_1,z_2)\Big|_{\substack{z_1 = \exp[j2\pi k/M] \\ z_2 = \exp[j2\pi l/N].}} \quad (14)$$

The raster of samples corresponding to a 20 × 10 point standard DFT is illustrated in Fig. 3. Computationally, both the standard and the modified DFT can be computed using an FFT algorithm using $(MN/2)\log_2 MN$ complex multiply and adds if both M and N are powers of 2. The modified DFT requires the computation of a single large DFT whereas the standard DFT requires a number of smaller DFT calculations.

Like the standard two-dimensional DFT the modified DFT can be used to perform convolutions. To demonstrate how this can be done let $F_M(k,l)$ and $G_M(k,l)$ be the $M \times N$ point modified DFT's of $f(m,n)$ and $g(m,n)$ respectively, and let $h_M(m,n)$ denote the inverse transform of $F_M(k,l)G_M(k,l)$. Then

$$h_M(m,n) = \frac{1}{MN}\sum_{k=0}^{M-1}\sum_{l=0}^{N-1} F_M(k,l)G_M(k,l)$$

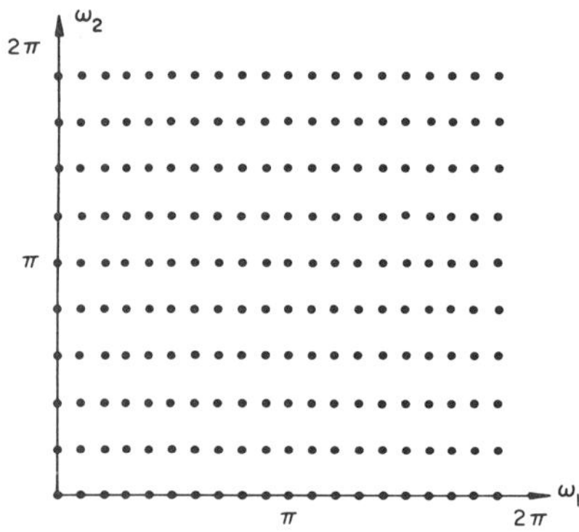

Fig. 3. The raster of samples in the two-dimensional Fourier plane which corresponds to the 20 × 10 point standard two-dimensional DFT, $F_S(k,l)$.

$$\cdot \exp\left[j \frac{2\pi}{MN} (Nm + n)(k + Ml) \right]$$

$$= \sum_{r=0}^{M-1} \sum_{s=0}^{N-1} f(r,s) g(u,v) \Bigg|_{\substack{Nu+v=((N(m-r)\\ +(n-s)))_{MN}}} \qquad (15)$$

where $((\cdot))_L$ means the argument should be considered modulo L.

By way of contrast, the inverse transform of $F_S(k,1)G_S(k,1)$, say $h_S(m,n)$, can be written as

$$h_S(m,n) = \sum_{r=0}^{M-1} \sum_{s=0}^{N-1} f(r,s) g(((m-r))_M, ((n-s))_N). \qquad (16)$$

The convolution represented by (15) will be referred to as a helical convolution, whereas that in (16) is normally referred to as a circular or toroidal convolution.

To perform a linear convolution of an $M \times N$ point array $f(m,n)$ with a $P \times Q$ point array $g(m,n)$ we can define two $(M + P - 1) \times (N + Q - 1)$ point arrays $\tilde{f}(m,n)$ and $\tilde{g}(m,n)$ in the following fashion:

$$\tilde{f}(m,n) = \begin{cases} f(m,n), & 0 \leq m \leq M-1, 0 \leq n \leq N-1 \\ & \text{otherwise, such that} \\ 0 & 0 \leq m \leq M+P-2, \\ & 0 \leq n \leq N+Q-2 \end{cases}$$

$$\tilde{g}(m,n) = \begin{cases} g(m,n), & 0 \leq m \leq P-1, 0 \leq n \leq Q-1 \\ & \text{otherwise, such that} \\ 0, & 0 \leq m \leq M+P-2, \\ & 0 \leq n \leq N+Q-2 \end{cases}$$

as illustrated in Fig. 4. The toroidal convolution of $\tilde{f}(m,n)$ and $\tilde{g}(m,n)$ over the range $0 \leq m \leq M + P - 2$, $0 \leq n \leq N + Q - 2$ is numerically equal to the linear convolutional of $f(m,n)$ and $g(m,n)$. The helical convolution of $\tilde{f}(m,n)$ and $\tilde{g}(m,n)$ is also numerically equal to the linear convolution over this range.

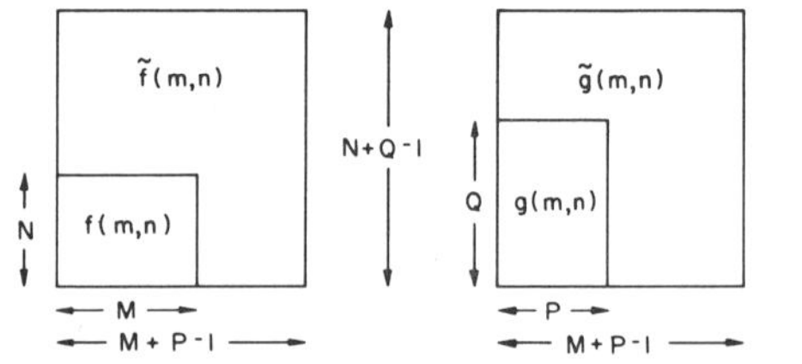

Fig. 4. The sequences $\tilde{f}(m,n)$ and $\tilde{g}(m,n)$ which are formed from $f(m,n)$ and $g(m,n)$. Either a toroidal or helical convolution of $\tilde{f}(m,n)$ and $\tilde{g}(m,n)$ is numerically equal to the linear convolution of $f(m,n)$ and $g(m,n)$.

IV. PROCESSING DRUM SCANNER OUTPUTS

One use for the modified DFT is in processing the output of a drum scanner. Such a scanner scans an image along a series of diagonal lines such as those illustrated in Fig. 5. This is strongly reminiscent of the manner in which a slice "scans" a two-dimensional Fourier transform. Therefore, it should be possible to calculate $F_S(k,l)$, the standard two-dimensional DFT of a two-dimensional function, which has been scanned using a drum scanner by considering our earlier derivations with the time and frequency domains reversed.

Assume that we have a continuous two-dimensional function $f(x,y)$ such that $f(x,y) = 0$ unless $0 \leq x \leq X$ and $0 \leq y \leq Y$. The function $f(x,y)$ can then be considered to represent one period of a periodic two-dimensional function $\tilde{f}(x,y)$ which has period X in x and period Y in y. In order that $f(x,y)$ can be represented by its DFT let us further assume that $\tilde{f}(x,y)$, when expressed in a Fourier series, have only MN nonzero Fourier coefficients where N is the number of scan lines as illustrated in Fig. 5.
Thus

$$f(x,y) = \frac{1}{XY} \sum_{k=0}^{M-1} \sum_{l=0}^{N-1} F\left(\frac{2\pi k}{X}, \frac{2\pi l}{Y}\right) \exp\left[j2\pi \left(\frac{kx}{X} + \frac{ly}{Y}\right)\right] \qquad (17)$$

for

$$0 \leq x \leq X, \qquad 0 \leq y \leq Y.$$

The output of the drum scanner is a function of a single variable and will be denoted as $g(u)$. It corresponds to the N scan lines arranged as a single continuous function. From Fig. 5 we see that

$$g(u) = f\left(\left(\left(\frac{XNu}{(N^2X^2 + Y)^{1/2}}\right)\right)_X, \frac{Yu}{(N^2X^2 + Y^2)^{1/2}}\right). \qquad (18)$$

Now suppose that $g(u)$ is sampled at MN evenly spaced points over the interval $0 \leq u \leq (N^2X^2 + Y^2)^{1/2}$. This corresponds to choosing a sampling interval

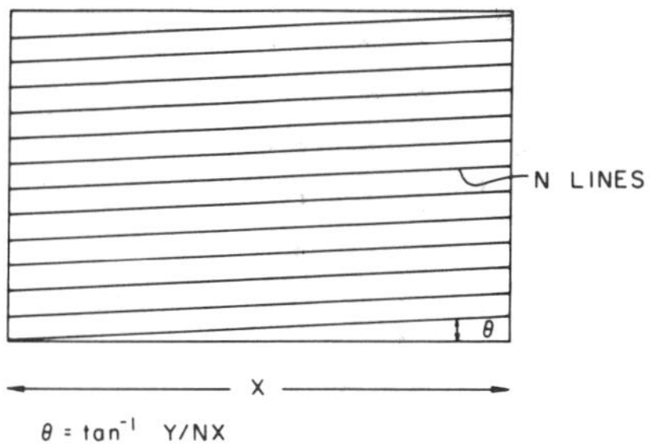

Fig. 5. The series of lines along which an image is scanned by a drum scanner.

$$\Delta u = \frac{(N^2X^2 + Y^2)^{1/2}}{MN}. \tag{19}$$

If we let $\hat{g}(p)$ denote the sequence of samples

$$\hat{g}(p) = g(p\Delta u) = f\left(\left(\left(\frac{pX}{M}\right)\right)_X, \frac{pY}{MN}\right). \tag{20}$$

Setting $p = m + Mn$,

$$\hat{g}(m + Mn) = f\left(\frac{mX}{M}, \frac{Y}{MN}(m + Mn)\right). \tag{21}$$

Taking the MN point DFT of $\hat{g}(p)$ it follows that

$$\hat{G}(Nk + l) = \frac{MN}{XY} F\left(\frac{2\pi k}{X}, \frac{2\pi l}{Y}\right). \tag{22}$$

Thus, by sampling the drum scan and computing the DFT of the resulting sequence, one can obtain the standard two-dimensional DFT of the original image. Thus, the modified DFT can be used in conjunction with the standard two-dimensional DFT in processing a drum scanned image exactly. Note that no constraints have been placed upon the dimensions of the scanning area or upon the number of scan lines except that the number of scan lines exceed the number of nonzero Fourier coefficients in the vertical direction.

V. FINITE IMPULSE RESPONSE (FIR) FILTER DESIGN USING PROJECTIONS

Projections may be useful for some two-dimensional FIR filter design problems. To design a filter which approximates an ideal two-dimensional function, we can first evaluate the ideal on a critical slice to obtain a one-dimensional ideal function. Then using any one-dimensional FIR filter design algorithm, we may obtain a one-dimensional FIR filter whose frequency response approximates the one-dimensional ideal. The unit sample response of this filter can then be back-projected to give a two-dimensional unit sample response, which hopefully approximates the two-dimensional ideal. This technique is outlined in Figs. 6–10. In Fig. 6 is the two-dimensional ideal. It is of the form of a low-pass filter with a cusp-shaped passband and the stopband is the exterior of a diamond. The region between the pass and stopbands is

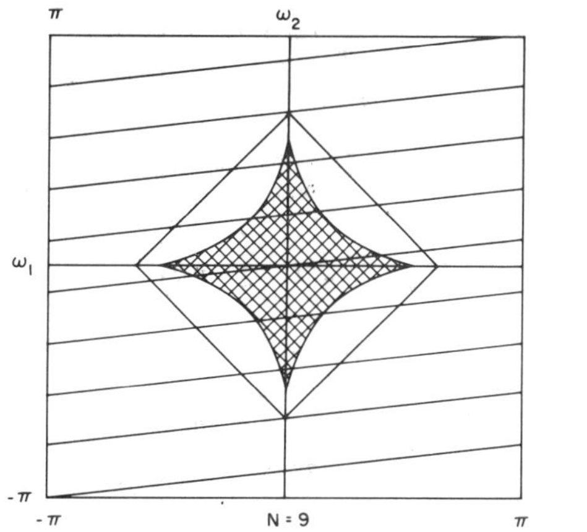

Fig. 6. An ideal filter characteristic which we wish to approximate by a 9 × 9 point FIR digital filter. The cross hatched region represents a passband, the area outside the diamond is the stopband, and the area between the two is the transition band. The $N = 9$ critical slice is superimposed.

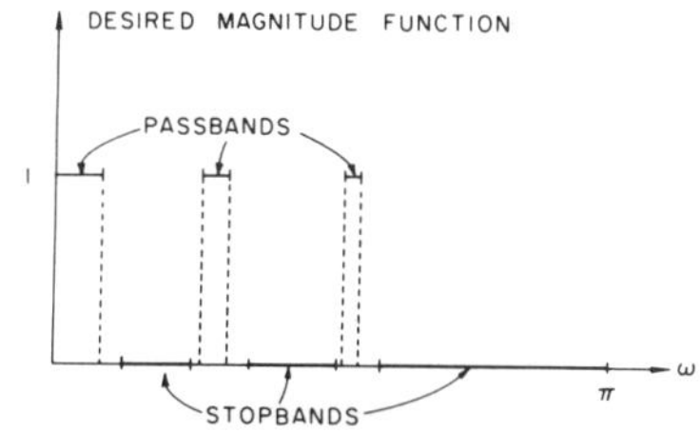

Fig. 7. The one-dimensional ideal response obtained by evaluating the two-dimensional ideal along the critical slice.

a transition band where the ideal response is not specified. In Fig. 7 is the one-dimensional ideal which is the two-dimensional ideal evaluated along the slice corresponding to $N = 9$. Note the one-dimensional ideal has multiple pass, stop, and transition bands. In Fig. 8 is shown an approximation to the one-dimensional ideal. The approximation was obtained by using a Parks–McClellan equiripple design [3] with an 81 point unit sample response. The frequency response of the 9 × 9 two-dimensional filter, which results from back-projecting the filter of Fig. 8, is shown in Figs. 9 and 10. In Fig. 9 is an intensity plot of the log magnitude of the frequency response and in Fig. 10 is a surface drawing of the frequency response. Note that the approximation does not possess all of the symmetry of the ideal. Work is continuing on ways to improve this design technique.

VI. GENERALIZATION TO SLICES OF RATIONAL FREQUENCY RESPONSES

If we are given a rational two-dimensional frequency response

$$H[\exp(j\omega_1), \exp(j\omega_2)] = \frac{A[\exp(j\omega_1), \exp(j\omega_2)]}{B[\exp(j\omega_1), \exp(j\omega_2)]} \tag{23}$$

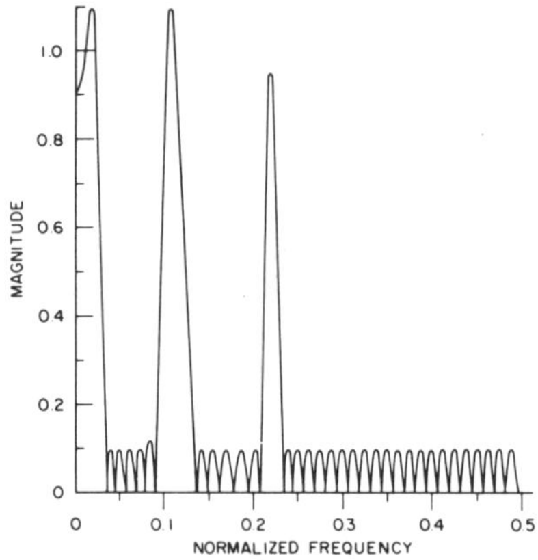

Fig. 8. The frequency response of an 81 point one-dimensional filter which approximates the ideal in Fig. 7.

Fig. 9. An intensity plot of the log-magnitude of the resulting two-dimensional filter.

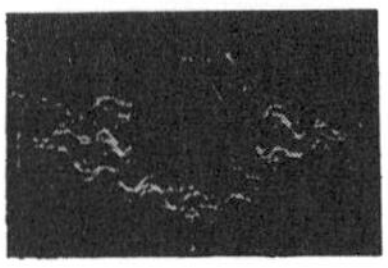

Fig. 10. A surface drawing of the filter of Fig. 9, which represents an approximation to the filter in Fig. 6.

then, in general, the inverse Fourier transform of H is not of finite extent. However, if A and B are polynomials of finite degree, then there exist critical slices from which $H[\exp(j\omega_1), \exp(j\omega_2)]$ can be recovered. If A is a polynomial of degree M_1 in $\exp[j\omega_1]$ and of degree N_1 in $\exp[j\omega_2]$ and if B is a polynomial of degree M_2 in $\exp[j\omega_1]$ and N_2 in $\exp[j\omega_2]$, then let

$$N = \max[N_1, N_2].$$

Letting $\hat{H}(e^{j\omega})$ denote the frequency response evaluated on the slice at angle $\tan^{-1} 1/N$

$$\hat{H}(e^{j\omega}) = H[\exp(jN\omega), \exp(j\omega)] = \frac{A[\exp(j\omega N), \exp(j\omega)]}{B[\exp(j\omega N), \exp(j\omega)]}. \tag{24}$$

If $\hat{h}(p)$ denotes the projection associated with this slice, then

$$\hat{h}(p) = \sum_{m=-\infty}^{\infty} h(m, p - Nm) \tag{25}$$

where $h(m,n)$ is the inverse Fourier transform of $H[\exp(j\omega_1), \exp(j\omega_2)]$.

The sequences $\hat{a}(p)$ and $\hat{b}(p)$, which are the inverse transforms of $A[\exp(jN\omega), \exp(j\omega)]$ and $B[\exp(jN\omega], \exp(j\omega)]$, can be recovered from $\hat{h}(p)$ by Prony's method, Shank's method, or any similar technique which can be used to simultaneously estimate the poles and zeros of a rational transfer function from samples of its unit sample response [4], [5]. Then $\hat{a}(p)$ and $\hat{b}(p)$ can be separately back-projected by the methods of the previous sections to give $a(m,n)$ and $b(m,n)$ from which $H[\exp(j\omega_1), \exp(j\omega_2)]$ can be easily reconstructed. The implications of this result for two-dimensional filter design and implementation have not been fully explored.

VII. CONCLUSIONS

A set of representations of two-dimensional sequences (arrays) as one-dimensional sequences has been presented and a few possible uses for such a representation have been considered, although many issues remain to be explored in depth. Two-dimensional arrays of finite length can be processed either two-dimensionally using standard techniques and standard DFT's or one-dimensionally using a modified two-dimensional DFT presented here. Some signals such as drum scans and television signals are perhaps most efficiently processed using both sets of techniques together.

REFERENCES

[1] R. M. Mersereau, "The digital reconstruction of multi-dimensional signals from their projections," in *Proc. 10th Annu. Allerton Conf. Circuit and System Theory*, 1972, pp. 326–334.
[2] ——, "Digital reconstruction of multi-dimensional signals from their projections," Sc.D. thesis, Dep. Elec. Eng., Mass. Inst. Tech., Cambridge, 1973.
[3] J. H. McClellan, T. W. Parks, and L. R. Rabiner, "A computer program for designing optimum FIR linear phase digital filters," *IEEE Trans. Audio Electroacoust.*, vol. AU-21, pp. 506–526, Dec. 1973.
[4] J. L. Shanks, "Recursive filters for digital processing," *Geophysics*, vol. 32, pp. 33–51, 1967.
[5] C. S. Burrus and T. W. Parks, "Time domain design of recursive digital filters," *IEEE Trans. Audio Electroacoust.* (*Special Issue on Digital Filtering*), vol. AU-18, pp. 137–141, June 1970.

31

Reprinted from *IEEE Trans. Geosci. Electron.* **GE-9**(1):10–27 (1971)

The Design of Multistage Separable Planar Filters

SVEN TREITEL AND JOHN L. SHANKS, MEMBER, IEEE

Abstract—A two-dimensional, or planar, digital filter can be described in terms of its planar response function, which is in the form of a matrix of weighting coefficients, or filter array. In many instances the dimensions of these matrices are so large that their implementation as ordinary planar convolutional filters becomes computationally inefficient. It is possible to expand the given coefficient matrix into a finite and convergent sum of matrix-valued stages. Each stage can be separated with no error into the product of an m-length column vector multiplied into an n-length row vector, where m is the number of rows and n is the number of columns of the original filter array. Substantial savings in computer storage and speed result if the given filter array can be represented with a tolerably small error by the first few stages of the expansion. Since each constituent stage consists of two vector-valued factors, further computational economies accrue if the one-dimensional sequences described by these vectors are in turn approximated by one-dimensional recursive filters. Two geophysical examples have been selected to illustrate how the present design techniques may be reduced to practice.

I. Introduction

A SUBSTANTIAL amount of geophysical data is recorded in the form of two-dimensional, or planar, arrays of discrete observations. A seismogram consists of a multiplicity of one-dimensional time sequences, or traces, each of which is registered at a different point in space. If the receivers are spaced along a straight line at uniform distances, and if the individual traces are sampled at a uniform temporal sampling increment, the resulting discrete seismogram is conveniently visualized as a planar array of observations, or weighting coefficients. Filters that are designed to operate explicitly on such planar input data are characterized as planar, or two-dimensional, in contradistinction to filters that act only on one-dimensional sequences, such as a single seismic trace. The velocity, or fan filter, is a familiar illustration of a planar digital operator that has found widespread use in the reduction of seismic data. Other geophysical examples also abound. For instance, potential field surveys are often recorded on an observation plane, rather than merely along a profile. If this is not done at uniform sampling increments, the data are reducible to this form by the application of two-dimensional interpolation techniques. More recently, the field implementation of the acoustical holography method has required the recording of ground motion over a plane at very fine sampling rates [5].

Manuscript received July 16, 1970; revised August 30, 1970. This paper was presented at the IEEE Workshop on Digital Filtering, Harriman, N. Y., January 13, 1970.

The authors are with the Research Center, Pan American Petroleum Corporation, Tulsa, Okla. 74102.

Planar digital filters can be obtained to satisfy a wide variety of optimization criteria [17]. The response functions of such operators must often be described by large arrays of weighting coefficients, so that computing times grow rapidly and computer-storage allocation problems become increasingly complex. These difficulties have stimulated the search for methods that reduce the total computational labor necessary for the implementation of a given planar digital filter. In the present treatment we assume that the two-dimensional response function of the filter is already known in the form of a uniformly sampled array, or matrix, of weighting coefficients. The problem of reducing the computational labor associated with the application of planar filters to planar input data can be attacked in several ways. One method that shows promise involves the approximation of the given planar response function by a two-dimensional recursive filter [14]. The current approach, on the other hand, is based on the expansion of the given coefficient matrix into a finite and converging sum of matrices having the same dimensions as the original filter array. Each constituent matrix of this sum is separable; that is, each matrix in the expansion can be expressed as the product of a column vector multiplied into a row vector. In many cases the given matrix can be represented with small error by only the first few terms of the expansion, and thus substantial gains in computing efficiency arise. Since each term, or filter stage, is expressible as a product of vectors, which are one-dimensional arrays, further speed and storage advantages result if each vector is itself approximated by an appropriate one-dimensional recursive filter.

The matrix expansion method presented here is based on the so-called "spectral representation" of an arbitrary rectangular matrix in terms of the eigenvalues and eigenvectors of two particular matrices associated with the given coefficient array. The salient features of this expansion technique are described in the main text, while several lengthy proofs are relegated to three appendices. A somewhat related approach has been given by Sanathanan [12], but the expansion he proposes does not in general converge in a finite number of matrix-valued stages. In addition, his method requires *a priori*, knowledge of one of the two vectors that comprise each stage of the expansion, and such information is not generally available in most applications. The present approach, on the other hand, determines these vectors uniquely in terms of the given planar-filter coefficient array, and no preliminary knowledge of their particular form is necessary.

Throughout this treatment we make extensive use of eigenvalue theory. Readers unfamiliar with this subject are referred to standard reference books such as Frazer *et al.* [7] and Hildebrand [8, ch. 1].

II. Expansion of a Planar Filter into a Sum of Individually Separable Planar Stages

A. Theoretical Background

Consider an arbitrary real and nonnull planar response function $\mathbf{D}$[1] which is described by the $(m \times n)$ rectangular matrix $\mathbf{D}$

$$\mathbf{D} = \begin{bmatrix} d_{11} & d_{12} & \cdots & d_{1n} \\ d_{21} & & & \cdot \\ \cdot & & & \cdot \\ \cdot & & & \cdot \\ d_{m1} & d_{m2} & \cdots & d_{mn} \end{bmatrix}. \tag{1}$$

We say that $\mathbf{D}$ is separable when it can be split into the product of an m-length column vector $\mathbf{c}$ and an n-length row vector $\mathbf{r}$

$$\mathbf{D} = \mathbf{cr} = \begin{bmatrix} c_1 \\ c_2 \\ \cdot \\ \cdot \\ c_m \end{bmatrix} [r_1 r_2 \cdots r_n] \tag{2}$$

or

$$\mathbf{D} = \mathbf{cr} = \begin{bmatrix} c_1 r_1 & c_1 r_2 & \cdots & c_1 r_n \\ c_2 r_1 & c_2 r_2 & \cdots & c_2 r_n \\ \cdot & & & \\ \cdot & & & \\ c_m r_1 & c_m r_2 & \cdots & c_m r_n \end{bmatrix}. \tag{3}$$

From the last of the preceding relations it is evident that a separable matrix has proportional rows and proportional columns; that is, all rows of $\mathbf{D}$ are of the form $c_1\mathbf{r}, c_2\mathbf{r}, \cdots, c_m\mathbf{r}$, and all columns of $\mathbf{D}$ are of the form $r_1\mathbf{c}, r_2\mathbf{c}, \cdots, r_n\mathbf{c}$.

The array $\mathbf{D}$ of (1) is described by mn elements, or weighting coefficients, but if $\mathbf{D}$ is a separable matrix, we see from (2) that it is equally well describable by the $(m+n)$ elements of its constituent vectors $\mathbf{c}$ and $\mathbf{r}$. As m and n become large, one sees that substantial savings in computer time and computer storage will be achieved if the planar-filter $\mathbf{D}$ is so separable.

Unfortunately, we cannot generally expect that an arbitrary planar-filter matrix satisfies such stringent conditions. Suppose, however, that we require only that $\mathbf{D}$ be approximately separable; that is, we seek two vectors $\mathbf{c}$ and $\mathbf{r}$ such that

$$\mathbf{D} = \mathbf{cr} + \mathbf{E} \tag{4}$$

where $\mathbf{E}$ is an $(m \times n)$ error matrix, whose elements we would like to be as small as possible in relation to the elements of $\mathbf{D}$. If this is not the case, we might then attempt to generalize the separation (4) in the form

$$\begin{aligned} \mathbf{D} &= \mathbf{c}_1\mathbf{r}_1 + \mathbf{c}_2\mathbf{r}_2 + \cdots + \mathbf{c}_i\mathbf{r}_i + \cdots + \mathbf{c}_k\mathbf{r}_k + \mathbf{E}_k \\ &= \mathbf{\Gamma}_1 + \mathbf{\Gamma}_2 + \cdots + \mathbf{\Gamma}_i + \cdots + \mathbf{\Gamma}_k + \mathbf{E}_k \end{aligned} \tag{5}$$

where each term,

$$\mathbf{\Gamma}_i = \mathbf{c}_i\mathbf{r}_i, \qquad i = 1, 2, \cdots, k \tag{6}$$

is thus an exactly separable $(m \times n)$ matrix, and where $\mathbf{E}_k$ is the $(m \times n)$ error matrix associated with a k-term expansion of $\mathbf{D}$ in the form (5). In keeping with nomenclature from filter theory, we call each term $\mathbf{\Gamma}_i$ a stage, and we say that (5) represents the "multistage separable" expansion of $\mathbf{D}$ into a sum of individually separable stages $\mathbf{\Gamma}_i$ plus a not necessarily separable error matrix $\mathbf{E}_k$. We demonstrate later that each stage is implementable as a member of a parallel configuration of individually separable planar filters.

Assuming that $\mathbf{D}$ is well approximated by k stages, we require $k(m+n)$ elements, or weighting coefficients, to describe the filter in this manner. Substantial savings in computer time and storage will result as long as $k(m+n) \ll mn$. Further advantages of this expansion will be discussed in later sections of the present treatment.

Up to this point we have merely postulated that $\mathbf{D}$ is representable as a sum of individually separable stages. We demonstrate in Appendix I that such an expansion can always be obtained and that in fact an exact representation of $\mathbf{D}$ can be guaranteed with a finite number of stages. Here we shall only state the salient features of the derivation—for detailed proofs and references the reader is referred to Appendix I.

Without loss of generality we may assume that m, the number of rows of $\mathbf{D}$, is equal to or greater than n, the number of columns of $\mathbf{D}$. We also assume first that the n columns of $\mathbf{D}$ are linearly independent.[2]

Then a solution to our problem is given by the spectral representation of $\mathbf{D}$ [see (54)],

$$\begin{aligned} \mathbf{D} &= \sqrt{\lambda_1}\,\boldsymbol{\kappa}_1\boldsymbol{\varrho}_1 + \sqrt{\lambda_2}\,\boldsymbol{\kappa}_2\boldsymbol{\varrho}_2 + \cdots + \sqrt{\lambda_n}\,\boldsymbol{\kappa}_n\boldsymbol{\varrho}_n \\ &= \sum_{i=1}^{n} \sqrt{\lambda_i}\,\boldsymbol{\kappa}_i\boldsymbol{\varrho}_i \end{aligned} \tag{7}$$

where

- $\lambda_1, \lambda_2, \cdots, \lambda_n$ real and positive eigenvalues of the n-order, symmetric and positive definite matrix $\mathbf{S} = \mathbf{D}^T\mathbf{D}$ (the superscript T denotes the matrix transpose);
- $\boldsymbol{\varrho}_1, \boldsymbol{\varrho}_2, \cdots, \boldsymbol{\varrho}_n$ normalized row eigenvectors, or eigenrows, associated with the corresponding eigenvalues $\lambda_1, \lambda_2, \cdots, \lambda_n$ of $\mathbf{S}$,

[1] A symbol in boldface type indicates that a matrix or a vector is implied.

[2] This restriction is relaxed at the end of the present section.

$\kappa_1, \kappa_2, \cdots, \kappa_n$ normalized column eigenvectors, or eigencolumns associated with the corresponding nonvanishing eigenvalues $\lambda_1, \lambda_2, \cdots, \lambda_n$ of the m-order, symmetric, and positive semidefinite matrix $Q = DD^T$.

Each term of the spectral representation (7) is the separable $(m \times n)$ matrix resulting from the multiplication of the m-length column vector $\sqrt{\lambda_i}\kappa_i$ into the n-length row vector ϱ_i. Furthermore, this representation is *exact*, provided that all n terms associated with the n eigenvalues $\lambda_1, \lambda_2, \cdots, \lambda_n$ are retained in the expansion. The matrices $Q = DD^T$ and $S = D^TD$ have an *identical* set of nonzero eigenvalues, namely, $\lambda_1, \lambda_2, \cdots, \lambda_n$.

We accordingly *identify* the spectral representation (7) with our desired multistage expansion of D, and hence,

$$D = \sum_{i=1}^{n} \sqrt{\lambda_i}\kappa_i\varrho_i = \sum_{i=1}^{n} \Gamma_i \tag{8}$$

where the ith stage Γ_i is given by

$$\Gamma_i = \sqrt{\lambda_i}\kappa_i\varrho_i. \tag{9}$$

Comparison of (6) and (7) yields the explicit forms of the desired column vectors c_i and of the desired row vectors r_i

$$c_i = \sqrt{\lambda_i}\kappa_i$$

$$r_i = \varrho_i$$

so that the c_i's are scale-factored versions of the normalized eigen columns κ_i, while the r_i's are identically equal to the normalized eigenrows ϱ_i.

The identification of the multistage expansion of D with the spectral representation of D does not constitute the only possible solution of our problem. We show in Appendix II that a least squares approach also leads to an expansion in terms of separable stages, but in general only an infinite number of stages computed in this manner will represent D with no error. It is further demonstrated in Appendix II that the spectral representation of D can be obtained as a special case of the least squares formulation.

The exact representation of D requires $(m+n)$ coefficients per stage, or a total of $n(m+n) = nm + n^2$ coefficients for the n separable-stages Γ_i. Since the original filter matrix D is described by only mn coefficients, one obviously does not obtain any direct computing speed advantages in this way.[3] On the other hand, an approximate multistage representation of D is always possible by truncating the exact expansion (8) to only k terms. Computing speed advantages will then still result as long as $k(m+n) < mn$.

In general the n real and positive eigenvalues λ_i of the symmetric positive definite n-order matrix

$$S = D^TD \tag{10}$$

will have distinct magnitudes, so that some will be larger than others. One may therefore suspect that the n terms of (8) do not contribute equally to the entire expansion. If this should turn out to be the case, one will naturally wish to retain only the k *dominant* terms.

Accordingly, we suppose that the eigenvalues λ_i are listed in order of decreasing magnitude,

$$\lambda_1, \lambda_2, \cdots, \lambda_{n-1}, \lambda_n \tag{11}$$

where λ_1 is the largest and λ_n the smallest eigenvalue. We suppose further that any repeated eigenvalues are labeled separately.

Let us note that the expansion (8) can be written in any one of the n forms,

$$D = \sum_{i=1}^{k} \Gamma_i + E_k, \qquad k = 1, 2, \cdots, n \tag{12}$$

where the $(m \times n)$ array E_k is the truncation error matrix arising from the retention of the first k terms in the expansion (8). Since the expansion is exact for $k = n$, we have

$$E_n = 0$$

where 0 is the $(m \times n)$ null matrix.

A convenient scalar-valued measure of the truncation error is given by the normalized error energy ϵ_k, which we define to be

$$\epsilon_k = \frac{J_k}{J_0}. \tag{13}$$

Here J_k is the sum of the squares of the elements of the truncation error matrix E_k, while J_0 is the sum of the squares of the elements of the given matrix D.

It is shown in Appendix I (66) that ϵ_k is uniquely determined by the eigenvalues λ_i of S,

$$\epsilon_k = 1 - \frac{\lambda_1 + \lambda_2 + \cdots + \lambda_k}{\lambda_1 + \lambda_2 + \cdots + \lambda_n}, \qquad k = 1, 2, \cdots, n. \tag{14}$$

The normalized error energy ϵ_k decreases as k increases, for the term

$$\frac{\lambda_1 + \lambda_2 + \cdots + \lambda_k}{\lambda_1 + \lambda_2 + \cdots + \lambda_n}$$

is less than unity for $k < n$ and becomes unity when $k = n$. The bounds on ϵ_k are then

$$0 \le \epsilon_k < 1.$$

Let us see how rapidly the normalized error energy approaches zero as the number of retained stages k in-

[3] Other advantages do, however, arise from the exact multistage representation of D. These are described in Section IV.

crease. From (14) we obtain an expression for the difference between successive values of ϵ_k,

$$\epsilon_k - \epsilon_{k+1} = \frac{\lambda_{k+1}}{\lambda_1 + \lambda_2 + \cdots + \lambda_n}, \quad k = 1, 2, \cdots, n-1. \tag{15}$$

Suppose for the moment that all eigenvalues λ_i are distinct. Then we have that $\lambda_1 > \lambda_2 > \cdots > \lambda_n$, and ϵ_k diminishes rapidly for the smaller values of k. As k increases further, the difference $\epsilon_k - \epsilon_{k+1}$ becomes smaller because λ_{k+1} becomes progressively smaller. In other words, the dominant terms of the expansion (8) are associated with the larger, or dominant, eigenvalues of the matrix $\boldsymbol{S} = \boldsymbol{D}^T\boldsymbol{D}$. The ordering (11) then guarantees that the leading terms of (8) are also the dominant ones. Knowledge of the eigenvalues of $\boldsymbol{S}$ is sufficient to establish the number of stages k required to reduce ϵ_k equal to or below some desired value in the range $0 \le \epsilon_k < 1$.

If all eigenvalues λ_i are equal, we deduce from (15) that no term of the expansion (8) is dominant; if some of the λ_i are equal, corresponding terms of (8) contribute equally to the expansion.

Thus far it has been assumed that the n columns of $\boldsymbol{D}$ are linearly independent; that is, no set of constants $a_1, a_2, \cdots, a_n$, at least one of which is not zero, exists such that

$$a_1\boldsymbol{d}_1 + a_2\boldsymbol{d}_2 + \cdots + a_i\boldsymbol{d}_i + \cdots + a_n\boldsymbol{d}_n = 0$$

where $\boldsymbol{d}_i$ $(i = 1, 2, \cdots, n)$ is the ith m-length column of $\boldsymbol{D}$ [8, p. 25]. If this is not the case, and say a certain number $p(<n)$ of the columns $\boldsymbol{d}_i$ are linearly dependent, the spectral representation for $\boldsymbol{D}$ given by (7) terminates with no error for $n-p$ stages. Then the expansion (12) becomes

$$\boldsymbol{D} = \sum_{i=1}^{k} \boldsymbol{\Gamma}_i + \boldsymbol{E}_k, \qquad k = 1, 2, \cdots, n-p \tag{16}$$

where

$$\boldsymbol{E}_{n-p} = \boldsymbol{0}.$$

The proof for this result is given in the discussion leading to (67) of Appendix I. In particular, if $p = n-1$ so that only one column of $\boldsymbol{D}$ is linearly independent, the expansion (16) terminates with no error for $(n - [n-1]) = 1$ stage,

$$\boldsymbol{D} = \boldsymbol{\Gamma}_1$$

where we have by (9) that

$$\boldsymbol{\Gamma}_1 = \sqrt{\lambda_1}\,\boldsymbol{\kappa}_1\boldsymbol{\varrho}_1 = \boldsymbol{c}_1\boldsymbol{r}_1.$$

In this instance the given array $\boldsymbol{D}$ is thus *itself* separable.

The numerical problem of computing the eigenvalues λ_i and their associated eigenvectors $\boldsymbol{\kappa}_i$ and $\boldsymbol{\varrho}_i$ is the subject of the subsequent section.

B. The Numerical Solution

We have previously seen that the most general form of the multistage expansion of $\boldsymbol{D}$ is given by (16),

$$\boldsymbol{D} = \sum_{i=1}^{k} \boldsymbol{\Gamma}_i + \boldsymbol{E}_k, \qquad k = 1, 2, \cdots, n-p \tag{17}$$

and where, in particular,

$$\boldsymbol{E}_{n-p} = \boldsymbol{0}.$$

The ith separable stage $\boldsymbol{\Gamma}_i$ is given by (9)

$$\boldsymbol{\Gamma}_i = \sqrt{\lambda_k}\,\boldsymbol{\kappa}_i\boldsymbol{\varrho}_i. \tag{18}$$

It may be deduced from the derivations of Appendix I that the m-length column vectors $\boldsymbol{\kappa}_i$ constitute the set of $(n-p)$ orthonormal eigencolumns of the symmetric, positive semidefinite, m-order matrix $\boldsymbol{Q}$,

$$\boldsymbol{Q} = \boldsymbol{D}\boldsymbol{D}^T.$$

Similarly, the n-length row vectors $\boldsymbol{\varrho}_i$ constitute the set of $(n-p)$ orthonormal eigenrows of the symmetric nth order matrix $\boldsymbol{S}$,

$$\boldsymbol{S} = \boldsymbol{D}^T\boldsymbol{D}$$

where $\boldsymbol{S}$ is positive definite when $p = 0$ and positive semidefinite otherwise. Furthermore, both $\boldsymbol{Q}$ and $\boldsymbol{S}$ have the identical set of nonvanishing eigenvalues,

$$\lambda_1, \lambda_2, \cdots, \lambda_{n-p}$$

where λ is the largest and λ_{n-p} the smallest in magnitude.

Throughout this treatment we assume that $m \ge n$, and it is shown in Appendix I that no loss of generality results from such a restriction. If $m > n$, it is obviously better to extract the $(n-p)$ nonvanishing λ_i's from $\boldsymbol{S}$, which is only of order n, rather than from $\boldsymbol{Q}$, which is of order m. If $m = n$, the eigenvalues may be extracted equally well either from $\boldsymbol{Q}$ or from $\boldsymbol{S}$, but for the sake of convenience we shall always extract these eigenvalues from $\boldsymbol{S}$.

Once the $(n-p)$ nonvanishing eigenvalues of $\boldsymbol{S}$ are known, we can compute the normalized error energy ϵ_k from the formula

$$\epsilon_k = 1 - \frac{\lambda_1 + \lambda_2 + \cdots + \lambda_k}{\lambda_1 + \lambda_2 + \cdots + \lambda_{n-p}}, \quad k = 1, 2, \cdots, n-p \tag{19}$$

which is the generalization of (14) for the case $p \ne 0$. It is convenient to compute the full set of $(n-p)$ normalized error energies at this point, because one may thus determine the number of stages k which are required to reduce ϵ_k to or below some desired value in the range $0 \le \epsilon_k < 1$. In this manner one avoids the calculation of unnecessary eigenvectors.

Next one computes the first k orthonormal eigenrows $\boldsymbol{\varrho}_i$ of $\boldsymbol{S}$, which are associated with the eigenvalues

$\lambda_1, \cdots, \lambda_k$, $(k \leq n-p)$. These vectors constitute the set of desired k row vectors of the multistage expansion (16).

We could now calculate the required first k orthonormal eigencolumns $\boldsymbol{\kappa}_i$ directely from $\boldsymbol{Q}$. The extraction of an eigenvector from a matrix is generally a laborious numerical problem. Although this must be done for the eigenrows $\boldsymbol{\varrho}_i$ of $\boldsymbol{S}=\boldsymbol{D}^T\boldsymbol{D}$, we may make use of the expression

$$\sqrt{\lambda_i}\,\boldsymbol{\kappa}_i = \boldsymbol{D}\boldsymbol{\varrho}_i^T, \qquad i = 1, 2, \cdots, k \tag{20}$$

which is (53a) of Appendix I. Indeed, it is shown there that this is one of the two relations that provide the coupling between the eigenvalue problems for $\boldsymbol{S}=\boldsymbol{D}^T\boldsymbol{D}$ and $\boldsymbol{Q}=\boldsymbol{D}\boldsymbol{D}^T$.

From (18) we see that the desired column vectors are actually of the scale-factored form

$$\boldsymbol{c}_i = \sqrt{\lambda_i}\,\boldsymbol{\kappa}_i. \tag{21}$$

Combination of (20) and (21) yields

$$\boldsymbol{c}_i = \boldsymbol{D}\boldsymbol{\varrho}_i^T.$$

In summary, if we express the ith stage $\boldsymbol{\Gamma}_i$ as the product

$$\boldsymbol{\Gamma}_i = \boldsymbol{c}_i\boldsymbol{r}_i \tag{22}$$

the vectors comprising this stage are

$$\left.\begin{aligned} \boldsymbol{r}_i &= \boldsymbol{\varrho}_i, \\ \boldsymbol{c}_i &= \boldsymbol{D}\boldsymbol{\varrho}_i^T, \end{aligned}\right\} \quad i = 1, 2, \cdots, k \tag{23}$$

and we see that the eigencolumns $\boldsymbol{\kappa}_i$ need not be computed explicitly. On the other hand, it will always be necessary to solve the eigenvalue problem associated with $\boldsymbol{S}=\boldsymbol{D}^T\boldsymbol{D}$; that is, one will have to make use of one of the many numerical schemes available for the extraction of the eigenvalues $\lambda_1, \cdots, \lambda_n$ and the desired associated first k eigenrows $\boldsymbol{\varrho}_1, \cdots, \boldsymbol{\varrho}_k$.

The choice of a particular eigenvalue algorithm will be governed by many factors, including desired computational speed and accuracy and the idiosyncracies of one's digital computer installation. The numerical examples described in the present treatment were calculated with the aid of a Fortran IV subroutine[4] based on the method of Jacobi. Readers interested in an exhaustive analysis of the numerical eigenvalue problem are referred to the excellent book by Wilkinson [18].

A small illustrative example will be helpful. Let $\boldsymbol{D}$ be the matrix

$$\boldsymbol{D} = \begin{bmatrix} 1 & 2 \\ 2 & 1 \\ 1 & 3 \end{bmatrix}$$

[4] Subroutine EIGEN, IBM System 360 Scientific Subroutine Package, H20-0205-3, 4th ed., IBM Corporation, New York, 1968.

so that $m=3$ and $n=2$. Then

$$\boldsymbol{S} = \boldsymbol{D}^T\boldsymbol{D} = \begin{bmatrix} 1 & 2 & 1 \\ 2 & 1 & 3 \end{bmatrix}\begin{bmatrix} 1 & 2 \\ 2 & 1 \\ 1 & 3 \end{bmatrix} = \begin{bmatrix} 6 & 7 \\ 7 & 14 \end{bmatrix}.$$

The eigenvalues of $\boldsymbol{S}$ are

$$\lambda_1 = 18.06226$$

$$\lambda_2 = 1.93774.$$

Since neither of these eigenvalues is zero, $p=0$; i.e., the columns of $\boldsymbol{D}$ are linearly independent, and therefore we expect $\boldsymbol{D}$ to be representable exactly with $(n-p)=2$ stages. From (19) we obtain the normalized error energies

$$\epsilon_1 = 1 - \frac{\lambda_1}{\lambda_1+\lambda_2} = 0.09689 \approx 0.1$$

$$\epsilon_2 = 1 - \frac{\lambda_1+\lambda_2}{\lambda_1+\lambda_2} = 0.$$

The associated orthonormal eigenrows are

$$\boldsymbol{\varrho}_1 = [0.50193 \quad 0.86491] = \boldsymbol{r}_1$$

$$\boldsymbol{\varrho}_2 = [0.86491 \quad -0.50193] = \boldsymbol{r}_2.$$

We compute the column vectors $\boldsymbol{c}_1$ and $\boldsymbol{c}_2$ from (23),

$$\boldsymbol{c}_1 = \begin{bmatrix} 1 & 2 \\ 2 & 1 \\ 1 & 3 \end{bmatrix}\begin{bmatrix} 0.50193 \\ 0.86491 \end{bmatrix} = \begin{bmatrix} 2.23175 \\ 1.86876 \\ 3.09666 \end{bmatrix}$$

$$\boldsymbol{c}_2 = \begin{bmatrix} 1 & 2 \\ 2 & 1 \\ 1 & 3 \end{bmatrix}\begin{bmatrix} 0.86491 \\ -0.50193 \end{bmatrix} = \begin{bmatrix} -0.13894 \\ 1.22789 \\ -0.64087 \end{bmatrix}.$$

The desired two-stage expansion of $\boldsymbol{D}$ is thus

$$\boldsymbol{D} = \boldsymbol{\Gamma}_1 + \boldsymbol{\Gamma}_2$$

where

$$\boldsymbol{\Gamma}_1 = \begin{bmatrix} 2.23175 \\ 1.86876 \\ 3.09666 \end{bmatrix}[0.50193 \quad 0.86491]$$

$$\boldsymbol{\Gamma}_2 = \begin{bmatrix} -0.13894 \\ 1.22789 \\ -0.64087 \end{bmatrix}[0.86491 \quad -0.50193].$$

Retention of the first and dominant stage $\boldsymbol{\Gamma}_1$ produces the truncation error matrix

$$\boldsymbol{E}_1 = \boldsymbol{D} - \boldsymbol{\Gamma}_1 = \begin{bmatrix} -0.12017 & 0.06974 \\ 1.06202 & -0.61631 \\ -0.55430 & 0.32167 \end{bmatrix}.$$

The elements of $\boldsymbol{E}_1$ are certainly not small in comparison to the elements of $\boldsymbol{D}$, and hence one would use both stages in order to obtain a satisfactory (in this instance, perfect) agreement between D and its multistage expansion. The single-stage normalized error energy ϵ_1 is as we have seen, of the order of 0.1. The preceding example illustrates that 0.1 is not "small" because the single-stage Γ_1 does not represent D adequately. In a subsequent section we shall find that a good multistage expansion is achievable whenever k is large enough to reduce ϵ_k to the order of 0.01 or less.

The skeptical reader has probably noticed that even the single-stage expansion requires $(m+n)=(3+2)=5$ coefficients in comparison to the $mn=6$ coefficients that describe D. The advantages of the multistage expansion method only become apparent for larger values of m and n, and the above example is merely meant to illustrate the principles of the technique.

III. The Implementation of Multistage Separable Planar Filters

In the preceding section we have demonstrated how an arbitrary planar response function can be expanded into a series of individually separable stages. Here we shall discuss how such filters may be reduced to practice.

It is convenient to represent the planar filtering operation in terms of the two-dimensional, or planar, z transform [9, p. 73].[5] Suppose that we have an $(M\times N)$ input matrix X that is to be filtered with an $(m\times n)$ filter matrix F. Both X and F are representable as two-dimensional, or planar, polynomials in the variables z_1 and z_2.

$$X(z_1, z_2) = \sum_{i=1}^{M}\sum_{j=1}^{N} x_{ij} z_1^{i-1} z_2^{j-1} \tag{24}$$

and

$$F(z_1, z_2) = \sum_{i=1}^{m}\sum_{j=1}^{n} f_{ij} z_1^{i-1} z_2^{j-1} \tag{25}$$

where x_{ij} is an element of X and f_{ij} is an element of F.

Let Y be the output matrix resulting from the action of the filter F on the input X. Then this operation can be represented by the planar polynomial product

$$Y(z_1, z_2) = X(z_1, z_2)F(z_1, z_2) \tag{26}$$

[14]. Alternatively, this filtering operation is expressible as the planar convolution

$$y_{kl} = \sum_{i=1}^{m}\sum_{j=1}^{n} f_{ij} x_{k-i+1,l-j+1} \tag{27}$$

where y_{kl} is an element of Y, and where we assume that the input elements x_{ij} vanish outside the range

$$i, j:\ i = 1, 2, \cdots, M$$

and

$$j = 1, 2, \cdots, N.$$

The convolution of an m-length one-dimensional filter with an M-length one-dimensional input sequence results in a one-dimensional output sequence of length $M+m-1$. Similarly, the planar convolution of an $(m\times n)$ filter F with an $(M\times N)$ input X yields an $(M+m-1)$ by $(N+n-1)$ output array Y, so that (27) holds for the subscript ranges

$$k = 1, 2, \cdots, \quad M+m-1$$
$$l = 1, 2, \cdots, \quad N+n-1.$$

Suppose that the filter F is exactly separable. Then its planar response function is given by (2).[6]

$$F = cr = \begin{bmatrix} c_1 \\ c_2 \\ \cdot \\ \cdot \\ \cdot \\ c_m \end{bmatrix} [r_1 r_2 \cdots r_n] \tag{28}$$

where c is an m-length column vector and r is an n-length row vector. In this case the elements of F are simply

$$f_{ij} = c_i r_j$$

and substitution of this relation into (27) yields

$$y_{kl} = \sum_{i=1}^{m}\sum_{j=1}^{n} c_i r_j x_{k-i+1,l-j+1}.$$

In terms of planar z transforms, the separation (28) is

$$F(z_1, z_2) = C(z_1)R(z_2) \tag{29}$$

where

$$c(z_1) = c_1 + c_2 z_1 + c_3 z_1^2 + \cdots + c_m z_1^{m-1}$$
$$R(z_2) = r_1 + r_2 z_2 + r_3 z_2^2 + \cdots + r_n z_2^{n-1}.$$

The planar z transform of the output $Y(z_1,z_2)$ is obtained by inserting (29) into (26),

$$Y(z_1, z_2) = X(z_1, z_2)C(z_1)R(z_2). \tag{30}$$

Since polynomial multiplication is commutative, $Y(z_1, z_2)$ may also be expressed in the form

$$Y(z_1, z_2) = X(z_1, z_2)R(z_2)C(z_1). \tag{31}$$

Let us define the intermediate outputs

$$W(z_1, z_2) = X(z_1, z_2)C(z_1)$$

and

$$V(z_1, z_2) = X(z_1, z_2)R(z_2).$$

Then (30) and (31) become

$$Y(z_1, z_2) = W(z_1, z_2)R(z_2) \tag{32a}$$

and

$$Y(z_1, z_2) = V(z_1, z_2)C(z_1). \tag{32b}$$

[5] Our z-transfom definition differs from Jury's [9] in that his z is our z_1^{-1} and his w is our z_2^{-1}.

[6] In the sequel we use the symbol F to represent the filter; the symbol D was used in Section II to denote the fact that either the filter or the input is expressible in terms of a series of separable stages.

The formulations (32a) and (32b) both produce the same output $Y(z_1,z_2)$, but they do not necessarily have the same computing efficiency. This efficiency can be established by determining the number of "multiply and add" operations (MADS) required by each formulation. Let us call the number of MADS required by the algorithms (32a) and (32b) MADS (a) and MADS (b), respectively. We show in Appendix III that

$$\begin{aligned} \mathrm{MADS}(a) &= MN(m+n) + n(m-1)N \\ \mathrm{MADS}(b) &= MN(m+n) + m(n-1)M. \end{aligned} \tag{33}$$

Since the first terms of these two expressions are identical, we will wish to pick the algorithm for which the second term is smallest. If we make the generally reasonable assumption that $n(m-1)$ is of the order of $m(n-1)$, the scheme (32a) will be more efficient when $N \ll M$, while the scheme (32a) will be more efficient when $N \gg M$. If $M \approx N$, both schemes have roughly similar efficiency. Furthermore, if $mn \ll MN$, so that the filter array is of much smaller size than the input array, the second terms of the two expressions (33) can be neglected. Then

$$\mathrm{MADS}(a) \approx \mathrm{MADS}(b) \approx MN(m+n) \tag{34}$$

and again both schemes exhibit comparable computational efficiency.

If the filter $F(z_1, z_2)$ is not separable, the original formulation (26) would have to be used. It is shown in Appendix III that the number of MADS required for the complete convolution of the $(m \times n)$ filter array $\boldsymbol{F}$ with the $(M \times N)$ input array $\boldsymbol{X}$ are

$$MNmn. \tag{35}$$

The computing efficiencies of the various filtering algorithms can be conveniently compared by the introduction of a quantity that we call the "effective length" L_{eff} of a given algorithm. We define the effective length to be the number of MADS required by a given filtering algorithm divided by the size MN of the input array. For standard planar convolution this quantity follows directly from (35) and is

$$L_{\mathrm{eff}} = \frac{MNmn}{MN} = mn. \tag{36}$$

If the filter $\boldsymbol{F}$ is separable and if $mn \ll MN$, the appropriate effective length is by (34)

$$L_{\mathrm{eff}} = \frac{MN(m+n)}{MN} = m+n. \tag{37}$$

Thus the effective length for a nonseparable filter grows as the product of the filter dimensions, whereas it grows only as their sum if the filter is separable.

In the previous section we showed how an arbitrary planar response function can be expanded into a sum of k separable stages $\boldsymbol{\Gamma}_i$, $(i=1, 2, \cdots, k)$. Assuming that $m \geq n$, we also established that this expansion is exact in, at most, $k=n$ stages, and that good approximations of the given filter response are often achievable when $k<n$. Let us represent the k-stage approximation of the planar filter $\boldsymbol{F}$ by the symbol $\boldsymbol{F}^{(k)}$, where

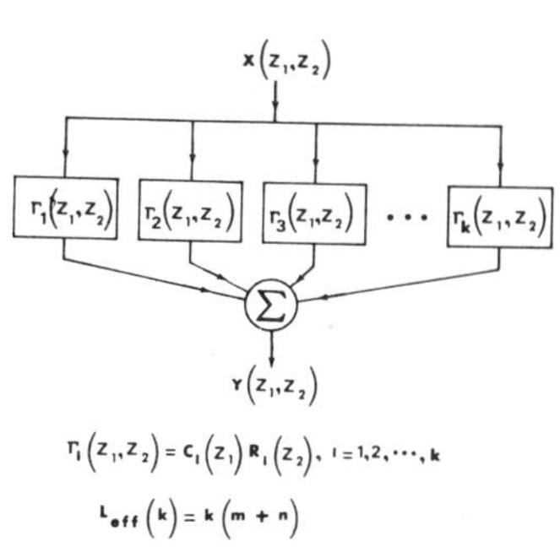

Fig. 1. Block diagram showing implementation of a k-stage planar filter [see (38)].

$$\boldsymbol{F}^{(k)} = \boldsymbol{\Gamma}_1 + \boldsymbol{\Gamma}_2 + \cdots + \boldsymbol{\Gamma}_k = \sum_{i=1}^{k} \boldsymbol{\Gamma}_i.$$

For every filter $\boldsymbol{F}^{(k)}$ there exists an associated normalized error energy ϵ_k which vanishes for some value of $k \leq n$, and whose explicit form is given by (14).

The planar z transform of the filter $\boldsymbol{F}^{(k)}$ is

$$F^{(k)}(z_1, z_2) = \sum_{i=1}^{k} \Gamma_i(z_1, z_2).$$

If we replace $F(z_1,z_2)$ by $F^{(k)}(z_1,z_2)$, in (26) and use the preceding expansion, we obtain

$$Y(z_1, z_2) = X(z_1, z_2) \sum_{i=1}^{k} \Gamma_i(z_1, z_2). \tag{38}$$

But every stage Γ_i is separable, and thus (29) generalizes to

$$\Gamma_i(z_1, z_2) = C_i(z_1) R_i(z_2) \tag{39}$$

where $C_i(z_1)$ and $R_i(z_2)$ are the one-dimensional z transforms of the m-length column vectors $\boldsymbol{c}_i$ and the n-length row vectors $\boldsymbol{r}_i$, respectively. Substitution of (39) into (38) yields

$$Y(z_1, z_2) = X(z_1, z_2) \sum_{i=1}^{k} C_i(z_1) R_i(z_2). \tag{40}$$

We thus obtain a sum of the k planar z transform products of the type (30). Equation (40) naturally reduces to (30) when $k=1$ if we make the identifications $C_1(z_1) = C(z_1)$ and $R_1(z_2) = R(z_2)$. But the effective length L_{eff} for a single stage of the filter $\boldsymbol{F}^{(k)}$ is given by (37), and hence the effective length $L_{\mathrm{eff}}(k)$ for the k-stage filtering operation represented by (38) or (40) is

$$L_{\mathrm{eff}}(k) = k(m+n). \tag{41}$$

The block diagram of Fig. 1 illustrates the implementation of the filter equation (38). We observe that each of the k separable stages $\boldsymbol{\Gamma}_i$ operates individually on the common input $\boldsymbol{X}$, so that k partial output arrays are produced. The sum of these output matrices then results in the overall output matrix $\boldsymbol{Y}$.

If the filter matrix $\boldsymbol{F}$ can be suitably approximated by a sum of k stages $\boldsymbol{\Gamma}_i$ such that

$$k(m+n) < mn \tag{42}$$

the expanded filter $\boldsymbol{F}^{(k)}$ is more efficient than the given filter $\boldsymbol{F}$. The inequality (42) suggests the introduction of the efficiency parameter α,

$$\alpha = \frac{mn}{k(m+n)} = \frac{mn}{L_{\text{eff}}(k)} \cdot \tag{43}$$

If α is greater than unity, the algorithm (38) is more efficient, while the standard filtering algorithm (26) is more efficient whenever α is less than unity. If $\alpha = 1$, both algorithms are equally efficient. Of course the smaller the acceptable value of the number of stages k the greater is the magnitude of α for a given set of filter dimensions m and n. The price one pays as k decreases is an increasing value of the normalized error energy ϵ_k (see Section II), so that one will want to pick a value of k large enough to render ϵ_k acceptably small. For example, suppose that $m = n = 100$. If five stages constitute a suitable approximation of $\boldsymbol{F}$, $k = 5$ and then

$$\alpha = \frac{10\,000}{1000} = 10$$

so that the expansion scheme (38) is ten times as efficient as the standard scheme (26). On the other hand, if as many as ten stages are required to yield a tolerably small value of ϵ_k, $k = 10$ so that

$$\alpha = \frac{10\,000}{2000} = 5$$

and the scheme (38) is only five times as efficient as the scheme (26).

We remark also that the efficiency parameter α always has a lower limit, since the multistage expansion of $\boldsymbol{F}^{(k)}$ is exact when k is at most equal to n. The minimum value of α, $\alpha_{\min}$, is obtainable from (43) by setting $k = n$

$$\alpha_{\min} = \frac{m}{m+n} \cdot$$

If $m = n$ (so that $\boldsymbol{F}$ is a square filter matrix), $\alpha_{\min} = 1/2$. Such a value of α does not warrant use of the expansion scheme unless further refinements of the method can be made.

One such refinement becomes possible once we recognize that the columns $C_i(z_1)$ and the rows $R_i(z_2)$ of all stages $\Gamma_i(z_1,z_2)$ are themselves one-dimensional convolutional filters. In many instances these filters can be well approximated by one-dimensional recursive filters [13], and thus further gains in computing efficiency result that are over and above those achievable by the multistage separation technique alone. All these matters will be illustrated by example in the following section.

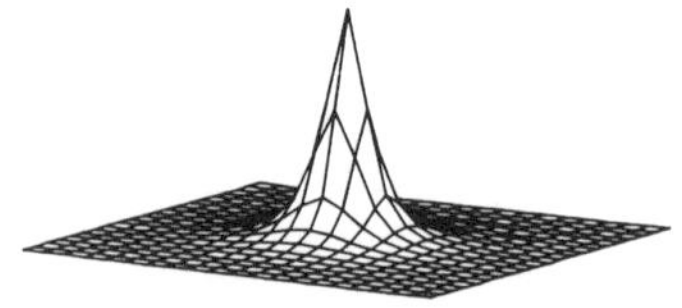

Fig. 2. Perspective view of discrete upward continuation operator. $L_{\text{eff}} = 441$.

IV. Illustrations of the Design Technique

The previously developed expansion method will here be illustrated with two examples. The first application is to a filter array that can be expanded with good accuracy into only a small number of individually separable stages $\boldsymbol{\Gamma}_i$, while the second involves a response function requiring a greater number of stages for a satisfactory approximation.

A. Upward Continuation Operator

The upward continuation operator is a planar filter that acts on gridded potential field data of one level and yields as output the field that would be measured at a distance h above the original recording plane. The continuous form of this filter in the space domain is

$$f(x, y) = \frac{1}{2\pi} \frac{h}{(h^2 + x^2 + y^2)^{3/2}}$$

where x and y are spatial coordinates [2]. If this response function is truncated in both x and y beyond a point for which the magnitude of $f(x, y)$ is negligibly small, and if the resulting function is then sampled in both directions at a uniform increment Δs, a planar array of weighting coefficients is obtained. Here we have

$$h = 1.5\Delta s$$

chosen to make[7] and to truncate $f(x, y)$ in such a manner that a square array $\boldsymbol{F}$ of size $(21) \times (21)$ is produced, and hence $m = n = 21$.

Fig. 2 shows a perspective view of the discrete upward continuation operator, where the magnitudes of the elements f_{ij} of $\boldsymbol{F}$ are plotted against the vertical axis. Use of the standard planar convolution algorithm (26) leads to the effective length L_{eff} given by (36), namely,

$$L_{\text{eff}} = (21)^2 = 441.$$

The matrix $\boldsymbol{F}$ was expanded into a single stage $\boldsymbol{\Gamma}_1 = \boldsymbol{c}_1\boldsymbol{r}_1$, with an associated normalized error energy (NEE)

$$\epsilon_1 = 0.01136.$$

The effective length for this single-stage ($k = 1$) expan-

[7] It can be shown that the aliasing error is negligible as long as $h/\Delta s \geq 1$. S. Treitel, unpublished note.

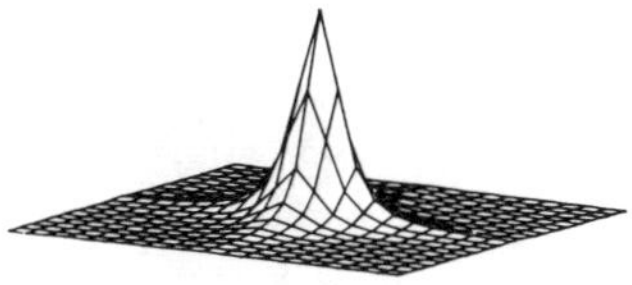

Fig. 3. Perspective view of single-stage expansion of discrete upward continuation operator. $\epsilon_k = 0.01136$ and $L_{eff} = 42$.

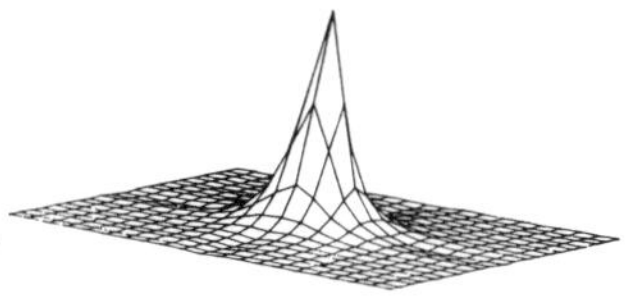

Fig. 4. Perspective view of two-stage expansion of discrete upward continuation operator. $\epsilon_k = 0.00057$ and $L_{eff} = 84$.

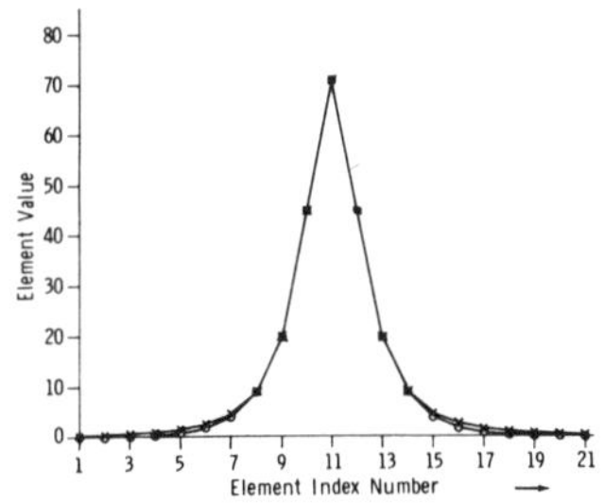

Fig. 5. The column vector C_1 (X) and its recursive approximation (O).

sion of F is then by (41)

$$L_{eff}(1) = 21 + 21 = 42.$$

The array Γ_1 is shown in perspective view in Fig. 3. Comparison with Fig. 2 clearly reveals that even the single separable stage Γ_1 constitutes a satisfactory though not perfect approximation of the given array F.

The array

$$F^{(2)} = \Gamma_1 + \Gamma_2$$

that results from the two-stage ($k=2$) expansion of F is depicted in Fig. 4. The value of ϵ_k has now decreased to

$$\epsilon_2 = 0.00057$$

while the effective length is by (41)

$$L_{eff}(2) = 2(21 + 21) = 84.$$

The two-stage expansion $F^{(2)} = \Gamma_1 + \Gamma_2$ is seen to be in excellent agreement with the given upward continuation filter array F. The efficiency parameter α is by (43) for $k=2$

$$\alpha = \frac{441}{84} = 5.25.$$

Therefore use of the filter $F^{(2)}$ in the expanded convolution scheme (38) is more than five times as efficient as use of the given filter F in the standard convolution scheme (26).

The planar filter $F^{(2)}$ consists of four individual one-dimensional filters, so that in terms of z transforms we have

$$\begin{aligned} F^{(2)}(z_1, z_2) &= \Gamma_1(z_1, z_2) + \Gamma_2(z_1, z_2) \\ &= C_1(z_1)R_1(z_2) + C_2(z_1)R_2(z_2). \end{aligned}$$

The computing efficiency of the present expansion method can be further increased by replacing the four one-dimensional convolutional filters $C_1(z_1)$, $C_2(z_1)$, $R_1(z_2)$, and $R_2(z_2)$ by corresponding one-dimensional recursive filters [13].

Fig. 5 shows a plot of the 21-length operator $C_1(z_1)$. We notice that this weighting coefficient sequence is symmetric with respect to the central; or eleventh coefficient. It was therefore decided to approximate $C_1(z_1)$ as a sum of two individually recursive stages that act in the forward and reverse directions, respectively [15]. In other words, given the sequence

$$C_1 = (c_1, c_2, \cdots, c_{11}, \cdots, c_{20}, c_{21})$$

we first design a "forward" recursive filter $C_1'(z_1)$ that approximates the sequence

$$0.5c_{11}, c_{12}, \cdots, c_{21}.$$

Then, since the sequence C_1 is symmetric about c_{11}, the "reverse" recursive filter is merely $C_1'(1/z_1)$; that is, the reverse filter is obtained by replacing z_1 by $1/z_1$ in the expression for $C_1'(z_1)$. The reverse filter thus approximates the sequence

$$c_1, c_2, \cdots, 0.5c_{11}$$

and evidently the sum

$$C_1(z_1) \approx C'(z_1) + C'(1/z_1) \tag{44}$$

will then constitute an approximation to the given filter $C_1(z_1)$. The method for computing the coefficients of a recursive filter as well as the algorithms required to apply them to digitized data are described by Shanks [13]. Fig. 5 also shows a plot of the approximation (44), which is seen to be quite satisfactory. In this particular case $C_1'(z_1)$ is given by

$$C_1'(z_1) = \frac{0.35447 + 0.29128z_1}{1.0 - 0.44040z_1}$$

so that three MADS per output value are required. Thus, whereas use of the convolutional filter $C_1(z_1)$ requires 21 MADS per output value, use of the recursive form (44) leads to only 2×3, or six MADS per output value.

The remaining one-dimensional convolution filters $R_1(z_2)$, $C_2(z_1)$, and $R_2(z_2)$ can likewise be approximated by sums of recursive filters of the form (44). For the present

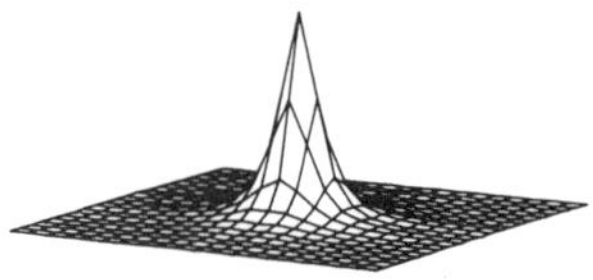

Fig. 6. Perspective of two-stage expansion of the discrete upward continuation operator. The constituent vectors have been approximated by one-dimensional recursive filters. $\epsilon_k = 0.002$ and $L_{eff} = 28$.

example satisfactory fits have been obtained by approximating $C_1(z_1)$ and $R_1(z_2)$ by sums of two recursive stages that each require three MADS per output value, while the stages corresponding to $C_2(z_1)$ and $R_2(z_2)$ each require four MADS per output value. As a result the effective length of the recursive approximation to the filter $\boldsymbol{F}^{(2)}$ is

$$L_{eff} = (4 \times 3) + (4 \times 4) = 28.$$

This effective length must be compared with $L_{eff} = 84$ for the original two-stage expansion of $\boldsymbol{F}$, and with $L_{eff} = 441$ for use of $\boldsymbol{F}$ in its original nonsuspended form. Fig. 6 shows a perspective view of the two-stage filter $\boldsymbol{F}^{(2)}$ after its constituent one-dimensional convolutional filters $C_i(z_1)$ and $R_i(z_2)$, $(i = 1, 2)$ have been approximated by one-dimensional recursive filters. The value of the normalized error energy ϵ_2, is now

$$\epsilon_2 = 0.002$$

which compares with $\epsilon_2 = 0.00057$ for the original two-stage expansion. Thus a price for this further increase in efficiency is paid, but comparison of Fig. 6 with Fig. 2 reveals good agreement beween the original filter $\boldsymbol{F}$ and its recursively approximated two-stage expansion.

Quite evidently the use of one-dimensional recursive filters can produce savings in computer time and storage beyond those achievable by the planar expansion method alone.

B. Fan-Pass Filter

The fan-pass operator is a planar filter that is designed to pass events on a seismogram whose apparent velocities fall within a given wedge-shaped region in the frequency–wavenumber, or (f, κ), plane [4], [6], and [15]. Here we have chosen a 22-trace fan filter, each of whose rows consists of 21 sampled values. This filter is therefore described by a matrix of weighting coefficients having $m = 22$ rows and $n = 21$ columns. Fig. 7 shows a contour map of the two-dimensional amplitude spectrum of the operator.[8] Also shown is a perspective view of the shaded portion of the contour map. The performance of the separation technique for the upward continuation operator was illustrated by successive views of the appropriate approximations to the given space-domain response function. A fan filter is more easily visualized in terms of its (f, κ)-domain transfer function. Accordingly, we illustrate the performance of the expansion technique by displaying the (f, κ) amplitude spectra of the successive approximations. These are, of course, computed by the expansion method from the exact representation of the fan filter in the space–time, or (x, t)-domain.

[8] The first and second quadrants of this spectrum are shown. The quadrants are bounded by the respective temporal and spatial Nyquist frequencies.

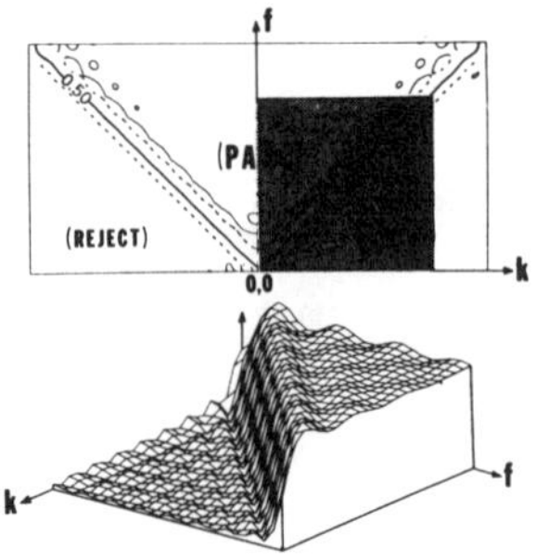

Fig. 7. Amplitude spectrum of the 22-trace 21 sampled values per trace fan filter. $L_{eff} = 462$.

TABLE I

EFFECTIVE LENGTHS AND NORMALIZED ERROR ENERGIES FOR THE EXPANSION OF THE FAN FILTER $(k = 1, 2, \cdots, 5)$

Number of Stages (k)	Effective Length $L_{eff}(k)$	Normalized Error Energy (ϵ_k)
1	43	0.173
2	86	0.081
3	129	0.048
4	172	0.031
5	215	0.021

The effective length of the given (22×21) operator $\boldsymbol{F}$ is

$$L_{eff} = (22) \times (21) = 462.$$

This operator was expanded into its first five dominant separable stages $\boldsymbol{\Gamma}_k (k = 1, 2, \cdots, 5)$. The effective lengths and associated normalized error energies for the approximations $\boldsymbol{F}^{(1)}, \boldsymbol{F}^{(2)}, \cdots, \boldsymbol{F}^{(5)}$ are given in Table I. The large value of the error energy for $k = 1$ indicates that the given filter is not readily separable, and thus the single-stage approximation $\boldsymbol{F}^{(1)}$ does not constitute an acceptable representation of $\boldsymbol{F}$.

Figs. 8–12 show contour maps of the two-dimensional amplitude spectra of the approximations $\boldsymbol{F}^{(k)}$, $(k = 1, 2, \cdots, 5)$. In all these figures perspective views of the shaded portions of the contour maps are also displayed. Comparison of Fig. 8 with Fig. 7 confirms that the single-stage approximation of $\boldsymbol{F}$ is inadequate. However, progressive comparison of Figs. 9–12 with Fig. 7 clearly reveals that the approximation improves in quality as k increases. For $k = 5$ the agreement is quite good; the efficiency parameter for the filter $\boldsymbol{F}^{(5)}$ is by (43)

$$\alpha = \frac{462}{215} \approx 2.15$$

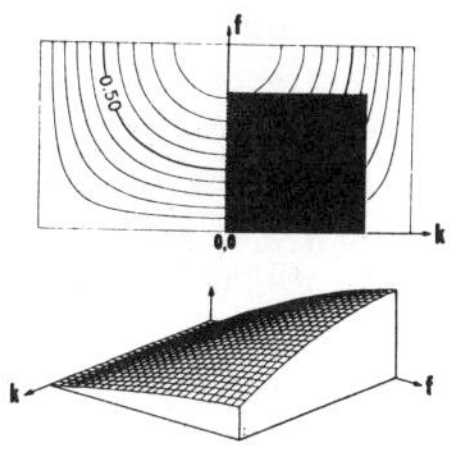

Fig. 8. Amplitude spectrum of single-stage approximation to fan-pass filter. $\epsilon_k = 0.173$ and $L_{eff} = 43$.

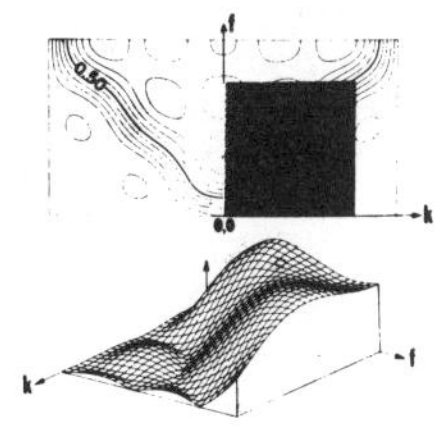

Fig. 10. Amplitude spectrum of three-stage approximation to fan-pass filter. $\epsilon_k = 0.048$ and $L_{eff} = 129$.

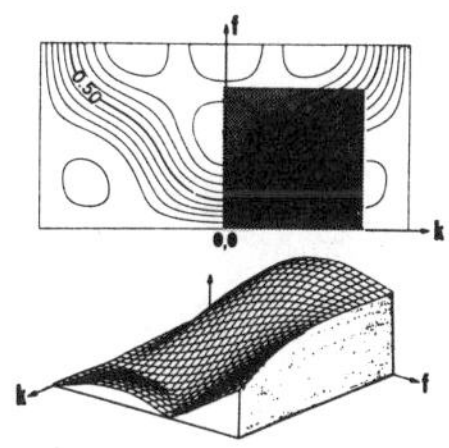

Fig. 9. Amplitude spectrum of two-stage approximation to fan-pass filter. $\epsilon_k = 0.081$ and $L_{eff} = 86$.

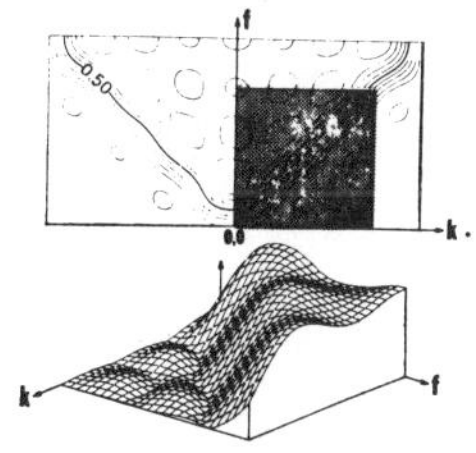

Fig. 11. Amplitude spectrum of four-stage approximation to fan-pass filter. $\epsilon_k = 0.031$ and $L_{eff} = 172$.

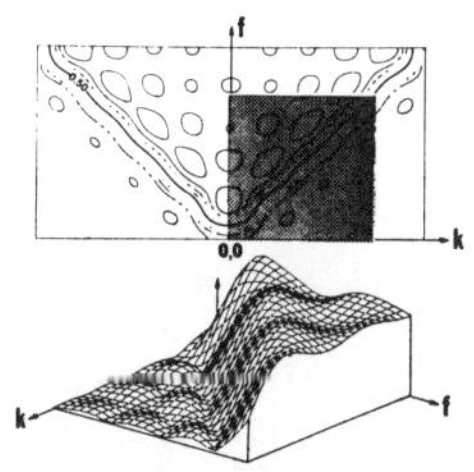

Fig. 12. Amplitude spectrum of five-stage approximation to fan-pass filter. $\epsilon_k = 0.021$ and $L_{eff} = 215$.

so that use of the filter $\boldsymbol{F}^{(5)}$ in the expanded convolution scheme (38) is more than twice as efficient as use of the given filter $\boldsymbol{F}$ in the standard convolution scheme (26). Further gains in computing efficiency could again be achieved if the convolutional components of each separable stage $\boldsymbol{\Gamma}_k$ were approximated by recursive filters. This refinement will be omitted here because its usefulness has already been demonstrated for the previous example.

V. Concluding Remarks

We have presented a method that permits the expansion of an arbitrary planar filter response function into a converging sum of individually separable planar stages. Judicious use of this technique can result in substantial savings in computer time and storage. These savings are further increased if the one-dimensional convolutional filters of each separable stage are approximated by appropriate one-dimensional recursive filters.

The advent of digital computers with parallel processing units renders the present scheme attractive even when the total number of weighting coefficients in the expanded form of the filter exceeds the number of such coefficients in the original filter response function array. One computer of this type is the ILLIAC IV [1], which will have the capacity to execute simultaneously a large number of identical algorithms. Thus each of the k planar stages of the expanded filter could operate *simultaneously* on the common input array $\boldsymbol{X}$, and in this manner the output is obtainable much more rapidly than if the original filter array is convolved directly with the input. Such a scheme will be particularly useful if the same planar filter is to be applied to a large number of different input arrays.

The expanded filter is expressible as a sum of products of column vectors and row vectors. The one-dimensional fast Fourier transform algorithm (FFT) can then be used to evaluate the Fourier transforms of each constituent vector, and the resulting complex-valued vector product sum will constitute an approximation to the two-dimensional Fourier transform of the original array. In many cases, particularly when the given array is expressible with small error by only a few separable stages, this scheme could result in faster computing speeds than if the two-dimensional FFT algorithm is applied directly to the original array. Such an approach to planar Fourier analysis certainly merits closer examination.

Appendix I

Eigenvector Expansion of a Planar Array

We desire to find an expansion of the given $(m\times n)$ matrix $\boldsymbol{D}$ into a series of individually separable $(m\times n)$ matrix stages $\boldsymbol{\Gamma}_i$ such that the sum

$$\boldsymbol{\Gamma}_1+\boldsymbol{\Gamma}_2+\cdots+\boldsymbol{\Gamma}_i+\cdots$$

gives a good approximation to $\boldsymbol{D}$ with a small number of stages. Each stage $\boldsymbol{\Gamma}_i$ can be written

$$\boldsymbol{\Gamma}_i = \boldsymbol{c}_i\boldsymbol{r}_i \tag{45}$$

where $\boldsymbol{c}_i$ is an m-length column vector and $\boldsymbol{r}_i$ is an n-length row vector.

Let us consider the two square matrices

$$\boldsymbol{Q} = \boldsymbol{D}\boldsymbol{D}^T$$
$$\boldsymbol{S} = \boldsymbol{D}^T\boldsymbol{D}$$

where the superscript $(^T)$ denotes the matrix transpose. We note that $\boldsymbol{Q}$ and $\boldsymbol{S}$ are real, symmetric, and of order m and n, respectively [7, p. 26]. Such matrices are expressible in terms of the finite matrix sums

$$\boldsymbol{Q} = \sum_{i=1}^{m} \mu_i\boldsymbol{\kappa}_i\boldsymbol{\kappa}_i^T \tag{46}$$

$$\boldsymbol{S} = \sum_{i=1}^{n} \lambda_i\boldsymbol{\varrho}_i^T\boldsymbol{\varrho}_i. \tag{47}$$

See [8, p. 38]. The numbers $\mu_i(i=1, 2, \cdots, m)$ are the eigenvalues of $\boldsymbol{Q}$, while the m-length vectors $\boldsymbol{\kappa}_i$ are the associated normalized column eigenvectors, or eigencolumns of $\boldsymbol{Q}$. Similarly, the numbers $\lambda_i(i=1, 2, \cdots, n)$ are the eigenvalues of $\boldsymbol{S}$, while the n-length vectors $\boldsymbol{\varrho}_i$ are the associated normalized row eigenvectors, or eigenrows of $\boldsymbol{S}$. The eigenvalues μ_i and λ_i are real because the matrices $\boldsymbol{Q}$ and $\boldsymbol{S}$ are symmetric [8, p. 32].

We observe that the term

$$\mu_i\boldsymbol{\kappa}_i\boldsymbol{\kappa}_i^T$$

of (46) is a separable matrix, since it results from the multiplication of the column vector $\mu_i\boldsymbol{\kappa}_i$ into the row vector $\boldsymbol{\kappa}_i^T$. Similarly, the term

$$\lambda_i\boldsymbol{\varrho}_i^T\boldsymbol{\varrho}_i$$

of (47) is a separable matrix because it results from the multiplication of the column vector $\lambda_i\boldsymbol{\varrho}_i^T$ into the row vector $\boldsymbol{\varrho}_i$.

The expansions for $\boldsymbol{Q}$ and $\boldsymbol{S}$ as given by (46) and (47) are called "spectral representations" [16, p. 49]. Quite evidently these representations at once enable one to split $\boldsymbol{Q}$ and $\boldsymbol{S}$ into a sum of separble matrix stages. The problem at hand, however, is to determine whether the matrix $\boldsymbol{D}$ can be split into such a sum of separable matrices or, equivalently, to determine whether there exists a spectral representation for $\boldsymbol{D}$. We shall show that such a representation can indeed be found from knowledge of the spectral representations for the associated matrices $\boldsymbol{Q}=\boldsymbol{D}\boldsymbol{D}^T$ and $\boldsymbol{S}=\boldsymbol{D}^T\boldsymbol{D}$.

We now assume that $m\geq n$ so that the number of rows of $\boldsymbol{D}$ are always equal to or greater than the number of columns of $\boldsymbol{D}$. This assumption does not result in a loss of generality, since any matrix $\tilde{D}$ for which $m<n$ can always be transposed, and then the resulting matrix $\tilde{\boldsymbol{D}}^T=\boldsymbol{D}$ does satisfy the postulated condition $m\geq n$. Let us assume further that the n columns of $\boldsymbol{D}$ are linearly independent.[9] The matrices $\boldsymbol{Q}=\boldsymbol{D}\boldsymbol{D}^T$ and $\boldsymbol{S}=\boldsymbol{D}^T\boldsymbol{D}$ are both products resulting from the multiplication of a rectangular matrix by its transpose. In view of these considerations, the mth-order matrix $\boldsymbol{Q}$ is positive semidefinite and of deficiency $(m-n)$; that is, $\boldsymbol{Q}$ is only of rank n. On the other hand, the n-order matrix $\boldsymbol{S}$ is positive definite and thus of full rank n [7, p. 20] and [3, p. 268].

Since $\boldsymbol{Q}$ is positive semidefinite and only of rank n, $(m-n)$ of its eigenvalues μ_i vanish, while its remaining n eigenvalues are real and positive. Since $\boldsymbol{S}$ is positive definite and of full rank n, none of its eigenvalues λ_i vanish, and these eigenvalues are all real and positive. Thus both $\boldsymbol{Q}$ and $\boldsymbol{S}$ have the same number n of real and positive eigenvalues.

An even closer relationship exists between the eigenvalue problems associated with $\boldsymbol{Q}$ and $\boldsymbol{S}$. This matter has been treated with great elegance by Lanczos [10, pp. 115–117], whose approach we use here with some modifications. Consider the eigenvalue problem

$$\boldsymbol{U}\boldsymbol{w} = \sqrt{\lambda}\,\boldsymbol{w} \tag{48}$$

where the $(m+n)$ by $(m+n)$ matrix $\boldsymbol{U}$ and the $(m+n)$-length eigenvector $\boldsymbol{w}$ are formed as shown in Fig. 13. Here 0_m and 0_n denote the null matrices of order m and n, respectively. The reason for writing an eigenvalue of $\boldsymbol{U}$ as the square root of a number λ will become apparent in a moment. Clearly, the system (48) can be written in the coupled form

$$\boldsymbol{D}\boldsymbol{\varrho}^T = \sqrt{\lambda}\,\boldsymbol{\kappa} \tag{49a}$$

$$\boldsymbol{D}^T\boldsymbol{\kappa} = \sqrt{\lambda}\,\boldsymbol{\varrho}^T. \tag{49b}$$

Premultiplying the first of the relations by $\boldsymbol{D}^T$ and the second by $\boldsymbol{D}$, we obtain

$$\boldsymbol{D}^T\boldsymbol{D}\boldsymbol{\varrho}^T = \sqrt{\lambda}\,\boldsymbol{D}^T\boldsymbol{\kappa} \tag{50a}$$

$$\boldsymbol{D}\boldsymbol{D}^T\boldsymbol{\kappa} = \sqrt{\lambda}\,\boldsymbol{D}\boldsymbol{\varrho}^T. \tag{50b}$$

[9] This restriction will be relaxed in the sequel.

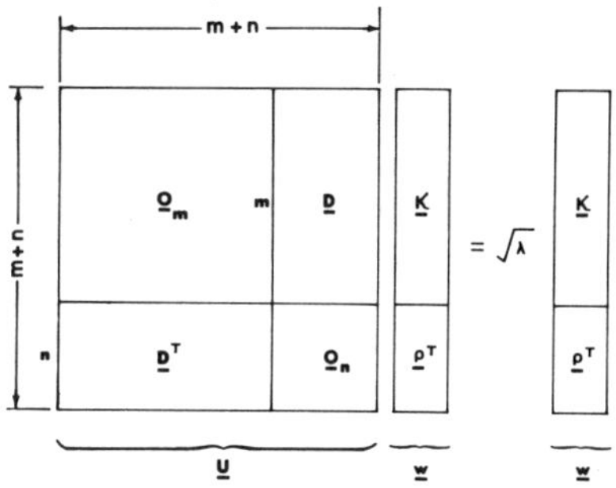

Fig. 13. The eigenvalue problem $Uw=\sqrt{\lambda}\,w$ in terms of D, μ, and ϱ.

Substitution of (49b) into (50a) and of (49a) into (50b) yields

$$D^T D \varrho^T = \lambda \varrho^T$$
$$D D^T \kappa = \lambda \kappa$$

and we see why we have have written the eigenvalue as $\sqrt{\lambda}$ in (48). Since $Q=DD^T$ and $S=D^TD$ these relations can be written

$$S\varrho^T = \lambda \varrho^T \quad (51a)$$
$$Q\kappa = \lambda \kappa. \quad (51b)$$

Now we have already established that $(m-n)$ of the eigenvalues of Q vanish, and that the remaining n eigenvalues of Q are real and positive. From (51a) and (51b) we then deduce that the n nonvanishing eigenvalues of Q coincide with the n eigenvalues of S, all of which are real and positive since S is positive definite. In other words, we have

$$\mu_i = \lambda_i, \qquad i = 1, 2, \cdots, n$$
$$\mu_i = 0, \qquad i = n+1, n+2, \cdots, m.$$

This means that the spectral representations for Q and S given by (46) and (47) become

$$Q = \sum_{i=1}^{n} \lambda_i \kappa_i \kappa_i^T \quad (52a)$$

$$S = \sum_{i=1}^{n} \lambda_i \varrho_i^T \varrho_i. \quad (52b)$$

Let us next realize that the coupling relations (49a) and (49b) can equally well be written in the subscripted form

$$D\varrho_i^T = \sqrt{\lambda_i}\,\kappa_i \quad (53a)$$
$$D^T\kappa_i = \sqrt{\lambda_i}\,\varrho_i^T \quad (53b)$$

for $i=1, 2, \cdots, n$. If we replace S by D^TD and express the eigenvalue λ_i as $\sqrt{\lambda_i}\,\sqrt{\lambda_i}$, (52b) can be rewritten in the form

$$D^T D = \sum_{i=1}^{n} \sqrt{\lambda_i}\,\varrho_i^T \sqrt{\lambda_i}\,\varrho_i.$$

Substitution of the coupling relation (53b) into the right-hand side of this equation produces

$$D^T D = \sum_{i=1}^{n} D^T \kappa_i \sqrt{\lambda_i}\,\varrho_i = D^T \sum_{i=1}^{n} \sqrt{\lambda_i}\,\kappa_i \varrho_i.$$

From this matrix identity we deduce that

$$D = \sum_{i=1}^{n} \sqrt{\lambda_i}\,\kappa_i \varrho_i \quad (54)$$

which is the spectral representation for the given matrix D. An identical result is obtained if we replace Q by DD^T in (52a) and substitute into the resulting equation the transpose of (53a).

Consider the ith term of the spectral representation for D,

$$\sqrt{\lambda_i}\,\kappa_i \varrho_i.$$

This term can be viewed as the *separable* $(m\times n)$-matrix resulting from the multiplication of the m-length column vector $\sqrt{\lambda_i}\kappa_i$ into the n-length row vector ϱ_i. We also see that the sum of n such separable matrices represents the given $(m\times n)$ matrix D with no error. But these are the very criteria that we ourselves had set for the implementation of the multistage expansion of D, and hence we *identify* the spectral representation (54) with this desired multistage expansion:

$$D = \Gamma_1 + \Gamma_2 + \cdots + \Gamma_i + \cdots + \Gamma_n = \sum_{i=1}^{n} \Gamma_i = \sum_{i=1}^{n} \sqrt{\lambda_i}\,\kappa_i \varrho_i. \quad (55)$$

The ith stage, or term, of this expansion is thus

$$\Gamma_i = \sqrt{\lambda_i}\,\kappa_i \varrho_i, \qquad i = 1, 2, \cdots, n. \quad (56)$$

Comparing (56) with (45), we see further that the desired series of column vectors c_i can be written

$$c_i = \sqrt{\lambda_i}\,\kappa_i$$

and that the desired series of row vectors r_i is then

$$r_i = \varrho_i.$$

It is to be particularly noted that the multistage expansion (55) is exact when the number of individually separable stages Γ_i is n, and we recall that $n \le m$. We remark that the spectral representation of D constitutes only one possible solution of the problem we have posed originally (see Appendix II). However, it has the distinct advantage of yielding an exact representation of D in a finite number of stages. Other advantages will become evident later.

In many applications (see Section IV) we will be satisfied with a multistage separation that will merely approximate $\boldsymbol{D}$. One way to do this is to retain only k stages in the expansion (55). Quite obviously we will wish to retain the dominant stages and neglect those that contribute little to the expansion. We accordingly suppose that the real and positive eigenvalues λ_i are listed in order of decreasing magnitude,

$$\lambda_1, \lambda_2, \cdots, \lambda_{n-1}, \lambda_n \tag{57}$$

where λ_1 is the largest and λ_n the smallest eigenvalue of the positive definite matrix $\boldsymbol{S}$.[10] We suppose also that any repeated eigenvalues are labeled separately.

Let us then assume that the sequence of terms of the multistage expansion (55) conforms with the ordering (57), so that $\boldsymbol{\Gamma}_1$ is the stage associated with the largest eigenvalue λ_1, $\boldsymbol{\Gamma}_2$ is the stage associated with the second largest eigenvalue λ_2, and so on. If we retain the first k terms in the expansion, we may write

$$\boldsymbol{D} = \sum_{i=1}^{k} \sqrt{\lambda_i}\, \boldsymbol{\kappa}_i \boldsymbol{\varrho}_i + \boldsymbol{E}_k, \qquad k = 1, 2, \cdots, n \tag{58}$$

where the $(m\times n)$ matrix $\boldsymbol{E}_k$ is the truncation error matrix resulting from the neglect of all terms beyond the kth. Since the expansion is exact for $k=n$, we have that

$$\boldsymbol{E}_n = \boldsymbol{0}$$

where 0 is the $(m\times n)$ null matrix. Furthermore, the truncation error matrix must be equal to the sum of all neglected terms in the expansion (55).

$$\boldsymbol{E}_k = \sum_{i=k+1}^{n} \sqrt{\lambda_i}\, \boldsymbol{\kappa}_i \boldsymbol{\varrho}_i. \tag{59}$$

Although we could inspect the individual magnitudes of the mn elements of $\boldsymbol{E}_k$ in order to determine how well the sum of the first k separable stages approximate $\boldsymbol{D}$, it is more practical to measure this performance with a single number. A convenient scalar-valued error indicator is simply the sum of the squares of the elements in the matrix $\boldsymbol{E}_k$, which we call the "error energy" in $\boldsymbol{E}_k$, and which we identify with the symbol J_k,

$$J_k = \sum_{i=1}^{m} \sum_{j=1}^{n} e_{k,ij}^2$$

where $e_{k,ij}$ is an element of $\boldsymbol{E}_k$. But J_k is also equal to the trace[11] of the nth-order matrix $\boldsymbol{S}_k$,

$$\boldsymbol{S}_k = \boldsymbol{E}_k^T \boldsymbol{E}_k$$

so that

$$J_k = \mathrm{Tr}\,(\boldsymbol{S}_k)$$

where the notation $\mathrm{Tr}(\cdot)$ denotes the trace of the matrix with parentheses.

[10] The usual convention is to label λ_n the largest eigenvalue and λ_1 the smallest eigenvalue, but here we will find it convenient to reverse standard practice.

[11] The trace of a square matrix is equal to the sum of the elements on its principal diagonal. It is easily verified that the energy in any rectangular matrix $\boldsymbol{X}$ is equal to the trace of the square matrices $\boldsymbol{X}^T\boldsymbol{X}$ or $\boldsymbol{X}\boldsymbol{X}^T$.

It will be desirable to express J_k as a function of the neglected terms in (55). Using (59) we have

$$\begin{aligned}\boldsymbol{E}_k^T \boldsymbol{E}_k &= \left[\sum_{i=k+1}^{n} \sqrt{\lambda_i}\, \boldsymbol{\kappa}_i \boldsymbol{\varrho}_i\right]^T \left[\sum_{j=k+1}^{n} \sqrt{\lambda_j}\, \boldsymbol{\kappa}_j \boldsymbol{\varrho}_j\right] \\ &= \sum_{i=k+1}^{n} \sum_{j=k+1}^{n} \sqrt{\lambda_i}\sqrt{\lambda_j}\, \boldsymbol{\varrho}_i^T \boldsymbol{\kappa}_i^T \boldsymbol{\kappa}_j \boldsymbol{\varrho}_j \end{aligned} \tag{60}$$

where j is a convenient dummy summation index. From (46) and (47) we recall that $\boldsymbol{\kappa}_i$ and $\boldsymbol{\varrho}_i$ are the normalized eigencolumns and eigenrows of the symmetric matrices $\boldsymbol{Q}$ and $\boldsymbol{S}$, respectively. But the normalized eigenvectors of a symmetric matrix are mutally orthonormal, that is,

$$\boldsymbol{\kappa}_i^T \boldsymbol{\kappa}_j = \delta_{ij} \tag{61}$$

$$\boldsymbol{\varrho}_i \boldsymbol{\varrho}_j^T = \delta_{ij} \tag{62}$$

where

$$\delta_{ij} = \begin{cases} 1, & i = j \\ 0, & i \neq j \end{cases}$$

[8, pp. 32–33]. Substitution of the orthonormality relation (61) into (60) produces

$$\begin{aligned}\boldsymbol{E}_k^T \boldsymbol{E}_k &= \sum_{i=k+1}^{n} \sum_{j=k+1}^{n} \sqrt{\lambda_i}\sqrt{\lambda_j}\, \boldsymbol{\varrho}_i^T \delta_{ij} \boldsymbol{\varrho}_j \\ &= \sum_{i=k+1}^{n} \sqrt{\lambda_i}\, \boldsymbol{\varrho}_i^T \sum_{j=k+1}^{n} \sqrt{\lambda_j}\, \delta_{ij} \boldsymbol{\varrho}_j \\ &= \sum_{i=k+1}^{n} \sqrt{\lambda_i}\, \boldsymbol{\varrho}_i^T \big[\sqrt{\lambda_{k+1}}\, \delta_{i,k+1} \boldsymbol{\varrho}_{k+1} \\ &\qquad + \sqrt{\lambda_{k+2}}\, \delta_{i,k+2} \boldsymbol{\varrho}_{k+2} + \cdots + \sqrt{\lambda_n}\, \delta_{i,n} \boldsymbol{\varrho}_n\big].\end{aligned}$$

Since $\delta_{ij}=1$ only when $i=j$ and vanishes otherwise, we obtain

$$\boldsymbol{S}_k = \boldsymbol{E}_k^T \boldsymbol{E}_k = \sum_{i=k+1}^{n} \lambda_i \boldsymbol{\varrho}_i^T \boldsymbol{\varrho}_i \tag{63}$$

which is the spectral representation of $\boldsymbol{S}_k$. Finally,

$$\begin{aligned} J_k &= \mathrm{Tr}\,(\boldsymbol{S}_k) \\ &= \mathrm{Tr}\,(\lambda_{k+1} \boldsymbol{\varrho}_{k+1}^T \boldsymbol{\varrho}_{k+1} + \cdots + \lambda_n \boldsymbol{\varrho}_n^T \boldsymbol{\varrho}_n). \end{aligned}$$

Now the trace of a sum of equal-order matrices is equal to the sum of the traces of the component matrices. Furthermore, each of these component matrices is the product obtained by multiplying the column vector $\lambda_i \boldsymbol{\varrho}_i^T$ into the row vector $\boldsymbol{\varrho}_i$.

For such matrices one easily verifies that

$$\mathrm{Tr}\,(\lambda_i \boldsymbol{\varrho}_i^T \boldsymbol{\varrho}_i) = \lambda_i \boldsymbol{\varrho}_i \boldsymbol{\varrho}_i^T = \lambda_i$$

where the last equality follows from the orthonormality relation (62). These considerations lead to the interesting result

$$\begin{aligned} J_k &= \lambda_{k+1} + \lambda_{k+2} + \cdots + \lambda_n \\ &= \sum_{i=k+1}^{n} \lambda_i \end{aligned} \tag{64}$$

and we see that the error energy resulting from the retention of the first k terms of the expansion (55) is equal to the sum of the "neglected" eigenvalues $\lambda_{k+1}, \lambda_{k+2}, \cdots \lambda_n$. In other words, the energy in the truncation error matrix $\boldsymbol{E}_k$ of (58) is equal to the sum of the eigenvalues associated with the neglected stages of this approximate multistage expansion.

The particular magnitude of the error energy J_k will be a function of the particular magnitudes of the elements of $\boldsymbol{D}$, and hence one finds it convenient to define the normalized energy ϵ_k in the form

$$\epsilon_k = J_k/J_0 \tag{65}$$

where J_0 is the energy in the matrix $\boldsymbol{D}$. The energy in $\boldsymbol{D}$ is equal to the trace of the matrix $\boldsymbol{S}=\boldsymbol{D}^T\boldsymbol{D}$,

$$J_0 = \sum_{i=1}^{m}\sum_{j=1}^{n} d_{ij}^2 = \text{Tr}(\boldsymbol{S}).$$

Then, using the spectral represention (52b) for $\boldsymbol{S}$ and proceeding as before, we find

$$J_0 = \sum_{i=1}^{n} \lambda_i.$$

Thus the energy in the given matrix $\boldsymbol{D}$ is equal to the sum of the eigenvalues of the matrix $\boldsymbol{S}=\boldsymbol{D}^T\boldsymbol{D}$. The normalized error energy ϵ_k now becomes

$$\epsilon_k = \frac{\lambda_{k+1} + \lambda_{k+2} + \cdots + \lambda_n}{\lambda_1 + \lambda_2 + \cdots + \lambda_n}$$

or

$$\epsilon_k = 1 - \frac{\lambda_1 + \lambda_2 + \cdots + \lambda_k}{\lambda_1 + \lambda_2 + \cdots + \lambda_n}, \quad k = 1, 2, \cdots, n. \tag{66}$$

In particular,

$$\epsilon_1 = 1 - \frac{\lambda_1}{\lambda_1 + \lambda_2 + \cdots + \lambda_n}$$

$$\epsilon_2 = 1 - \frac{\lambda_1 + \lambda_2}{\lambda_1 + \lambda_2 + \cdots + \lambda_n}$$

$$\cdots\cdots\cdots$$

$$\epsilon_n = 1 - 1 = 0.$$

Since the λ_1 are all real and positive, the bounds on ϵ_i are

$$1 > \epsilon_1 > \epsilon_2 > \cdots > \epsilon_{n-1}, \qquad \epsilon_n = 0$$

and thus the normalized error energy is always less than unity in magnitude no matter what the particular values taken on by the elements d_{ij} of the given matrix $\boldsymbol{D}$ may be.

Up to this point we have assumed that the n columns of $\boldsymbol{D}$ are linearly independent. This means that $\boldsymbol{S} = \boldsymbol{D}^T\boldsymbol{D}$ is positive definite, and hence has no vanishing eigenvalues. Suppose, however, that p of the columns of $\boldsymbol{D}$ are not linearly independent. Then $\boldsymbol{S}$ is positive semidefinite, and p of its n eigenvalues λ_i vanish [7, p. 29]. This means that the ordering (57) now becomes

$$\lambda_1, \lambda_2, \cdots, \lambda_{n-p}$$

so that λ_{n-p} is the smallest nonvanishing eigenvalue of $\boldsymbol{S}$. As a result the multistage expansion (55) is exact for only $(n-p)$ stages. Equation (58) is rewritten in the form

$$\boldsymbol{D} = \sum_{i=1}^{k} \sqrt{\lambda_i}\,\boldsymbol{\kappa}_i\boldsymbol{\varrho}_i + \boldsymbol{E}_k, \qquad k = 1, 2, \cdots, n-p \tag{67}$$

where

$$\boldsymbol{E}_{n-p} = \boldsymbol{0}.$$

If $p=n-1$, the expansion (67) terminates with no error after the first stage. Therefore only one of the n columns of $\boldsymbol{D}$ is linearly independent, so that all the columns (and rows) of $\boldsymbol{D}$ are proportional to each other [7]. This is just the condition that $\boldsymbol{D}$ must satisfy in order to be exactly separable with a single stage,

$$\boldsymbol{D} = \boldsymbol{\Gamma}_1 = \boldsymbol{c}_1\boldsymbol{r}_1 = \sqrt{\lambda_1}\,\boldsymbol{\kappa}_1\boldsymbol{\varrho}_1$$

for then the rows and columns of $\boldsymbol{\Gamma}_1$ and hence of $\boldsymbol{D}$ are proportional (compare with Section II). The present method yields the gratifying result of an error-free single-stage separation when $\boldsymbol{D}$ is so separable by construction.

Finally, if p of the columns of $\boldsymbol{D}$ are not linearly independent, the normalized error energy (66) becomes

$$\epsilon_k = 1 - \frac{\lambda_1 + \lambda_2 + \cdots + \lambda_k}{\lambda_1 + \lambda_2 + \cdots + \lambda_{n-p}}, \qquad k = 1, 2, \cdots, n-p \tag{68}$$

Appendix II

Least Squares Expansion of a Planar Array

Suppose that we are given the $(m \times n)$ matrix $\boldsymbol{D}$ and any nonnull *a priori* estimate of the n-length row vector $\boldsymbol{r}_1$. We first seek that m-length column vector $\boldsymbol{c}_1$ that minimizes the error energy J_1 in the error matrix $\boldsymbol{E}_1$

$$\boldsymbol{E}_1 = \boldsymbol{D} - \boldsymbol{c}_1\boldsymbol{r}_1$$

where

$$J_1 = \sum_{i=1}^{m}\sum_{j=1}^{n} e_{1,ij}^2 \tag{69}$$

and $e_{1,ij}$ is an element of $\boldsymbol{E}_1$. Since we can also write

$$\boldsymbol{D} = \boldsymbol{c}_1\boldsymbol{r}_1 + \boldsymbol{E}_1 \tag{70}$$

we see that the least squares solution $\boldsymbol{c}_1$ and the *a priori* estimate $\boldsymbol{r}_1$ will constitute a possible single-stage separation of $\boldsymbol{D}$ in the least squares sense.

Equation (70) may be written in the alternate form

$$d_{ij} = c_i r_j + e_{1,ij} \tag{71}$$

where d_{ij}, c_i, and r_j are elements of $\boldsymbol{c}_1$, $\boldsymbol{r}_1$, and $\boldsymbol{E}_1$, respectively.

Substitution of (71) into (69) allows us to express the error energy J_1 in the form

$$J_1 = \sum_{i=1}^{m} \sum_{j=1}^{n} (d_{ij} - c_i r_j)^2. \tag{72}$$

In order to find the elements c_i that minimize J_1 we differentiate this expression with respect to each of the elements $c_l (l=1, 2, \cdots, m)$ and set the resulting system of m normal equations equal to zero,

$$\begin{aligned} \frac{\partial J_1}{\partial c_l} &= \frac{\partial}{\partial c_l} \sum_{i=1}^{m} \sum_{j=1}^{n} (d_{ij}^2 - 2d_{ij}c_i r_j + c_i^2 r_j^2) \\ &= -2 \sum_{j=1}^{n} d_{ij} r_j \frac{\partial}{\partial c_l} \sum_{i=1}^{m} c_i + \sum_{j=1}^{n} r_j^2 \frac{\partial}{\partial c_l} \sum_{i=1}^{m} c_i^2 \\ &= -2 \sum_{j=1}^{n} d_{lj} r_j + 2c_l \sum_{j=1}^{n} r_j^2 \\ &= 0, \qquad l = 1, 2, \cdots, m. \end{aligned} \tag{73}$$

This system is linear in the unknown elements c_l, and its solution can be found directly from the last of the preceding relations,

$$c_l = \sum_{j=1}^{n} d_{lj} r_j \Big/ \sum_{j=1}^{n} r_j^2, \qquad l = 1, 2, \cdots, m \tag{74}$$

or using matrix notation,

$$\mathbf{c}_1 = \frac{\boldsymbol{D}\boldsymbol{r}_1^T}{\boldsymbol{r}_1\boldsymbol{r}_1^T}. \tag{75}$$

This column vector thus minimizes the error energy in the single-stage expansion of $\boldsymbol{D}$ for a given known *a priori* estimate of the row vector $\boldsymbol{r}_1$.

If the elements of $\boldsymbol{E}_1$ are not small in comparison to the elements of $\boldsymbol{D}$, we may apply this technique once more, but this time to $\boldsymbol{E}_1$ rather than to $\boldsymbol{D}$ itself. Let $\boldsymbol{r}_2$ be a nonnull *a priori* estimate in the separation

$$\boldsymbol{E}_1 = \mathbf{c}_2 \boldsymbol{r}_2 + \boldsymbol{E}_2 \tag{76}$$

where $\mathbf{c}_2$ will be computed such that the error energy J_2 in the error matrix $\boldsymbol{E}_2$ is minimized. If we compare (76) with (70), we see that the least squares problems connected with these relations have identical structure, and hence the solution for $\mathbf{c}_2$ may be written down by appropriately extending the solution for $\mathbf{c}_1$ given by (75),

$$\mathbf{c}_2 = \frac{\boldsymbol{E}_1\boldsymbol{r}_2^T}{\boldsymbol{r}_2\boldsymbol{r}_2^T}. \tag{77}$$

Finally, if we desire a k-stage expansion in this sense, we deduce inductively that

$$\mathbf{c}_i = \frac{\boldsymbol{E}_{i-1}\boldsymbol{r}_i^T}{\boldsymbol{r}_i\boldsymbol{r}_i^T}, \qquad i = 1, 2, \cdots, k. \tag{78}$$

For $i=1$, this solution reduces to (75) if we set $\boldsymbol{E}_0 = \boldsymbol{D}$. We are thus able to expand $\boldsymbol{D}$ in the form

$$\boldsymbol{D} = \mathbf{c}_1\boldsymbol{r}_1 + \mathbf{c}_2\boldsymbol{r}_2 + \cdots + \mathbf{c}_k\boldsymbol{r}_k + \boldsymbol{E}_k \tag{79}$$

where $\boldsymbol{E}_k$ is the error matrix resulting from this k-stage least squares expansion of $\boldsymbol{D}$.

The minimum error energy $\hat{J}_1$ for the single-stage expansion of $\boldsymbol{D}$ is obtained by substituting the solution (74) into the expression for the error energy (72). After some algebra this gives

$$\hat{J}_1 = J_0 - \frac{\boldsymbol{r}_1\boldsymbol{D}^T\boldsymbol{D}\boldsymbol{r}_1^T}{\boldsymbol{r}_1\boldsymbol{r}_1^T}$$

where $J_0 = \sum_{i=1}^{m} \sum_{j=1}^{n} d_{ij}^2$ = energy in the given matrix $\boldsymbol{D}$ [11]. It is convenient to deal with the normalized minimum square error (NMSE), which we define to be

$$\begin{aligned} \epsilon^{(1)} &= \hat{J}_1/J_0 \\ &= 1 - \frac{1}{J_0} \frac{\boldsymbol{r}_1\boldsymbol{D}^T\boldsymbol{D}\boldsymbol{r}_1^T}{\boldsymbol{r}_1\boldsymbol{r}_1^T}. \end{aligned} \tag{80}$$

But $\epsilon^{(1)}$ is a sum of squares and cannot hence be negative The matrix $\boldsymbol{D}$ and the *a priori* vector estimates $\boldsymbol{r}_i$ are nonnull by hypothesis. Therefore the bounds on $\epsilon^{(1)}$ are

$$0 \le \epsilon^{(1)} \le 1$$

[11].

It can be similarly shown that the NMSE for the least squares separation of $\boldsymbol{E}_1$ (76), namely $\epsilon^{(2)}$, also satisfies these bounds, and in general

$$0 \le \epsilon^{(i)} \le 1, \qquad (i = 1, 2, \cdots, k) \tag{81}$$

where $\epsilon^{(i)}$ is the NMSE for the least squares separation of $\boldsymbol{E}_{i-1}$,

$$\boldsymbol{E}_{i-1} = \mathbf{c}_i\boldsymbol{r}_i + \boldsymbol{E}_i. \tag{82}$$

Moreover, (80) can be generalized in the form

$$\epsilon^{(i)} = \hat{J}_i/\hat{J}_{i-1}, \qquad i = 1, 2, \cdots, k \tag{83}$$

where $\hat{J}_i$ is the energy in the error matrix $\boldsymbol{E}_1$, which is minimum for a given *a priori* choice of the vector $\boldsymbol{r}_i$ and where for $i=1$, we set $\hat{J}_0 = J_0$, the energy in $\boldsymbol{D}$.

But (83) can be written

$$\hat{J}_i = \epsilon^{(i)}\hat{J}_{i-1}$$

and in particular,

$$\begin{aligned} \hat{J}_1 &= \epsilon^{(1)}J_0 \\ \hat{J}_2 &= \epsilon^{(2)}\hat{J}_1 = \epsilon^{(2)}\epsilon^{(1)}J_0 \\ &\cdots\cdots\cdots \\ \hat{J}_k &= \epsilon^{(k)}\hat{J}_{k-1} = \epsilon^{(k)}\epsilon^{(k-1)} \cdots \epsilon^{(1)}J_0. \end{aligned}$$

The last of these relations can be written,

$$\hat{J}_k = J_0 \prod_{i=1}^{k} \epsilon^{(i)}.$$

Now $\hat{J}_k/J_0$ is the ratio of the energy in $\boldsymbol{E}_k$ to the energy in $\boldsymbol{D}$, and hence we have

$$\epsilon_{(k)} = \hat{J}_k/J_0 = \prod_{i=1}^{k} \epsilon^{(i)} \tag{84}$$

where $\epsilon_{(k)}$ is the *cumulative* NMSE for a k-stage expansion of $\boldsymbol{D}$. But the bounds (81) guarantee that $\epsilon_{(k)}<1$ unless all $\epsilon^{(i)}$ is of unit magnitude.[12] If no $\epsilon^{(i)}$ is of unit magnitude for $i=1, 2, \cdots, k$, we conclude from (84) that the least squares multistage expansion (79) converges uniformly to $\boldsymbol{D}$ as k increases. If for any stage $\epsilon^{(i)}=0$, the least squares expansion represents $\boldsymbol{D}$ exactly with a finite number of stages.

Although this result is gratifying at first thought, it becomes less so when one reflects that an inordinately large number of stages may be required to bring $\epsilon_{(k)}$ down to an acceptable small level. Up to this point we have assumed that the least squares computation of the vector $\mathbf{c}_i$ is always carried out with an arbitrary, nonnull estimate of the row vector $\boldsymbol{r}_i$ (82).

Instead, let us now go back to the solution for $\mathbf{c}_1$ given by (75), and let us there *identify* the vector $\boldsymbol{r}_1$ with $\boldsymbol{\varrho}_1$, where $\boldsymbol{\varrho}_1$ is the normalized eigenrow associated with the largest eigenvalue λ_1 of the matrix $\boldsymbol{S}=\boldsymbol{D}^T\boldsymbol{D}$. Then (75) becomes

$$\mathbf{c}_i = \boldsymbol{D}\boldsymbol{\varrho}_1^T \tag{85}$$

since $\boldsymbol{r}_1\boldsymbol{r}_1^T=\boldsymbol{\varrho}_1\boldsymbol{\varrho}_1^T=1$. If we compare (85) with (23), we obtain the remarkable result that the single-stage least squares solution for $\mathbf{c}_1$ is now identical to the solution for $\mathbf{c}_1$ obtained via the spectral representation of $\boldsymbol{D}$.

Next, let us identify the vector $\boldsymbol{r}_2$ of (77) with $\boldsymbol{\varrho}_2$, where $\boldsymbol{\varrho}_2$ is the normalized eigenrow associated with the second-largest eigenvalue, λ_2, of $\boldsymbol{S}$. Then (77) becomes

$$\mathbf{c}_2 = \boldsymbol{E}_1\boldsymbol{\varrho}_2^T. \tag{86}$$

But we have previously seen that $\boldsymbol{E}_1$ can be expressed as the difference,

$$\boldsymbol{E}_1 = \boldsymbol{D} - \mathbf{c}_1\boldsymbol{r}_1$$

which by the result just obtained is

$$\boldsymbol{E}_1 = \boldsymbol{D} - \mathbf{c}_1\boldsymbol{\varrho}_1.$$

Substituting this expression in (86) we obtain

$$\begin{aligned}\mathbf{c}_2 &= (\boldsymbol{D} - \mathbf{c}_1\boldsymbol{\varrho}_1)\boldsymbol{\varrho}_2^T\\ &= \boldsymbol{D}\boldsymbol{\varrho}_2^T - \mathbf{c}_1\boldsymbol{\varrho}_1\boldsymbol{\varrho}_2^T.\end{aligned} \tag{87}$$

It has been previously state (62) that the n normalized eigenrows $\boldsymbol{\varrho}_i$ $(i=1, 2, \cdots, n)$ of the nth order symmetric matrix $\boldsymbol{S}$ form an orthonormal set[13]

$$\boldsymbol{\varrho}_i\boldsymbol{\varrho}_j^T = \delta_{ij}, \qquad \delta_{ij} = \begin{cases}1, & i=j\\ 0, & i\neq j.\end{cases}$$

[12] From (80) we deduce that $\epsilon^{(1)}=1$ if and only if $\boldsymbol{r}_1$ is orthogonal to all rows of $\boldsymbol{D}$. Similarly, $\epsilon^{(2)}=1$ if and only if $\boldsymbol{r}_2$ is orthogonal to all rows of $\boldsymbol{E}_1$, and so on. One could at each stage avoid a unit value for the NMSE by chosing an $\boldsymbol{r}_i$ that is not orthogonal to all rows of $\boldsymbol{E}_{i-1}$. This entire question is obviated in view of our subsequent results.

[13] We assume for the sake of simplicity that $\boldsymbol{S}$ has no vanishing eigenvalues, i.e., that $\boldsymbol{S}$ is positive definite. The present results are easily generalized for the case when λ_i is a p-fold null eigenvalue of $\boldsymbol{S}$ (see also Appendix I).

In particular, this implies

$$\boldsymbol{\varrho}_i\boldsymbol{\varrho}_2^T = 0$$

and thus the second term of (87) vanishes so that

$$\mathbf{c}_2 = \boldsymbol{D}\boldsymbol{\varrho}_2^T. \tag{88}$$

If we compare (88) with (23) we find that the second stage of the least squares expansion of $\boldsymbol{D}$ is now identical to the second stage as. determined from the spectral representation of $\boldsymbol{D}$. This result, together with the previous result (85), then implies that the two-stage least squares expansion of $\boldsymbol{D}$ is equivalent term by term to the expansion obtained via its spectral representation.

It is evident that if in the general least squares solution (78) we identify $\boldsymbol{r}_i$ with $\boldsymbol{\varrho}_i$, there results

$$\begin{aligned}\mathbf{c}_i &= \boldsymbol{E}_{i-1}\boldsymbol{\varrho}_i^T\\ &= (\boldsymbol{D} - \mathbf{c}_1\boldsymbol{\varrho}_1 - \mathbf{c}_2\boldsymbol{\varrho}_2 - \cdots - \mathbf{c}_{i-1}\boldsymbol{\varrho}_{i-1})\boldsymbol{\varrho}_i^T\\ &= \boldsymbol{D}\boldsymbol{\varrho}_i^T, \qquad i = 1, 2, \cdots, n\end{aligned}$$

where we have made use of (79) to express $\boldsymbol{E}_{i-1}$ in terms of $\boldsymbol{D}$ and the previously computed separable stages, and where we have again used the fact that the $\boldsymbol{\varrho}_i$ are orthonormal.

Therefore the identification $\boldsymbol{r}_i=\boldsymbol{\varrho}_i$ for $i=1, 2, \cdots, n$ renders the least squares expansion for $\boldsymbol{D}$ term by term equivalent to the spectral representation for $\boldsymbol{D}$. Moreover, the least squares expansion of $\boldsymbol{D}$ then terminates with no error for a total of $k=n$ stages (see Section II-A and Appendix I).

It may be of interest to note that the authors arrived at the spectral representation of $\boldsymbol{D}$ only after having obtained the least squares expansion of $\boldsymbol{D}$ in an arbitrary, not necessarily finite number of stages [Compare with (79) and (84))]. If for no other reason, this Appendix has value because it illustrates that research rarely proceeds in the chronological sequence in which it is finally presented!

Finally, we remark that an identical result will be achieved if we pose the original problem in such a way that the column vectors $\mathbf{c}_i$ constitute *a priori* estimates; the row vectors $\boldsymbol{r}_i$ are then the solutions to appropriate least squares problems.

Appendix III

Number of "Multiply and Add" Operations Required for the Implementation of Planar Filtering Algorithms

We begin by computing the number of "multiply and adds" MADS required for the one-dimensional convolution of an m-length sequence a_i with an M-length sequence b_j. In terms of one-dimensional z-transforms, we have

$$\begin{aligned}C(z) &= A(z)B(z)\\ &= \left\{\sum_{i=1}^{m} a_i z^{i-1}\right\}\left\{\sum_{j=1}^{M} b_j z^{j-1}\right\}\end{aligned} \tag{89}$$

where $C(z)$ is the z transform of the $(m+M-1)$-length output sequence c_k. When we carry out the preceding polynomial multiplication, we observe that each of the m coefficients a_i is multiplied by each of the M coefficients b_j. Thus the number of MADS required to convolve the two sequences is simply mM, and the total length of the output sequence is $m+M-1$.

Let us next determine the number of MADS required for the implementation of the algorithm given by (32a)

$$Y(z_1, z_2) = W(z_1, z_2)R(z_2) \tag{90}$$

where $W(z_1, z_2)$ is the intermediate output,

$$W(z_1, z_2) = X(z_1, z_2)C(z_1). \tag{91}$$

We first carry out the operations associated with (91). The m-length column vector $\boldsymbol{C}$ is convolved with each of the N M-length columns of $\boldsymbol{X}$. Since each such one-dimensional convolution requires mM MADS, and since $\boldsymbol{X}$ has N columns, the total number of MADS needed for the implementation of (91) is mMN. In order to carry out the operations associated with (90), we notice that the intermediate output array $\boldsymbol{W}$ contains $m+M-1$ rows, since every column of $\boldsymbol{W}$ resulted from the one-dimensional convolution of the m-length column vector $\boldsymbol{C}(z_1)$ with each M-length column of $\boldsymbol{X}$. Each N-length row of W is convolved with the n-length row vector $\boldsymbol{R}$, so that nN MADS are required. But since W contains $(m+M-1)$ rows, the total number of MADS needed for the implementation of (89) is $(m+M-1)$ (nN). Therefore the number of MADS necessary for the combined evaluation of (90) and (91) is

$$mMN + (m + M - 1)(nN)$$

or after rearrangement,

$$MN(m + n) + n(m - 1)N$$

which is the first of the relations (33).

The number of MADS required for the implementation of the filtering algorithm (32b) can be obtained in a similar manner, and this analysis then leads to the second of the relations (33).

Finally, we determine the number of MADS needed for the implementation of the standard planar convolution algorithm (26),

$$Y(z_1, z_2) = X(z_1, z_2)F(z_1, z_2). \tag{92}$$

The planar z transform of the filter $F(z_1, z_2)$ can be written

$$F(z_1, z_2) = \sum_{j=1}^{n} F_j(z_1)z_2^{j-1}$$

where $F_j(z_2)$ is the one-dimensional z transform of the jth column of $\boldsymbol{F}$. Similarly, $X(z_1, z_2)$ can be written

$$X(z_1, z_2) = \sum_{k=1}^{N} X_k(z_1)z_2^{k-1}$$

where $X_k(z_1)$ is the one-dimensional z transform of the kth column of $\boldsymbol{X}$. Thus (92) becomes

$$Y(z_1, z_2) = \left\{\sum_{k=1}^{N} X_k(z_1)z_2^{k-1}\right\} \left\{\sum_{j=1}^{n} F_j(z_1)z_2^{j-1}\right\}. \tag{93}$$

We observe that (93) is structurally similar to (89), except that now the "coefficients" X_k and F_j are polynomials in z_1, rather than mere constants. Then by analogy with our analysis of (89) we find that the evaluation of (93) requires the computation of nN polynomial products of the form

$$X_k(z_1)F_j(z_1).$$

But each such product represents the one-dimensional convolution of the M-length column vector $\boldsymbol{X}_k$ with the m-length column vector $\boldsymbol{F}_j$, for whose evaluation mM MADS are needed. Since there are nN convolutions of this kind, we deduce that the standard planar convolution algorithm (26) requires a total of $mn\ MN$ MADS, which is the result (35).

References

[1] G. M. Ackins and D. J. Kuck, "Seismic signal processing via the ILLIAC IV computer," *IEEE Trans. Geosci. Electron.*, vol. GE-7, Jan. 1969, pp. 34–41.

[2] W. C. Dean, "Frequency analysis for gravity and magnetic interpretation," *Geophysics*, vol. 23, 1958, pp. 97–127.

[3] P. M. De Russo, R. J. Roy, and C. M. Close, *State Variables for Engineers*. New York: Wiley, 1965.

[4] J. P. Fail and G. Grau, "Les filtres en eventail," *Geophys. Prospect.*, vol. 11, 1963, pp. 131–163.

[5] J. B. Farr, "Acoustical holography experiments using digital processing," in *Acoustical Holography*, vol. 2. New York: Plenum, 1970, pp. 225–242.

[6] M. R. Foster, R. L. Sengbush, and R. J. Watson, "Suboptimum filter systems for seismic data processing," *Geophys. Prospect.*, vol. 12, 1964, pp. 173–191.

[7] R. A. Frazer, W. J. Duncan, and A. R. Collar, *Elementary Matrices*. New York: Cambridge, 1938.

[8] F. B. Hildebrand, *Methods of Applied Mathematics*, 2nd ed. Englewood Cliffs, N. J.: Prentice-Hall, 1965.

[9] E. I. Jury, *Theory and Application of the Z-Transform Method*. New York: Wiley, 1964.

[10] C. Lanczos, *Linear Differential Operators*. New York: Van Nostrand-Reinhold, 1961.

[11] E. A. Robinson and S. Treitel, "Principles of digital Wiener filtering," *Geophys. Prospect.*, vol. 15, 1967, pp. 311–333.

[12] C. K. Sanathanan, "On the synthesis of space dependent transfer functions," *IEEE Trans. Automat. Contr.*, vol. AC-11, Oct. 1966, pp. 724–729.

[13] J. L. Shanks, "Recursion filters for digital processing," *Geophysics*, vol. 32, 1967, pp. 33–51.

[14] J. L. Shanks, "Two-dimensional recursive filters," *SWIEEECO Rec.*, 1969, pp. 19E1–19 E8.

[15] S. Treitel, J. L. Shanks, and C. W. Frasier, "Some aspects of fan filtering," *Geophysics*, vol. 32, 1967, pp. 789–800.

[16] P. Whittle, *Prediction and Regulation*. London: The English Universities Press, 1963.

[17] R. A. Wiggins, "ω-k filter design," *Geophys. Prospect.*, vol. 14, 1966, pp. 427–440.

[18] J. H. Wilkinson, *The Algebraic Eigenvalue Problem*. New York: Oxford, 1965.

Editors' Comments on Papers 32 Through 37

32 **EKLUNDH**
A Fast Computer Method for Matrix Transposing

33 **TWOGOOD and EKSTROM**
An Extension of Eklundh's Matrix Transposition Algorithm and Its Application in Digital Image Processing

34 **ONOE**
A Method for Computing Large-Scale Two-Dimensional Transform Without Transposing Data Matrix

35 **DeLOTTO and DOTTI**
Two-Dimensional Transforms by Minicomputers Without Matrix Transposing

36 **HARRIS et al.**
Vector Radix Fast Fourier Transform

37 **BUIJS et al.**
Implementation of a Fast Fourier Transform (FFT) for Image Processing Applications

TWO-DIMENSIONAL IMPLEMENTATIONS USING THE FFT

Because of the availability of effective design algorithms and absence of stability considerations, most practical implementations of 2-D digital filtering have involved nonrecursive or FIR forms. In this case, the implementation issues are those associated with realizing the 2-D linear convolution in Eq. (1.3).

The standard procedures used to realize 1-D linear convolutions, i.e., lagged product and FFT approaches, can be suitably generalized to the 2-D case. In contrast to the 1-D case, however, the computations required may place severe demands on even the largest scientific computers. This is because of the large-size data arrays usually encountered in practical applications. Typically, these data arrays may be on the order of 1024x1024.

The complications in implementation occur because the filter and data arrays are too large to be stored in the primary (random access) storage of the computer. Thus, they must be stored, at least partially, in the secondary (sequential access) storage. The effect of employing multilevel storage in filter implementation is profound. It turns out that the characterization of arithmetic operations (adds, multiplies, delays) is not sufficient to adequately describe the computational requirements involved with the filter implementation. This is because the data manipulation between primary and secondary storage (I/O) constitutes a substantial amount of the computational effort involved with filtering the data.

This situation has had a major effect on development of procedures based on the 2-D DFT. The basic nonrecursive filter implementation is quite straightforward: the 2-D DFTs' of the filter and data arrays are computed, multiplied together, and the result inverse transformed. The 2-D DFTs' are classically computed by a repeated application of a 1-D FFT. The computation proceeds as follows. First each row in the array is transformed with the 1-D FFT. The array is then transposed and the row transform process is repeated (which computes the column transforms of the original array). This three-step procedure yields the transpose of the 2-D transform array.

Assuming the data array to be filtered is stored in a row sequential format on the secondary storage device, the first step of computing the row transforms can be performed straightforwardly. However, the data manipulations associated with performing the matrix transpose of the second step, if directly implemented, would require "random" accessing of the secondary storage device. This is grossly inefficient, particularly when transposing large arrays.

An efficient approach for resolving this problem was presented in a true benchmark paper by Eklundh (Paper 32). His matrix transposition algorithm is designed to transpose a $2^n \times 2^n$ matrix (n is an integer) in n steps. Each of these steps involves a specific sequence of matrix element interchanges between rows of the matrix taken two at a time. The efficiency of his scheme derives from its minimization of data transfer between the primary and secondary storage. A similar approach for transposing nonsquare matrices has also been proposed (Schumann, 1973).

Eklundh's approach involves only two rows of the matrix simultaneously resident in core. Twogood and Ekstrom (Paper 33) have shown that if 2^j rows (j is an interger) are available in core, a more efficient algorithm may be devised. Their algorithm is an extension of Eklundh's and has been shown to require less computational time. A high speed filtering algorithm which uses their transposition method is also described. Specifically, Twogood and Ekstrom quote times of 37 sec. (15 sec. CPU, 27 sec. I/O) to filter a 1024x1024 data array.

Onoe (Paper 34) approaches the problem of transform computation by essentially integrating the FFT and transposition steps together. The row transforms are computed and subsequent I/O is arranged such that the data elements required for the butterflies of the column transform calculation are simultaneously resident in core. While the manipulations required to perform a 16X16 FFT are presented, the general case of an $N \times N$ transform calculation is not described. Somewhat more detail is provided in a subsequent paper of Onoe (1975).

This integration of transposition and FFT calculation is also characteristic of Delotto and Dotti's technique (Paper 35). They describe the order of the data on the secondary storage device via an "order key" which codes the ordering of the data in terms of a binary expansion of addresses. The I/O required at any point in the transform calculation is a function of the current order key and the calculations to be performed (in this case, FFT butterflies). As in Paper 35, their 2-D FFT is computed by 1-D row and column transforms. They avoid performing an explicit matrix transpose by accessing/storing the partially transformed data in a prescribed sequence. An apparent limitation of the approach is that it restricts the record length used in the data transfers to be the physical record length of the secondary storage device. On most modern digital computers, however, it is possible to read many sequential records with one secondary storage access. It is this transfer of large blocks of contiguous data which results in the efficient data manipulation in Papers 32–34. An extension of their technique to incorporate variations in record length which are independent of the physical record length should yield an algorithm of sizable practical significance.

While the previous papers deal with issues arising from the computation of a 2-D DFT using iterated 1-D FFTs', the contribution of Harris et al. (Paper 36) describes a direct calculation of a 2-D FFT. Hoyer and Berry (1977) independently derived this transform also. As in the 1-D development, the 2-D FFT involves decomposing the 2-D transform into successively smaller 2-D transforms involving appropriate twiddle factors. This 2-D decimation allows the elimination of certain redundant operations (multiplies) occuring with use of the iterated 1-D FFT. While no "direct" transpose is performed when the array to be transformed overflows the primary storage, very similar I/O manipulations (as described above) are required. A consideration is that the final transform result does occur in bit reversed order (in both variables) and may need to be unscrambled for some applications.

The use of sectioning for the implementation of nonrecursive filters has also been extended to two dimensions (Hunt, 1972, and Twogood et al., 1977). This approach is computationally efficient when the filter is of much smaller dimension than the data array to be filtered. It remains true that the data manipulation (I/O) required in sectioning is an important part of its total computational effort involved and must be considered in deriving optimal sectioning procedures (Twogood et al., 1977).

While the majority of effort has involved filter implementation performed off-line using general purpose computers, Buijs et al. (Paper 37) have studied the development of real-time processing with a dedicated hardware processor. They have simulated a hardware design utilizing a serial input-parallel output transformer with an expected processing rate of 5 MHz.

REFERENCES

Hoyer, E. A., and W. R. Berry. 1977. "An Algorithm for the Two-Dimensional FFT," *Proc. 1977 IEEE Internat. Conf. Acoust., Speech, Signal Process.*, 552–555.

Hunt, B. R. 1972. "Minimizing the Computation Time for Using the Technique of Sectioning for Digital Filtering of Pictures," *IEEE Trans. Computers* **C-21**: 1219–1222.

Onoe, M. 1975. "A Fast Algorithm for Two-Dimensional Transform and Its Application to Digital Image Processing," *Proc. 2nd USA-JAPAN Computer Conf.*, 97–101.

Schumann, U. (1973). "Comments on 'A Fast Computer Method for Matrix Transposing' and Application to the Solution of Poisson's Equation," *IEEE Trans. Computers* **C-22**:542–544.

Twogood, R. E., M. P. Ekstrom, and S. K. Mitra. 1977. "On the Implementation of Two-Dimensional FIR Digital Filters by Sectioning," *Proc. 1977 IEEE Internat. Symp. Circuits and Syst.*, 670–673.

32

Reprinted from *IEEE Trans. Computers* **C-21**(7):801–803 (1972)

A FAST COMPUTER METHOD FOR MATRIX TRANSPOSING

J. O. Eklundh

***Abstract*—A method is given for transposition of $2^n \times 2^n$ data matrices, larger than available high-speed storage. The data should be stored on an external storage device, allowing direct access. The performance of the algorithm depends on the size of the main storage, which at least should hold 2^{n+1} points. In that case the matrix has to be read in and written out n times.**

***Index Terms*—Externally stored matrices, fast Fourier transform, fast Hadamard transform, matrix transposition, two-dimensional transformations.**

Introduction

The purpose of this correspondence is to present a fast computer method for matrix transposing. This method makes it possible to apply fast transforms such as the fast Fourier transform (FFT) and the fast Hadamard transform (FHT) on matrices of size exceeding available main storage. The algorithm presented is useful provided that each row (or column) of the matrix can be addressed, e.g., by use of direct access. When the data records have to be accessed sequentially, the computational and input–output time required will be excessive. In fact, the number of backspace and read operations will then be of the same magnitude as the number of matrix elements.

Manuscript received December 9, 1971; revised March 1, 1972.

The author is with the Research Institute of National Defence, Stockholm, Sweden.

A characteristic feature of the FFT, as presented by Cooley and Tukey [1], is the assumption that all the data points in the array to be transformed are kept in the main storage simultaneously. This may be a serious restriction considering two-dimensional applications, such as image processing.

To state the problem explicitly, let

$$f = (f(m, n)), \qquad m = 0, 1, \cdots, M-1,\ n = 0, 1, \cdots, N-1 \quad (1)$$

be the matrix to be transformed. The transform is defined by

$$F(p, q) = \sum_{m=0}^{M-1} \sum_{n=0}^{N-1} f(m, n) \exp\left(-2\pi j\left(\frac{mp}{M} + \frac{nq}{N}\right)\right),$$
$$p = 0, 1, \cdots, M-1,\ q = 0, 1, \cdots, N-1. \quad (2)$$

Suppose that f is stored on an external unit, e.g., a disk or a tape, row by row, that is in the order $f(0, 0), f(0, 1), \cdots, f(0, N-1), f(1, 0), f(1, 1), \cdots, f(M-1, N-1)$. We may then read f row by row and calculate the sums

$$G(m, q) = \sum_{n=0}^{N-1} f(m, n) \exp\left(-2\pi j \frac{nq}{N}\right), \qquad q = 0, 1, \cdots, N-1. \tag{3}$$

To minimize required storage, $G(m, q)$ is stored in the location that earlier contained $f(m, q)$. If we write the resulting arrays on an external data set—that may or may not be identical with the one containing f—we still have a matrix stored row by row. Now the next operation is to calculate

$$F(p, q) = \sum_{m=0}^{M-1} G(m, q) \exp\left(-2\pi j \frac{mp}{M}\right), \qquad p = 0, 1, \cdots, M-1, \tag{4}$$

that is, to form the sum of the elements in the matrix G multiplied by their corresponding exponential factor *columnwise*. But since G is stored row by row, we cannot obtain the elements in one column without reading through the whole matrix, picking out for each row the proper element. Doing this straightforwardly, we shall have to go through the matrix M times, which will result in excessive input–output time, even in the case of moderate matrix sizes. One way to avoid this is evidently to transpose the matrix G, which, however, by no means is trivial. The problem can be formulated as the one of how to construct a (quite specific) permutation of a given sequence when only part of the sequence can be kept in the main storage and operated on at the same time, minimizing the number of times the matrix is "worked through," and each element is taken into the core.

Singleton [2] and Brenner [3] have presented procedures for computation of the Fourier transform of externally stored one-dimensional data. Their methods may be modified to apply also in two dimensions (see [3]). In both cases, matrices of a size exceeding the available main storage have to be transposed.

In this correspondence we shall demonstrate a fast method for transposition of matrices of the order $2^n \times 2^n$, that is, matrices appearing in the two-dimensional application of the FFT and FHT. Simple generalizations to arbitrary matrices will be discussed later.

Theoretically, the algorithm to be described requires that the matrix—that we suppose is stored on an external data set unit—is brought into the main storage and returned again n times. However, this number can be reduced substantially by efficient programming. In fact, two times suffice in most implementations. This implies that we have to work through the matrix only one extra time, since any linear transformation requires that the data are taken into the main storage and that the result is written out on some unit at least once.

The Algorithm

The idea of the method is that a $2^n \times 2^n$ matrix can be transposed, taking two rows at the same time into the main storage, interchanging certain elements in the two rows, and finally returning the resulting two rows to the very same locations.

Consider the matrix

$$A = (\alpha_{IK}), \qquad I = 0, 1, \cdots, 2^n - 1, \; K = 0, 1, \cdots, 2^n - 1. \tag{5}$$

We can write I and K in the form

$$I = \sum_{p=0}^{n-1} i_p 2^p = b(i_{n-1}, i_{n-2}, \cdots, i_1, i_0) \tag{6}$$

$$K = \sum_{p=0}^{n-1} k_p 2^p = b(k_{n-1}, k_{n-2}, \cdots, k_1, k_0) \tag{7}$$

where i_p and k_p are binary integers.

Let us also write $\alpha(I, K)$ instead of α_{IK}.

We transpose A in n steps, forming pairs of rows (or columns) and interchanging elements according to the following rules.

In step j, $0 \leq j \leq n-1$,

row I' and row I'' form a pair iff

$$i_j' \neq i_j'' \text{ and } i_p' = i_p'', \qquad \text{for all values of } p \neq j. \tag{8}$$

In each pair of rows I' and I'', the elements $\alpha(I', K')$ and $\alpha(I'', K'')$ change places iff

$$k_j' \neq k_j'', \; k_j' \neq i_j', \text{ and } k_p' = k_p'', \qquad \text{for all values of } p \neq j. \tag{9}$$

It follows from (8) and (9) that

$$k_j'' = i_j' \quad \text{and} \quad k_j' = i_j''. \tag{10}$$

Hence if $\alpha(I', K')$ is the element in row I and column K before step j is performed, the new element in this place will be $\alpha(I'', K'')$ where

$$I'' = b(i_{n-1}', \cdots, i_{j+1}', k_j', i_{j-1}', \cdots, i_0') \tag{11}$$

$$K'' = b(k_{n-1}', \cdots, k_{j+1}', i_j', k_{j-1}', \cdots, k_0'), \tag{12}$$

the primes referring to I' and K'.

Consider an arbitrary (but fixed) element $\alpha(I, K)$ of A. Let I_j and K_j, respectively, be the indices of the row and the column containing $\alpha(I, K)$ when step j is performed. Using the notation of (6) and (7), we get from (11) and (12) that

$$\begin{aligned}
(I_0, K_0) &= (b(i_{n-1}, \cdots, i_2, i_1, k_0), b(k_{n-1}, \cdots, k_2, k_1, i_0)) \\
(I_1, K_1) &= (b(i_{n-1}, \cdots, i_2, k_1, k_0), b(k_{n-1}, \cdots, k_2, i_1, i_0)) \\
&\vdots \\
(I_{n-1}, K_{n-1}) &= (b(k_{n-1}, \cdots, k_2, k_1, k_0), b(i_{n-1}, \cdots, i_2, i_1, i_0)).
\end{aligned} \tag{13}$$

(Incidentally, these formulas also apply to the case when $k_j' = i_j'$ in (9).)

We conclude that after n steps the element $\alpha(I, K)$ is found in row K and column I, which means that the matrix is transposed.

A simple example will illustrate how the algorithm works.

Example: Suppose that the 4×4 matrix

$$A = \begin{pmatrix} \alpha_{11} & \alpha_{12} & \alpha_{13} & \alpha_{14} \\ \alpha_{21} & \alpha_{22} & \alpha_{23} & \alpha_{24} \\ \alpha_{31} & \alpha_{32} & \alpha_{33} & \alpha_{34} \\ \alpha_{41} & \alpha_{42} & \alpha_{43} & \alpha_{44} \end{pmatrix} \tag{14}$$

is stored row by row. We transpose it in the following two steps.

Step 1: First read row 1 and row 2 and let α_{12} and α_{14} change places with α_{21} and α_{23}, respectively. This means that the elements 2 and 4 in row 1 change places with the elements 1 and 3, respectively, in row 2. Then write row 1 and row 2 in their former locations and repeat the procedure with rows 3 and 4. We get

$$A_1 = \begin{pmatrix} \alpha_{11} & \alpha_{21} & \alpha_{13} & \alpha_{23} \\ \alpha_{12} & \alpha_{22} & \alpha_{14} & \alpha_{24} \\ \alpha_{31} & \alpha_{41} & \alpha_{33} & \alpha_{43} \\ \alpha_{32} & \alpha_{42} & \alpha_{34} & \alpha_{44} \end{pmatrix}. \tag{15}$$

Step 2: Read row 1 and row 3 (of A_1), interchange element 3 in row 1 and element 1 in row 3, element 4 in row 1 and element 2 in row 3, and finally write row 1 and row 3 in their proper locations. Repeat the procedure with row 2 and row 4. We obtain

$$A_2 = \begin{pmatrix} \alpha_{11} & \alpha_{21} & \alpha_{31} & \alpha_{41} \\ \alpha_{12} & \alpha_{22} & \alpha_{32} & \alpha_{42} \\ \alpha_{13} & \alpha_{23} & \alpha_{33} & \alpha_{43} \\ \alpha_{14} & \alpha_{24} & \alpha_{34} & \alpha_{44} \end{pmatrix} = A^t. \tag{16}$$

Hence A being a $2^2 \times 2^2$ matrix, we get A^t, the transpose of A, by twice reading through the matrix, swapping the elements, and writing it out again.

It may be noted that the rows are put together in couples in the same manner as the elements are interchanged.

Programming Considerations

Evidently, the optimal way to implement the algorithm is to take more than two rows into the main storage simultaneously. For instance, during the first three steps, the rows numbered 0, 1, 2, $\cdots$, 7 form a group in which the elements are interchanged independently of the other rows. Suppose that 2^j rows can be held in the high-speed storage. Then, if $n = j_1 + \cdots + j_k$, where $j_p \leq j$ for all

TABLE I

NUMBER OF EXTRA I/O PASSES WHEN 2^j ROWS OUT OF 2^n CAN BE HELD IN THE MAIN STORAGE

j \ n	10	9	8	7
5	1	1	1	1
4	2	2	1	1
3	3	2	2	2
2	4	4	3	3

TABLE II

CENTRAL PROCESSOR (LEFT) AND ESTIMATED INPUT–OUTPUT (RIGHT) TIME IN SECONDS FOR TRANSPOSITION OF $2^n \times 2^n$ COMPLEX MATRIX WHEN 2^j ROWS FIT INTO CORE

j \ n	10	9	8	7	j \ n	10	9	8	7
5	23.0	5.8	2.2	0.8	5	134	54	24	12
4	25.2	6.9	2.2	0.8	4	268	108	24	12
3	27.4	6.9	2.8	1.0	3	402	108	48	24
2	29.6	9.1	3.3	1.3	2	536	216	71	34

values of p, k input–output passes are needed, e.g., the matrix has to be taken into the main storage and returned again k times. Hence the number of input–output passes caused by the transposition algorithm is $k-1$. Table I lists $k-1$ as a function of n and j.

Required computing and input–output times can be expressed as

$$T_c = (k-1)(2^n T_{\text{read}} + 2^n T_{\text{write}}) + n2^{2n-2}T_{\text{change}} \tag{17}$$

and

$$T_{I/O} = (k-1)(2^{n+1}T_{\text{acc}} + 2^{2n+1}T_{\text{tran}}). \tag{18}$$

Here T_{read} and T_{write} are the central processor times to read and write one data record, and T_{change} is the time to perform a permutation in the core of two complex elements. Moreover, T_{acc} and T_{tran} are the input–output times to access one data record and read or write one complex number, respectively.

The algorithm has been implemented on an IBM 360/75, with 2314 disk packs. The routine is written in Fortran and utilizes the so-called BSAM routines for input and output. Its performance as a function of n and j is exemplified in Table II.

It should be pointed out that T_c is dominated by the second term, which is independent of the working storage area. (Typical values for the time constants are $T_{\text{read}} = T_{\text{write}} = 1$ ms, $T_{\text{change}} = 8$ μs.) Consequently, even a drastic reduction of high-speed storage will affect the computing time but little. This may be compared with the method suggested by Brenner [3], which for transposition requires 2^{2n-j} records to be read and written. Computing time, then, is dominated by the time to read and write, and hence is approximately proportional to working storage.

Generalization to arbitrary square matrices seems feasible if we use the approach of Bergland [4] for the FFT case. Obviously, there exists no generalization to nonsquare matrices, as row and column length must coincide when the elements are changed. However, if enough external storage is accessible, an arbitrary matrix could be imbedded in a square matrix, filling out the latter one with zero elements. Of course, this trick can also be used to obtain a $2^n \times 2^n$ matrix from an arbitrary one.

REFERENCES

[1] J. W. Cooley and J. W. Tukey, "An algorithm for the machine calculation of complex Fourier series," *Math. Comput.*, vol. 19, pp. 297–301, Apr. 1965.
[2] R. C. Singleton, "A method for computing the fast Fourier transform with auxiliary memory and limited high-speed storage," *IEEE Trans. Audio Electroacoust.*, vol. AU-15, pp. 91–98, June 1967.
[3] N. M. Brenner, "Fast Fourier transform of externally stored data," *IEEE Trans. Audio Electroacoust.*, vol. AU-17, pp. 128–132, June 1969.
[4] G. D. Bergland, "The fast Fourier transform recursive equations for arbitrary length records," *Math. Comput.*, vol. 21, pp. 236–238, Apr. 1967.

33

Reprinted from *IEEE Trans. Computers* C-25 (9):950–952 (1976)

AN EXTENSION OF EKLUNDH'S MATRIX TRANSPOSITION ALGORITHM AND ITS APPLICATION IN DIGITAL IMAGE PROCESSING

R. E. Twogood and M. P. Ekstrom

***Abstract*—Eklundh's fast matrix transposition algorithm for transposing a $2^n \times 2^n$ matrix is extended here to the general case where $2^j (j \geq 1)$ of its rows will fit into primary storage. The performance characteristics of the extension are described and subsequently compared with those of Eklundh, and its use in a two-dimensional image filtering routine is described.**

***Index Terms*—Digital image processing, externally stored matrices, fast Fourier transform, matrix transposition, two-dimensional filtering.**

I. Introduction

In a recent correspondence, Eklundh [1] presented a very fast algorithm for transposing matrices which overflow the primary or core storage of a computer. His method is specifically applicable to transposing matrices of dimension $2^n \times 2^n$ where n is an integer. It requires a minimum of 2^{n+1} words of primary storage and the matrix available in secondary storage in either a row—or column—sequential format. A total of n steps are required, each of which involves a selected sequence of element interchanges between rows of the matrix taken two at a time (i.e., a pair). The principal advantage of the method over other transposition schemes is its minimization of primary storage to/from secondary storage data transfer.

The Eklundh algorithm treats the specific case where only two rows of the matrix are resident in core at any one time. This case is of particular interest to those involved with implementing image processing algorithms on minicomputers having small-capacity core memories. When using large-scale scientific computers for such applications, however, it is of interest to consider the more general case of allowing 2^j rows simultaneously resident in core, where $j \geq 1$. In this correspondence we present a more efficient extension of Eklundh's basic method to this general problem of transposing a matrix when only $2^j (j \geq 1)$ of its rows can be simultaneously resident in primary storage. Our algorithm requires the same number of input/output (I/O) passes for the matrix (see [1, table I]), but is more efficient in terms of computational time. In addition to describing the general algorithm and characterizing its performance, we also discuss its use in a fast two-dimensional linear filtering routine for processing digital imagery [2].

Manuscript received August 23, 1974; revised June 30, 1975. This work was performed under the auspices of the U.S. Energy Research and Development Administration and was partially supported by the Advanced Research Projects Agency of the Department of Defense under ARPA Order 2436.

The authors are with the Lawrence Livermore Laboratory, University of California, Livermore, CA 94550.

II. The Matrix Transposition Algorithm

As mentioned above, our extension of Eklundh's method for transposing a $2^n \times 2^n$ matrix considers the general case where 2^j rows of the matrix fit into central memory. The algorithm consists of $\lceil n/j \rceil = K + 1$ steps[1] (a table of the values of K as a function of n and j is given in [1, table I]), the first K of which can be described by the following procedure.

In the kth step, $0 \leq k \leq K - 1$, (i.e., in all but the last step), every 2^{kj}th row is read from secondary storage beginning with the first row until 2^j rows have been read into core memory. At this point it is useful to partition the $2^j \times 2^n$ array in core into a matrix of $2^{kj} \times 1$ vectors $V_{p,q}$, where $0 \leq p \leq 2^j - 1$ and $0 \leq q \leq 2^{n-kj} - 1$. The rule in this kth step is then to interchange the vector $V_{p,q}$ with $V_{q \bmod 2^j, p+q-q \bmod 2^j}$

$$\text{iff } q \bmod 2^j < p, \qquad \text{for } 0 \leq p \text{ and } 0 \leq q \leq 2^{n-kj} - 1. \tag{1}$$

The rows of the array are then written back to secondary storage to their original places.

Next, a new set of 2^j rows, spaced 2^{jk} apart, is read from secondary storage to core. This new set begins with the row having the smallest index which has not been previously accessed in the present step. The sequence of vector interchanges is then repeated as above. Note that in this step the process of transferring 2^j rows and performing vector interchanges occurs a total of 2^{n-j} times. Consequently, every row in the $2^n \times 2^n$ matrix is read and rewritten only once per step.

The last step in the transposition (i.e., $k = K$) is handled differently. Here, $2^{j(K-1)-n}$ consecutive rows of the matrix are read from addresses 2^{jk} rows apart until a total of 2^j rows have been read. The resulting $2^j \times 2^n$ array in core is then partitioned into submatrices $M_{p,q}$ of dimension $2^{j(K+1)-n} \times 2^{jK}$ The rule in this Kth step is to interchange the submatrix $M_{p,q}$ with $M_{q,p}$

$$\text{iff } q < p, \qquad \text{for } 0 \leq p \leq 2^{n-jk} - 1 \text{ and } 0 \leq q \leq 2^{n-jK} - 1. \tag{2}$$

Following these interchanges, the array is written back to secondary storage. The process of reading in 2^j new rows, interchanging the submatrices, and rewriting is repeated until all the rows of the matrix have been read and rewritten once in this step.

The above relations completely describe our matrix transposition algorithm. As we previously mentioned, this algorithm is more efficient than Eklundh's in terms of CPU time. In demonstrating this we first note that the computational time required for our method is on the order of

[1] In our notation, $\lceil n/j \rceil$ is the result of dividing n by j and rounding up to the nearest integer.

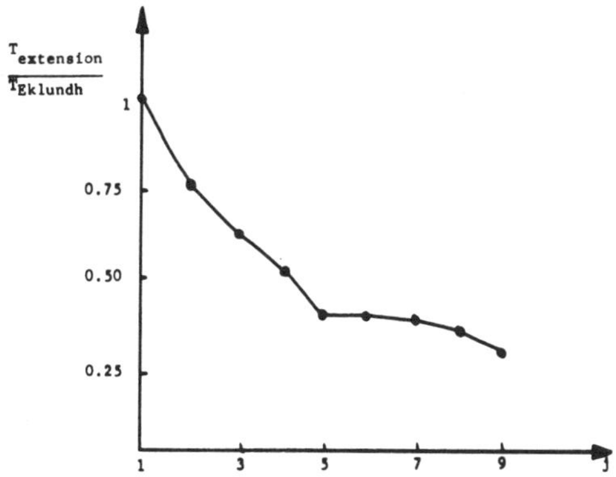

Fig. 1. Ratio of CPU time required by the transposition algorithm described above to that required by Eklundh's algorithm for $n = 10$.

$$T_{CPU} \approx \frac{1}{2} K(2^{2n} - 2^{2n-j})T_{change} + \frac{1}{2}(2^{2n} - 2^{n+jK})T_{change}. \quad (3)$$

In this expression, the first term accounts for the interchanges required in the first K steps and the second term accounts for the interchanges in the last step. While it can be shown that T_{CPU} in (3) is always less than T_{CPU} in [1, eq. (17)], quantitative comparisons are apparently best performed by considering specific cases.[2] In Fig. 1 we have such a comparison of the CPU times required in the two algorithms for the case of transposing a 1024 × 1024 matrix (i.e., $n = 10$). These speed ratios are, of course, calculated as a function of core capacity. Interestingly, with only 16 rows simultaneously resident in core, i.e., $j = 4$, our algorithm has a 2:1 speed advantage. For smaller dimension matrices, this improvement is less dramatic.

The I/O time required for our algorithm is given by

$$T_{I/O} = (2^{n-j}\, T_{acc} + 2^{2n+1}\, T_{tran}) + (K-1)(2^{n+1}\, T_{acc} + 2^{2n+1}\, T_{tran}) + (2^{2n-j(K+1)}\, T_{acc} + 2^{2n+1}\, T_{tran}). \quad (4)$$

Here, the first term is the time required in the first step, the second term is the time required in the $K - 1$ intermediate steps, and the third term is the time required in the last step.

III. Application in Two-Dimensional Linear Filtering

Having fully developed our transposition algorithm, we now discuss its use in an application of interest in digital image processing, that of filtering images. Our two-dimensional (2-D) nonrecursive filtering program uses 2-D FFT's to perform high-speed convolutions. The digital image is Fourier transformed, multiplied by the filtering function and inverse transformed to give the filtered image. The transforms are performed by repeated application of a 1-D FFT to all the rows and then columns of the image. This requires a transposition of the image matrix in the forward FFT and inverse FFT steps. Because images of interest generally overflow available core, our algorithm is directly applicable to this transposition.

[2] Notice that for $j = 1$, (3) becomes identical to [1, eq. (17)].

The program we describe was designed to filter a 1024 × 1024 digital image and runs on our CDC 7600 computer. This machine has 64K of 275-ns memory [small-core memory (SCM)], 512K of 1-μs memory [large-core memory (LCM)], and a very fast intermemory blockcopy instruction. Secondary storage is provided by over 150M words of disk memory. With this sizable core capacity, we are able to store in core 256 rows of the image matrix in performing this transposition. The structure of the program is outlined below:

LOOP 1 (executed 4 times):
- Read in 256 consecutive rows, 1024 each, from disk to LCM.
- Do row FFT's, two rows at a time (in SCM).
- Perform first step of transposition (resulting in four 256 × 256 transposed submatrices in LCM).
- Write the 256 rows from LCM to disk.

LOOP 2 (executed 4 times):
- Read in 256 rows, in four 64K chunks, from sections of disk spaced by 256 rows.
- Perform last step of transposition.
- Do column FFT's, since columns are now sequential.
- Bring in filter, multiply by data.
- Do inverse column FFT's.
- Perform last step of transposition algorithm (this starts the second transposition, using commutativity of the steps since these columns are already present in the core memory).
- Write out 256 rows, in four 64K chunks, from LCM to appropriate sections of disk.

LOOP 3 (executed 4 times):
- Read in 256 consecutive rows, 1024 each, from disk to LCM.
- Perform first step of transposition (four 256 × 256 transpositions, finishing the 1024 × 1024 transposition)
- Do inverse row transforms, two at a time.
- Write the 256 rows from LCM to disk.

Note that $n = 10$ and $j = 8$ in this implementation, and there are two steps (and two I/O passes) required per transposition. The entire filtering program, however, requires only three passes through the data despite the fact that two transpositions are performed. This reduction is accomplished by finishing the first transposition and starting the second one in the same I/O pass. This is possible because the order in which the steps of the transposition algorithm are performed is unimportant, i.e., the steps are commutative.

The linear filtering program described above requires only 15-s CPU time and 22 s I/O time on the 7600. (Approximately 2 s of the CPU time and 7 s of the I/O time are used in performing the transpositions with limited concurrency, the remainder of the time being spent to do FFT's and the multiplication of the data and filter.) These times represent a significant improvement (about a factor of 2) over any other image filtering program reported in the literature [3], [4]. As a figure of merit for this filtering routine, the CPU time required to perform only the 2-D FFT for the 1024 × 1024 array is ≈ 6.3 s.

IV. Summary and Conclusions

We have presented in this correspondence an extension of an algorithm proposed by Eklundh for transposing matrices which are larger than available core memory. The performance of our transposition method was characterized and shown to be more efficient in terms of computational time.

While the direct relation between our transposition algorithm and Eklundh's may be somewhat obscure, the methods are fundamentally similar. For the case of $j = 1$ (i.e., only two rows fit into core at any one time), the methods are in fact identical. In the more general case of $j > 1$, their difference lies in the structure of their element interchanges and intrastep data transfers. By minimizing the number of interchanges required in each pass through the data, our algorithm realizes its savings in CPU time. As we have shown, for large-scale problems, this savings will increase significantly with increases in core capacity.

It should be noted that in minicomputer applications, I/O considerations are generally more important than CPU considerations. This is because the number of passes through the data increases as the core capacity decreases, and the number of times the secondary storage must be accessed may become very large. With large-scale scientific computers such as the CDC 7600, however, the CPU time represents a much larger percentage of the time required to perform the matrix transposition, and the gains in CPU time illustrated above are of considerable practical importance.

References

[1] J. O. Eklundh, "A fast computer method for matrix transposing," *IEEE Trans. Comput.* (Corresp.), vol. C-21, pp. 801–803, July 1972.
[2] R. E. Twogood, "A fast two-dimensional linear filtering algorithm using the finite Fourier transform and multi-level storage," Lawrence Livermore Lab., Livermore, CA, Rep. UCRL-51845, June 1975.
[3] M. P. Ekstrom and O. L. Rater, "Computer processing of radiographic images with the finite Fourier transform," *IEEE Trans. Nucl. Sci.*, vol. NS-19, pp. 40–45, Feb. 1972.
[4] B. R. Hunt and H. J. Trussell, "Recent data on image enhancement programs," *Proc. IEEE* (Lett.), vol. 61, p. 466, Apr. 1973.

34

Reprinted from *IEEE Proc.* **63**(1):196–197 (1975)

A METHOD FOR COMPUTING LARGE-SCALE TWO-DIMENSIONAL TRANSFORM WITHOUT TRANSPOSING DATA MATRIX

M. Onoe

***Abstract*—A fast algorithm is presented for a two-dimensional transform of a data matrix such as Fourier or Hadamard transforms. It can be applied to a matrix, which is too large for the main storage and is stored sequentially rowwise in an auxiliary storage. The size of the matrix may be of R^m rows and of arbitrary number of columns. Whenever R rows of working area are available in the main storage, the matrix is read out, processed, and stored back m times. This new algorithm does not require the presently used method of transposing the data matrix.**

Introduction

In digital image processing, two-dimensional transforms, such as Fourier and Hadamard transforms, are indispensable tools for filtering, enhancement, convolution, correlation, and data compression of images. These two-dimensional transforms can be decomposed into two one-dimensional transforms: namely, rowwise transforms of data matrix followed by columnwise transforms.

When the whole data matrix is stored within a main random access storage, there is a trivial difference between rowwise and columnwise transforms. Unfortunately, the size of data matrix of an image is usually large, say 1024 by 1024, and exceeds the capacity of the main storage of most computers. Hence the data have to be stored, say row by row in sequential access storages, such as disks or tapes.

The rowwise transform is easily done: a row is read, transformed, and written back in the same or in another sequential storage. The columnwise transform, however, is incompatible with the sequential-row access storages. A conventional approach to solve this difficulty is first transposing the intermediate data matrix and then performing another rowwise transform [1].

Drawbacks of this approach are as follows.

1) Transposing requires a separate program and a considerable processing time, even though several ingenious fast algorithms have been proposed [2]–[4].

2) The final result is given in a transposed form.

3) The fast transposing algorithms in place are available only for a square matrix, and hence a nonsquare matrix has to be imbedded in a square one by adding meaningless elements before transposing.

The new algorithm, presented in this letter, does not require the transposing for the two-dimensional transform of a data matrix which is sequentially stored in auxiliary storages.

Manuscript received June 3, 1974; revised July 12, 1974.
The author is with the Institute of Industrial Science, University of Tokyo, Roppongi, Tokyo, Japan.

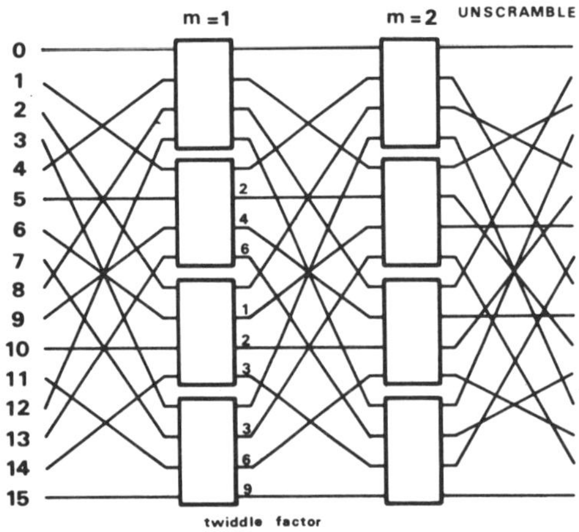

Fig. 1. Signal flow graph for a 16-point one-dimensional Fourier transform.

The Algorithm

The fast algorithm for one-dimensional Fourier transform has been implemented in various forms [5], [6]. Its essence is in the decomposition of original transform of size $N = (R^m)$ into m steps of N/R transforms of size R. At each step, intermediate results are corrected by twiddle factors and regrouped into N/R sets of R points, which become inputs for transforms in the next step. Since each set is needed only once, transforms may be done in place, that is, the output may be stored back into the same location as the corresponding input.

Based on these facts, this paper shows that the columnwise transform of two-dimensional data, stored row by row, can be performed in m steps.

At each step, a set of R rows is read into the main storage and columnwise transforms are performed over the entire row. The resultant R rows are written back in the auxiliary storage. Then the next set of R rows is read into the main storage.

During one step this process is repeated N/R times, and the total number of required steps is m. Hence, the whole matrix has to be read into the main storage and returned m times.

If transforms are performed in place, the final result is in the digit reversed order along column. Thus in this case one more step is needed for unscrambling.

Similar algorithms exist for any two-dimensional transform, such as Hadamard or Haar transform, which is decomposable into one-dimensional transforms.

The following simple example shows how the present algorithm works. Fig. 1 is a signal flow graph for an $N = 16$ one-dimensional Fourier transform, which is decomposed into $m = 2$ steps of $N/R = 4$ transforms of size $R = 4$. According to this pattern, the two-dimensional transform is performed as follows.

1) The first step ($m = 1$):

a) the rows, 0, 4, 8, and 12, are read into the main storage and a 16-point one-dimensional transform is performed for each row;

b) a 4-point transform in place is performed on the first element of each row; next, the transform is repeated on the second element, then on the third, and so on; the result is returned to the previous location in the auxiliary storage;

c) the two preceding procedures are applied on the rows 1, 5, 9, and 13, then on the rows 2, 6, 10, and 14, and finally on the rows 3, 7, 11, and 15; the multiplication of twiddle factors is done either separately or jointly with transforms.

2) The second step ($m = 2$): the same procedures as 1-b) and 1-c) are applied on the rows 0–4, 5–8, 9–12, and finally on 13–15.

3) The unscrambling step: the rows 1, 2, 4, and 8 are read out and stored back in the unscrambled order. The same procedure is applied on the rows 3, 5, 10, and 12 and later on the rows 7, 11, 13, and 14.

Various modifications of this basic algorithm are possible. For example, mixed radix can be used in the transform. The unscrambling step may be eliminated if enough capacity is available in the auxiliary storage.

Instead of performing the rowwise transformation in the first step, it may be broken down evenly into m steps. This may be advantageous in programming and also in designing a dedicated hardware because all the operation will be almost identical for both rows and columns.

In conclusion, the proposed algorithm yields a fast two-dimensional transform of nonsquare, as well as square, data matrix which sequentially stored in auxiliary storage.

Acknowledgment

The author wishes to thank Dr. S. Nishikawa for his careful reading of this manuscript.

References

[1] B. R. Hunt, "Data structures and computational organization in digital image enhancement," *Proc. IEEE*, vol. 60, pp. 884–887, July 1972.

[2] J. O. Eklundh, "A fast computer method for matrix transposing," *IEEE Trans. Comput.*, vol. C-21, pp. 801–803, July 1972.

[3] U. Schumann, "Comments on 'A fast computer method for matrix transposing' and application to the solution of Poisson's equation," *IEEE Trans. Comput.*, vol. C-22, pp. 542–543, May 1973.

[4] M. Takagi and Y. Yokoi, "A fast algorithm for transposing matrix," *Rep. Inst. Electron. Commun. Eng. Jap.*, C 73-26, Sept. 1973. (In Japanese.)

[5] J. W. Cooley and J. W. Tukey, "An algorithm for the machine calculation of complex Fourier series," *Math. Comput.*, vol. 19, pp. 297–301, Apr. 1965.

[6] W. M. Gentleman and G. Sande, "Fast Fourier transforms—for fun and profit," *AFIPS Proc.*, vol. 29, pp. 563–578, 1966.

35

Reprinted from *Computer Graphics and Image Process.*
4(3):271–278 (1975)

Two-Dimensional Transforms by Minicomputers without Matrix Transposing*

I. DeLotto

CIOC–CNR and Pavia University, Pavia, Italy

AND

D. Dotti

CISE, P.O. Box 3986, Milan, Italy

Communicated by T. S. Huang

Received July 5, 1974; revised November 29, 1974

A new procedure for two-dimensional transforms without matrix transposing is briefly presented. Its structure makes it suitable for minicomputers that have a small core capacity compared with the picture dimensions.

1. Introduction

Fast transform (FT) algorithms are widely used in picture processing techniques [1]. As an actual quantized picture requires a storage capacity on the order of 10^7 bits, an interesting problem arises in the management of the data and the structure of the FT algorithm in order to use the mass storage and processor of the computer system efficiently.

Standard techniques for transforming a large matrix, stored on an external memory device, usually perform the fast transform (FT) of the columns first, then the matrix transposition, and then the FT of the rows; sometimes a final transposition is required so as to get the results in the same order as the data. Moreover the matrix is assumed to be square, with row and column length equal to record length in the mass memory; rectangular matrices are dealt with by squaring them using additional zeroes [2].

The procedure presented here does not impose the above restrictions, and performs a two-dimensional transform without transposing the matrix. These features make it suitable for minicomputers with small core capacity and for multiprogrammed systems. The mass memory devices under consideration are both sequential (magnetic tapes) and random access (disks, drums). Greater attention will be paid to the latter, because they are more suitable for the task described.

* Work sponsored by CNR, Rome, Italy.

The distribution of items in the mass memory will be described in terms of an "order key," as illustrated in two recent papers [3,5]. This is briefly outlined in the next section.

2. *The Order Key* [5]

The address of a given datum in an ordered sequence of 2^n data[1] is:

$$A_0 = \sum_{j=0}^{n-1} b_j 2^j, \qquad b_j = 0,1. \tag{1}$$

In this paper, we deal with order variations corresponding to address changes from A_0 to A with:

$$A = \sum_{j=0}^{n-1} b_j 2^{k(j)}, \tag{2}$$

where the $k(j)$ set is a permutation of the j set; the $k(j)$ set will be called the "order key."

This class of order changes, although limited, embraces important situations; for instance:

(a) scrambling of frequency components performed by an FT algorithm, where $k(j) = n - j - 1$ (Fig. 1a);

(b) transposition of a $2^r \times 2^c$ data matrix, with the matrix stored in columns, where the order key is $k(j) = j + r$ modulo $(c + r)$ (Fig. 1b).

Item exchanges between mass storage and main memory are assumed to be performed by means of "read/write routines," each of which transfers to and from the core memory the entire data set in blocks of a size equal to the available computer memory. The order key can thus be divided into three areas A, B, and

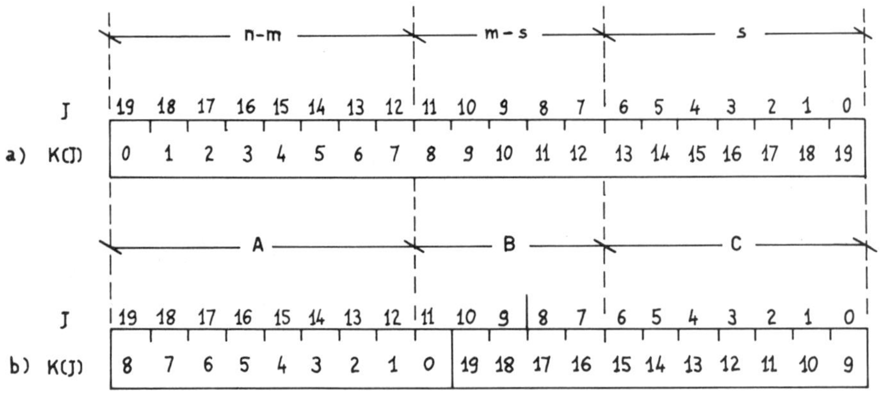

FIG. 1. Order keys: (a) $k(j) = n - j - 1$, scrambling performed by an FFT algorithm; (b) $k(j) = j + r$, transposition of a $2^r \times 2^c$ matrix stored by columns with $r = 9$ and $c = 11$. Order keys are divided into three areas: A, B, and C. Area A contains the block address in the data set, area B the record address in the block, and area C the datum address in the record.

[1] A sequence length which can be factored in a different way can also be dealt with. This aspect will, however, be neglected, owing to its relatively limited significance.

C (see Fig. 1): area C (s bits) contains the datum address in the record, area B ($m-s$ bits) the record address in the block, and area A ($n-m$ bits) the block address in the data set; clearly 2^s and 2^m are the length of the record and of the available fast memory, respectively.

Changes in the order key can take place in the course of a read/write routine:

(a) during the reading phase, bits can be exchanged between A and B;

(b) in the main memory, bits can be exchanged between B and C;

(c) during the writing phase, bits can be exchanged between A and B, but only with random access mass memories.

Since each writing phase except the last is followed by a reading phase, from this viewpoint disks and tapes differ only as far as result writing is concerned.

3. *Fast Transforms*

It is well known that the FT processing of a vector of 2^n data may easily be performed in n steps [4], in each of which simple calculations are carried out on pairs of complex items, generating other pairs of complex items, to be stored in the same memory positions. The pair chosen must be such that their addresses do not differ from each other except in one bit, which in the first step is the most significant one, in the second step the next, and so on. The resulting transform components are placed in memory positions whose addresses, as far as Fourier transforms are concerned, are related through bit reversal to the frequency number; thus the jth transforming step can be interpreted as converting the $(n-j)$th bit of the data address into the $(j-1)$th bit of the transform address. In the case of two-dimensional transforms the above considerations apply irrespectively to both columns and rows, that is, the FT on columns may be merged with the FT on rows, with the restriction that one must proceed from the most significant bit to the least significant bit for both variables.

When the data are stored on external devices, the item blocks to be read into the memory must contain the pairs to be processed. In terms of the order key, the corresponding bits must therefore lie in the B or C area. In practice it is convenient to have more than one bit in these areas, so that many FT steps can be carried out on the same data block. Considering that s bits are initially in area C, and $m-s$ bits can be brought from A to B in each block reading, the number α_t of read/write sequences required to transform two-dimensional (and also one-dimensional) data is:

$$\alpha_t = I[(n-s)/(m-s)], \tag{3}$$

where $I(x)$ is the smallest integer greater than or equal to x. A procedure will now be described which allows fast and standardized execution of the transform and result ordering without any transposition of the matrix. Simple rules for generation of the order keys will be given, and the corresponding record selection will be briefly outlined.

4. *Result Ordering*

The transform components obtained are seldom in their natural order, because of the FT scrambling mentioned previously and the possible mixing of column and row bits performed by the chosen sequence of transforming steps. The ordering problem involves both in-record and off-record ordering for a tape unit, but only in-record ordering for a disk, because of the random access capabilities of the latter device.

Assuming that ordering begins in the course of the last FT read–write routine, the number α_0 of extra "read/write routines" for ordering depends on the order key both of the last writing phase and of the desired result arrangement; this dependence is given by the following equation:

$$\alpha_0 = I[t/(m-s)],$$

where t is the number of order key bits to be transferred from A to B or C areas in tape units, and from A to C area in a disk, after completion of the last transforming read/write routine.

5. *Order Keys for Two-Dimensional Transforms of Large Matrices*

In Sections 3 and 4 the minimum number of read/write routines needed for transforming large matrices of data are given, without giving any particular procedures which allow the performance quoted to be obtained. In this section read/write routines are described in terms of order keys for both the reading and writing phase, taking into account the restrictions on order key changes mentioned in Section 2.

Let a $2^c \times 2^r$ matrix be stored on the mass storage, and assume $s \leqslant r \leqslant m$, conditions which cover most actual cases for two-dimensional transforms. Thus the transform may be carried out on the column bits sooner than on the row bits, bringing the key elements in their proper order into the B area. This task can be achieved by rotating to the left the $(n-s)$ bits of the A and B areas by $(m-s)$ places, and transforming on the bits of the B area. This operation is repeated $I[(n-s)/(m-s)]$ times, with possible reduced rotation at the last step, so that one complete revolution is undergone by the A and B area elements. During the last step the bits of the C area, each of which pertains to the row address, are also transformed. Furthermore, the last read/write routine allows the transform components to be ordered suitably in core, exchanging bits between the B and C areas, and then between the A and B areas during the writing phase, at least when one can take advantage of the random access property of disk units.

All the operations described are illustrated in Fig. 2, where $k(j)$ is given by the column bits, indicated by y_i (Y_i after FFT), and the row bits, indicated by x_i (X_i after FFT). The correspondence between these symbols and $k(j)$ is given at the bottom of the figure.

6. *Implementation of the Procedure*

In the order key sequence in Fig. 2 the $(n-s)$ off-record bits (A and B areas) are rotated to the left by $(m-s)$ bits in each step but the last one. This means that in each reading phase data blocks are collected, taking one record every 2^{n-m} records from the mass storage, beginning from the first, the second, and so on up to the (2^{n-m})th record. Using this technique, records whose addresses have the same $(m-s)$ most significant bits are brought together into the main

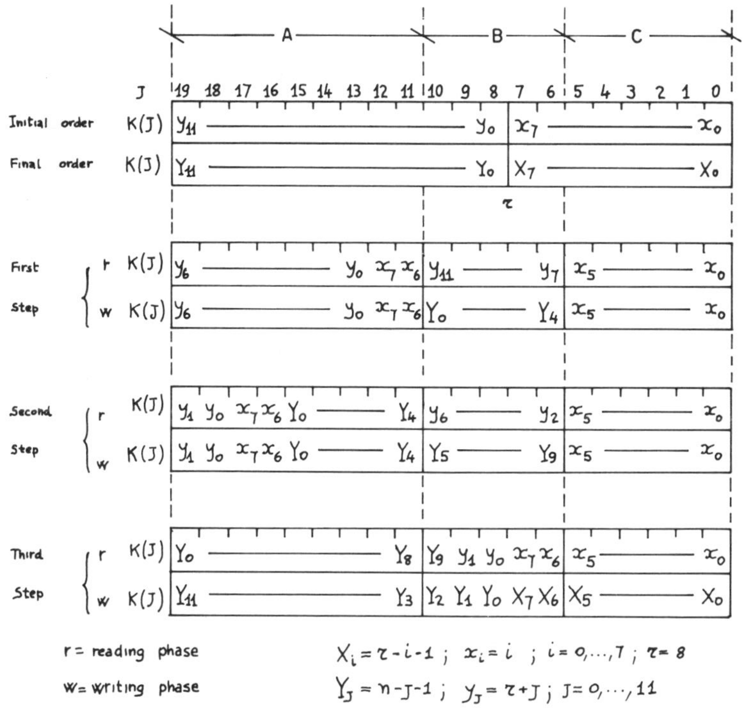

FIG. 2. An order key sequence for a two-dimensional transform of a $2^8 \times 2^{12}$ data matrix. Note that the final result gives the correct sequence of frequency components without transposition and ordering of the matrix. The last writing step, in which the record sequence ordering is performed, requires a random access mass memory.

memory, as required by the order keys. Record writing is sequential, so that this procedure is also suitable for magnetic tapes, in a mass storage area which does not coincide with the reading area.

In the last step, the order key is obtained by left-rotating the $(n-s)$ off-record bits by $z=(n-s)-(\alpha_t-1)\cdot(m-s)$, that is, the remainder of the ratio $(n-s)/(m-s)$, if any, or otherwise $(m-s)$. This means that during this step data blocks are written in core by reading from the mass storage groups of $2^{m-(s+z)}$ records every $2^{n-(s+z)}$ records, beginning from the first, the $(2^{m-(s+z)}+1)$th, and so on. The writing phase is in general not sequential in this last step; it is accomplished by writing groups of 2^{r-s} records every $2^{n-m+r-s}$ records. The address of the initial record of each group is calculated by scrambling the address bits of the columns.

Finally, it should be mentioned that the order key of each transforming step determines the FT operations to be performed. Indeed, it defines the appropriate sequence of elementary transforming steps, i.e., not only the data addresses but also the set of coefficients required by the FT algorithm.

To better explain the procedure given above let us refer to a small matrix $F(2\times 4)$ with the values $r=1$, $c=2$, $s=1$, $m=2$, $n=3$. That is, the matrix is stored by columns, with record length equal to column length;

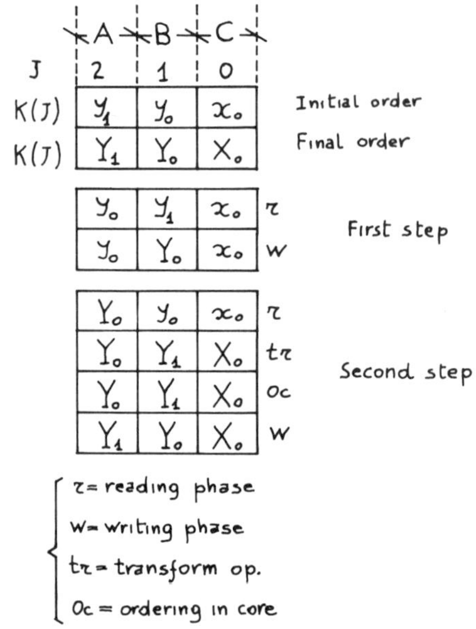

FIG. 3. Order key sequence for transforming a matrix $F(2 \times 4)$.

$$F = \begin{vmatrix} f_{11} & f_{12} & f_{13} & f_{14} \\ f_{21} & f_{22} & f_{23} & f_{24} \end{vmatrix}.$$

The sequence of the order keys is given in Fig. 3. In fact, $\alpha^t = I[(n-s)/(m-s)] = 2$, that is, two steps are needed. The rotation in the reading phase of the former step is by $(m-s) = 1$ bit and in the latter step by $z = 1$ bit. In the latter step four order keys have been drawn; the first one gives the order in which records are read in, the second indicates the transform operation, the third the ordering performed in core between areas C and B (note that in this case no ordering is done in core), and the last the order in which data are written in the mass memory. This last writing operation requires a random access memory; in fact some ordering is performed between areas A and B and consequently writing cannot be sequential.

Let us now follow step by step the whole procedure on a data flow diagram which may be compared to that of a standard FT algorithm. For the standard procedure, the three transform steps and final ordering are shown in Fig. 4a. Figure 4b gives the data flow diagram for all the elementary operations indicated by the order keys in Fig. 3. Symbols show the correspondence with the diagram in Fig. 4a and the order key sequences in Fig. 3. Dot–dashed lines separate the operations to be carried out on blocks of data which fit into available core capacity.

7. Efficiency of the Procedure

To give an example of the efficiency of the procedure, let us refer to a case such as that in Fig. 2. Since $m \geq r \geq s$, the total transformation time, with the final data ordered, is

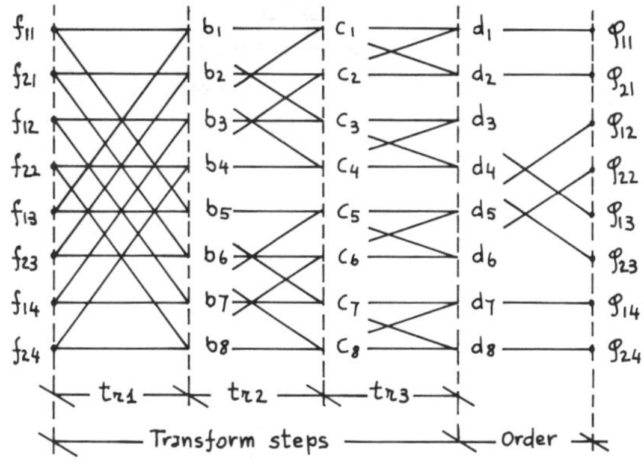

a

FIG. 4a. Flow diagram for a standard FT algorithm on a matrix $F(2 \times 4)$ with a final ordering. f_{ij} are the coefficients of the original matrix, ϕ_{ij} are the coefficients of the transformed matrix.

$$T = \alpha_t T_t + T_c + T_m, \tag{4}$$

where $\alpha_t = I[(n-s)/(m-s)]$. T_t, the transfer time to and from the mass storage for each transforming step, is

$$T_t = 2[t_{tr} + (t_{ac}/2^s)]2^n,$$

where t_{tr} is datum transfer time, t_{ac} record access time, and the factor 2 is needed to account for reading and writing of data at each step; T_c, the in-core computation time for an FT algorithm, is given by

$$T_c = n2^n(t_p + t_s)$$

with t_p the "product time" and t_s the "addition time" for in-core transform operations with address time included; T_m, the core ordering time, is

$$T_m = 2^n t_m,$$

where t_m is the move time.[2]

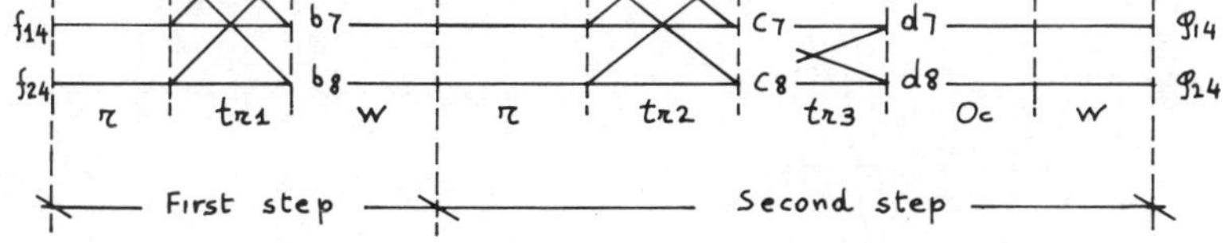

b

FIG. 4b. Flow diagram of the proposed procedure in the case of a matrix $F(2 \times 4)$.

[2] All times evaluated for complex data.

Equation (4) is written for a serial organization of the operations needed in the procedure presented here. However, with a given size M of core space available for data, better results (down to $\simeq 0.5T$) are obtained by interleaving transfer operations and transform computations using size $M/2$ of core space for both. In fact it often happens that

$$T_c + T_m > T_t(\alpha_t^{(M/2)} - \alpha_t^{(M)}) = T_t \Delta \alpha_t \tag{5}$$

where $\alpha_t^{(M/2)}$ and $\alpha_t^{(M)}$ are the number of steps, given by Eq. (2), needed for data spaces of sizes $M/2$ and M. The total transform time is consequently the greater of $\alpha_t^{(M/2)}\, T_t$ and $(T_c + T_m)$.

In the example in Fig. 2, for instance, where $s = 6$, $r = 9$, $m = 12$, and $n = 20$, $\Delta\alpha_t$ is 0. In a standard minicomputer the following times can be assumed. $t_m = t_s = 10$ μsec, $tp = 60$ μsec, $t_{tr} = 20$ μsec, $t_{ac} = 20$ msec. With these computer features Eq. (4) gives $T = 3300$ sec for a serial organization of the operations, and $T = 1900$ sec when computation and transfer operations are interleaved.

REFERENCES

1. H. C. Andrews, *Computer Techniques in Image Processing,* Academic Press, New York, 1970.
2. I. O. Eklundh, A fast computer method of matrix transposing, *IEEE Trans.* **C-21,** 1972, 801.
3. I. DeLotto and D. Dotti, A new procedure for optimum mass storage use in FT algorithms, *Alta Frequenza* **42,** 1973, 379.
4. J. W. Cooley and J. W. Tukey, An algorithm for the machine calculation of complex Fourier series, *Math. Comp.* **19,** 1965, 297.
5. I. DeLotto and D. Dotti, Large matrix ordering technique with applications to transposition, *Electronics Letters* **9,** 1973, 374.

36

Reprinted from pp. 548–551 of the *1977 IEEE Internat. Conf. on Acoust., Speech, Signal Process. Rec.*, 879 pp.

VECTOR RADIX FAST FOURIER TRANSFORM

D. B. Harris, J. H. McClellan, D. S. K. Chan, and H. W. Schuessler
M.I.T. Research Laboratory of Electronics, Cambridge, Mass.

ABSTRACT

A new radix-2 two-dimensional direct FFT developed by Rivard is generalized in this paper to include arbitrary radices and non-square arrays. It is shown that the radix-4 version of this algorithm may require significantly fewer computations than conventional row-column transform methods. Also, the new algorithm eliminates the matrix transpose operation normally required when the array must reside on a bulk storage device. It requires the same number of passes over the array on bulk storage as efficient matrix transpose routines, but produces the transform in bit-reversed order. An additional pass over the data is necessary to sort the array if normal ordering is desired.

INTRODUCTION

Since the appearance of the original Cooley-Tukey algorithm in 1965, the standard methods of computing the two-dimensional (2-D) discrete Fourier transform (DFT) of an array have capitalized on the separability of the 2-D DFT [1]. Using a 1-D FFT algorithm, row-wise and columnwise, 1-D DFT's can be computed to yield the 2-D transform. This scheme amounts to decimating and transforming the array first in one index and then in the other.

A new algorithm which performs the decimation in both indices simultaneously has been derived by Rivard using a holor algebra formalism [2]. Rivard demonstrated that his radix-2 direct 2-D FFT eliminates 25% of the multiplies required by the conventional row-column approach.

The purpose of this paper is to present an alternate derivation of the new algorithm and to extend it to rectangular arrays and arbitrary radices. Further, we show that even larger savings in multiplies are obtained when the algorithm is generalized to operate with larger radices and on higher dimensional arrays. We refer to the general algorithm as the vector radix algorithm, since to specify the decimation of the array, multiple radices are required, one for each index of the array.

An additional, machine dependent, implication of the new algorithm is explored here as well. When the DFT computation is implemented on a computer with an insufficient amount of core memory to contain the entire array, a matrix transpose operation is a necessary component of the row-column approach. This fact has occasioned a literature on fast transpose techniques, led by Eklundh [3]. However, the vector radix algorithm requires no transpose. The transpose is, in effect, incorporated into the transform. A vector radix transform requires the same number of passes over the array on secondary storage (e.g. disk) as the Eklundh transpose algorithm, when the resulting DFT can be tolerated in bit reversed order. If the DFT must be in correct order, an additional transfer of the array is required to perform bit reversed sorting. On a machine which is I/O intensive, this extra pass over the array may compromise the computational advantages of the new algorithm.

DERIVATION

As with the one-dimensional FFT algorithm, the new direct two-dimensional FFT is derived by decomposing the DFT into sums of smaller DFT's multiplied by "twiddle" factors. We derive here a single stage of the general vector radix algorithm for the decimation in time case. This is all that is necessary, since the complete algorithm is obtained by recursive application of this basic decomposition.

We suppose the 2-D DFT

$$X(k,\ell) = \sum_{m=0}^{M-1} \sum_{n=0}^{N-1} x[m,n]\, W_M^{km} W_N^{\ell n} \qquad k=0,\ldots,M-1 \quad \ell=0,\ldots,N-1 \tag{1}$$

of the MxN array $x[m,n]$ is desired, where

$$W_M = \exp(-j2\pi/M)$$

This work was supported in part by the Advanced Research Projects Agency monitored by ONR under Contract N00014-75-C-0951-NR 049-308 and in part by the National Science Foundation under Grant ENG71-02319-A02.

Provided radix r_1 divides M and radix r_2 divides N:

$$M/r_1 = P \qquad N/r_2 = Q \quad \text{integers,} \tag{2}$$

the DFT may be computed with an $r_1 x r_2$ stage. Using the change of variables:

$$\begin{array}{ll} m = r_1 p+u & n=r_2 q+v \\ p = 0,\ldots,P-1 & q=0,\ldots,Q-1 \\ u = 0,\ldots,r_1-1 & v=0,\ldots,r_2-1 \end{array} \tag{3}$$

the DFT (1) is decomposed into $r_1 r_2$ PxQ DFT's. Performing the substitution of variables of (3) in equation (1), we obtain:

$$X(k,\ell) = \sum_{p=0}^{P-1} \sum_{u=0}^{r_1-1} \sum_{q=0}^{Q-1} \sum_{v=0}^{r_2-1} x[r_1 p+u, r_2 q+v] \cdot W_M^{k(r_1 p+u)} \cdot W_N^{\ell(r_2 q+v)} \tag{4}$$

From equation (2):

$$W_M^{k(r_1 p+u)} = W_P^{kp} W_M^{ku} ; W_N^{\ell(r_2 q+v)} = W_Q^{\ell q} W_N^{\ell v} . \tag{5}$$

Substituting the relations of (5) into equation(4), we obtain the desired result:

$$X(k,\ell) = \sum_{u=0}^{r_1-1} \sum_{v=0}^{r_2-1} W_M^{ku} W_N^{\ell v} X_{PQ}^{uv} (k,\ell) \tag{6}$$

where

$$X_{PQ}^{uv}(k,\ell) = \sum_{p=0}^{P-1} \sum_{q=0}^{Q-1} x[r_1 p+u, r_2 q+v] W_P^{kp} W_Q^{\ell q} \cdot$$

is the 2-D DFT of a PxQ element array.

We introduce a second change of variables to split the computation of $X(k,\ell)$ into $r_1 x r_2$ element butterfly groupings:

$$\begin{array}{ll} k = a+bP & \ell = c+dQ \\ a = 0,\ldots,P-1 & c = 0,\ldots,Q-1 \\ b = 0,\ldots,r_1-1 & d = 0,\ldots,r_2-1 \end{array} \tag{7}$$

Performing the change of variables (7) on equation (6) yields:

$$X(a+bP,\ c+dQ) = \sum_{u=0}^{r_1-1} \sum_{v=0}^{r_2-1} W_M^{(a+bP)u} W_N^{(c+dQ)v} X_{PQ}^{uv} (a+bP,\ c+dQ) . \tag{8}$$

Again:

$$W_M^{(a+bP)u} = W_M^{au} W_{r_1}^{bu} ; W_N^{(c+dQ)v} = W_N^{cv} W_{r_2}^{dv} \tag{9}$$

The PxQ point DFT's are periodic:

$$X_{PQ}^{uv}(a+bP,\ c+dQ) = X_{PQ}^{uv}(a,c) . \tag{10}$$

Using substitutions (9) and 10) in equation (8) produces:

$$X(a+bP,\ c+dQ) = \sum_{u=0}^{r_1-1} \sum_{v=0}^{r_2-1} W_{r_1}^{bu} W_{r_2}^{dv} [(W_M^{au} W_N^{cv}) X_{PQ}^{uv}(a,c)] . \tag{11}$$

Equation (11) defines the basic structure of the $r_1 x r_2$ butterfly. A few observations will clarify this. First, there is one butterfly for each combination of values of a and c, giving a total of PQ butterflies. Thus, a and c are the butterfly indices. Each butterfly computes $r_1 r_2$ values of $X(\cdot,\cdot)$ (indexed by b and d) from $r_1 r_2$ values of the smaller PxQ element DFT's (indexed by u and v). As input to the butterfly, one element is taken from each PxQ point transform. These elements are not used by any other butterflies in the stage, so that the $r_1 r_2$ output values of the butterfly may be stored in the memory locations occupied by the input elements. Therefore, as with the 1-D FFT algorithm, the vector radix transform may be computed in place.

The inputs to the butterfly are premultiplied by the twiddle factors (in parentheses), then an $r_1 x r_2$ 2-D DFT of the resulting array (in square brackets) is computed. The fact that the butterfly can be interpreted as a 2-D DFT calculation suggests several structures for performing the computation. The conventional row-column FFT structure can be employed or the vector radix approach can be used. The latter is generally more efficient.

One very important observation is that the product of the twiddle factors W_M^{au} and W_N^{cv} may be precomputed and stored in a table. The computational advantage enjoyed by the vector radix algorithm arises from the fact that such products are not computed, whereas they are computed implicitly in the row-column approach.

As an example we consider the radix 2x2 case, where $M=N=2^{\nu}$, ν an integer, and $r_1=r_2=2$. Equation (11) reduces to:

$$X(a+b\tfrac{N}{2},\ c+d\tfrac{N}{2}) = \sum_{u=0}^{1} \sum_{v=0}^{1} (-1)^{bu+dv} [(W_N^{au+cv}) X_{\frac{N}{2}\frac{N}{2}}^{u\ v} (a,c)] \tag{12}$$

$$a,c=0,\ldots,N/2-1 \qquad b,d=0,1$$

One possible butterfly structure for the computation of equation (12) is shown in Figure 1. This structure has a minimum number of adds and multiplies, and is the same as an efficient radix four one-dimensional butterfly structure with the internal multiplies by j removed.

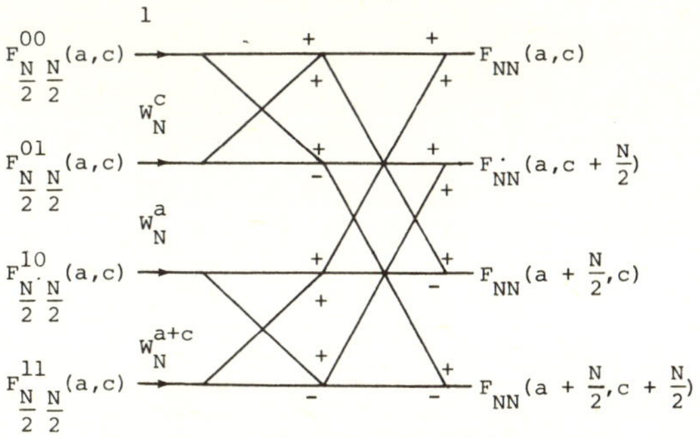

Figure 1: Radix 2x2 Butterfly

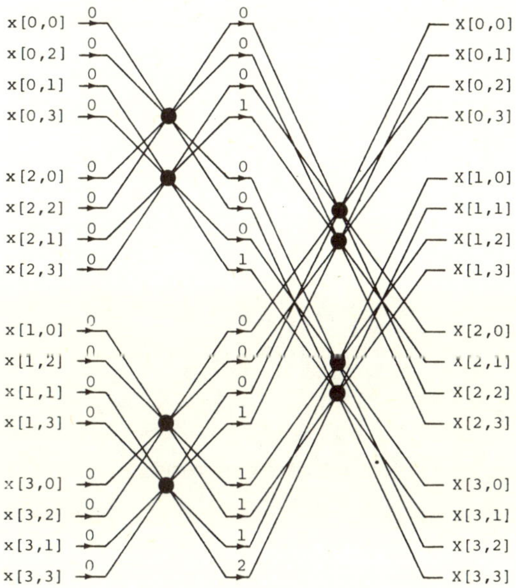

Figure 2: 4x4 DFT with radix 2x2 algorithm. Solid circles indicate 2x2 butterflies and twiddle factor multiplies by W_4^k are indicated by the exponents k.

In Figure 2 we display the radix 2x2 structure for computation of a 4x4 ($\nu=2$) element DFT. Rows of the bit-reversed array are concatenated and arranged vertically on the left hand side of the figure. As is generally true for the 2x2 algorithm, only 2 rows of the array are accessed simultaneously, so that 2 rows worth of core storage are sufficient to implement the algorithm.

COMPARISON AND DISCUSSION

Introduction of the vector radix FFT provides some flexibility in trading CPU speed against I/O transfer rate. In an attempt to quantify this tradeoff, statistics for the two algorithms are presented in a table at the end of this section. The total number of complex multiplies and butterflies required for NxN transforms are tabulated for both algorithms and various radices. The number of butterflies is included as a partial measure of the amount of "overhead" that the algorithms might be expected to have.

In counting the number of multiplies we adopt the philosophy that all butterflies are treated identically. This implies that all twiddle factor multiplies are counted for the second and subsequent stages. Twiddle factors are unity in the first stage, so these multiplies are not counted. Multiplies by ±1 and ±j internal to the butterflies are not counted.

As a figure of relative computational speed, the ratio of the total number of multiplies required by each algorithm to the total number of multiplies required by the radix 2 row-column algorithm (for N=4096) is also tabulated.

The left-hand side of the table contains information on the number of times that the entire array must be transfered from bulk memory to core memory and back. This information is presented as a function of the number of rows of the array (2, 4,8,16) that can be stored in core memory. It is assumed that the Eklundh transpose algorithm is used in the row-column approach. This requires $\log_2 N/\log_2 R$ transfers, where R is the available number of rows of core storage and we assume that N is a power of R. No extra transfers are necessary to compute the row and column FFT's, since these may be computed during the first and last stages of the Eklundh algorithm. The vector radix algorithm requires $\log_2 N/\log_2 R + 1$ passes over the array, since (in the radix 2x2 case) $\log_2 R$ stages of the FFT may be computed per array transfer. The extra pass over the array in the vector radix algorithm is necessitated by the 2-D bit-reversed sorting of the data.

It is readily apparent that the vector radix rxr algorithm has fewer multiplies and butterflies than the row-column radix r algorithm. In the case where r=2, there is a 25% reduction in number of multiplies and fewer, albeit more complicated, butterflies with the 2x2 algorithm. Experiments with FORTRAN codings of the two algorithms have shown that a 25% reduction in computation time is achieved in practice for N=64.

On the basis of relative number of multiplies and butterflies, we conclude that the radix 4 row-column algorithm has about the same computational complexity as the vector radix 2x2 algorithm. The fact that the radix 4 1-D algorithm requires one less array transfer argues strongly in its favor. Its disadvantages are less flexibility in transform size (N must be a power of 4 as opposed to a power of 2) and the fact that it produces the transpose of the DFT instead of the normally-ordered DFT. This last, rather minor, problem is characteristic of all radices in the row-column approach.

The larger radix algorithms are even more attractive as far as the relative numbers of multiplies and butterflies are concerned. The radix 4x4 and 8x8 algorithms appear to be computationally advantageous even when compared with the 1-D radix 4, 8 and 16 algorithms. However, the coding complexity will increase dramatically with increases in the radix size for the vector radix approach. The rather marginal decrease in multiplies in the radix 8x8 case as compared to the radix 4x4 case probably does not justify the extra coding effort. With these facts in mind, we speculate that the radix 4x4 algorithm will be the algorithm of choice for most applications involving machines limited by computational speed or machines with sufficient core storage to hold the entire array.

In situations where the DFT is implemented on an I/O intensive machine and must not be in scrambled order, the row-column approach may be preferable. The row-column algorithm of any radix requires one fewer transfer than the vector radix 2x2 algorithm. When the amount of core storage is sufficient for a number of rows which is not a power of 4, the row-column algorithms require substantially fewer transfers than the vector radix 4x4 algorithm. In such cases, the row-column approach makes better use of increased core memory size.

Transforms of non-square arrays can be accomodated in the vector radix approach by means of mixed radix algorithms. For example, Nx2N transforms may be computed with one stage of radix 2x4 butterflies, followed by $\log_2 N-1$ stages of radix 2x2 butterflies.

In conclusion, we note that the vector radix algorithm can be generalized to higher dimensional transforms. In three dimensions, decimation in all three indices of the array simultaneously leads to a radix 2x2x2 algorithm. This algorithm requires 7/12 the number of multiplies that a conventional algorithm using 1-D FFT's requires. This ratio improves with higher dimensionality and is equal to

$$\frac{2^d - 1}{d \cdot 2^{d-1}}$$

for d-dimensional transforms. For large values of d, this ratio asymptotically approaches 2/d.

REFERENCES

[1] Cooley, J. W., and J. W. Tukey, "An Algorithm for the Machine Calculation of Complex Fourier Series," Math. of Comput., vol. 19, pp. 297-301, April 1965.

[2] Rivard, G. E., "Algorithm for Direct Fast Fourier Transform of Bivariant Functions," paper presented at the 1975 Annual Meeting of the Optical Society of America, Boston, Massachusetts, October 1975.

[3] Eklundh, J. O., "A Fast Computer Method for Matrix Transposing," IEEE Trans. on Computers, vol. C-21, pp. 801-803, July 1972.

TABLE OF OPERATION COUNTS ($N=2^\nu$)

FFT TYPE	Radix	Disk Accesses 2	4	8	16	Multiplies per Butterfly*	Number of Butterflies	Total Number of Multiplies	Ratio
VR	2x2	$\nu+1$	$\nu/2+1$	$\nu/3+1$	$\nu/4+1$	3+0	$1/4\ N^2\nu$	$3/4\ N^2(\nu-1)$	.75
VR	4x4	-	$\nu/2+1$	$\nu/2+1$	$\nu/4+1$	15+0	$1/32\ N^2\nu$	$15/32\ N^2(\nu-2)$	.426
VR	8x8	-	-	$\nu/3+1$	$\nu/3+1$	63+24	$1/192\ N^2\nu$	$29/64\ N^2(\nu-63/29)$	.405
RC	2	ν	$\nu/2$	$\nu/3$	$\nu/4$	1+0	$N^2\nu$	$N^2(\nu-1)$	1.00
RC	4	ν	$\nu/2$	$\nu/3$	$\nu/4$	3+0	$1/4\ N^2\nu$	$3/4\ N^2(\nu-2)$	.682
RC	8	ν	$\nu/2$	$\nu/3$	$\nu/4$	7+2	$1/12\ N^2\nu$	$3/4\ N^2(\nu-7/3)$	.659
RC	16	ν	$\nu/2$	$\nu/3$	$\nu/4$	15+9	$1/32\ N^2\nu$	$3/4\ N^2(\nu-5/2)$	.648

*Twiddle factor multiplies (first) and internal butterfly multiplies (second) are tabulated separately.

37

Reprinted from *IEEE Trans. Acoust., Speech, Signal Process.*
ASSP-22(6):420-424 (1974)

Implementation of a Fast Fourier Transform (FFT) for Image Processing Applications

HENRY L. BUIJS, ANDRÉ POMERLEAU, STUDENT MEMBER, IEEE, MARTIN FOURNIER, SENIOR MEMBER, IEEE, AND W. G. TAM

***Abstract*—Different fast Fourier transform (FFT) algorithms for hardware implementation have been considered. We propose an implementation whereby two radix-$N^{1/2}$ passes are carried out in parallel and in which each $N^{1/2}$-point transform is carried out via a serial input parallel output transform circuit. The processing rate is one clock cycle per input point for the N-point transform regardless of the value of N chosen. The circuit is being implemented with TTL logic and will be used to perform spatial frequency domain filtering on two dimensional infrared camera images in real time; real time meaning processing between frame display.**

INTRODUCTION

IN ORDER to treat a sampled two-dimensional function—or image—in the Fourier transform domain, it is necessary to have a means to transform into and out of this domain. An $N \times M$ two-dimensional Fourier transform is most conveniently carried out by means of factoring the summations into a set of one-dimensional row transformations of N samples each, followed by a set of column transformations of M samples each. One then requires an $N \times M$ word storage medium to retain some intermediate results, and later the transform, and a means to compute many N-point or M-point Fourier transforms, successively. For most image processing applications, N and M may vary from 50 to 1000.

If one-dimensional transforms can be carried out in a very short time, then it may be possible to treat relatively rapidly varying images on a continuous basis. The introduction of the fast Fourier transform (FFT) algorithm by Cooley and Tukey [1] in 1965 allowed a tremendous saving in calculating effort in carrying out a discrete complex Fourier transformation. However, in the majority of implementations, the computations are performed in a strictly serial fashion in which case if $N = r^n$, the computation time is proportional to $n \times N/r \times r^2$. Here we imply that each r-point transform of each radix r pass requires r^2 computations where, in practice, one can usually take advantage of symmetries in the trigonometric transformation matrix. As mentioned by a number of authors [2]–[5], when one considers a special circuit for the implementation of an FFT algorithm there are numerous possibilities for reducing the computation time through such considerations as higher circuit performance and particularly by means of parallel operation of various parts of the circuit. Veenkant [2] has introduced a possible implementation which makes use of two types of parallel operation. Firstly, he reduces the computation time of each r-point Fourier transform from r^2 computation cycles to r-computation cycles by means of an "arithmetic unit (AU)" circuit that must effectively attribute the contribution of all r-input points simultaneously to one of r-output frequency channels. (The twiddle factor is applied as soon as each channel is completed.) Secondly, he suggests that one could introduce more than one AU circuit so that a number of r-point transforms can be carried out within each radix r-pass. If m is the number of AU's, then the computation time is proportional to $n \times (1/m) \times (N/r) \times r$. Corinthios [3], [4] proposes an implementation with a very high degree of parallelism in the computation of each r-point transform and "twiddle" operation so that the reading of the r-input points, the r-point transform, and the writing of the r results is done in one "clock cycle." In this case the computation time is proportional to the number of r-point transforms required, i.e., $n \times (N/r)$. Both implementations mentioned above make use of a "source" and "sink" memory which is addressed differently in the read mode as compared to the write mode in order to reorder the data between radix passes. These memories consist of short chains of serial memories, the number of which increases with r or r^2. Gold and Bially [5] implement essentially the same parallelism as Corinthios, but the memory organization is perhaps somewhat simpler and the manner in which the r-point transform circuit effectively computes one r-point transform per clock cycle is more evident through the well described technique of "pipe lining." (Meaning that at each clock cycle one starts a new r-point transform while partial results of former data advances through later sections of the arithmetic unit.)

The parallel implementations of Corinthios and of Gold and Bially imply that for the best speed performance, r should be as large as possible so that the time $n \times (N/r)$ is reduced both by r directly and by the reduction in n (note that $N = r^n$). It appears that an increase in r results in a very rapid increase in circuit complexity.

We would like to present here yet another type of parallel implementation which appears to be relatively advantageous for small N. This consists of carrying out more than one radix pass simultaneously. This principle

Manuscript received October 25, 1973. This work was supported in part by the Defense Research Board under Grant 9931-38 and the National Research Council of Canada under Grant A-4398.

H. L. Buijs is with the Centre de Recherches sur les Atomes et Molecules, Laval University, Quebec, P. Q., Canada.
A. Pomerleau and M. Fournier are with the Department of Electrical Engineering, Laval University, Quebec, P. Q., Canada.
W. G. Tam is with the Defense Research Establishment, Valcartier, P. Q., Canada.

was suggested by Corinthios [4] as a means of further speeding up his implementation, but because of the particular mode of operation for each pass, this means he must pipe line the different passes by multiplying his circuit complexity by a factor n. We have restricted ourselves to carrying out only a two-pass FFT algorithm regardless of the size of N so that $N = r^2$. In particular, if one carries out each small r-point Fourier transform in a "classical" manner, then each completed calculation in the first pass can be propagated via the twiddle operation and treatment in the second pass to all the outputs of the second pass. In the circuit implementation we also suggest to compute the r-point transforms in a parallel fashion so that only r clock cycles are required to complete each transform. The computing time in this case becomes simply the time that is required to complete the first radix-r pass and is proportional to $r \times N/r$. The greatest complexity of this implementation lies in the realization of the two "classical parallel" implementations of the r-point Fourier transform. However, it has been found that this implementation is simpler than suggested by the general r-point transformation equation. This arises from the many symmetries in the trigonometric matrix, particularly when r contains a factor 4. Furthermore, it will be shown that a considerable saving in hardware can be realized when we consider real data as the input to the first pass, and obtain only the corresponding nonredundant part of the spectrum at the output of the second pass or vice versa. It should be recognized that we require a source and sink memory, but these do not reverse roles as in other implementations. This makes both the address gating and the data entry and retrieval to the transform unit simpler.

THEORY

Consider

$$F(j) = \sum_{n=0}^{N-1} f(n) \exp(-2\pi i/N)nj. \tag{1}$$

For a two-radix pass FFT calculation, let $N = pq$, then j and n may be uniquely represented by

$$j = pj_1 + j_0 \qquad \text{where } 0 \le j_0 < p, \quad 0 \le j_1 < q$$

$$n = qn_1 + n_0 \qquad 0 \le n_1 < p, \quad 0 \le n_0 < q$$

and

$$F(pj_1 + j_0) = \sum_{n_0=0}^{q-1} \sum_{n_1=0}^{p-1} f(qn_1 + n_0) \cdot \exp(-2\pi i/pq)(qn_1 + n_0)(pj_1 + j_0). \tag{2}$$

Expanding the indices of the complex exponential and simplifying

$$F(pj_1 + j_0) = \sum_{n_0=1}^{q-1} \sum_{n_1=0}^{p-1} f(qn_1 + n_0) \exp(-2\pi i/p)n_1 j_0 \cdot \exp(-2\pi i/pq)n_0 j_0 \cdot \exp(-2\pi i/q)n_0 j_1. \tag{3}$$

The above equation can be decomposed into successive partial results,

$$F'(j_0,n_0) = \sum_{n_1=0}^{p-1} f(qn_1 + n_0) \exp(-2\pi i/p)n_1 j_0 \tag{4}$$

$$F''(j_0,n_0) = F'(j_0,n_0) \exp(-2\pi i/pq)n_0 j_0 \tag{5}$$

$$F(pj_1 + j_0) = \sum_{n_0=0}^{q-1} F''(j_0,n_0) \exp(-2\pi i/q)n_0 j_1. \tag{6}$$

In the above derivation constants have been neglected and the input function and its transform and all intermediate results are complex. Equations (4) and (6) constitute sets of p- and q-point Fourier transforms, respectively. In carrying out the two radix passes in parallel, as soon as some values of F' have been evaluated in (4) (such as completing the summation over n_1 for say $n_0 = 0$), these values may be weighted by the "twiddle" factors and processed in (6) in order to contribute to the final result. Logistically, this is possible only when the summation in (6) is sequential (classical) and when partially completed output points can be upgraded frequently. Fig. 1 shows a block diagram of an implementation incorporating the above considerations. In order to perform the inverse Fourier transform one simply changes the sign of the index of each complex exponential. With respect to circuit realization, it is important to attempt minimization of arithmetic operations involving complex numbers since for example the product of two complex numbers requires four multiplications and the addition of two pairs of numbers. Also, one must always set aside two storage cells for each complex number.

In the case where either the input data or the output data are real, one often inserts two real vectors of data into the corresponding complex vector in order to process two sets of real data with one application of the Fourier transform procedure. We have considered to adapt our implementation to use real input data and to compute only the nonredundant part of the output function for the forward transform, and the reverse for the inverse transform. This procedure has the advantage that one obtains the same computing efficiency (or computing time per real input point) as in the case of processing two data vectors in a complex transform, but with the added advantage of requiring approximately one half the components. We adapt (3) to indicate real input data and complex output data.

$$F_R(pj_1 + j_0) + iF_I(pj_1 + j_0) = \sum_{n_0=0}^{q-1} \sum_{n_1=0}^{p-1} f(qn_1 + n_0) \cdot \exp(-2\pi i/p)n_1 j_0 \cdot \exp(-2\pi i/pq)n_0 j_0 \cdot \exp(-2\pi i/q)n_0 j_1 \tag{7}$$

where now $0 \le j_1 < q/2$, and all other coefficients are as before. Decomposing (7) into successive partial results gives

$$F_R'(j_0,n_0) = \sum_{n_1=0}^{p-1} f(qn_1 + n_0) \cos(2\pi/p)n_1 j_0$$

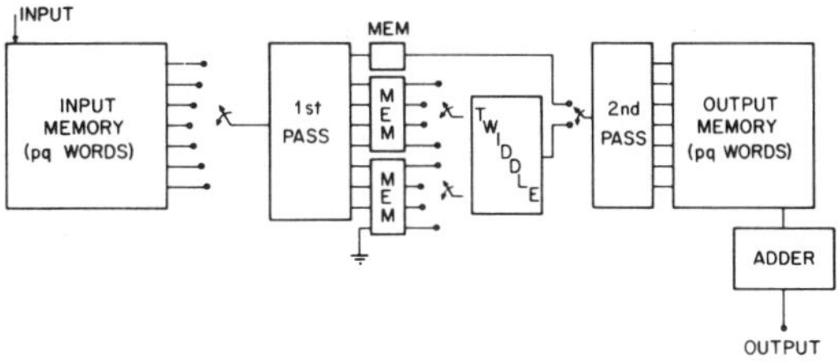

Fig. 1. Block diagram of system implementation.

$$F_I'(j_0,n_0) = -\sum_{n_1=0}^{p-1} f(qn_1 + n_0)\sin(2\pi/p)n_1j_0. \quad (8)$$

One needs to evaluate the above terms only for $0 \leq j_0 < p/2$ because of the following symmetries:

$$\begin{aligned} F_R'(j_0,n_0) &= F_R'(p - j_0,n_0) \\ F_I'(j_0,n_0) &= -F_I'(p - j_0,n_0). \end{aligned} \quad (9)$$

Then the following step will be

$$\begin{aligned} F_R''(j_0,n_0) &= F_R'(j_0,n_0)\cos(2\pi/N)n_0j_0 \\ &\quad + F_I'(j_0,n_0)\sin(2\pi/N)n_0j_0 \\ F_I''(j_0,n_0) &= -F_R'(j_0,n_0)\sin(2\pi/N)n_0j_0 \\ &\quad + F_I'(j_0,n_0)\cos(2\pi/N)n_0j_0. \end{aligned} \quad (10)$$

Again we have the symmetry relations

$$\begin{aligned} F_R''(p - j_0,n_0) &= F_R''(j_0,n_0)\cos(2\pi/q)n_0 \\ &\quad - F_I''(j_0,n_0)\sin(2\pi/q)n_0 \\ F_I''(p - j_0,n_0) &= -F_R''(j_0,n_0)\sin(2\pi/q)n_0 \\ &\quad - F_I''(j_0,n_0)\cos(2\pi/q)n_0. \end{aligned} \quad (11)$$

These symmetries imply that one may develop the entire final result using only 50 percent of the possible intermediate results.

$$\begin{aligned} F_R(pj_1 + j_0) &= \sum_{n_0=0}^{q-1} F_R''(j_0,n_0)\cos(2\pi/q)n_0j_1 \\ &\quad + F_I''(j_0,n_0)\sin(2\pi/q)n_0j_1 \\ F_R(pj_1 + p - j_0) &= \sum_{n_0=0}^{q-1} F_R''(j_0,n_0)\cos(2\pi/q)n_0(j_1 + 1) \\ &\quad - F_I''(j_0,n_0)\sin(2\pi/q)n_0(j_1 + 1) \\ F_I(pj_1 + j_0) &= \sum_{n_0=0}^{q-1} - F_R''(j_0,n_0)\sin(2\pi/q)n_0j_1 \\ &\quad + F_I''(j_0,n_0)\cos(2\pi/q)n_0j_1 \\ F_I(pj_1 + p - j_0) &= \sum_{n_0=0}^{q-1} - F_R''(j_0,n_0)\sin(2\pi/q)n_0(j_1 + 1) \\ &\quad - F_I''(j_0,n_0)\cos(2\pi/q)n_0(j_1 + 1). \end{aligned} \quad (12)$$

In order to avoid evaluating redundant terms one may store the following:

$$\begin{aligned} F_{CR}''(j_0,j_1) &= \sum_{n_0=0}^{q-1} F_R''(j_0,n_0)\cos(2\pi/q)n_0j_1 \\ F_{SR}''(j_0,j_1) &= \sum_{n_0=0}^{q-1} F_R''(j_0,n_0)\sin(2\pi/q)n_0j_1 \\ F_{CI}''(j_0,j_1) &= \sum_{n_0=0}^{q-1} F_I''(j_0,n_0)\cos(2\pi/q)n_0j_1 \\ F_{SI}''(j_0,j_1) &= \sum_{n_0=0}^{q-1} F_I''(j_0,n_0)\sin(2\pi/q)n_0j_1 \end{aligned} \quad (13)$$

and obtain the final result

$$\begin{aligned} F_R(pj_1 + j_0) &= F_{CR}''(j_0,j_1) + F_{SI}''(j_0,j_1) \\ F_R(pj_1 + p - j_0) &= F_{CR}''(j_0,j_1 + 1) - F_{SI}''(j_0,j_1 + 1) \\ F_I(pj_1 + j_0) &= -F_{SR}''(j_0,j_1) + F_{CI}''(j_0,j_1) \\ F_I(pj_1 + p - j_0) &= -F_{SR}''(j_0,j_1 + 1) - F_{CI}''(j_0,j + 1). \end{aligned} \quad (14)$$

One may also start with complex input data for which the real part is symmetric and the imaginary part is antisymmetric so that the Fourier transform results in a set of real data

$$\begin{aligned} f(qn_1 + n_0) &= \sum_{j_0=0}^{p-1}\sum_{j_1=0}^{q-1} (F_R(pj_1 + j_0) + iF_I(pj_1 + j_0)) \\ &\quad \cdot \exp(-2\pi i/q)j_1n_0 \exp(-2\pi i/pq)j_0n_0 \\ &\quad \cdot \exp(-2\pi i/p)j_0n_1. \end{aligned} \quad (15)$$

Here it is found that it is convenient to compute the intermediate results $f_1(qn_1 + n_0)$ and $f_2(qn_1 + n_0)$ so that

$$f(qn_1 + n_0) = f_1(qn_1 + n_0) + if_2(qn_1 + n_0). \quad (16)$$

Because of the symmetries in F_R and F_I, f_1 is found to be real and symmetric and f_2 is imaginary and antisymmetric. We calculate $f_1(qn_1 + n_0)$ and $f_2(qn_1 + n_0)$ independently and according to the same procedure as for the direct transform. To derive f_1 from F_R, however, requires only half the computing effort compared to the earlier described direct transform. This is because F_R is not only considered real, but also symmetric. Likewise, to derive f_2 from F_I requires half the effort of the regular direct transform. Hence, it can be seen that the computation of f from $F_R + iF_I$ requires the same computing effort as the computation of $F_R + iF_I$ from f. It is evident that with a change in sign of the complex exponentials, this latter procedure corresponds to the inverse Fourier transform as compared to the earlier derivation.

The timing diagram illustrating the sequence of operations of the above formalism is shown in Fig. 2 for the case of real input data.

IMPLEMENTATION

The features of implementing a two-pass FFT in a hardwire circuit according to the above formalism are

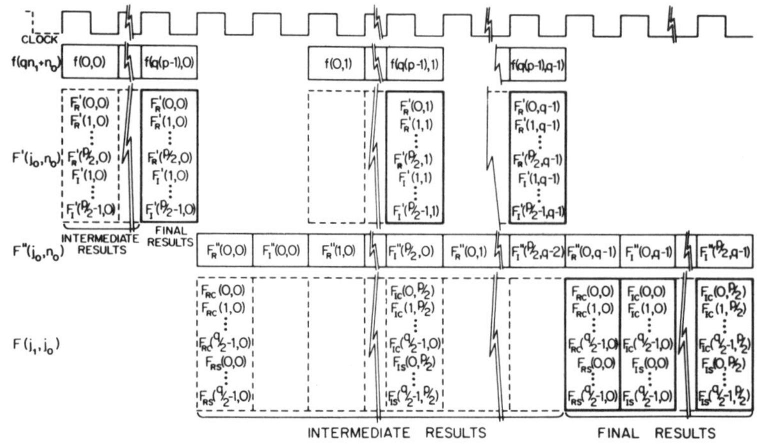

Fig. 2. Timing diagram.

the following: 1) one needs to implement two small Fourier transform circuits which are connected via a complex multiplier circuit where one transform circuit is fed from an input memory and the other deposits partial results into a parallel set of circulating output memories; 2) there is a read-out section that must prepare sums and differences of pairs of numbers via an adder.

The small transforms are carried out in a manner such that each input point is processed sequentially and its contribution is simultaneously attributed to all frequency channels. For an r-point transform, this requires r accumulators to add successive partial results. The weighting of the input points before addition into the accumulators is done in parallel since the absolute values of the real and imaginary parts of the rth roots of 1 take only $(r - 4)/4$ values different from 0 and 1. The direct input and the $(r - 4)/4$ multiplier outputs are simultaneously available and need to be distributed to the r accumulators. Since a number of accumulators always require access to the same weighted value, one needs only a relatively small number of multiplexers.

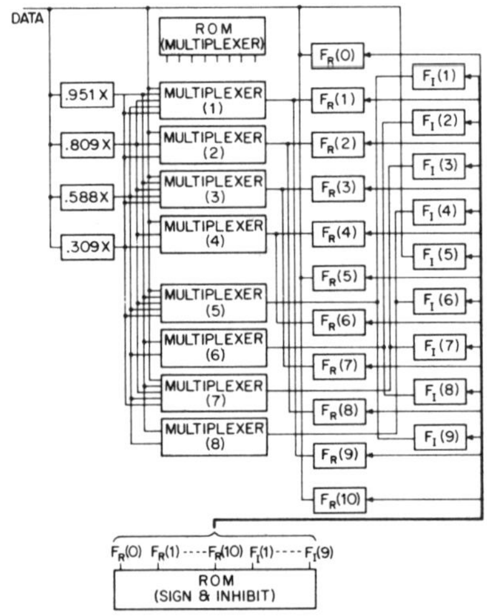

Fig. 3. 20-point Fourier transform unit.

As an example, when N consists of 400 real samples, one requires two 20-point transform units as shown in Fig. 3, each of which requires $\frac{1}{4}(r - 4) = 4$ multiplier circuits, with fixed multiplicands, four 5-input word multiplexers, two 3-input word multiplexers, two 2-input word multiplexers, and 20 adders. These transform units are linked by two parallel multipliers and a ROM with a sine table for the twiddle operation. Finally there is a 400-word input memory, a 20-word intermediate memory and an output memory. The output memory consists of twenty 20-word memories in which case each 20-word memory is connected to an accumulator belonging to the second 20-point transform unit. If a circuit is realized in which $N = 1024 = 32^2$, then arrays with $N = 784$, 576, 400, 256, 64, and 16 can be easily computed requiring only a change in procedure with no change in circuit layout.

The advantages of this circuit implementation are: 1) the parallelism introduced here provides an effective processing rate of one input point per clock cycle for whatever value of N chosen (i.e., per point processing time can be about 200 ns for TTL logic circuitry); 2) the delay between the entry of the last input word and the completion of the transform is only $N^{1/2}$ clock cycles and hence one obtains a high through-put rate not only for continuous groups of input vectors, but also for isolated input vectors; 3) the realization of the transform unit for real input data which calculates nonredundant output data or vice versa results in a reduction of operations and a reduction of components of nearly 50 percent as compared to a complex transform unit and hence appears more favorable than treating two real vectors in a complex transform unit; 4) the accuracy obtained with finite size integers is maximized by having accumulators of sufficient size (i.e., $B + \log_2 N^{1/2}$ bits where B is the input word size) which then require scaling only when data is propagated from the first pass to the second and from the second pass into the final results, hence only two truncation operations. This latter advantage is particularly evident for a small word-size implementation.

SIMULATION

The exact sequence of operations which has been proposed for the hardwire implementation has been coded into a Fortran program for simulation on a general-purpose computer. This was done to verify the feasibility of the implementation for both the direct and inverse transforms and to analyze the effect of truncation errors. This simulation has indicated that the accuracy is better than that of a radix-2 FFT algorithm. A detailed study is presently under way to evaluate computation accuracy by means of a number of different criteria and for various numbers of input samples and word sizes. This study is particularly directed towards evaluating the procedure of image modification via spatial frequency domain digital processing.

CONCLUSION

Partly on the basis of the results of the simulation and mostly on the basis of availability of suitable TTL integrated circuits, the construction of a prototype Fourier transform unit has been undertaken. This circuit operates on 64 8-bit words of input data and evaluates 32 complex pairs of 8-bit words or vice versa. The accumulators at the output of each small Fourier transform unit have 12-bit words in order to accumulate groups of 8 points without the necessity of rescaling in case of overflows. For subsequent processing the most useful 8 bits are propagated only. The processing rate is expected to be nearly 5 MHz however the unit will be used to transform a 64×64 element image that is stored in a digital scan converter associated with an infrared camera. The scan converter displays 30 frames per second. It is expected that images may be transformed directly and then inversely in the time interval between two consecutive frames. This would enable image processing in the spatial frequency domain in real time on this scale. Further possibilities of scaling up this operation while retaining real time processing are being considered.

REFERENCES

[1] J. W. Cooley and J. W. Tukey, "An algorithm for the machine computation of complex Fourier series," *Math. Comput.*, vol. 19, pp. 297–301, Apr. 1965.

[2] R. L. Veenkant, "A serial-minded FFT," *IEEE Trans. Audio Electroacoust.*, vol. AU-20, pp. 180–185, Aug. 1972.

[3] M. J. Corinthios, "The design of a class of fast Fourier transform computers," *IEEE Trans. Comput.*, vol. C-20, pp. 617–623, June 1971.

[4] ——, "A fast Fourier transform for high-speed signal processing," *IEEE Trans. Comput.*, vol. C-20, pp. 843–846, Aug. 1971.

[5] B. Gold and T. Bially, "Parallelism in fast Fourier transform hardware," *IEEE Trans. Audio Electroacoust.*, vol. AU-21, pp. 5–16, Feb. 1973.

Editors' Comments on Papers 38 Through 41

38 CHAKRABARTI, BOSE, and MITRA
Sum and Product Separabilities of Multivariable Functions and Applications

39 MITRA, SAGAR, and PENDERGRASS
Realizations of Two-Dimensional Recursive Digital Filters

40 ROESSER
A Discrete State-Space Model for Linear Image Processing

41 MECKLENBRÄUKER and MERSEREAU
McClellan Transformations for Two-Dimensional Digital Filtering: II—Implementation

2-D RECURSIVE AND NONRECURSIVE STRUCTURES

As mentioned previously, there is no general factorization scheme for arbitrary 2-D polynomials. However, under certain conditions, two variable polynomials can be expressed as a product of special types of two-variable polynomials of lower degree. These conditions and the algorithm for factorization are described in by Chakrabarti et al. (Paper 38). The authors also treat the case of expressing two-variable functions as a sum of such functions of lower order. These types of separabilities can be used in developing classical cascade or parallel type realizations of both 2-D recursive and nonrecursive transfer functions. While this work was being carried out, the general problem of factorization of multivariable polynomials over the integers was also receiving considerable attention by researchers in other fields. As a result, general algorithms for determining the reducibility and the irreducible factors of any multivariable polynomial over the integers are now available. The first such algorithm has been advanced by Musser (1975). Another such algorithm is due to Wang and Rothschild (1975). Recently, Chakrabarti and Mitra (1977) have improved Musser's algorithm.

In the more general (and typical) case, the type of polynomial decomposition used in one dimension to generate the profuse variety of structures is not possible in two dimensions. This condition has severely limited the extent to which 2-D filters can be realized in anything but a composite form. Realization methods advanced invariably involve implementing the complete transfer function as a single structure.

A recursive filter realization based on a direct implementation of Eq. (1.4) was originally proposed by Shanks et al. (Paper 1). Because, in the final configuration, the multiplier coefficients are identical to the coefficients of the difference equation, this type of realization is commonly known as a *direct form* realization. Three other direct form realizations of 2-D transfer functions are outlined by Mitra et al. (Paper 39). In addition, they propose several other realizations based on a two-variable continued-fraction expansion. As indicated in this paper, such expansions may not always exist.

Realizations can also be developed from the state-space descriptions of the 2-D filters. To this end, the state-space model of 2-D structures proposed by Roesser (Paper 40) can be utilized. From a known 2-D realization, a state-space characterization in a form proposed by Roesser can be easily developed by considering the input variables to all delay elements as state variables. Then, by applying linear transformations to the state variables, many equivalent state-space models (not all) and 2-D structures can be developed. Proceeding along this line, Kung et al. (1977) proposed new implementation schemes for 2-D filters which are 2-D equivalents of the controllable—and observable—forms of standard 1-D theory.

A new realization scheme for 2-D IIR filters, based on linear algebraic techniques, has also been outlined by Mitra and Chakrabarti (1977). Their scheme is based on realizing multi-input, multi-output subnetworks of multipliers and adders and interconnecting 2-D digital subnetworks containing z_1- and z_2- delay elements (and possibly multipliers and adders) in known configurations. By choosing different subnetworks, one may obtain different realizations of the same transfer function.

For the case of 2-D nonrecursive transfer functions, direct realizations can be obtained from the convolution sum description of the filter as given by Eq. (1.3). Alternate structures are directly derivable from this case where there exist symmetries in the filter weights $\{h_{m,n}\}$. In particular, Paper 41 by Mecklenbräuker and Mersereau describes the computational savings which derive from the special structure of 2-D linear phase filters designed using generalized McClellan transformations (Paper 19). McClellan and Chan (1977) present alternate realizations based on this approach which achieve identical savings while avoiding previous coefficient round-off and sensitivity problems.

REFERENCES

Chakrabarti, S., and S. K. Mitra. 1977. "An Algorithm for Multivariable Polynomial Factorization," *Proc. 1977 IEEE Internat. Symp. Circuits Syst.*, 678–683.

Kung, S. Y., B. Levy, M. Morf, and T. Kailath. 1977. "New Results in 2-D Systems Theory, Part II," *IEEE Proc.* **65**(6):945–961.

McClellan, J. H., and D. S. K. Chan. 1977. "A 2-D FIR Filter Structure Derived From the Chebyshev Recursion," *IEEE Trans. Circuits Syst.,* **CAS-24**(7):372–378.

Mitra, S. K., and S. Chakrabarti. 1977. "A General Realization Scheme for 2-D Digital Transfer Functions," *Proc. 1977 IEEE Internat. Conf. Acoust., Speech, Signal Process.* 539–542.

Musser, D. R. 1975. "Multivarate Polynomial Factorization," *J. Ass. Comp. Mach.* **22**(2):291–308.

Wang, P. W., and L. P. Rothschild. 1975. "Factoring Multivariable Polynomials Over the Integers," *Math. Comp.* **29**:935–950.

38

Reprinted from *Franklin Inst. J.* **299**(1):53-66 (1975)

*Sum and Product Separabilities of Multivariable Functions and Applications**

by S. CHAKRABARTI

Department of Electrical Engineering
Concordia University, Montreal, Quebec, Canada

N. K. BOSE

Department of Electrical Engineering
University of Pittsburgh, Pittsburgh, Pennsylvania

and S. K. MITRA

Department of Electrical Engineering
University of California, Davis, California

ABSTRACT: *Necessary and sufficient conditions for reducibility of certain classes of multivariable polynomials are considered. The special case when a polynomial in that class can be factored into a finite product of single variable polynomials is investigated. A set of necessary and sufficient conditions for decomposing a multivariable p.r.f. into a sum of single variable p.r.f.'s is given. Examples to illustrate applications of sum and product separabilities of functions are given.*

I. Introduction

Research in the areas of multidimensional filtering and multivariable realizability theory is continuously expanding, due to potential applications in subjects ranging from image processing (**1**), seismic, gravity and magnetic data processing (**2**), to lumped-distributed network synthesis (**3**). A fundamental trouble in multidimensional problems results from the nonfeasibility of factorization of an arbitrary multivariable polynomial into a product of linear factors, as in the single variable case. This has made the problem of synthesis as well as implementation of multidimensional stability tests more difficult. One approach in the field of synthesis of recursive filters has been the transformation of the problem, whenever possible, into that of the synthesis of one-dimensional filters. Multidimensional filters of this type have been called *separable filters* (**2**). In Section II, the necessary and sufficient conditions when a polynomial, $f(p_1, p_2, \ldots, p_n)$ can be finitely factored into polynomials of the form $f = \prod_{i=1} f_i$, where the set of indeterminates corresponding to each f_i are mutually disjoint, are given. The special case of

$$f(p_1, p_2, \ldots, p_n) = \prod_{i=1}^{n} f_i(p_i)$$

is also discussed.

* This work was partly supported by the National Science Foundation under Grant GK-38205, while N. K. Bose was a visiting faculty at the University of California, Berkeley, U.S.A.

Sum type of separability, whenever possible, is useful in network synthesis, because the difficult problem of multivariable synthesis is then converted into the synthesis of single variable functions, for which well-known techniques are available. A necessary and sufficient condition for the decomposition of a multivariable reactance function into the sum of several single variable reactance functions, was advanced (**4**). In Section III, a necessary and sufficient condition for the decomposition of a multivariable positive real function into a sum of several single variable positive real functions, is given. Some consequences and extensions of this result are also considered.

In Section IV, some examples illustrating scopes for applications of the separability results are given.

II. Reducibility Criteria for Some Classes of Multivariable Polynomials

In general, the multivariable polynomial $f(p_1, \ldots, p_n)$ admits the representation

$$f(p_1, \ldots, p_n) = \sum a_{i_1 i_2 \cdots i_n} p_1^{i_1} p_2^{i_2} \cdots p_n^{i_n},$$

where the i's assume nonnegative integer values. Each product term is called a *monomial* and if the coefficients of the monomials are replaced by unity, the monomials are called *primitive* monomials (**5**).

Definition 1: *Form of a multivariable polynomial*

The set of primitive monomials including 1 corresponding to a multivariable polynomial f will be called the "form" of the polynomial, denoted by $[f]$. A constant is a trivial form. All other forms are nontrivial.

For example, for $f = 3p_1 p_2 + 7$, $[f] = \{p_1 p_2, 1\}$.

Suppose $[A]$ and $[B]$ are two forms. The product of $[A]$ and $[B]$, denoted by $[A][B]$, is defined as follows:

$$[A][B] \triangleq \left\{ \bigcup_{\forall b \in [B]} [A]b \colon [A]b = b[A] \triangleq \{ab(=ba) \colon \forall a \in [A]\} \right\},$$

where ab correspond to the usual product of primitive monomials and the union is the ordinary union of sets.

Subject to this definition, it is easy to observe that if $[f]$ is the form of a reducible multivariable polynomial $f = f_1 f_2$ and $[f_1]$ and $[f_2]$ are the corresponding forms of f_1 and f_2, respectively, then $[f] = [f_1][f_2]$ are "factors" of $[f]$.

Definition 2: *Reducible and irreducible forms*

A form is called "reducible" if and only if it has more than one nontrivial factors. Otherwise, the form is said to be "irreducible".

Lemma 1

If the form of a multivariable polynomial is irreducible, the polynomial itself is irreducible.

The proof is straightforward and hence omitted. It must be observed, however, that if the form of a multivariable is reducible, the polynomial itself is not necessarily reducible. For example, the polynomial

$$f(p_1, p_2) = p_1 p_2 + \alpha p_2 + \beta p_1 + \gamma$$

has the form $[f] = \{p_1 p_2, p_1, p_2, 1\}$ with factors $\{p_1, 1\}$ and $\{p_2, 1\}$. However, unless $\gamma = \alpha\beta$, $f(p_1, p_2)$ is not reducible.

Definition 3: *p-Reducible polynomials and p-factors*

A multivariable polynomial will be called "p-reducible" if and only if it can be expressed as a product of nontrivial factors, so that the sets of indeterminates in any two factors are mutually disjoint. The factors themselves will be called p-factors, and need not be themselves irreducible.

Since the factorization of a p-reducible polynomial into such factors induces a partitioning in the set of indeterminates, this type of reducibility is termed partition–reducibility or p-reducibility.

A multivariable polynomial $f(p_1, ..., p_n)$ will be called "1-reducible" if and only if f can be expressed as $f = \prod_{i=1}^{n} f_i(p_i)$ where each $f_i(p_i)$ has a degree ≥ 1.

The testing procedure of p-reducibility and the actual determination of p-factors are presented next.

p-Reducibility conditions

In order to check the p-reducibility of a multivariable polynomial, one first notes the following lemma which follows from the definition of p-reducibility and may be thought of as a corollary to Lemma 1.

Lemma 2

If the form of a multivariable polynomial is not p-reducible, the polynomial itself is not p-reducible.

Consider a n-variable polynomial $f(p_1, ..., p_n)$ with total degree d. Let $M \subseteq [f]$ denote the set of elements of $[f]$ each of which has total degree d. If $[f]$ is reducible, one can write $[f] = [f_1][f_2]$ for appropriate nontrivial $[f_1]$ and $[f_2]$. Let the sets M_1 and M_2 be the counterparts of the set M for $[f_1]$ and $[f_2]$, respectively. Let $\operatorname{card} M = m$, $\operatorname{card} M_1 = m_1$ and $\operatorname{card} M_2 = m_2$, where "card" denotes the cardinality. Clearly, $M = M_1 M_2$, where the product operation is applied to the relevant forms; and $m = m_1 m_2$. Let X, X_1, X_2, X_M, X_{M_1} and X_{M_2} denote the sets of indeterminates corresponding to $[f]$, $[f_1]$, $[f_2]$, M, M_1 and M_2, respectively. Then $X = X_1 \cup X_2$, $X_M = X_{M_1} \cup X_{M_2}$ and $X_M \subseteq X$. If $[f]$ is p-reducible, by definition $X_1 \cap X_2 = \phi$ and hence $X_{M_1} \cap X_{M_2} = \phi$.

Let γ denote the largest common factor of all elements of M. Let x denote an indeterminate in X or the product of several indeterminates in X (raised to respective exponents, if any) occurring in the elements of $[f]$. Let Π_x denote the set of all elements of $[f]$ in which x occurs, and $x^{-1}\Pi_x$ denote the form-product of $\{x^{-1}\}$ and Π_x.

In studying the p-reducibility conditions of $[f]$, two possible cases are investigated; (i) $m > 1$, m is prime; (ii) $m \geqslant 1$, m is a composite number for $m > 1$.

Case (i): $m > 1$, *m is prime*

The following lemma will be proved.

Lemma 3

The form $[f]$ of a multivariable polynomial f, for which $m > 1, m$ is prime holds, will be p-reducible, if and only if, (i) there exists a unique nontrivial (i.e. of total degree $\geqslant 1$) largest common factor γ for all elements of M such that $M = \gamma\bar{M} \Rightarrow X_\gamma \cap X_{\bar{M}} = \Phi$: (ii) the sets $\beta^{-1}\Pi_\beta$ are identical to one another, $\forall \beta \in \gamma^{-1}\Pi_\gamma$ and $(\gamma^{-1}\Pi_\gamma) \cap (\beta^{-1}\Pi_\beta) = \Phi$, where Φ denotes the null set.

Proof: *Sufficiency*

Since γ is unique, $M = \gamma\bar{M} \Rightarrow \bar{M}$ is unique. Also, $X_\gamma \cap X_{\bar{M}} = \phi$, which implies that if $[f]$ is reducible, then $[f]$ may be p-reducible. Now $\bar{M} \subseteq \gamma^{-1}\Pi_\gamma$ and $X_\gamma \subseteq X - X_{\bar{M}}$. From the construction of the sets, it then follows that $[f] = \bigcup_{\forall\beta \in \gamma^{-1}\Pi_\gamma} \beta[\beta^{-1}\Pi_\beta] = [\gamma^{-1}\Pi_\gamma][\beta^{-1}\Pi_\beta]$ which follows from hypothesis (ii). Since $X_{\gamma^{-1}\Pi_\gamma} \cap X_{\beta^{-1}\Pi_\beta} = \phi$, $[f]$ is p-reducible.

Necessity

$[f]$ is p-reducible, i.e. M is p-reducible, and $M = M_1 M_2 \Rightarrow X_{M_1} \cap X_{M_2} = \phi$ and $m = m_1 m_2$. Since m is a prime > 1, either $m_1 = 1$ or $m_2 = 1$.

Assume that there is no such γ as stated in (1). Then M is irreducible. This being a contradiction, at least one such γ must exist.

Suppose γ is not unique. This implies $m_1 > 1$ and $m_2 > 1$ which is again a contradiction. Hence γ is unique. Since M is p-reducible, by definition of p-reducibility $M = \gamma\bar{M} \Rightarrow X_\gamma \cap X_{\bar{M}} = \phi$.

Suppose the sets $\beta^{-1}\Pi_\beta$ are not identical to one another, $\forall \beta \in \gamma^{-1}\Pi_\gamma$. Since $[f] = \bigcup_{\forall\beta}\Pi_\beta$, this implies $[f]$ is not reducible. This being a contradiction, $\beta^{-1}\Pi_\beta$ must be the same $\forall \beta \in \gamma^{-1}\Pi_\gamma$. Hence, clearly,

$$[f] = [\gamma^{-1}\Pi_\gamma][\beta^{-1}\Pi_\beta],$$

and p-reducibility would imply that $(\gamma^{-1}\Pi_\beta) \cap (\beta^{-1}\Pi_\beta) = \Phi$.

Example I

Let

$$[f] = \{p_1^2p_2^2p_4^3, p_1^2p_2^2p_4^2, p_1^2p_2^2p_5, p_1p_2p_3^2p_4^3, p_1p_2p_3^2p_4^2, p_1p_2p_3^2p_5, p_3^4p_4^3, p_3^4p_4^2, p_3^4p_5\}.$$

In this case

$$M = \{p_1^2p_2^2p_4^3, p_1p_2p_3^2p_4^3, p_3^4p_4^3\}, \quad m = 3, \quad \gamma = p_4^3,$$

$$\gamma^{-1}\Pi_\gamma = \{p_1^2p_2^2, p_1p_2p_3^2, p_3^4\}$$

which can be checked easily to be a factor of $[f]$. The other factor

$$\beta^{-1}\Pi_\beta = (p_1^2 p_2^2)^{-1}\Pi_{p_1^2 p_2^2} = (p_3^4)^{-1}\Pi_{p_3^4}$$
$$= (p_1 p_2 p_3^2)^{-1}\Pi_{(p_1 p_2 p_3^2)}$$
$$= \{p_4^3, p_4^2, p_5\}.$$

Case (ii): $m \geqslant 1$, $m > 1 \Rightarrow m$ *is a composite number*

In this case the following lemma is proved.

Lemma 4

The form $[f]$ with $m \geqslant 1$, $m > 1 \Rightarrow m$ is composite, will be p-reducible if and only if, for any p_i^k occurring in any element of M, the sets $\beta^{-1}\Pi_\beta$ are identical to one another $\forall \beta \in \hat{\Pi}_{p_i^k}$ where $\hat{\Pi}_{p_i^k}$ is obtained from $\Pi_{p_i^k}$ after extracting the largest factor common to all elements of $\Pi_{p_i^k}$ with $X_{\hat{\Pi}_{p_i^k}} \cap X_{\beta^{-1}\Pi_\beta} = \phi$. If (i) $\hat{\Pi}_{p_i^k}$ is irreducible, $[f] = \hat{\Pi}_{p_i^k}(\beta^{-1}\Pi_\beta)$; if (ii) $\hat{\Pi}_{p_i^k}$ is reducible and $\hat{\Pi}_{p_i^k} = \hat{\Pi}_1 \hat{\Pi}_2$, say, then $[f] = \hat{\Pi}_1[\alpha^{-1}\Pi_\alpha]$, where $\hat{\Pi}_1$ is a factor of $[f]$, $\hat{\Pi}_2$ is not, $\alpha \in \hat{\Pi}_1$ where $\alpha^{-1}\Pi_\alpha$ is identical to one another for all such α's, and $\hat{\Pi}_1 \cap \alpha^{-1}\Pi_\alpha = \Phi$.

The proof of this lemma is along somewhat similar lines as the previous lemma, and is omitted for brevity. Instead, a simple example, presented below, is used to illustrate the idea.

Example II

Let $[f] = \{x^2yp^2v, xy^2p^2v, w^2zp^2v, x^2yp^2r, xy^2p^2r, w^2zp^2r, x^2yr^2s, xy^2r^2s, w^2zr^2s\}$. In this case, $M = [f]$, $m = 9$. Form Π_{p^2}, where it is noted that every indeterminate-raised-to-power which occurs in one element of $[f]$, occurs in at least one other element of $[f]$. $\hat{\Pi}_{p^2} = \{x^2yv, xy^2v, w^2zv, x^2yr, xy^2r, w^2zr\}$ Let $\hat{\Pi}_{p^2} = [g]$. It is required to test for the reducibility of $[g]$, where $M_{[g]} = [g]$ and $m_{[g]} = 6$. Form Π_{x^2} in $[g]$. Then $\Pi_{x^2} = \{x^2yv, x^2yr\} \Rightarrow \Pi_{x^2} = \{v, r\}$ which leads to the irreducible decomposition of $\hat{\Pi}_{p^2}$ as $\hat{\Pi}_{p^2} = \{x^2y, xy^2, w^2z\}\{v, r\}$. Since there are terms in $[f]$ which are free of p^2v and p^2r, while p^2 never occurs in a term without one of v or r also occurring simultaneously, $\{x^2y, xy^2, w^2z\}$ must be one of the factors of $[f]$. The other factor, in that case, comes out to be $\{p^2v, p^2r, r^2s\}$. It is seen that Lemma 3 cannot be used here to obtain the factorization. However, in special cases, it is possible to use Lemma 3, even when m is non-prime.

At this stage, therefore, one can check whether a particular form is p-reducible or not. From the result of p-reducibility of forms, the next step, viz. the p-reducibility of the polynomials follows directly. For convenience, the following definition is introduced.

Definition 4: *Monic multivariable polynomials*

A multivariable polynomial with a total degree d will be called "monic" if and only if the coefficient of at least one of the monomials with a total degree d, contained in the sum of monomials expression of the polynomial, is unity.

The coefficients of the p-factors of f_σ can be found by inspection, as the example below illustrates. (f_σ is the monic polynomial, related to f by $f = \sigma f_\sigma$.)

Example III

Let

$$f = 24x_1^2x_4x_6 + 16x_2x_3x_4x_6 + 12x_1^2x_5x_6 + 8x_2x_3x_5x_6 + 42x_1^2x_4x_7^2 + 28x_2x_3x_4x_7^2 + 21x_1^2x_5x_7^2 + 14x_2x_3x_5x_7^2,$$

$$[f] = \{x_1^2x_4x_6, x_2x_3x_4x_6, x_1^2x_5x_6, x_2x_3x_5x_6, x_1^2x_4x_7^2, x_2x_3x_4x_7^2, x_1x_5x_7^2, x_2x_3x_5x_7^2\}.$$

Using Theorem I, $[f]$ is found to be reducible. A particular decomposition is given below, obtained by use of Lemma 4, or in this case, also by Lemma 3,

$$[f] = \{x_1^2x_4, x_2x_3x_4, x_1^2x_5, x_2x_3x_5\}\{x_6, x_7^2\}.$$

Obtain for convenience, a monic f_σ from f, with $\sigma = 42$ being chosen to be the coefficient of $x_1^2x_4x_7^2$.

$$\begin{aligned}\tfrac{1}{42}f = f_\sigma &= x_1^2x_4x_7^2 + \tfrac{2}{3}x_2x_3x_4x_7^2 + \tfrac{1}{2}x_1^2x_5x_7^2 + \tfrac{1}{3}x_2x_3x_5x_7^2 + \tfrac{4}{7}x_1^2x_4x_6 \\ &\quad + \tfrac{8}{21}x_2x_3x_4x_6 + \tfrac{2}{7}x_1^2x_5x_6 + \tfrac{4}{21}x_2x_3x_5x_6 \\ &= (x_1^2x_4)(x_7^2) + \tfrac{2}{3}(x_2x_3x_4)(x_7^2) + \tfrac{1}{2}(x_1^2x_5)(x_7^2) + \tfrac{1}{3}(x_2x_3x_5)(x_7^2) \\ &\quad + \tfrac{4}{7}(x_1^2x_4)(x_6) + \tfrac{8}{21}(x_2x_3x_4)(x_6) + \tfrac{2}{7}(x_1^2x_5)(x_6) + \tfrac{4}{21}(x_2x_3x_5)(x_6).\end{aligned}$$

Since

$$\tfrac{2}{3}\cdot\tfrac{4}{7} = \tfrac{8}{21}, \quad \tfrac{1}{2}\cdot\tfrac{4}{7} = \tfrac{2}{7}, \quad \tfrac{1}{3}\cdot\tfrac{4}{7} = \tfrac{4}{21},$$

f_σ is p-reducible and

$$\begin{aligned}&f_\sigma(x_1^2x_4 + \tfrac{2}{3}x_2x_3x_4 + \tfrac{1}{2}x_1^2x_5 + \tfrac{1}{3}x_2x_3x_5)(x_7^2 + \tfrac{4}{7}x_6) \\ &= (x_1^2 + \tfrac{2}{3}x_2x_3)(x_4 + \tfrac{1}{2}x_5)(x_7^2 + \tfrac{4}{7}x_6).\end{aligned}$$

Hence

$$f = 42(x_1^2 + \tfrac{2}{3}x_1x_3)(x_4 + \tfrac{1}{2}x_5)(x_7^2 + \tfrac{4}{7}x_6).$$

1-*Reducibility criterion*

As a special case of the above discussion, the conditions of 1-reducibility are presented.

Definition 5: *Full and scarce multivariable polynomial*

If $f(p_1, \ldots, p_n)$ is of degree d_i in the indeterminate p_i and contains every exponent of p_i from 1 to d_i, f is said to be "full" in p_i. Otherwise, f is "scarce in p_i for $i_{\alpha_1}, i_{\alpha_2}, \ldots, i_{\alpha_\beta}$" where $\beta =$ the total number of absent exponents of p_i, $1 \leqslant \beta \leqslant d_i$ and i_{α_j}, $j = 1, 2, \ldots, \beta$ are the absent exponents. If $\beta = d_i$, f is simply said to be "scarce" in p_i. The same definitions can also be used for $[f]$.

The 1-reducibility conditions for $[f]$ is given by the following lemma:

Lemma 5

Let $f(p_1, \ldots, p_n) = \sum_{j=0}^{d_n} f_j(p_1, \ldots, p_{n-1}) p_n^j$. The form $[f]$ will be 1-reducible if and only if any one of the following conditions hold: (i) $[f_j]$ full in p_m, $m = 1, 2, \ldots, n-1$ for a particular $j \Rightarrow [f_j]$ is full in p_m, $m = 1, 2, \ldots, n-1$, $\forall j$; (ii) $[f_j]$ scarce in p_m for $m_{\alpha_1}, m_{\alpha_2}, \ldots, m_{\alpha_\beta} \Rightarrow [f_j]$ is scarce in p_m for $m_{\alpha_1}, m_{\alpha_2}, \ldots, m_{\alpha_\beta}, \forall j$.

Proof: *Necessity*

Given $[f]$ is 1-factorable. Write

$$[f] = \prod_{i=1}^{n} [f_i(p_i)] = [f_n(p_n)] \prod_{i=1}^{n-1} [f_i(p_i)] = \bigcup_{\forall \beta_n \in [f_n(p_n)]} \beta_n \prod_{i=1}^{n-1} [f_i(p_i)].$$

Evidently, if $\prod_{i=1}^{n-1} [f_j]$ is scarce or full in an indeterminate for a particular β_n, it has to be so for all $\beta_n \in [f_n(p_n)]$.

Sufficiency

The hypothesis for sufficiency implies that $\prod_{i=1}^{n-1} [f_i(p_i)]$ is same $\forall \beta_n \in [f_n(p_n)]$. The rest is obvious. For 1-reducible f, obviously,

$$\operatorname{card} M = m = 1.$$

Assertion

A 1-reducible multivariable polynomial is uniquely 1-reducible so fas as the locations of the roots of individual $f_i(p_i) = 0$ are concerned, even when f is not monic.

As before, the coefficients of 1-factors are obtained by inspection.

Example IV

Let

$$\begin{aligned} f(p_1, p_2, p_3) = {} & 6p_1^2 p_2^3 p_3^2 + 14\sqrt{(3)}\, p_2^3 p_3^2 + 9p_1^2 p_2 p_3^2 + 21\sqrt{(3)}\, p_2 p_3^2 + 3p_1^2 p_3^2 \\ & + 7\sqrt{(3)}\, p_3^2 + 2\sqrt{(3)}\, p_1^2 p_2^3 p_3 + 14 p_2^3 p_3 + 3\sqrt{(3)}\, p_1^2 p_2 p_3 + 21 p_2 p_3 \\ & + \sqrt{(3)}\, p_1^2 p_3 + 7p_3 + 2\sqrt{(3)}\, p_1^2 p_2^3 + 14 p_2^3 + 3\sqrt{(3)}\, p_1^2 p_2 + 21 p_2 \\ & + \sqrt{(3)}\, p_1^2 + 7. \end{aligned}$$

Note that $[f] = \{p_3^2, p_3, 1\}\{p_1^2 p_2^3, p_2^3, p_1^2 p_2, p_2, p_1^2, 1\}$. Again

$$\{p_1^2 p_2^3, p_2^3, p_1^2 p_2, p_2, p_1^2, 1\} = \{p_1^2, 1\}\{p_2^3, p_2, 1\}.$$

Though $[f]$ is scarce in both p_1 and p_2, the conditions of Lemma 5 are satisfied for 1-reducibility.

The polynomial f is not monic. This f is converted to monic polynomial f_6, using the coefficient 6 of the only element of M_f as the scale factor. It is easily checked that

$$f = 6\left(p_1^2 + \frac{7}{\sqrt{3}}\right)\left(p_2^3 + \tfrac{3}{2} p_2 + \tfrac{1}{2}\right)\left(p_3^2 + \frac{1}{\sqrt{3}} p_3 + \frac{1}{\sqrt{3}}\right).$$

Before concluding this section, another method for testing the 1-reducibility determination of 1-factor of f will be discussed.

Let $f(p_1, p_2, ..., p_n)$ be a multivariable polynomial:

$$f(p_1, p_2, ..., p_n) = \sum_{i=0}^{d_j} a_{ij} p_j^{dj-i},$$

where d_j is the degree of f in the variable p_i, and a_{0j} is not identically zero. Then the following assertion is made:

Assertion

A necessary and sufficient condition for a polynomial

$$g(p_1, p_2, ..., p_{j-1}, p_{j+1}, ..., p_n),$$

independent of the variable p_i, to be a factor of $f(p_i, ..., p_n)$ in (2) is that it be a factor of all a_{ij}'s $0 \leqslant i \leqslant d_j$.

The proof of the above assertion can be given along the lines indicated by Bôcher (**6**) for two or three variable cases. Then the following theorem applies.

Theorem I

A multivariable polynomial, $f(p_1, ..., p_n)$ with all coefficients of like sign is 1-reducible if and only if the polynomials $f(1, p_2, p_3, ..., p_n)$, $f(1, 1, p_3, ..., p_j, ..., p_n)$, $f(1, 1, 1, p_4, ..., p_j, ..., p_n)$, ..., $f(1, 1, 1, ..., p_j, 1, ..., 1)$ are factors of $f(p_1, p_2, ..., p_n)$.

Proof: The proof will be given for any j, arbitrarily chosen to be $j = n$. The necessity is obvious. The sufficiency is proved by using successively the foregoing assertion in the manner indicated next. If $f(1, p_2, p_3, ..., p_j, ..., p_n)$ is a factor it directly follows using the foregoing assertion that

$$f(p_1, p_2, ..., p_n) = f_1(p_1) f(1, p_2, ..., p_j, ..., p_n),$$

where $f_1(p_1)$ is a 1-factor. Again if $f(1, 1, p_3, ..., p_j, ..., p_n)$ is a factor of $f(p_1, ..., p_n)$, it must be a factor of $f(1, p_2, ..., p_n)$. Using the foregoing assertion once again,

$$f(p_1, p_2, ..., p_n) = f_1(p_1) f_2(p_2) f(1, 1, p_3, ..., p_j, ..., p_n).$$

The procedure is repeated till finally all 1-factors are obtained.

The results of the theorem can be extended to the case when not all coefficients are of like sign, after recognizing that in this case, some of the polynomials, $f(1, p_2, p_3, ..., p_j, ..., p_n)$, $f(1, 1, p_3, ..., p_j, ..., p_n)$, etc. could be identically equal to zero.

Example V

Let

$$f(p_1, p_2, p_3) = p_1^2 p_2^2 p_3 + p_1^2 p_2 p_3 + p_1^2 p_2^2 + p_1^2 p_3 + p_2^2 p_3 + p_1^2 p_2 + p_1^2 + p_2^2 + p_2 p_3 + p_2 + p_3 + 1.$$

It is required to determine whether $f(p_1, p_2, p_3)$ can be 1-factored. Using Theorem I and a result to determine whether two multivariable polynomials are relatively prime (**7**), it is seen that, for $j = 2$,

$$f(1, p_2, p_3) = 2(p_2^2 p_3 + p_2^2 + p_2 p_3 + p_2 + p_3 + 1)$$

and $f(1, p_2, 1) = 4(p_2^2 + p_2 + 1)$ are factors of $f(p_1, p_2, p_3)$, consistent with the fact that, indeed, $f(p_1, p_2, p_3)$ can be factorized in the form

$$f(p_1, p_2, p_3) = (p_1^2 + 1)(p_2^2 + p_2 + 1)(p_3 + 1).$$

III. Sum Separability

No algorithm is available to implement the synthesis of multivariable positive real functions and matrices. Koga (**8**) considers the existence problem based on a matrix factorization technique but fails to advance a procedure by which the factorization can be carried out, in general, and lead to a network, characterized by the prescribed open circuit impedance matrix. In the absence of any general constructive synthesis technique, specialized approaches exploiting known results in the single variable case, are helpful. For example, a necessary and sufficient condition for a multivariable reactance function to be expressed as a sum of single variable reactance functions, was obtained (**4**). Given below is a decomposition theorem for multivariable positive real functions that is likely to find use in multivariable synthesis problems.

Theorem II

A multivariable p.r.f., $F(p_1, p_2, \ldots, p_n)$ can be decomposed as

$$F(p_1, p_2, \ldots, p_n) = F_1(p_1) + F_2(p_2) + \ldots + F_n(p_n) + k, \quad k \geqslant 0,$$

where $F_i(p_i)$ are each minimum p.r.f.'s, for $i = 1, 2, \ldots, n$ (a p.r.f. $F(p)$ is minimum if and only if there exists at least an ω, such that the real part of $F(j\omega) = 0$) if and only if with $F_i(p_i) = F(1, 1, \ldots, 1, p_i, 1, \ldots, 1) - k_i$ and $k_i = \min \quad \operatorname{Re} F(1.1, \ldots, 1, j\omega_i, 1, \ldots, 1), \; -\infty \leqslant \omega_i \leqslant \infty, \quad i = 1, 2, \ldots, n$ (where "min" and "Re" denote "minimum" and "real part of", respectively; also $F_i(p_i)$'s are readily seen to be minimum p.r.f.'s).

(i) $F^{(1)}(p_2, p_3, \ldots, p_n) = F(p_1, p_2, \ldots, p_n) - F_1(p_1)$, is a $(n-1)$ variable function

$F^{(2)}(p_3, p_4, \ldots, p_n) = F^{(1)}(p_2, p_3, \ldots, p_n) - F_2(p_2)$, is a $(n-2)$ variable function

. . . . . .

$F^{(n-1)}(p_n) = F^{(n-2)}(p_{n-1}, p_n) - F_{n-1}(p_{n-1})$, is a *single variable positive real function*

and

(ii) $k = \sum_{i=1}^{n} k_i - (n-1) F(1, 1, \ldots, 1)$.

Proof: Only if part:
Suppose,

$$F(p_1, p_2, \ldots, p_n) = \sum_{i=1}^{n} F_i(p_i) + k, \quad k \geqslant 0$$

and the $F_i(p_i)$'s are minimum p.r.f.'s for $i = 1, 2, \ldots, n$. Then, the fact that (i) is satisfied follows directly. Also,

$$F_i(1) = F(1, 1, \ldots, 1) - k_i, \quad i = 1, 2, \ldots, n$$

or

$$\sum_{i=1}^{n} F_i(1) = nF(1, 1, \ldots, 1) - \sum_{i=1}^{n} k_i$$

or

$$k = \sum_{i=1}^{n} k_i - (n-1) F(1, 1, \ldots, 1),$$

satisfying (ii).

If part:

As noted, because $F(p_1, p_2, \ldots, p_n)$ is multivariable p.r.f., $F_i(p_i)$ is a single variable p.r.f., for $i = 1, 2, \ldots, n$. Therefore, if $F^{(n-1)}(p_n)$ is a p.r.f.,

$$F^{(n-2)}(p_{n-1}, p_n) = F^{(n-1)}(p_n) + F_{n-1}(p_{n-1})$$

$$F^{(n-3)}(p_{n-2}, p_{n-1}, p_n) = F^{(n-2)}(p_{n-1}, p_n) + F_{n-2}(p_{n-2})$$

$$\cdot \quad \cdot \quad \cdot \quad \cdot \quad \cdot \quad \cdot$$

$$F^{(1)}(p_2, p_3, \ldots, p_n) = F^{(2)}(p_3, p_4, \ldots, p_n) + F_2(p_2)$$

are each p.r.f.'s. Therefore, $F^{(1)}(p_2, p_3, \ldots, p_n)$ is a $(n-1)$ variable p.r.f., obtained after extraction of $F_1(p_1)$ from $F(p_1, p_2, \ldots, p_n)$. Similarly, $F^{(2)}(p_3, p_4, \ldots, p_n)$ is a $(n-2)$ variable p.r.f., obtained after extraction of $F_2(p_2)$ from $F^{(1)}(p_2, p_3, \ldots, p_n)$ and so on. Finally, $F^{(n-1)}(p_n)$ is a single variable p.r.f., obtained after extraction of $F_{n-1}(p_{n-1})$ from $F^{(n-1)}(p_{n-1}, p_n)$, leaving $F^{(n-1)}(p_n)$ to be decomposed as

$$F^{(n-1)}(p_n) = F_n(p_n) + k, \quad k \geqslant 0,$$

where $F_n(p_n)$ is a minimum p.r.f. Note that condition (ii) serves as a check after the k_i's are calculated.

The above theorem is, in principle, simple to apply, and the algorithm suggested provides a simple synthesis technique at least for a class of multivariable p.r.f.'s. Note that the theorem only involves *one* test for *single* variable positive realness, besides the routine extraction from *each* of *n single variable p.r.f.'s* the maximum possible nonnegative constant without making the remaining function nonpositive real.

IV. Application of the Multiplicative Reducibility Criteria

The technique of multivariable multiplicative reducibility, developed in Section II, has an immediate application in the synthesis of certain types of multivariable network functions. For the purpose of application it is necessary to extend the definition of reducibility to the multivariable network functions which may be expressed as a ratio of two nontrivial multivariable polynomials.

Definition 6: *Reducibility and* 1-*reducibility of multivariable network functions*

Let $Q(p_1,\ldots,p_n) = N(p_1,\ldots,p_n)/D(p_1,\ldots,p_n)$ be a multivariable network function where each of $N(p_1,\ldots,p_n)$ and $D(p_1,\ldots,p_n)$ has total degree > 1. The network function Q will be called "upper (lower)-semirreducible (semi-1-reducible)" if and only if $N(p_1,\ldots,p_n)$ [resp. $D(p_1,\ldots,p_n)$] is reducible (1-reducible) and $D(p_1,\ldots,p_n)$ [resp. $N(p_1,\ldots,p_n)$] is not.

The function Q will be called "reducible (1-reducible)" if and only if, both N and D are reducible (1-reducible).

It is evident that if a multivariable driving-point function is semi-reducible or if a transfer function is reducible, a simplification and reduction of computational complexities can be expected. This is particularly so when one replaces the term "reducible" by "1-reducible" in the last sentence. As an illustration, two examples, the first dealing with 1-reducible transfer functions and the second with a special type of lower-semi-1-reducible driving-point functions, are presented. Also a third example will be used to illustrate the algorithm in Theorem II.

a. *Synthesis of* 1-*reducible transfer functions.* Evidently, if a transfer function is 1-reducible, the synthesis problem trivially reduces to that of a single-variable transfer functions. As an illustration, consider the synthesis of 1-reducible transfer functions in terms of cascaded sections of constant-resistance lattice. The lattice section is chosen for the simplicity of illustration only and, in fact, one can use any other constant-resistance section instead of a lattice. Extending the conditions of single-variable realizability, for a 1-reducible transfer function $z_{12}(p_1,\ldots,p_n)$ in order that a cascaded constant resistance realization exists in terms of lattice sections terminated in the prescribed resistance, each 1-factor of z_{12} must have a magnitude $\leqslant 1$ and should have no poles in the right-half plane or on the imaginary axis (**9**). Consider the following 3-variable transfer function:

$$z_{12}(p_1,p_2,p_3) = \frac{p_1p_2^2 - p_1p_2 + p_1}{\begin{array}{r}(p_1p_2^2p_3 + p_1p_2^2 + p_3p_2^2 + p_1p_2p_3 + p_2^2 + p_1p_2 + p_2p_3\\ +p_1p_3 + p_1 + p_2 + p_3 + 1)\end{array}}.$$

For simplicity, z_{12} is selected such that both the numerator and the denominator are monic; however, the same technique is applicable in the non-monic cases as well.

Observe that z_{12} is 1-reducible

$$z_{12}(p_1,p_2,p_3) = \frac{p_1}{p_1+1}\,\frac{p_2^2-p_2+1}{p_2^2+p_2+1}\,\frac{1}{p_3+1}.$$

All three 1-factors are passive lattice-realizable network functions. Synthesizing z_{12} as the cascade of three constant-resistance lattice sections terminated in 1Ω, the network configuration of Fig. 1 is obtained.

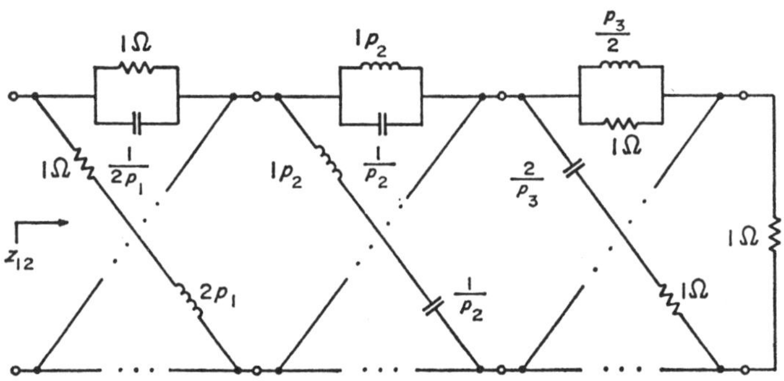

FIG. 1. Reducible transfer function synthesis, product separable.

b. *Synthesis of a special type of lower-semi-1-reducible driving-point function.* In this case, consider a very special type of a multivariable driving-point functions which can be expressed as

$$I(p_1, \ldots, p_n) = \frac{N(p_1, \ldots, p_n)}{D(p_1, \ldots, p_n)} = \sum_{i=1}^{n} \frac{A_i}{(p_i + \alpha_i)},$$

where α_i's and A_i's *are* real constants.

The necessary and sufficient condition that a multivariable function $I(p_1, \ldots, p_n) = N(p_1, \ldots, p_n)/D(p_1, \ldots, p_n)$ can be expressed as above is that I is lower-semi-1-reducible, the polynomials $N(p_1, \ldots, p_n)|_{p_i=-\alpha_i}$, $i = 1, 2, \ldots, n$ are 1-reducible with the 1-factors being constant multiples of the corresponding 1-factors of $D(p_1, \ldots, p_n)$.

A simple driving-point impedance function of above type is given below:

$$F(p_1, p_2, p_3) = \frac{3p_1p_2 + 2p_1p_3 + p_2p_3 + 8p_1 + 4p_2 + 4p_3 + 10}{p_1p_2p_3 + 2p_1p_3 + p_2p_3 + p_1p_2 + 2p_1 + p_2 + 2p_3 + 2}$$

$$= \frac{3p_1p_2 + 2p_1p_3 + p_2p_3 + 8p_1 + 4p_2 + 4p_3 + 10}{(p_1+1)(p_2+2)(p_3+1)}$$

$$= \frac{1}{p_1+1} + \frac{2}{p_2+2} + \frac{3}{p_3+1}.$$

The trivial realization follows immediately as in Fig. 2. As the above positive real function is sum separable, Theorem II must be satisfied. It is seen that $k_1 = \min \operatorname{Re} F(j\omega_1, 1, 1) = \frac{13}{6}$, $k_2 = \min \operatorname{Re} F(1, j\omega_2, 1) = 2$ and

$$k_3 = \min \operatorname{Re} F(1, 1, j\omega_3) = \tfrac{7}{6}.$$

Whence

$$k = k_1 + k_2 + k_3 - (3-1)F(1, 1, 1) = 0.$$

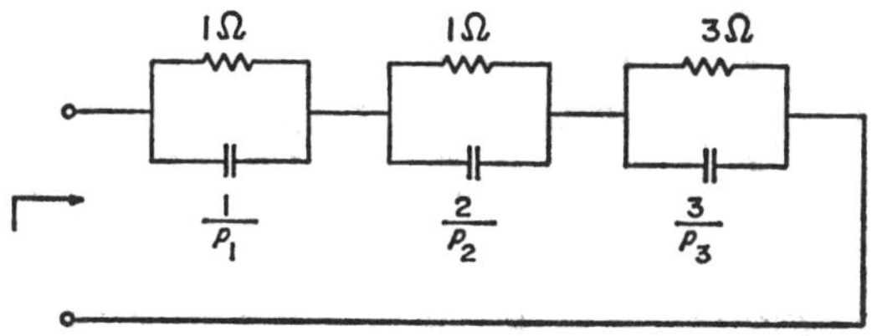

FIG. 2. Sum separable synthesis.

Also,

$$F_1(p_1) = F(p_1, 1, 1) - k_1 = \frac{1}{p_1+1},$$

$$F_2(p_2) = F(1, p_2, 1) - k_2 = \frac{2}{p_2+2},$$

$$F_3(p_3) = F(1, 1, p_3) - k_3 = \frac{3}{p_3+1}.$$

Also, $F^{(1)}(p_2, p_3)$, $F^{(2)}(p_3)$ are positive real as expected, and the theorem is indeed satisfied.

The next example is given specifically to illustrate step-by-step the algorithm contained in Theorem II.

Example VI

Consider the feasibility of decomposition of the three variable p.r.f.,

$$F(p_1, p_2, p_3) = \frac{5p_1p_2p_3 + 2p_1p_3 + 14p_2p_3 + 3p_1p_2 + 6p_3 + 12p_2 + p_1 + 4}{3p_1p_2p_3 + p_1p_3 + 72p_2p_3 + 4p_3},$$

$$F(p_1, 1, 1) = \frac{11p_1+36}{4p_1+16}, \quad k_1 = \min \operatorname{Re} F(j\omega_1, 1, 1) = \tfrac{9}{4},$$

$$F_1(p_1) = F(p_1, 1, 1) - k_1 = \frac{\frac{1}{2}p_1}{p_1+4},$$

$$F^{(1)}(p_2, p_3) = F(p_1, p_2, p_3) - F_1(p_1) = \frac{7p_2p_3 + 3p_3 + 6p_2 + 2}{6p_2p_3 + 2p_3},$$

$$F(1, p_2, 1) = \frac{34p_2+13}{15p_2+5}, -k_2 = \min \operatorname{Re} F(1, j\omega_2, 1) = \tfrac{34}{15},$$

$$F_2(p_2) = F(1, p_2, 1) - k_2 = \frac{\frac{1}{3}}{3p_2+1},$$

$$F^{(2)}(p_3) = F^{(1)}(p_2, p_3) - F_2(p_2) = \frac{6+7p_3}{6p_3},$$

which is obviously a p.r.f. $F^{(2)}(p_3)$ can be further decomposed as

$$F^{(2)}(p_3) = F_3(p_3) + k = \frac{1}{p_3} + \frac{7}{6}.$$

It is seen that

$$k = \frac{7}{6} = \sum_{i=1}^{3} k_i - (3-1)F(1,1,1).$$

The relevant decomposition is

$$F(p_1, p_2, p_3) = \frac{\frac{1}{2}p_1}{p_1+4} + \frac{\frac{1}{3}}{3p_2+1} + \frac{1}{p_3} + \frac{7}{6}.$$

V. Conclusion

First, the necessary and sufficient conditions for p-reducibility of forms are considered. This is followed by a discussion on actual construction of the p-reducible factors, whenever possible. An attempt has been made to supply necessary rigor in presentation without destroying clarity or obscuring the feasible engineering applications of the result. Examples are used to elucidate the discussions. The special case of 1-reducibility is considered from two different angles, especially because of applications of this result in separable filter design. A set of necessary and sufficient conditions for decomposing a multivariable positive real function into a sum of single variable positive real functions, plus a nonnegative constant, is given. Some examples dealing with scopes for applications of the result are included.

References

(1) T. S. Huang, W. F. Schreiber and O. J. Tretiak, "Image processing", *Proc. IEEE*, Vol. 59, pp. 1586–1609, Nov. 1971.

(2) J. L. Shanks, S. Treitel and J. H. Justice, "Stability and synthesis of two-dimensional recursive filters", *IEEE Trans. Audio Electroacoust.*, Vol. AU-20, pp. 115–128, June 1972.

(3) N. K. Bose, K. Zaki and R. W. Newcomb, "A multivariable rounded reality criterion", *J. Franklin Inst.*, Vol. 297, No. 6, pp. 479–484, June 1974.

(4) A. M. Soliman and N. K. Bose, "A decomposition theorem for multivariable reactance functions", *Proc. IEEE*, Vol. 59, pp. 309–311, Feb. 1971.

(5) S. Lang, "Algebra", Addison-Wesley, Reading, Mass., 1965.

(6) M. Bôcher, "Introduction to Higher Algebra", pp. 203–217, Macmillan, New York, 1907.

(7) N. K. Bose, "A criterion to determine if two multivariable polynomials are relatively prime", *Proc. IEEE*, Vol. 60, pp. 134–135, Jan. 1972.

(8) T. Koga, "Synthesis of finite passive n-ports with prescribed positive real matrices of several variables", *IEEE Trans. Circuit Theory*, Vol. CT-15, pp. 2–24, March 1968.

(9) E. S. Kuh and D. O. Pederson, "Principles of Circuit Synthesis", Chap. 12, McGraw-Hill, New York, 1959.

39

Reprinted from *IEEE Trans. Circuits Syst.* **CAS-22**(3):177–184 (1975)

Realizations of Two-Dimensional Recursive Digital Filters

S. K. MITRA, FELLOW, IEEE, A. D. SAGAR, AND N. A. PENDERGRASS, STUDENT MEMBER, IEEE

***Abstract*—Algorithms are derived for realization of two-dimensional recursive digital filters using direct forms and continued fraction expansions. Included is a test for continued fraction expansion which at the same time determines the expansion coefficients. Transposed structures are also discussed, and a comparison is made of the various realizations.**

INTRODUCTION

TWO-DIMENSIONAL linear digital filtering can be described in terms of two-dimensional z-transforms. Consider the signal to be a function of two orthogonal variables, say τ_1 and τ_2, which span a two-dimensional space. Let us introduce a uniform mesh on this space so that the coordinates of any knot of the mesh will be (mT_1,nT_2) where T_1 and T_2 are the mesh spacings along the τ_1 and τ_2 axis, respectively. Any knot can be identified by the coordinates (m,n), and the signal at that knot can be denoted by S_{mn}. Thus the discretized signal can be written as a two-dimensional array

$$S = \begin{bmatrix} & \vdots & \vdots & \vdots & \\ \cdots & S_{m-1,n-1} & S_{m-1,n} & S_{m-1,n+1} & \cdots \\ \cdots & S_{m,n-1} & S_{m,n} & S_{m,n+1} & \cdots \\ \cdots & S_{m+1,n-1} & S_{m+1,n} & S_{m+1,n+1} & \cdots \\ & \vdots & \vdots & \vdots & \end{bmatrix}$$

The two-dimensional z-transform of this array is given by

$$S(z_1,z_2) = \sum_{m=-\infty}^{\infty} \sum_{n=-\infty}^{\infty} S_{mn} z_1^{m-1} z_2^{n-1}.$$

We can write the transfer function of a two-dimensional recursive filter as a ratio of two-dimensional polynomials in the following manner [1]:

$$H(z_1,z_2) = \frac{A(z_1,z_2)}{B(z_1,z_2)} = \frac{\sum_{i=1}^{M_a} \sum_{j=1}^{N_a} a_{ij} z_1^{i-1} z_2^{j-1}}{\sum_{k=1}^{M_b} \sum_{l=1}^{N_b} b_{kl} z_1^{k-1} z_2^{l-1}}. \quad (1)$$

For an input signal $X(z_1,z_2)$, the output of the filter is given by

$$Y(z_1,z_2) = H(z_1,z_2)X(z_1,z_2).$$

Manuscript received April 1, 1974; revised June 25, 1974. This work was supported in part by the National Science Foundation under Grant GK 38205.
S. K. Mitra is with the Department of Electrical Engineering, University of California, Davis, Calif. 95616.
A. D. Sagar is with the Department of Electrical Engineering, Rice University, Houston, Tex. 77001.
N. A. Pendergrass is with the Department of Electrical Engineering and Computer Sciences, and the Electronics Research Laboratory, University of California, Berkeley, Calif. 94720.

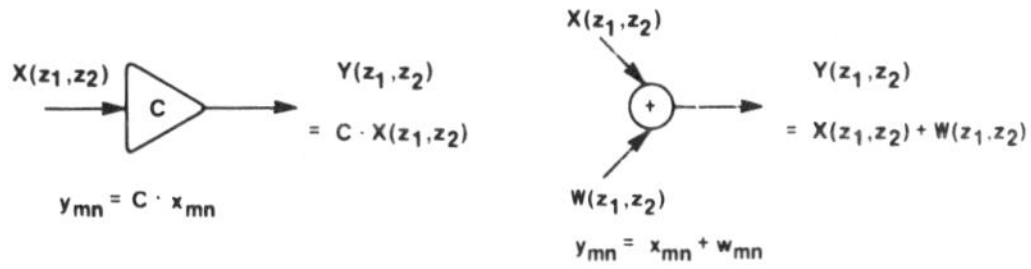

Fig. 1. Block elements to multiply input signal by a constant and to add input signals.

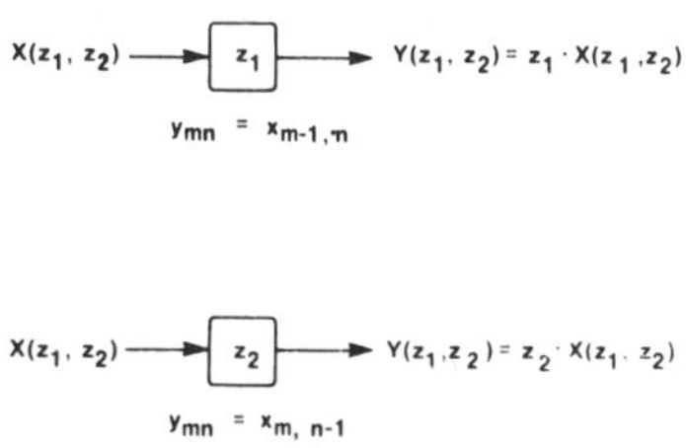

Fig. 2. z_1 and z_2 block elements.

The question of how to realize (1) is an important one. Different realizations of the same transfer function can have widely diverse performance in terms of roundoff errors and coefficient sensitivity. Furthermore, some realizations may offer significant advantages over others by requiring less memory or fewer multipliers. Up to this time, however, only one method of realization has been advanced [1] for two-dimensional recursive digital filters. In this present paper, several alternative configurations will be proposed and discussed.

ELEMENTARY BLOCK ELEMENTS

In order to represent alternative realizations clearly, it is desirable to show them in block diagram forms which are made up of basic realizable elements.

The block elements used to represent two-dimensional digital filters are very similar to those for one-dimension. The elements that multiply by a constant or add input signals are the same as for one-dimension (Fig. 1). In two-dimensions, analogous to the delay element in one-dimension, there are elements with transfer functions z_1 (Fig. 2(a)) and z_2 (Fig. 2(b)). As shown in the figure, the signal value occurring at knot (m,n) of the input of a z_1 block will occur at knot $(m + 1, n)$ of the block output. Similarly, the value at knot (m,n) of the input to a z_2 block will occur at knot $(m, n + 1)$ of the block output.

Direct Realizations

A variety of algorithms for realizing two-dimensional filters will now be introduced using direct forms or continued fraction expansions. In the development of these structures, it has been assumed that it is not possible to perform z_1^{-1} or z_2^{-1} operations. This is the reason that block elements of these types were not introduced previously. This assumption is analogous to requiring causality in the one-dimensional case but need not be a condition for realization in two-dimensions.

Let us assume that a two-dimensional filter is described by (1). In addition, let us assume for convenience that

$$b_{11} = 1.$$

A direct realization follows from the algorithm given by Shanks *et al.* [1]. There it is shown that (1) can be written as

$$Y(z_1,z_2) = \left[\sum_{i=1}^{M_a}\sum_{j=1}^{N_a} a_{ij} z_1^{i-1} z_2^{k-1}\right] X(z_1,z_2) - \left[\sum_{\substack{k=1 \\ k\cdot l\neq 1}}^{M_b}\sum_{l=1}^{N_b} b_{kl} z_1^{k-1} z_2^{l-1}\right] Y(z_1,z_2). \quad (2)$$

Another direct form can be derived as follows; let

$$W(z_1,z_2) = \frac{1}{B(z_1,z_2)} X(z_1,z_2).$$

Then,

$$B(z_1,z_2)W(z_1,z_2) = X(z_1,z_2).$$

Hence,

$$\left[1 + \sum_{\substack{k,l=1 \\ k\cdot l\neq 1}}^{M_b}\sum^{N_b} b_{kl} z_1^{k-1} z_2^{l-1}\right] W(z_1,z_2) = X(z_1,z_2)$$

$$W(z_1,z_2) = X(z_1,z_2) - \sum_{\substack{k,l=1 \\ k\cdot l\neq 1}}^{M_b}\sum^{N_b} b_{kl} z_1^{k-1} z_2^{l-1} W(z_1,z_2).$$

By comparing coefficients, we get

$$w_{pq} = x_{pq} - \sum_{\substack{k,l=1 \\ k\cdot l\neq 1}}^{M_b}\sum^{N_b} b_{kl} w_{p-k+1,q-l+1}. \quad (3)$$

Now

$$Y(z_1,z_2) = A(z_1,z_2)W(z_1,z_2)$$

which gives

$$y_{mn} = \sum_{i=1}^{M_a}\sum_{j=1}^{N_a} a_{ij} w_{m-i+1,n-j+1} = \sum_{i=1}^{M_a}\sum_{j=1}^{N_a} a_{ij} \cdot \left[x_{m-i+1,n-j+1} - \sum_{\substack{k,l=1 \\ k\cdot l\neq 1}}^{M_b}\sum^{N_b} b_{kl} w_{m-i-k+2,n-j-l+2}\right]. \quad (4)$$

From the equations of (3) and (4), we can obtain a configuration realizing the filter. This is simply illustrated by considering an example. Let

$$H(z_1,z_2) = \frac{a_{11} + a_{12}z_2 + a_{21}z_1 + a_{22}z_1z_2}{1 + b_{12}z_2 + b_{13}z_2^2 + b_{21}z_1 + b_{22}z_1z_2 + b_{23}z_1z_2^2}. \quad (5)$$

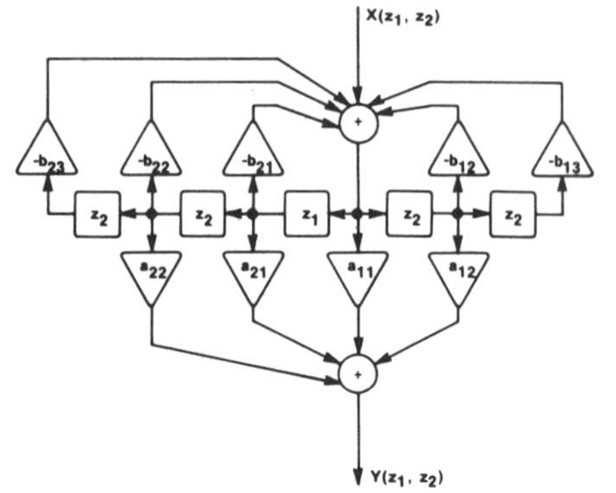

Fig. 3. Direct form using (3) and (4) with $H(z_1,z_2)$ given by (5).

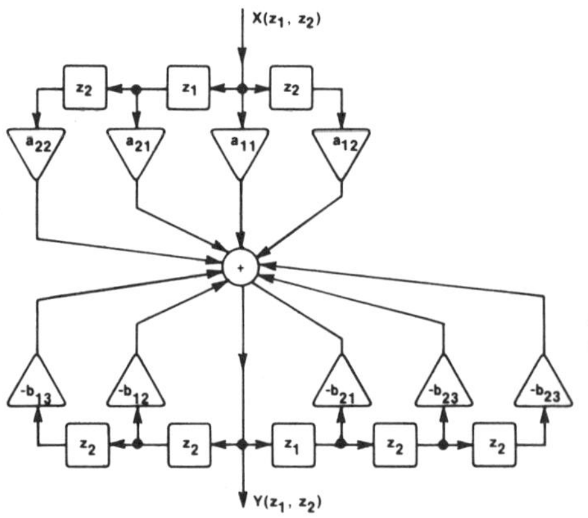

Fig. 4. Direct form using (2) with $H(z_1,z_2)$ given by (5).

Fig. 3 shows the realization using (3) and (4). Fig. 4 shows the realization using (2). Fig. 5 shows the realization of (3) and (4) in a way which graphically illustrates the computation and storage required. Fig. 6 shows the realization of (2) in this same form.

A comparison of the figures immediately shows that the number of z_1 and z_2 blocks does not represent the memory required for computation of the output values as would be the case in one dimension. For example, consider the case of Fig. 5 with input and output arrays of N rows and M columns. If the output is calculated row by row, the input must be read in row by row and elements of the preceding row of w's must be stored in order to calculate each new w value and hence each new output value. After the w values in the preceding row have been used for the last time, they

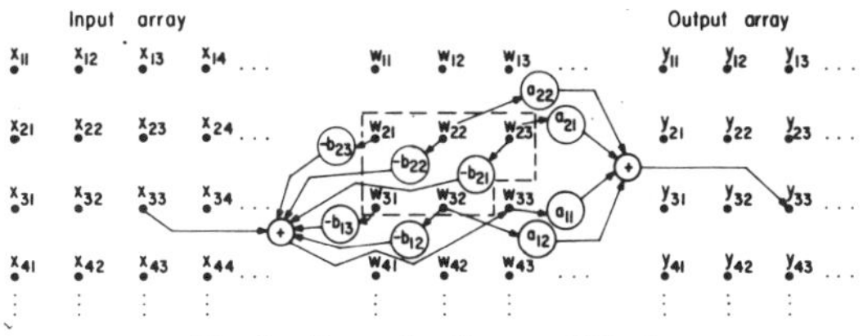

Fig. 5. Alternative diagram of Fig. 3.

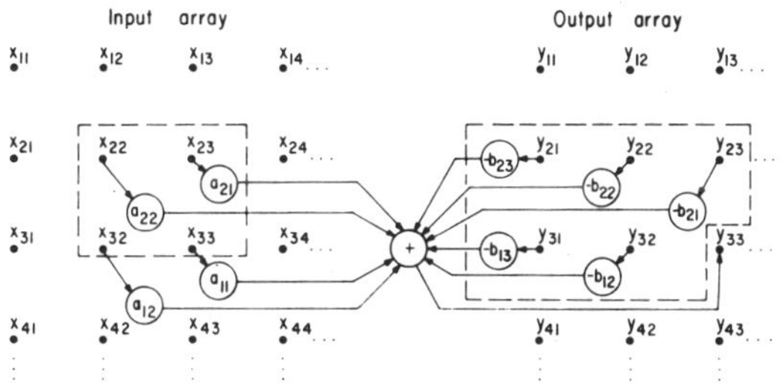

Fig. 6. Alternative diagram of Fig. 4.

may be replaced by stored values from the new row. Therefore, the memory required will be $M + 2$ elements in Fig. 5. Similarly, if the output is computed column by column then the memory required is $2N + 1$ elements. Other choices of computation sequence could be made as long as the necessary values have been calculated before being required for each new computation. Memory requirements will always depend, however, on the output array dimensions and the computation sequence in addition to the interconnection of the z_1 and z_2 blocks.

Continued Fraction Expansion Realizations

Depending on the particular coefficients in (1), it may be possible to express the transfer function as one of many continued fraction expansion forms and thereby to arrive at a new structure. A check for the existence of these expansions will be discussed in the next section.

The expansions that follow can all be altered easily to arrive at a second set of expansions which have similar form and block diagram. The additional expansion types are obtained by simply interchanging the roles of z_1 and z_2 in each expansion. The corresponding block diagrams result from a similar interchange.

Additional expansions are also possible since hybrids of different types could be introduced. It is not the purpose of this paper, however, to provide an exhaustive list of all possible configurations. It is rather to suggest some approaches.

In the expansions that follow, it must be noted that coefficients in a given expansion may or may not be the same as those in another.

Type IA

If we can expand the transfer function as

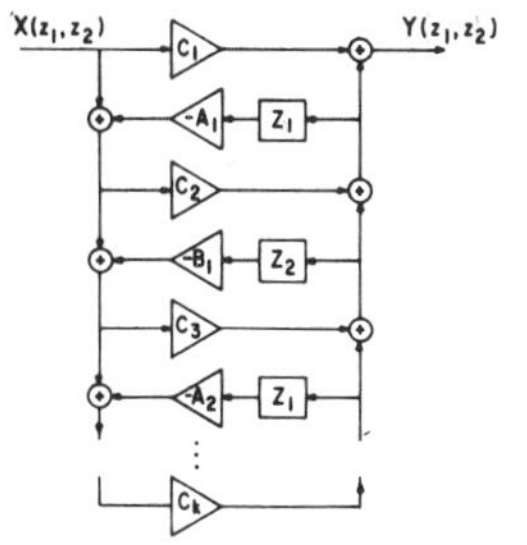

Fig. 7. Type IA realization.

$$H(z_1,z_2) = C_1 + \cfrac{1}{A_1 z_1 + \cfrac{1}{C_2 + \cfrac{1}{B_1 z_2 + \cfrac{1}{\ddots}}}}$$

then we can obtain a structure (Fig. 7) realizable under the constraint of our assumption that the expansion terminates in $1/C_K$ where K is any integer.

In order to show how Fig. 7 would be implemented, consider the following example:

$$H(z_1,z_2) = C_1 + \cfrac{1}{A_1 z_1 + \cfrac{1}{C_2 + \cfrac{1}{B_1 z_2 + \cfrac{1}{C_3 + \cfrac{1}{A_2 z_1 + \cfrac{1}{C_4}}}}}} \tag{6}$$

The implementation of this transfer function is illustrated in Fig. 8.

Type IB

If we can expand $H(z_1^{-1},z_2^{-1})$ as in Type IA, then we get after rearrangement

$$H(z_1,z_2) = C_1 + \cfrac{1}{A_1 z_1^{-1} + \cfrac{1}{C_2 + \cfrac{1}{B_1 z_2^{-1} + \cfrac{1}{\ddots}}}}.$$

If $C_2,C_3,\cdots,A_1,A_2,\cdots,B_1,B_2,\cdots$, are not zero, then we obtain the configuration given by Fig. 9. If any coefficients are zero, the structure can be modified accordingly.

Type IIA

We may be able to expand $H(z_1,z_2)$ as

$$H(z_1,z_2) = C_1 + \cfrac{1}{A_1 z_1 + C_2 + \cfrac{1}{B_1 z_2 + C_3 + \cfrac{1}{\ddots}}}$$

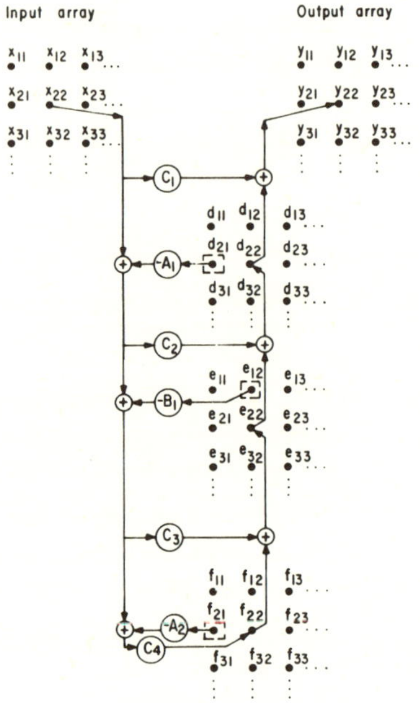

Fig. 8. Alternative diagram of Type IA with $H(z_1,z_2)$ given by (6).

In this case, it is not possible to obtain a realizable configuration given our constraint that we cannot implement z_1^{-1} or z_2^{-1} blocks. We note, however, that in the absence of this constraint, this structure (and others) would be possible.

Type IIB

Another configuration is possible if $H(z_1,z_2)$ can be expanded to become

$$H(z_1,z_2) = C_1 + \cfrac{1}{A_1 z_1^{-1} + C_2 + \cfrac{1}{B_1 z_2^{-1} + C_3 + \cfrac{1}{\ddots}}}.$$

Providing that $A_1, A_2, \cdots, B_1, B_2, \cdots, C_2, C_3, \cdots$, are nonzero, the configuration is given in Fig. 10. Otherwise, the structure may be modified.

Type IIIA

Another realization results if we can expand $1/H(z_1,z_2)$ as a type IA. Then,

$$H(z_1,z_2) = \cfrac{1}{C_1 + \cfrac{1}{A_1 z_1 + \cfrac{1}{C_2 + \cfrac{1}{B_1 z_2 + \cfrac{1}{\ddots}}}}}.$$

For this to yield a structure realizable under our assumption, the expansion must terminate in $1/A_m z_1$ or $1/B_m z_2$

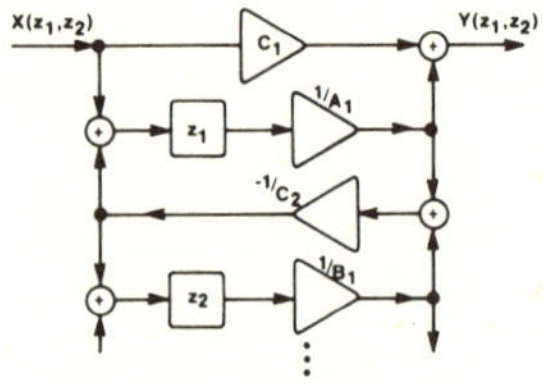

Fig. 9. Type IB realization.

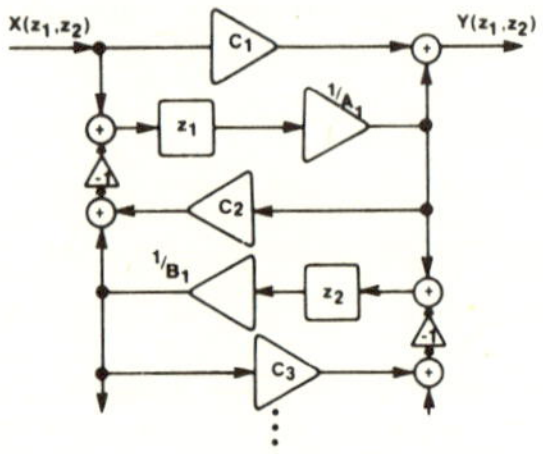

Fig. 10. Type IIB realization.

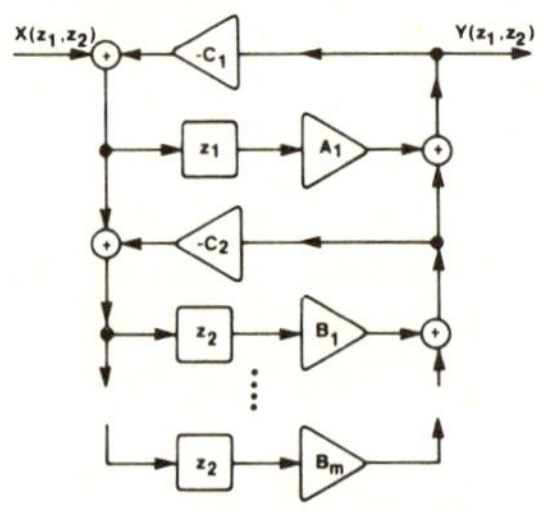

Fig. 11. Type IIIA realization.

where m is any integer. Then, we obtain the block diagram of Fig. 11.

Type IIIB

If we expand $H(z_1,z_2)$ as follows with $C_1 \neq 0$,

$$H(z_1,z_2) = \cfrac{1}{C_1 + \cfrac{1}{A_1 z_1^{-1} + \cfrac{1}{C_2 + \cfrac{1}{B_1 z_2^{-1} + \cfrac{1}{\ddots}}}}}$$

then we can obtain the realizable configuration given in Fig. 12.

Type IVA

Assume we can write $H(z_1,z_2)$ as

$$H(z_1,z_2) = \cfrac{1}{C_1 + \cfrac{1}{A_1 z_1 + C_2 + \cfrac{1}{B_1 z_2 + C_3 + \cfrac{1}{\ddots}}}}.$$

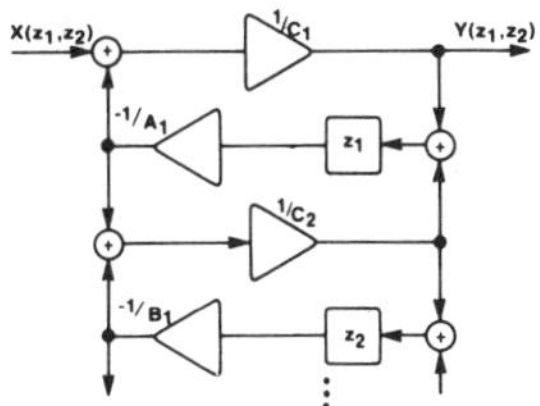

Fig. 12. Type IIIB realization.

In this case, as in type IIA, there is no realizable expansion under our constraint.

Type IVB

If we find that

$$H(z_1,z_2) = \cfrac{1}{C_1 + \cfrac{1}{A_1 z_1^{-1} + C_2 + \cfrac{1}{B_1 z_2^{-1} + C_3 + \cfrac{1}{\ddots}}}}$$

then the configuration in Fig. 13 results if $A_1, A_2, \cdots, B_1, B_2, \cdots, C_1, C_2, \cdots$, are nonzero. If any of these terms are zero, the configuration must be modified.

Existence of Continued Fraction Expansions

A procedure for checking for the existence of any of the continued fraction expansions for a given transfer function could be established in a manner similar to that outlined by Ramachandran and Rao [4]. However, a simpler method can be devised which provides the expansion coefficients while at the same time checking existence of the expansion. This approach will also systematize the computations and provide a structure which gives insight into the expansion process. Furthermore, it allows one to easily check alternative paths of expansion should one path fail.

First consider the one-dimensional problem

$$\frac{a_n z^n + a_{n-1} z^{n-1} + \cdots + a_0}{b_n z^n + b_{n-1} z^{n-1} + \cdots + b_0}.$$

The first step of division will produce a value

$$\frac{a_n}{b_n}$$

with remainder

$$\frac{\left(a_{n-1} - \frac{a_n}{b_n} b_{n-1}\right) z^{n-1} + \left(a_{n-2} - \frac{a_n}{b_n} b_{n-2}\right) z^{n-2} + \cdots + \left(a_0 - \frac{a_n}{b_n} b_0\right)}{b_n z^n + b_{n-1} z^{n-1} + \cdots + b_0}. \tag{7}$$

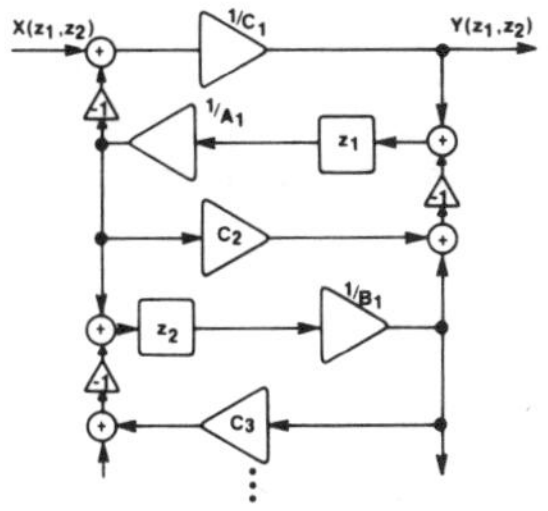

Fig. 13. Type IVB realization.

But we notice that the coefficients in the numerator are exactly the coefficients that would be obtained in generating the third row of a Routh array [7], [9], using

$$\begin{array}{ccccc} a_n & a_{n-1} & \cdots & a_1 & a_0 \\ b_n & b_{n-1} & \cdots & b_1 & b_0 \\ c_{n-1} & c_{n-2} & \cdots & c_0 & \end{array}$$

with

$$c_k = \frac{-\begin{vmatrix} a_n & a_k \\ b_n & b_k \end{vmatrix}}{b_n}; \qquad k = 0,1,\cdots,n-1.$$

Since we are interested in a continued fraction expansion, (7) is inverted in the next step. Hence, we can see immediately that continuing the table will give all of the expansion coefficients as ratios of the terms in the first column as follows:

$$\frac{a_n}{b_n} + \cfrac{1}{\cfrac{b_n}{c_{n-1}} z + \cfrac{1}{\ddots}}.$$

Extension to two-dimensions only requires careful ordering of the numerator and denominator terms. Let us assume that we wish to extract a constant term first followed by a z_1 term. Then, we can set up a Routh array in decreasing powers of z_1 as follows (the array is filled out with no missing terms).

The constant coefficient is given by a_{mn}/b_{mn}. Row c is computed as in the one-dimensional case except that each new term is aligned directly below the elements used in its computation, i.e.,

$$c_{kl} = \frac{-\begin{vmatrix} a_{mn} & a_{kl} \\ b_{mn} & b_{kl} \end{vmatrix}}{b_{mn}}.$$

$z_1^m z_2^n$	$z_1^m z_2^{n-1}$	$\cdots$	$z_1^m z_2^0$	$z_1^{m-1} z_2^n$	$\cdots$	$z_1^{m-1} z_2^0$	$\cdots$	$z_1^0 z_2^n$	$z_1^0 z_2^{n-1}$	$\cdots$	$z_1^0 z_2^0$
a_{mn}	$a_{m,n-1}$	$\cdots$	$a_{m-1,0}$	$a_{m-1,n}$	$\cdots$	$a_{m-1,0}$	$\cdots$	$a_{0,n}$	$a_{0,n-1}$	$\cdots$	a_{00}
b_{mn}	$b_{m,n-1}$	$\cdots$	$b_{m-1,0}$	$b_{m-1,n}$	$\cdots$	$b_{m-1,0}$	$\cdots$	$b_{0,n}$	$b_{0,n-1}$	$\cdots$	b_{00}
0	0		0	$c_{m-1,n}$	$\cdots$	$c_{m-1,0}$	$\cdots$	$c_{0,n}$	$c_{0,n-1}$	$\cdots$	c_{00}

Of course the first term will be zero in row c. The next n terms must also be zero in row c. If they are not, further steps in the expansion will not be able to reduce the degree of resulting terms. This would indicate that this expansion path fails.

Now form row c' by shifting row c to the left $n + 1$ terms. The z_1 coefficient in the expansion is then given by $b_{mn}/c_{m-1,n}$. Row d is then computed in the next row as given. Again the first $n + 1$ terms must be zero for the expansion to continue.

Row e is next computed using rows c and d. First, however, we note that the expansion can be continued in any direction desired at this point. If the next two terms in the expansion are to be a constant followed by a z_1 term, we simply continue the table as given. If the next two terms are to be a constant followed by a z_2 term, then we reorder rows c and d as decreasing powers of z_2 and use these to continue the table.

Because of the connection between the actual division and the table as pointed out previously, special cases that might occur in the table are easily accounted for. Examples that follow will demonstrate the procedure and illustrate a special case to clarify the method.

Example 1

Consider

$$H(z_1,z_2) = \frac{48z_1^2z_2 + 12z_1^2 + 144z_1z_2 + 33z_1 + 81z_2 + 15}{48z_1^2z_2 + 12z_1^2 + 120z_1z_2 + 27z_1 + 27z_2 + 2}.$$

We form the array

Rows	$z_1^2z_2$	z_1^2	z_1z_2	z_1	z_2	c	
a	48	12	144	33	81	15	$C_1 = \frac{48}{48} = 1$
b	48	12	120	27	27	2	
c	0	0	24	6	54	13	$A_1 = \frac{48}{24} = 2.$
c'	24	6	54	13			
d	0	0	12	1	27	2	

Reordering rows c and d in powers of z_2, we get

Rows	z_1z_2	z_2	z_1	c	
c	24	54	6	13	$C_2 = \frac{24}{12} = 2$
d	12	27	1	2	
e	0	0	4	9	$B_1 = \frac{12}{4} = 3.$
e'	4	9			
f	0	0	1	2	

Next reordering rows e and f in powers of z_1, we get

Rows	z_1	c	
e	4	9	$C_3 = \frac{4}{1} = 4$
f	1	2	
g	0	1	$A_2 = \frac{1}{1} = 1.$
g'	1		

From row g, the remainder is $1/(z_1 + 2)$. Hence, the continued fraction expansion for $H(z_1,z_2)$ is given as

$$H(z_1,z_2) = 1 + \cfrac{1}{2z_1 + \cfrac{1}{2 + \cfrac{1}{3z_2 + \cfrac{1}{4 + \cfrac{1}{z_1 + 2}}}}}.$$

Example 2

Consider

$$H(z_1,z_2) = \frac{z_1^2z_2^2 + 9z_1z_2 + 3z_1z_2^2 + 3z_1^2z_2 + 3z_2^2 + 7z_2 + 2z_1^2 + 7z_1 + 3}{z_1^2z_2^2 + 3z_2^2 + 3z_1z_2^2 + 2z_1^2z_2 + 6z_1z_2 + 4z_2 + z_1^2 + 3z_1 + 1}.$$

Then,

Rows	$z_1^2z^2$	$z_1z_2^2$	z_2^2	$z_1^2z_2$	z_1z_2	z_2	z_1^2	z_1	c	
a	1	3	3	3	9	7	2	7	3	$C_1 = 1$
b	1	3	3	2	6	4	1	3	1	
c	0	0	0	1	3	3	1	4	2	$A_1 = 1.$
c'	1	3	3	1	4	2				
d	0	0	0	1	2	2	1	3	1	

Now reordering rows c and d in powers of z_2 gives

Rows	$z_1^2z_2$	z_1^2	z_1z_2	z_1	z_2	c	
c	1	1	3	4	3	2	$C_2 = 1$
d	1	1	2	3	2	1	
e	0	0	1	1	1	1	$B_1 = 1.$
e'	1	1	1	1			
f	0	0	1	2	2	1	

Next reordering rows e and f in powers of z_2 gives

Rows	z_1z_2	z_2	z_1	c
e	1	1	1	1
f	1	2	2	1
g	0	−1	−1	0

↗ expansion fails.

In this case, the expansion fails since row g should have two zeros leading the row. The continued fraction expansion is thus

$$H(z_1,z_2) = 1 + \cfrac{1}{z_2 + \cfrac{1}{1 + \cfrac{1}{z_1 + \cfrac{z_1z_2 + 2z_1 + 2z_2 + 1}{z_1z_2 + z_1 + z_2 + 1}}}}$$

We can easily verify that the last term cannot be expanded further.

Example 3

Let

$$H(z_1,z_2) = \frac{48z_1^2z_2 + 72z_1^2 + 24z_1z_2^2 + 78z_1z_2 + 60z_1 + 16z_2^2 + 30z_2 + 7}{48z_1^2z_2 + 72z_1^2 + 24z_1z_2^2 + 54z_1z_2 + 24z_1 + 4z_2^2 + 7z_2 + 1}.$$

Filling out the table and including the missing $z_1^2z_2^2$ term, we get

Rows	$z_1^2z_2^2$	$z_1^2z_2$	z_1^2	$z_1z_2^2$	z_1z_2	z_1	z_2^2	z_2	c	
a	0	48	72	24	78	60	16	30	7	
b	0	48	72	24	54	24	4	7	1	$C_1 = 1$
c	0	0	0	0	24	36	12	23	6	
c'	0	24	36	12	23	6				$A_1 = 2.$
d	0	0	0	0	8	12	4	7	1	

Reordering rows c and d in powers of z_2, we get

Rows	$z_2^2z_1$	z_2^2	z_2z_1	z_2	z_1	c	
c	0	12	24	23	36	6	
d	0	4	8	7	12	1	$C_2 = 3$
e	0	0	0	2	0	3	$B_1 = 2.$
e'	0	2	0	3			
f	0	0	8	1	12	1	

Now reorder e, f, in powers of z_1 to get

	z_1z_2	z_1	z_2	c
e	0	0	2	3
f	8	12	1	1.

We note that no constant term can be extracted at this step. However, we can take a zero constant term and continue the expansion by noting that if this is done, row g is equal to row e. We can then continue without difficulty, as seen by

Rows	z_1z_2	z_1	z_2	c	
e	0	0	2	3	
f	8	12	1	1	$C_3 = 0$
g	0	0	2	3	$A_2 = 4$
g'	2	3	0	0	
h	0	0	1	1	

Rows	z_2	c	
g	2	3	
h	1	1	$C_4 = 2$
i	0	1	$A_3 = 1.$
i'	1		

Note that the expansion coefficients are determined by the first nonzero column and that this column is used to calculate the new rows. The column of zeros is added simply to complete the table with no missing terms so that the rule for shifting rows is maintained. The expansion obtained is thus

$$H(z_1,z_2) = 1 + \cfrac{1}{2z_1 + \cfrac{1}{3 + \cfrac{1}{2z_2 + \cfrac{1}{0 + \cfrac{1}{4z_1 + \cfrac{1}{2 + \cfrac{1}{z_2 + 1}}}}}}}$$

which can be rewritten as

$$H(z_1,z_2) = 1 + \cfrac{1}{2z_1 + \cfrac{1}{3 + \cfrac{1}{2z_2 + 4z_1 + \cfrac{1}{2 + \cfrac{1}{z_2 + 1}}}}}$$

An examination of these examples and of the test procedure shows that any of the expansion forms (or hybrids of the different types) can be obtained through simple modifications of the method. Furthermore, decisions as to which expansion to pursue can be made easily at each stage as the expansion progresses.

A special case of interest occurs when the numerator and denominator are not relatively prime, i.e., when they contain a common factor. In this case, the continued fraction expansion will prematurely terminate and produce a whole row of zeros in the test for existence. This provides valuable information since the factor can be cancelled out of $H(z_1,z_2)$ simplifying the realization. It should be noted, however, that the existence of a common factor in the numerator and denominator of $H(z_1,z_2)$ cannot be detected if the expansion fails for any reason other than a whole row of zeros. In that case, a method such as in [8] could be used to determine the existence of the common factor.

Transpose Structures

The transpose of a given signal-flow network is formed by reversal of the signal-flow direction in each branch. Since the network topology is not changed, application of Masons rule [6] immediately shows that the transpose of any of the structures introduced will realize the same transfer function. Only the transposes of the two direct forms, however, will give essentially different configurations. The continued fraction structures can be drawn in such a way that they are the same as their transposes.

Discussion and Conclusion

Some comparisons of the different structures are useful. For example, when they can be used, the continued fraction expansions will usually require fewer multiplications. The direct forms will require one less multiplication than the total number of coefficients in the numerator and denominator. This number is less than or equal to $M_aN_a + M_bN_b - 1$ where M_a, N_a, M_b, and N_b are given in (1). The continued fraction expansions, however, require only a number of multipliers equal to the number of terms in the continued fraction expansion. The number required is less than or equal to $2N + 1$ where N is the order of $H(z_1,z_2)$. This follows since the number of z_1 and z_2 terms in the expansion is equal to N [4]. Therefore, we can see that the savings in the number of multiplications may be substantial if a continued fraction expansion realization can be used. Of course the number of additions required is related directly to the number of multiplications so savings will be obtained here also.

A comparison of memory requirements can also be made. As pointed out earlier, the number of z_1 or z_2 blocks does not directly indicate the amount of memory that will be needed by a given algorithm in two dimensions. However, for a given realization, the storage requirements can be calculated. This is done by first formulating the new arrays that could be generated in addition to the input and the output arrays. The minimum storage is then found by totaling the number of rows or columns and single elements of storage required in each of these arrays. The storage requirement is determined for a given array by choosing the computation sequence and then noting the elements that must be available from the array at each computation.

For an example, consider the Type IA realization of Fig. 8. Arrays d, e, and f could be generated during computation. Others could also be generated after other summations but no z_1 or z_2 blocks operate on them so they need not be considered. If computation is row by row, then arrays d and f would require only a single element of storage each. Array e would require one row of storage. Thus the total storage required by Fig. 8 is a minimum of $M + 2$ where M is the length of a row. Similarly, if the calculation is done column by column, the memory requirement is $2N + 1$ where N is the length of the output columns.

The comparison of memory requirements can now be made based on this discussion. For a given $H(z_1,z_2)$ and sequence of calculations and output array dimensions, the direct realizations using (3) and (4) will require storage space less than or equal to that required by a realization using (2). Furthermore, the continued fraction expansion based realizations will need an amount of storage equal to that required by the realizations based on (3) and (4).

It may be noted from the structure of the test for continued fraction expansion that strong relationships must exist between coefficients of terms in $H(z_1,z_2)$ in order for an expansion to be possible. This situation indicates that a useful design technique might be to assume that a particular expansion structure is desired and then to evaluate the coefficients of $H(z_1,z_2)$ to meet other design criteria subject to the expansion constraint. This would insure that the resulting filter could be implemented in the desired continued expansion form with the resulting savings in the number of multipliers and adders that must be used.

Acknowledgment

The authors wish to thank E. I. Jury and P. S. Kamat for their assistance and helpful comments during the preparation of this paper.

References

[1] J. L. Shanks, S. Treitel, and J. H. Justice, "Stability and synthesis of two-dimensional recursive filters," *IEEE Trans. Audio Electroacoust.*, vol. AU-20, pp. 115–128, June 1972.

[2] S. K. Mitra and R. J. Sherwood, "Canonic realizations of digital filters using the continued fraction expansion," *IEEE Trans. Audio Electroacoust.*, vol. AU-20, pp. 185–194, Aug. 1972.

[3] S. K. Mitra and A. D. Sagar, "Additional canonic realizations of digital filters using the continued fraction expansion," *IEEE Trans. Circuits and Systems*, vol. CAS-21, pp. 135–136, Jan. 1974.

[4] V. Ramachnadran and A. S. Rao, "A multivariable array and its applications," *IEEE Trans. Circuit Theory*, vol. CT-20, pp. 511–517, Sept. 1973.

[5] B. O. Watkins, *Introductions to Control Systems.* New York: Macmillan, 1969.

[6] S. J. Mason and H. J. Zimmermann, *Electronic Circuits, Signals and Systems.* New York: Wiley, 1960.

[7] J. L. Willems, *Stability Theory of Dynamical Systems.* New York: Wiley, p. 67, 1970.

[8] N. K. Bose, "A criterion to determine if two multivariable polynomials are relatively prime," *Proc. IEEE*, vol. 60, pp. 134–135, Jan. 1972.

[9] W. D. Fryer, "Applications of Routh's algorithm to network theory problems," *IRE Trans. Circuit Theory*, vol. CT-6, pp. 144–149, June 1959.

Reprinted from *IEEE Trans. Autom. Control* **AC-20**(1):1-10 (1975)

A Discrete State-Space Model for Linear Image Processing

ROBERT P. ROESSER, MEMBER, IEEE

***Abstract*—The linear time-discrete state-space model is generalized from single-dimensional time to two-dimensional space. The generalization includes extending certain basic known concepts from one to two dimensions. These concepts include the general response formula, state-transition matrix, Cayley–Hamilton theorem, observability, and controllability.**

I. Introduction

IMAGE processing by nonoptical means has been receiving extensive attention in the last few years. Several books, e.g., [1]–[3] and many papers, e.g., [4]–[6], have been published that have established nonoptical image processing as a viable area of research. A large portion of this research emphasizes the linear processing of images for two main reasons: 1) Many image processing tasks are linear in nature. These tasks include image enhancement, image restoration, picture coding, linear pattern recognition, and TV bandwidth reduction. 2) There are many known linear techniques that may be brought to bear in the treatment of linear image processing, and therefore simplify such treatment. These techniques include transform theory, matrix theory, superposition, etc.

Several ways are commonly used to represent the operations involved in image processing. These include transfer functions, partial difference (recursive) equations, and convolution summations. For example, VanderLugt [7], [11] has presented an extensive development of linear optics based on transfer functions. The transfer functions relate the two-dimensional Fourier transform of an output image to that of the input image. Complex optical systems are easily described by combinations of transfer functions that correspond to individual components of the optical system.

Partial difference equations are used by Habibi [6] to describe a model for estimating images corrupted by noise. The model corresponds to a two-dimensional extension of Kalman filters.

Convolution summations are discussed by Fryer and Richmond [5] in work that involves simplifying a two-dimensional filter to a single-dimensional filter.

The time-discrete state-space model offers great utility in the formulation and analysis of linear systems. Linear systems that are described by transfer functions, difference equations, or convolution summations are easily formulated into a state-space representation. Once so formulated, many known techniques may be applied to systematically analyze the model. Consequently, the state-space model is a general and powerful tool that is used to unify the research and study of time-discrete linear systems.

This paper develops a discrete model for linear image processing that closely parallels the well-known state-space model for time-discrete systems. Because of this parallel many of the concepts that are known for the temporal model may be carried over to the spatial model. This is done by generalizing from a single coordinate in time to two coordinates in space. The spatial model will hopefully have some of the same utility in unifying the study of two-dimensional linear systems as does the temporal model for one-dimensional linear systems.

Temporal systems are inherently nonanticipatory and are often treated as such for the sake of physical realizability in real time; whereas spatial systems do not have causality as an inherent limitation. That is, an image processor may have right to left dependency as well as left to right dependency. Causality is built into the temporal state-space model if an initial state is assumed to be fully specified. In order to establish a close parallel for the spatial model the same built-in causality will be intentionally assumed despite the fact that causality is not necessary for physical realizability in real space. Such an image processor is said to be unilateral. If the constraint of causality is removed, then an image processor is said to be bilateral.

Concepts that are developed in this paper include 1) formulation of the state-space model, 2) the definition of a state-transition matrix, 3) the derivation of a general response formula, 4) a two-dimensional parallel to the Cayley–Hamilton theorem, 5) observability and controlability, and 6) computation of the state-transition matrix. Some of these concepts are based upon an extension of published material on linear iterative circuits coauthored by this writer [8], [9]. A finite field is assumed in the case of iterative circuits, whereas a real field is assumed for image processing. One particular concept, the two-dimensional Cayley–Hamilton theorem, is treated in [9]. An interesting alternative proof to the theorem is the topic of a paper published by Vilfan [10].

Manuscript received February 5, 1974. Paper recommended by A. S. Morse, Chairman of the IEEE S-CS Linear Systems Committee.
The author is with the Department of Electrical Engineering, Wayne State University, Detroit, Mich. 48202.

II. The Model

An image is a generalization of a temporal signal in that it is defined over two spatial dimensions instead of a single temporal dimension. Consequently, two space coordinates i and j take the place of time t. Also two-state sets are introduced to replace the single-state set. The following definitions are made for the model.

i	An integer-valued vertical coordinate.
j	An integer-valued horizontal coordinate.
$\{R\}$	A set of real n_1-vectors which convey information vertically.
$\{S\}$	A set of real n_2-vectors which convey information horizontally.
$\{u\}$	A set of real p-vectors that act as inputs.
$\{y\}$	A set of real m-vectors that act as outputs.

A specific image processor is then defined as a 6-tuple $<\{R\}, \{S\}, \{u\}, \{y\}, f, g>$ where f is the next state function:

$$f:\{R\}\times\{S\}\times\{u\}\rightarrow\{R\}\times\{S\}$$

and g is the output function;

$$g:\{R\}\times\{S\}\times\{u\}\rightarrow\{y\}.$$

Now, since f and g are to be linear functions they may be represented by the following matrix equations:

$$R(i+1,j)=A_1R(i,j)+A_2S(i,j)+B_1u(i,j)$$

$$S(i,j+1)=A_3R(i,j)+A_4S(i,j)+B_2u(i,j)$$

$$y(i,j)=C_1R(i,j)+C_2S(i,j)+Du(i,j), \qquad i,j\geqslant 0.$$

A_1, A_2, A_3, A_4, B_1, B_2, C_1, C_2, D are matrices of appropriate dimensions. Boundary conditions $R(0,j)$ and $S(i,0)$ and also the input $u(i,j)$ are externally specified. In the next section a computational rule is obtained that uniquely determines the states $R(i,j)$ and $S(i,j)$ and also the output $y(i,j)$ (for $i,j\geqslant 0$) from the boundary conditions and inputs. Thus given values for the boundary conditions (such as all zero) the equations produce an output vector image from an input vector image. This formulation is general so that any discrete linear image process may be so represented. Notation is condensed somewhat by introducing the following matrices and vectors:

$$A=\begin{bmatrix} A_1 & A_2 \\ A_3 & A_4 \end{bmatrix} \quad B=\begin{bmatrix} B_1 \\ B_2 \end{bmatrix} \quad C=[C_1\ C_2]$$

$$T(i,j)=\begin{bmatrix} R(i,j) \\ S(i,j) \end{bmatrix} \quad T'(i,j)=\begin{bmatrix} R(i+1,j) \\ S(i,j+1) \end{bmatrix}.$$

Then

$$T'(i,j)=AT(i,j)+Bu(i,j)$$

$$y(i,j)=CT(i,j)+Du(i,j).$$

III. General Response Formula

A state-transition matrix $A^{i,j}$ is defined as follows.

Definition: For

$$A=\begin{bmatrix} A_1 & A_2 \\ A_3 & A_4 \end{bmatrix}$$

$$A^{i,j}=A^{1,0}A^{i-1,j}+A^{0,1}A^{i,j-1}, \qquad (i,j)>(0,0)$$

$$A^{0,0}=I$$

$$A^{-i,j}=A^{i,-j}=0, \qquad \text{for } j\geqslant 1,\ i\geqslant 1.$$

Examination of this definition bears out that it is an effective recursive definition for integer values of i and j, such that either $i>0$ or $j>0$ or $(i,j)=(0,0)$. It parallels the definition of the time-discrete state-transition matrix $A^t=A^1A^{t-1}$.

It is now to be shown that this state-transition matrix, $A^{i,j}$, may be used in an expression for the response of the model in terms of the inputs and boundary conditions. The term boundary conditions is used here to refer to the states along the edges of the model. Specifically, the set of boundary conditions consist of $R(0,j)$ for $j\geqslant 0$ and $S(i,0)$ for $i\geqslant 0$.

Definition: The following partial ordering is used for integer pairs:

$$(h,k)\leqslant(i,j), \qquad \text{iff } h\leqslant i \text{ and } k\leqslant j$$

$$(h,k)=(1,j), \qquad \text{iff } h=i \text{ and } k=j$$

$$(h,k)<(i,j), \qquad \text{iff } (h,k)\leqslant(i,j) \text{ and } (h,k)\neq(i,j).$$

Lemma: Let the input, $u(i,j)$, for all (i,j) and the boundary conditions, $R(0,j)$ and $S(i,0)$, for $(i,j)\neq(0,0)$ be equal to zero. Then $T(i,j)=A^{i,j}T(0,0)$.

Proof: The proof is accomplished by induction.

First, $T(0,0)=IT(0,0)=A^{0,0}T(0,0)$. This implies the hypothesis is true for $(i,j)=(0,0)$. Now assume the hypothesis is true for all (h,k) such that $(0,0)\leqslant(h,k)<(i,j)$, and show that it is true for (i,j).

$$T(i,j)=\begin{bmatrix} R(i,j) \\ S(i,j) \end{bmatrix}=\begin{bmatrix} A_1R(i-1,j)+A_2S(i-1,j)+B_1\cdot 0 \\ A_3R(i,j-1)+A_4S(i,j-1)+B_2\cdot 0 \end{bmatrix}$$

$$=\begin{bmatrix} A_1 & A_2 \\ 0 & 0 \end{bmatrix}T(i-1,j)+\begin{bmatrix} 0 & 0 \\ A_3 & A_4 \end{bmatrix}T(i,j-1)$$

$$=A^{1,0}A^{i-1,j}T(0,0)+A^{0,1}A^{i,j-1}T(0,0)$$

$$=A^{i,j}T(0,0).$$

This is an effective inductive proof because an enumeration can be found such that all (i,j) are reached but not before all $(h,k)<(i,j)$. Such an enumeration is the diagonal enumeration (0,0), (0,1), (1,0), (0,2), (1,1), (2,0), ⋯ Q.E.D.

It is to be noted that since the matrices are not functions of (i,j) the model is spatially invariant. The effect then of $T(h,k)$ on $T(i,j)$ is $A^{i-h,j-k}T(h,k)$. The superposition property of linear systems may be used to obtain a more general expression for $T(i,j)$ in which $u(i,j)$ and the other boundary conditions are not assumed to be zero.

Effect of u(h,k): Assume $u(h,k)$ for some $(h,k)<(i,j)$ is the only nonzero input and all boundary conditions are zero. Then

$$T(h+1,k)=\begin{bmatrix} A_1\cdot 0+A_2\cdot 0+B_1\cdot u(h,k) \\ 0 \end{bmatrix}=\begin{bmatrix} B_1 \\ 0 \end{bmatrix}u(h,k)$$

and

$$T(h,k+1)=\begin{bmatrix} 0 \\ A_3\cdot 0+A_4\cdot 0+B_2\cdot u(h,k) \end{bmatrix}=\begin{bmatrix} 0 \\ B_2 \end{bmatrix}u(h,k).$$

Then

$$T(i,j)=A^{i-(h+1),j-k}T(h+1,k)+A^{i-h,j-(k+1)}T(h,k+1)$$

$$=\left(A^{i-h-1,j-k}\begin{bmatrix} B_1 \\ 0 \end{bmatrix}+A^{i-h,j-k-1}\begin{bmatrix} 0 \\ B_2 \end{bmatrix}\right)u(h,k).$$

Effect of R(0,k): Assume that $R(0,k)$ is the only nonzero boundary condition and that all inputs are zero.

$$T(0,k)=\begin{bmatrix} R(0,k) \\ 0 \end{bmatrix}.$$

Then

$$T(i,j)=A^{i,j-k}T(0,k)$$

$$=A^{i,j-k}\begin{bmatrix} R(0,k) \\ 0 \end{bmatrix}.$$

Effect of S(h,0): Similarly to $R(0,k)$ the effect of $S(h,0)$ is

$$T(i,j)=A^{i-h,j}\begin{bmatrix} 0 \\ S(h,0) \end{bmatrix}.$$

We thus have the following theorem.

Theorem: For all $(i,j)\geqslant 0$.

$$T(i,j)=\sum_{k=0}^{j}A^{i,j-k}\begin{bmatrix} R(0,k) \\ 0 \end{bmatrix}+\sum_{h=0}^{i}A^{i-h,j}\begin{bmatrix} 0 \\ S(h,0) \end{bmatrix}$$

$$+\sum_{(0,0)\leqslant(h,k)<(i,j)}\sum$$

$$\cdot\left(A^{i-h-1,j-k}\begin{bmatrix} B_1 \\ 0 \end{bmatrix}+A^{i-h,j-k-1}\begin{bmatrix} 0 \\ B_2 \end{bmatrix}\right)u(h,k).$$

Proof: By superposition of the effects of all inputs and boundary conditions. Q.E.D.

An expression for $y(i,j)$, called the *general response formula*, may now be written

$$y(i,j)=[C_1\ C_2]\left(\sum_{k=0}^{j}A^{i,j-k}\begin{bmatrix} R(0,k) \\ 0 \end{bmatrix}\right.$$

$$+\sum_{h=0}^{i}A^{i-h,j}\begin{bmatrix} 0 \\ S(h,0) \end{bmatrix}$$

$$+\sum_{(0,0)\leqslant(h,k)<(i,j)}\sum\left(A^{i-h-1,j-k}\begin{bmatrix} B_1 \\ 0 \end{bmatrix}\right.$$

$$\left.\left.+A^{i-h,j-k-1}\begin{bmatrix} 0 \\ B_2 \end{bmatrix}\right)u(h,k)\right)+Du(i,j).$$

IV. Properties of $A^{i,j}$

Some properties of $A^{i,j}$ are

1)

$$A=\begin{bmatrix} A_1 & A_2 \\ A_3 & A_4 \end{bmatrix}=\begin{bmatrix} A_1 & A_2 \\ 0 & 0 \end{bmatrix}+\begin{bmatrix} 0 & 0 \\ A_3 & A_4 \end{bmatrix}.$$

Thus $A=A^{1,0}+A^{0,1}$.

2)

$$A^{i,0}=A^{1,0}A^{i-1,0}+A^{0,1}A^{i,-1}=A^{1,0}A^{i-1,0}.$$

Thus $A^{i,0}=(A^{1,0})^i$. Similarly $A^{0,j}=(A^{0,1})^j$.

3)

$$I=\begin{bmatrix} I & 0 \\ 0 & I \end{bmatrix}$$

where I is the identity matrix with appropriate dimensions. Thus

$$I^{1,0}=\begin{bmatrix} I & 0 \\ 0 & 0 \end{bmatrix} \text{ and } I^{0,1}=\begin{bmatrix} 0 & 0 \\ 0 & I \end{bmatrix}.$$

4)

$$I^{1,0}A=\begin{bmatrix} I & 0 \\ 0 & 0 \end{bmatrix}\begin{bmatrix} A_1 & A_2 \\ A_3 & A_4 \end{bmatrix}=\begin{bmatrix} A_1 & A_2 \\ 0 & 0 \end{bmatrix}=A^{1,0}.$$

Briefly $I^{1,0}A=I^{1,0}A^{1,0}=A^{1,0}$. Similarly $I^{0,1}A=I^{0,1}A^{0,1}=A^{0,1}$.

5)

$$I^{0,1}A^{1,0}=\begin{bmatrix} 0 & 0 \\ 0 & I \end{bmatrix}\begin{bmatrix} A_1 & A_2 \\ 0 & 0 \end{bmatrix}=0.$$

Briefly $I^{0,1}A^{1,0}=0$. Similarly $I^{1,0}A^{0,1}=0$.

V. Characteristic Function of a Matrix

If the primary inputs and outputs are neglected in the model equations, a representation arises for the state behavior of the circuit, having the form

$$R(i+1,j)=A_1R(i,j)+A_2S(i,j)$$

$$S(i,j+1)=A_3R(i,j)+A_4S(i,j).$$

These equations are useful in the development of a form for a two-dimensional characteristic matrix of A. Operators are first introduced that advance a particular coordinate of their operand.

Definition: Let E be an operator that has the effect of advancing the vertical coordinate or the first subscript of the function upon which it is operating. Likewise, let F be an operator that has the effect of advancing the horizontal coordinate or second subscript of the function upon which it is operating.

The effect of these operators on the state vectors is

$$R(i+1,j)=ER(i,j)$$

$$S(i,j+1)=FS(i,j).$$

The state equations can be rewritten using these advance operators.

$$(EI-A_1)R(i,j)-A_2S(i,j)=0$$

$$-A_3R(i,j)+(FI-A_4)S(i,j)=0.$$

These equations are equivalently represented in the overall matrix form.

$$\begin{bmatrix} (EI-A_1) & -A_2 \\ -A_3 & (FI-A_4) \end{bmatrix}T(i,j)=0.$$

The above equation represents a system of homogeneous equations in the elements of $T(i,j)$. If the system is to have a nontrivial solution for $T(i,j)$, then the transformation represented by the matrix must be singular.

The above matrix is said to be the two-dimensional characteristic matrix of the partitioned matrix A where

$$A=\begin{bmatrix} A_1 & A_2 \\ A_3 & A_4 \end{bmatrix}.$$

The characteristic matrix of A is denoted by $cm(A)$ and may be represented as

$$cm(A)=EI^{1,0}+FI^{0,1}-A.$$

Now since $cm(A)$ must be singular, its determinate must be equal to zero.

$$|cm(A)|=0.$$

If E and F are placed in the above by general indeterminates x and y, respectively, the result is an expression called the two-dimensional characteristic equation for A. The determinate of $cm(A)$ with x and y replacing E and F is called the two-dimensional characteristic function of the matrix X, and is denoted by $f(x,y)$.

$$|cm(A)|=f(x,y)=0.$$

$f(x,y)$ will be a monic multinomial in x and y with degree n_1 in x and degree n_2 in y; where n_1 is the dimension of R and n_2 is the dimension of S. $f(x,y)$ has the form

$$f(x,y)=\sum_{(0,0)<(i,j)<(n_1,n_2)}\sum a_{i,j}x^iy^j, \qquad \text{where } a_{n_1,n_2}=1.$$

Comparing these concepts to the one-dimensional case, it is observed that they are correspondingly analogous to the one-dimensional characteristic matrix, equation and function of a matrix. $xI-A$ is the one-dimensional characteristic matrix of A and $f(x)=|xI-A|=0$ is the one-dimensional characteristic equation.

The Cayley–Hamilton theorem in the one-dimensional case states that a matrix A satisfies its own characteristic equation, i.e., $f(A)=0$. The following theorem extends this to the two-dimensional case.

Definition: $E^iF^jA=F^jE^iA=A^{i,j}$ for any 2×2 partition of A.

Two-Dimensional Cayley–Hamilton Theorem: Every partitioned matrix

$$A=\begin{bmatrix} A_1 & A_2 \\ A_3 & A_4 \end{bmatrix}$$

satisfies its own characteristic equation. That is $f(E,F)A = 0$.

Proof: Let $B=(xI^{1,0}+yI^{0,1}-A)$, so that $f(x,y)=\det B$. Cramer's rule for computing the inverse of a matrix, in this case matrix B, states

$$\operatorname{adj} B\cdot(xI^{1,0}+yI^{0,1}-A)=\det B\cdot I=f(x,y)I \tag{1}$$

where adj B is the transpose of the cofactor matrix of B. The elements of adj B transpose are computed by taking the determinate of the matrix formed by deleting the row and column containing the corresponding element in B. The elements of adj B will consequently be multinomials in x and y having degrees in x and y not greater than n_1 and n_2, respectively, where n_1 is the rank of $I^{1,0}$ and n_2 is the rank of $I^{0,1}$. Therefore, adj B may be written in the form of a matrix multinomial.

$$\operatorname{adj} B=\sum_{i=0}^{n_1}\sum_{j=0}^{n_2} B_{i,j}x^iy^j. \tag{2}$$

Represent the characteristic multinomial $f(x,y)$ as

$$f(x,y)=\sum_{i=0}^{n_1}\sum_{j=0}^{n_2} b_{i,j}x^iy^j, \qquad \text{where } b_{n_1,n_2}=1. \tag{3}$$

Substituting (2) and (3) into (1), we have,

$$\sum_{i=0}^{n_1}\sum_{j=0}^{n_2} B_{i,j}x^iy^j(xI^{1,0}+yI^{0,1}-A)=\sum_{i=0}^{n_1}\sum_{j=0}^{n_2} b_{i,j}x^iy^jI.$$

Expand the left side and adjust i and j.

$$\sum_{i=1}^{n_1+1}\sum_{j=0}^{n_2} B_{i-1,j}I^{1,0}x^iy^j+\sum_{i=0}^{n_1}\sum_{j=1}^{n_2+1} B_{i,j-1}I^{0,1}x^iy^j$$

$$-\sum_{i=0}^{n_1}\sum_{j=0}^{n_2} B_{i,j}Ax^iy^j=\sum_{i=0}^{n_1}\sum_{j=0}^{n_2} b_{i,j}x^iy^jI.$$

The coefficients of each term x^iy^j on both sides of the equation must be equal. Equating these coefficients yields the following:

$$B_{n_1,j}I^{1,0}=0, \qquad \text{for } 0\leqslant j\leqslant n_2 \tag{4}$$

$$B_{i,n_2}I^{0,1}=0, \qquad \text{for } 0\leqslant i\leqslant n_1 \tag{5}$$

$$B_{i-1,j}I^{1,0}+B_{i,j-1}I^{0,1}-B_{i,j}A=b_{i,j}I, \qquad \text{for } i\neq 0\; j\neq 0 \tag{6}$$

$$B_{0,j-1}I^{0,1}-B_{0,j}A=b_{0,j}I, \qquad \text{for } j\neq 0 \tag{7}$$

$$B_{i-1,j}I^{1,0}-B_{i,0}A=b_{i,0}I, \qquad \text{for } i\neq 0 \tag{8}$$

$$-B_{0,0}A=b_{0,0}I. \tag{9}$$

Multiply each of the equations (6)–(8) on the right by $A^{i,j}$ and sum equations (6)–(9) over all i,j.

$$\sum_{i=1}^{n_1}\sum_{j=1}^{n_2}(B_{i-1,j}I^{1,0}+B_{i,j-1}I^{0,1}-B_{i,j}A)A^{i,j}$$
$$+\sum_{j=1}^{n_2}(B_{0,j-1}I^{0,1}-B_{0,j}A)A^{0,j}$$
$$+\sum_{i=1}^{n_1}(B_{i-1,0}I^{1,0}-B_{i,0}A)A^{i,0}-B_{0,0}A$$
$$=\sum_{i=0}^{n_1}\sum_{j=0}^{n_2} b_{i,j}IA^{i,j}.$$

Collect coefficients of $B_{i,j}$ on the left side of the equation.

$$B_{i,j}(I^{1,0}A^{i+1,j}+I^{0,1}A^{i,j+1}-AA^{i,j}), \qquad \text{for } 1\leqslant i\leqslant n_1\; 1\leqslant j\leqslant n_2 \tag{10}$$

$$B_{n_1,j}(I^{0,1}A^{n_1,j+1}-AA^{n_1,j}), \qquad \text{for } j\neq 0\; j\neq n_2 \tag{11}$$

$$B_{i,n_2}(I^{1,0}A^{i+1,n_2}-AA^{i,n_2}), \qquad \text{for } i\neq 0\; i\neq n_1 \tag{12}$$

$$B_{0,j}(I^{1,0}A^{1,j}+I^{0,1}A^{0,j+1}-AA^{0,j}), \qquad \text{for } j\neq 0\; j\neq n_2 \tag{13}$$

$$B_{i,0}(I^{0,1}A^{i,1}+I^{1,0}A^{i+1,0}-AA^{i,0}), \qquad \text{for } i\neq 0\; i\neq n_1 \tag{14}$$

$$B_{n_1,n_2}(AA^{n_1,n_2}) \tag{15}$$

$$B_{0,0}(I^{0,1}A^{0,1}+I^{1,0}A^{1,0}-A). \tag{16}$$

Expressions (10)–(16) exhaust all i,j combinations. We will now evaluate each in turn.

$$I^{1,0}A^{i+1,j}=I^{1,0}A^{1,0}A^{i,j}+I^{1,0}A^{0,1}A^{i+1,j-1}$$
$$I^{0,1}A^{i,j+1}=I^{0,1}A^{1,0}A^{i-1,j+1}+I^{0,1}A^{0,1}A^{i,j} \tag{10}$$

but $I^{1,0}A^{0,1}=0$ and $I^{0,1}A^{1,0}=0$.

Thus,

$$I^{1,0}A^{i+1,j}+I^{0,1}A^{i,j+1}-AA^{i,j}=A^{1,0}A^{i,j}+A^{0,1}A^{i,j}-AA^{i,j}$$
$$=AA^{i,j}-AA^{i,j}=\mathbf{0}.$$

$$I^{0,1}A^{n_1,j+1}=I^{0,1}A^{1,0}A^{n_1-1,j+1}+I^{0,1}A^{0,1}A^{n_1,j}$$
$$=A^{0,1}A^{n_1,j}. \tag{11}$$

Thus,

$$B_{n_1,j}(I^{0,1}A^{n_1,j+1}-AA^{n_1,j})$$
$$=B_{n_1,j}(A^{0,1}A^{n_1,j}-A^{0,1}A^{n_1,j}-A^{1,0}A^{n_1,j})$$
$$=-B_{n_1,j}A^{1,0}A^{n_1,j}=-B_{n_1,j}I^{1,0}A^{1,0}A^{n_1,j}=\mathbf{0}$$

since $B_{n_1,j}I^{1,0}=0$ from (4).

$$\text{Similarly}=\mathbf{0} \text{ as in (11).} \tag{12}$$

$$AA^{0,j}=A^{1,0}A^{0,j}+A^{0,1}A^{0,j}=A^{1,0}A^{0,j}+A^{0,j+1}$$
$$I^{1,0}A^{1,j}=I^{1,0}A^{1,0}A^{0,j}+I^{1,0}A^{0,1}A^{1,j-1}=A^{1,0}A^{0,j}. \tag{13}$$

Thus,

$$I^{1,0}A^{1,j}+I^{0,1}A^{0,j+1}-AA^{0,j}$$
$$=A^{1,0}A^{0,j}+A^{0,j+1-A1,0}A^{0,j}-A^{0,j+1}=\mathbf{0}.$$

$$\text{Similarly}=\mathbf{0} \text{ as in (13).} \tag{14}$$

$$B_{n1,n2}=\mathbf{0}, \text{ since adj } B \text{ can have no term in } x^{n1}y^{n2}. \tag{15}$$

$$A=A^{1,0}+A^{0,1}, \text{ thus, } I^{0,1}A^{0,1}+I^{1,0}A^{1,0}-A=\mathbf{0}. \tag{16}$$

Briefly, the left side of the equation being considered is zero, giving

$$0=\sum_{i=0}^{n_1}\sum_{j=0}^{n_2} b_{i,j}A^{i,j}=f(E,F)A$$

which completes the proof.

VI. Observability and Controllability

The notions of observability and controllability for time-discrete systems carry over to parallel notions for discrete-space image processors.

Definition: A state T_0 is observable iff whenever it appears as the initial state and all other boundary conditions are zero, there exists a pattern of inputs and a pair $(i,j) \geqslant (0,0)$ such that $y(i,j)$ is not the same as when the initial state is zero, and the same pattern of inputs is applied.

Definition: A state T_0 is controllable iff when all boundary conditions are zero there exists some pair $(i,j) > 0$ and some input pattern such that $T(i,j) = T_0$.

An image processing model is said to be observable (controllable) iff all states are observable (controllable).

It is often desirable to reduce a model to an equivalent model that is observable and controllable. Here equivalence between models will be taken to mean no pattern of inputs exist so that the output from one model is different at some pair $(i,j) \geqslant (0,0)$ than the output from the other model at (i,j) when the boundary conditions of both models are zero.

To test for observability the output $y(i,j)$ for each possible initial state is compared with the output $y_0(i,j)$ with zero initial state, for all $(i,j) \geqslant (0,0)$ and all input patterns. If $y(i,j) = y_0(i,j)$ for all $(i,j) \geqslant 0$ and all input conditions, then the model is not observable. Using the general response formula this condition reduces to

$$CA^{i,j}T(0,0) = 0, \qquad \text{for all } (i,j) \geqslant 0.$$

The two-dimensional Cayley–Hamilton theorem implies that any two-tuple power of A is linearly dependent upon those $A^{i,j}$ for which $(0,0) \leqslant (i,j) < (n_1,n_2)$ so that the condition for nonobservability can be limited to those (i,j) such that $(0,0) \leqslant (i,j) < (n_1,n_2)$. The condition may then be put into matrix form $KT = 0$, where K is the diagnostic matrix defined as follows:

$$K = \begin{bmatrix} C \\ CA^{0,1} \\ CA^{0,2} \\ \vdots \\ CA^{0,n_2} \\ CA^{1,0} \\ CA^{1,1} \\ \vdots \\ CA^{n_1,0} \\ CA^{n_1,1} \\ \vdots \\ CA^{n_1,n_2-1} \end{bmatrix}.$$

If $KT = 0$ then the model is not observable, but may be reduced to an observable model. Let K_1 be a matrix consisting of the first n_1 columns of K and let K_2 be a matrix consisting of the last n_2 columns of K. The conditions $KT = 0$ may then be split into the two conditions $K_1R = 0$ and $K_2S = 0$. A reduced model may then be formed by using the equivalence classes of $\{R\}$ modulo the null space of K_1 as the new vertical state set, and the equivalence classes of $\{S\}$ modulo the null space of K_2 as the new horizontal state set. As a result only the new zero vector will satisfy $K_1R = 0$ and $K_2S = 0$, implying that the new model is observable. It remains to find the characterizing matrices for the reduced model. First vector representation for the equivalence classes of vertical and horizontal states is found, then the original characterizing matrices are modified so that the behavior of the model remains the same. K_1 and K_2 may be reduced so that their new dimension agrees with their rank. This is done by forming matrices G_1 and G_2 from the first complete set of linearly independent rows of K_1 and K_2, respectively. $G_{(\,)}$ and $K_{(\,)}$ have the same null space. Furthermore each equivalence class of $\{R\}$ and $\{S\}$ will correspond to a single vector G_1R and G_2S, respectively. The equivalence classes of $\{T\}$ may then be represented as

$$\overline{T} = GT, \qquad \text{where } G = \begin{bmatrix} G_1 & 0 \\ 0 & G_2 \end{bmatrix}.$$

Now let

$$H = \begin{bmatrix} H_1 & 0 \\ 0 & H_2 \end{bmatrix}$$

be a right inverse to G, i.e., $GH = I$. The reduced model will then have the following characterizing matrices:

$$\overline{A} = GAH \qquad \overline{B} = GB \qquad \overline{C} = CH \qquad \overline{D} = D.$$

To test for controllability the state $T(i,j)$ is examined with all boundary conditions equal to zero for all input patterns. If $T(i,j)$ doesn't equal T_0 for some T_0, for any $(i,j) \geqslant (0,0)$ and any input pattern, then the model is uncontrollable. Using the general formula for $T(i,j)$ this condition becomes

$$\sum_{(0,0) \leqslant (h,k) < (i,j)} \sum \left(A^{i-h-1,j-k} \begin{bmatrix} B_1 \\ 0 \end{bmatrix} + A^{i-h,j-k-1} \begin{bmatrix} 0 \\ B_2 \end{bmatrix} \right) u(k,k) \neq T_0$$

for all (i,j) and all $u(h,k)$. As for observability, the two-dimension Cayley–Hamilton theorem allows us to limit the condition for noncontrollability to those (i,j) such that $(0,0) \leq (i,j) < (n_1,n_2)$. The condition may then be put into matrix form.

$QU \neq T_0$ for all U and some T_0 where U and Q are defined as follows:

$$U = \begin{bmatrix} u(0,0) \\ u(0,1) \\ \vdots \\ u(0,n_2) \\ u(1,0) \\ \vdots \\ u(n_1,n_2-1) \end{bmatrix}$$

$$Q = [M(0,0), M(0,1), \ldots, M(0,n_2), \\ M(1,0), \ldots, M(n_1,n_2-1)]$$

$$M(i,j) = A^{i-1,j}\begin{bmatrix} B_1 \\ 0 \end{bmatrix} + A^{i,j-1}\begin{bmatrix} 0 \\ B_2 \end{bmatrix}.$$

The maximum rank of Q is equal to n_1+n_2 (the number of rows), which is equal to the dimension of $\{T\}$. If Q has n_1+n_2 as its rank, then its range space must equal $\{T\}$ so that there would be no T_0 such that $QU \neq T_0$ for all U. The model would then be controllable. If however the rank of Q is less than n_1+n_2, then the model may be reduced to a controllable model. Let Q_1 be formed from the first n_1 rows of Q and Q_2 be formed from the last n_2 rows of Q.

Q_1 and Q_2 may be reduced so that their column dimension agrees with their rank. This is done by forming matrices G_1 and G_2 from the first complete set of linearly independent columns of Q_1 and Q_2, respectively.

The controllable states T will be those formed from the direct sum of the vectors in the range spaces of G_1 and G_2. A state set for the reduced model may now be specified as a set of vectors

$$\left\{ \bar{T} = \begin{bmatrix} \bar{R} \\ \bar{S} \end{bmatrix} \right\}$$

which are mapped into the controllable states under the direct sum of G_1 and G_2. That is $\{\bar{T}\}$ is the domain of the linear mapping

$$T = \begin{bmatrix} G_1 & 0 \\ 0 & G_2 \end{bmatrix}\begin{bmatrix} \bar{R} \\ \bar{S} \end{bmatrix} = G\bar{T}.$$

Let

$$H = \begin{bmatrix} H_1 & 0 \\ 0 & H_2 \end{bmatrix}$$

be a left inverse of G, i.e., $HG = I$. Then,

$$\bar{T} = HG\bar{T} = HT.$$

Note that the states in the range space of G are controllable and that the dimension of $\{\bar{T}\}$ is equal to the rank of G. This implies that each $\bar{T}$ is mapped by G uniquely into a controllable T. Therefore, each $\bar{T}$ of the reduced model will be controllable. The characterizing matrices of the reduced model can now be found by noting the effect of the mappings G and H on the original characterizing matrices

$$\bar{A} = HAG \qquad \bar{B} = HB$$

$$\bar{C} = CG \qquad \bar{D} = D.$$

VIII. Computation of the Transition Matrix

Computing the transition matrix $A^{i,j}$ using the recursive definition becomes quite tedious as i and j become large. It is therefore desirable to extend the techniques that are known for computing a single power of a matrix, such as the Cayley-Hamilton technique or Sylvester's theorem, to parallel techniques for $A^{i,j}$. The Cayley–Hamilton technique will be treated in what follows. Other methods may be extended in a similar fashion.

From the two-dimensional Cayley–Hamilton theorem we have

$$f(E,F)A = 0.$$

Letting $a_{h,k}$ be the coefficient of E^hF^k in $f(E,F)$ the last equation becomes

$$\sum_{(0,0) \leq (h,k) \leq (n_1,n_2)} \sum a_{h,k} A^{h,k} = 0.$$

If both sides of this equation are operated upon by E^iF^j it becomes

$$\sum_{(0,0) \leq (h,k) \leq (n_1,n_2)} \sum a_{h,k} A^{h+i,k+j} = 0.$$

Consider this equation for the set of pairs $\{(i,j) | (-n_1 \leq i \leq 0 \text{ and } j = 1) \text{ or } (-n_2 \leq j \leq 0 \text{ and } i = 1)\}$. These pairs have the property that whenever one exponent of $A^{h+i,k+j}$ is negative the other will be positive for $(0,0) \leq (h,k) \leq (n_1,n_2)$. From the definition of $A^{i,j}$ each such $A^{h+i,k+j}$ will be equal to zero. The following set of equations is produced, each corresponding to a pair (i,j) in the set above:

$$\sum_{(i,0)\leqslant(h,k)\leqslant(n_1,n_2)}\sum a_{h,k}A^{h-i,k+1}=0, \qquad 0\leqslant i\leqslant n_1$$

and

$$\sum_{(0,j)\leqslant(h,k)\leqslant(n_1,n_2)}\sum a_{h,k}A^{h+1,k-j}, \qquad 0\leqslant j\leqslant n_2.$$

Each of these equations can be written in terms of a function of E and F obtained by modifying the two-dimensional characteristic function $f(E,F)$. The modification in the first case consists of multiplying $f(E,F)$ by $E^{-i}F$ and then deleting all terms involving negative exponents of E. The modification for the second case consists of multiplying $f(E,F)$ by EF^{-j} and then deleting all terms involving negative exponents of F. Let

$$f_{i,j}(E,F)=E^{-i}F^{-j}f(E,F)$$

(with deletion of terms having negative exponents). Matrix A then must satisfy the following set of equations:

$$f_{i,-1}(E,F)A=0, \qquad 0\leqslant i\leqslant n_1-1$$

$$f_{-1,j}(E,F)A=0, \qquad 0\leqslant j\leqslant n_2-1.$$

That is, not only must A satisfy the two-dimension characteristic equation, but it must also satisfy the above set of n_1+n_2 equations as well. It will be shown that these equations may be used to reduce $A^{i,j}$ to a linear combination of those $A^{h,k}$ where $(0,0)\leqslant(h,k)<(n_1,n_2)$. This leads to the definition of two-dimensional eigenvalues.

Definition: The two-dimensional eigenvalues of a partitioned matrix

$$A=\begin{bmatrix} A_1 & A_2\\ A_3 & A_4\end{bmatrix}$$

are the pairs (x,y) that simultaneously solve the following set of equations:

$$f_{i,-1}(x,y)=0, \qquad 0\leqslant i\leqslant n_1-1$$

$$f_{-1,j}(x,y)=0, \qquad 0\leqslant j\leqslant n_2-1$$

$$f(x,y)=0.$$

The equations in the above definition may be used to reduce a multinomial in x,y to one of degree $(h,k)<(n_1,n_2)$. The next theorem establishes this result.

Theorem: Two-dimensional division algorithm. Any multinomial $g(x,y)$ of degree $\geqslant(0,0)$ may be expressed as follows:

$$g(x,y)=\sum_{j=0}^{n_2}p_j(x)f_{-1,j}(x,y)+\sum_{i=0}^{n_1}q_i(y)f_{i,-1}(x,y)+m(x,y)f(x,y)+r(x,y)$$

where the degree (h,k) of $r(x,y)$ is less than (n_1,n_2).

Proof: Each term in $g(x,y)$ of degree $(i,j)\geqslant(n_1,n_2)$, having the coefficient $b_{i,j}$, may be reduced to a sum of terms of degree less than (k,j) by subtracting $b_{i,j}x^{i-n_1}y^{j-n_2}f(x,y)$ from it. This is repeated until there are no terms of degree greater than (n_1,n_2). The remainder will have the form

$$g(x,y)-m(x,y)f(x,y).$$

Each term of this remainder of degree $(i,n_2-1)\geqslant(n_1,n_2-1)$ having the coefficient $c_{i,j}$, may then be reduced to a sum of terms of degree less than (i,n_2-1) by subtracting $c_{i,j}f_{-1,n_2-1}(x,y)$ from it. This is repeated until there are no terms $(h,n_2-1)\geqslant(n_1,n_2-1)$. A similar process is used for the other functions $f_{-1,j}$ and $f_{i,-1}$. The result will be a remainder of the form

$$r(x,y)=g(x,y)-\sum_{j=0}^{n_2}p_j(x)f_{-1,j}(x,y)+\sum_{i=0}^{n_1}q_i(y)f_{i,-1}(x,y)+m(x,y)f(x,y)$$

where the degree of $r(x,y)$ is less than (n_1,n_2). Q.E.D.

If the equation $f_{i,1}(E,F)\ A=0$ is operated upon by F^j no additional terms are generated so that the result will be a valid equation. Likewise, $E^if_{1,j}(E,F)A=0$ is a valid equation. Consequently, the two-dimensional division algorithm may be applied to matrix multinomials and in particular to $A^{i,j}$. Thus,

$$A^{i,j}=\sum_{j=0}^{n_2}p_j(E)f_{-1,j}(E,F)A+\sum_{i=0}^{n_1}q_i(F)f_{i,-1}(E,F)A+M(E,F)f(E,F)A+r(E,F)A$$

where the degree of $r(E,F)A$ is less than (n_1,n_2). The first three terms are zero because A satisfies the previously-mentioned equations. Then $A^{i,j}=r(E,F)A$. Likewise, if (x,y) is an eigenvalue of A, then from the division algorithm $g(x,y)=r(x,y)$.

Since the operations are the same for obtaining $r(x,y)$ and $r(E,F)A$ they will have the same coefficients. $A^{i,j}$ may then be computed by determining $r(x,y)$ from x^iy^j. $r(x,y)$ has $(n_1+1)(n_2+1)-1$ coefficients. These are determined by solving the set of simultaneous equations resulting from $x^iy^j=r(x,y)$, using different eigenvalues (x,y). The same number of eigenvalues as coefficients are necessary.

The determination of two-dimensional eigenvalues of a matrix A is not, in general, obvious. However, if the characteristic function $f(x,y)$ is factorable into linear factors then the two-dimensional eigenvalues are easily identified.

Theorem: Suppose the two-dimensional characteristic function of a matrix A is factorable into linear factors as follows:

$$f(x,y)=(x-a_1)(x-a_2)\cdots(x-a_{n_1})(y-b_1)(y-b_2)\cdots(y-b_{n_2}).$$

Let $a_0=b_0=0$. The set of two-dimensional eigenvalues is

$$\{(a_i,b_j)|(0,0)<(i,j)\leqslant(n_1,n_2)\}.$$

Proof: From the hypothesis $f(x,y)$ may be expressed as the product $p(x)q(y)$. Consequently, $f_{i,-1}(x,y)$ may be expressed as $p_i(x)yq(y)$, where $p_i(x)=x^{-i}p(x)$ (with negative powers of x deleted). Thus, $f_{i,-1}(x,y)$ contains each $(y-b_j)$ as a factor, which implies that $f_{i,-1}(x,b_j)=0$. Likewise $f_{-1,j}(a_i,y)=0$. Thus, for $(x,y)=(a_i,b_j)$ all functions, including f, will be zero. Q.E.D.

If the factors of $f(x,y)$ in the previous theorem are distinct then there will be a total of $(n_1+1)(n_2+1)-1$ eigenvalues, which is the same as the number of coefficients in $r(x,y)$. There is, therefore, just enough equations $a_h^i b_k^j=r(a_h,b_k)$ to solve for the coefficients of $r(x,y)$.

Example: Let

$$A=\left[\begin{array}{c|c} A_1 & A_2 \\ \hline A_3 & A_4 \end{array}\right]=\left[\begin{array}{c|c} a & b \\ \hline 0 & d \end{array}\right];$$

then

$$cm(A)=\begin{bmatrix} x-a & -b \\ 0 & y-d \end{bmatrix}$$

$$f(x,y)=|cm(A)|=(x-a)(y-d).$$

The two-dimensional eigenvalues are $(0,d)$, $(a,0)$, and (a,d).

$$A^{i,j}=r(E,F)A \quad \text{where deg } r(E,F)<(1,1)$$
$$=a_{0,0}I+a_{1,0}A^{1,0}+a_{0,1}A^{0,1}.$$

To find the coefficients substitute the eigenvalues into

$$x^iy^j=r(x,y)=a_{1,1}xy+a_{1,0}x+a_{0,1}y$$

$$0^id^j=a_{0,0}+a_{1,0}0+a_{0,1}d$$

$$a^i0^j=a_{0,0}+a_{1,0}a+a_{0,1}0$$

$$a^id^j=a_{0,0}+a_{1,0}a+a_{0,1}d.$$

The solution to these equations is

$$a_{0,0}=-a^id^j \quad a_{1,0}=a^{i-1}d^j \quad a_{0,1}=a^id^{j-1}$$

Note that

$$a^{1,0}=\begin{bmatrix} a & b \\ 0 & 0 \end{bmatrix}$$

$$a^{0,1}=\begin{bmatrix} 0 & 0 \\ 0 & d \end{bmatrix}.$$

Then

$$a^{i,j}=-a^id^j\begin{bmatrix} 1 & 0 \\ 0 & 1 \end{bmatrix}+a^{i-1}d^j\begin{bmatrix} a & b \\ 0 & 0 \end{bmatrix}+a^id^{j-1}\begin{bmatrix} 0 & 0 \\ 0 & d \end{bmatrix}$$

$$=\begin{bmatrix} 0 & a^{i-1}d^jb \\ 0 & 0 \end{bmatrix}.$$

To check:

$$A^{1,1}=A^{1,0}A^{0,1}+A^{0,1}A^{1,0}$$

$$=\begin{bmatrix} a & b \\ 0 & 0 \end{bmatrix}\begin{bmatrix} 0 & 0 \\ 0 & d \end{bmatrix}+\begin{bmatrix} 0 & 0 \\ 0 & d \end{bmatrix}\begin{bmatrix} a & b \\ 0 & 0 \end{bmatrix}$$

$$=\begin{bmatrix} 0 & bd \\ 0 & 0 \end{bmatrix}+\begin{bmatrix} 0 & 0 \\ 0 & 0 \end{bmatrix}=\begin{bmatrix} 0 & bd \\ 0 & 0 \end{bmatrix}.$$

This checks with the previous result.

VIII. Conclusion

This paper is an attempt to establish a parallel of the linear discrete-time state-space model for linear discrete-space image processing. However, it can only be assumed to be an initial attempt. Only the more basic and well-known concepts have been extended. Thus, there is much room for future research to be done along this line. Specifically this research should include:

1) Generalization to bilateral models.

2) Methods for programming a spatial transfer function into a state-space model.

3) Discovery of a general method for factoring multinomials and a method for finding two-dimensional eigenvalues.

4) Finding methods for obtaining canonical forms.

5) Establishing criteria for stability.
6) Application of estimation theory.

Finally, of a more general nature, the techniques and concepts of optimal control could be extended to the spatial model.

References

[1] A. Rosenfeld, *Picture Processing by Computer*. New York: Academic, 1969.
[2] H. C. Andrews, *Computer Techniques in Image Processing*. New York: Academic, 1970.
[3] B. S. Lipkin and A. Rosenfeld, *Picture Processing and Psychopictorics*. New York: Academic, 1970.
[4] T. S. Huang, W. F. Schreiber, and O. J. Tretiak, "Image processing," *Proc. IEEE*, vol. 59, pp. 1586–1609, Nov. 1971.
[5] W.D. Fryer and G.E. Richmond, "Two-dimensional spatial filtering and computers," in *Proc. Nat. Electron. Conf.*, Oct. 19, 1962.
[6] A. Habibi, "Two-dimensional Bayesian estimate of images," *Proc. IEEE (Special Issue on Digital Picture Processing)*, vol. 60, pp. 878–883, July 1972.
[7] A. Vander Lugt, "Operational notation for the analysis and synthesis of optical data-processing systems," *Proc. IEEE*, vol. 54, pp. 1055–1063, Aug. 1966.
[8] D. D. Givone and R. P. Roesser, "Multidimensional linear iterative circuits—general properties," *IEEE Trans. Comput.*, vol. C-21, pp. 1067–1073, Oct. 1972.
[9] ——, "Minimization of multidimensional linear iterative circuits," *IEEE Trans. Comput.*, vol. C-22, pp. 673–678, July 1973.
[10] B. Vilfan, "Another proof of the two-dimensional Cayley–Hamilton theorem," *IEEE Trans. Comput.* (Corresp.), vol. C-22, p. 1140, Dec. 1973.
[11] F. T. S. Yu, *Introduction to Diffraction, Information Processing, and Holography*. Cambridge, Mass. : M.I.T. Press, 1973.

41

Reprinted from *IEEE Trans. Circuits Syst.* **CAS-23**(7):414–422 (1976)

McClellan Transformations for Two-Dimensional Digital Filtering: II — Implementation

WOLFGANG F. G. MECKLENBRÄUKER AND RUSSELL M. MERSEREAU, MEMBER, IEEE

***Abstract*—A family of 2-D structures is presented for implementing 2-D FIR digital filters designed by means of a transformation of a 1-D design. These implementations are computationally efficient for filters up to degree 50×50 and are straightforward to program or build in hardware. Many details of the implementations are discussed such as its susceptibility to coefficient quantization and arithmetic roundoff. A comparison is made between these implementations and other implementations for 2-D FIR filters.**

Manuscript received August 8, 1975; revised January 22, 1976. This work was supported in part by the Advanced Research Projects Agency monitored by ONR under Contract N00014-75-C-0951-NR049-308 and in part by the National Science Foundation under Grant ENG71-02319-A02.

W. F. G. Mecklenbräuker was a Visiting Scientist in the Department of Electrical Engineering and Computer Science, Massachusetts Institute of Technology, Cambridge, MA. He is now with the Philips Research Laboratories, Eindhoven, The Netherlands.

R. M. Mersereau was with the Research Laboratory of Electronics and Department of Electrical Engineering and Computer Science, Massachusetts Institute of Technology, Cambridge, MA. He is now with the School of Electrical Engineering, Georgia Institute of Technology, Atlanta, GA 30332.

I. Introduction

TRADITIONALLY there have been two difficulties inherently associated with two-dimensional (2-D) digital filtering. First, 2-D digital filters are difficult to design because they must possess many degrees of freedom. Secondly, they are expensive to implement because their realization typically requires many arithmetic operations. In [1] it was shown that an extension of an FIR design technique, originally due to McClellan [2], could overcome the first of these difficulties. In this paper it is shown that filters designed by this method also possess efficient implementations which can partially overcome the second.

After first reviewing the transformation which forms the basis for the design method, we present this new realization in Section II and discuss computational details in Section III. This structure suggests a number of related variations, which are discussed in Section IV.

Although not all 2-D filters can be designed using transformations (the technique is only truly valuable for designing linear phase filters with quadrilateral symmetry) the authors feel that the McClellan transformation is the best technique available for designing high precision FIR digital filters. The algorithms are fast, filters of arbitrary order can be designed, and for filters of moderate degree, there is a powerful implementation.

II. The Implementation

It can be recalled from [1, eqs. (3) and (4)] that the frequency response of a 2-D FIR filter designed by a transformation is given by

$$H(e^{j\omega_1}, e^{j\omega_2}) = \sum_{n=0}^{N} b(n)\left[F(\omega_1,\omega_2)\right]^n \tag{1}$$

where

$$F(\omega_1,\omega_2) = \sum_{p=0}^{P}\sum_{q=0}^{Q} t(p,q)\cos p\omega_1 \cos q\omega_2. \tag{2}$$

The impulse response of this filter is $(2NP+1)\times(2NQ+1)$ points in extent. From (1) and (2) it can be noted that the entire $(2NP+1)\times(2NQ+1)$ point filter is completely specified by $(P+1)(Q+1)+N+1$ free parameters. It is thus reasonable to expect that there should be specific realizations for these filters which have an implementation complexity that is proportional to N rather than to N^2. Such implementations will be described in this paper.

As an aid to understanding this 2-D structure, let us first consider a related one-dimensional (1-D) filter structure for 1-D zero-phase filters of length $2N+1$. The frequency response of such a filter can be expressed as

$$H(e^{j\omega}) = \sum_{n=0}^{N} b(n)\left[\cos\omega\right]^n. \tag{3}$$

It is a well-known property of linear shift-invariant systems that if two such systems are arranged in cascade the frequency response of the cascade is the product of the individual frequency responses. If $h_f(m)$ represents the impulse response of a linear shift-invariant system which has frequency response

$$H_f(e^{j\omega}) = \cos\omega$$

the cascade of n such identical systems will have frequency response $[\cos\omega]^n$. A tapped cascade of such systems with tap coefficients $b(n)$ will thus have the frequency response depicted in (3).

The system $h_f(m)$ corresponds to a noncausal FIR filter with an impulse response length of 3

$$h_f(m) = \begin{cases} \frac{1}{2}, & m=\pm 1 \\ 0, & \text{otherwise} \end{cases}. \tag{4}$$

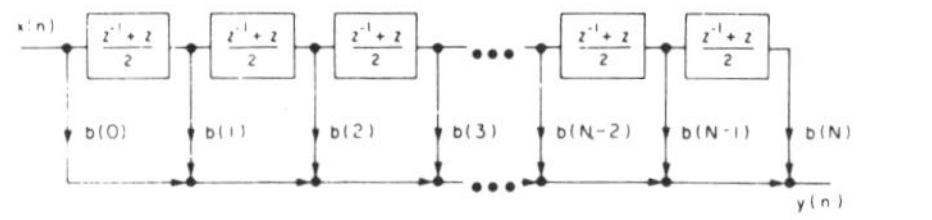

Fig. 1. Implementation of 1-D zero-phase filter in terms of zero-phase operators $(z+z^{-1})/2$.

The z-transform corresponding to this system is

$$H_f(z) = \frac{1}{2}(z+z^{-1}) \tag{5}$$

and the overall network which implements $h(n)$ is diagrammed in Fig. 1.

The McClellan transformation can be implemented by replacing the 1-D linear shift-invariant subnetwork $h_f(n)$ by a 2-D linear shift-invariant subnetwork which has impulse response $h_f(m,n)$ and frequency response $H_f(e^{j\omega_1}, e^{j\omega_2})$ where

$$H_f(e^{j\omega_1}, e^{j\omega_2}) = F(\omega_1,\omega_2). \tag{6}$$

If this substitution is made, then by reference to the 2-D network in Fig. 2(a), the frequency response of the whole 2-D filter is identical to (1). The filter $h_f(m,n)$ is a zero-phase $(2P+1)\times(2Q+1)$ point noncausal FIR filter and its impulse response is related to the transformation coefficients by the relation

$$t(|m|,|n|) = (2-\delta_{m0})(2-\delta_{0n})h_f(m,n), \qquad \text{for all } m,n$$

where δ_{mn} is the Kronecker delta.

To implement a transformed design it is necessary to first realize the prototype 1-D zero-phase filter $h_f(n)$ using zero-phase operators and then to physically implement the transformation by replacing those operators by transformation filters (operators) $h_f(m,n)$. Clearly three-dimensional filters could be designed and implemented the same way by replacing $h_f(n)$ by a three-dimensional transformation filter $h_f(m,n,p)$. Oppenheim, Mecklenbräuker, and Mersereau [3] have used the technique to design 1-D filters by replacing $h_f(n)$ by another 1-D filter $h_f'(n)$. For filters of dimension two or greater this implementation allows some computational savings, which will be elaborated in the next section. As the dimension increases the savings become greater.

There are many ways of representing the 1-D prototype filter in terms of zero-phase operators. Each of these yields, in turn, a different 2-D implementation after the substitution is made. In Fig. 2(b) we show the transposed implementation and in Section V we consider still different implementations. As an attempt to adopt some kind of standardized nomenclature, the structure depicted in Fig. 2(a) will be called the direct transformed structure and the structure in Fig. 2(b) will be called the transposed direct transformed structure.

The tap coefficients in the filter implementation are the $b(n)$'s which are given by the 1-D prototype filter. The subnetworks $h_f(m,n)$ depend only upon the specific transformation being used. If we are thus only working with nearly circularly symmetric filters ($P=Q=1$, $-t(0,0) = t(0,1) = t(1,0) = t(1,1) = 1/2$), to change say the cutoff frequency of the filter, we only need to change the tap coefficients $\{b(n)\}$.

If a single input array is to be filtered by several 2-D filters, each of which was designed by the same transformation, the chain of subnetworks $h_f(m,n)$ can be shared among the different filters with a considerable savings in computational effort. Such a fact could be useful in implementing the 2-D equivalent of a filter bank.

III. Discussion

A. Computational Organization

There are a variety of ways to turn the networks of Fig. 2 into viable computer code or functioning digital hardware. It is our goal in this section to address many of these computational issues—the first of which concerns how the computation should be organized. We shall describe the direct transformed realization first. To this end it will be beneficial to refer to Fig. 2(a).

The array $x(m,n)$ denotes the input array to the filter and the array $y(m,n)$ represents its output. The nodes labeled $x_i(m,n)$, $i=1,2,\cdots,N$ and $g_i(m,n)$, $i=1,2,\cdots,N$ are intermediate arrays which will be used in the calculation. The algorithm consists of N identical iterations, at each one of which the arrays $x_{i-1}(m,n)$ and $g_{i-1}(m,n)$ are used to produce arrays $x_i(m,n)$ and $g_i(m,n)$ by means of the following array operations:

$$x_i(m,n) = x_{i-1}(m,n) * h_f(m,n) \tag{7}$$

$$g_i(m,n) = g_{i-1}(m,n) + b(i)x_i(m,n) \tag{8}$$

($*$ denotes 2-D linear convolution). These computations are done for $i=1, 2,\cdots,N$ and they are initialized by

$$x_0(m,n) = x(m,n) \tag{9}$$

$$g_0(m,n) = b(0)x(m,n). \tag{10}$$

$g_N(m,n)$ corresponds to the output array $y(m,n)$. The convolution indicated in (7) can be implemented directly from the convolution sum in two dimensions

$$x_i(m,n) = \sum_{k=-P}^{P} \sum_{l=-Q}^{Q} h_f(k,l)x_{i-1}(m-k,n-l). \tag{11}$$

From the foregoing it is clear that once the array $x_i(m,n)$ has been computed, $x_{i-1}(m,n)$ is no longer needed and that x_i can thus be stored in place of x_{i-1}. Similarly, $g_i(m,n)$ can be stored in place of $g_{i-1}(m,n)$.

The transpose configuration in Fig. 2(b) can be implemented in a very similar fashion. It too requires two auxiliary arrays—one, $x(m,n)$, to contain the input array and one, $g_i(m,n)$, which will ultimately contain the output.

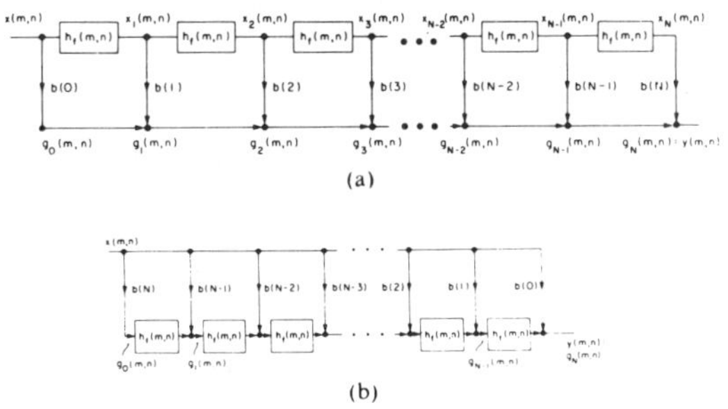

Fig. 2. Two implementations of 2-D digital filter which has been designed using transformation. (a) Direct transformed implementation. (b) Transpose direct transformed implementation.

At each of the N iterations we set

$$g_i(m,n) = b(N-i)x(m,n) + g_{i-1}(m,n)*h_f(m,n), \qquad i=1,2,\cdots,N \tag{12}$$

with

$$g_0(m,n) = b(N)x(m,n).$$

As before, the array $g_i(m,n)$ can be stored in place of the array $g_{i-1}(m,n)$.

It should be noted that each of these structures is highly modular which makes each easy to program on a digital computer or straightforward to build as a hardware device.

B. Complexity of the Implementations

It was claimed earlier that the complexity of these implementations was proportional to N where the 1-D prototype filter is of length $2N+1$ and the order of the 2-D design is $(2NP+1)\times(2NQ+1)$.

For the derivation of this fact let us assume that $P=Q$ (the usual case), that N is the parameter defined above, and that the input array is of finite extent and measures M points by M points. The output array $y(m,n)$ is then $(M+2PN)$ points by $(M+2PN)$ points.

As we proceed along the chain of transformation filters $h_f(m,n)$ each output $x_i(m,n)$ will be larger in extent than its predecessor $x_{i-1}(m,n)$. If, in fact, common storage arrays are to be used to store the arrays $\{x_i(m,n)\}$ and $\{g_i(m,n)\}$, then these arrays must be large enough to hold $(M+2PN)^2$ samples. The $M\times M$ input array will thus appear as an array of nonzero values encircled with P square rings of zeros. In a software implementation it is undoubtedly advantageous to recognize these zero values and thus save some processing time. This has the effect of not making the different stages of the implementation identical to each other which would probably be undersirable in a hardware implementation where for purposes of logic timing, pipelining, etc., it is advantageous to keep the computations identical from stage to stage. In the balance of this discussion, we shall distinguish between a nonmodular (software) and a slightly less efficient modular (hardware) implementation.

First consider the convolution which must be performed by each filter $h_t(m,n)$ in the chain. As written in (11), this operation will require $(2P+1)^2$ real multiplies for each output point (m,n). In fact, we can reduce this number by roughly a factor of four by exploiting the fact that $h_t(m,n)$ is a zero-phase filter and that thus

$$h_t(m,n)=h_t(-m,n)=h_t(m,-n)=h_t(-m,-n). \quad (13)$$

Therefore, the direct convolution can be performed using $\alpha=(P+1)^2$ multiplies for each point (m,n) and approximately $(2P+1)^2$ adds. If we are making the assumption that the structure is completely modular and the size of the output of the filter is $(M+2NP)\times(M+2NP)$ points, a single convolution by $h_t(m,n)$ thus requires

$$(M+2NP)^2(P+1)^2 \quad \text{real multiplies}$$

and

$$(M+2NP)^2(2P+1)^2 \quad \text{real adds.}$$

Actually these numbers represent upper bounds. The true number of multiplies could be smaller depending upon the specific transformation. For example, if the transformation is symmetric

$$t(m,n)=t(n,m) \quad (14)$$

then the number of multiplies (but not the number of adds) can be further halved. If the transformation was derived using the contour matching approach described in [1], then the presence of equality constraints among the coefficients can again be exploited to reduce the number of multiplies required in the implementation. Furthermore, for some transformations, such as McClellan's original transformation for circular symmetry [2], ($P=1$, $-t(0,0)=t(1,0)=t(0,1)=t(1,1)=1/2$), or for fan filters [4], ($P=1$, $t(0,0)=t(1,1)=0$, $t(1,0)=-t(0,1)=1/2$) the filter coefficients are all powers of two and thus can be implemented as single bit shifts rather than as multiplies with further computational savings. In these cases we have $\alpha=0$.

The operation of forming the array $g_i(m,n)$ from $x_i(m,n)$ requires an additional

$$(M+2PN)^2 \quad \text{real multiplies}$$

and

$$(M+2PN)^2 \quad \text{real adds.}$$

Summing the number of operations required for all stages we get

$$M_M=(M+2PN)^2\left[(P+1)^2N+N+1\right] \quad \text{multiplies} \quad (15)$$

TABLE I

COMPARISON OF NUMBERS OF MULTIPLIES NEEDED FOR TWO IMPLEMENTATIONS OF FILTER DESIGNED USING FIRST-ORDER TRANSFORMATION, AND FOR IMPLEMENTATION BY DIRECT CONVOLUTION (INPUT ARRAY HAS EXTENT $M\times M$ POINTS, 1-D PROTOTYPE FILTER IS OF LENGTH $2N+1$)

M	N	MODULAR IMPLEMENTATION	NONMODULAR IMPLEMENTATION	DIRECT CONVOLUTION (ZERO PHASE)
256	5	1.84×10^6	1.74×10^6	2.36×10^6
	10	3.88×10^6	3.49×10^6	7.93×10^6
	15	6.22×10^6	5.31×10^6	1.68×10^7
	20	8.85×10^6	7.21×10^6	2.89×10^7
	25	1.18×10^7	9.20×10^6	4.43×10^7
1024	5	2.78×10^7	2.74×10^7	3.78×10^7
	10	5.56×10^7	5.40×10^7	1.27×10^8
	15	8.44×10^7	8.09×10^7	2.68×10^8
	20	1.14×10^8	1.08×10^8	4.62×10^8
	25	1.45×10^8	1.36×10^8	7.09×10^8

$$A_M=(M+2PN)^2\left[(2P+1)^2N+N\right] \quad \text{adds} \quad (16)$$

for the modular implementation and

$$\hat{M}_M=\left[(P+1)^2+1\right]N\left(M^2+MP(N+1)+2P^2(N+1)(2N+1)/3\right)+M^2 \quad (17)$$

$$\hat{A}_M=\left[(2P+1)^2+1\right]N\left(M^2+MP(N+1)+2P^2(N+1)(2N+1)/3\right) \quad (18)$$

for the nonmodular implementation. For $M\gg N$ each of these figures is proportional to N. These figures are tabulated in Table I for typical values of M, N, and P. For those data $\hat{M}_M$ is between 5 and 20 percent smaller than M_M.

The number of multiplies and adds is not the only figure of merit. In addition, the amount of storage needed must be considered. Some of the savings in arithmetic that are afforded by this implementation are at the expense of increased storage. We must provide storage for the arrays $x_i(m,n)$ and $g_i(m,n)$. $x(m,n)$ can be stored in place of $x_0(m,n)$ and the final array $y(m,n)$ can be stored in place of $g_N(m,n)$. Thus storage for two arrays of size $(M+2N)^2$ is sufficient. The total storage required is thus

$$S_M=2(M+2NP)^2 \quad (19)$$

registers.

The above figures all refer to the direct transformed implementation. For the transpose version since the input array $x(m,n)$ is used repeatedly and this array does not grow from stage to stage, these figures are slightly smaller. In this case

$$M_M=\left[M+2PN\right]^2\left[(P+1)^2N\right]+M^2(N+1) \quad (20)$$

$$A_M = \left[M + 2PN \right] \left[(2P+1)^2 N \right] + M^2 (N+1) \quad (21)$$

$$S_M = (M + 2NP)^2 + M^2 \quad (22)$$

for the modular version.

Now we shall show that these implementations for transformed filters compare favorably with other implementations, at least for moderate chain lengths.

For filters of low order the direct implementation can be used. This corresponds to performing the convolution sum directly using the impulse response $h(m,n)$ of the transformed filter. This is the technique which was outlined in Section II for the implementation of the transformation filters $h_f(m,n)$. As in that case, the fact that $h(m,n)$ is zero phase and symmetric about the m and n axes, implies that some multiplies can be saved. Therefore, $M_D = (M + 2NP)^2 \; (NP+1)^2$ real multiplies, $(M + 2NP)^2 (2NP+1)^2$ real adds, and $(M+2NP)(2NP+1)$ words of storage are required. For $M \gg N$ the number of multiplies and additions in the direct implementation is proportional to N^2, and this number is higher than that for the transformed implementation. The storage, however, is smaller than that required for the transformed implementation.

FIR filters designed by any means including transformations can also be implemented using the 2-D discrete Fourier transforms (DFT's). This can be an efficient procedure if the transforms are computed using the fast Fourier transform (FFT) algorithm. The details of the technique are available in a variety of sources including [5] and will not be repeated here. It is necessary to compute two $(M+2NP)$ by $(M+2NP)$ point DFT's (it is generally assumed that the DFT of the filter itself has been precomputed and stored) and to multiply the transform of the data (which is complex) and the transform of the filter (which is real) together. If $M+2NP$ is a power of two this requires approximately

$$M_{\text{FFT}} = 4(M+2NP)^2 \log_2 (M+2NP) + (M+2NP)^2 \text{ real multiplies}$$

a similar number of adds and $2(M+2NP)^2$ words of storage. In the DFT implementation the storage required is therefore comparable to that needed for the transformed implementation. For filters of moderate degree, however, the number of multiplies is higher than for the transformed implementation. In Table II the multiplies required in the DFT, direct, and transformed implementations are compared.

C. Block Processing

The idea of partitioning the input array into blocks and processing each block separately which is widely used in FFT implementations can also be effectively used with these transformed implementation to reduce the required storage drastically at a small increase in the number of adds and multiplies.

TABLE II
COMPARISON OF COMPUTATION EFFORT REQUIRED FOR DIRECT, FFT, AND TRANSFORMED IMPLEMENTATION OF TRANSFORMED FILTER (FIGURES ARE RATIOS OF NUMBERS OF MULTIPLIES PER OUTPUT POINT)

Order of 2-D Filter	M = 256, α = 0		M = 1024, α = 4	
	M_D/M_{FFT}	M_M/M_{FFT}	M_D/M_{FFT}	M_M/M_{FFT}
5 x 5	.27	.09	.22	.27
9 x 9	.76	.15	.61	.51
11 x 11	1.09	.18	.88	.63
13 x 13	1.48	.21	1.20	.76
17 x 17	2.45	.27	1.98	1
33 x 33	8.76	.52	7.05	1.98
65 x 65	33	.1	26.56	3.93

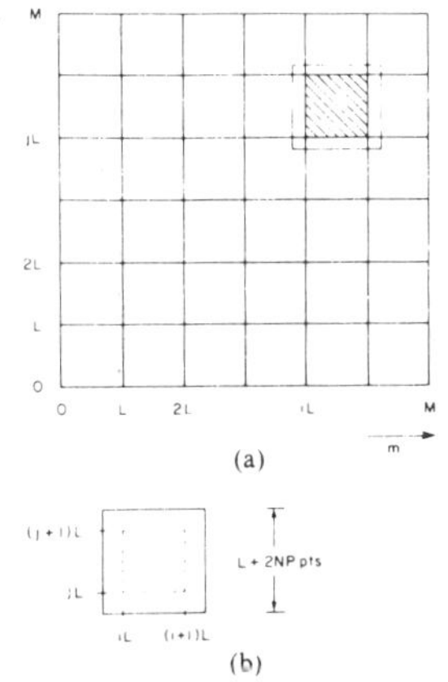

Fig. 3. Use of partitioning to conserve storage. (a) Array $x(m,n)$ partitioned into subarrays. The subarray $x_{ij}(m,n)$ is shown shaded. (b) Representation of contribution to output of single section.

Assume that the input array $x(m,n)$ is partitioned into subarrays $\{x_{ij}(m,n)\}$ such that

$$x_{ij}(m,n) = \begin{cases} x(m,n), & iL \leqslant m < (i+1)L, jL \leqslant n < (j+1)L \\ 0, & \text{otherwise} \end{cases} \quad (23)$$

It then follows that

$$x(m,n) * h(m,n) = \sum_{i=0}^{[M/L]} \sum_{j=0}^{[M/L]} x_{ij}(m,n) * h(m,n) \quad (24)$$

where [ ] denotes the greatest integer function.

The output subarrays $y_{ij}(m,n) = x_{ij}(m,n) * h(m,n)$ overlap one another and from (24) where they overlap they must be added. In Fig. 3 we have shown how the overlapping will take place. This decomposition is similar to an overlap-add decomposition as described in [5]. We could also have proceeded according to an overlap-save decomposition and similar savings would ensue.

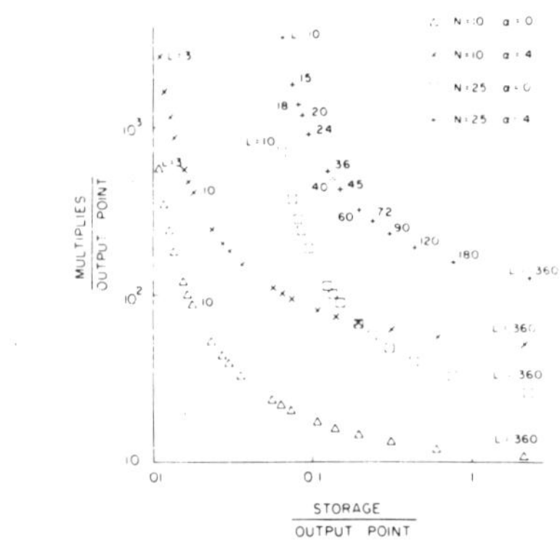

Fig. 4. Number of multiplies per output point versus storage per output point as function of block size, L, for different filter parameters.

If we use (15) as a guide to how many multiplies will be required per block of size $L\times L$, the number of real multiplies is thus

$$M_L=\left[\frac{M}{L}+1\right]^2(L+2PN)^2((P+1)^2N+N+1). \quad (25)$$

In computing the number of adds we must remember that some are required in combining the subarrays $y_{ij}(m,n)$.

$$A_L=\left[\frac{M}{L}+1\right]^2((M+2PN)^2((2P+1)^2+1)N+4NPL+4N^2P^2). \quad (26)$$

The total storage required is

$$S_L=2[L+2NP]^2+(2NP)(2L+2NP). \quad (27)$$

In Fig. 4, we have plotted $M_L/(M+2NP)^2$, the number of multiplies per output point, versus $S_L/(M+2NP)^2$, the storage per output point, for various values of L for $M=360$, $P=1$, and $N=10$ and 25. Two sets of curves are plotted for different α, the number of multiplies required per output point for a single transformation filter. $\alpha \doteq 4$ is a worst case value for a first-order transformation and $\alpha=0$ is a best case. It should be noted that if we are willing to increase the number of multiplies by, say a factor of 2, the amount of storage can be reduced severalfold. M_L dictates the speed of the implementation and S_L in some sense dictates the size of the computer. In all of this discussion it is assumed that the number of multiplies, adds, and storage registers is a reliable figure of merit. In practice the exact form the implementation should take will be tempered by the nature of the computing hardware available.

D. Coefficient Sensitivity and Computational Noise

The savings in arithmetic operations that are possible with this implementation do not come for free. Here the price that must be paid is that the direct transformed implementation is generally more numerically sensitive than the direct convolution. On a general purpose computer with wordlengths on the order of 32 bits or more, these problems will probably not be apparent for chains of modest length. For hardware or minicomputer implementations, however, the issue of computational error should be examined. Variations of the transformed structures exist which have less computational error and these are discussed in the next section.

The source of the numerical difficulties is the size of the coefficients $\{b(n)\}$. The coefficient $b(N)$, where N is the length of the chain, grows as 2^N, while the overall gain of the system is of the order of unity. Thus, in the implementation large, nearly equal numbers must be subtracted from one another. This implies that extra bits might be necessary to guarantee a given level of precision.

1) The Direct Transformed Implementation: Let us address the issue of coefficient sensitivity first. Since

$$H(e^{j\omega})=\sum_{n=0}^{N} b(n)[\cos\omega]^n$$

if Δ_n represents the error in the representation of $b(n)$, and $\Delta H(e^{j\omega})$ represents the error in the frequency response

$$\Delta H(e^{j\omega})=\sum_{n=0}^{N}\Delta_n[\cos\omega]^n \quad (28)$$

$$|\Delta H(e^{j\omega})|\leqslant\sum_{n=0}^{N}|\Delta_n|. \quad (29)$$

The total error in the frequency response is thus, in the worst case, proportional to the absolute deviation in each coefficient. To guarantee that the error in the frequency response be less than 2^{-b} will require a precision in the coefficients $b(n)$ of roughly $N+b+\log_2 N$ bits.

The next source of computational error to consider is arithmetic roundoff error caused by quantizations after each of the multiplies. Let us assume that the filter is implemented in fixed point arithmetic and that the error produced by each multiplier is independent both of the input and also of the other errors in the system. Let us make the further assumption that each error sequence introduced is white and that each sample is a random variable uniformly distributed between $-1/2\cdot 2^{-b}$ and $1/2\cdot 2^{-b}$ where b is the number of bits to be used in the representation of the data which are to the right of the binary point.

Fig. 5(a) shows the locations of a series of equivalent noise inputs for a chain of length 6. The noises $\eta_i(m,n)$ are due to the computations by the transformation filters $h_f(m,n)$ while those labeled $\epsilon_i(m,n)$ are due to the multiplications by $b(i)$. Each of these noises has a sample variance $1/12\cdot 2^{-2b}$. For convenience let $h_i(m,n)$ denote the impulse response of a chain of length i. Thus

$$h_i(m,n)\leftrightarrow[F(\omega_1,\omega_2)]^i. \quad (30)$$

Further set

$$R_i=\sum_{m=-\infty}^{\infty}\sum_{n=-\infty}^{\infty}h_i^2(m,n). \quad (31)$$

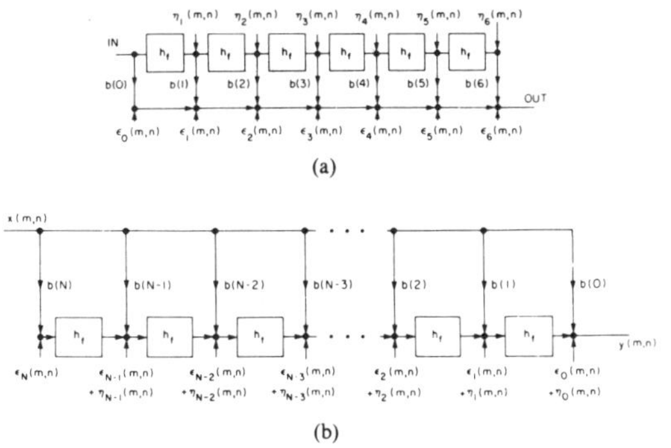

Fig. 5. Sources of arithmetic quantization error for (a) direct transformed realization, (b) transpose direct transformed realization.

TABLE III
PARAMETER R_i WHICH REPRESENTS NOISE GAIN AS FUNCTION OF i, CHAIN LENGTH FOR PARTICULAR FIRST-ORDER TRANSFORMATION

i	R_i	i	R_i
0	1.0000	15	.2084
1	.5625	16	.2031
2	.4463	17	.1983
3	.3858	18	.1939
4	.3468	19	.1897
5	.3189	20	.1859
6	.2976	21	.1823
7	.2805	22	.1790
8	.2664	23	.1759
9	.2545	24	.1729
10	.2443	25	.1701
11	.2354	26	.1674
12	.2275	27	.1649
13	.2204	28	.1625
14	.2141		

If we make the (invalid) assumption that the noises η_i are independent of one another then a straightforward 2-D extension of the analysis technique presented in [6, ch. 9] can be used to estimate the variance of the error signal at the output. Denoting that variance by σ_e^2 we observe

$$\sigma_e^2=\frac{2^{-2b}}{12}\left\{\sum_{i=1}^{N} b^2(i)\sum_{j=0}^{i-1} R_j+1\right\}. \tag{32}$$

In Table III the values of R_i are tabulated for $i=1$ to 28 for the transformation $-t(0,0)=t(1,0)=t(0,1)=t(1,1)=0.5$. In Table IV we present σ_e for several different values of N. The agreement between the simulated results and the theoretical values predicted by the simple model represented by (32) is very good. The simulations of the finite wordlength effects were made by filtering a finite area 2-D random sequence by a variety of transformed low-pass filters. The low-pass filters did not all have

TABLE IV
STANDARD DEVIATION OF OUTPUT NOISE DUE TO ARITHMETIC QUANTIZATION FOR LOW-PASS FILTERS OF VARIOUS ORDERS

chain length	FILTER NOISE $\sigma_e \times 2^b$: Transpose Direct Transformed Structure — theoretical	actual	Direct Transformed Structure — theoretical	actual
3	.4467	.4017	.3445	.3178
4	.4780	.4128	.5337	.3966
5	.5050	.4150	.9736	.6086
6	.5283	.4619	1.4983	.7797
7	.5506	.4338	1.5016	.7396
8	.5704	.7333	11.1511	5.2249
9	.5887	1.3846	22.9456	9.4785
10	.6058	1.3046	26.3047	10.1859
11	.6218	2.1383	92.1070	63.1761
12	.6368	3.6492		
13	.6511	8.3521		
15	.6776	2.9198		

exactly the same cutoff frequency, but generally the cutoff was between $\pi/3$ and $\pi/2$. The random input sequence was never larger than 28×28, so there is some statistical uncertainty in the measured values.

What is clear from the simulation and the model is that, due to the growth of the filter coefficients, the standard deviation of the output roundoff noise grows as $2^{(N-b)}$. Thus for every transformation filter that is added to the chain, the number of bits to the right of the binary point must be increased by one.

2) The Transpose Direct Transformed Implementation: Since the coefficients used in the transpose direct transformed structure are identical to those used in the direct transpose implementation, the coefficient sensitivity difficulties for the two structures are identical. The effects of arithmetic quantization, however, will be different. In fact, the transpose implementation is better behaved with respect to computational noise.

In Fig. 5(b) we show the equivalent noise sources for the transpose configuration. Under the same assumptions as with the direct structure the variance of the output noise is given by

$$\sigma_e^2=\frac{1}{12}2^{-2b}\sum_{i=0}^{N-1} R_i. \tag{33}$$

In Table IV we have included these theoretical and simulated noise figures for comparison with the corresponding figures for the direct transformed realization.

Thus for long chains the transpose implementation is clearly to be preferred over the direct. Both realizations, however, require long wordlengths to accommodate the size of the coefficients and in general with either of these realizations the wordlength required for the implementation must grow linearly with N.

IV. Generalizations and Alternate Forms

A. The Cascade Transformed Implementation

The direct transformed implementation affords a savings in computational effort at the expense of coefficient sensitivity and computational noise. The noise could be reduced, but not the sensitivity problems, by recourse to the transpose transformed structure. In this section we present a third structure which appears to be a reasonable compromise while still affording the same computational efficiency as the direct transformed structure. This structure is the cascade transformed structure.

This structure is possible because the transfer functions of filters of any dimensionality designed by a transformation must always be factorable. This is readily seen by reference to (1) where the 2-D transfer function is expressed as a 1-D polynomial in the variable $F(\omega_1,\omega_2)$. The z-transform of the prototype filter can be written as

$$H(z)=\sum_{n=0}^{N} b(n)\left(\frac{z+z^{-1}}{2}\right)^{n}. \tag{34}$$

Letting $z+z^{-1}/2=F$, this can be factored as

$$H(z)=\prod_{k=1}^{N_1}\left(d_k(2)F^2+d_k(1)F+d_k(0)\right) \cdot \prod_{l=1}^{N_2}\left(c_l(1)F+c_l(0)\right) \tag{35}$$

where the coefficients $d_k(i)$, $c_l(i)$ are all real. Here $2N_1+N_2=N$. The coefficients in (35) can be readily computed from the zeros of the prototype filter. For each zero on the unit circle and its complex conjugate we get one first-order factor in (35). For each zero not on the unit circle we get one second-order factor by combining that zero, its reciprocal, and their complex conjugates. Due to the dynamic range of the coefficients $b(n)$ this approach is to be preferred over actually factoring equation (35). From (35) a 2-D filter now results by replacing F by its 2-D equivalent as indicated in (2). One structure which results is shown in Fig. 6.

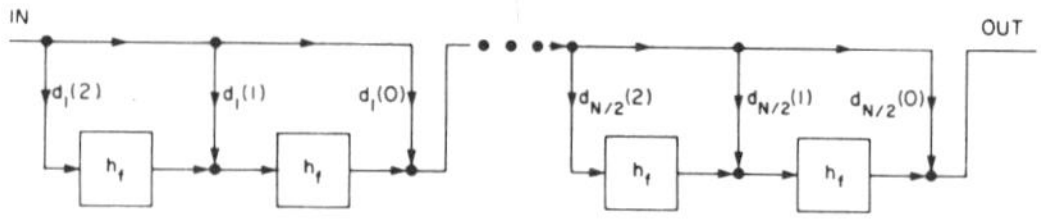

Fig. 6. Cascade transformed realization.

The coefficients needed for this implementation are those labeled $d_k(i)$ and $c_l(j)$ which have always been observed to be on the order of unity for typical low-pass filters. Although we have not carried out detailed calculations of coefficient sensitivity, it is believed that this structure should be far less sensitive than the implementations of Fig. 2, based on limited simulations.

The effects of arithmetic quantization will vary in their severity depending upon how the sections are ordered. Once again detailed studies have not been carried out, but we have found the guidelines suggested by Chan and Rabiner [7] for ordering the sections in a 1-D FIR cascade implementation to be useful. The gain from any point in the filter to the output should be kept as close to unity as possible. In Table V we present noise figures for a 16-bit simulation of a 17×17-point transformed filter. The orderings indicated in that table can perhaps be made clearer by reference to Fig. 7 where the zeros in the transfer function are labeled. The effect of quantization error in the best of the cascade implementations appears to lie between that of the transpose and direct transformed realizations.

TABLE V
Arithmetic Roundoff Noise for Several Orderings of Sections in Cascade Transformed Structure

ORDERING	NOISE x 2^b
1 2 3 4	7.1936
1 2 4 3	6.1700
1 3 2 4	6.1467
1 3 4 2	5.3179
1 4 2 3	3.7183
1 4 3 2	4.0040
2 1 3 4	7.3773
2 1 4 3	6.2632
2 3 1 4	5.8480
2 3 4 1	5.8426
2 4 1 3	3.2616
2 4 3 1	3.4676
3 1 2 4	5.8766
3 1 4 2	5.1312
3 2 1 4	6.0744
3 2 4 1	5.8288
3 4 1 2	13.5344
3 4 2 1	14.0742
4 1 2 3	7.1996
4 1 3 2	7.3126
4 2 1 3	5.2097
4 2 3 1	7.1876
4 3 1 2	12.2807
4 3 2 1	12.3597
transpose direct transformed	.7333
direct transformed	5.2249

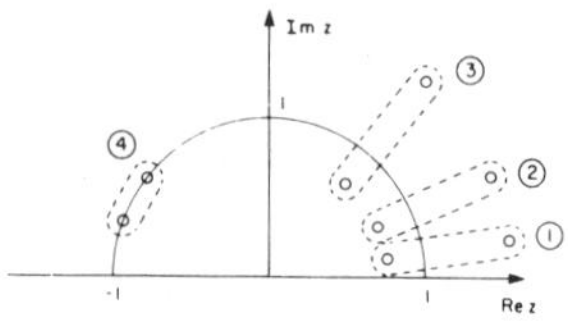

Fig. 7. Zeros of 1-D prototype filter used to produce Table V, labeled for convenience.

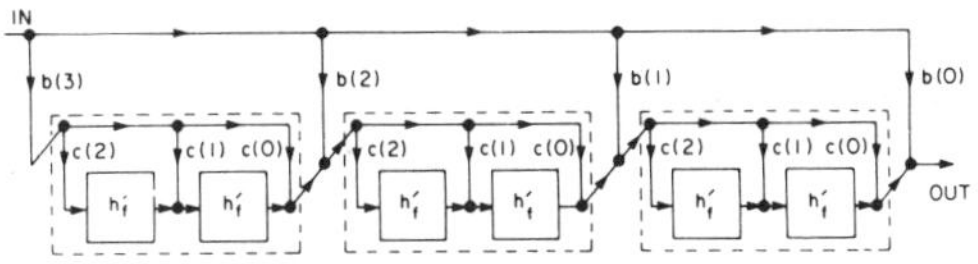

Fig. 8. Implementation of iterated transform design.

B. Iterated Transformations

Suppose the transformation filter $h_f(m,n)$ was itself designed by a transformation. If $H(e^{j\omega_1},e^{j\omega_2})$ is the frequency response of the filter then

$$H(e^{j\omega_1},e^{j\omega_2}) = \sum_{n=0}^{N-1} b(n)\left[H_f(e^{j\omega_1},e^{j\omega_2})\right]^n$$

$$= \sum_{n=0}^{N-1} b(n)\left[\sum_{m=0}^{M-1} c(m)\left[H_f'(e^{j\omega_1},e^{j\omega_2})\right]^m\right]^n. \quad (36)$$

The structure implied by (36) (in the transposed form) is shown in Fig. 8.

It follows from (36) that the contours

$$H(e^{j\omega_1},e^{j\omega_2}) = \text{constant} \quad (37)$$

are equivalent to the contours

$$H_f'(e^{j\omega_1},e^{j\omega_2}) = \text{constant}. \quad (38)$$

The contour shapes that result are thus no different than if the filter had been designed using $h_f'(m,n)$ as a transformation filter. The advantage to doing the design in two steps is that the coefficients $\{b(n), c(m)\}$ are now much smaller than in the case of a noniterated design and thus the coefficient sensitivity should be less. The price paid by using this technique is that the effective 1-D prototype filter will have a larger passband and stopband ripple than would be the case with a one step design and this will be reflected in the 2-D filter. The number of arithmetic operations that must be performed in an iterated implementation is effectively the same as with the comparable one-step implementation of the same filter.

The design problem for an iterated transformation is straightforward using the algorithms of [1, sec. III-3].

C. Other Structures Which Have Similar Implementations

Filters designed using McClellan transformations are not the only filters whose arithmetic complexity increases linearly with N. For example, it is not necessary that all filters in the chain of Fig. 2 have the same transfer function. If all of these low-order filters have different impulse responses all of the statements made to this point concerning implementation efficiency still apply. The FIR cascade realization is a special case of this class of filters.

Unfortunately, an arbitrary 2-D FIR filter cannot necessarily be implemented as such a chain and the authors know of no design algorithms for designing filters in the general case for a chain length greater than two or three. Maria and Fahmy [8] have considered the design of recursive filters in cascade form but only for filters of low order.

V. Summary

In this paper we have shown that 2-D FIR filters designed using a transformation due to McClellan possess a series of highly efficient implementations which are more efficient than the direct implementations and are even more efficient than an FFT implementation for filters of order up to 50×50 or so. The implementation exploits the structure of the 1-D prototype filter. We have also presented techniques for overcoming potential difficulties inherent in these realizations such as storage and effects of finite register length.

References

[1] R. M. Mersereau, W. F. G. Mecklenbräuker, and T. F. Quatieri, Jr., "McClellan transformations for two-dimensional digital filtering: I-design," *IEEE Trans. Circuits and Systems*, this issue, pp. 405–414.
[2] J. H. McClellan, "The design of two-dimensional digital filters by transformations," in *Proc. 7th Annual Princeton Conference on Information Sciences and Systems*, pp. 247–251, 1973.
[3] A. V. Oppenheim, W. F. G. Mecklenbräuker, and R. M. Mersereau, "Variable cutoff linear phase digital filters," *IEEE Trans. Circuits and Systems*, vol. CAS-23, Apr. 1976.
[4] J. H. McClellan and T. W. Parks, "Equiripple approximation of fan filters," *Geophysics*, vol. 37, pp. 573–583, 1972.
[5] R. M. Mersereau and D. E. Dudgeon, "Two-dimensional digital filtering," *IEEE Proceedings*, vol. 63, pp. 610–623, Apr. 1975.
[6] A. V. Oppenheim and R. W. Schafer, *Digital Signal Processing*. Englewood Cliffs, N J: Prentice-Hall, 1975.
[7] D. S. K. Chan and L. R. Rabiner, "An algorithm for minimizing roundoff noise in cascade realizations of finite impulse response digital filters," *Bell System Technical Journal*, vol. 52, no. 3, pp. 347–385, Mar. 1973.
[8] G. A. Maria and M. M. Fahmy, "An l_p design technique for two-dimensional digital recursive filters," *IEEE Trans. Acoust., Speech, and Signal Processing*, vol. ASSP-22, pp. 15–22, Feb. 1974.

Editors' Comments on Papers 42 and 43

42 SHIPP and AGGARWAL
An Error Analysis for a Vector Model of Two Dimensional Recursive Filters

43 NI and AGGARWAL
Error Analysis of Two-Dimensional Recursive Digital Filters Employing Floating-Point Arithmetic

ANALYSIS OF QUANTIZATION EFFECTS

As a consequence of the finite word length of any digital machine, unavoidable errors occur in the implementation of a digital filter transfer function. These errors are due to the effects of quantizing filter coefficients and the results of arithmetic operations. Because for most applications the 2-D filter implementations are carried out on a general purpose computer with reasonably large wordlength, these quantization errors have not created a problem comparable to that occurring in one-dimensional digital filtering. As a result, very little work has been reported on the error analysis aspect of 2-D filter implementations. The first paper to consider this problem was written by Ni and Aggarwal (1974), who have directly generalized the error analysis procedures used for 1-D filters to the 2-D filter case. However, these results are difficult to apply for the practical evaluation of quantization effects. Paper 42 by Shipp and Aggarwal offers an alternate, more feasible approach. Here the authors develop a bound for the rms value of the quantization error in the output of a fixed point implementation of an all-pole 2-D recursive filter. Their analysis approach is based on a vector model of the recursive filter obtained by sectioning the output sequence. The bounds derived by the authors are valid for "transient" output sequences as well as limit cycles for constant inputs. It should be possible to extend their approach to other cases.

Paper 43 considers the analysis of quantization effects for 2-D direct form structures implemented using floating-point arithmetic. It provides a method for estimating the mean-squared error due to round-off, input and multiplier coefficient quantizations, and also gives expressions for computing error-norm bounds. The approach followed here is similar to that used for the error analysis of 1-D structures.

In 2-D stable digital filters, the output may be of an oscillatory nature for constant (zero and nonzero) inputs. Such oscillations, known as limit cycles, occur due to rounding the results of arithmetic operations. Publications dealing with limit cycle oscillations in 2-D filters have been contributed by Maria and Fahmy (1975a, 1975b) and Chang (1977). It is expected that more attention to the error analysis problem (and associated limit cycle problem) will occur in the future when dedicated hardware implementations of 2-D filters are considered for practical applications.

In the case of 2-D nonrecursive filters, the analysis of quantization effects is somewhat similar to that used for the 2-D recursive filters. In Paper 41 by Mecklenbräuker and Mersereau, some discussion concerning quantization effects of various implementations of nonrecursive transfer functions (designed via generalized McClellan transformations) can be found.

REFERENCES

Chang, T-L. 1977. "Limit Cycles in a Two-Dimensional First-Order Digital Filter," *IEEE Trans. Circuits Syst.* **CAS-24**(1):15–19.

Maria, G. A., and M. M. Fahmy. 1975a. "Limit Cycle Oscillations in First-order Two-Dimensional Digital Filters," *IEEE Trans. Circuits Syst.* **CAS-22**:246–251.

Maria, G. A., and M. M. Fahmy. 1975b. "Limit Cycle Oscillations in a Cascade of Two-Dimensional Digital Filters," *IEEE Trans. Circuits Syst.* **CAS-22**:826–830.

Ni, M-D., and J. K. Aggarwal. 1974. "Two-Dimensional Digital Filtering and Its Error Analysis," *IEEE Trans. Computers* **C-23**:942–954.

42

Reprinted from *IEEE Trans. Acoust., Speech, Signal Process.*
ASSP-24(4):339–341 (1976)

AN ERROR ANALYSIS FOR A VECTOR MODEL OF TWO-DIMENSIONAL RECURSIVE FILTERS

K. O. Shipp, Jr. and J. K. Aggarwal

***Abstract*—The problem of bounding the output quantization errors in linear two-dimensional recursive filters is examined. A two-dimensional filter is analyzed by developing a model for the filter based on a vector formed from the two-dimensional output sequence. A deterministic bound is obtained and then applied to a special class of second-order filters.**

This correspondence offers a new approach for analyzing two-dimensional filters. An equation, using a vector formed from a subset of previous output points, is used to bound the rms value of the quantization error in the output of a fixed-point implementation of a two-dimensional recursive filter. We assume that the quantization used is roundoff and the quantization step size is $q = 2^{-t+1}$, where t represents the finite register length of the implementation in bits. The bounds obtained are valid for transient output arrays as well as limit cycles for both zero and nonzero inputs.

We will consider filters implemented by the difference equation

$$y(i, k) = -\sum_{\substack{m=0 \\ m+n \neq 0}}^{M} \sum_{n=0}^{N} a_{mn} y(i-m, k-n) + u(i, k), \tag{1}$$

where $\{y(i, k)\}$ and $\{u(i, k)\}$ are the output and input arrays, respectively. For convenience, we assume that $M = N$.

Consider the following N^2 column vector:

$$x(m, n) = \{y(m-N+i, n-N+k)\}_{i,k=1}^{N}. \tag{2}$$

Using the above vector, our filter may also be represented by the following set of equations:

$$x(m, n) = \sum_{\substack{r=0 \\ r+s \neq 0}}^{1} \sum_{s=0}^{1} A_{rs} x(m-r, n-s) + b\,u(m, n) \tag{3a}$$

$$y(m, n) = c^t x(m, n), \tag{3b}$$

where the A_{rs} are $N^2 \times N^2$ matrices and b and c are N^2 vectors.

The set of coefficient matrices in the above representation is not unique. One possible set for $N = 2$ is

$$A_{10} = \begin{bmatrix} 0 & 0 & 0 & 0 \\ 0 & 0 & 0 & 1 \\ 0 & 0 & 0 & 0 \\ -a_{21} & -a_{20} & 0 & -a_{10} \end{bmatrix}, \quad A_{01} = \begin{bmatrix} 0 & 0 & 0 & 0 \\ 0 & 0 & 0 & 0 \\ 0 & 0 & 0 & 1 \\ -a_{12} & 0 & -a_{02} & -a_{01} \end{bmatrix}$$

$$A_{11} = \begin{bmatrix} 0 & 0 & 0 & 1 \\ 0 & 0 & 0 & 0 \\ 0 & 0 & 0 & 0 \\ -a_{22} & 0 & 0 & -a_{11} \end{bmatrix}, \text{ and } b = c = \begin{bmatrix} 0 \\ 0 \\ 0 \\ 1 \end{bmatrix}. \tag{4}$$

The fact that the above matrices are singular is not crucial for the purpose of this correspondence, as will be shown later.

Following Yakowitz and Parker [1], let $[\,\cdot\,]_q$ denote the roundoff of the multiplications of each of the elements of a vector to a quantization level of q and let primes, $(\,\cdot\,)'$, denote rounded quantities. Also assume that the initial conditions of the states and that the input are previously rounded quantities. Therefore, using the above notation and taking into account the roundoff errors, (3a) becomes

$$x'(m, n) = \sum_{\substack{r=0 \\ r+s \neq 0}}^{1} \sum_{s=0}^{1} [A_{rs} x'(m-r, n-s)]_q + [b\,u(m, n)]_q. \tag{5}$$

Define the following error vectors:

$$e(m, n) \triangleq \sum_{\substack{r=0 \\ r+s \neq 0}}^{1} \sum_{s=0}^{1} (A_{rs} x'(m-r, n-s) - [A_{rs} x'(m-r, n-s)]_q) + b\,u(m, n) - [b\,u(m, n)]_q, \tag{6a}$$

$$v(m, n) \triangleq x(m, n) - x'(m, n), \text{ and} \tag{6b}$$

$$z(m, n) \triangleq y(m, n) - y'(m, n). \tag{6c}$$

Manuscript received May 20, 1975; revised October 3, 1975. This research was sponsored in part by the Joint Services Electronics Program under JSEP Contract F44620-70-C-0091 and in part by the National Science Foundation under Grant GK 42790.

The authors are with the Department of Electrical Engineering, University of Texas at Austin, Austin, TX 78712.

Therefore, from (3a) and (5) we obtain

$$v(m,n)=\sum_{\substack{r=0\\r+s\neq 0}}^{1}\sum_{s=0}^{1}A_{rs}v(m-r,n-s)+e(m,n). \tag{7}$$

The z transform of some vector, $w(m,n)$, is defined as

$$W(z_1,z_2)\triangleq Z\{w(m,n)\}=\sum_{m=0}^{\infty}\sum_{n=0}^{\infty}w(m,n)z_1^{-m}z_2^{-n}. \tag{8}$$

Taking the z transform of (7) and manipulating we obtain

$$V(z_1,z_2)=G(z_1,z_2)\cdot E(z_1,z_2)$$

where

$$G(z_1,z_2)=[I-A_{10}z_1^{-1}-A_{01}z_2^{-1}-A_{11}z_1^{-1}z_2^{-1}]^{-1}. \tag{9}$$

Let $g(m,n)=Z^{-1}\{G(z_1,z_2)\}$; then from (7) and (9) we obtain

$$v(m,n)=\sum_{i=0}^{\infty}\sum_{k=0}^{\infty}g(i,k)\,e(m-i,n-k)$$

$$=\sum_{i=-\infty}^{m}\sum_{k=-\infty}^{n}g(m-i,n-k)\,e(i,k). \tag{10}$$

Letting $x=\{x_i\}_{i=1}^{N^2}$, define the following norm on vectors:

$$\|x\|\triangleq(x^*x)^{1/2}=\left(\sum_{i=1}^{N^2}|x_i|^2\right)^{1/2}. \tag{11}$$

Assuming that $z(m,n)$ and $v(m,n)$ are periodic, define the rms values of $z(m,n)$ and $v(m,n)$, respectively as

$$\langle z\rangle\triangleq\left[\frac{1}{(K_1+1)(K_2+1)}\sum_{m=0}^{K_1}\sum_{n=0}^{K_2}|z(m,n)|^2\right]^{1/2} \quad\text{and} \tag{12a}$$

$$\langle v\rangle\triangleq\left[\frac{1}{(K_1+1)(K_2+1)}\sum_{m=0}^{K_1}\sum_{n=0}^{K_2}\|v(m,n)\|^2\right]^{1/2} \tag{12b}$$

where $z(m,n)=z(m+K_1+1,n+K_2+1)$ and $\|v(m,n)\|=\|v(m+K_1+1,n+K_2+1)\|$. The following two theorems are straightforward and are given without proof.

Theorem 1: $\langle z\rangle=(1/N)\langle v\rangle$, where N is the order of the filter.

Theorem 2:

$$\min_{k,l,r}\left|\lambda_r\left(\exp\left(j\frac{2\pi k}{K_1+1}\right),\exp\left(j\frac{2\pi l}{K_2+1}\right)\right)\right|\langle e\rangle\leqslant\langle v\rangle$$

$$\leqslant\max_{k,l,r}\left|\lambda_r\left(\exp\left(j\frac{2\pi k}{K_1+1}\right),\exp\left(j\frac{2\pi l}{K_2+1}\right)\right)\right|\langle e\rangle$$

where $\lambda_r(\exp(j(2\pi k/K_1+1)),\exp(j(2\pi l/K_2+1)))$; $r=1,2,\cdots,N^2$ are the eigenvalues of $G(\exp(j2\pi k/K_1+1),\exp(j2\pi l/K_2+1))$.

From Theorems 1 and 2, it is easy to show that

$$\langle z\rangle\leqslant\frac{1}{N}\langle e\rangle\max_{r,\varphi_1,\varphi_2}|\lambda_r(\exp(j\varphi_1),\exp(j\varphi_2))|$$

$$=\frac{\langle e\rangle}{N\min_{r,\varphi_1,\varphi_2}|\xi_r(\exp(j\varphi_1),\exp(j\varphi_2))|}, \tag{13}$$

where $\xi_r(\exp(j\varphi_1),\exp(j\varphi_2))$; $r=1,2,\cdots,N^2$ are the eigenvalues of $G^{-1}(\exp(j\varphi_1),\exp(j\varphi_2))$.

Now assume that $N=2$. For the representation by (7), the coefficient matrices are given by (4). The fact that the A_{rs} are singular as we have chosen them is not crucial as long as $G^{-1}(\exp(j\varphi_1),\exp(j\varphi_2))$ is nonsingular, i.e.,

$$\det[G^{-1}(\exp(j\varphi_1),\exp(j\varphi_2))]\neq 0,\qquad\text{for all }\varphi_1,\varphi_2.$$

This will be true for all stable filters since upon computing the determinant of $G^{-1}(\exp(j\varphi_1),\exp(j\varphi_2))$ we find that

$$H(\exp(j\varphi_1),\exp(j\varphi_2))=\frac{1}{\det[G^{-1}(\exp(j\varphi_1),\exp(j\varphi_2))]}, \tag{14}$$

where $H(z_1,z_2)$ is the transfer function of our second-order filter, i.e.,

$$H(z_1,z_2)=\frac{1}{1+\sum_{\substack{m=0\\m+n\neq 0}}^{2}\sum_{n=0}^{2}a_{mn}z_1^{-m}z_2^{-n}}. \tag{15}$$

It may be shown that $\langle e\rangle\leqslant(q/2)(9+\alpha^2)^{1/2}$, where α is the number of coefficients (a_{mn}, $m,n=0,1,2,m+n\neq 0$) which are neither 0 nor 1. Upon computing $\min_{r,\varphi_1,\varphi_2}|\xi_r(\exp(j\varphi_1),\exp(j\varphi_2))|$, the bound of (13) becomes

$$\langle z\rangle\leqslant\frac{\frac{q}{4}(9+\alpha^2)^{1/2}}{\min_{\varphi_1\varphi_2}\left|1+\sum_{\substack{m=0\\m+n\neq 0}}^{2}\sum_{n=0}^{2}a_{mn}\exp(-jm\varphi_1)\exp(-jn\varphi_2)\right|}. \tag{16}$$

Maria and Fahmy [3] extended a theorem by Sandberg and Kaiser [2] to bounding the rms value of limit cycles for first-order two-dimensional filters. Upon inspection it is easily seen that the theorem also applies to any order filter. The theorem is stated below.

Theorem 3:

$$\min_{r,s}\Phi(r,s)\langle\hat{s}\rangle\leqslant\langle z\rangle\leqslant\max_{r,s}\Phi(r,s)\langle\hat{s}\rangle,$$

where

$$\Phi(r,s)=\left|H\left(\exp\left(j\frac{2\pi r}{K_1+1}\right),\exp\left(j\frac{2\pi s}{K_2+1}\right)\right)\right|$$

and $\hat{s}(m,n)$ is an error array which is bounded by α, the number of coefficients which are neither 0 nor 1; the quantization step size is q. Therefore, $\langle\hat{s}\rangle\leqslant\alpha q/2$. Thus the bound of Theorem 3 for $N=2$ becomes

$$\langle z\rangle\leqslant\frac{\frac{\alpha q}{2}}{\min_{\varphi_1,\varphi_2}\left|1+\sum_{\substack{m=0\\m+n\neq 0}}^{2}\sum_{n=0}^{2}a_{mn}\exp(-jm\varphi_1)\exp(-jn\varphi_2)\right|}. \tag{17}$$

If $\alpha\geqslant 2$, then the bound of (16) will be smaller than the bound of (17).

Calculating the denominator of (16) and (17) is not easy, in general. At present, numerical techniques are the only available methods. Consider the following special case; letting

$$W(\exp(j\varphi_1)\exp(j\varphi_2)) = 1 + \sum_{\substack{m=0\\ m+n\neq 0}}^{2}\sum_{n=0}^{2} \cdot a_{mn}\exp(-jm\varphi_1)\exp(-jn\varphi_2), \tag{18}$$

assume that $W(\exp(j\varphi_1), \exp(j\varphi_2))$ is "almost" separable, i.e.,

$$W(\exp(j\varphi_1), \exp(j\varphi_2)) = W_1(\exp(j\varphi_1))\, W_2(\exp(j\varphi_2)) + E(\exp(j\varphi_1), \exp(j\varphi_2)), \tag{19}$$

where

$$\max_{\varphi_1,\varphi_2} |E(\exp(j\varphi_1), \exp(j\varphi_2))| < \min_{\varphi_1,\varphi_2} |W_1(\exp(j\varphi_1))\, W_2(\exp(j\varphi_2))|.$$

Hence,

$$|W(\exp(j\varphi_1), \exp(j\varphi_2))| \geqslant \min_{\varphi_1,\varphi_2} |W_1(\exp(j\varphi_1)) \cdot W_2(\exp(j\varphi_2))| - \max_{\varphi_1,\varphi_2} |E(\exp(j\varphi_1), \exp(j\varphi_2))| = \widetilde{W} - \widetilde{E}. \tag{20}$$

There are many ways to write $W(\exp(j\varphi_1), \exp(j\varphi_2))$ in the form of (19). One way is to let

$$W_1(\exp(j\varphi_1)) = 1 + a_{10}\exp(-j\varphi_1) + a_{20}\exp(-j2\varphi_1), \tag{21a}$$

$$W_2(\exp(j\varphi_2)) = 1 + a_{01}\exp(-j\varphi_2) + a_{02}\exp(-j2\varphi_2). \tag{21b}$$

It is easy to show that for stable second-order 2-D filters,

$$|a_{20}| < 1,\ |a_{02}| < 1,\ |a_{10}| < 1 + a_{20},\ \text{and}\ |a_{01}| < 1 + a_{02}. \tag{22}$$

Thus, if the 2-D filter is stable then the separable components, W_1 and W_2, must also be stable. Therefore, $\widetilde{W}$ can be easily bounded from Sandberg and Kaiser [2]. Since $|E(\exp(j\varphi_1), \exp(j\varphi_2))|$ can be reduced to a first-order section of a 2-D filter, $\widetilde{E}$ can be easily bounded following a method used by Maria and Fahmy [4]. Thus, the rms bound of (16) for this special case becomes

$$\langle z\rangle \leqslant \frac{\frac{q}{4}(9+\alpha^2)^{1/2}}{\widetilde{W} - \widetilde{E}}. \tag{23}$$

Two examples are given below in which zero-input limit cycles were simulated for several sets of initial conditions, $y(m, -1)$, $y(m, -2)$, $y(-1, n)$, $y(-2, n)$. The initial conditions were assumed to be zero for $m, n > 20$ to allow the output array to reach a steady state. We limited the output array to 200 X 200 points. The exact rms value was not found since the entire array had not reached a steady state; however, for our examples most of the array had reached steady state outside the first quadrant of 50 X 50 points. Therefore, our estimate of the rms value was taken from the points outside this first quadrant. Even if we had extended our array out until the entire array had reached a steady state, visually detecting the exact period, $(K_1 + 1, K_2 + 1)$, of the limit cycle would have been very difficult.

Example 1:

$$\begin{aligned} y(m,n) = &-0.7y(m,n-1) - 0.6y(m,n-2)\\ &- 0.8y(m-1,n) - 0.53y(m-1,n-1)\\ &- 0.47y(m-1,n-2) - 0.7y(m-2,n)\\ &- 0.48y(m-2,n-1) - 0.4y(m-2,n-2). \end{aligned}$$

Maximum rms value of the limit cycle	3.477
rms bound of (16)	20.703[1]
rms bound of (17)	38.770.[1]

Example 2:

$$\begin{aligned} y(m,n) = &\ 1.2y(m,n-1) - 0.5y(m,n-2) + 1.5y(m-1,n)\\ &- 1.8y(m-1,n-1) + 0.75y(m-1,n-2)\\ &- 0.6y(m-2,n) + 0.72y(m-2,n-1)\\ &- 0.29y(m-2,n-2). \end{aligned}$$

Maximum rms value of the limit cycle	10.813
rms bound of (23)	129.789.

In conclusion, this correspondence has developed a vector model of a two-dimensional filter by using a vector formed from the two-dimensional output sequence. This model is then used to produce a bound on the errors produced due to the roundoff of multiplications in a fixed-point implementation of a two-dimensional filter. The bound theorem is applied to a special class of filters called "almost" separable filters. Examples are given bounding a type of roundoff errors called limit cycles. The rms bounds developed appear to be about one order of magnitude larger than the estimated rms value of the observed limit cycles.

REFERENCES

[1] S. Yakowitz and S. R. Parker, "Computation of bounds for digital filter quantization errors," *IEEE Trans. Circuit Theory*, vol. CT-20, pp. 391-396, July 1973.

[2] I. W. Sandberg and J. F. Kaiser, "A bound on limit cycles in fixed-point implementations of digital filters," *IEEE Trans. Audio Electroacoust.*, vol. AU-20, pp. 110-112, June 1972.

[3] G. A. Maria and M. M. Fahmy, "Limit cycle oscillations in first order two-dimensional digital filters," *IEEE Trans. Circuits and Systems (Special Issue on Digital Filtering and Image Processing)* vol. CAS-22, pp. 246-251, Mar. 1975.

[4] —, "Limit cycle oscillations in a cascade of two-dimensional digital filters," *IEEE Trans. Circuits and Systems*, vol. CAS-22, pp. 826-830, Oct. 1975.

[1] These bounds were found by a numerical method.

Reprinted from *IEEE Trans. Computers* **C-25**(7):755–759 (1976)

ERROR ANALYSIS OF TWO-DIMENSIONAL RECURSIVE DIGITAL FILTERS EMPLOYING FLOATING-POINT ARITHMETIC

M.-D. Ni and J. K. Aggarwal

***Abstract*—In this correspondence, error analysis for two-dimensional direct form recursive digital filters employing floating-point arithmetic is carried out. A systematic way of estimating the mean-squared errors due to roundoff, coefficient, and input quantizations is discussed. Norm error bounds are also given. Extensive numerical experiments have shown that the present approach leads to satisfactory results.**

***Index Terms*—Digital filter, error analysis of digital filters, floating-point arithmetic digital filters, two-dimensional digital filters.**

I. Introduction

The three common sources of digital filter error, due to finite word length, are roundoff error accumulation, inaccuracies in the representation of filter coefficients, and quantizations of input signal and initial conditions. The effects of these errors depend on the type of arithmetic employed by the digital filter. There have been extensive studies on digital filters employing fixed-point arithmetic. In this correspondence, we investigate the error properties of 2D (two-dimensional) recursive digital filters employing floating-point arithmetic. There are quite a few papers on floating-point 1D (one-dimensional) digital filters ([3], [4], [6], [8], [9], etc.). Frequency domain calculation is generally a suitable way for estimating the output errors in 1D digital filters. Generally, the methods (e.g., [4], [6], [8], [9]) and the expressions (e.g., [4], [6], [8]) for roundoff errors are quite readily extended from 1D to 2D recursive digital filters. However, a fundamental difficulty of 2D digital filters is that we generally cannot factorize a 2D recursive digital filter into a multiplication of lower order digital filters. This fundamental difficulty has caused serious problems in the designing and analyzing of 2D recursive digital filters. For example, we can obtain integral results in closed form for the expressions given by Kaneko and Liu [8], [9], but when the expressions are extended to two dimensions, we generally cannot get exact evaluations, although numerical approximations are possible. For this reason and the fact that we generally encounter high-order filters in the 2D case (in contrast to the 1D case where digital filters are often realized by cascading first- and second-order filters), space domain calculations of output errors may be more suitable for most practical 2D recursive digital filters. In the following, we discuss and formulate a space domain method of estimating the output mean-squared errors and norm error bounds due to the three sources of error for 2D recursive digital filters. Extensive numerical experiments simulated on a general-purpose computer (CDC 6600) support the validity of the method. It can also be used to estimate errors for 1D and multidimensional digital filters.

We concentrate our analysis on the following equation which represents a direct form of realization:

$$w_{mn} = \sum_{j=0}^{M_b} \sum_{k=0}^{N_b} b_{jk} x_{m-j,n-k} - \sum_{\substack{j=0\\ j+k\neq 0}}^{M_a} \sum_{k=0}^{N_a} a_{jk} w_{m-j,n-k} \tag{1}$$

where

1) $m \geq 0, n \geq 0$,

2) x_{mn} is the ideal input signal,

$$x_{ij} = 0 \qquad \text{for } i < 0 \text{ or } j < 0,$$

3) w_{mn} is the ideal output signal,

$$w_{ij} = W(i,j), \qquad j = 0, -1, \cdots, -N_a,$$
$$i = 0, -1, \cdots, -M_a,$$
$$i + j \neq 0,$$
$$w_{ij} = 0, \qquad \text{for } i < -M_a \text{ or } j < -N_a.$$

The digital transfer function corresponding to (1) is

$$G(z_1,z_2) = \frac{N(z_1,z_2)}{D(z_1,z_2)} \tag{2}$$

$$N(z_1,z_2) = \sum_{j=0}^{M_b} \sum_{k=0}^{N_b} b_{jk} z_1^{-j} z_2^{-k} \tag{3}$$

$$D(z_1,z_2) = 1 + \sum_{\substack{j=0\\ j+k=0}}^{M_a} \sum_{k=0}^{N_a} a_{jk} z_1^{-j} z_2^{-k}. \tag{4}$$

II. Mean-Squared Error Analysis

A. *Mean-Squared Estimations for Roundoff Errors*

Output Error System Equations: By having the same assumptions as in [4, p. 1777] and following the same concept as in [5]–[8] with some appropriate arrangements, we can obtain the following output error system equations for roundoffs:

$$e_{mn} = -\sum_{\substack{j=0\\ j+k\neq 0}}^{M_a} \sum_{k=0}^{N_a} a_{jk} e_{m-j,n-k} + e_{mn}^{(d)} \tag{5}$$

where

1) e_{mn} is the output error, i.e., the difference between ideal output and the actual output with finite word length,

2) $$e_{mn}^{(d)} = \sum_{j=0}^{M_b} \sum_{k=0}^{N_b} b_{jk} (\theta_{mn,jk} - 1) x_{m-j,n-k} - \sum_{\substack{j=0\\ j+k\neq 0}}^{M_a} \sum_{k=0}^{N_a} a_{jk} (\varphi_{mn,jk} - 1) w_{m-j,n-k}, \tag{6}$$

Manuscript received June 5, 1975; revised September 19, 1975. This work was supported in part by the National Science Foundation under Grant GK 42790 and by the Air Force Office of Scientific Research under Grant 72-2371.

The authors are with the Department of Electrical Engineering and the Electronics Research Center, University of Texas, Austin, TX 78712.

3) $\theta_{mn,jk} = (1 + \xi_{mn})(1 + \delta_{mn,jk}) \prod_{i=j,l=k}^{M_b,N_b} (1 + \gamma_{mn,il}),$

with $\gamma_{mn,00} = 0, j = 0,\cdots,M_b$, and $k = 0,\cdots,N_b$,

4) $\varphi_{mn,jk} = (1 + \xi_{mn})(1 + \epsilon_{mn,jk}) \left(\prod_{i=j,l=k}^{M_a,N_a} (1 + \eta_{mn,il})\right),$

with $j + k \neq 0, j = 0,\cdots,M_a, k = 0,\cdots,N_a$, and $\eta_{mn,01} = 0$,(7)

5) ξ_{mn}, $\delta_{mn,jk}$, $\gamma_{mn,il}$, and $\epsilon_{mn,jk}$ are relative errors taking values in $(-2^{-t}, 2^{-t})$ with t being the number of binary bits. A more detailed explanation of these quantities is found in [8] and [9].

The numbers of factors in the expressions for $\theta_{mn,jk}$ and $\varphi_{mn,jk}$ are, respectively, as follows:

$$\beta'_{jk} = \begin{cases} (M_b + 1)(N_b + 1) + 1, & \text{for } j = 0, k = 0 \quad (8) \\ (M_b - j + 1)(N_b - k + 1) + 2, & \text{otherwise} \quad (9) \end{cases}$$

$$\alpha'_{jk} = \begin{cases} (M_a + 1)(N_a + 1), & \text{for } j = 0, k = 1 \quad (10) \\ (M_a + 1)(N_a + 1) - k + 2, & \text{for } j = 0, k = 2,\cdots, N_a, \quad (11) \\ (M_a - j + 1)(N_a - k + 1) + 2, & \text{otherwise.} \quad (12) \end{cases}$$

Notice that if some of the filter coefficients are 0 (no addition and multiplication) or 1 (no multiplication), the numbers in (8)–(12) will be correspondingly reduced.

Basic Assumptions: As usual, we assume that the relative errors $\delta_{mn,ij}$, $\epsilon_{mn,ij}$, $\gamma_{mn,ij}$, ξ_{mn}, etc., are independent random variables and are uniformly distributed in the range $(-2^{-t}, 2^{-t})$. Thus, they are zero mean and have a variance of $\sigma^2 = E_0{}^2/3$ with $E_0 = 2^{-t}$, and they are uncorrelated with each other or with any signals. We also assume that the input signal x_{mn} is zero mean and wide sense stationary. Since our system is a linear spatially invariant system, the output signal is also zero mean and wide sense stationary.

Mean-Squared Error: Expanding (7) and taking the first-order approximation, we have

$$\varphi_{mn,jk} - 1 \doteq \xi_{mn} + \epsilon_{mn,jk} + \sum_i \sum_l \eta_{mn,il}. \tag{13}$$

The right-hand side of (13) has a total of α'_{jk} [see (12)] independent identically distributed random variables. We then consider $(\varphi_{mn,jk} - 1)$ to be an independent random variable consisting of the sum of α'_{jk} independent identically distributed random variables. Likewise, we consider each $(\theta_{mn,jk} - 1)$ to be an independent random variable which is the sum of β'_{jk} independent identically distributed random variables.

In (6), let

$$e^{(a)}_{mn,jk} = (\varphi_{mn,jk} - 1)a_{jk}w_{m-j,n-k} \tag{14}$$

$$e^{(b)}_{mn,jk} = (\theta_{mn,jk} - 1)b_{jk}x_{m-j,n-k}. \tag{15}$$

The means of the random variables $e^{(a)}_{mn,jk}$ and $e^{(b)}_{mn,jk}$ are zero. Let $\sigma^2 e^{(a)}_{jk}$ and $\sigma^2 e^{(b)}_{jk}$ represent their variances, respectively; then

$$\sigma^2 e^{(a)}_{jk} = \frac{E_0^2}{3} \cdot \alpha'_{jk} \cdot \overline{(a_{jk}w_{m-j,n-k})^2} \tag{16}$$

$$\sigma^2 e^{(b)}_{jk} = \frac{E_0^2}{3} \cdot \beta'_{jk} \cdot \overline{(b_{jk}x_{m-j,n-k})^2}. \tag{17}$$

The bar denotes expected value. We approximate it by the following operation:

$$\overline{A^2} = \lim_{\substack{n\to\infty \\ m\to\infty}} \frac{\sum_{i=0}^{m}\sum_{j=0}^{n} A_{ij}^2}{mn}. \tag{18}$$

Considering $e^{(d)}_{mn}$ [see (6)] to be composed of $(M_b + 1)(N_b + 1)$ of the $e^{(b)}_{mn,jk}$'s and $[(M_a + 1)(N_a + 1) - 1]$ of the $e^{(a)}_{mn,jk}$'s, then the steady-state output error is found to be

$$\sigma^2 e = \left(\sum_{\substack{j=0 \\ j+k\neq 0}}^{M_a}\sum_{k=0}^{N_a} \sigma^2 e^{(a)}_{jk} + \sum_{j=0}^{M_b}\sum_{k=0}^{N_b} \sigma^2 e^{(b)}_{jk}\right) \sum_{m=0}^{\infty}\sum_{n=0}^{\infty} h_{mn}^2, \tag{19}$$

where $\{h_{mn}\}$ is the impulse response of the filter $1/A(z_1,z_2)$.

B. Mean-Squared Error Estimation for Coefficient Quantizations

The output error due to coefficient quantization is

$$\hat{\sigma}^2 e = \left(\sum_{j=0}^{M_a}\sum_{k=0}^{N_a} \hat{\sigma}^2 e^{(a)}_{jk} + \sum_{\substack{j=0 \\ j+k\neq 0}}^{M_b}\sum_{k=0}^{N_b} \hat{\sigma}^2 e^{(b)}_{jk}\right) \sum_{m=0}^{\infty}\sum_{n=0}^{\infty} h_{mn}^2 \tag{20}$$

with

$$\hat{\sigma}^2 e^{(a)}_{jk} = \overline{(a_{jk}\delta_{jk}w_{m-j,n-k})^2} = \delta_{jk}^2 \overline{(a_{jk}w_{m-j,n-k})^2} \tag{21}$$

$$\hat{\sigma}^2 e^{(b)}_{jk} = \overline{(b_{jk}\epsilon_{jk}x_{m-j,n-k})^2} = \epsilon_{jk}^2 \overline{(b_{jk}x_{m-j,n-k})^2} \tag{22}$$

where δ_{jk} and ϵ_{jk} are relative errors for a_{jk} and b_{jk}, respectively, and are fixed numbers.

C. Mean-Squared Error Estimations for Quantizations of Input Signal and Initial Conditions

The output error due to the quantization of the input signal can be calculated as

$$\bar{\sigma}^2 e = \left(\frac{E_0^2}{3}\overline{x_{mn}^2}\right) \sum_{m=0}^{\infty}\sum_{n=0}^{\infty} g_{mn}^2 \tag{23}$$

where $\{g_{mn}\}$ is the impulse response of the filter $G(z_1,z_2)$.

As for the output error due to the quantization of the initial conditions, it usually can be neglected if the impulse response dies out fast enough. The output error is governed by the following equation:

$$\bar{e}_{mn} = -\sum_{\substack{j=0 \\ j+k\neq 0}}^{M_a}\sum_{k=0}^{N_a} h^{(jk)}_{mn}\epsilon_{jk}W(-j,-k) \tag{24}$$

where ϵ_{jk} is the relative error for the initial conditions $W(-j,-k)$ and

$$\sum_{m=0}^{\infty}\sum_{n=0}^{\infty} h^{(jk)}_{mn} z_1^{-m} z_2^{-n} = \frac{\sum_{p=j}^{M_a}\sum_{q=k}^{N_a} a_{pq} z_1^{-(p-j)} z_2^{-(q-k)}}{D(z_1,z_2)},$$

$$j = 0,\cdots,M_a;\ k = 0,\cdots,N_a;\ j + k \neq 0. \tag{25}$$

It can be seen that

$$\langle\bar{e}\rangle^{K_2,K_1} \leq \sum_{j=0}^{M_a}\sum_{k=0}^{N_a} \langle h^{(jk)}\rangle^{K_2,K_1} |\epsilon_{jk}W(-j,-k)| \tag{26}$$

where $\langle\cdot\rangle^{K_2,K_1}$ is the "2D sequence mean-squared average norm" defined by

$$\langle x\rangle^{K_2,K_1}_{p,q} \triangleq \left(\frac{1}{(K_2 - p + 1)(K_1 - q + 1)} \sum_{m=p}^{K_2}\sum_{n=q}^{K_1} |x_{mn}|^2\right)^{1/2}$$

The value of the right-hand side dies out as K_2 and K_1 increase.

D. Discussion

We have formulas for the steady-state output mean-squared error for each of three sources of error. If two or more sources of error are present, the output mean-squared errors due to their combined effects can be approximated by the summations of $\sigma^2 e$ [see (19)], $\hat{\sigma}^2 e$ [see (20)], and/or $\bar{\sigma}^2 e$ [see (23)].

For nonstationary or deterministic inputs, we cannot obtain similar results to those in (19), (20), and (23). As an illustration, for roundoff error with deterministic inputs, but with the same assumption for the relative errors, we have the following equation instead of (19):

$$\sigma^2 e_{mn} = \frac{E_0^2}{3}\Bigg[\sum_{\substack{j=0\\ j+k\neq 0}}^{M_a}\sum_{k=0}^{N_a} a_{jk}^2 \cdot \alpha'_{jk} \cdot \Big(\sum_{\mu=0}^{m}\sum_{\nu=0}^{n} w_{\mu-j,\nu-k}^2 h_{m-\mu,n-\nu}^2\Big) + \sum_{j=0}^{M_b}\sum_{k=0}^{N_b} b_{jk}^2 \cdot \beta'_{jk} \cdot \Big(\sum_{\mu=0}^{m}\sum_{\nu=0}^{n} x_{\mu-j,\nu-k}^2 h_{m-\mu,n-\nu}^2\Big)\Bigg]. \quad (27)$$

III. Norm Bounds

In this section, we give norm bounds for output errors due to the three sources of error. We still assume that the relative errors $\gamma_{mn,jk}$, $\delta_{mn,jk}$, ⋯, etc., are independent random variables uniformly distributed in $(-2^{-t}, 2^{-t})$. We make no assumption for the input signal.

For roundoff errors, by the application of the lemma in [5] on (5), (6), (14), and (15), we can have

$$E\{{}^2\langle e\rangle^{K_2,K_1}\} \leq \max_{\substack{0\leq\omega_1\leq 2\pi\\ 0\leq\omega_2\leq 2\pi}} |D^{-1}(e^{i\omega_1},e^{i\omega_2})|^2 \cdot \frac{E_0^2}{3} \cdot \Bigg[\sum_{\substack{j=0\\ j+k\neq 0}}^{M_a}\sum_{k=0}^{N_a} a_{jk}^2\alpha'_{jk} \cdot {}^2\langle w\rangle_{j,k}^{K_2,K_1} + \sum_{j=0}^{M_b}\sum_{k=0}^{N_b} b_{jk}^2\beta'_{jk} \cdot {}^2\langle x\rangle_{j,k}^{K_2,K_1}\Bigg] \quad (28)$$

where $E\{\cdot\}$ denotes the expected value, ${}^2\langle\cdot\rangle_{j,k}^{K_2,K_1}$ is the squared value of $\langle\cdot\rangle_{j,k}^{K_2,K_1}$, and we have assumed that the initial conditions of the 2D digital filters are zero.

Similarly, for quantization errors of the filter coefficients, we have

$$E\{{}^2\langle \hat{e}\rangle^{K_2,K_1}\} \leq \max_{\substack{0\leq\omega_1\leq 2\pi\\ 0\leq\omega_2\leq 2\pi}} |D^{-1}(e^{i\omega_1},e^{i\omega_2})|^2 \cdot \Bigg[\sum_{\substack{j=0\\ j+k\neq 0}}^{M_a}\sum_{k=0}^{N_a} a_{jk}^2\delta_{jk}^2 \cdot {}^2\langle w\rangle_{j,k}^{K_2,K_1} + \sum_{j=0}^{M_b}\sum_{k=0}^{N_b} b_{jk}^2\epsilon_{jk} \cdot {}^2\langle x\rangle_{j,k}^{K_2,K_1}\Bigg]. \quad (29)$$

For quantization errors of the input signal,

$$E\{{}^2\langle \bar{e}\rangle^{K_2,K_1}\} \leq \max_{\substack{0\leq\omega_1\leq 2\pi\\ 0\leq\omega_2\leq 2\pi}} |G(e^{i\omega_1},e^{i\omega_2})|^2 \cdot \frac{E_0^2}{3} \cdot \Bigg[\sum_{j=0}^{M_b}\sum_{k=0}^{N_b} b_{jk}^2\beta'_{jk} \cdot {}^2\langle x\rangle_{j,k}^{K_2,K_1}\Bigg]. \quad (30)$$

IV. Numerical Results

A decimal simulation program has been written to simulate, decimally, all the filtering operations on a CDC 6600 computer. The program includes subroutines which simulate floating-point additions, floating-point multiplications, and quantizations. All the floating-point variables are double precision. For a CDC 6600 computer, the mantissa part of a floating-point number is represented by 96 binary bits when double precision is used. The maximum possible error for the mantissa is $2^{-97} \simeq 10^{-30}$, which is fairly small, and we can count on the results the program provides.

The following examples are among those we used to test the validity of the method in (19), (20), (23), and (28)–(30). For each example, with a given input signal, the following quantities are calculated: let $\Delta = {}^2\langle w\rangle^{K_2,K_1}$ where w_{mn} is the ideal output sequence,

1) *RE*: $\sigma^2 e/\Delta$, where $\sigma^2 e$ is the theoretical roundoff error [see (19)],
2) *RA*: ${}^2\langle e\rangle^{K_2,K_1}/\Delta$, where e_{mn} is the actual output error due to roundoffs,
3) *RNE*: Norm bound for output error due to roundoffs [see (28)] divided by Δ,
4) *CE*: $\hat{\sigma}^2 e$ [see (20)]/Δ,
5) *CA*: ${}^2\langle \hat{e}\rangle^{K_2,K_1}/\Delta$, where $\hat{e}_{mn}$ is the actual output error due to coefficient quantizations,
6) *CNE*: Norm bounds for output error due to coefficient quantizations [see (29)] divided by Δ,
7) *IE*: $\bar{\sigma}^2 e$ [see (23)]/Δ,
8) *IA*: ${}^2\langle \bar{e}\rangle^{K_2,K_1}/\Delta$, where $\bar{e}_{mn}$ is the actual output error due to input quantizations,
9) *INE*: Norm bound for output error due to input quantizations [see (20)] divided by Δ,
10) *RCE* = *RE* + *CE*,
11) *RCA*: ${}^2\langle e^{rc}\rangle^{K_2,K_1}/\Delta$, where e^{rc}_{mn} is the actual output error due to roundoffs and coefficient quantizations,
12) *RCNE* = *RNE* + *CNE*,
13) *RCIE* = *RE* + *CE* + *IE*,
14) *RCIA*: ${}^2\langle e^{rci}\rangle^{K_2,K_1}/\Delta$, where e^{rci}_{mn} is the actual output error due to the combined effects of all the three sources of error,
15) *RCINE* = *RNE* + *CNE* + *INE*.

For Examples 1 and 2, we let $N(z_1,z_2) = \bar{z}_1\bar{B}\bar{z}_2^T$ and $D(z_1,z_2) = \bar{z}_1\bar{A}\bar{z}_2^T$, where $\bar{z}_1 = (1,z_1^{-1},z_1^{-2},z_1^{-3})$, $\bar{z}_2 = (1,z_2^{-1},z_2^{-2},z_2^{-3})$, and $\bar{z}_2^T$ is the matrix transpose of $\bar{z}_2$. The input signals and operating conditions are the same for both examples.

Example 1: Two-Dimensional Digital Filter Designed by Shanks' Rotation Method and Modified by a Planar Least Squares Inverse Algorithm [2]:

$$\bar{B} = \begin{bmatrix} 4.39\times 10^{-3} & 1.317\times 10^{-2} & 1.317\times 10^{-2} & 4.39\times 10^{-3} \\ 1.317\times 10^{-2} & 3.951\times 10^{-2} & 3.951\times 10^{-2} & 1.317\times 10^{-2} \\ 1.317\times 10^{-2} & 3.951\times 10^{-2} & 3.951\times 10^{-2} & 1.317\times 10^{-2} \\ 4.39\times 10^{-3} & 1.317\times 10^{-2} & 1.317\times 10^{-2} & 4.39\times 10^{-3} \end{bmatrix}$$

$$\bar{A} = \begin{bmatrix} 1. & 9.72193\times 10^{-2} & 1.35846\times 10^{-2} & 6.245431\times 10^{-4} \\ 2.105095\times 10^{-2} & -1.785671 & -0.1398205 & -1.197869\times 10^{-2} \\ 7.184771 & -8.055841\times 10^{-2} & 1.262808 & 5.927248\times 10^{-2} \\ -7.236237\times 10^{-4} & -5.415724\times 10^{-3} & 5.705888\times 10^{-2} & -3.541561\times 10^{-1} \end{bmatrix}.$$

The following three input signals are used:

$2I1$: $x_{jk} = \cos(j)\cos(k)$, $j = 0,\cdots,127$, and $k = 0,\cdots,127$

$2I2$: $x_{jk} = \cos(j\cdot k)$, $j = 0,\cdots,127$, and $k = 0,\cdots,127$

$2I3$: $x_{jk} = \text{noise}\,(j,k)$, $j = 0,\cdots,127$, and $k = 0,\cdots,127$

where noise (j,k) is generated by a random number generator.

The operations are from $m = 0,\cdots,127$ and $n = 0,\cdots,127$. The impulse responses for $1/D(z_1,z_2)$ at $(m,n) = (0,0)$ and $(m,n) =$

TABLE I

Errors	Inputs 2I1	2I2	2I3	Powers of 10
RE	2.14062	2.14409	2.15345	10^{-8}
RA	2.76639	1.75509	1.55308	10^{-8}
RNE	1.47309	1.47548	1.48192	10^{-7}
CE	1.18103	1.19667	1.19465	10^{-9}
CA	2.18830	1.73785	1.57992	10^{-9}
CNE	8.12738	8.23503	8.22112	10^{-9}
IE	2.73986	8.64918	7.066	10^{-11}
IA	1.28770	3.15256	3.55109	10^{-11}
INE	0.681206	2.15043	1.75680	10^{-9}
RCE	2.25872	2.26376	2.27292	10^{-8}
RCA	2.02213	1.57048	1.40364	10^{-8}
RCNE	1.55437	1.55783	1.56413	10^{-7}
RCIE	2.26146	2.27241	2.27998	10^{-8}
RCIA	2.08492	1.60492	1.41122	10^{-8}
RCINE	1.56118	1.57934	1.58170	10^{-7}

TABLE II

Errors	Inputs 2I1	2I2	2I3	Power of 10
RE	6.70882	5.97688	6.89292	10^{-7}
RA	8.27063	1.54102	5.85148	10^{-7}
RNE	8.91590	7.94316	9.16056	10^{-6}
CE	5.36286	4.55883	5.56276	10^{-8}
CA	0.979651	1.53522	1.00613	10^{-8}
CNE	7.12713	6.05960	7.39280	10^{-7}
IE	0.429636	0.761428	1.072392	10^{-10}
IA	0.380303	0.322345	0.66187	10^{-10}
INE	1.37214	1.03698	1.46045	10^{-10}
RCE	7.2451	3.91965	3.95618	10^{-7}
RCA	8.57491	1.54024	6.69687	10^{-7}
RCNE	9.62861	8.54902	9.89984	10^{-6}
RCIE	7.2503	3.92726	3.95725	10^{-7}
RCIA	7.94085	1.62868	5.98533	10^{-7}
RCINE	9.62875	8.54913	9.89998	10^{-6}

(127,127) are 1.0 and 5.732×10^{-16}, respectively. The number of bits for simulations is $t = 16$. Table I shows the numerical results.

Example 2: Two-Dimensional High-Pass Digital Filter:

Table II gives the numerical results.

Example 3: One-Dimensional Bandpass Digital Filter Designed by Linear Programming Method [10]:

$$\overline{B} = \begin{bmatrix} 1.2 & -3.6 & 3.6 & -1.2 \\ -3.6 & 10.8 & -10.8 & 3.6 \\ 3.6 & -10.08 & 10.08 & -3.6 \\ -1.2 & 3.6 & -3.6 & 1.2 \end{bmatrix}$$

$$\overline{A} = \begin{bmatrix} 1. & -1.529578 & 9.6912 & -2.14664 \\ -1.529578 & 2.339608 & -1.482344 & -0.208035 \\ 0.96912 & -1.482344 & 0.939193 & 4.6081 \times 10^{-2} \\ -0.214644 & -3.28345 & -0.208035 & 4.6081 \times 10^{-2} \end{bmatrix}.$$

TABLE III

t = 15 bits

Errors	Inputs I1	I2	I3	I4	I5	I6	I7	I8	I9	Powers of 10
RE	6.73	7.05	6.51	1.64	1.60	6.15	1.73	1.60	1.61	10^{-5}
RA	2.25	3.00	2.05	0.75	0.52	2.25	0.71	0.52	0.91	10^{-5}
RNE	7.33	7.68	7.01	1.79	1.74	6.70	1.89	1.74	1.76	10^{-4}
CE	1.51	1.58	1.46	0.30	0.29	1.37	0.32	0.29	0.29	10^{-5}
CA	1.56	1.42	0.85	1.57	1.66	0.84	2.52	1.66	1.59	10^{-5}
CNE	16.4	17.2	15.8	3.29	3.18	14.9	3.53	3.18	3.22	10^{-5}
IE	1.21	1.29	1.15	0.033	0.023	1.08	0.058	0.023	0.026	10^{-8}
IA	0.47	0.26	0.39	0.012	0.014	0.26	0.027	0.014	0.011	10^{-8}
INE	13.7	14.6	13.1	0.37	0.26	12.2	0.62	0.26	0.29	10^{-8}
RCE	8.23	8.63	7.97	1.94	1.89	7.52	2.06	1.89	1.90	10^{-5}
RCA	5.51	9.60	3.38	3.23	2.97	3.02	2.52	2.97	1.72	10^{-5}
RCNE	8.97	9.4	8.7	2.11	2.06	8.2	2.24	2.06	2.08	10^{-4}
RCIE	8.23	8.63	7.97	1.94	1.89	7.52	2.06	1.89	1.90	10^{-5}
RCIA	6.51	8.18	2.26	2.22	2.49	2.90	2.46	2.50	2.23	10^{-5}
RCINE	8.97	9.41	8.7	2.11	2.06	8.2	2.24	2.06	2.08	10^{-4}

$$N(z) = 0.10202 - 0.3031065z^{-1} + 0.3986085z^{-2} - 0.2786962z^{-3} + 0.0812557z^{-4}$$

$$D(z) = 1. - 2.98247z^{-1} + 3.95745z^{-2} - 2.5993z^{-3} + 0.758117z^{-4}$$

$I1$: $x_j = \cos(10j)$, $I2$: $x_j = \cos(0.1j)$,
$I3$: $x_j = \cos(j)$, $I4$: $x_j = \text{noise}(j)$,
$I5$: $x_j = \text{noise}(j)\cdot\cos(j)$, $I6$: $x_j = \cos(j + \text{noise}(j))$,
$I7$: $x_j = \cos(\text{noise}(j)\cdot j)$,
$I8$: $x_j = \text{noise}(j)\cdot\cos(j) + \text{noise}^1(j)\cdot\sin(j)$,
$I9$: $x_j = \text{noise}^2(j)\cdot\cos(j) + \text{noise}^3(j)\cdot\sin(j)$.

where $j = 0,\cdots,255$, and the signals noise (j), noise^1 (j), noise^2 (j), and noise^3 (j) are generated by a random number generator, and they have different variances and different starting values. The operations are from $n = 0$ to $n = 255$. Table III lists the numerical results.

V. Conclusion

A systematic way of estimating the mean-squared errors due to roundoff errors, coefficient inaccuracies, and quantizations of input signals has been presented. Norm bounds for the three sources of error are also given. The numerical experiments have shown that the estimated errors are within an order of magnitude of the actual errors.

References

[1] T. S. Huang, "Stability of two-dimensional recursive filters," *IEEE Trans. Audio Electroacoust.*, vol. AU-20, June 1972.
[2] J. L. Shanks, S. Treitel, and J. H. Justice, "Stability and synthesis of two-dimensional recursive filters," *IEEE Trans. Audio Electroacoust.*, vol. AU-20, June 1972.
[3] A. V. Oppenheim , and C. J. Weinstein, "A comparison of roundoff noise in floating point and fixed point digital filter realization," *Proc. IEEE*, vol. 57, pp. 1181–1183, June 1969.
[4] I. W. Sandberg, "Floating-point-roundoff accumulation in digital filter realizations," *Bell Syst. Tech. J.*, vol. 46, pp. 1775–1791, Oct. 1967.
[5] M. D. Ni and J. K. Aggarwal, "Two-dimensional digital filtering and its error analysis," *IEEE Trans. Comput.*, vol. C-23, Sept. 1974.
[6] E. P. F. Kan and J. K. Aggarwal, "Error analysis of digital filter employing floating-point arithmetic," *IEEE Trans. Circuit Theory*, vol. CT-18, Nov. 1971.
[7] ——, "Correction to 'Error analysis of digital filters employing floating-point arithmetic,' " *IEEE Trans. Circuit Theory*, vol. CT-20, Sept. 1973.
[8] B. Liu, "Effect of finite word length on the accuracy of digital filters—A review," *IEEE Trans. Circuit Theory*, vol. CT-18, Nov. 1971.
[9] B. Liu and T. Kaneko, "Error analysis of digital filters realized with floating-point arithmetic," *Proc. IEEE*, vol. 57, pp. 1735–1747, Oct. 1969.
[10] P. Thajchayapong and P. J. W. Rayner, "Recursive digital filter design by linear programming," *IEEE Trans. Audio Electroacoust.*, vol. AU-21, Apr. 1973.
[11] B. Gold and C. M. Rader, *Digital Processing of Signals*. New York: McGraw-Hill, 1969.
[12] E. I. Jury, *Theory and Application of the z-Transformation Method.* New York: Wiley, 1964.
[13] A. V. Oppenheim and C. J. Weinstein, "Effects of finite register length in digital filtering and the fast Fourier transform," *Proc. IEEE*, vol. 60, Aug. 1972.

AUTHOR CITATION INDEX

SUBJECT INDEX

About the Editors

SANJIT K. MITRA received the Ph.D. in electrical engineering from the University of California, Berkeley (1962). He has been on the faculty of the University of California since 1967, and currently is a professor of electrical engineering and computer science at the Santa Barbara campus. He is also a consultant to the Lawrence Livermore Laboratory of the University of California.

Dr. Mitra has published over 100 papers in analog networks and digital filters, and is author of *Analysis and Synthesis of Linear Active Networks,* coeditor of *Modern Filter Theory and Design* (with G. C. Temes), and editor of *Active Inductorless Filters.*

He is a fellow of the IEEE, a member of ASEE, Sigma Xi, and the New York Academy of Sciences. He was an associate editor of the *IEEE Transactions on Circuit Theory,* a member of the Administrative Committee of the IEEE Circuits and Systems Society, general chairman of the 1974 IEEE International Symposium on Circuits and Systems, a member of the editorial board of the IEEE Press, and is presently a member of the editorial board of the Proceedings of the IEEE. He is the recipient of the 1973 F. E. Terman Award of the American Society of Engineering Education as an outstanding young electrical engineering educator.

MICHAEL P. EKSTROM received the Ph.D. in Electrical Engineering from the University of California, Davis (1971). He is presently an associate division leader, Engineering Research Division, Lawrence Livermore Laboratory, specializing in two-dimensional digital signal processing involving both deterministic and stochastic formalisms. He is responsible for the R & D activities of the Lawrence Laboratory in digital image processing and has published numerous papers in the field. These activities involve applications in radar signal processing, radiographic image processing, computerized-axial tomography, and stochastic aperture encoding.

Dr. Ekstrom is a lecturer in the Department of Electrical Engineering and Computer Science at the University of Santa Clara, teaching courses in advanced digital signal processing. He was a co-recipient of the 1976 Senior Award of the IEEE Acoustics, Speech and Signal Processing Society, general chairman and technical program chairman of the Asilomar Conference on Circuits, Systems and Computers, organizer and general chairman of the Lawrence Symposium on Systems and Decision Sciences, and is a member of the IEEE Circuits and Systems, Technical Committee on Digital Signal Processing. He is also a member of Pi Mu Epsilon and Phi Kappa Phi.